# Crop Post-Harvest:
## Science and Technology

### Volume 1

### Principles and Practice

**Edited by**
**Peter Golob, Graham Farrell and John E. Orchard**



Blackwell Science

Natural Resources Institute

Editorial Offices:
Osney Mead, Oxford OX2 0EL, UK
Tel: +44 (0)1865 206206
Blackwell Science, Inc., 350 Main Street, Malden, MA 02148-5018, USA
Tel: +1 781 388 8250
Iowa State Press, a Blackwell Publishing Company, 2121 State Avenue, Ames, Iowa 50014-8300, USA
Tel: +1 515 292 0140
Blackwell Publishing Asia Pty, 550 Swanston Street, Carlton South, Melbourne, Victoria 3053, Australia
Tel: +61 (0)3 9347 0300
Blackwell Wissenschafts Verlag, Kurfürstendamm 57, 10707 Berlin, Germany
Tel: +49 (0)30 32 79 060



First published 2002 by Blackwell Science Ltd

Library of Congress
Cataloging-in-Publication Data
is available

ISBN 0-632-05723-8

A catalogue record for this title is available from the British Library

Set in 9/11.5 pt Times
by Sparks Computer Solutions Ltd, Oxford
http://www.sparks.co.uk
Printed and bound in Great Britain by
Ashford Colour Press Ltd, Gosport, UK

For further information on
Blackwell Science, visit our website:
www.blackwell-science.com

This series of volumes is dedicated to the memory of
Dr Philip C. Spensley, Director of the Tropical Products Institute 1966–1982
for his leadership in a period of great expansion of post-harvest research and development

Part of a three-volume set from Blackwell Publishing and the Natural Resources Institute, University of Greenwich:

*Crop Post-Harvest: Science and Technology Volume 1: Principles and Practice*
Edited by P. Golob, G. Farrell and J.E. Orchard
0 632 05723 8

*Crop Post-Harvest: Science and Technology Volume 2: Durables*
Edited by R.J. Hodges and G. Farrell
0 632 05724 6

*Crop Post-Harvest: Science and Technology Volume 3: Perishables*
Edited by D. Rees, J.E. Orchard and G. Farrell
0 632 05725 4

# Contents

# Contributors

Roger D. Bancroft, Steven R. Belmain, Lucy A. Birkinshaw, Robert Black, Mick E. Blowfield, Robin A. Boxall, John R. Brice, John A. Conway, H. Claire Coote, Andrew D. Devereau, Graham Farrell, Peter Golob, Ann E. Gordon, Peter Greenhalgh, Lynda Hammond, Paul S. Hindmarsh, Richard J. Hodges, Heather M. Kindness, Ulrich K. Kleih, Neil K. Marsland, Adrian N. Meyer, Linda Nicolaides, John E. Orchard, Deborah Rees, Tanya E. Stathers, Anthony A. Swetman, Robert W.D. Taylor, Sarah J. Taylor, David J. Walker, Peter W. Wareing, John F. Wood: at the time of writing, all on the staff of the Natural Resources Institute, University of Greenwich, Chatham ME4 4TB, UK.

Savitri Abeyasakera, University of Reading, Reading RG6 6AH, UK.

Campbell Anderson, Campden and Chorleywood Food Research Association, Chipping Campden GL55 6LD, UK.

Francesco Goletti, President, Agrifood Consulting International, 2715 Harmon Road, Silver Spring, MD 20902, USA.

Adrian N. Meyer, The Acheta Partnership, Garden Cottage, Horsemoor, Chieveley, Newbury RG20 8XD, UK.

Robert Myhara, Food Industry Consultant, 12-5 Cheryl Road, Ottawa, Ontario, Canada K2G-0V5.

John H. New, Stubbings, Weavering Street, Maidstone ME14 5JH, UK.

Emma Samman, St Antony's College, Oxford OX2 6JF, UK.

Ian M. Wilson, University of Reading, Reading RG6 6AH, UK.

# Foreword

In our world of six billion people, almost 800 million are hungry or starving, a quarter of whom are children under five years old. More than an eighth of all humans, or one-sixth of the developing world's population, are chronically food-insecure and do not have access to sufficient safe food at all times to lead healthy, active, productive lives. Food insecurity brings with it the vicious cycles of incapacity to work, increased susceptibility to communicable disease, permanent debilitation from childhood malnutrition, withdrawal from education, and social or political exclusion – and, ultimately, starvation. Yet all authorities agree that, globally, humankind harvests sufficient food to meet the needs of all, and most analysts believe that production trends will continue to keep ahead of demand in the medium-term future. In many parts of the world, yield productivity has increased dramatically in recent decades due in part to changes in agronomic practice and, especially, to the improved crop varieties arising from international investment in the Green Revolution. Nevertheless, vast numbers of people are malnourished, notably in sub-Saharan Africa where one in three people are food-insecure.

Efforts to improve harvest yield and quality continue apace, including contentious research on biotechnology and genetic modification in plant and animal breeding. Increasingly, however, international attention in agricultural development is focused on issues, not of food production *per se*, but of people's access to safe and nutritious food, which is compromised – and not just in developing countries – by poverty, war or civil insecurity, social or political exclusion, ignorance, ill-health, ineffective markets, and inadequate food quality management. Self-evidently, several of these constraints to access are only amenable to social and political solutions. However, many barriers to access could be solved by technological and economic improvements in commodity marketing, processing, packaging, storage and distribution, that is, by actions to reduce losses and to add value and quality across the post-harvest (or post-production) sector.

Public concern about food safety and quality is increasing both in industrialised countries and in urban centres in the developing world. This has been fuelled by long-term anxieties about pesticides and other contaminants in food, by growing awareness of the prevalence of food-borne disease, by the media profile of specific 'food scares', and by the impact of globalisation on quality management in international trade. While some elements of food safety and quality have their origins in the pre-harvest sector, many of the problems occur or develop post-harvest, and food quality management falls squarely in the post-harvest sector, whatever the origin of the problem.

In spite of its importance to food security, food safety and food quality, the post-harvest sector has long been the poor relation of agricultural development. One possible reason for this is that many of its proponents have focused exclusively on their specific commodity sub-sectors, with separation of teams working on grains, roots and tubers, tree crops, oilseeds, fruit and vegetables. In recent years there has been greater convergence and interaction of these sub-sectors, and thus the development of cross-commodity post-harvest sectoral thinking. This three-volume series brings together the results of these interactions for the first time in what should become the standard text on post-harvest technology.

Chris Haines
*Professor of Post-Harvest Technology*
*Natural Resources Institute*
*University of Greenwich*

# Preface

In much of the world without artificial irrigation, the agricultural year can be divided into a crop production period followed by a post-harvest period. Crop production lasts for three months or more but the post-harvest period may stretch from the end of one growing season to the next, often at least six months. If there is only one rainy season then it may last for as long as ten months. In the European Union, intervention storage may last for several years.

When one considers the pre- and post-harvest phases of crops, technical information on crop production far outweighs that available on crop harvesting, storage, processing and marketing. Every country has undergraduate and postgraduate university courses dedicated to teaching agronomy and crop protection. However, post-harvest science is taught on a very few specialised courses, mostly at postgraduate diploma or MSc levels.

The UK Natural Resources Institute (NRI), of the University of Greenwich, has taught courses in post-harvest science, technology, economics and development for more than 30 years. Many other institutions of further and higher education are seeking to teach post-harvest science as further development in the post-harvest sector offers substantial opportunities for improving rural livelihoods. Furthermore, the globalisation of trade, presenting opportunities for every country to export agricultural commodities, will require quality and safety standards to be maintained and unified. Government and private institutions will need to know how to conform to the requirements of importing countries and they will need educated and trained personnel capable of providing advice and information.

To cope with this demand, NRI undertook the compilation of a three-volume standard text on post-harvest science and technology. The intention of the editors was to mine the extensive historical literature, often unreferenced but based on accumulated wisdom gained over many years of practical experience, and combine this with the latest research findings, applications and methodologies. Although basic concepts remain constant the subject has been changing rapidly, as a result, for example, of the ozone depletion effects of methyl bromide, the process of agricultural market liberalisation and the increasing emphasis on adaptive and participatory approaches to investigations in the field. To address the changing nature of the subject and to bring it up to date, information is included that describes the most recent knowledge and developments.

The work is designed primarily for university students, lecturers and researchers. However, the text can be regarded as a standard work of reference in the subject so that it can be used by anyone wishing to acquire information on post-harvest issues, in tropical and temperate regions.

The series comprises three volumes. Volume 1 describes the principles and practices applying to all post-harvest issues related to both durable and perishable agricultural products. Volume 2 contains case studies of how durable commodities are stored around the globe and Volume 3 does the same for perishables.

In Volume 1 the subject is dealt with in a holistic fashion so that, as well as natural science and engineering, other disciplines including economics, trade and international regulation are covered. The first chapter puts the subject into perspective and gives an indication as to how globalisation might affect post-harvest systems in the future. Chapter 2 considers the basic structure of crops, including their physiology and biochemistry, and how these systems influence post-harvest characteristics. Chapter 3 describes physical factors influencing commodity deterioration, particularly temperature and relative humidity. Biological factors affecting quality, for example, insects, rodents and fungi are described in Chapter 4. The types of losses that occur in commodities once they have matured in the field are described in Chapter 5, which includes descriptions of good management practices to prevent deterioration from occurring. Management practices to prevent pest attack, including the use of conventional and non-chemical methods, are described in Chapter 6 and methods employed to disinfest commodities, using fumigation, controlled atmospheres and radiation, are described in Chapter 7. Processing of durables into flours, oils and animal feed is covered in Chapter 8. Food security issues are dealt with in Chapter 9 and this leads into Chapter 10, which considers the most effective way of researching the subject and disseminating the information obtained. Finally, Chapter 11 describes the international practices and agreements that regulate trade and associated health and safety issues.

The editors would like to thank the UK Department for International Development, which in various guises has, over the past 40 years, funded much of the post-harvest work reported in these pages. We also greatly appreciate being granted time by the University of Greenwich for writing, compiling and editing. Colleagues at the Natural Resources Institute and its predecessors gave generously of their expertise and advice. We are also indebted to the many overseas collaborators who have shared their insights into post-harvest systems and from whom we have learnt so much.

Every effort has been made to contact copyright holders of photographs and diagrams. However, the editors would be pleased to make amendments in future editions.

# Chapter 1
# Post-Harvest Systems in World Agriculture

**F. Goletti and E. Samman**

The post-harvest system is a crucial and dynamic component of the agricultural complex. Increasingly, agricultural products are not consumed in their raw form, and post-harvest activities such as transport, storage, processing and marketing account for a growing part of their final value. In poorer countries, these activities often provide the first step for the development of manufacturing. Processed products can generate export earnings, and even if consumed domestically, can free up foreign exchange to be used for other imports and time traditionally spent preparing food for more productive uses. An efficient post-production system can help to ensure a safe and sufficient food supply in growing cities, and can give impetus to the development of infrastructure and institutions that are vital for further economic growth. Moreover, if the post-harvest chain linking producers and consumers does not function properly, investment in production may be relatively more costly, risky and susceptible to wastage, particularly in the case of perishables. An efficient post-production system can contribute toward environmental sustainability by minimising unnecessary production, thereby saving on scarce land and water resources, and by providing alternatives to the heavy application of chemical inputs with potentially harmful side-effects.

Further, in connecting farmers to domestic industry and international markets, post-harvest activities can engender rural development. It can utilise surplus labour, particularly in less favoured areas, thereby slowing rural–urban migration. The integration of small farmers as suppliers to processors can promote their incorporation into the larger marketing chain and reduce the risk of engaging in market activity for both. Demand for agroprocessing in remote rural areas provides a crucial boost to small-enterprise development. Post-harvest development in rural areas can also result in increased demand for agricultural raw materials for processing. If post-harvesting develops in a broad-based manner, it has the potential to improve rural income distribution.

In richer countries, post-harvest activities account for a higher share of value; they are key to meeting emergent forms of consumer demand, and ensuring food safety and quality – as well as a significant source of employment. In the context of globalisation, companies in developed countries are becoming enmeshed with those in developing countries, with implications for farmers and firms involved there.

Ensuring a positive outcome for all groups involved will depend crucially on a combination of technological innovation and adaptation, institutional and policy frameworks, and appropriate management systems. Several trends suggest that the potential for post-harvest will continue to increase, yet at the same time they present a number of challenges that will need to be overcome to realise dynamic and broad-based systems, particularly in developing countries.

This chapter aims to provide a holistic overview of the post-harvest sector by examining its economic importance, the trends shaping its future development and emerging challenges to development. The first section seeks to highlight the economic importance of post-harvest to growth, exports, employment, food security and rural welfare. The second part turns to qualitative trends which show that the sector has strong growth potential and that the structure of the sector is changing dramatically. The third section draws upon these structural changes to highlight the future challenges the post-harvest sector will face. Finally, the chapter closes with a discussion of the role policy can play in supporting post-harvest agriculture, and ensuring it develops in a broad-based and sustainable manner.

We define post-harvest to involve all activities that occur after production of agricultural commodities, including storage, packaging, procurement, transportation, processing and marketing of agricultural products from the farmgate to the distributors, either for domestic consumption or export. By agroindustry, which we will sometimes use to roughly gauge levels of post-harvest activity, we refer to processing and marketing of agricultural products. Although this chapter deals with all forms of agroprocessing, it is especially concerned with food products, which are discussed in the most depth. However, the distinction between fresh and processed foods is not a clear one. Even 'fresh' foods actually undergo sophisticated forms of cleaning, quality control, packaging, storage, refrigeration and transport, so that 'the development of agroindustry is increasingly becoming one with the development of industrial agriculture' (FAO 1997). Therefore, we also expand our scope to consider agricultural production and the effects of post-harvest activity on commercial farmers.

## The importance of post-harvest agriculture

The global food industry, with a value of US$2.2 trillion annually, is the single most important industry in the world economy (Huang 2000). Agroindustry is expected to be worth $10 trillion by 2028 (*The Economist* 23 May 2000), and most of this growth will come from the developing world. Clearly, the direct impact of the subsector on growth and indirect stimulus to other types of economic activity carry important implications for employment, exports, food security and living standards.

Throughout the world, the importance of post-harvest activity relative to farming is increasing. In 1900, US farmers received about 60% of money spent on food, but this share fell to about 20% by the mid-1990s (Austin 1995). Globally, farmers received over one-third of the value of food products in 1950, and it is predicted that this share will drop to 10% by 2028 (*The Economist* 23 May 2000). These shares differ by product type: US farmers received a relatively high 35% of the value of meat and dairy products in 1995, but under 10% of the value of cereal and bakery products (USDA data). Even for relatively unprocessed foods, such as fresh fruits and vegetables, post-harvest contributes about 80% of final value, mostly through sophisticated storage, transport and marketing systems. In developing countries, farmers receive a higher share of product value, but the importance of post-harvest is growing as well.

A few figures highlight the huge gulf in the importance of agroprocessing between developed and developing countries. On average, agroindustry accounts for about 2% of GDP in developing regions but 9% in developed countries (UNIDO data). The value of agroprocessing is about three to four times that of agriculture in developed countries, while it is typically a fraction of the value of agriculture in developing countries. Another indicator is the huge gap in consumption of certain processed products between rich and poor countries. For example, China's soft drink consumption averaged just 4 litres per person per year in 1996, compared with 200 litres in the United States. Further, considering that China contains more than 20% of the world's population – while the US has 4% – it signals vast future growth potential.[1]

The economic contribution of post-harvest is difficult to quantify with any precision because its components do not fit neatly into traditional economic categories. The ideal measure would include appropriate elements of agriculture, the whole of agroprocessing, as well as other relevant manufactured goods (such as fertiliser and other inputs) and related services. However, even such a measure would neglect the indirect effects of the subsector in terms of its effects on agricultural production, and on spending and investment in the broader community. Here we rely on measures of agroprocessing to give a very rough approximation of the size, growth and distribution of the post-harvest sector, using principally data gathered by the United Nations Industrial Development Organization (UNIDO).[2]

### Contribution to GDP, exports and employment

To measure the economic contribution of agroindustry to output, we examine its share in manufacturing, its growth rate and its distribution among regions in the world. Table 1.1 contrasts the shares of agroprocessing in total manufacturing value-added[3] (MVA) in 1980 and 1998. The first striking fact is the very high share of agroprocessing in MVA in both rich and poor countries; in 1998, the share ranged from 24% in North America to 45% in the category of 'other' developing countries (those excluding the newly industrialising countries (NICs) and second generation NICs). In Africa, agroprocessing is especially important; the food products subsector in particular is the most important subsector in most of its economies, sometimes accounting for as much as 50% of MVA (UNIDO 1990). The table also shows considerable decline in the importance of agroindustry relative to total manufacturing over the

**Table 1.1** Share[1] of agroindustries in total manufacturing value-added in selected country groups, 1980 and 1998. Source: UNIDO 2001.

| | Food products | | Beverages | | Tobacco | | Textiles | | Wearing apparel | | Leather and fur | | Footwear | |
|---|---|---|---|---|---|---|---|---|---|---|---|---|---|---|
| Country groups | 1980 | 1998 | 1980 | 1998 | 1980 | 1998 | 1980 | 1998 | 1980 | 1998 | 1980 | 1998 | 1980 | 1998 |
| Industrialised countries | 9.9 | 8.1 | 2.2 | 1.8 | 1.4 | 1.0 | 4.7 | 2.4 | 2.6 | 1.3 | 0.4 | 0.2 | 0.6 | 0.3 |
| European Union | 8.0 | 8.9 | 2.2 | 2.3 | 1.6 | 1.3 | 4.3 | 3.1 | 2.8 | 1.4 | 0.5 | 0.3 | 0.9 | 0.5 |
| Japan | 9.5 | 7.9 | 1.6 | 1.1 | 0.3 | 0.2 | 4.9 | 1.9 | 2.0 | 1.2 | 0.3 | 0.1 | 0.2 | 0.1 |
| North America | 10.0 | 6.5 | 1.9 | 1.2 | 2.2 | 1.0 | 3.2 | 2.0 | 2.7 | 1.2 | 0.3 | 0.1 | 0.4 | 0.1 |
| Eastern Europe & CIS | 16.5 | 13.3 | 3.3 | 4.7 | 1.0 | 1.4 | 9.3 | 2.2 | 3.3 | 1.5 | 1.0 | 0.2 | 0.9 | 0.5 |
| Developing countries[2] | 11.2 | 10.2 | 3.7 | 3.6 | 3.5 | 3.4 | 8.6 | 6.2 | 3.9 | 2.5 | 0.9 | 0.4 | | |
| Newly industrialising countries (NICs)[3] | 9.2 | 8.2 | 3.1 | 2.8 | 2.6 | 2.0 | 8.5 | 5.7 | 3.8 | 1.8 | 1.0 | 0.5 | | |
| Second generation NICs[4] | 15.1 | 12.1 | 4.4 | 4.2 | 5.7 | 6.2 | 8.4 | 6.6 | 6.2 | 4.8 | 0.4 | 0.2 | | |
| Others[5] | 14.6 | 15.5 | 5.0 | 5.8 | 5.0 | 5.3 | 9.0 | 7.6 | 2.7 | 1.9 | 1.0 | 0.6 | | |

**Table 1.1** (*Continued.*)

| | Wood products | | Furniture | | Paper | | Printing | | Rubber | | All agroprocessing | |
|---|---|---|---|---|---|---|---|---|---|---|---|---|
| | 1980 | 1998 | 1980 | 1998 | 1980 | 1998 | 1980 | 1998 | 1980 | 1998 | 1980 | 1998 |
| Industrialised countries | 2.0 | 1.5 | 1.6 | 1.2 | 3.1 | 3.1 | 4.9 | 4.8 | 1.2 | 1.1 | 34.6 | 26.8 |
| European Union | 1.9 | 1.6 | 1.9 | 1.5 | 2.8 | 3.1 | 4.0 | 4.2 | 1.2 | 1.3 | 32.1 | 29.5 |
| Japan | 2.7 | 1.1 | 1.4 | 0.7 | 2.4 | 2.8 | 6.4 | 6.5 | 1.4 | 1.3 | 33.1 | 24.9 |
| North America | 1.7 | 1.3 | 1.2 | 1.0 | 4.7 | 3.4 | 7.2 | 4.8 | 0.9 | 0.9 | 36.4 | 23.5 |
| Eastern Europe & CIS | 1.4 | 2.0 | 1.3 | 1.4 | 1.8 | 2.1 | 0.4 | 1.1 | 1.4 | 1.3 | 41.6 | 31.7 |
| Developing countries[2] | 1.5 | 1.1 | 1.4 | 0.7 | 2.2 | 2.3 | 2.1 | 2.2 | 1.5 | 1.7 | 42.0 | 34.9 |
| Newly industrialising countries (NICs) | 1.2 | 0.6 | 1.2 | 0.7 | 2.3 | 2.5 | 2.2 | 2.4 | 1.6 | 1.7 | 38.3 | 29.6 |
| Second generation NICs[4] | 2.6 | 2.5 | 0.7 | 0.7 | 1.8 | 1.9 | 1.6 | 1.2 | 1.9 | 2.3 | 49.5 | 43.1 |
| Others[5] | 1.9 | 1.1 | 2.2 | 0.6 | 2.1 | 2.1 | 2.2 | 2.3 | 0.9 | 1.0 | 48.1 | 44.5 |

1 % share at constant 1990 prices.
2 1998 column is 1997 for developing countries.
3 Includes Argentina, Brazil, India, Hong Kong, Mexico, Singapore, South Korea, Taiwan and the former Yugoslavia.
4 Includes Colombia, Indonesia, Malaysia, Philippines, Thailand and Turkey.
5 Includes all but the NICs, 2nd generation NICs and least developed countries.

18-year period throughout the world but especially in industrialised countries, with the notable exception of the European Union.

Table 1.2 reveals the rate of growth in value-added by product and country groupings. In general, growth rates fell in the industrialised countries over the 1980s and 1990s, and were sometimes negative. In contrast, in the developing countries, a number of product groups experienced higher growth in the 1990s than the 1980s, with tobacco, food and beverages appearing particularly dynamic. The increases were most notable in the second generation NICs but also significant in the 'other' devel-

**Table 1.2** Annual growth of value-added in agroindustry by selected groups, 1980–90 and 1990–98. Source: UNIDO 2000.

| | Industrialised countries | | Eastern Europe & CIS | | Developing countries | | NICs | | 2nd generation NICs | | Other developing countries | |
|---|---|---|---|---|---|---|---|---|---|---|---|---|
| Branch | 1980–90 | 1990–98 | 1980–90 | 1990–98 | 1980–90 | 1990–98 | 1980–90 | 1990–98 | 1980–90 | 1990–98 | 1980–90 | 1990–98 |
| Food | 1.8 | 1.3 | 1.7 | | 2.4 | 3.8 | 2.8 | 2.8 | 2.9 | 6.9 | 1.6 | 2.4 |
| Beverages | 1.7 | 0.8 | –1.4 | | 2.4 | 4.2 | 2.5 | 3.1 | 3.4 | 7.5 | 1.5 | 3.9 |
| Tobacco | 0 | 0.4 | 0.7 | | 1.7 | 5.7 | 1.6 | 1.9 | 2.9 | | 0.9 | |
| Textiles | 0.1 | –0.9 | 1.1 | | 2.5 | 1.3 | 2.1 | 0.7 | 6.5 | 3.1 | 0.7 | 1.2 |
| Wearing apparel | –0.6 | –1.4 | 1.7 | | 2.6 | –0.3 | 2.2 | –3.5 | 5.5 | | | |
| Leather & fur | –1.5 | –2.9 | 0 | | 0.8 | –1.5 | 2.0 | –3.4 | 0.5 | 3.0 | | |
| Footwear | –2.8 | –2.4 | 2.4 | | –0.6 | –2.8 | 0.6 | –3.3 | 1.9 | 1.8 | | |
| Wood products | 1.5 | 0.7 | 2.2 | | 2.0 | | 1.0 | | 6.9 | 2.7 | | |
| Furniture | 1.6 | 0.3 | 3.0 | | –0.3 | | 2.5 | | 5.3 | 4.4 | | |
| Paper | 3.2 | 1.8 | 1.2 | | 4.5 | 4.1 | 5.0 | 3.6 | 6.5 | 7.6 | 1.5 | 3.2 |
| Printing | 3.2 | 1.1 | 2.9 | | 3.2 | 4.2 | 4.4 | 3.9 | 3.0 | 4.7 | | |
| Rubber | 3.1 | 2.4 | 1.6 | | 5.2 | 4.2 | | 2.7 | 6.3 | 8.4 | 2.2 | 3.3 |
| All manufacturing | 2.8 | 1.6 | 2.6 | | 5.1 | 6.7 | 4.0 | 4.3 | 7.1 | 6.7 | 2.3 | 3.2 |

See notes to Table 1.1.

oping countries. In terms of product groups, typically paper and printing were the most dynamic categories across country groupings.

Of course these aggregate figures conceal a wide variation of country-level experiences. Agroindustry has had a disproportionate impact on some countries and areas within countries while bypassing others. It has been particularly dynamic in Latin America, certain African countries (notably Republic of South Africa, Côte d'Ivoire and Kenya), as well as in China, India and some of East Asia (especially Indonesia, South Korea and the Philippines). As shown in Fig. 1.1, ten developing countries accounted for 74% of processed food production in 1997.

For all product groups, agroprocessing activity is overwhelmingly concentrated in the industrialised countries (Table 1.3). Yet there has been a shift from developed to developing regions over the last two decades. During this time, in the industrialised countries, the share of every product group has fallen. Conversely, the share of developing countries has increased during this period.[4] However, the distribution of agroprocessing activity within the developing world is relatively unequal (Table 1.4). In 1997, about 40% of agroprocessing value-added in developing countries was concentrated each in South and East Asia and in Latin America. Africa accounted for just 8% of the total, while the group of least developed countries contained just 3% of value-added. Moreover, over the 1990s, the share of agroindustrial value-added grew only in South and East Asia, and, marginally, in Latin America.

Processed agricultural products have provided a crucial source of export revenue to developing countries, particularly in light of the long-term trend towards declining prices for traditional agricultural commodity exports.[5] Trade in food products increased ten-fold between 1961 and 1990 (Giovannucci & Satin 2000). Trade in processed products was much more dynamic than trade in raw materials, climbing 9.4% each year over the three decades, compared with an annual average increase of 2.1% for agricultural commodities (Traill 2000). Growth has been concentrated among the richer countries, suggesting that developing countries have yet to realise the potential gains of growth. About 85% of EU food exports are processed (Traill 2000), while 60% of African exports are primary commodities (UNIDO 2000a). Figure 1.2 gives a more disaggregated picture by product group, showing that while developed countries dominate some areas – such as fresh foods, alcohol and paper – developing areas are

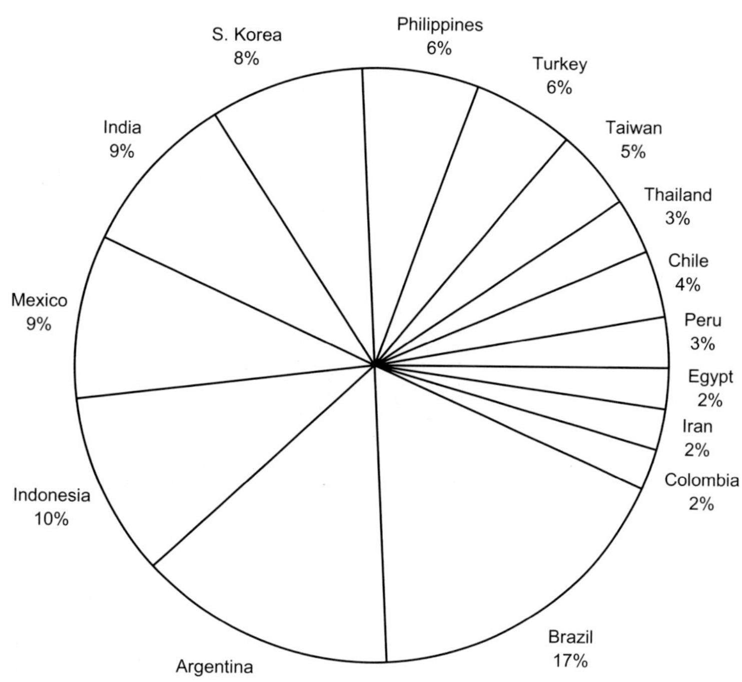


**Fig. 1.1** Leading developing countries in production of processed foods. Source: UNIDO 2000. Note: Refers to ISIC-2 revision categories 3.11 and 3.12. Sum of countries' contributions is 74% (of developing countries' total production).

**Table 1.3** Distribution of world value-added by branch of agroindustry, 1980 and 1998. Source: UNIDO 2000.

| Branch | Year | Industrialised countries | European Union | Japan | North America | Eastern Europe & CIS | Developing countries | NICs | Others |
|---|---|---|---|---|---|---|---|---|---|
| Food | 1980 | 85.5 | 28.0 | 14.8 | 22.3 | 16.3 | 14.5 | 7.4 | 7.1 |
| | 1998 | 80.1 | 33.1 | 12.9 | 24.4 | 5.5 | 20.0 | 9.4 | 10.6 |
| Beverages | 1980 | 79.3 | 32.6 | 10.4 | 18.6 | 14.0 | 20.7 | 11.1 | 9.6 |
| | 1998 | 72.1 | 35.3 | 7.1 | 17.9 | 7.9 | 27.9 | 13.1 | 14.8 |
| Tobacco | 1980 | 73.7 | 33.7 | 3.2 | 29.3 | 5.8 | 26.3 | 12.2 | 14.1 |
| | 1998 | 59.1 | 29.2 | 2.3 | 22.6 | 3.4 | 40.9 | 13.0 | 27.9 |
| Textiles | 1980 | 78.1 | 29.3 | 14.4 | 14.0 | 17.9 | 21.9 | 13.2 | 8.7 |
| | 1998 | 67.9 | 32.6 | 9.0 | 21.2 | 2.6 | 32.1 | 17.7 | 14.4 |
| Wearing apparel | 1980 | 81.5 | 34.2 | 11.1 | 21.7 | 11.6 | 18.5 | 10.9 | 7.6 |
| | 1998 | 73.5 | 29.3 | 11.5 | 26.0 | 3.5 | 26.5 | 11.2 | 15.3 |
| Leather | 1980 | 76.7 | 34.6 | 9.9 | 12.0 | 18.9 | 23.3 | 15.3 | 8.0 |
| | 1998 | 68.4 | 43.6 | 8.6 | 10.3 | 3.4 | 31.6 | 19.1 | 12.5 |
| Footwear | 1980 | 74.1 | 42.1 | 4.4 | 13.1 | 11.7 | 25.9 | 17.6 | 8.3 |
| | 1998 | 69.7 | 47.0 | 5.0 | 8.4 | 5.9 | 30.3 | 20.3 | 10.0 |
| Wood products | 1980 | 89.6 | 33.5 | 22.1 | 19.4 | 7.6 | 10.4 | 4.9 | 5.5 |
| | 1998 | 88.2 | 36.3 | 10.6 | 28.9 | 5.1 | 11.8 | 3.9 | 7.9 |
| Paper | 1980 | 90.4 | 33.1 | 12.7 | 35.0 | 6.1 | 9.6 | 6.3 | 3.3 |
| | 1998 | 87.7 | 32.9 | 13.0 | 36.0 | 2.5 | 12.3 | 8.0 | 4.3 |
| Rubber | 1980 | 84.9 | 36.3 | 17.2 | 17.6 | 11.3 | 15.1 | 10.1 | 5.0 |
| | 1998 | 78.7 | 33.6 | 14.5 | 14.5 | 3.8 | 21.3 | 12.8 | 8.5 |
| All manufactures | 1980 | 87.1 | 35.7 | 14.2 | 23.9 | 9.5 | 12.9 | 8.2 | 4.7 |
| | 1998 | | | | | | | | |

See notes to Table 1.1. Totals do not equal 100 because non-EU countries in Western Europe are not included.

**Table 1.4** Distribution of value-added among developing regions.

| | Africa | | Latin America | | South and East Asia & Europe | | West Asia countries | | Least developed | |
|---|---|---|---|---|---|---|---|---|---|---|
| Branch | 1990 | 1997 | 1990 | 1997 | 1990 | 1997 | 1990 | 1997 | 1990 | 1997 |
| Food | 9.8 | 8.7 | 43.8 | 43.5 | 34.6 | 39.0 | 11.8 | 8.8 | 5.0 | 4.8 |
| Beverages | 12.9 | 12.4 | 54.9 | 52.8 | 23.7 | 28.1 | 8.5 | 6.7 | 5.4 | 3.9 |
| Tobacco | 8.6 | 6.5 | 44.3 | 38.2 | 37.0 | 44.6 | 10.1 | 10.7 | 5.3 | 3.5 |
| Textiles | 8.5 | 7.1 | 23.3 | 21.1 | 54.3 | 59.4 | 13.9 | 12.4 | 3.9 | 3.7 |
| Wearing apparel | 6.1 | 7.1 | 27.2 | 25.6 | 51.4 | 54.2 | 15.3 | 13.1 | 2.2 | |
| Leather | 6.0 | 6.8 | 41.8 | 51.4 | 38.1 | 30.0 | 14.1 | 11.8 | 2.9 | |
| Footwear | 8.0 | 8.0 | 48.6 | 57.7 | 26.2 | 21.4 | 17.2 | 12.9 | 2.7 | |
| Wood products | 12.2 | 9.6 | 22.3 | 26.3 | 53.5 | 55.0 | 12.0 | 9.1 | 2.9 | 2.7 |
| Paper | 4.8 | 3.6 | 46.9 | 43.6 | 37.5 | 44.8 | 10.8 | 8.0 | 1.3 | 0.7 |
| Rubber | 4.2 | 3.7 | 31.0 | 28.3 | 55.1 | 58.3 | 9.7 | 9.7 | 0.6 | 0.4 |
| Average agroprocessing | 8.1 | 7.4 | 38.4 | 38.9 | 41.1 | 43.5 | 12.3 | 10.3 | 3.2 | 2.8 |

responsible for close to half the supply of goods such as man-made fabric and footwear. Developed countries account for the overwhelming majority of processed food imports: in 1992, 30 developed and newly industrialised countries (NICs) accounted for 90% of processed food imports, of which the NIC share was just 6% (Traill 2000).

The last 40 years have seen higher growth of food product exports in Asia and to some extent in Latin America, spurred by liberalisation, but such growth declined in much of Africa, owing to war and other economic factors.[6] At the same time there has been a shift in Asian and Latin American processed exports towards oilseeds, fruits and vegetables, relative to

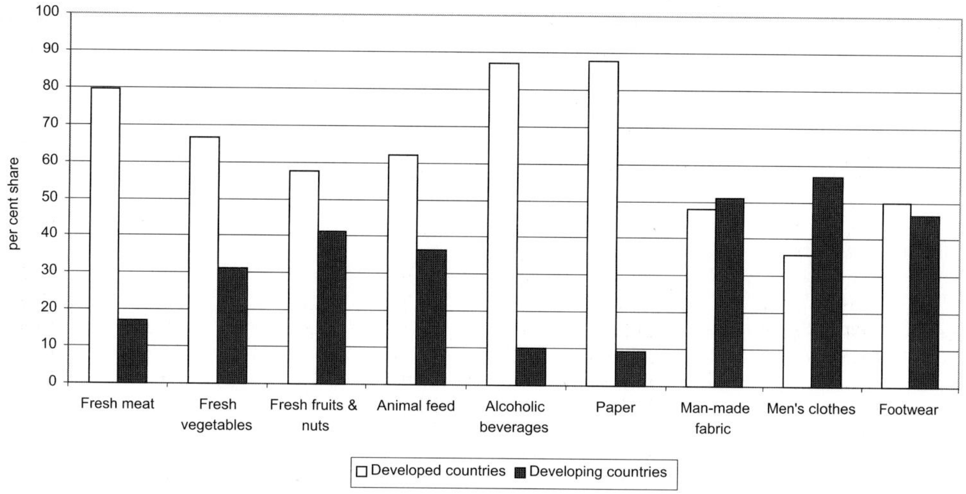


**Fig. 1.2** Share of developed and developing countries in selected agroindustrial exports, 1997–98. Source: UNCTAD 2000.

traditional products such as sugar, coffee and cocoa. The share of oilseed, fruit and vegetable exports jumped from about 20% of total agricultural exports by developing countries in the 1960s to slightly more than 35% in the 1990s, while that of traditional exports fell from about 35–40% of exports (1960s–1980s) to about 25% in the 1990s. In Brazil, the relative share of traditional commodities in agricultural exports fell from 75% at the beginning of the 1970s to less than 40% at the beginning of the 1990s.

The growth of non-traditional food exports in Latin America in recent years has been very high. In Central America their value increased at an annual average rate of 17% between 1985 and 1992. In South America – excluding Brazil – these exports grew at an average of 48% annually over the same period. In Guatemala, Costa Rica and Chile, annual average growth rates between 1984 and 1989 were 78%, 348% and 222%, respectively (Thrupp 1995).

The dramatic decline in world market share for aggregate African exports stands in stark contrast to the recent sharp growth of certain products in some of its countries. Between 1989 and 1997, the value of fresh vegetable exports from sub-Saharan Africa to the European Union increased 150% (Dolan & Humphrey 2000). Exports of fresh cut flowers from Kenya have also grown at a high rate; the Kenya Flower Council estimates real growth of 65% between 1995 and 1999 alone (Kenya Flower Council, pers. comm.). The increase in processed wood exports from Ghana in the decade after 1983 was also significant (Owusu 2001).

Employment in agroindustry is significant, both as a source of permanent labour, and in providing labour during the slack season in agriculture. In the poorest countries, the relative share of employment generated by agroindustry is highest, and it generally falls with the level of development. As would be expected, trends in employment roughly conform to those in total output, with the employment share declining between 1980 and 1999 in developed countries and in some developing ones, but growing in a number of developing countries (e.g. see Hong Kong, Sri Lanka, Peru, Kenya, among others). While in developed countries agroindustrial wages tend to be lower than those in the rest of manufacturing, in developing countries they often compare favourably (FAO 1997).

These figures do not include the significant employment spillover effects on raw material production. In agriculture too, the aggregate impact on employment masks disproportionately large effects on certain groups and areas. Employment effects tend to be crop specific, but commercialisation of agriculture generally increases demands on labour, and many non-traditional crops have higher labour requirements than traditional commodities (von Braun & Kennedy 1994). In Guatemala the net returns to non-traditional crops per unit of family labour were about twice as high as those from maize and 60% higher than from traditional vegetable production (von Braun & Immink 1994).[7]

However, the broader community level employment effects of increases in agroprocessing are not always so clearly positive. In their study of asparagus and cotton production in Chincha, Peru, Escobar *et. al.* (2000) show that the emergence of asparagus farming based on contracting excluded small farms, thus lowering employment levels, while reinforcement of smallholder cotton production boosted employment. They judged that the net effect was ambiguous but tended to be negative.

### Contribution to food security

Post-harvest development contributes to food security in several ways. Improved storage technologies such as biological pest control or controlled atmosphere storage reduce post-harvest losses, thereby increasing the amount of food available for consumption. For example, control of the larger grain borer greatly reduced maize lost in on-farm storage among smallholders in a number of African countries, heightening their food security (Goletti & Wolff 1998). Further, post-harvest can provide income-generating opportunities for farmers in rural areas. Studies of the commercialisation of smallholder producers in a number of developing countries show that producers' nutritional status is typically not compromised; income gains generally lead to higher spending on food in absolute terms (von Braun & Kennedy 1994).

### Contribution to living standards

Developing post-harvest systems has the potential to raise living standards in urban and rural areas. In urban areas, it makes food available more efficiently and at a lower cost. In rural areas, post-harvest activities can benefit the poorest members of society in particular, through its contribution to farm and non-farm income.

The overwhelming majority of the world's poor and indigenous populations live in rural areas, and therefore stand to benefit from increased development of post-harvest and consequent increases in agricultural production to provide raw material inputs. In Africa, some 80% of those living below the poverty line are in rural areas. In Asia, where the share of the rural population is smaller and where urban concentrations are higher, about 75% of those living in absolute poverty are in rural areas. In Latin America, the comparable range is 50 to 60%. Moreover much urban poverty represents a flight from rural poverty (Mellor 1995).

Post-harvest can have a strong positive impact on agricultural production. It can increase the availability of inputs while lowering their prices, increase market demand for agricultural produce and reduce output price variability (Giovannucci 2000a; Lamb 2000). Local input provision may allow for reduced foreign imports, saving foreign exchange and potentially allowing for the development of inputs that are more closely adapted to local circumstances (UNIDO 2000a). Agroenterprises may provide services that promote agriculture – such as land preparation and levelling, irrigation system design and other types of technical consulting (Lamb 2000). Moreover, they can enable farmers to obtain inputs and services associated with economies of scale – such as costly agricultural machinery, arranging for the transport of produce, or providing processing facilities (Lamb 2000). They may also supply organisational capital needed to connect farmers with traders and customers (Lamb 2000; Escobar *et al.* 2000). Countries typically go through a structural shift from producing goods intended mainly for consumers to producing raw materials for processors.[8]

Development of post-harvest clearly has effects on surrounding communities, particularly in poorer countries. These effects, known as linkages, can occur through the impetus post-harvest gives to production, consumption and spending. A small body of work seeks to quantify the linkages between agriculture and non-agriculture in rural communities (see Reardon *et al.* 1998 for a succinct review), but there is none we are aware of that focuses on linkages stemming from post-harvest development.

Agroindustry may have indirect employment effects too, relating to forward and backward production linkages and consumption and expenditure linkages from both firms and farms (Escobar *et al.* 2000). In their case study of Peru, the impact of these linkages on local communities tended to be stronger if small firms and farms were involved, since larger ones tended to 'leapfrog' local suppliers to take advantage of markets in Lima.

## Global trends bolstering the importance of post-harvest science

In the section above, we discussed the important contribution the post-harvest sector makes to economic growth, particularly in poorer countries, but in richer ones too. We examined its impact on output, exports, employment and overall rural welfare, and documented considerable recent growth in some developing countries over the past two decades. Here we build on this argument by outlining several qualitative trends that reveal dynamic prospects for continued growth of post-harvest activity, but are also rapidly changing its structure, with particular relevance for developing countries. These trends include the contraction of agriculture, urbanisation, growing concern over gender issues and the environment, and globalisation and food safety. Each is addressed in turn. Although treated as analytically distinct, in reality they are interconnected, with the links between them either reinforcing or opposing one another. An effort is made to trace some of these links. Further, as above, we seek to distinguish between aggregate and distributional effects of the trends where appropriate; again, the recurring theme is that while aggregate effects are generally positive, they often mask some adverse distributional consequences.

## Contraction of agriculture

The first trend we consider is the contraction of the agricultural sector. As countries develop, the share of agriculture in GDP typically contracts more quickly than its share of the labour force, creating a labour surplus in rural areas (Fig. 1.3). Those displaced by shrinking agricultural employment generally migrate to urban areas in search of better opportunities, however cities are increasingly unable to absorb this influx, or policies are in place to discourage or prevent such migration. This suggests that policies are needed to boost productive rural employment. Agroprocessing is generally labour-intensive (Boeh-Ocansey 1988; FAO 1995), generates higher value-added than agriculture, and is usually located in rural areas where most of the poor live. Development of rural agroenterprise can directly contribute to strengthening the rural economy despite agricultural contraction.

It is argued that the only sustainable increases in GDP and employment in rural areas come from rural off-farm activity (Mellor 1995). Reardon *et al.* (1998) support this argument through a comprehensive review of about 100 surveys of rural households conducted all over the developing world in the 1980s and 1990s. They find that shares of non-agricultural activity (mostly agricultural or agribusiness related) to farm household income averaged roughly 42% in Africa, 40% in Latin America and 32% in Asia. Comparing results over time, they conclude that the importance of non-agricultural income to rural households has increased, and argue that this is inevitable since traditional agriculture is increasing unable to absorb a growing rural labour force, and that non-agricultural income increases with population and infrastructure density (Anderson & Leiserson 1980, cited in Reardon *et al.* 1998).

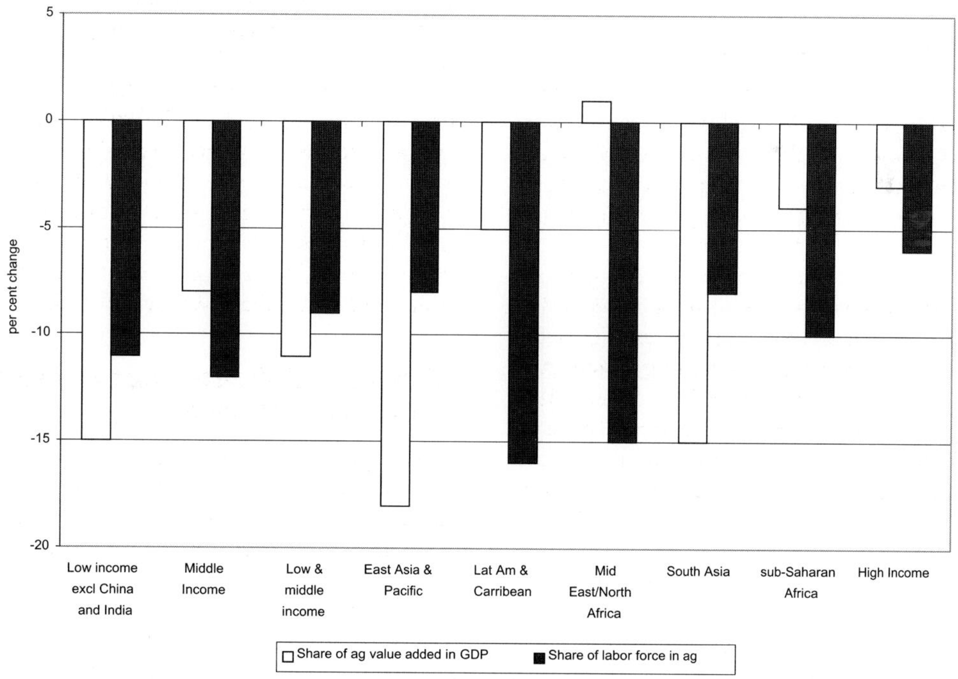


**Fig. 1.3** Share of agriculture in value-added and labour force (% change, 1970–98).

The development literature provides numerous concrete demonstrations of these points. Examples come from countries as diverse as China, where the growth of township village enterprises (TVEs) in rural areas has been very strong in recent years (Huang 2000), Ethiopia and Mali (Toulmin *et al.* 2000). Of course there is a need to ensure balance in rural–rural migration and in the regional distribution of production. Unbalanced rural–rural migration can force up local rents, overload sewage systems, and cause heath problems in overcrowded slums (Blowfield 1999), while numerous observers have commented on the unequal distribution of rural industry in China (OECD 2000).

### Demographic shifts

A second set of explanations relates to increased income growth and urbanisation in developing countries, and changing family structures and demand patterns in developed countries. It is well established that income growth in poor countries triggers a shift towards consumption of higher-value foods (Bennett's Law). For instance, Huang and Bouis (1996) found that over the course of Thailand's rapid development between 1960 and 1990, per capita rice consumption fell by half, while fish consumption doubled, meat consumption quadrupled, and fruit consumption jumped five-fold. They observed similar patterns in the development experiences of Japan and Korea.

Urbanisation is proceeding quickly, with implications for the amounts and variety of products consumed, and for the structure of post-harvest activity. The share of the urban population in developing countries grew from 20% in 1960 to more than 40% in 2000, and it is expected to reach nearly 60% by 2030 (Fig. 1.4). In high-income economies, the urban share of the population has stabilised at about 75%, indicating that this trend is likely to continue in developing regions. Urbanisation affects the post-harvest chain in two ways. First, when people live further away from where food is produced, they rely increasingly on transport, storage, processing and marketing systems for a secure and safe food supply. The complexity of the post-harvest chain is compounded by the rapid growth of some cities. Indeed, it is expected that by 2015, some 26 mega-cities will have populations of 10 million or more, mostly in developing countries (Reardon *et al.* 1998). Feeding a city of this size requires importing at least 6000 tonnes of food per day (Reardon *et al.* 1998).

Second, structural changes associated with urbanisation – namely changes in tastes, lifestyles, occupations and marketing systems – lead to a shift in consumption towards higher-value items even when income and prices are held constant. For instance, in China, Huang and Bouis (1996) found that after controlling for income and price effects, urban residents ate 6 to 9 kg more of meat, fish and dairy products per capita per year than rural residents. Because of higher incomes and changing preferences in cities, demand for more processed and highly differentiated goods rises (Jaffee & Gordon 1993; Poleman & Thomas 1994).[9] Higher value food – such as fresh and processed fruits, vegetables, meats, fish, dairy products and vegetable oils – tends to have a shorter shelf-life than traditional staples and thus requires a well-organised post-harvest chain to ensure freshness. Increased demand and reduced time for food preparation[10] highlight the need to develop healthy, affordable food products and appropriate processing systems to

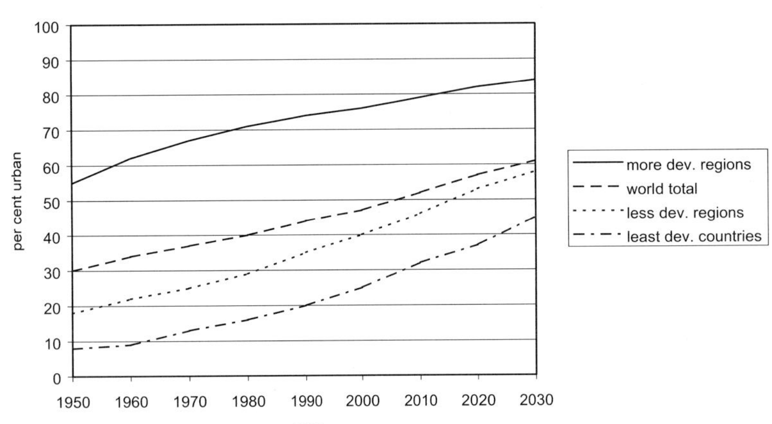


**Fig. 1.4** World urbanisation trends. Source: United Nations. World Urbanization Report (The 1996 Revision).

provide food for the rapidly growing urban population in developing countries (Kennedy & Reardon 1994). Advances in post-harvest can also reduce food prices for the urban poor as well as reduce interseasonal variation.

In developed countries too, although overall demand for food is inelastic, demand patterns are changing as a result of changes in lifestyle, health consciousness and family structures. The recent upsurge in dual income households, female-headed households and single-person households has had a significant effect (Rattray *et al.* 2000). Processors stand to capture huge gains by taking simple steps to make foods more convenient.[11] The market for some processed products – such as speciality vegetables, pre-washed salads and organic foods – has grown considerably in recent years. Further, consumers have become accustomed to buying fresh fruits and vegetables throughout the year, creating a market for suppliers in the southern hemisphere during the winter 'off season' in the North.

### Gender issues

Growing interest in redressing gender inequality has also contributed to post-harvest development (Fleischer *et al.* 1996). Female participation in agroindustry is growing rapidly, and particularly in poorer countries (see Fig. 1.6). The sector has made a huge contribution to female employment: according to UNIDO, agroprocessing accounts for more than 70% of female manufacturing labour in Latin America and Africa, and nearly 90% in South Asia (see Fig. 1.5). Moreover, women dominate many subsectors of agroindustry: over 80% of textile, clothing and leather production; 75% of food, beverage and tobacco production; and over 60% of wood production and processing (UNIDO 2000b). They are extensively involved in producing non-traditional export crops as well;[12] indeed in sub-Saharan Africa, women typically provide most of the labour that goes into producing cash crops such as tea in Tanzania (Mbilinyi 1988), green beans in Kenya (Dolan 2001), shea butter in Guinea (UNIDO 2000a) and fruit in Senegal (UNIDO 2000c).

The availability of processed food may reduce demands on female labour in preparing food, thus releasing this labour for other purposes. However, a large part of women's work in post-production is often in the informal sector such as the preparation of traditional foods, and small-scale production of fish,

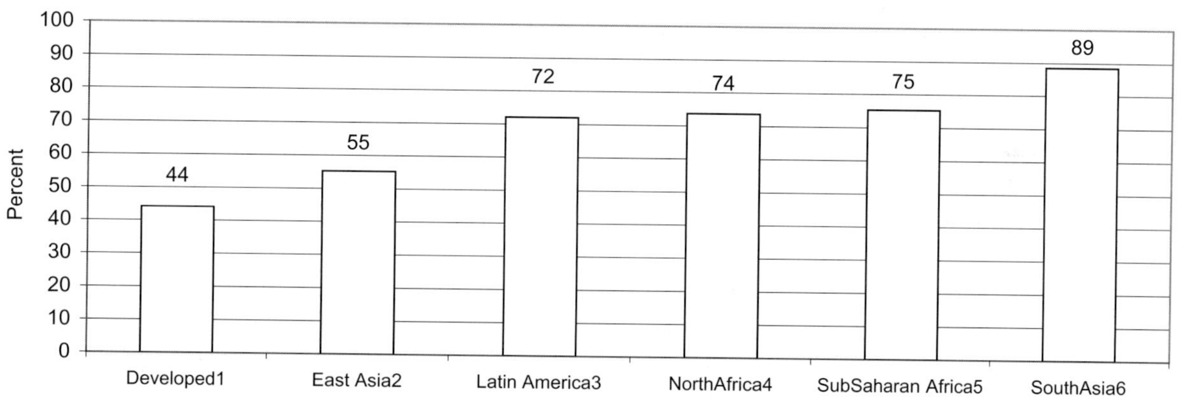


**Fig. 1.5** Female labour force in agroindustry as share of total female labour in manufacturing, 1990–95.
Notes: 1 Australia (1990); Austria (1990–94); Canada (1990–91); Denmark (1990–92); West Germany (1990); Ireland (1990–91); Italy (1991–94); New Zealand (1990–93); Portugal (1990–95); Sweden (1990); United Kingdom (1990, 1992, 1995); United States (1992–95).
2 Indonesia (1993–96); Korea (1990–95); Malaysia (1990–95); Philippines (1992–95); Taiwan (1992–96); Thailand (1990–91, 1993–94).
3 Argentina (1994); Chile (1990–95); Colombia (1991, 1993–94); El Salvador (1993–96); Panama (1990).
4 Algeria (1990–92); Egypt (1991–94); Morocco (1992–96).
5 Angola (1992–93); Ethiopia (1991–96); Kenya (1990–93); Nigeria (1991–92); Tanzania (1990–91).
6 Bangladesh (1990–92); India (1993–94); Sri Lanka (1990–93).
Source: UNIDO ISIC-3 digit dataset, 1999.

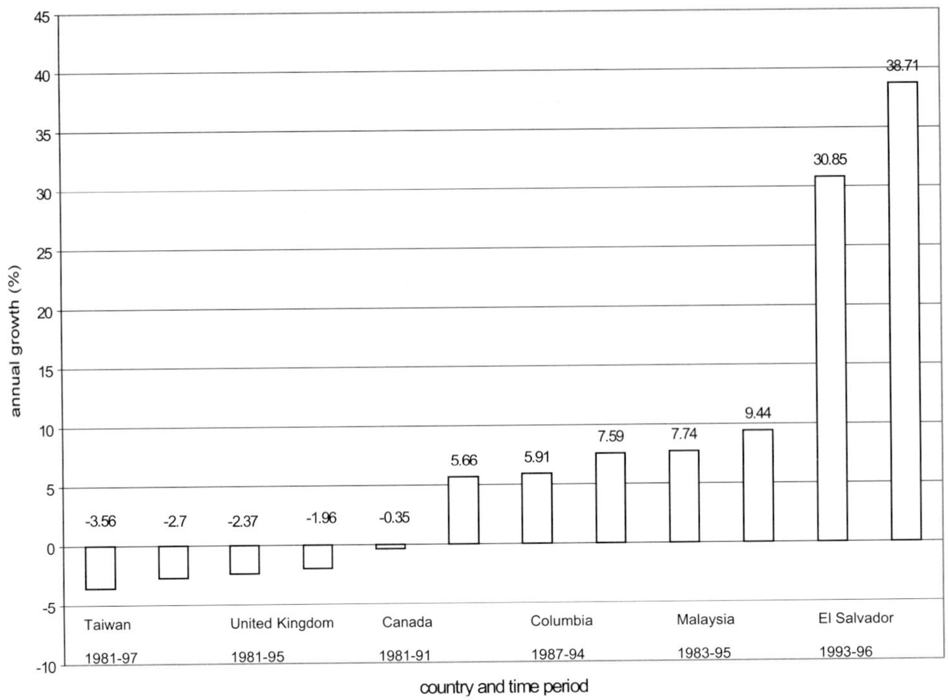


**Fig. 1.6** Growth of female employment in agroindustry.

palm oil, cassava and dairy products (Petritisch 1985). The development of post-production systems should explicitly take into account the role of women in these activities, and allow them to move into the formal sector (Jaffee & Morton 1995).

In view of the fact that agroprocessing is often seasonal, that wages are typically lower than those in the rest of the manufacturing sector, and that case studies suggest this labour tends to be insecure and sometimes dangerous, its effects on women and their traditional roles should be explored further. Women receive less pay than men for agroprocessing. While figures are only available for a few countries, they show that women's earnings in the subsector range from 55% (South Korea) to 92% (Sweden) of men's wages, and the disparity seems to be larger within agroprocessing than for manufacturing on the whole (Fig. 1.7). Among farming households, unequal power relations often result in the expropriation of cash crop income by men, although women are typically providing most of the labour. In Meru, Kenya, women provide 72% of the labour for French bean production and receive 38% of the income. In Papua New Guinea, returns to labour for coffee production accruing to women are one-third those received by men (Overfield 1998). The literature contains numerous similar examples (see, for instance, von Braun & Kennedy 1994).

### Environmental sustainability

Concern over environmental issues and sustainable development is growing. An effective post-harvest system minimises unnecessary production, thus saving on scarce land and water resources, and providing alternatives to the heavy application of chemicals that could be harmful to human health and the environment. Examples include the replacement of blanket spraying of pesticides with seed dressing (UNIDO 2000a), and the evolution of integrated pest management (IPM) techniques.

Estimates of post-harvest loss from disease, inadequate storage, or suboptimal handling vary greatly, depending on the crop, country, climate and estimation method (Giovannucci 2000a). For some staples, losses can be less than 10% (most US grains are under 3%), while certain horticultural crops can suffer losses higher than 50% (Giovannucci 2000a). Okezie (1998) finds that post-harvest loss typically ranges from 15% to 50% for horticultural products, and from 10% to 20% for grains

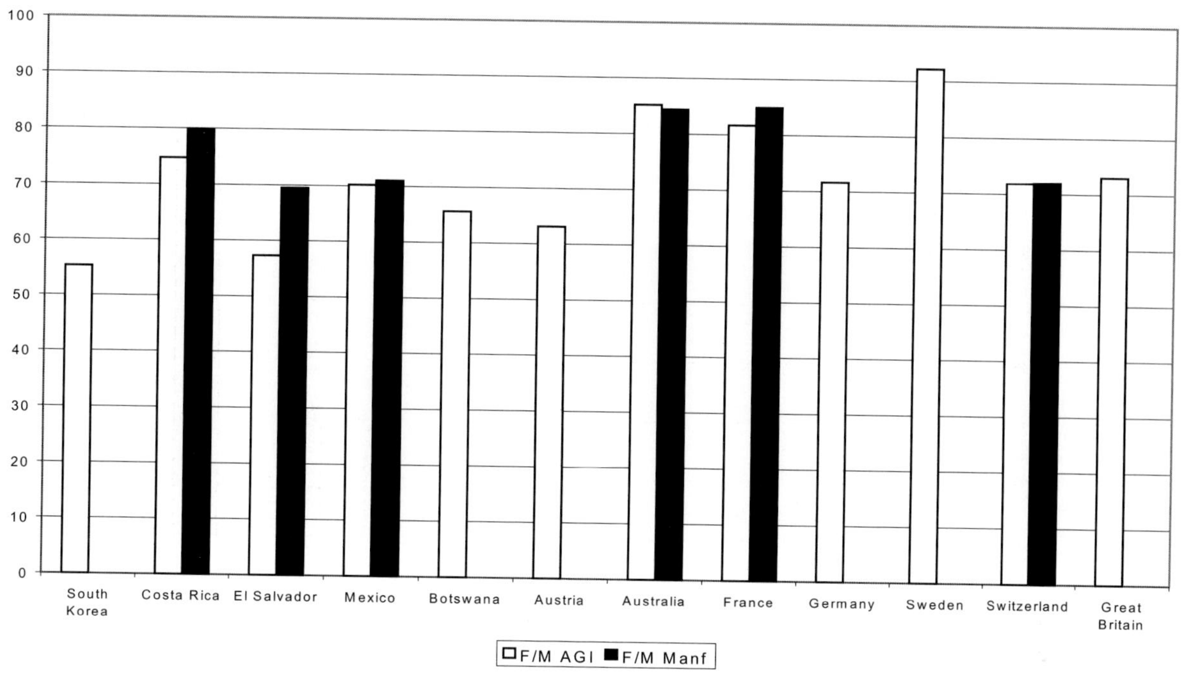


**Fig. 1.7** Ratio of female/male wages in food processing and total manufacturing, 1998. Note: Based on ISIC-3 Revision, category 01, which denotes food and beverage manufacturing. Source: International Labour Organization Laborsta database, 2000.

and oil seeds (cited in Hofman 2000). In Africa, post-harvest losses are estimated at 30% to 40% (UNIDO 2000c). These figures are probably underestimates because they do not account for income loss resulting from quality deterioration when wastage is partial (Hofman 2000). In contrast, Greeley (1991) finds that loss estimates from traditional post-harvest systems are generally much lower than experts predict for the simple reason that poor farmers cannot afford to waste food (cited in Goletti & Wolff 1998). In addition to reducing loss, post-harvest advances can also improve production technologies or generate useful by-products such as paper pulp, fuel, fertiliser or animal feed, reducing pressure on natural sources (FAO 1997; Reardon & Barrett 2000).

In industrialised countries, consumers are demanding reductions in pesticide use, more organic food products and biodegradable packaging. Prompted by several food safety scares, the market for organic foods is growing remarkably, from a low base (Table 1.5). Growth in consumption tripled in Europe from 1990 to 1997 (AgraEurope 2000), and has grown at about 25% each year since 1991 in the United States, despite price premiums of between 10% and 100% for organic products (USDA, various). It is predicted that US consumption will quadruple by 2010 (ABC News 2000), while in European countries, consumption will grow at annual rates of between 20 and 40% (AgraEurope 2000). The European Union has made a concerted effort in recent years to expand its organic production; one result is that land area under organic production more than tripled between 1993 and 1998 (Foster & Lampkin 2000). On current trends, almost 20–30% of Western Europe's farmland could meet organic standards by 2010 (*The Economist* 23 May 2000). In developing countries too, production and consumption are growing. In Argentina and Brazil, production increased 25% and 20%, respectively, from 1995 to 2000 (USDA), most of which was exported; Argentina increased its organic hectarage from 5500 ha in 1992 to 3 000 000 ha in 2000, a 500-fold increase (Willer & Youssefi 2001). In several richer Asian countries, notably Japan, Singapore, Hong Kong and Taiwan, domestic demand is growing sharply.[13]

However, while post-harvest development stands to mitigate some environmental concerns, it also raises new ones, particularly in developing countries where agroindustrial growth has been rapid and environmental rules are scarce. A major problem is the run-off from

**Table 1.5** International demand for organic products. Source: Willer & Youssefi 2001.

| Country | % of total food sales (1999) | Yearly expected growth (%) | Forecast retail sales (2000) ($m) |
|---|---|---|---|
| Germany | 1.2 | 10 | 2 500 |
| Italy | 0.6 | 20 | 1 100 |
| France | 0.5 | 20–25 | 1 250 |
| Great Britain | 0.4 | 25–30 | 900 |
| Switzerland | 2 | 20–30 | 700 |
| Netherlands | 1 | 15–20 | 600 |
| Denmark | 2.5 | 30–40 | 600 |
| Austria | 2 | 15 | 400 |
| Sweden | 0.6 | 30–40 | 400 |
| Other Europe | – | – | 500 |
| USA | 1.25 | 15–20 | 8 000 |
| Japan | – | – | 2 500 |
| Australia | – | – | 170 |
| New Zealand | – | – | 58.8 |
| Argentina | – | – | 20.0 |
| China | – | – | 12.0 |
| Taiwan | – | – | 10.0 |
| Philippines | – | – | 6.2 |
| Total | | | 19 727 |

agricultural enterprises contaminating water supplies, as with sugar refining in China, fishmeal production in Chile, starch production in Vietnam, and tanneries and flower farming in Kenya. For instance, the explosion of flower farming around Kenya's Lake Naivasha, the centre of the country's flower growing industry, has caused water levels to sink quickly as producers siphon the water for irrigation, and has led to fertiliser and pesticide contamination, with implications for biodiversity and human health (BBC 2001). Intensification of farming in many areas – to supply agroprocessors – has increased use of chemicals and water with adverse consequences for soil and water resources, as well as expansion of cultivated area into fragile forest, rangeland and watersheds (see Pingali 2000 and Barbier 2000, both cited in Reardon & Barrett 2000).

These concerns present opportunities and challenges for post-harvest research, to develop alternative technologies for storage pest and disease control, waste treatment from processing plants, aquaculture, livestock feedlotting and improved safety (Austin 1995; Arnold 1996; Johnson 1998).

## Globalisation

Globalisation is very quickly restructuring the production, processing and distribution of agrifood products, with dramatic implications for developing countries, as they become increasingly integrated with global supply chains in the North. Here we consider two aspects of this transformation: trade liberalisation and the emergence of global value chains.

### *Liberalisation of international trade*

The last decade has witnessed a dramatic move towards a liberalised international trading system, along with an increasing economic orientation towards export markets as a source of growth. This orientation is attributable to increased surpluses arising from the introduction of green revolution technology, to policy changes wrought by structural adjustment programmes and to a proliferation of global and regional trade agreements.

As described above, processed agricultural products make up a significant share of exports, especially in developing countries. A recent study by Athukorala and Sen (1998) found that an open trade policy is crucial to the success of export growth in developing countries in processed foods (cited in Díaz-Bonilla & Reca 2000). Participation in export markets, however, also requires relatively sophisticated marketing, information and transportation networks. Successful competition requires quality control and product standardisation, and improved storage and trade facilities (Jaffee & Gordon 1993; Johnson 1998). Developing industries may require support in exploring new methods, institutions and technologies to make them more competitive in international markets. At the same time, policymakers should account for possible adverse effects of opening markets on vulnerable populations, such as rising prices for basic staples (Islam & Valdés 1990; Ingco 1997).

### *Emergence of global value chains*

Possibly the most important qualitative trend goes beyond increased trade to cover the way in which production processes are being restructured on a global level to take advantage of reduced trade barriers, lower costs and seasonal differences. This phenomenon is inextricably linked with the increase in trade and foreign direct investment (FDI), the drive to capture new economies of scale and scope, and a host of food safety issues. It is evident in the multinationalisation, consolidation and increasing concentration of market share among firms, particularly in developed countries, but in developing countries too, and in the changing role suppliers and processors in developing countries must play. This restructuring has unfolded at an unprecedented rate over the past decade and carries dramatic implications for post-harvest science.

Total net flows of FDI to developing countries grew from \$24.5 bn in 1990 to about \$105 bn in 1996. Few data exist on the specific direction of these flows, but they suggest that in 1993, for instance, about 10% of US FDI and 6% of UK FDI went into the food, beverage and tobacco sectors. Numerous observers have recognised that the value of FDI has exceeded that of trade, and is growing more quickly. Indeed, multinational activity has been described as a final leap in the process of internationalisation, a process which begins in the export phase (FAO 1997). Foreign production by subsidiaries of multinational food companies is 4.3 times the value of processed food exports in the United States and the figure is slightly higher when taking into account the United Kingdom, Germany, France, Italy and Holland (Traill 2000). Many countries' trade/GDP share did not change much over the course of the twentieth century (Kaplinsky 2000). However, he notes the key difference that in the earlier period, trade was 'largely in arm's length relationships, with final products being largely manufactured in a particular country and then exported', whereas by the end of the century, in contrast, trade was 'increasingly in sub-components and services and was considerably more complex'.

Industrial concentration has occurred along all stages of the production and post-harvest chain, and for numerous products.[14] Data on the market shares of major companies is difficult to obtain. Globally it is estimated that the top ten multinational groups in the agrifood sector accounted for 32% of the market in 1990, with a trend towards further concentration (FAO 1997). In the US, few data are available, and their availability has worsened since the mid-1980s.[15] However, the figures that exist reveal both substantial four-firm concentration and its increase over time (Table 1.6). Four firms control over 40% of the processing of the major commodities produced in the mid-west, and a few firms reappear as the top processors for several commodities (Heffernan 1999).[16] Obtaining concentration ratios on global market shares is even more difficult.

Case studies of specific industries help to fill this gap. Seed companies have become tremendously concentrated in recent years as a rash of mergers and acquisition activity accompanied the rush to develop genetically engineered seeds. Processors have become more concentrated too. In the EU, Viaene and Gellynck (1995) report the increasing consolidation of the food, drink and tobacco industry since the early 1980s, while Giles (1999) foresees its further concentration so that as few as 20 major groups will dominate by about 2010. In the US, Koontz *et al.* (1993) find a large increase in concentration of meat packing; this is the case for lamb (Brester & Musick 1995), beef (Azzam 1998) and pork (Hayenga 1998, 2000).

Similar trends characterise food retailing. By 1996, the top five supermarket chains in the EU had a total market share of more than 50% in all but three countries (Baas *et al.* 1998, cited in Dolan & Humphrey 2000). In the UK, Dolan *et al.* (1999) describe the dramatic con-

**Table 1.6** US four-firm concentration ratios, earliest and latest years available. Source: Heffernan 1999.

| Commodity | Year: CR-4* | Year: CR-4 |
|---|---|---|
| Beef packing | 1990: 72% | 1998: 79% |
| Pork packing | 1987: 37% | 1998: 57% |
| Broilers | 1986: 35% | 1998: 49% |
| Turkeys | 1988: 31% | 1996: 40% |
| Flour milling | 1982: 40% | 1997: 62% |
| Wet corn milling | 1977: 63% | 1987: 74% |
| Soybean crushing | 1977: 54 % | 1997: 80% |

* Market share of the four largest firms.
Note: Hayes (2000) writes that since these figures come from trade publications, they may be slightly dated and probably underestimate current shares.

centration of food retailing over three decades so that four retailers accounted for nearly 75% of food sales in 1998. In the US, supermarket consolidation began more slowly but is proceeding apace (*The Economist* 23 May 2000); by 2000, six supermarket chains controlled 53% of supermarket sales (Cotterill 2001).

There is also evidence of this trend in developing countries. Between 1991 and 2000, 12% of Brazil's food and beverage firms (269 out of 2300) were either taken over or merged (Agrafood Latin America 2001). A remarkable 143 of these mergers and acquisitions took place in the three years following the 1994 Real Plan. In both Kenya and Zimbabwe, five exporters have come to control over 75% of all fresh vegetable exports (Dolan *et al.* 1999). In China, most agroenterprises are small-scale but their average scale grew 1.84 times between 1989 and 1996, and average output per firm climbed from Rmb3.72m to Rmb10.45m (Du 2000). Industrial concentration from the 1970s onward has been reported in South Africa (Kaplinsky & Manning 1998) and in Indonesia (APO 1992).

Multinationals have structured their global supply chains in two main ways: (1) by vertically integrating production and (2) by vertically co-ordinating production through outsourcing various stages of the production process. Vertical integration allows companies far greater control over various stages of the production and post-harvest process. Global sourcing, in contrast, permits greater flexibility but also requires that firms be able to procure goods from suppliers as needed, be it through spot markets or contracting with independent producers (FAO 1997; Schejtman 1998).

Vertical specialisation (defined as the use of imported inputs in producing goods for export) may account for up to 30% of the world's exports, and grew as much as 40% over the last 25 years (Hummels *et al.* 1999). They argue that since trade barriers (tariffs and transport costs) are incurred repeatedly as goods in process cross multiple borders, even small reductions in barriers can lead to extensive vertical specialisation, large growth in trade and large gains from trade. While they find that most specialisation has occurred in the chemical and machinery industry, their findings also have clear implications for recent developments in agroprocessing.

A practical example clearly illustrates this process.[17] The multinational company Cargill is the major producer of phosphate fertiliser in Florida. It ships this fertiliser elsewhere in the US and to Argentina to grow soybeans. The beans are processed into meal and oil, at which point the meal is shipped to Thailand to feed chickens at a lower cost than if they were fed only from Thai sources. Cargill then buys the chickens from Thai farmers and takes them to its facility outside Bangkok for processing, cooking and packaging. Finally, it ships the processed meat to Europe and Japan to be sold to consumers there.

These shifts carry profound implications for the post-harvest chain, and raise both new possibilities for and new challenges to broad-based and dynamic development. The closer integration of various stages of the post-harvest chain suggests new possibilities for more efficient co-ordination and growth. At the same time, the concentration of agroindustry and their increased presence in developing countries is vastly increasing the vulnerability of small farmers and firms who cannot meet the stringent standards multinational processors and retailers demand (T. Reardon, pers. comm.). For instance, while in 1992, smallholders provided about 75% of fruits and vegetables to exporters in Kenya, by 1998 their share had fallen to 18% (Dolan & Humphrey 2000). Similar changes are taking place in other developing areas as well.[18] Though the net economic effect of multinational activity in developing countries is ambiguous and context-specific (Reardon & Barrett 2000), the clear tendency towards exclusion of small-scale farms and firms is an important challenge for post-harvest. We return to these challenges below.

### Food quality and safety

The growth in urbanisation and changes connected with globalisation and new technologies have called attention to and amplified problems with food safety and quality. While the former is vital for public health, both have a significant impact on economic growth and trade. Assurance of food safety is a prerequisite for trade, while evidence of food quality is increasingly essential in reaping higher margins from the sale of fresh and processed foods.

Food is the major source of exposure to biological and chemical pathogens. While the true incidence of food-borne disease and its economic costs are difficult to quantify with any certainty,[19] surveys suggest that up to 30% of industrial populations become infected annually, and that up to 20 per million die (WHO 2001). In the United States, the most recent estimates suggest that food-borne illnesses cause approximately 76 million cases of illness, resulting in 325 000 hospitalisations and 5000 deaths each year (Mead *et al.* 1999).[20] In developing countries, it is estimated that contaminated food causes an estimated 70% of global episodes of diarrhoea (resulting in 2.2 million deaths per year, 1.8 million of which are children), and is key to the epidemiology of cholera (WHO 2001). It is estimated that 2% to 3% of cases of food-borne disease result in long-term ill-health.

Relatively few resources are devoted to the detection and prevention of food-borne illness in sub-Saharan Africa and Asia (excepting Japan) and so data are scarce and few episodes of food-based illness are documented. Tests of food quality, however, can reveal severe problems.[21] In Latin America, monitoring systems are more developed. An outbreak of cholera, the first to afflict the Western Hemisphere in nearly a century, infected over 1 million people and claimed nearly 10 000 lives between 1991 and 1994 (WHO 1996).

In addition to illness, unsafe food exacts substantial economic costs in terms of health care, lost work effort and international trade and tourism. Estimates of the annual economic cost of disease in the US (including hospital care and lost productivity) are as high as $110 bn (Buzby & Roberts 1997, cited in WHO 2001). The cholera epidemic in Peru in 1991 cost over $700m in fish and fishery product exports, and $70m from closed restaurants and a decline in tourism. The 1999 chemical contamination of animal feed cost Belgian farmers $600m in sales (*The Economist* 23 May 2000). The UK lost more than $6 bn in culled herds and lost trade through its bovine spongiform encephalopathy (BSE) crisis and trade in beef products was all but decimated; in 2001, exports were still down 99% from 1995 (Brundtland 2001).

Frequent outbreaks of food-borne illness in the last two decades have focused renewed attention on food safety. Recent scares include the 1992 outbreak of *Escherichia coli* food poisoning in the United States and the 1996 outbreak in Japan (the largest ever recorded for this pathogen); in Europe, BSE or 'mad cow' disease, with its connection to variant Creutzfeldt-Jakob disease in humans; the huge increase in the incidence of salmonella in many countries (a 1994 outbreak in the US affected an estimated 224 000 persons); and the rapid growth of food-borne trematodes in southeast Asia, linked to under-processing or insanitary processing of aquaculture.[22]

Numerous episodes of chemical poisoning have occurred, too. Recent episodes include the discovery of the pesticide Alar in apples, toxins in Belgian poultry in 1999, toxic mustard seed oil in India in 1998 and toxic cooking oil in Spain in 1999 (this last incident left 800 dead and 20 000 ill). In developing countries, 'where an overwhelming majority of acute pesticide poisoning occurs', the few available data indicate 'significant exposure of the general population to pesticide residues in food' (WHO 2001). Chemical contaminants can be even more insidious than biological pathogens since they may be chronic and build up over time.

Data problems notwithstanding, the evidence suggests food-borne disease has grown since the early 1980s,[23] owing to the emergence of new microbiological pathogens, greater awareness of the role of food in transmitting disease, and social changes relating to globalisation.[24] New pathogens include *Escherichia coli* O157: H7 (identified in 1982 in ground meat), *Vibrio vulnificus* (in shellfish) and *Cyclospora cayetanesis* (found in Guatemalan raspberries in 1996) (Giovannucci & Satin 2000). Scientists recently identified the role of food in transmitting listeria and the increased antibiotic resistance of pathogens.

Food safety problems also have social causes, stemming from changes in production and marketing channels for food, increased movement of people and pathogens across national borders, rapid urbanisation and changes in food handling (Unnevehr & Hirschhorn 2000).

> The WHO writes:
> 'Globalisation of the food trade presents a transnational challenge to food safety control agencies in that food can become contaminated in one country and cause outbreaks of food-borne illness in another. The intensification of food production and the consolidation of the food industries present opportunities for food-borne pathogens to infect large numbers of consumers… The strain on limited resources (in developing countries) will be further compounded by demographic changes, such as expanding urbanization, the increased dependence on stored foods, and insufficient access to safe water and essential facilities for safe food preparation.'

Increased movement of food and people can be dangerous not only because it allows new pathogen variants to spread but also it lets familiar pathogens into new areas that are less 'immune':

> 'Ordinarily, pathogenic organisms come to some type of ecological balance with other forms of life in their native environment. However, when these organisms are transferred to a new environment, they will often demonstrate much greater virulence simply because natural immunity or appropriate

social or medical management practices have not been established'

*Giovannucci & Satin (2000)*

Demographic changes have also led to an expansion in the number of people who are highly susceptible to disease, owing to ageing, malnutrition, HIV infection or other conditions, while lifestyle changes entail more dining out in establishments where food preparation may be unhygienic (WHO 1996).

Lastly, advances in post-harvest technology have been implicated. According to WHO (2001), 'microorganisms have the ability to adapt and change, and hence changing modes of food production, preservation and packaging have resulted in altered food safety hazards'. For instance, the re-emergence of organisms such as *Listeria monocytogenes* and to a lesser extent *Clostridium botulinum* has been traced to changes in the way high-risk foods are packaged and purchased (WHO 2001).

The increasing importance of food safety and quality has led to a dramatic reshaping of quality control systems in developed and developing countries in recent years, with a renewed focus on the links between them. Most notably, attentions have focused on grades and standards (vital in conducting long-distance trade[25]), the emergence of international standards (CODEX Alimentarius, which became the international standard for food safety in 1993), the shift from a focus on outcome to process (e.g. through Hazard Analysis Critical Control Point (HACCP)) and towards a holistic farm-to-table perspective. These developments pose a particular barrier to developing countries, many of which lack the resources to develop and enforce adequate safety and quality systems. Referring back to Fig. 1.2, this may play a role in explaining the very small share of developing countries in exports of meat, and fresh fruits and vegetables.

The main implication for post-harvest is that while increased development of post-harvest systems has led to some food safety scares, renewed attention to post-harvest activity is crucial to overcoming them. In response to concerns over food safety and amid continued global restructuring, post-harvesting technology and policy is needed to strengthen quality control systems and their management, as well as to facilitate their adoption by poor countries and small suppliers.

### Technological advance

Scientific progress has provided vital impetus to the growth of post-harvest. The initial focus of much research was on reducing post-harvest loss of staple commodities through improved drying and storage technologies. This emerged as a serious problem following the Green Revolution, when traditional storage systems proved ill-equipped to deal with vastly increased production, particularly during the rainy season (Goletti & Wolff 1998). Research also led to advances in processing techniques for food and feed, and to improvements in safety and quality (Goletti & Wolff 1998). Examples include the mechanisation of drying and milling technologies, reduction of toxins, increase in the micronutrient content of staples, biological pest control and tropical fruit preservation (Goletti & Samman 1999a).

New industrial applications evolved, such as new forms of heat processing, low energy production, pasteurisation, semi-finished production techniques (filtration, extraction, centrifugation), chilling and freezing (UNIDO 1986). More recently, computers have been used to develop sophisticated monitoring systems; for instance, horticultural scientists have developed computer sensors that continually measure plants' 'vital signs' such as tissue temperature and swelling, and regulate irrigation and atmospheric gas concentrations accordingly (*The Economist* 23 May 2000). Packaging technologies – like vacuum packing of milk – have also improved (Reardon & Barrett 2000). Such developments have served to produce foods with better taste, appearance, shelf-life and pest resistance.

In turn, marketing opportunities have expanded. For instance, better grain storage techniques and post-harvest pest management allow developing countries with humid tropical climates to compete in world grain markets with virtually insect-free exports from temperate zones (Goletti & Wolff 1998). Improved processing techniques for palm oil allow it to compete with vegetable oils for a wide variety of uses benefiting South-east Asian producers (Wolff 1998), while the development of controlled atmosphere storage, improved transport and chemical fungicides facilitates marketing of tropical fruits over long distances (OECD 1996). Marketing technologies like bar coding allow for product differentiation and tracing, increasing value-added.

Recent developments in biotechnology are fostering more concentrated seed production, vertical integration of production and processing, and the need for segregated handling systems to preserve the identity of distinct products. Chemical and seed companies have combined forces to develop transgenic seed varieties engineered to provide certain 'input' traits – such as resistance to insects (*Bt* toxin) and weeds – or 'output' traits – mainly nutritional enhancements as in vitamin A rice. Research suggests that such modifications may increase the efficiency of production and processing (especially storage), and could potentially boost the nutritional quality and safety of foods (WHO 2001). For instance, it is cheaper to genetically modify plants for disease and pest resistance than to spray them during production (Hayes 2000). At the same time, doubts persist over ethics, safety and environmental effects, as reflected in consumer resistance in Europe and Japan particularly, and more research is needed to assuage critics. Nonetheless, such technologies are transforming bulk commodities into 'tailor-made' products that target distinct groups of consumers. The aversion of many consumers to genetically modified (GM) foods creates the need for handling systems that allow the identity of the commodity to be maintained from farmer to customer (Hayes 2000).[26]

Despite their potential, current GM technologies remain clumsy and ill-adapted to deal with a host of agricultural problems, particularly those facing developing countries:

> 'Sticking a *Bt* gene in rice and expecting it to solve the problem of insect predation is a "silver bullet" approach to a highly complex ecological problem. More useful are plants bred to be better at extracting nutrients from the ground or tolerating salinity, making more effective use of existing land and bringing new land into cultivation.'
>
> *The Economist 23 May 2000*

In addition, better tests are needed to detect the presence of GM material in food products – currently a time-consuming process – and to gauge the quantities present.

The internet is rapidly emerging as a means of disseminating market information – such as news, agronomic advice and risk management tools – and facilitating market exchange. It has been estimated that by 2003, up to 10% of the world's $4 trillion market in agricultural goods with be traded on-line (*The Economist* 23 May 2000). The internet has transformed goods marketing in many developed countries, particularly where long distances might otherwise impede trade, such as in Canada and the mid-western United States. For instance, in Canada, one of the main systems used to market hogs is an electronic auction that operates as a price discovery system with prices bid for hog lots (Rattray *et al.* 2000). In Iowa, an internet company arranges for producers in particular areas to deliver specified quantities of identity-preserved grain to particular elevators on pre-arranged dates. All business transactions, including the signing of contracts, are conducted on-line (Hayes 2000). The upsurge of electronic communications has also led to the emergence of 'virtual co-ordination' systems whereby purchasers can monitor farmers' production and management practices in return for agreeing to purchase output, usually with a minimum revenue guarantee (Hayes 2000). Further, the internet allows retailers to track consumer preferences precisely, and become more responsive to them.

Finally, the emergence of precision agriculture in recent years, based on a combination of global positioning systems (GPS), yield monitors and information on input use, allows farmers to obtain very precise information about land productivity. In practice it does not always work so well. A key implication for post-harvest is that precision agriculture allows decisions regarding input use and cropping to be made away from the farm, facilitating processor involvement (Heffernan 1999).

## Future challenges for post-harvest science

So far, we have discussed the importance of post-harvest and its high future growth potential. At the same time, we pointed to a number of trends that are dramatically affecting its structure – with implications for post-harvest systems' development and for the various actors involved. Here we focus on what we see as the main challenges to the development of dynamic and broad-based post-harvest systems: implementation of a systems approach, clear delineation of public and private goods and careful consideration of the scale issues involved, particularly for small-scale farmers and firms in poor countries.

### Operationalisation of systems approaches and multi-disciplinarity

Traditional post-harvest science focused on reducing post-harvest loss by devising better storage and processing technologies, and on improving the transit of goods from producers to consumers. Researchers focused their efforts on making technological improvements to discrete components of the supply chain in isolation from one another. Over the last decade, spurred by the rise of global supply chains, and their renewed emphases on efficiency and product safety, there has been a 'paradigmatic shift'[27] in the way in which post-harvest is conceptualised from a series of individual components to an integrated chain linking the producer and consumer.[28] Adoption of this approach and the opportunities it presents for an interdisciplinary perspective can lead to greater systemic efficiency, food safety and quality and a clearer picture of various actors along the chain and the benefits they are deriving from their participation.

The systems approach reveals new ways of adding value to post-harvest by shifting the focus of attention from point to systemic efficiency (Kaplinsky 2000), thereby allowing for better co-ordination of the system as a whole. Studying specific supply chains allows the identification of bottlenecks and constraints to efficiency while heightening the impact of research in a specific area (Young 1991; GTZ 1998). In reference to hog production in Canada, Rattray *et. al.* (2000) write that the pressures of global competition led to massive structural change in the industry, involving significant rationalisation of the sector. This involved:

> 'increased interest in capturing system efficiencies associated with improved coordination of product flow and information along the chain. This attention to system efficiencies represents a significant change from traditional efforts to improve efficiencies at each horizontal stage of the supply chain in relative isolation from vertical chain considerations.'

A practical example illustrates this point. Tesco, one of Europe's major retailers, was able to make considerable efficiency gains by restructuring its own operations – reducing its inventories, and ensuring a process of just-in-time delivery from its warehouses and those of its main suppliers to its retail stores. However, it increasingly came to realise that these activities accounted for only a slight share of its products' total value-added and that 'without "governing" its chain to achieve broader levels of systemic integration, little more could be done to achieve competitive advantage'. The company then undertook to streamline the value chains of particular products. In so doing, it discovered that while under optimal conditions it would take a minimum of 3 hours to produce and cool a tin of cola, it was actually taking 319 hours, leading to a significant cost in working capital throughout the chain. Accordingly, Tesco came to realise that it should strive to improve efficiency throughout its value chains since 'the activities which it (was) directly responsible for in its internal operations (accounted) for only a small share of product costs' (Womack & Jones 1996, cited in Kaplinsky 2000).

An integrated approach to production and post-harvest chains is also a prerequisite for implementing effective food safety and quality control systems. The 'first wave' of food safety research, in the 1950s, focused on abattoirs, factories and supermarkets as sources of contamination, neglecting the role of farms and farm-level processes. More recently, attention has widened to include the farm level and the entire farm to table process. The privatisation of grades and standards propelled this development (T. Reardon, pers. comm.), as did legislation in some countries making downstream firms legally responsible for the actions of upstream suppliers (Loader & Hobbs 1999). Producers and processors need to be able to guarantee that the products they sell are safe, while firms increasingly must be able to trace the produce they source along the agro-food chain, to maintain its quality and identity (Rattray *et al.* 2000).

Renewed attention to food safety has propelled a shift from an outcome (or inspection-based) to a process orientation (Giovannucci & Reardon 2000). The HACCP system, which is emblematic of this change, is predicated on ensuring quality throughout the supply chain as a particular product moves from farm to retailer (USFDA 1999). It operates by identifying critical control points where hazards might occur then putting in place processes to rectify potential hazards. The theory behind this approach is that tracing food throughout the system rather than through a single component is a more preventative and systematic way of avoiding contaminants.

A practical example, focusing on Indian groundnut production, gives an illustration of the systemic approach and the shift to HACCP. Aflatoxin contamination is a common problem in nut production with serious implications for food safety. A group of smallholders producing groundnuts in Gujarat, India faced especially severe problems in this regard. In domestic markets, the problem was not as severe, because most nuts were sold to oil millers who were not very concerned about aflatoxins. However, the toxicity posed a severe barrier to smallholder attempts to market their produce internationally; about three quarters of consignments tested were found to be poisonous.

Researchers sought to overcome this problem by conducting a thorough assessment of the Indian groundnut supply chain to determine at which points contaminants were introduced. On farms, they identified the method of seed treatment before planting, the need for annual crop rotation and proper soil treatment, and techniques for drying after harvest and on-farm storage. At sheller mills, they found unacceptably high moisture levels and unhygienic practices. The producers introduced new methods of dealing with production and storage, and set about persuading millers to adopt new techniques. In addition to technological changes, new management practices were introduced to review and monitor the quality control system set in place (Dietz 2000).

The systems approach does not solely have technical implications, but rather allows for a broader focus that harnesses the insights of diverse disciplines for the design of complementary institutions, policies and management systems. It thus presents new challenges for researchers and requires much interdisciplinary cooperation (Ferris *et al.* 1997).

Management theory can be used to strengthen co-ordination along the chain (Cook & Chaddad 2000) and to shape the culture within which systems are operating (Wong *et al.* 2000). Hofman (2000) writes of the need for more integrated approaches to production and marketing:

> 'More integrated approaches to production and marketing will be developed. No longer can the artificial "separation" of production, post-harvest processes and marketing be sustained, and approaches to address issues will need to consider the potential for solutions to exist in other parts of the system. Management of quality in the future will have to be more integrated to achieve the increasing consumer demands.'

Meanwhile, the social sciences are concerned with efficiency, equity and the broader economic and social impact of post-harvest systems' development. Research examines the broader economic and political context within which supply chains are operating to suggest policies and institutions that might make them function more successfully. Examples could include supportive policies directed at building infrastructure, providing credit, opening trade, or more general efforts to strengthen the macroeconomic climate. Further, social scientists consider the broader picture of who is included and excluded from these chains, power relations between actors, and the relative distribution of gains. They can use this information to devise policies and institutions to ensure that post-harvest systems develop in a broad-based and sustainable manner, to the furthest extent possible, and design policies to promote the involvement of disadvantaged groups and areas.

Kaplinsky (2000) observes that while economists determine the basis of *comparative advantage*, other disciplines identify the determinants of *competitive advantage* (the factors that explain why some firms are able to appropriate these economic rents). He writes:

> 'Because value chain enquiry spans different economic branches and sectors, effective analysis requires the participation of different disciplines. This is most clearly the case in relation to the focus on agricultural and manufacturing production systems, but the focus on dynamics of rent also requires inputs from management studies and engineering. Moreover, since power is a key component of governance, and trust is critical to enhanced inter-firm cooperation and new forms of work organisation, there is a simultaneous need to draw on the insights of political science and sociology.'

The challenge for post-harvest science is to integrate various disciplines to ensure their total contribution is greater than the contribution of each single discipline in understanding the post-harvest system.

### Public versus private goods in post-harvest agriculture

Public goods refer to those made available by governments or by international organisations while private goods are those provided by private companies. Conventional theory holds that government should supply public goods only when private companies lack an incentive to do so because they cannot sufficiently appropriate the gains of their investment. Distinguishing between public and private goods, and forming hybrid public–private organisations where appropriate, poses another challenge for post-harvest science.

Considering the wider social implications of post-harvest development, a strong case can be made for government intervention to provide needed support services that fall beyond the purview of any individual company, such as providing physical and communications infrastructure, suitable institutions, credit, education and training. The targeting of such assistance towards small and medium enterprises and small firms, and disadvantaged areas, can be decisive in fostering their involvement in lucrative markets.

With regard to grades and standards, it is argued that the state has a particularly important role to play. At minimum it must establish a baseline set of standards to ensure consumer safety, communicate these standards to suppliers, and enforce compliance (Giovannucci & Satin 2000). Moreover, it is argued that government has a significant role to play in co-ordinating with international organisations to harmonise standards and regulations, in surveying food-borne diseases, and in designing programmes for consumer education, training, extension and research (World Bank 1999, cited in Giovannucci

& Satin 2000). The private sector then is typically responsible for food quality management at the level of the firm or trade association, in conformity with government requirements (Henson & Caswell 1999).

In poor countries, there is an especially compelling case for government assistance to firms to help them survive in competitive markets, for instance by supporting compliance with grades and standards, providing information regarding potential markets, and helping with transport. In such cases, the public and private sectors can often work in concert until companies become capable of sustainably providing services for themselves (Giovannucci & Reardon 2000).

However, developing country governments may not have the capacity to undertake the full range of public goods deemed necessary for adequate investment in the sector. For instance, most developing country governments do not have the capability to cope with the costs, logistics, and technical expertise required to effectively control quality, even within a specific commodity subsector (Giovannucci & Satin 2000). At such times, the assistance of international organisations may be warranted, particularly when the programmes they develop could be replicated in a number of countries. Again in the arena of food safety, Giovannucci and Reardon (2000) give the example of a 'public or semi-public body that interfaces with national and international CODEX and ISO organisations, and has links with committees or bodies from export-oriented agriculturalists and farmers'. Alternatively, some international organisations (like FAO) are working directly with producers and processors in developing countries to improve their quality control systems. The challenge for post-harvest science is to identify which goods are best provided under public and private auspices respectively, when some sort of hybrid structure is called for, and when the broader assistance of the international community is warranted.

### *Scale issues*

Although more than 90% of agroenterprises are small-scale and rural (Giovannucci 2000b), multinationals account for a disproportionately large share of output. A 1985 study of the food industry in Latin American found that large companies made up 3.5% of the total but earned 85% of income, while micro-enterprises accounted for 88% of the total but just 2.8% of income (Resende 1985, cited in UNIDO 1986). As stated previously, there is a clear tendency towards increased scale in production and processing, and towards the rapid growth of horizontally and vertically integrated global supply chains. Under the pressure of changing demand patterns and competition from large domestic or foreign enterprises, there has been a clear tendency towards the exclusion of small farmers and small firms.

A key implication of this concentration is that supermarkets are very quickly crowding out traditional neighbourhood markets in urban and rural areas, and in a wide range of developing countries at very different income levels (T. Reardon, pers. comm.).[29] These groups cannot meet the stringent quantity and quality demands made by large firms, or overcome their tendency to use large suppliers in order to overcome the transaction costs involved in dealing with a multitude of dispersed producers.[30] As a result, there are serious concerns about the ability of small and medium suppliers to survive in the short and medium term.

Several facets of globalisation tend to discriminate against small and medium enterprises. In the case of fruits and vegetables, for example, the need to control for high perishability requires specialised production, packing techniques and refrigerated transport. Computer-controlled deep irrigation systems in production, the intensive use of fertiliser and pesticides, sophisticated packing plants that resemble large modern factories, and temperature and atmospherically controlled storage and transport all contribute to 'cool chain' supply systems which allow fresh produce to be supplied to major supermarkets around the world. More generally, marketing products requires highly sophisticated and well-integrated information and transportation networks. The need to comply with aesthetic, hygiene, and health requirements involves investment in research, development, and marketing that small and medium enterprises cannot easily afford. The privatisation of grades and standards by large companies has clear exclusionary implications (T. Reardon, pers. comm.).

This exclusionary tendency is particularly worrying in the context of developing countries, because post-harvest systems and agroindustry are typically dominated by a multitude of small and medium enterprises (see Fig. 1.8, panel A). In the long term (see Fig. 1.8, panel B) the distribution of firms engaged in post-harvest and agroindustry might well be dominated by large enterprises. As the economy moves from a predominantly rural to a predominantly industrial structure, the consolidation of agroindustry might be optimal on both growth and efficiency grounds. However, in the medium term, there are important reasons why the promotion of a broad-based approach including small, medium and large enterprises might be more favourable to growth than a bias towards

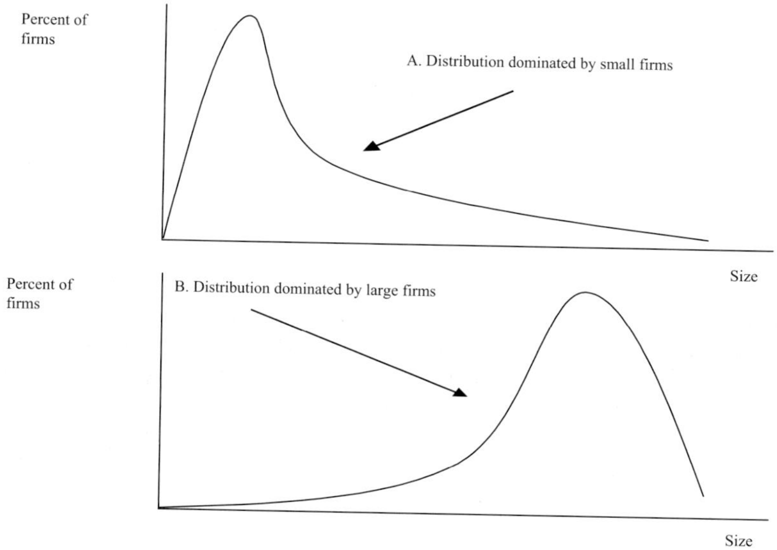


**Fig. 1.8** Firm size distribution.

large enterprises. The key challenge, therefore, is how to involve small farmers and firms in large value chains.

There are two possibilities for their involvement. The first is that in some poor rural areas, there may be diseconomies of scale in production and marketing that render smallholders and small enterprises even more competitive than their larger counterparts, as in the case of starch firms in Vietnam (see Box 1.1 and Goletti *et al.* 1998, 2001). The other is the use of technology and institutional mechanisms to help small actors surmount the difficulties posed by their size.

In certain developing regions, particularly in Asia, smallholder production is so prevalent that agrofood processors must secure their produce in order to ensure themselves an adequate supply of raw materials. Elsewhere, where processors have a choice, smallholders can be more efficient for a number of reasons:

- They tend to use labour more efficiently because they generally rely upon family labour, thereby avoiding costs associated with finding, screening and supervising hired labour (Lipton 1993). For labour-intensive crops, this is a clear advantage.
- Because of market imperfections and higher risk adversity, smallholders are typically willing to pay a higher premium for credit and for insurance (Key & Runsten 1999).

Also, firms may prefer to deal with diffuse and fragmented smallholders than larger and more powerful farms. Examples in which firms have contracted with smallholders suppliers include palm oil production in Malaysia (Kajisa *et al.* 1997) and fruit production in Guatemala (Key & Runsten 1999). However, such cases are not very common.

For small firms, diseconomies of scale could arise from:[31]

- the greater flexibility of small firms in adapting to disruptive circumstances or interruptions in the supply of inputs (e.g. Sandee 1999 on Indonesia);
- an input supply that is insufficient to permit substantial economies of scale (e.g. large fish plants are operating below capacity in Tanzania (McCormick 1998) and medium and large wood processing factories are well below capacity in Ghana (Owusu 2001[32]);
- more flexibility in small firm use of labour or by combining several business activities that can be frozen or expanded according to market fluctuations;
- production for domestic markets that are too small or inconstant to absorb the output of a large plant (e.g. demand for rattan furniture in Indonesia (Smyth 1992), coupled with high production costs that decrease export competitiveness;
- the fact that managerial problems involved in large-scale production can be very complex and lack of effective co-ordination can result in input loss;
- diseconomies of scale that are outside of the plant itself, e.g. if adequate infrastructure is lacking, new roads my be needed, large concentrations of estate workers may require housing, and poor industrial linkages may make the maintenance of sophisticated large-scale plant difficult; and
- production for niche markets.

**Box 1.1 Scale issues in starch processing in Vietnam.**

In Vietnam, food processing accounts for about 9% of GDP and 36% of industrial value-added; in the 1991–97 period, value-added in this area grew 14% yearly and provided 47% of exports. Data from the 1995 census suggest food processing employed about 1.5% of the labour force or perhaps 8% of formal sector labour (Minot 1998).

Over the past decade, the starch industry has undergone rapid development. In light of increasing demand for starch-related foods – expected to grow over 10% per year – the processing of starch, a high value item, from cassava, a low value item, was promoted to reduce the country's reliance on starch imports and as a potential export. Although the sector is relatively small, it has high potential in terms of demand growth, poverty reduction and income diversification in rural areas, particularly less favoured ones (Goletti *et al.* 1998).

Growth in starch processing has been dramatic. Between 1988 and 1998, the share of cassava used for starch grew 250% (from 10% to 24%) while average investment increased 78% each year. Industrial capacity grew eight-fold between 1994 and 1997 alone. In 1998, starch processing employed 20 000 rural households, most of which derived half their income from farming. The total income of these households (from agriculture and processing) was more than twice the rural per capita income in Vietnam.

The industry is largely dominated by small and micro enterprises, which together account for three-quarters of processors. The last few years have witnessed the arrival of a few large-scale foreign and domestic enterprises with significant potential to develop the starch industry in Vietnam. Nonetheless, a recent study finds decreasing returns to scale in starch production, owing to high transaction costs, induced by low levels of infrastructure development and market integration, and poor productivity of cassava (Goletti & Rich 1998). Small firms utilised more of their capacity, made more efficient use of capital and faced fewer constraints in terms of access to raw materials and markets.

Modelling work simulating the effects of directing capital to firms of various sizes showed that much higher gains would accrue from exclusively targeting small firms; under such a scenario, a 10% increase in capital would increase processors' income by 10%, exports by 137% and farm income by 10%. The study recommended policies to support small enterprises, in a continuum of firms that includes large enterprises and multinational companies. More generally, this research suggests that a broad-based approach to post-harvest development may be more adequate than a large enterprise model of development in responding to the challenges of rural poverty in developing countries.

Although growth potential is considerable, it may be limited by serious constraints relating to sustainability and quality. The most serious environmental by-product is water pollution, particularly for small enterprises. In the absence of sewage and water purification systems, starch processors have no choice but to dump the residues and polluted water into the village streets and inadequate sewage systems. Further, the export of starch requires quality improvement in line with international standards. Overcoming these problems would benefit not just starch producers but other processing industries as well.

Kaplinsky (1990) shows how these problems have led to large-scale sugar processing plants being supplanted by smaller-scale models in China, India and Kenya, where production at levels originally envisioned would have doubled sugar production costs.

The second way in which small farms and firms may be able to compete with their larger counterparts – where scale economies do exist – is through technological or institutional innovations designed to enable them to surmount the barriers posed by their size.

Technological innovations should enable small-scale suppliers to compete by developing appropriate technologies that do not require extensive capital outlay. For instance, the development of dehullers increased chances for small firms while changes in dairy technology excluded small suppliers, leading to 'blindingly fast concentration' (T. Reardon, pers. comm.). The exclusionary tendency of grades and standards referred to above also highlights the need for remedial technologies and processes suitable for adoption by small suppliers. More broadly, there is a need to consider the distribution of gains and losses arising from technological advances. For instance, Greeley (1987) studies the impact of the introduction of pedal threshers and rice mills in Bangladesh, and concludes that while both had adverse effects on income distribution, the welfare losses from the introduction of rice mills exceeded the production gains.

A variety of institutional arrangements have emerged in an attempt to integrate small firms and processors to meet quality requirements, lower risks and overcome transaction costs. These often involve:

- relying on external rather than internal economies of scale through some sort of network (e.g. Schmitz 1995), i.e. farmer co-operatives or firm clusters;
- encouraging small firms to adopt more efficient production techniques that are suited to small-scale production, following the premise that in certain industries mass production is becoming inefficient and being replaced by techniques requiring smaller-scale methods; and
- linking farmers and firms through contracting relationships, in which processors arrange with producers to provide a steady and timely supply of a particular crop for less than it would cost for them to operate a plantation, while supplying farmers with credit, inputs, information and other support services.

A growing body of literature focuses on contract farming in particular and finds that contracting with smallholders tends to raise their income (Watts *et al.* 1988; Glover 1990; Porter & Phillips-Howard 1997) and promote rural development. Singh (2000a, b) provides a more critical perspective, arguing that it may lead to economic and environmental degradation in the long term.

A recent case study of Peru shows the endogenous emergence of an institution – private management companies – which served as an intermediary between cotton farmers and buyers. The company provided management services (marketing and negotiation expertise, and organisational capital) in return for labour supervision through a share tenancy agreement (Escobar *et al.* 2000). Smallholders benefited through easier access to credit and through the bulk purchase of inputs. Whereas they had previously sold their raw cotton to ginners, they were now able to subcontract ginning services before selling directly to textile firms. Under this arrangement, smallholder profits were 80% higher than if they worked alone, and 50% higher than if they took advantage of assistance from non-governmental organisations.

However, it is also important to bear in mind that 'small is not always beautiful' (Lamb 2000). He points out that a preponderance of small-scale suppliers can make production costly and jeopardise international competitiveness by increasing transaction costs and margins within a supply chain. Moreover, shifts from worker to small-scale entrepreneur may entail risk, while fragmentation within a food supply chain may heighten quality control problems.

## The role of policy in post-harvest agriculture

### Regulation

It is commonly acknowledged that governments should implement a regulatory framework that allows post-harvest activity to be competitive, while safeguarding consumer health and fostering broad-based participation in the sector. Policy also has a role in assigning costs to negative environmental and social consequences of post-harvest activity, for instance, by enacting legislation to protect against unsustainable resource use and to safeguard worker rights, and by ensuring laws are enforced. Assisting domestic suppliers to comply with international conventions and agreements is also important. Finally, government has a broader duty to create a stable and dynamic macroeconomic climate free from any clear distortions that may dampen post-harvest activity, such as excessive subsidies and taxes or restrictions on trade.

### Information

By collecting and disseminating market information where the private sector lacks the capacity to do so itself, governments can significantly bolster domestic and international trade. Lack of information and information asymmetries pose significant barriers to market entry (Giovannucci 2000b; Lamb 2000), while knowledge of market information can reduce risks and lower transaction costs of market participation. It can also foster more efficient production processes, since suppliers are able to plan according to market demand and choose optimal marketing channels and traders can make more efficient decisions about when and how to allocate resources (Giovannucci 2000b).

In extending the reach of such information to disadvantaged areas, governments can also foster more broad-based market participation. Relevant information would include prices, quality requirements and trade regulations. In particular, participants could also benefit from 'current and accurate information about potential markets: how they are structured, how they operate, who are the competitors, what are the product specifications, what are the formal grades and standards, what are typical prices and marketing costs, what are the import procedures' (Lamb 2000).

Surveys show that market information systems often function poorly in developing countries, giving added impetus to the case for more government intervention and for international assistance. A recent FAO study of 120 countries found only 53 functioning national market information systems and only 13 that offered daily price disseminations (Giovannucci 2000b). FAO attributed the lack of widespread systems to poor initial planning and a lack of operational resources. Most of these systems were operated by the public sector. Guidelines for how countries should go about developing more effective systems, where possible in concert with private enterprises, are given by Giovannucci (2000b).

### Investment in research

More investment in post-harvest research would be beneficial for several reasons. First, post-harvest research has demonstrated high rates of return. Second, while much research with such high rates of return is carried out by the private sector, many types of post-harvest research have public good characteristics. Therefore, the private sector will not engage in those activities and it is generally left to governments and international organisations to supply them.

The development of methods to evaluate the impact of the post-harvest system is quite recent and very few studies have sought to quantify the effects (Goletti & Wolff 1998). However, a recent review of the limited literature on the impact of post-harvest research found rates of return ranging from 13% to 14 000% (Goletti & Samman 1999b). In their very extensive study of rates of return to agricultural R&D in general, Alston *et al.* 1999 found an average rate of return of 50–60%, which would roughly indicate that returns to post-harvest research may be as high as returns to all agricultural research. Furthermore, many improvements – e.g. in human health, environmental sustainability and gender roles – have important non-monetary value which is excluded from internal rate of return calculations. The economic impact of post-harvest research investments is encouraging and does not warrant a continued discrimination against such activities in funding allocations.

Post-harvest research can have a public good character for two main reasons. The first relates to private sector investments. Under certain circumstances, the private sector should and will participate in post-harvest research. For example, agribusiness firms find it necessary and worthwhile to engage in final product development, market studies and the search for improved processing technologies. However, the private sector will not fund research when it does not stand to appropriate the economic gains. Such research can nevertheless yield widespread benefits, but they are not easily reduced to only the clients of a commercial firm.

In circumstances in which the gains from research tend to be more diffuse, national agricultural research systems (NARS) or international institutes should be involved. Generally, NARS should focus on local post-harvest systems for certain commodities where market failures are common. International institutes should mainly come into play when post-harvest research has the character of a public good and stands to benefit a variety of countries.

Under what circumstances will post-harvest research have public good character? Where markets do not work well, or are absent (such as in the case of land or credit markets), smallholder farmers and rural enterprises do not have access to the same technology, information, assets, input supplies and market outlets. Under these conditions, smallholder households and small-scale enterprises are subject to significant transaction costs for producing and selling the same output mix (Akerlof 1970; Lopez 1984; de Janvry *et al.* 1991). Transaction costs include marketing costs (storage, transportation, handling, packaging) and intangibles such as searching, monitoring and enforcing (Hoff *et al.* 1993; Jaffee & Morton 1995), bargaining and lags in production. They vary by products, type of agent in the marketing chain and individual agent within a category of agents. This case is common to many less developed countries.

Most high value-added products in agriculture are characterised by a high ratio of transaction costs to final value, and include such commodities as animal and horticultural products, which are prime candidates for improvement through post-harvest activities. Since these products tend to have high income elasticities, they also tend to offer better prospects for long-term growth. Due to high transaction costs and lower assets, poor households and small-scale enterprises will have more difficulty than wealthier ones in engaging in the production of these products if an adequate post-harvest system is set in place.

Policies for growth and poverty reduction should then focus on reducing high transaction cost barriers separating the poor from markets and increasing access to information and assets for improving the adjustment of agricultural output and non-farm income mix to major changes in relative prices. These policies stress the development of the post-harvest system and are of strategic significance for growth in those countries with a large share of agriculture in GDP and exports and with large regions or sectors of the economy characterised by a low level of commercialisation and poorly functioning markets.

Examples of international public goods in the post-harvest area abound. The International Center for Tropical Agriculture's cassava project is a case in which a methodology for rural enterprise development was applied in several Latin American countries, and is being adapted for other regions too. The International Rice Research Institute's simple rice drying technology has been copied and modified by small manufacturers. The International Food Policy Research Institute's research on policy and institutions can be applied in a variety of countries that experience similar constraints to the development of post-harvest systems. Private funding of this type of research is marginal.

Sometimes, public institutions are able to invent a technology and then let the private sector control its development. Thus the International Potato Center, for instance, has stopped investing in potato processing because private companies can easily appropriate the gains from this type of research.

The second reason is that post-harvest research stands to achieve a number of goals that do not benefit private companies directly but have large social externalities. As we have seen, research in this area can benefit poorer populations, thus contributing to food security, poverty alleviation and income growth. When less food is lost, resources are used more efficiently and value is added at the community level, thus rendering development more sustainable. Therefore, such research benefits not only a specific population but has the potential to benefit policymakers and producers in general.

As the significant social contribution of post-harvest research becomes clear, and in light of high rates of return, the very skewed allocation of funds to production versus post-harvest topics cannot be justified. So far, relatively little has been invested in post-harvest research, despite its large potential as constraints and bottlenecks are removed. It would thus be desirable to re-examine current funding priorities and to allocate a larger proportion of resources to the post-harvest system.

### International agreements

International agreements have had an important and growing effect on countries' post-harvest systems, as countries must enact prescribed trade liberalisation measures while striving to adhere to international standards. The principal international agreements with a bearing on post-harvesting are those pertaining to agricultural trade and to quality standards under the World Trade Organization (WTO). Three of these are especially pertinent: the agreements on agriculture, on sanitary and phytosanitary standards (SPS) and on technical barriers to trade (TBT).

The movement to liberalise trade has a special bearing upon agriculture, since the sector has traditionally been heavily subsidised, particularly in developed countries. The 1995 Agreement on Agriculture acknowledged that trade in agricultural goods should be brought under clear effective rules that applied equally to all countries (Van Dyke 2000). Member countries agreed to convert non-tariff barriers to tariffs, and then to bind these at a maximum rate. Further, developed countries agreed to reduce tariffs over six years by 36%, with a minimum drop of 15% per tariff. For developing countries, the reduction was a less sharp 24%, with a minimum of 10% per tariff. Despite these measures, tariffs on agricultural goods still average 40%, compared with well under 10% for manufactures – and import quotas remain tight (*The Economist* 23 May 2000). Export subsidies are also a contentious issue; under the WTO Uruguay round, developed countries reduced these subsidies by 36% of their 1986–88 value for most commodities, but appear unwilling to take any further action. The current tariff structure discriminates against processed foods in particular since they escalate according to the value-added to the product, and against certain types of processed food products (Van Dyke 2000).

The SPS agreement strives to ensure safety of products being traded internationally. It allows countries to refuse to import goods from countries if it believes they fail to comply with its quality and safety standards, and therefore constitute a threat to human, animal or plant health. Countries are encouraged to adopt international standards, such as CODEX Alimentarius, as a benchmark, but may enact their own instead. This had led to the fear that countries may resort to this agreement to unfairly deny entry to others through false claims that their goods are not safe enough, thereby crossing a 'fine line between protection and protectionism' (*The Economist* 23 May 2000). It also places a heavy emphasis on risk assessment processes; while SPS does not specify which assessment procedures should be adopted, it does require that they be justified (Henson & Caswell 1999). Moreover, it is argued that the agreement puts excessive pressure on developing countries' food control systems, despite the exemptions it allows them (FAO 1997, cited in Hooker 1999).

The TBT agreement is intended to encourage countries to adopt standards and assessment procedures developed by international standard-setting bodies. It covers such aspects as terminology, packaging, marking and labelling requirements for products, processes and production methods (*The Economist* 23 May 2000).

## Conclusion

This chapter has sought to provide an overview of the current and likely future importance of the post-harvesting sector to the global economy, while emphasising its particular relevance for developing countries. The first section highlights its contribution to output, exports and employment in various parts of the world, and its vital role in promoting food security and rural welfare. It showed that throughout the world, agroprocessing accounts for a high share of manufacturing value, employment and trade, and that it is particularly significant in poorer countries. Further, the aggregate figures conceal particular dynamism for some countries and product groups. Within developing countries, agroprocessing activity is particularly important and within the poorest ones in particular; nonetheless, growth rates are higher and most agroprocessing activity is located in the relatively richer countries.

Next, the chapter examined what we believe to be the key qualitative trends which are propelling growth in the post-harvest sector and changing its structure in dramatic ways – with important implications for farmers, firms and consumers. In particular, this section considered the contraction of agriculture, the drive towards urbanisation, increased awareness of gender roles and the environment, technological advance, and perhaps most importantly, the massive changes being wrought by globalisation and increased attention to food safety. We stressed that while the overall impact on post-harvest development should be positive, the aggregate trends might mask adverse distributional consequences.

Having highlighted the key influences on the sector, the chapter then turned to the challenges firms and governments will face in the coming years, in seeking to ensure that the continued growth of post-harvesting is dynamic and broad-based. We identified three challenges as most important. First, the need for a systems approach that integrates sciences, management and social science theory in order to harness potential efficiency gains, promote food safety and direct attention to distributional outcomes throughout the value chain. Second, the importance of delineating between public and private goods in crafting policy that ensures the most efficient use of government and private sector resources. Third, the vital importance of scale issues and of taking steps to ensure that small firms are integrated into the new economic structures arising from global restructuring.

As they seek to cope with these challenges, governments face a number of pressing policy issues. The chapter closes by outlining those that appear most relevant on a national level: the need for a regulatory framework that promotes growth while safeguarding welfare; for adequate market information to be given to all actors involved; for further investment in post-harvest research; and for participation in international agreements that promote trade and food safety.

There can be little doubt that the post-harvest sector will continue to grow, in the face of buoyant demand for numerous agroprocessed products and rapid technological advance. What is not so certain is the ultimate outcome of these changes. Countries seeking to develop their post-harvest sectors are at a crucial juncture where they face the opportunity to craft policies that will allow the sector to develop in a manner that integrates poor rural farmers and firms, and allows for an equitable distribution of the gains from participation. Careful consideration of the likely outcomes of post-harvest sector development, and of policy options, will be crucial in realising such goals.

## Notes

1 This example is taken from Coquillard (2000).

2 These figures are beset by several problems. UNIDO collects the data from country governments without attempting to ensure their accuracy or consistency between countries. Countries may submit different types of data, derived from various sources, or none at all. In most countries, firms are typically recorded only if they employ five or more people; this effectively excludes many micro-enterprises or family-based enterprises, and may lead to underestimates of agroindustry and its employment effects in developing countries. No effort is made to estimate the size of the informal sector (UNIDO,

pers. comm.). To measure agroprocessing, we combined several branches of manufacturing activity using ISIC-2 and ISIC-3 revisions. Using ISIC-2 revisions, we followed FAO (1997) in including categories 3.1, 3.2, 3.3, 3.4 and 3.5. For ISIC-3, which several countries switched to in the 1990s, we tried to construct as similar a grouping as possible, using categories 15, 16, 17, 18, 19, 20, 21, 22 and 25. However, the two sets of categories are not identical, complicating any comparison of changes over time.

3 Value-added refers to the value of the output minus the intermediate inputs.

4 Note that lack of data for Eastern Europe and CIS in the 1990s precludes including them in this discussion.

5 Commodity prices since the 1970s show a strong downward trend, which has been estimated at about –0.5% per year in real terms. In 1994, real commodity prices were half those of their post-World War II 1970s peak level (ODI 1994).

6 If not otherwise indicated, information in this paragraph is from Díaz-Bonilla and Reca (2000).

7 In the six communities they studied, von Braun and Immink (1994) estimated that employment in production, input supply and output marketing grew 21%.

8 For Latin America see UNIDO (1986); for China see OECD (2000).

9 For instance, Greeley (1991) estimates that more than 50% of the value of cereal products consumed in urban areas was added by post-harvesting, in contrast to about 25% in rural areas.

10 Surveys have documented sharp drops in time allocated to food preparation. A recent study found that in Australia the current average is about 15 minutes, and is expected to drop further still (Hofman 2000).

11 A UK supermarket charged 15 times the per kilo price of loose carrots for carrots that were peeled, chopped, washed and ready-to-eat (Dolan & Humphrey 2000).

12 Barrientos *et al.* (1999) discuss women's role in producing grapes in Brazil's Sao Francisco valley, while Thrupp (1995) discusses their role in producing non-agricultural exports in Latin America more broadly.

13 In Japan and Singapore, FAO predicts 20% yearly growth in the coming years.

14 The few exceptions are commodities for which there are no obvious scale economies, e.g. calf raising and grain farming (*The Economist* 23 May 2000).

15 Heffernan (1999) reports that market shares were easy to obtain in the mid-1980s, but more recently have come to be considered proprietary information, with pressure being put on trade journals not to publish them.

16 The 40% figure is significant because it was once thought to be the four-firm concentration ratio at which a market was no longer competitive (Heffernan 1999).

17 This example is given by Coquillard (2000).

18 See, for instance, Key and Runsten (1999) on Latin America.

19 Even in developed countries, measurement suffers from poor surveillance systems and severe under-reporting: most cases of food poisoning do not result in a visit to a medical facility, are not precisely diagnosed and are not reported to public health authorities. WHO (2001) estimates that in industrialised countries, less than 10% of total cases are reported, and that in developing countries, the figure falls to less than 1%.

20 Note that this includes illness from known and unknown food-borne pathogens. See Mead *et al.* (1999) for a discussion of how and why these estimates differ from earlier ones.

21 For example, a 1999 survey of animal products in markets in Hanoi, Vietnam, found aerobic bacteria in 100% of sampled meats and milks; *Escherichia coli* in 100% of beef; salmonella in 60% of the beef and 40% of pork; antibiotic residues that were 22% higher than the norm in 100% of chicken; DDT in 28% of pork, 33% of kidneys, 25% of animal liver, 100% of chicken eggs and 60% of duck eggs; and heavy metals (lead, cadmium and mercury) in meat and dairy products that were 2 to 10 times higher than the norm (FAO 1999).

22 Animal diseases, such as the 2001 outbreak of foot and mouth disease in Europe, have also caused widepread alarm regarding the food supply system, even if not epizootic (Brundtland 2001).

23 Data problems seriously hinder the identification of trends in the incidence of disease. Because of ongoing efforts to strengthen and expand surveillance, and to improve methodologies used to analyse the results, surveys are not strictly comparable over time. Indeed, Mead *et. al.* (1999) write that differences between their estimates in the US and those previously published appear to be primarily attributable to the availability of better data and new analytical techniques rather than real changes in disease

frequency over time. However, a survey of global findings concludes that methodological problems not withstanding, there is enough evidence to demonstrate that the incidence of food-borne disease trended upwards in the 1980s and 1990s (WHO 2001).

24 Note that the majority of food-borne pathogens have yet to be identified. Mead *et al.* (1999) report that known agents cause just 19% of food-borne illnesses and account for 36% of deaths.

25 Giovannucci and Reardon (2000) observe that as long as buyers and sellers can meet and bargain with products at hand, there is little need for standardisation. But as soon as products are handled in greater volumes and over greater distances, grades and standards become crucial in conveying vital information about quality and measures.

26 Note that separate handling systems are primarily an issue for commodities with modified 'output' traits.

27 Giovannucci (2000a).

28 It was recently suggested that that the impact of global competition and new technologies are transforming the chain to a complex web (*The Economist* 23 May 2000).

29 T. Reardon (pers. comm.) gives the examples of Peru, Honduras, Mexico, Chile, Brazil and Argentina. FAO (1999) points to this trend in Hanoi, Vietnam, while WHO discusses it in the context of Chinese cities.

30 In Zimbabwe, it was estimated that in order to break even, a horticultural exporter would have to pay its smallholder suppliers less than 30% the price per kilo paid to commercial farmers who delivered their goods directly to a packing plant (Coulter *et al.* 1999).

31 The first three arguments adapted from Rasmussen *et al.* (1992), while the last four are taken from Kaplinsky (1990).

32 Estimated rates of capacity utilisation in Ghana for wood processing plants between 1978 and 1993 hit a low of 20% in 1983 and were never above 70%; the manufacturing sector as a whole reached its highest level of capacity utilisation – 46% – in 1993 (Owusu 2001).

## References


ABC News (2000) Organizing Organics. 21 December 2000. www.abcnews.com.

AgraEurope (2000) *Organic Food in the European Union. An AgraEurope Special Study.* AgraEurope, London, UK.

AgraFood Latin America (2001) Food and Drink Sector Tops Merger and Acquisition List, **49**.

Akerlof, G.A. (1970) The market for 'lemons': quality uncertainty and the market mechanism. *Quarterly Journal of Economics,* **84**, 488–500.

Alston, J.M., Pardey, P.G. & Smith, V.H. (1999) *Paying for Agricultural Productivity.* Food Policy Statement No. 30. Johns Hopkins University Press/International Food Policy Research Institute, Washington DC, USA.

APO (1992) Seminar on Promoting Agribusiness for Higher Productivity. 4–14 August 1992, Jakarta, Indonesia. Asian Productivity Organization, Tokyo, Japan.

Arnold, M. (1996) *Harvest and Post-harvest Problems in Agriculture, Forestry and Fisheries: The CGIAR Contribution to Research.* CGIAR Document No. DR/TAC:IAR/96/5. Consultative Group on International Agricultural Research.

Athukorala, P.C. & Sen, K. (1998) Processed food exports from developing countries: patterns and determinants. *Food Policy,* **23**, 41–54.

Austin, J.E. (1995) Reflections on IDB Strategy for Agricultural and Rural Development. Mimeo. World Bank, Washington, DC, USA.

Azzam, A.M. (1998) Competition in the US meatpacking industry: is it history? *American Journal of Agricultural Economics,* **80**, 96–106.

Barrientos, S., Bee, A., Matear A. & Vogel, I. (1999) *Women and Agribusiness: Working Miracles in the Chilean Fruit Export Sector.* Macmillan Press, Basingstoke, UK.

BBC (2001) Kenya's flower power. Radio 4, 5 April 2001. http://news.bbc.co.uk/hi/english/audiovideo/programmes/crossing_continents/africa/newsid_1260000/1260248.stm.

Blowfield, M. (1999) Ethical trade: a review of developments and issues. *Third World Quarterly,* **20**, 753–770.

Boeh-Ocansey, O. (1988) Small-scale food industries: improving performance in developing countries. *Food Policy,* November, 334–337.

Brester, G.W. & Musick, D.C. (1995) The effect of market concentration on lamb marketing margins. *Journal of Agricultural and Applied Economics,* **27**, 172–183.

Brundtland, Gro Harlem (2001) Food Safety: a World-wide Challenge. Speech delivered to Food Chain 2001 Meeting in Uppsala, Sweden, 14 March 2001. World Health Organization, Geneva, Switzerland. http://www.who.int/director-general/speeches/2001/english/20010314_foodchain2001 uppsala.en.html.

Cook, M.L. & Chaddad, F.R. (2000) Agroindustrialization of the global agrifood economy: bridging development economics and agribusiness research. *Agricultural Economics,* **23**, 207–218.

Coquillard, J.N. (2000) Determinants of Commercial Viability of Private Enterprises in the OECD Food Processing Sector. In: *The Agro-food Processing Sector in China: Developments and Policy Challenges.* Proceedings of a Conference held in October 1999 in Beijing. Organization for Economic Cooperation and Development, Paris, France.

Cotterill, R.W. (2001) Neoclassical explanations of vertical organisation and performance of food industries. *Agribusiness,* **17**, 33–57.

Coulter, J., Goodland, A., Tallontire, A. & Stringfellow, R.

(1999) Marrying farmer cooperation and contract farming for agricultural service provision in sub-Saharan Africa. World Bank, Washington, DC, USA.www.worldbank.org.

De Janvry, A., Fafchamps, M. & Sadulet, E. (1991) Peasant household behavior with missing markets: some paradoxes explained. *Economic Journal,* **10**, 1400–1417.

Díaz-Bonilla, E. & Reca, L. (2000) Trade and agroindustrialisation in developing countries: trends and policy impacts. *Agricultural Economics,* **23**, 219–229.

Dietz, H.M. (2000) Quality of Agricultural Products: The Challenge for Agricultural Trade and for Producers. *Proceedings of PhAction Seminar 12: Linking Farms to Markets*, 7 June 2000, Eschborn, Germany. GTZ and P*h*Action.

Dolan, C. (2001) The 'good wife': struggles over resources in the Kenyan horticultural sector. *Journal of Development Studies,* **37**, 39–70.

Dolan, C. & Humphrey, J. (2000) Governance and trade in fresh vegetables: the impact of UK supermarkets on the African horticulture industry. *Journal of Development Studies,* **37**, 147–176.

Dolan, C., Humphrey, J. & Harris-Pascal, C. (1999) *Horticultural Commodity Chains: the Impact of the UK Market on the African Fresh Vegetable Industry*. IDS Working Paper 96. Institute for Development Studies, Brighton, UK.

Du, Ying (2000) Development and policy options for China's agroprocessing sector. In: *The Agro-food Processing Sector in China: Developments and Policy Challenges*. Proceedings of a Conference, October 1999, Beijing. Organization for Economic Cooperation and Development, Paris, France.

Escobar, J., Agreda, V. & Reardon, T. (2000) Endogenous institutional innovation and agroindustrialization on the Peruvian coast. *Agricultural Economics,* **23**, 267–277.

FAO (1995) *Food for Consumers: Marketing, Processing and Distribution.* WFS 96/TECH/4: Provisional Version. United Nations Food and Agriculture Organization, Rome, Italy.

FAO (1997) *The State of Food and Agriculture.* United Nations Food and Agriculture Organization, Rome, Italy.

FAO (1999) Final Statement of Workshop on Food Supply and Distribution to Hanoi, 17–18 September 1999. United Nations Food and Agriculture Organization, Rome, Italy.

Ferris, R.S.B., Best, R., Goletti, F. & Wheatley, C. (1997) Towards closer collaboration and coordination in the post-harvest sector. In: *Integrated Control of Insect Pests in Rural Maize Stores, with Particular Reference to the Larger Grain Borer* Prostephanus truncatus*, and the Future Development of the Post-harvest Sector in sub-Saharan Africa.* Proceedings of an IITA/GTZ Conference, 13–15 October 1997, Cotonou, Benin. A CGIAR Strategy Perspective.

Fleischer, G., Waibel, H. & Dirksmeyer, W. (1996) Future priorities in post-harvest systems development – the role of GASGA. In: *Systems Approach in Post-harvest Activities.* GASGA Executive Seminar 8, 12–13 June 1996, CIRAD, Montpelier, France.

Foster, C. & Lampkin, N. (2000) *Organic and In-conversion Land Area, Holdings, Livestock and Crop Production in Europe*. Final report for EU project FAIR3-CT96–1979 'Effects of the CAP-reform and possible further developments on organic farming in the EU'. Welsh Institute of Rural Studies, University of Wales, Aberystwyth, Wales.

Giles, J.A. (1999) New Structures, New Customers and New Strategies: Implications for Food Safety. Paper presented at the Post-harvest Convention, 24 March 1999, Silsoe College, Cranfield University, Bedford, UK.

Giovannucci, D. (2000a) Introduction. In: D. Giovannucci (Ed.) *A Guide to Developing Agricultural Markets and Agro-enterprises.* World Bank, Washington, DC, USA. http://www.worldbank.org.

Giovannucci, D. (2000b) Market Information Services. In: D. Giovannucci (Ed.) *A Guide to Developing Agricultural Markets and Agro-enterprises.* World Bank, Washington, DC, USA http://www.worldbank.org.

Giovannucci, D. & Reardon, T. (2000) Understanding Grades and Standards and How to Apply Them. In: D. Giovannucci (Ed.) *A Guide to Developing Agricultural Markets and Agro-enterprises.* World Bank, Washington, DC, USA. http://www.worldbank.org.

Giovannucci, D. & Satin, M. (2000) Food Quality Issues: Understanding HACCP and Other Quality Management Techniques. In: D. Giovannucci (Ed.) *A Guide to Developing Agricultural Markets and Agro-enterprises.* World Bank, Washington, DC, USA. http://www.worldbank.org.

Glover, D. (1990) Contract farming and outgrower schemes in East and Southern Africa. *Journal of Agricultural Economics,* **41**, 303–315.

Goletti, F. & Rich, K. (1998) Policy Issues and Policy Analysis for Post-harvest Research. In: *GASGA Seminar 10. CGIAR Post-harvest Research and GASGA*, pp. 67–80.

Goletti, F. & Samman, E. (1999a) Globalisation and the benefits of a broad-based approach to post-harvest systems development. In: G.I. Johnson, Le Van To, Nguyen Duy Duc & M.C. Webb (Eds) *Quality Assurance and Agricultural Produce.* Proceedings of the 19th ASEAN/1st APEC Seminar on Post-harvest Technology, 9–12 November 1999, Ho Chi Minh City, Vietnam. ACIAR Proceedings **100**. Australian Centre for International Agricultural Research, Canberra, Australia.

Goletti, F. & Samman, E. (1999b) The Importance of Postproduction to Sustainable Rural Livelihoods. In: GASGA Seminar 11. *The Importance of Post-production to Sustainable Rural Livelihoods.* Natural Resources Institute, University of Greenwich, Chatham, UK.

Goletti, F. & Wolff, C. (1998) *The Impact of Post-harvest Research.* Markets and Structural Studies Discussion Paper **29**. International Food Policy Research Institute, Washington, DC, USA.

Goletti, F., Rich, K., Wheatley, C. & Nguyen, K.V. (1998) *Starch Industry Development as a Strategy for Agro-food based Rural Industrialization.* Donor report submitted to the International Development Research Center, December. International Food Policy Research Institute, Washington, DC, USA. www.ifpri.org.

Goletti, F., Rich, K. & Wheatley, C. (2001) The cassava starch industry in Vietnam: can small firms survive and prosper? *International Food and Agribusiness Management Review,* **1**, 345–357.

Greeley, M. (1987) *Post-harvest Losses, Technology and Employment: The Case of Rice in Bangladesh.* Westview Press, Boulder, CO, USA.

Greeley, M. (1991) *Post-harvest Technologies: Implications for Food Policy Analysis.* EDI Development Policy Case Series. World Bank, Washington, DC, USA.

GTZ (1998) *'Nacherntesystem', was ist das?* Deutsche Gesellschaft für Technische Zusammenarbeit, Eschborn, Germany. www.gtz.de/post_harvest/nes.htm.

Hayenga, M.L. (1998) Cost structures of pork slaughter and processed firms: behavioral and performance implications. *Review of Agricultural Economics,* **20**,574–583.

Hayenga, M.L. (2000) Structural Changes in the Pork Production and Processing Industry of the United States and other OECD Countries: Major Trends and Issues. In: *The Agro-food Processing Sector in China: Developments and Policy Challenges.* Proceedings of a Conference, October 1999, Beijing. Organization for Economic Cooperation and Development, Paris, France.

Hayes, D. (2000) Structural Developments and Policies in Agricultural Processing in OECD Countries. In: *The Agro-food Processing Sector in China: Developments and Policy Challenges.* Proceedings of a Conference, October 1999, Beijing. Organization for Economic Cooperation and Development, Paris, France.

Heffernan, W. (1999) *Consolidation in the Food and Agriculture System.* Report to the National Farmers Union. Department of Rural Sociology, University of Missouri-Columbia, Columbia MO, USA.

Henson, S. & Caswell, J. (1999) Food safety regulation: an overview of contemporary issues. *Food Policy,* **24**, 589–603.

Hoff, K., Braverman, A. & Stiglitz, J.E. (1993) *The Economics of Rural Organization: Theory, Practice and Policy*. Oxford University Press for the World Bank, Oxford, UK.

Hofman, P.J. (2000) Future research and development in agricultural produce quality. In: G.I. Johnson, Le Van To, Nguyen Duy Duc and M.C. Webb (Eds) *Quality Assurance and Agricultural Produce.* Proceedings of the 19th ASEAN/1st APEC Seminar on Post-harvest Technology, 9–12 November 1999, Ho Chi Minh City, Vietnam. ACIAR Proceedings **100**. Australian Centre for International Agricultural Research, Canberra, Australia.

Hooker, N.H. (1999) Food safety regulation and trade in food products. *Food Policy*, **24**, 653–668.

Huang, Jikun & Bouis, H. (1996) *Structural Changes in the Demand for Food in Asia.* IFPRI 2020 Discussion Paper **11**. International Food Policy Research Institute, Washington, DC, USA.

Huang, Shingming (2000) Present Situation, Policies and Outlook for China's Food Industry Development. In: *The Agro-food Processing Sector in China: Developments and Policy Challenges.* Proceedings of a Conference, October 1999, Beijing. Organization for Economic Cooperation and Development, Paris, France.

Hummels, D., Jun Ishii & Kei-Mu Yi (1999) *The Nature and Growth of Vertical Specialization in World Trade.* Federal Reserve Bank of New York, NY, USA. http://www.frb.ny.org.

Ingco, M. (1997) *Has agricultural trade liberalization improved welfare in the least developed countries? Yes.* Policy Research Working Paper **1748**. World Bank, Washington, DC, USA. http://www.worldbank.org/research/trade/wp1748.html.

Islam, N. & Valdés, A. (Eds) (1990) *The GATT, Agriculture and the Developing Countries.* International Food Policy Research Institute, Washington, DC.

Jaffee, S. & Gordon, P. (1993) *Exporting High-value Food Commodities.* World Bank Discussion Papers. World Bank, Washington DC, USA.

Jaffee, S. & Morton, J. (Eds) (1995) *Marketing Africa's High-value Foods: Comparative Experiences of an Emergent Private Sector.* World Bank, Washington DC, USA.

Johnson, G.I. (1998) *Post-harvest Technology in Asia: ACIAR's Framework for Collaborative Research and Development.* Australian Centre for International Agricultural Research, Canberra, Australia.

Kajisa, K., Maredia, M. & Boughton, D. (1997) *Transformation Versus Stagnation in the Oil Palm Industry: A Comparison Between Malaysia and Nigeria.* Staff Paper **97–5**, Michigan State University, Lancing MI, USA.

Kaplinsky, R. (1990) *The Economics of Small: Appropriate Technology in a Changing World.* Short Run Press, Exeter, UK.

Kaplinksy, R. (2000) Globalisation and unequalization: what can be learned from value chain analysis? *Journal of Development Studies,* **37**, 117–146.

Kaplinksy, R. & Manning, C. (1998) Concentration, competition policy and the role of small and medium enterprises in South Africa's industrial development. *Journal of Development Studies,* **35**,139–161.

Kennedy, E. & Reardon, T. (1994) Shift to non-traditional grains in the diets of East and West Africa: role of women's opportunity cost of time. *Food Policy,* **19**, 45–56.

Key, N. & Runsten, D. (1999) Contract farming, smallholders and rural development in Latin America: the organisation of agroprocessing firms and the scale of outgrower production. *World Development,* **27**, 386–401.

Koontz, S., Garcia, P. & Hudson, M. (1993) Meatpacker conduct in fed cattle pricing: an investigation of oligopsony power. *American Journal of Agricultural Economics,* **75**, 537–548.

Lamb, J.E. (2000) Non-farm Rural Enterprise Development. In: D. Giovannucci (Ed.) *A Guide to Developing Agricultural Markets and Agro-enterprises.* World Bank, Washington DC, USA www.worldbank.org.

Lipton, M. (1993) Land reform as commenced business: the evidence against stopping. *World Development,* **21**, 641–657.

Loader, R. & Hobbs, J.E. (1999) Strategic responses to food safety legislation. *Food Policy,* **24**, 685–706.

Lopez, R.E. (1984) Estimating substitution and expansion effects using a profit function framework. *American Journal of Agricultural Economics,* **33**, 358–367.

Mbilinyi, M. (1988) Agribusiness and women peasants in Tanzania. *Development and Change,* **19**, 549–583.

McCormick, D. (1998) *Enterprise Clusters in Africa: on the Way to Industrialization?* IDS Discussion Paper **366**. Institute of Development Studies, Brighton, UK.

Mead, P.S., Slutsker, L., Dietz, V., McCaig, L.F., *et al.* (1999) Food-Related Illness and Death in the United States. *Emerging Infectious Diseases,* **5**, 607–25. http://www.cdc.gov/ncidod/eid/vol5no5/mead.htm.

Mellor, J. (Ed.) (1995) *Agriculture on the Road to Industrialisation.* Johns Hopkins University Press for the International

Food Policy Research Institute, Baltimore OH, USA.

Minot, N. (1998) *Competitiveness of Food Processing in Vietnam: A Study of the Rice, Coffee, Seafood, and Fruit and Vegetable Sub-sectors.* Report to the Ministry of Planning and Development, Vietnam and the UNDP. International Food Policy Research Institute, Washington, DC, USA.

ODI (1994) *Commodity Markets: Options for Developing Countries.* Briefing Paper **5/95**. Overseas Development Institute, London, UK.

OECD (1996) *Export Fruit Boom From the South: A Threat for the North?* Organization for Economic Cooperation and Development, Paris, France.

OECD (2000) *The Agro-food Processing Sector in China: Developments and Policy Challenges.* Proceedings of a Conference, October 1999, Beijing. Organization for Economic Cooperation and Development, Paris, France.

Overfield, D. (1998) An investigation of the household economy: coffee production and gender relations in Papua New Guinea. *Journal of Development Studies,* **34**, 52–70.

Owusu, J.H. (2001) Determinants of export-oriented industrial output in Ghana: the case of formal wood processing in an era of economic recovery. *The Journal of Modern African Studies,* **39**, 51–80.

Petritisch, M. (1985) *Women's Participation in Manufacturing in Developing Countries, with Emphasis on Agro-industries.* Paper Prepared for Expert Group Meeting on Women in Agro-Industries, 14–18 October 1985, Vienna. United Nations Industrial Development Organization, Vienna, Austria.

Poleman, T.T. & Thomas, L. (1994) *Income and Dietary Change: International Comparisons Using Purchasing-Power-Parity Conversions.* Department of Agricultural, Resource and Managerial Economics Working Paper 94–1, Cornell University, Ithaca NY, USA.

Porter, G. & Phillips-Howard, K. (1997) Comparing contracts: an evaluation of contract farming schemes in Africa. *World Development,* **25,** 227–238.

Rasmussen, J., Schmitz, H. & van Dijk, M.P. (1992) Introduction: exploring a new approach to small-scale industry. *IDS Bulletin,* **23**, 1–7.

Rattray, B., Bamford, J. & Srivastava, R. (2000) An overview of the Canadian agri-food system: a case study involving the Canadian hog-pork sector. In: *The Agro-food Processing Sector in China: Developments and Policy Challenges.* Proceedings of a Conference, October 1999, Beijing. Organization for Economic Cooperation and Development, Paris, France.

Reardon, T. & Barrett, C.B. (2000) Agroindustrialization, globalisation and international development: an overview of issues, patterns and determinants. *Agricultural Economics,* **23**, 195–205.

Reardon, T., Stamoulis, K., Balisacan, A., Cruz, M.E., Berdegue, J. & Banks, B. (1998) Rural Nonfarm Income in Developing Countries. In: *The State of Food and Agriculture 1998.* United Nations Food and Agriculture Organization, Rome, Italy.

Sandee, H. (1999) *Clustered small enterprise under economic distress: the impact of the Indonesian crisis on small industry clusters, with reference to agro-industries.* Paper presented at the American Agricultural Economics Association Pre-conference on Agro-industrialisation, Globalisation and Economic Development, 6–7 August 1999, Nashville TE, USA.

Schejtman, A. (1998) Agroindustria y pequeña agricultura: experiencias y opciones de transformación. In: CEPAL/GTZ/FAO (Eds) *Agroindustria y Pequeña Agricultura: Vínculos, Potencialidades y Oportunidades Comerciales.* Santiago de Chile. Naciones Unidas.

Schmitz, H. (1995) Collective efficiency: growth path for small-scale industry. *Journal of Development Studies,* **31**, 529–566.

Singh, S. (2000a) Theory and practice of contract farming: a review. *Journal of Social and Economic Development,* **3**, 228–246.

Singh, S. (2000b) Contract farming for agricultural diversification in the Indian Punjab: a study of performance and problems. *Indian Journal of Agricultural Economics,* **55**, 283–294.

Smyth, I. (1992) Collective efficiency and selective benefits: the growth of the rattan industry in Tegalwangi (Indonesia). *IDS Bulletin,* **23**, 51–56.

Thrupp, L. (1995) *Bittersweet Harvests for Global Supermarkets: Challenges in Latin America's Export Boom.* World Resources Institute, Washington DC, USA.

Toulmin, C., Leonard, R., Brock, C., Coulibaly, N., Carswell, G. & Dea, D. (2000) Diversification of Livelihoods: Evidence from Mali and Ethiopia. *IDS Research Report,* **47**. Institute of Development Studies, Brighton, UK.

Traill, W.B. (2000) Trends and Outlook for the Agroprocessing Sector. In: *The Agro-food Processing Sector in China: Developments and Policy Challenges.* Proceedings of a Conference, October 1999, Beijing. Organization for Economic Cooperation and Development, Paris, France.

UNIDO (1986) *Agro-food Industrial Development in Latin America.* Sectoral Studies Series No. 25. United Nations Industrial Development Organization, Vienna, Austria.

UNIDO (1990) *Profiles of Key Branches of Agro-industries in Africa.* Studies on the Rehabilitation of African Industry No. 13. United Nations Industrial Development Organization, Vienna, Austria.

UNIDO (2000a) *Women, Industry and Technology.* United Nations Industrial Development Organization, Vienna, Austria.

UNIDO (2000b) *Women, Industry and Entrepreneurship.* United Nations Industrial Development Organization, Vienna, Austria.

UNIDO (2000c) *Women, Industry and the Environment.* Brochure. United Nations Industrial Development Organization, Vienna, Austria.

Unnevehr, L. & Hirschhorn, N. (2000) Developing Effective Food Safety Interventions in Developing Countries. In: D. Giovannucci (Ed.) *A Guide to Developing Agricultural Markets and Agro-enterprises.* World Bank, Washington DC, USA. http://www.worldbank.org.

USFDA (1999) HACCP: A State-of-the-Art Approach to Food Safety. FDA Backgrounder. United States Food and Drug Administration, Washington DC, USA. http://vm.cfsan.fda.gov/~lrd/bghaccp.html.

Van Dyke, J. (2000) The Role of Grades and Standards – A Case Study: The Role of USDA's Beef Grading Programme in the

Marketing of Beef. In: *The Agro-food Processing Sector in China: Developments and Policy Challenges*. Proceedings of a Conference, October 1999, Beijing. Organization for Economic Cooperation and Development, Paris, France.

Viaene, J. & Gellynck, X. (1995) Structure, conduct and performance of the European food sector. *European Review of Agricultural Economics,* **22**, 282–295.

Von Braun, J. & Immink, M. (1994) Nontraditional Vegetable Crops and Food Security among Smallholder Farmers in Guatemala. In: J. von Braun and E. Kennedy (Eds) *Agricultural Commercialization, Economic Development and Nutrition.* Johns Hopkins University Press for the International Food Policy Research Institute, Baltimore OH, USA.

Von Braun, J. & Kennedy, E. (Eds) (1994) *Agricultural Commercialization Economic Development and Nutrition.* Johns Hopkins University Press, International Food Policy Research Institute, Baltimore OH, USA.

Watts, M., Little, P.D., Mock, C., Billings, M. & Jaffee, S. (1988) *Contract farming in Africa: executive summary.* Institute for Development Anthropology Working Paper **87**. United States Agency for International Development, Washington DC, USA.

WHO (1996) *Emerging Food-borne Diseases*. Fact Sheet No. 124. World Health Organization, Geneva, Switzerland. http://www.who.int/inf-fs/en/fact124.html.

WHO (2001) *Background Paper: Developing a Food Safety Strategy.* Draft for WHO Strategic Planning Meeting. World Health Organization, Geneva, Switzerland. http://www.who.int/fsf/BACKGROUND%20PAPER.pdf.

Willer, H. & Youssefi, M. (2001) *Organic Agriculture Worldwide: Statistics and Future Prospects.* Foundation of Ecology and Agriculture, Bad Dürkheim, Germany. http://www.soel.de/inhalte/publikationen/s_74_03.pdf.

Wolff, C. (1998) *Oil Palm in Chiapas – Economic Impact, Profitability and Linkages to the Processing Industry.* MSc Thesis. Cornell University, Ithaca NY, USA.

Wong, W.C., Mansour-Nahra, P., Ogle, I. & Sorby-Adams, N. (2000) Quality assurance: the concept and its evolution. In: G.I. Johnson, Le Van To, Nguyen Duy Duc & M.C. Webb (Eds) *Quality Assurance and Agricultural Produce.* Proceedings of the 19th ASEAN/1st APEC Seminar on Post-harvest Technology, 9–12 November 1999, Ho Chi Minh City, Vietnam. ACIAR Proceedings **100**. Australian Centre for International Agricultural Research, Canberra, Australia.

Young, R.H. (1991) *Addressing the Human Dimension in Nutrition Sciences, Agro-industries and International Agricultural Research.* International Development Research Centre, Ottawa, Canada.

# Chapter 2
# Biology of Plant Commodities

**D. Rees and L. Hammond**

In this book almost all of the food products considered are derived from plants, and in most cases the plant products remain alive after harvest. Thus the maintenance of product quality depends on keeping the products metabolically healthy and unstressed. The objective of this chapter is to provide a summary of the biology necessary to understand the behaviour of commodities after harvest, and also to understand the effects both of post-harvest treatments and storage technologies considered in other chapters. The section starts by introducing the smallest constituents, the molecules of which all organisms are made. A brief description of a cell and the main components within a cell is followed by a description of the main tissue types found in plant products. These sections provide the necessary background for the later parts of the chapter which consider the physiology and biochemistry of harvested plant-based commodities.

The information in this chapter has largely been derived from a number of excellent textbooks. Rather than referencing these throughout the chapter, they are listed at the end as Recommended further reading.

## Molecular components of living tissue

Life on Earth is carbon-based. With only a few exceptions, all living organisms – animals, plants and bacteria – are composed primarily of molecules with carbon skeletons. These molecules are classified depending on their structures and functions into carbohydrates, lipids and proteins. In addition to these three classes of molecule, small amounts of other chemicals, known collectively as vitamins, are needed for organisms to survive. Given that within this book we are considering plants primarily as a source of food for humans, the role of each nutrient is summarised in animals and plants.

### Carbohydrates

Carbohydrates consist of the elements carbon (C), hydrogen (H) and oxygen (O). They are formed by green plants from carbon dioxide ($CO_2$) and water ($H_2O$) using the energy of absorbed sunlight through the process of photosynthesis. Their main function in the human diet is as a source of energy, which is released when the molecules are broken down by the process of respiration.

Carbohydrates can exist in many forms, as given below.

#### *Monosaccharides*

These are simple sugars. The most common ones in food contain six carbon atoms and have the general formula $C_6H_{12}O_6$. Common examples are glucose, fructose and galactose (Fig. 2.1). As well as being the form of carbohydrate used to provide energy through respiration, monosaccharides form the building blocks for the larger, more complex carbohydrates.

#### *Oligosaccharides*

Oligosaccharides are molecules made up of between two and ten monosaccharides. Common examples are the following disaccharide sugars:

- sucrose, which consists of a glucose molecule attached to one of fructose;
- maltose, which consists of two glucose units;
- lactose, which consists of a glucose and galactose unit.

#### *Polysaccharides*

Polysaccharides are linear or branched chains of many monosaccharide units. They are relatively compact mol-

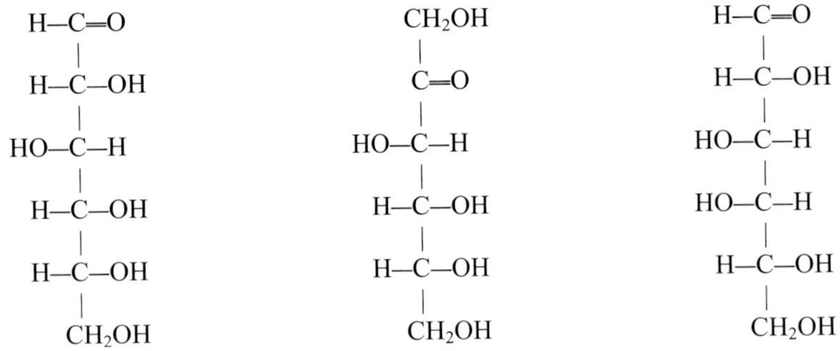

a. D-glucose b. D-fructose c. D-galactose

**Fig. 2.1** The structures of monosaccharides commonly found in plant products.

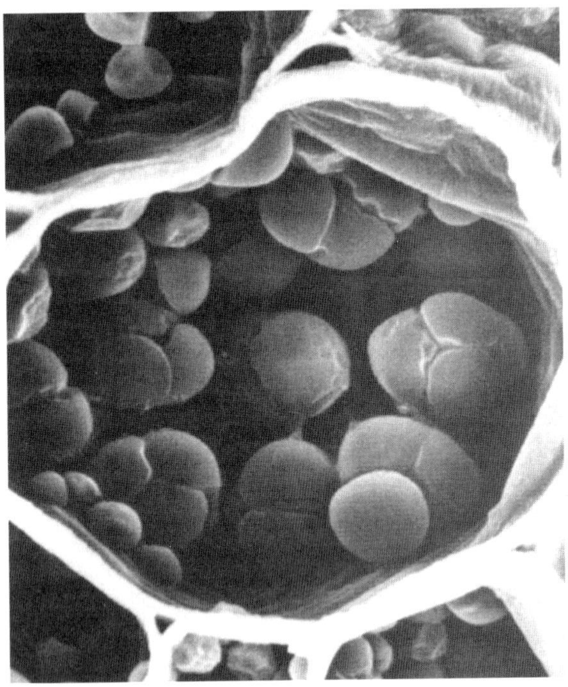

**Fig. 2.2** Scanning electron micrograph of starch grains in amyloplasts of a rice stem parenchyma cell. Source: Salisbury & Ross 1992 (courtesy of P. Dayanandan).

ecules, especially when the chains are folded, and so are used as an energy store both in plants (starch) and in animals (glycogen). In plants, but not in animals, polysaccharides also have a structural function (e.g. cellulose, hemicellulose, pectin).

Starch is a mixture of two different polysaccharides: amylose, a linear molecule, and amylopectin which is branched. These are tightly folded to form starch granules, which are stored within the cytoplasm of plant cells bound by a membrane in a structure called an amyloplast (Fig. 2.2). Starch is an important component in many of the commodities considered in this book. It has great commercial importance as it is used in the manufacture of many food and non-food products. The cooking and pasting characteristics of starch are critical for many uses. The differences in the characteristics of starches from different commodities are very complex. Among other things, though, these characteristics depend on the ratio of amylopectin and amylose.

Like starch, cellulose consists of long chains of sugars. In this case the molecules are unbranched and are bound together in long cylindrical fibres called microfibrils, as illustrated in Fig. 2.3. The resulting structure has great tensile strength, and cellulose is the major component of plant cell walls. The other components of cell walls are hemicelluloses, which have a more complex branched structure, and pectins, which are smaller acidic molecules which bind readily to ions and are important for cell cohesion.

### *The role of carbohydrates in plants*

When glucose is formed in the green leaves of plants by photosynthesis, it is either used to form starch granules and stored in the leaf, or it is converted to sucrose. This disaccharide is a water-soluble sugar, so that it can be transported to other parts of the plant to provide energy by respiration, to be stored as starch, or to be used to synthesise structural components such as cellulose for growth.

Commodities eaten by humans may contain any, or all, of the forms of carbohydrate mentioned above.

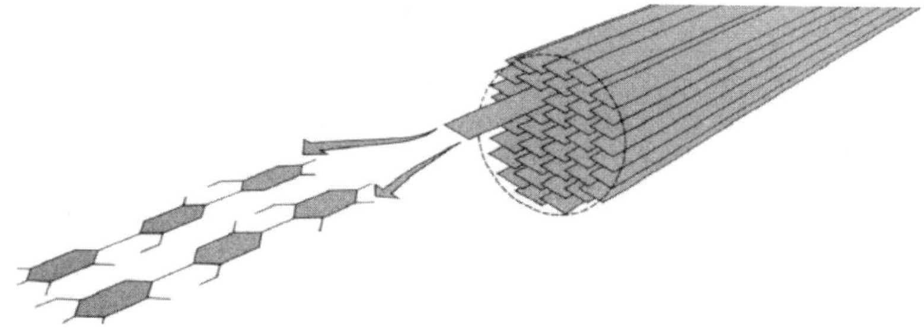

**Fig. 2.3** A schematic drawing of the arrangement of cellulose molecules to form a microfibril of cellulose. The hexagons represent glucose molecules in the long-chain cellulose molecules. Source: Salisbury & Ross 1992.

Many ripe fruits contain high levels of simple sugars – hence their sweet flavour. Some fruit, such as plantain and breadfruit, contain carbohydrate stored in the form of starch, as do plant storage organs such as cassava and sweet potato. All plant tissue contains considerable amounts of structural polysaccharides.

### *The role of carbohydrates in humans*

When eaten, starch and disaccharides must be broken down by digestive enzymes into monosaccharides before they can be absorbed into the body. Starch granules can be very difficult for humans to digest. In most cases starchy foods are cooked before being eaten in order to break the structure of the starch granule so that it is more accessible to the enzymes that break it down and thus more digestible. After absorption, monosaccharides are converted to glucose. The glucose is then either used to provide immediate energy by respiration or used to synthesise the polysaccharide glycogen, or fat for storage. Cellulose, on the other hand, cannot be broken down by humans. It is, however, important as a source of dietary fibre which is necessary to keep the digestive process moving. Cows and other ruminant animals are able to break down and utilise cellulose for energy due to the presence of cellulose-digesting micro-organisms in the rumen.

## Lipids

Fats and oils are together classed as lipids. Lipids, like carbohydrates, are composed of the elements C, H and O but in different proportions. Lipids are made primarily of triglycerides (Fig. 2.4), which are esters of glycerol and long chain (12–20 carbon) fatty acids. Glycerol has three hydroxyl groups, each of which may be esterified

```
        CH2-OOC-R1
            |
    R2--COO--CH
            |
            CH2-OOC-R3
```

Triglyceride

```
 H  H  H  H  H  H  H
 |  |  |  |  |  |  |
--C--C--C--C--C--C--C--
 |  |  |  |  |  |  |
 H  H  H  H  H  H  H
```

Saturated fatty acid

```
 H  H   H   H  H   H  H
 |  |   |   |  |   |  |
--C--C==C--C--C==C--C--
 |          |         |
 H          H         H
```

Unsaturated fatty acid

**Fig. 2.4** The structure of a triglyceride, saturated fatty acid and unsaturated fatty acid.

(a reaction which binds the two molecules), with a different fatty acid. There is a wide range of fatty acids. In saturated fatty acids, the hydrocarbon chain is saturated with hydrogen and the chain is straight. In unsaturated fatty acids, the hydrocarbon chain is not saturated with hydrogen and therefore has one double bond or more. In this case the chain is bent. When all the fatty acids are saturated, the straight chains give a very ordered structure for the triglyceride mixture, which therefore tends to be solid at room temperature, and is termed a fat. When the fatty acids are unsaturated, the bent chains give the triglyceride mixture a more mobile structure, so that it tends to be liquid at room temperature, and is termed an oil.

Triglycerides have a very important function in providing the main structure of cell membranes. The extent of saturation of the fatty acids has an important effect on the molecular packing and, therefore, on the fluidity of the membrane, which is vital for cell function.

### *The role of lipids in plants*

Most lipids within the human diet are derived from animal products. Plants can, however, be an important source of unsaturated lipids, usually in the form of oils. Coconuts, palms, soybean, sunflower seed, cottonseed and many nuts contain appreciable amounts of oil. Only certain fleshy fruits such as avocado and olive can compare with these. Leaf and stem vegetables and most fruits are normally low in oil content.

### *The role of lipids in humans*

In addition to their vital role within cell membranes, lipids or fats also function in humans for energy storage and insulation. Fat has more than twice the calorific value (energy content) per weight than carbohydrate. Thus, it is useful for people who have a particularly high energy requirement, as it reduces the bulk of the food that must be eaten. Excess fat, not immediately required for energy, is stored in adipose tissue. This not only acts as an energy store but also insulates against heat loss and can protect delicate organs from physical damage.

*Essential fatty acids:* Many of the fatty acids can be synthesised within the human body, either by conversion from other fatty acids or from carbohydrates. Certain fatty acids, however, must be ingested in small amounts for the proper functioning of the body. They must be supplied by fat in the diet, as they cannot be synthesised by humans. These essential fatty acids are linoleic, linolenic and arachidonic acids, which are needed for the formation of cell membranes and for the synthesis of certain hormones (Fig. 2.5).

Linoleic acid

$CH_3(CH_2)_4\,(CH{=}CHCH_2)_2(CH_2)_6COO$-

Linolenic acid

$CH_3CH_2\,(CH{=}CHCH_2)_3(CH_2)_6COO$-

Arachidonic acid

$CH_3(CH_2)_4\,(CH{=}CHCH_2)_4(CH_2)_2COO$-

**Fig. 2.5** Essential fatty acids.

It has been estimated that dietary requirements will be met if about 2% of total energy supply is in the form of fat. However, to ensure unrestricted absorption of fat-soluble vitamins it has been proposed that the amount of fat in the human diet should not fall below 10% (Jequier 1999).

In many developed countries there is concern that fat intake is too high, and many measures are being taken to reduce it. This is not usually such a problem in poorer communities.

## Proteins

Proteins are a very important group of nutrients, as they are needed to synthesise many structural and control components of the body. They contain the elements carbon, hydrogen, oxygen, nitrogen, sulphur and sometimes phosphorus. They are made up of small chemical units called amino acids, which are linked in chains. About 20 different amino acids occur in the proteins found in foods.

### *Amino acids*

The generalised structure of an amino acid, together with two examples, is shown in Fig. 2.6. Proteins are large, complex molecules with a typical molecule containing about 500 amino acids. The three-dimensional shape of the protein depends on the sequence of the amino acids and may be coiled, globular or fibrous.

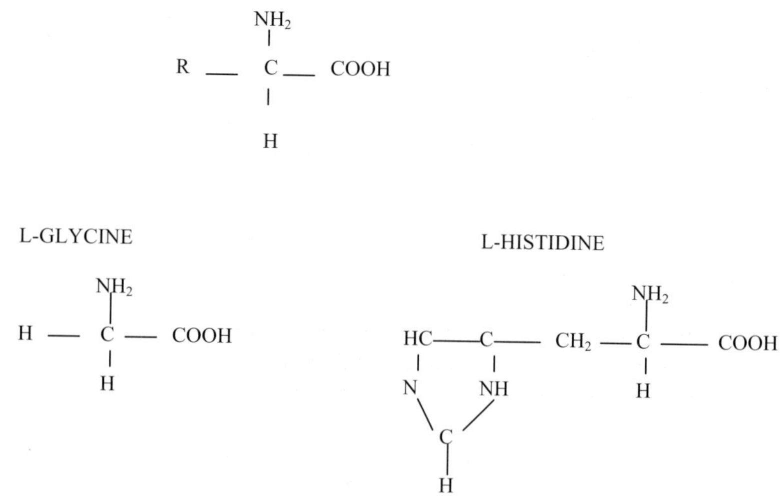


**Fig. 2.6** The structure of a generalised amino acid, with two examples.

### *The role of proteins in plants*

Proteins have a wide range of essential functions in plants. For example, proteins function as enzymes, which are vital for controlling metabolic reactions in all living systems. They also form an important part of cell membranes.

Plants are able to synthesise amino acids using carbohydrates formed by photosynthesis, and absorbing nitrogen (in the form of nitrates and nitrites), sulphur and phosphorus from the soil. Leguminous plants can absorb nitrogen from the air with the aid of bacteria.

### *The role of proteins in humans*

In animals, proteins also function as enzymes and have a vital role within cell membranes. In addition, in higher animals, proteins are involved as structural components of bones, cartilage and tendons, ligaments, hair and nails. Blood proteins are vital for disease resistance and for carrying oxygen.

Unlike plants, animals cannot synthesise amino acids from the original elements and therefore rely directly or indirectly on plant-derived proteins. As an approximation, the human body requires between 0.7 g and 0.9 g of protein to every kilogram of body weight per day.

Digestion of proteins from animal or plant origin releases the constituent amino acid sub-units. The amino acids are rapidly taken up by the tissues. After absorption the amino acids may be rearranged to form new proteins required in the body. Amino acids are also involved in a whole range of other important metabolic reactions.

For many of the amino acids, the human body is able to convert one that is in excess into another which is in short supply. However, there are eight amino acids for which this is not possible. These are called essential amino acids and must be supplied by the protein in the diet. One additional amino acid, histidine, must be supplied in the diets of rapidly growing infants. Essential and non-essential amino acids are listed in Table 2.1.

The proteins in animal products generally contain the whole range of required amino acids, whereas plant protein sources are frequently deficient in one or more amino acids. Cassava, for instance, is deficient in most of the essential amino acids. Consequently, for those people who derive their major or sole protein requirement from plants, it is important that they eat a variety of different plant products to avoid problems of certain amino acid deficiencies.

**Table 2.1** Essential and non-essential amino acids.

| Essential | Non-essential |
|---|---|
| Isoleucine | Alanine |
| Leucine | Arginine |
| Lysine | Asparagine |
| Methionine | Aspartic acid |
| Phenylalanine | Cysteine |
| Threonine | Glutamic acid |
| Tryptophan | Glycine |
| Valine | Ornithine |
| Histidine (essential for infants) | Proline |
| | Serine |
| | Tyrosine |

Excess protein cannot be stored by the body in the same way that fat and carbohydrate can, but excess amino acids are broken down. Nitrogen is removed by the liver, and excreted as urea, while the remaining part, now a carbohydrate, is used for energy. This means that it is important to maintain a constant supply of protein in the diet. Although proteins are poor sources of energy, in extreme cases of energy malnutrition proteins are oxidised to release energy, thus depriving the body of what few essential amino acids it has received.

The main functions of carbohydrates, lipids and proteins in plants and humans are listed in Table 2.2.

### Vitamins

The term vitamin does not refer to a molecule of particular structure, but covers a range of diverse compounds needed for the synthesis of certain vital components of the body, usually enzymes, that cannot be made by the body, and must therefore be supplied by the diet. While many vitamins are essential for good health and bodily functions, the quantities required are much smaller than for carbohydrates, lipids and proteins. Insufficient intake of vitamins leads to a disruption of the body biochemistry, which is expressed as deficiency diseases. Fruit, vegetables and root crops provide a major role in human nutrition as a source of many vitamins and minerals.

Over 80% of vitamin C needed in the diet is derived from fruits and vegetables, leafy vegetables and tropical fruit being important sources. Citrus fruits, such as oranges and grapefruit, are particularly high in vitamin C. Most animal products are deficient in this vitamin. The carotenoid pigments such as beta-carotene present in many plant products (carrots, peppers, pumpkin, squash) are the major precursors of vitamin A. These provitamins present in fruit and vegetables contribute significantly to the supplies of vitamin A in many diets. Plant tissues are generally a poorer source of B vitamins than animal products, but nevertheless in certain parts of the world they contribute not insignificant proportions of thiamine, riboflavin, niacin and in particular folic acid.

Certain vitamins, notably vitamin C (ascorbic acid), are liable to break down during processing, cooking or storage. Thus vitamin content can be an important quality characteristic which it is necessary to maintain during storage.

### Minerals

The human body requires a wide range of mineral elements to function healthily – potassium, sodium, chlorine, calcium, phosphorus, sulphur and iron being the major ones. Fruit and vegetables make an essential contribution to the supply of these elements, but in particular of calcium and iron. Leafy vegetables, particularly spinach, contain some calcium although there are certain factors in spinach and other commodities that render the calcium largely unavailable. Calcium deficiency leads to

**Table 2.2** A summary of the main functions of carbohydrates, lipids and proteins in plants and humans.

| | Main function in plants | Main function in humans |
|---|---|---|
| *Carbohydrates* (composed of carbon, hydrogen and oxygen) | | |
| Simple sugars (monosaccharides, disaccharides) | Energy to drive metabolism | Energy to drive metabolism (Calories) |
| Polysaccharides (large carbohydrates made from simple sugars) | Storage, e.g. starch<br>Structure, e.g. cellulose, lignin | Storage, e.g. glycogen<br>Not very important for structure |
| *Lipids* (composed of carbon, hydrogen and oxygen) | Constituents of cell membranes<br>Storage of energy (but less important than carbohydrates) | Constituents of cell membranes<br>Storage of energy (more important than carbohydrates)<br>Cushioning of organs<br>Heat insulation |
| *Proteins* (composed of carbon, hydrogen, oxygen, nitrogen, sulphur and some phosphorus) | Enzymes control metabolic processes<br>Important functional component of cell membranes | Enzymes control metabolic processes<br>Important functional component of cell membranes<br>Important in structural tissues such as bones, cartilage and tendons |

the diseases of rickets and osteomalacia. Iron is essential for the functioning of the blood protein haemoglobin as a transporter of oxygen. Deficiencies of the element lead to anaemia. Leafy vegetables also supply a significant amount of iron in many diets.

## The structure and components of a cell

Living tissues are made up of cells. Cells can be divided into two main types: eukaryotic cells and prokaryotic cells. All multicellular animals and plants, including all commodities that we consider here, are made up of eukaryotic cells, whereas certain unicellular organisms, such as bacteria and blue-green bacteria, consist of prokaryotic cells. Here we will consider only eukaryotic cells.

Plant and animal cells differ in many aspects. For example, plant cells, unlike animal cells, have a rigid external cell wall. This provides much of the structure for plant products. However, the cell also requires the internal pressure of water (cell turgor) to maintain its structure. After forming, cells differentiate into many different types that have different functions within the overall plant structure. However, cells also have many aspects in common. Figure 2.7 shows a generalised plant cell.

The main components of the plant cell are the cell wall, cell membrane, cytoplasm and nucleus. Within the cytoplasm there are many components with different functions, and some of these components are separated from the main cytoplasm by membranes. Thus, a cell may be considered to consist of several compartments in which the various different metabolic reactions essential for life occur. The correct functioning of a cell depends on these compartments being kept separate, but with the ability for chemicals to move selectively between them. The structure of the membranes separating the various parts of the cell is therefore critical. The term compartmentation is used to describe the controlled separation of different parts of the cell.

### Cell wall

The cell wall is composed primarily (90–95%) of large complex carbohydrates: cellulose, hemicellulose and pectin. The rest of the structure is glycoproteins, which are proteins with carbohydrate side chains.

The cell wall consists of different layers; the middle lamella (intercellular layer), the primary cell wall and the secondary cell wall. The middle lamella can be considered as the cement that holds the individual cells together. Although the detailed structure of the cell wall is little understood, it is known that pectins are located in the middle lamellae, and are very important in holding adjacent cells together. Softening of fruit during ripening

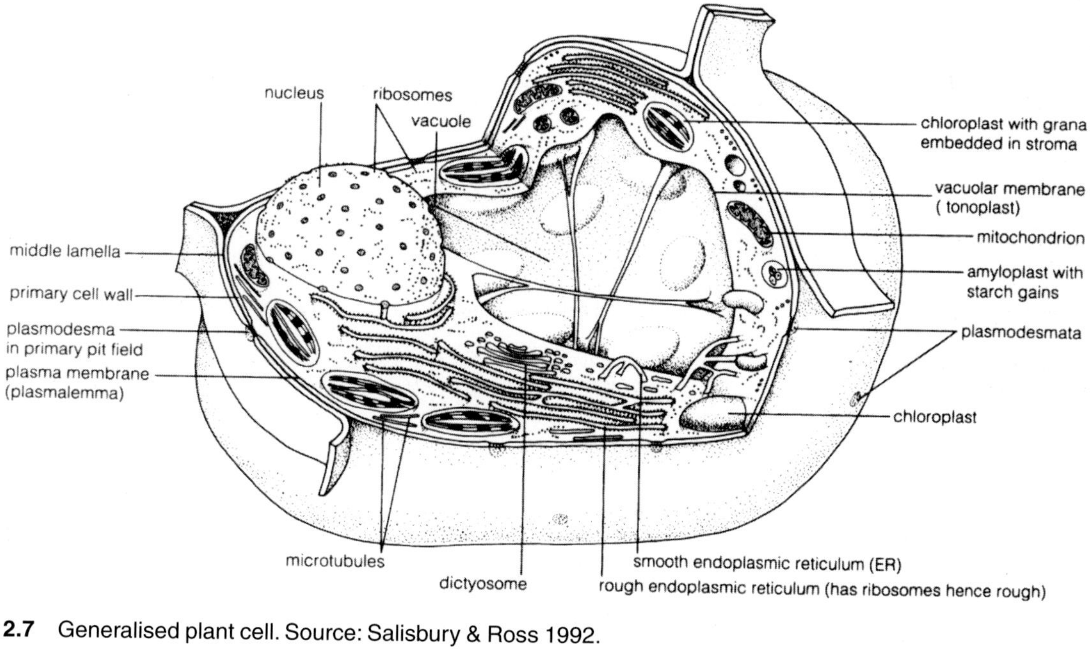


**Fig. 2.7** Generalised plant cell. Source: Salisbury & Ross 1992.

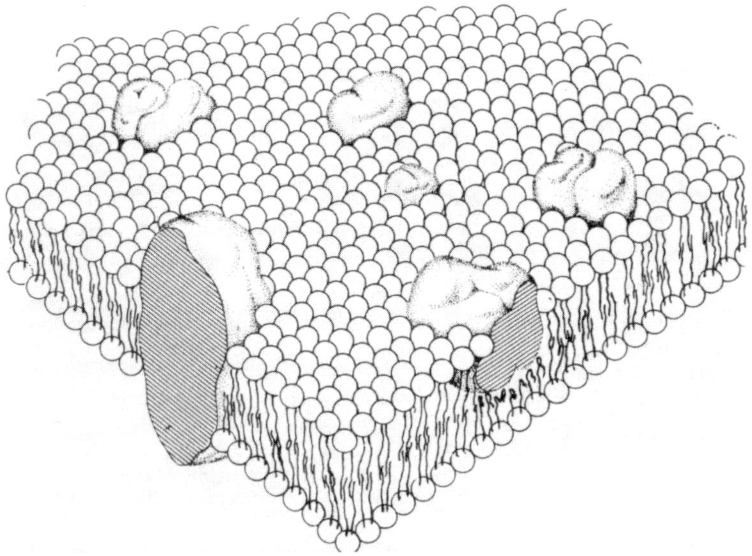

**Fig. 2.8** A three-dimensional drawing of a cell membrane. The phospholipid bilayer consists of: (a) ionic and polar head groups (the small circular structures in the drawing) of the phospholipid molecules which make contact with water, and (b) fatty acid chains (wavy lines). The large bodies with stippled cut surfaces represent the globular proteins. Source: Singer & Nicolson 1972.

is associated with the breakdown of pectins, so that cells detach from one another. The primary cell wall is the first true wall that develops on a new cell. The secondary cell wall consists of additional layers underneath the primary cell wall which only develop in cells which have ceased to grow. This often results in the tissue becoming increasingly tough and fibrous or woody (a secondary cell wall is not shown in Fig. 2.7).

In living cells there are areas where the cell wall is absent, known as pits. This allows communication between cells via plasmodesmata.

### Cell membranes

Cell membranes surround the cell beneath the cell wall and also provide important barriers between different parts of the cell. They are able to selectively control the movement of different substances between cells and between compartments within cells. This is vital for the correct functioning of the cell.

Membranes consist primarily of a double layer of triglycerides, with embedded proteins. The structure is illustrated in Fig. 2.8.

Triglycerides consist of phospholipids attached to three fatty acids. Phospholipids are polar, which means that they interact (or dissolve) easily with water, whereas fatty acids are hydrophobic and therefore tend to separate from water. Because of this, the triglycerides form into a double later with phospholipid on the outside and the fatty acid components on the inside, to avoid the surrounding water. The resulting structure is a bilayer which is generally fluid. Proteins are embedded in the membrane, such that the whole structure is often referred to as a fluid mosaic. The proteins have a range of very important functions. Many of them control the movement of different substances across the cell membrane. The mobility or fluidity of the membrane is very important for the correct functioning of these proteins.

Saturated fatty acids tend to be straight and form very rigid structures, whereas unsaturated fatty acids are bent and form more mobile structures. Thus, the inclusion of unsaturated fatty acids in the cell membrane is important for maintaining its fluidity and correct functioning (Fig. 2.9).

If the functioning of the cell membranes is impaired in any way, the movement of substances between com-

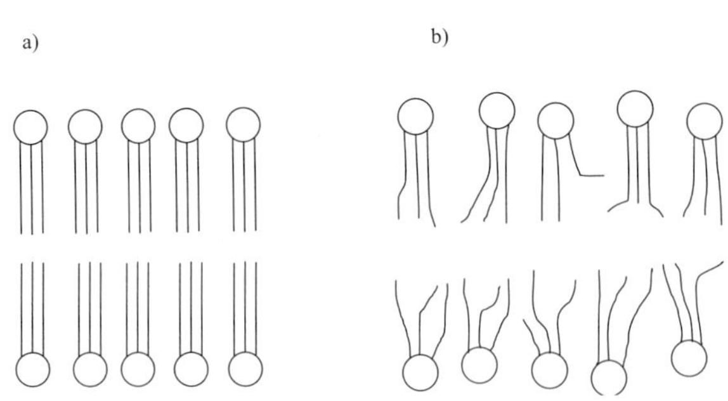


**Fig. 2.9** (a) Rigid structure when fatty acids are saturated, (b) fluid structure with many unsaturated fatty acids.

partments is affected, which can disrupt the metabolism of the whole cell. Low temperatures reduce membrane fluidity, and this is considered to be one of the causes of the damage to commodities, known as chilling injury, which occurs when commodities are stored at suboptimal temperatures.

### Cytoplasm

The cytoplasm (or cytosol) is a viscous solution that makes up the main part of the cell. Many important metabolic processes occur here, and the structures described below are all located within the cytoplasm. It is becoming apparent that rather than being an amorphous solution there is a great deal of structure within the cytoplasm. For example, there is a cytoskeleton composed of a network of protein filaments (microfilaments and microtubules) that gives the cell shape and enables the movement and rearrangement of cellular components.

#### *Nucleus*

The nucleus contains the genetic material of the cell in the form of deoxyribonucleic acid (DNA). This acts as the 'control centre' of the cell, providing most of the information for and control of synthesis of new components.

#### *Mitochondria*

Mitochondria are membrane bound structures or organelles in which the processes of respiration take place.

#### *Chloroplasts and other plastids*

Whereas mitochondria are also found in animal cells, plastids are membrane bound structures or organelles which are only found in plant cells.

Chloroplasts are plastids in which the processes of photosynthesis take place. They are green because they contain chlorophyll.

Chromoplasts are coloured, usually yellow, orange or red because they contain carotenoid pigments. They are responsible for the colour of petals, ripe fruits and some roots such as carrots. Their function, other than to store colour pigments, is little understood.

Amyloplasts are plastids used for the storage of starch grains. Likewise proteinoplasts can store protein and elaioplasts can store lipids.

#### *Vacuoles*

Vacuoles are compartments within the cell that basically act as a store for substances which the cell wants to keep out of the cytoplasm. Thus, they may be a store for waste material and also function to keep potentially destructive enzymes away from the cytoplasm. Vacuoles also help to maintain turgor pressure. In sugary fruit, for example, most of the sugar is within the vacuole. The vacuole often takes up the main part of the volume of the cell (more than 90%).

#### *Ribosomes and endoplasmic reticulum*

Ribosomes are proteins which control the synthesis of proteins as coded within nucleic acid sequences in the nucleus. Endoplasmic reticulum is a structure where protein synthesis occurs.

#### *Dictyosomes or Golgi bodies*

Dictyosomes or Golgi bodies are structures involved in the synthesis of substances to be secreted out of the cell.

## Plant tissues

There are many different types of cell, within different plant tissues. Some of these plant tissues are important for the understanding of post-harvest physiology.

### Surface (dermal) tissues

Surface tissues have the function of protecting the plant from water loss and invasion of pathogens. The two principal types are epidermis and periderm. Most above-ground parts of the plant are covered by the epidermis, whereas plant parts such as roots and stems that increase in thickness due to secondary growth are covered by a periderm. The epidermis is usually one cell thick, although a layer two to three cells thick is possible. A covering of wax, secreted by the epidermal cells, is usually present to reduce water loss. The periderm, on the other hand, is usually many cells thick, and may also be covered by wax. Another important difference between epidermis and periderm is that, for the cells of a periderm, secondary cell walls have developed and the cells are no longer living.

As the plant tissues are dependent on gas exchange for photosynthesis and respiration, pores exist in both types of dermal tissue; stomata in the epidermis and lenticels

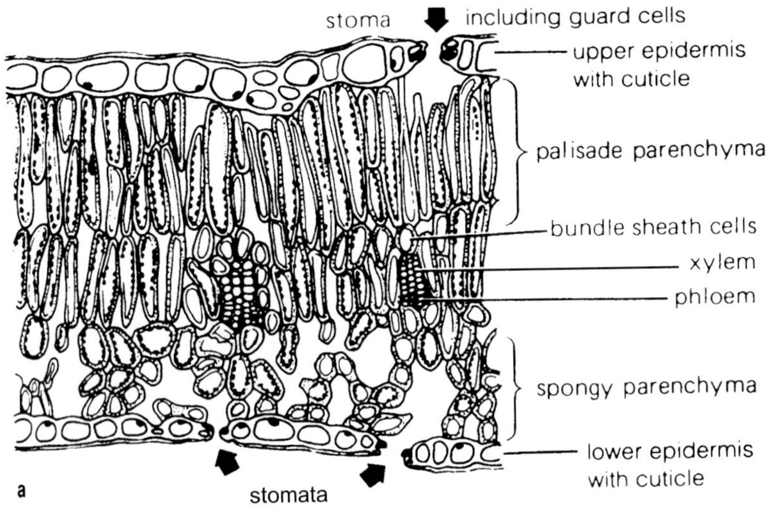


**Fig. 2.10** Cross-section through a leaf showing cell structure, including epidermis and stomata. Source: Salisbury & Ross 1992.

in the periderm. Stomata are bordered by cells called guard cells. They can be actively opened or closed. Lenticels, on the other hand, are openings which cannot be controlled.

Figure 2.10 shows the structure of a leaf, with a single-layer epidermis and stomata, while Fig. 2.11 shows the structure of a periderm. Within the periderm there are three different types of cell. The phellem is the water-resistant outer layer of cells which have secondary cell walls and are no longer living. The phellogen is the underlying layer of cells which are actively growing and dividing to form the phellem. The phelloderm are the cells under the phellem. Figure 2.12 shows a lenticel. Note that it is much larger than a stoma.

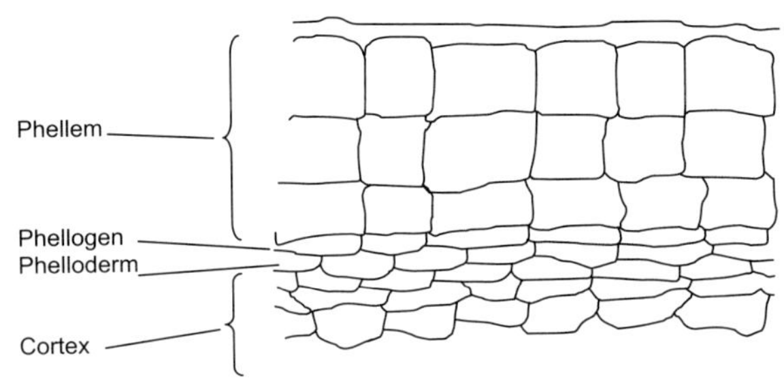


**Fig. 2.11** The structure of a periderm layer. Modified from Fahn 1990.

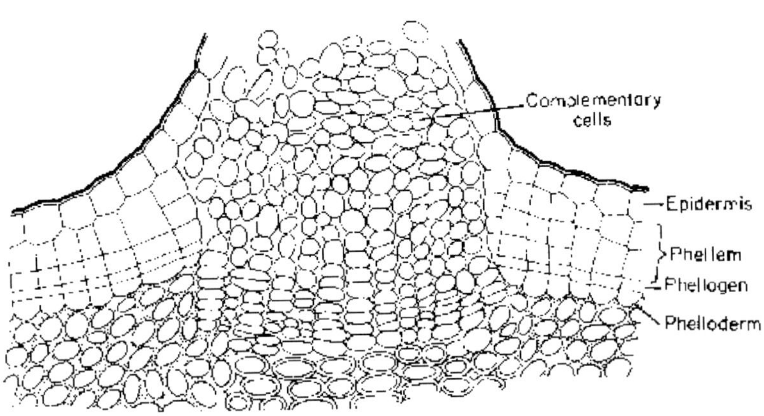


**Fig. 2.12** A lenticel. Source: Fahn 1990.

### Ground tissue, parenchyma

Parenchyma cells form the ground tissue of most plant products, in many cases acting as storage sites for carbohydrates, lipids or proteins. Parenchyma can be very diverse. In Fig. 2.10, parenchyma cells are shown in leaf tissue. In this case they contain chloroplasts for photosynthesis. In Fig. 2.12, which is typical of the structure of a storage root, the underlying cells are also parenchyma, but in this case are primarily stores of starch. In leaves, fruits, roots and tubers parenchyma tissues tend to contain many air spaces. Parenchyma cells do not usually have secondary cell walls.

### Support tissue

Collenchyma and sclerenchyma cells are similar to parenchyma but have secondary cell walls. Their main function is to provide greater support to the plant. These cell types are apparent as fibres in many commodities.

### Vascular tissue

Vascular tissue is involved in the transport of substances through the living plant. Xylem vessels transport water and minerals from the plant roots to the rest of the plant. They consist mainly of elongated cells with secondary cell walls that are no longer living. Xylem vessels may also contribute significantly to the plant structure. Phloem vessels transport carbohydrates (mainly sucrose but also organic acids) produced by photosynthesis in the leaves, to other parts of the plant. Although the component cells of phloem vessels are similar to xylem vessels in that they are elongated, they have less secondary thickening and are alive.

### Meristematic tissue

Meristematic cells are those that retain the ability for cell division, and they are essential for the formation of new tissues. They are found in the growing tip of shoots and roots. They are also active in the formation of periderm and for wound healing. In Fig. 2.11, the phellogen is meristematic tissue that is forming the phellem of the periderm.

## Plant structures and life cycles

Man uses plant products that correspond to many different parts of the plant and many different stages in plant development as a source of food. Different plant parts have different physical and physiological characteristics, and it is important to appreciate these in order to know how best to store and handle produce. Equally, it is very important to have an understanding of how the commodity in question fits into the developmental life of the plant, because when produce is harvested it will tend to try to fulfil its natural function. If the harvested produce does not first become unacceptable because of forms of deterioration such as wilting, microbial decay or insect attack, it will eventually become unacceptable because of the natural developmental, maturation and senescence processes.

Figure 2.13 shows a generalised plant indicating the different plant structures, and the food products that correspond to these different plant structures. This generalised plant bears buds, flowers and fruits simultaneously, but for most plants these would occur at different times. Figure 2.14 shows a simplified life cycle of a plant. Examples of commodities, corresponding to different developmental stages and plant parts, are given in Table 2.3.

The ease with which the various plant parts shown in Table 2.3 can be stored depends on what part of the plant they are, as some plant parts are more suited to storage than others. Commodities such as seed, nuts or storage roots store better than fruit and vegetables, as their purpose within the plant life cycle is preservation. Plant parts such as fleshy fruits and vegetables are not designed for preservation and will therefore tend to continue development and die (e.g. fruits ripen and senesce).

On the basis of origin and suitability for storage, plant products can be separated into durables and perishables. The main differences between durables and perishables are summarised in Table 2.4.

The plant products first stored and transported by man were all durables. It is only relatively recently that man has been able to keep and transport the more perishable products. It is important to appreciate, though, that those products defined as perishables can vary considerably in their keeping qualities, with normal storage periods ranging from hours to weeks.

### Vegetative growth

During the vegetative stage, the plant is concerned primarily with photosynthesising to grow. Photosynthesis is the process by which plants use the energy of

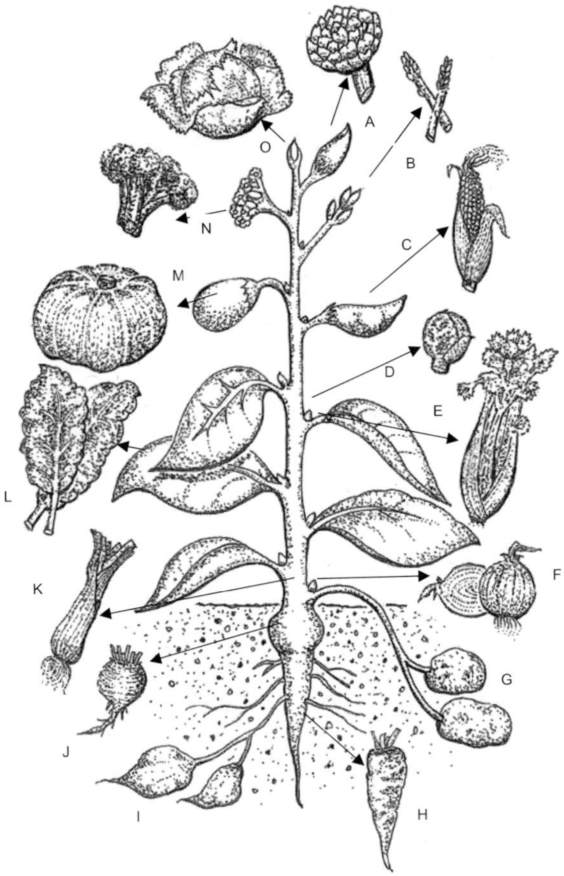


**Fig. 2.13** Plant structures: plant parts utilised by man are derived from virtually every portion of the plant. The diagram illustrates examples of various parts from a cross-section of plants used as vegetables. The letters indicate the principal origins of representative vegetables as follows: (A) flower bud, e.g. artichoke; (B) lateral sprout, e.g. asparagus; (C) seeds, e.g. corn; (D) axillary bud, e.g. Brussels sprout; (E) petiole, e.g. celery; (F) bulb (underground bud), e.g. onion; (G) stem tuber, e.g. potato; (H) swollen root, e.g. carrot; (I) swollen root, e.g. sweet potato; (J) swollen hypocotyl, e.g. beetroot; (K) swollen leaf base, e.g. leek; (L) leaf blade, e.g. spinach; (M) fruit, e.g. pumpkin; (N) swollen inflorescence, e.g. broccoli; (O) apical bud, e.g. lettuce. Source: Wills *et al.* 1998.

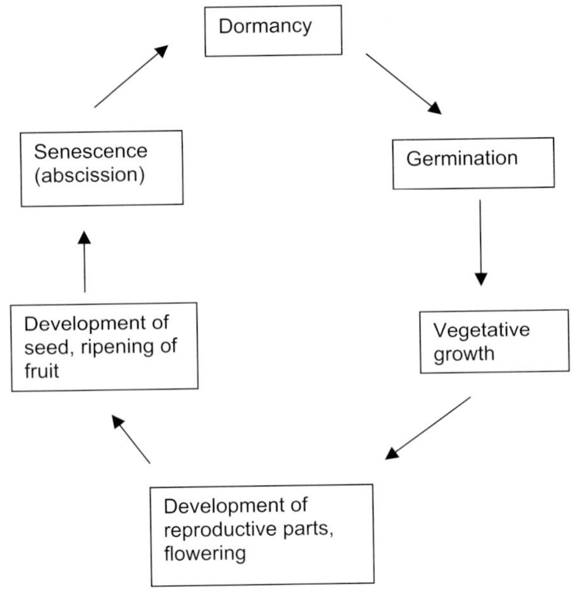


**Fig. 2.14** The plant life cycle.

absorbed light to synthesise carbohydrates from water and carbon dioxide. The main plant parts produced at this stage are roots, stems and leaves. Roots are needed to absorb water from the soil and anchor the plant. Other than storage roots, these plant parts are not usually an important food crop. The leaves are produced to absorb sunlight and carbon dioxide from the atmosphere, while stems and petioles are necessary to hold these leaves in the light, and for the transport of water to the leaves and carbohydrates away from the leaves. These structures are produced rapidly and have large intercellular spaces. They have a very high content of structural molecules such as cellulose and a very high water content. With respect to nutrition, these plant parts have low calorific value but may provide many vitamins and roughage.

The shape of the stems and leaves reflects their function. In order to photosynthesise efficiently, leaves must have a large surface area to absorb light and so plants tend to produce flat thin leaves. This gives them a large surface area to volume ratio, which means they have a tendency to lose moisture rapidly and therefore perish rapidly. Leaves must also be able to absorb carbon dioxide in order to provide the carbon that is built into larger molecules during the process of photosynthesis. For this, leaves and green stems have pores, called stomata, to allow gas exchange. While carbon dioxide is absorbed through the stomata, water vapour is continuously lost. This must be replaced by a continuous supply of water from the roots.

**Table 2.3** Examples of plant produce corresponding to different stages of development.

| Plant development stage | Plant part | Examples of produce |
|---|---|---|
| Dormancy | Seed | Wheat, rice, beans |
| | Tuber | Potato |
| Germination | Sprouting seed | Bean sprouts |
| Vegetative growth | Shoot | Asparagus |
| | Stem | Celery |
| | Leaves | Lettuce, cabbage |
| Flowering | Flower | Artichoke, broccoli, cauliflower |
| Development of fruit and seed | Immature fruit | Cucumber, green beans |
| Ripening of fruit | Ripe fruit | Tomato, banana, orange |

**Table 2.4** Differences between durable and perishable commodities.

| Durables | Perishables |
|---|---|
| Designed for preservation | Not designed for preservation |
| Low moisture content, usually 10–15% | High moisture content, usually 50–90% |
| Small unit size, less than 1 g | Large unit size, typically 5 g to 6 kg |
| Often symmetrical in shape | Often asymmetrical in shape |
| Hard texture | Soft texture |
| Stable – inherent storage life of years | Perishable – natural storage life of a few days to months depending on type |
| Losses mainly caused by external factors, e.g. mould, insects and rodents | Losses caused by external factors, mainly moulds and bacteria, and internal factors, e.g. respiration, sprouting, ripening, etc. |

Once the vegetative parts of plants are harvested, and therefore separated from the roots, they become very susceptible to water loss, because:

- leaves have a high surface area to volume ratio;
- leaves have thin skins (epidermis) with a translucent waxy surface, and many stomata, which make the epidermis permeable to water vapour; and
- once the above ground parts have been separated from the roots, there is no longer any replenishing water supply.

An additional important characteristic of vegetative tissue is that it often exhibits responses such as phototropism (growing towards the light) or negative geotropism (growing upwards). These responses have been developed to help the plant grow into a form that maximises light absorption. The fact that vegetative tissue may continue growing after harvest and may maintain these reactions has implications with respect to the way that they should be stored. For example, if asparagus (a growing shoot) is stored horizontally, it will have a tendency to bend upwards. The same response may be seen in cut flowers.

### Development of reproductive parts: flowering

After a period of vegetative growth, the plant enters a stage of reproductive growth initiated by the development of flowers. These are usually conspicuous, delicate structures that are morphologically similar to leaves. Many flowers are very important commercially for their ornamental value. However, there are also several edible plant products which are flowers, in botanical terms, such as broccoli and cauliflower, though cauliflowers no longer function biologically as flowers, but have been bred as an edible vegetable.

The role of flowers is to bring about fertilisation of the female ovule by the male pollen. For some flowers, such as grasses and cereal crops, the pollen from one flower is delivered to the ovule of another flower by the wind. For other flowers, the pollen is moved by insects. In the latter case, the flowers tend to be brightly coloured and are often scented to attract insects. Flowers marketed for ornamental purposes fall into this category. As pollination occurs fairly rapidly, the function of the flower is short-lived, so that the flower is not designed to survive for long. The flower is an example of a plant structure

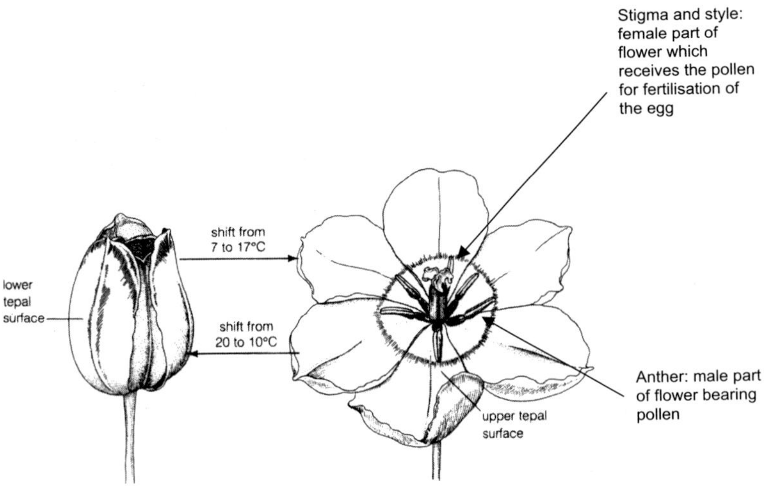


**Fig. 2.15** Example of an insect-pollinated flower. Source: Fahn 1990.

that is programmed to die or senesce, so that the next stage in reproduction can occur.

Figure 2.15 shows the structure of an insect-pollinated flower, while Fig. 2.16 shows examples of wind pollinated flowers of cereals. The flowers of all cereals, except maize, consist of an ovary, three stamens (six in rice) and two lodicules which are all enclosed in a pair of bracts called the lemma and palea. In maize the male and female flowers are located in separate inflorescences on the same plant. The male flowers are in an apical panicle (the tassel) while the female flowers are borne in the axils of the lower leaves. The female inflorescence takes the form of a central rachis (the cob) which bears rows of sessile spikelets. Each spikelet has a pair of glumes and two florets, of which only the upper one is fertile. The whole cob is enclosed by a number of overlapping bracts that will later form the husk.

Flowering may be induced by many factors. It may occur solely as a consequence of the stage of morphological or physiological development, or alternatively, especially at higher latitudes, it may be controlled by seasonal factors such as day length (photoperiodism) or temperature fluctuations (thermoperiodism). For some temperate crops, flowering will only occur if the seed of the plant has been subjected to low temperatures during the previous winter. This is called vernalisation. An understanding of the control of flower development can be especially important if the flower is the economically important part of the crop.

**Fig. 2.16** Examples of wind-pollinated flowers.

### *Senescence*

Once fertilisation has occurred, the function of the flower is fulfilled. At this point the flower tissues deteriorate and die. This process is termed senescence. Senescence describes a series of endogenously controlled deteriorative changes that result in the natural death of cells, tissues, organs or organisms. Senescence should not be thought of as cell death resulting from an inability to maintain function, but rather as a programmed cell death. This occurs throughout plant development as cells or organs become redundant. Many changes that occur after harvest, especially in highly perishable products, are part of senescence. As well as being used to describe the deterioration of flowers the term is also used to describe ripening of fruit.

### Development of seed: ripening of fruit

Following pollination of the flower comes fruit development. A fruit is a structure which contains a seed receptacle developed from the ovary of the flower. Botanically, fruits include grains, nuts, fleshy fruits such as bananas, apples and many commodities that are consumed as 'vegetables' (peppers, runner beans, mangetout).

These commodities differ greatly. Due to their biological function grains and nuts are very durable. They are of enormous economic importance. Conversely, the fleshy fruits are not designed to keep at all. They evolved to be eaten by animals as a mechanism for distributing the seed (the only part of the fruit designed to be preserved). Fleshy fruits are specifically designed to be attractive to animals in terms of taste, smell and appearance. Decomposing fruit may also act as a food supply for seed to grow. For both reasons these fruits are often high in sugars. There are many cases where the commercially important plant product is actually an immature, non-ripened fruit. In these cases, such as cucumber and green peppers, the fruits tend to be less sweet and less brightly coloured.

The ripening of fleshy fruit describes a process by which, while the seed is completing its development, the rest of the fruit changes in terms of structure and composition into its edible form. As ripening of the fruit is an immediate precursor to the biological death of the fruit tissue, it is often considered as a special case of senescence.

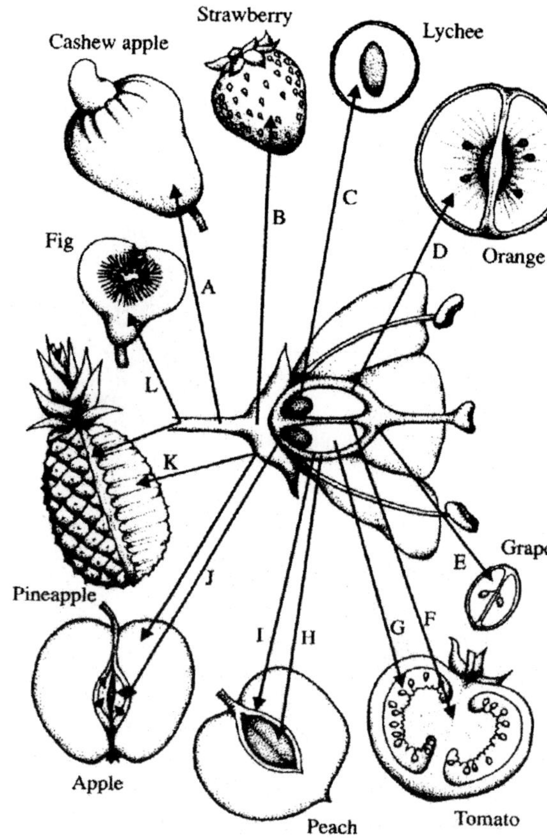


**Fig. 2.17** Derivation of some fruits from plant tissue. The letters indicate the tissues that comprise a significant portion of the fruit illustrated as follows: (A) pedicel, e.g. cashew apple; (B) receptacle, e.g. strawberry; (C) aril, e.g. lychee; (D) endodermal intralocular tissue, e.g. orange; (E) pericarp, e.g. grape; (F) septum, e.g. tomato; (G) placental intralocular tissue, e.g. tomato; (H) mesocarp, e.g. peach; (I) endocarp, e.g. peach; (J) carpel, e.g. apple; (K) accessory tissue, e.g. apple and pineapple; (L) peduncle, e.g. pineapple and fig. Source: Wills *et al.* 1998.

#### *The physiological origin of fleshy fruit*

Fleshy fruits differ greatly in their structure. This is related to the fact that, although the seed or stone must develop from the ovary of the flower, the associated fleshy edible tissues may arise from many different parts of the original flower. The post-harvest behaviour of fruit often differs depending on the origin of the structures. Figure 2.17 illustrates the structure of a flower, and the origin of a range of fruit with respect to the tissues from which they arise.

### Dormancy and germination

New plants may germinate from seeds or sprout from vegetative material. Examples of harvested commodities that have the potential to germinate and form new plants are seeds (such as grains, nuts and beans) and reproductive storage organs, such as tubers (e.g. potatoes) or storage roots (e.g. carrots and sweet potatoes). The physiological purpose of these plant parts is to survive for a period of time, often through environmentally difficult conditions such as drought or extreme cold, before sprouting to form new plants. In many cases (e.g. grains, beans, potatoes) these organs may be in a state of suspended animation, termed dormancy. These plant parts are usually less perishable than produce arising from other parts of the plant.

#### *Seeds*

Seeds, usually termed grain when edible, are particularly

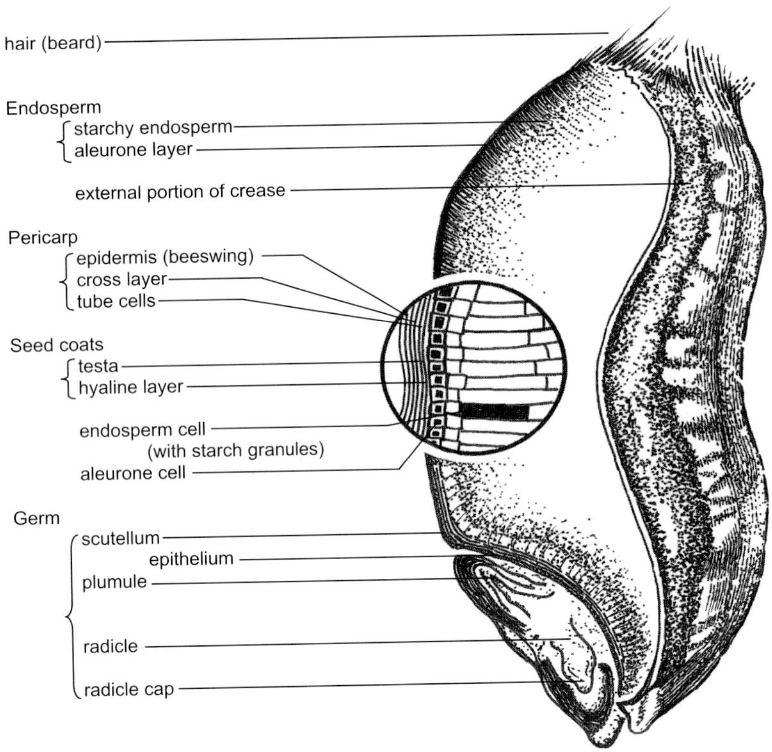


**Fig. 2.18** The naked structure of a grain. Redrawn from E.E. McDermott.

durable. The seed develops from the ovule of the flower. Seed durability is related to a very low water content (generally about 15%) and a very low metabolic rate (indicated by a low respiration rate). Seeds are in a state of dormancy prior to germination. Figure 2.18 shows the structure of a typical cereal grain. In wheat, rye, maize, pearl millet, finger millet and sorghum, the lemma and palea are loose and are easily removed from the grain by threshing, leaving a naked caryopsis. They are, therefore, known as naked grains. In hulled forms of barley, oats and rice, the lemma and palea fuse to form the hull or husk of the grain which is not easily separated – these grains are said to have a covered caryopsis.

### *Reproductive storage organs*

Reproductive storage organs include tubers, such as potatoes, and storage roots, such as carrots and sweet potatoes. These form due to the accumulation of storage products (usually starch) from the parent plant. The parent plant often dies down during the non-growing season and the storage organ is capable of producing a whole new plant by sprouting and rooting.

All root crops are storage organs. Although more perishable than grain, they tend to be less perishable than other horticultural produce as a result of the following attributes:

- relatively low respiration rate;
- low surface area relative to volume due to possessing a near spherical or cylindrical shape;
- hard texture (tightly packed cells with a low water content, the water having been replaced by starch); and
- thick skin with few natural openings (lenticels) allowing little gaseous exchange and water loss and providing protection from microbial decay.

### *The control of dormancy*

The end, or breaking of dormancy usually occurs when the external environment becomes more suitable for growth. Availability of water is often an important signal. Thus, in the case of beans, it is important to keep them dry during storage in order to prevent sprouting. In some plant species dormancy may be broken by more indirect signals, such as light levels, day length or temperature fluctuations. In potatoes, low levels of light are used during storage to prevent sprouting. This is because potatoes would normally only produce sprouts when they are under the soil, and therefore light acts as a signal that prevents the breaking of dormancy.

It is usually the case that the value of this type of produce will decrease if there is sprouting, such that it is important to maintain the dormant state. There are exceptions to this, such as bean sprouts, which are eaten (and often marketed) after germination has been initiated.

### *Germination*

Germination is accompanied by the breakdown of storage compounds (e.g. starch or fats) into sugars to be used for respiration to provide energy for growth, and for the synthesis of the larger structural molecules (e.g. structural polysaccharides and proteins) needed for that growth to occur. In some cases, therefore, germination is used by humans as a means for making a commodity more easily digestible. For example, germinated grain can be used as the basis of baby food and bean sprouts can be eaten without cooking, whereas the unsprouted beans must be cooked.

## Cereal grains

The anatomical structure of all grains is similar. Grains with a naked caryopsis consist of a fruit coat (pericarp), seed coat (testa), germ and endosperm. The covered grains also have a husk which surrounds the entire grain. Each part of the grain is further divided into various layers or regions, as illustrated by the diagram of a wheat grain (Fig. 2.18).

The husk or hull is the external, fibrous part of the grain that protects the grain during its formation and is totally indigestible.

The bran consists of several different layers and generally makes up around 15% of the kernel weight. The outer layers, or pericarp, are fibrous and waxy and create a protective barrier against water and mould penetration. The innermost layer of the bran is the testa or seed coat. It is in this layer that many of the pigments are located.

The aleurone layer lies between the bran and the floury part of the grain. This is a single layer of cells rich in fat, proteins and vitamins. In some milling processes the aleurone layer is removed as part of the bran while in others it is retained by the endosperm. It is not possible to isolate aleurone layers by milling.

The germ or embryo is the part of the grain from which a new plant could develop if given the right conditions (temperature and moisture). It is rich in protein, fat, vitamins and minerals. If separated from the remainder of the grain during milling it may be utilised in livestock feed or for oil extraction.

The endosperm or starchy part constitutes between 80% and 90% of the grain and is the food store of the grain, providing nourishment for the germinating plant. It is also the most important part of the grain for the miller since it is from here that flour is obtained. In some plant species and varieties the starch is more densely packed. This is associated with the protein bodies situated between the individual starch granules and determines the overall hardness of the grain. The starchy material in the centre of the grain is usually more loosely packed than that in the outer parts. This hard, outer layer helps to protect the grain from insect attack.

Table 2.5 shows the proportions of each structural component within the grain of four major cereals.

Each species of cereal has unique structural and chemical properties which influence the way it is processed and used and may have an effect on the nutritional content of the food produced. For example, sorghum grains are very similar in size to those of wheat and barley but are almost spherical in shape, while wheat and barley are more elongated and have a ventral crease. Maize has a much larger grain which is flattened and has a narrow basal part and broad apex. The germ of the maize kernel is very large and is situated at the base of the grain. In wheat and barley the germ is very small and is prominently situated at one end of the grain. Sorghum has a medium-sized germ that is partially surrounded by the endosperm.

The endosperm in both sorghum and maize is usually a mixture of both hard and soft (vitreous and floury). Wheat and barley kernels consist of either vitreous or floury endosperm and can thus be classified as either hard or soft varieties. The harder wheats have a higher protein content and are more suitable for breadmaking, while softer wheats are used in biscuits and cakes.

**Table 2.5** Composition of some cereal grains (%).

| | Endosperm | Germ | Bran and aleurone layers | Husk |
|---|---|---|---|---|
| Maize | 82 | 13 | 5 | |
| Wheat | 82 | 3 | 15 | |
| Rice | 73 | 10 | 5 | 20 |
| Sorghum | 82 | 2 | 8 | |

Sorghum is often highly pigmented and occurs in a wide range of colours from white through to very dark red. Wheat is also pigmented while maize is usually not, though some older varieties exhibit orange, black and red kernels. Sorghum has the highest polyphenol content of the grains, that may impart a bitter flavour to foods produced from it.

### Nutrient distribution

The nutrients are unevenly distributed in the grain, as indicated in Table 2.6. Processing can therefore have a very significant effect on the nutritional value of the grain.

#### *Protein*

Protein occurs throughout the grain but is concentrated in the germ and the aleurone layer. Cereal protein is deficient in the essential amino acid, lysine, and so people, especially children, whose diet consists almost exclusively of cereals are likely to suffer from protein deficiency. Supplementing the diet with other protein sources can prevent this. Refined cereals, such as polished rice, have had almost all of the protein removed during processing.

#### *Fat (oil)*

Oil from cereal germ is of good quality and can be extracted for commercial use. This is only possible when the germ (or other fat-rich fraction) can easily be separated from the remainder of the grain, for example, in corn (maize) oil and rice bran oil. Fat content is also important as it is the carrier for the oil-soluble vitamins. Without fat these vitamins would not be absorbed by the body.

#### *Carbohydrate*

Indigestible carbohydrates such as cellulose are located in the outer layers of the grain. Fibre is very important in the diet, and societies which consume highly refined foods are more prone to cancers of the digestive system. For this reason many people in the developed world now prefer to eat wholegrain products such as wholemeal flour and wholegrain breakfast cereals.

#### *Vitamins and minerals*

As can be seen from Table 2.6, the majority of minerals are located in the germ, bran and husk, as are most of the vitamins. Milling (refining) cereals, therefore, reduces the vitamin and mineral content of the grain.

## Processes occurring during ripening of fleshy fruit

While cereal grains are important as staples, fleshy fruits are of increasing commercial importance. Fleshy fruits are by their nature very perishable. An understanding of the processes that occur during ripening and how they are controlled is a very important area of post-harvest physiology. The changes that occur during fruit development and ripening can be divided into three categories: compositional changes, colour changes and textural changes.

**Table 2.6** Distribution of nutrients within some grains (%).

| Fraction | Cereal | Soluble carbohydrates | Protein | Fat | Sugar | Fibre | Ash/minerals |
|---|---|---|---|---|---|---|---|
| Endosperm | Wheat | 85.0 | 11.4 | 1.2 | 0.8 | 0.1 | 0.4 |
| | Maize | 86.4 | 9.4 | 0.8 | 0.6 | 0.1 | 0.3 |
| | Sorghum | 86.4 | 9.5 | 1.0 | 0.4 | 1.0 | 0.8 |
| | Rice | 88.9 | 9.8 | 0.5 | — | 0.3 | 0.6 |
| Germ | Wheat | 20.0 | 29.4 | 10.0 | 20.0 | 8.9 | 8.9 |
| | Maize | 19.0 | 18.8 | 34.5 | 10.8 | 4.6 | 10.1 |
| | Sorghum | 21.0 | 15.1 | 20.0 | 12.0 | 2.6 | 8.2 |
| | Rice | — | — | — | — | — | — |
| Bran + aleurone layers | Wheat | 1.0 | 11.1 | 3.5 | 0.3 | 13.5 | 6.1 |
| | Maize | 7.0 | 7.1 | 1.0 | 0.3 | 14.0 | 5.8 |
| | Sorghum | 5.0 | 8.9 | 5.5 | 0.3 | 8.6 | 2.4 |
| | Rice | 10.0 | 14.1 | 18.8 | 0.5 | 5.7 | 11.2 |
| Husk | Rice | 34.0 | 1.2 | 0.3 | 0.3 | 34.0 | 24.0 |

## Compositional changes

### *Carbohydrate (and lipid) changes*

A general pattern for fruit development and ripening is that fruits accumulate carbohydrate during development, often in the form of starch, and that this is then converted into sugars and acids during ripening. The final levels of sugars and acids are very important for taste. The most important sugars are generally sucrose, glucose and fructose, while the most important organic acids are malate and citrate.

Some fruits accumulate all the assimilate prior to the onset of ripening (e.g. banana, tomato). For other fruits, accumulation of sugar from the plant can continue during ripening (e.g. strawberry, grape). In the latter case it is extremely important for flavour that fruit is harvested when ripe, that is, the fruit is vine ripened, whereas for the former the fruit can be harvested at a mature but unripe stage. In the case of banana, the fruit can be harvested as soon as accumulation of starch is complete. In practice, this is achieved by determining the maturity of the banana by the cross-sectional shape, which becomes more rounded as the starch accumulates. For other fruits the extent of accumulation can be determined by fruit size. Note that there are a few fruit (e.g. avocado and olive) that do not accumulate carbohydrate but instead accumulate lipids.

The energy released by the breakdown of energy reserves such as polysaccharides, sugars or acids is used to propel and co-ordinate the many compositional changes associated with ripening. In addition to gross changes in starch, sugars, acids and proteins, etc. (Table 2.7), changes also often occur in carotenoid pigments and volatile aroma constituents.

**Table 2.7** Compositional changes during fruit ripening for banana and mango.

| | Banana | | Mango | |
|---|---|---|---|---|
| | Unripe | Ripe | Unripe | Ripe |
| Dry matter (%) | 29 | 27 | 19 | 19 |
| % (dry wt) | | | | |
| Starch | 83 | 10 | 28 | 1 |
| Sugar | 4 | 77 | 50 | 74 |
| Acid | 0.5 | 1 | 7 | 1 |
| Protein | 3 | 3 | 3 | 3 |
| Fibre | 5 | 5 | 9 | 9 |

### *Phenolics and flavour compounds*

Phenolics are responsible for bitter astringent tastes. Oxidation of phenolics can cause browning during processing. For these reasons, even when present at low concentrations, phenolics can have an important effect on taste and acceptability of fruit, usually detrimental. Phenolics tend to decline during ripening.

Flavour is a very complex issue, which is determined by a range of flavour volatiles which appear as a fruit ripens. Chemically, flavour compounds are very diverse. Their only common property is that they are volatile. They can be very complex and distinct for different fruits. This is illustrated by the fact that so far 230 flavour compounds have been found for apple and 330 for orange.

## Colour changes

Many fruits change colour during ripening. Most commonly, unripe fruit are green, but lose their green colour and develop additional pigmentation during ripening. The process usually involves chlorophyll breakdown with synthesis of other colour compounds, for example, carotenoids (orange, yellow, red) and anthocyanins (very diverse colour range from red to blue).

Chlorophyll, in unripe fruits as in leaves, is present within chloroplasts which are able to carry out photosynthesis. Loss of chlorophyll involves both the loss of photosynthetic capacity and the enzymatic breakdown of chlorophyll. Colour changes are not always completely connected to other ripening events. Although most fruit changes colour naturally during ripening, there are many cases (usually where the fruit is grown in a different environment from where it originally evolved) where this does not occur. Examples include oranges and pineapples, both of which may ripen without losing their green colour. In some cases, for consumer acceptability the colour change is artificially induced by treating with ethylene.

## Texture changes

Most fruit soften during ripening. This usually occurs due to starch breakdown and cell wall softening. The latter is generally the more important process. The detailed structures of fruit cell walls are still not completely understood and much of the information has been obtained from studies on cell walls of other plant structures. However, it is known that the fruit cell wall is composed primarily (90–95%) of carbohydrates (cellulose, hemicellulose and pectin), while the remainder of the structure is glycoprotein (Salisbury & Ross 1992).

Cellulose, hemicellulose and pectins are large complex molecules, held together by a variety of covalent and non-covalent bonds. The levels of pectins are particularly high in fruit. They are located primarily in the middle lamellae of the walls and appear to play a central role in the structure. Figure 2.19 illustrates a model of cell wall structure, while Fig. 2.20 illustrates the current 'egg-box model' of the binding between adjacent pectins (Grant *et al.* 1973; Seymour *et al.* 1993). This model emphasises the importance of calcium ions for pectin binding. Calcium is known to be important for the maintenance of fruit hardness (Seymour *et al.* 1993).

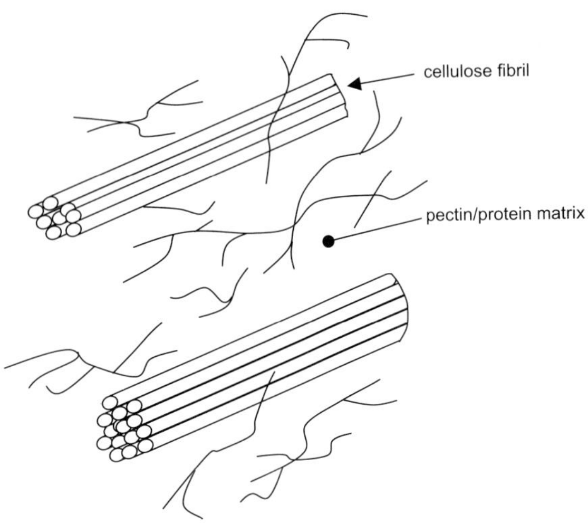


**Fig. 2.19** Idealised cell wall model. Source: Seymour *et al.* 1993.

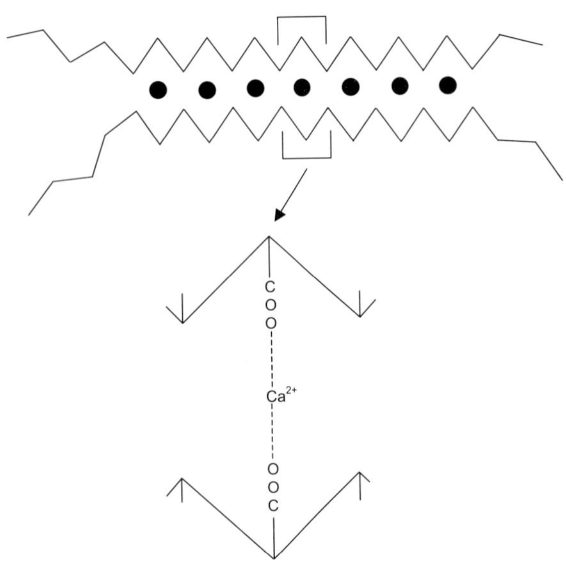


**Fig. 2.20** The 'egg-box' model for the non-covalent linkage of adjacent acidic pectin polymers in cell walls. Source: Seymour *et al.* 1993.

A wide range of cell wall hydrolases have been studied in order to determine their role in the solubilisation of pectin. The literature is very contradictory and a mechanism for softening in fruit has not been fully elucidated. It is unlikely that a single enzyme is responsible for textural change, and this probably involves a complex interaction of enzyme activity with physicochemical changes in the wall. Softening probably occurs through different mechanisms for different fruits.

## Metabolism and respiration

Energy is required to drive the biochemical reactions that enable growth and development to take place. The energy used by cells, both plant and animal, is derived from the breakdown of relatively complex molecules into simpler ones. This process releases energy stored in chemical bonds, which is then available as 'fuel' for living processes, including construction of structural molecules, synthesis of food reserves and transport of metabolites. The series of biochemical reactions which provides energy for cells is called respiration. The rate of respiration provides a very good indication of the rate of overall metabolism of a commodity. Likewise, it is often possible to control the rate of metabolism of commodities by controlling the rate of respiration.

Respiration consists of a number of complex reactions but can be summarised as follows:

$$C_6H_{12}O_6 + 6\,O_2 \rightarrow 6\,CO_2 + 6\,H_2O + \text{energy (chemical and heat)}$$

$$\text{glucose} + \text{oxygen} \rightarrow \text{carbon dioxide} + \text{water} + \text{energy}$$

The processes involved in respiration are essentially the same in animals and in plants. They occur in the cytoplasm and mitochondria. Respiration results in the release of energy, some of which is preserved chemically in the form of the high-energy molecule, adenosine triphosphate (ATP). The breakdown of ATP can, in turn, be coupled to and provide energy for essential metabolic reactions. This is illustrated in Fig. 2.21. Here ATP is shown as adenosine (represented as a square, A) with three phosphate groups (each represented as a circle, P).

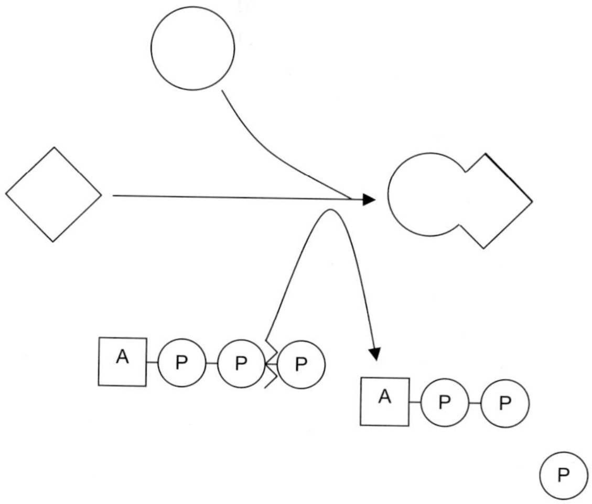


**Fig. 2.21** The breakdown of ATP into ADP and phosphate can provide the energy to drive metabolic reactions.

The energy to drive the synthetic reaction shown (creation of one larger molecule from two smaller molecules) is provided by the breakdown of ATP to adenosine diphosphate (ADP).

The respiration of sugar (glucose) is summarised in Fig. 2.22. Sugars are broken down (oxidised) to carbon dioxide ($CO_2$) and water ($H_2O$) in a two-step process, involving three interacting metabolic pathways: glycolysis, the tricarboxylic acid (TCA) cycle and the electron transport system. Often starch or sucrose is broken down into glucose or fructose as a substrate for respiration. Although monosaccharide sugars are considered here as the substrate for respiration, other molecules such as organic acids (e.g. citric acid, malic acid) or fatty acids (e.g. palmitic acid) can also be used.

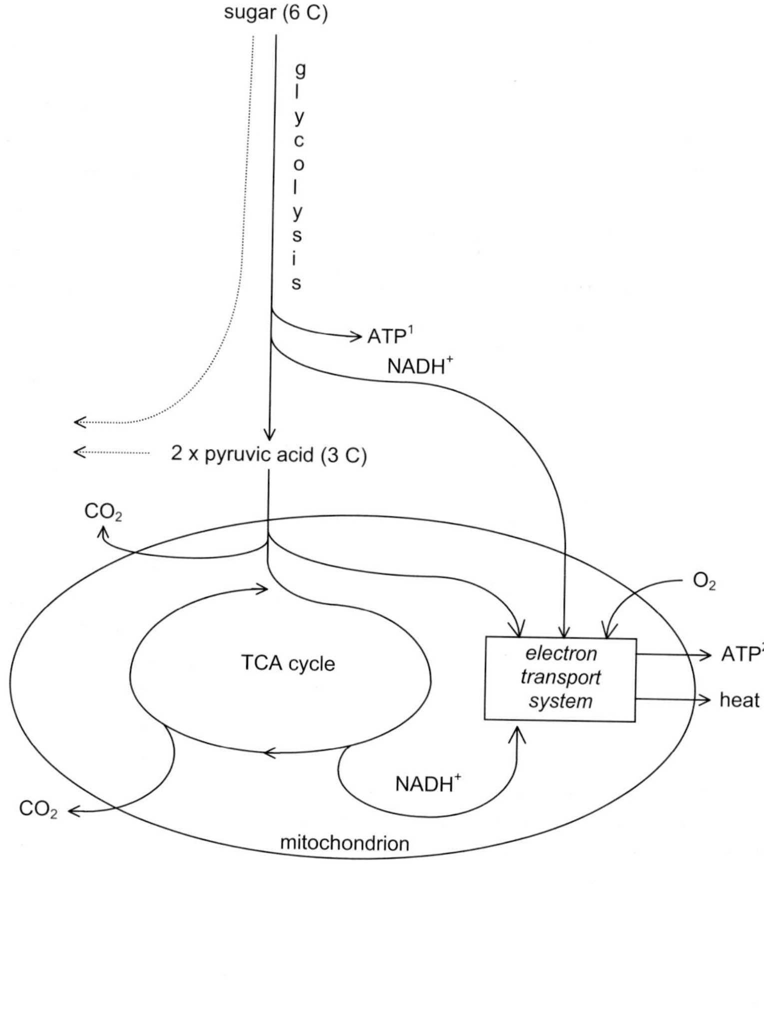


**Fig. 2.22** The process of respiration.

## Glycolysis

During the first stage of respiration, glycolysis, sugar is broken down into pyruvic acid which is a three-carbon molecule. The reactions of glycolysis occur in the cytoplasm. Figure 2.23 shows the reactions in detail. Some ATP (one molecule per sugar metabolised) is formed directly as part of the biochemical reactions of glycolysis. The remainder (two molecules per sugar metabolised) is formed as a result of the oxidation of a molecule called NADH, which is formed by the reduction of NAD during glycolysis. This occurs through the process of electron transport.

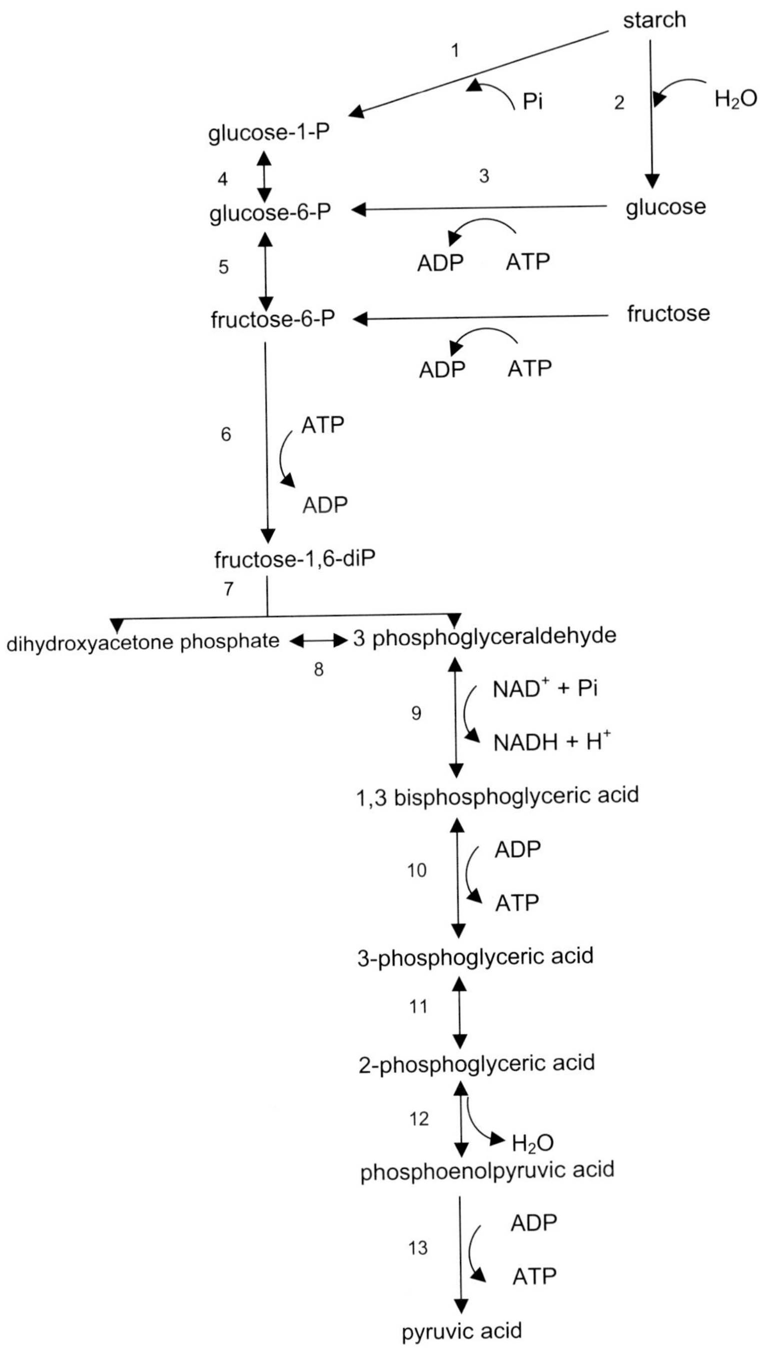


**Fig. 2.23** The reactions of glycolysis. The numbers refer to the enzymes which catalyse the reactions. 1, phosphorylase; 2, α- and β-amylases, oligo-1,6-glucosidase and α-glucosidase; 3, hexokinase; 4, phosphoglucomutase; 5, glucosephosphate isomerase; 6, phosphofructokinase; 7, fructose bisphosphate aldolase; 8, triosephosphate isomerase; 9, glyceraldehyde-phosphate dehydrogenase; 10, phosphoglycerate kinase; 11, phosphoglycerate phosphomutase; 12, enolase; 13, pyruvate kinase.

## The electron transport system

This process by which the oxidation of NADH is used to synthesise ATP is called the electron transport system and takes place within the mitochondria. Figure 2.24 shows the three-dimensional structure of a mitochondrion, partially cut away to reveal the inner structure.

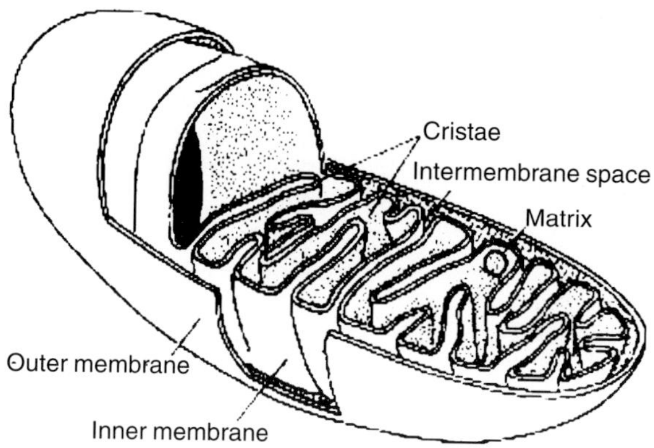


**Fig. 2.24** A mitochondrion, where the processes of electron transfer take place. Source: Wolfe 1972. Reprinted with permission of Brooks/Cole, an imprint of the Wadsworth Group, a division of Thomson Learning, Fax 800 730 2215.

The mitochondrion is a membrane-bound organelle, which has a second inner membrane that is extensively folded to form structures called cristae. The components which allow the electron transport system to function are protein complexes embedded in the membranes of the cristae. On arriving in the mitochondrion, the NADH reduces (i.e. donates an electron) to certain of these protein complexes. This initiates a series of reactions in which the electron is passed from one complex to another, and finally to oxygen. Thus oxygen is reduced to water. By a very complicated mechanism which is still not fully understood, the chain of reactions in the membrane result in the movement of hydrogen ions (protons) across the membrane, resulting in the concentration of protons in the intermembrane space. A specialised enzyme called ATP synthase, which is embedded in the same membrane, is able to use the movement of protons back across the membrane to provide the energy to drive the synthesis of ATP from ADP and phosphate (Fig. 2.25).

The process of respiration is not completely efficient, and results in the release of some of the energy of the breakdown of sugars as heat. This occurs primarily at the stage of the electron transport system. As the electron transport system occurs in the mitochondrial membranes it is particularly sensitive to low temperature, which affects membrane structure.

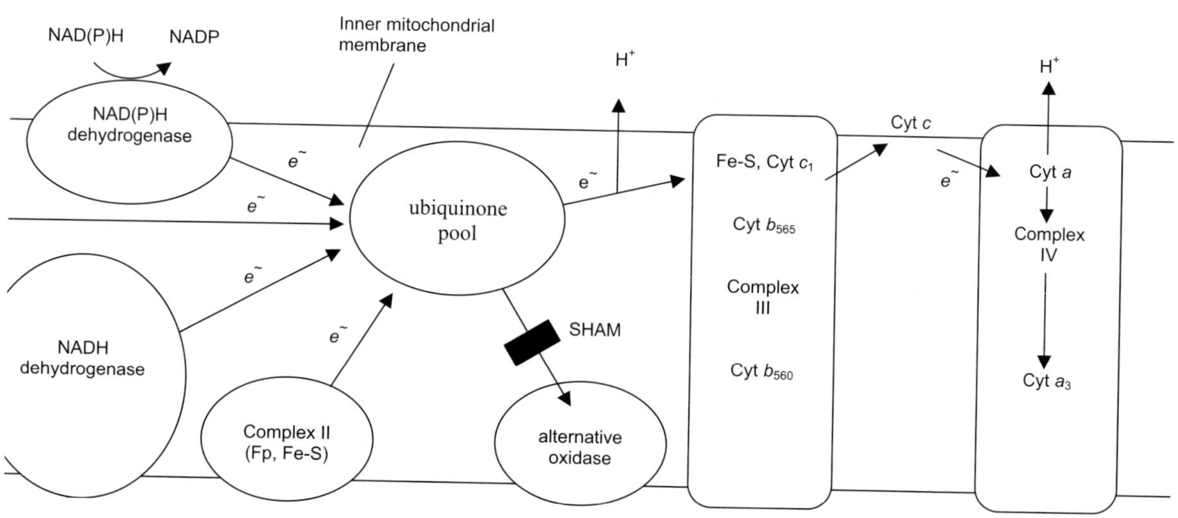


**Fig. 2.25a** Two-dimensional representation of the electron transport chain. Vertical arrows indicate proton movements. The function of the alternative oxidase (inhibited by salicylhydroxamic acid (SHAM)) is not clear. Source Taiz & Zeiger 1991.

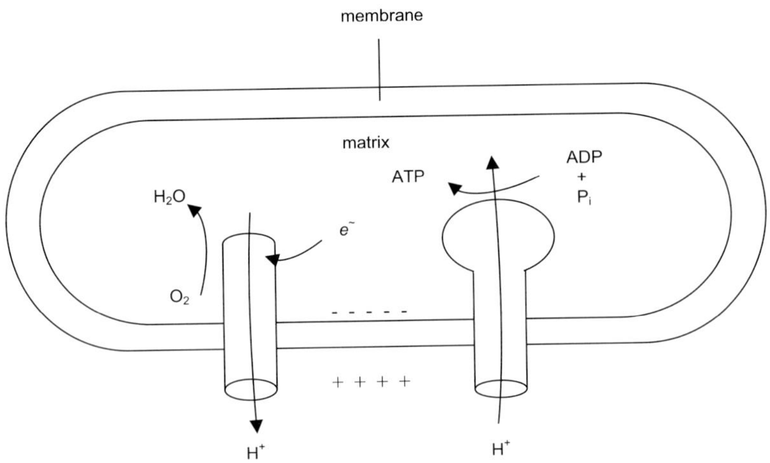


**Fig. 2.25b** The proton gradient created by the action of the electron transport chain is coupled to the synthesis of ATP by the ATP synthase located in the mitochondrial membrane. Source: Stryer 1995.

### Tricarboxylic acid cycle

During the tricarboxylic acid cycle (TCA cycle) (Fig. 2.26) the pyruvic acid formed during glycolysis is also taken into the mitochondria where it is broken down into $CO_2$ with the formation of many more molecules of NADH, which are again used to make ATP through the electron transport system. This process, therefore, also requires the uptake of oxygen.

### Aspects of respiration important in post-harvest systems

In summary, respiration is necessary for the production of ATP which provides the energy for development and maintenance of good cell health. It results in the uptake of oxygen and the production of carbon dioxide, heat and water. Some of these factors, and others, are particularly important to post-harvest technologists.

#### *Rates of respiration differ by commodity*

The origin of the commodity in terms of plant part – young shoot, leaf, flower, fruit, shoot, seed – has a significant effect on the inherent level of respiration. For example, young, rapidly developing, active plant organs tend to have higher rates of respiration than storage organs. As a general rule, the post-harvest life of a given commodity is directly related to the natural lifespan of that particular plant organ, and both are generally inversely proportional to its rate of respiration. Table 2.8 shows the rates of respiration for a range of commodities and indicates how they vary between plant parts.

As well as plant origin, respiration rates also vary with species. For example, fruits of avocado have a maximum respiratory rate some eight times that of apples. However, even within a species there can be significant differences in respiration rates between varieties.

There are many fruits which, once they reach an adequate state of development, can easily be picked before they are ready to eat and allowed to ripen after harvest. At an early stage in the development of post-harvest technology, it was discovered that apples and other fruits of this type showed an increase in respiration after harvest which continued until they approached eating ripeness and then once again declined. This respiratory change could be a steady drift extending over a period of weeks with a doubling of respiration, or an intense change with perhaps a quadrupling of respiration in three days (Fig. 2.27). This respiratory rise, which occurs in many fruits, is called the climacteric, and fruits which exhibit this behaviour are called climacteric fruit. In some fruits, such as oranges, which only ripen while attached to the tree, a contrasting respiratory pattern is shown in which respiration decreases steadily during growth and continues to decrease after harvesting, irrespective of the stage of development at which they were harvested (Fig. 2.27). These fruits are called non-climacteric. Climacteric fruits often, but not always, have a starch reserve that is hydrolysed to sugar during the climacteric and the occurrence of the climacteric is usually essential for the fruit to be palatable. Non-climacteric fruit do not improve in eating quality after harvest (through sugar or acidity metabolism) though some slight softening and loss of green colour may occur.

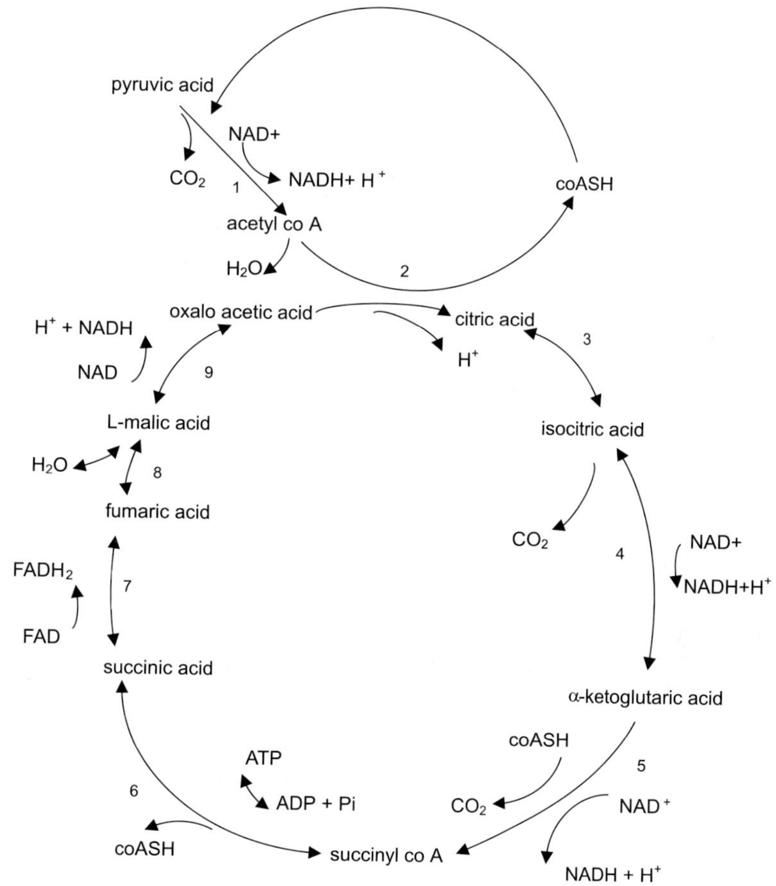


**Fig. 2.26** The reactions of the TCA cycle. The numbers refer to the enzymes which catalyse the reactions. 1, pyruvic acid dehydrogenase; 2, citrate synthase; 3, aconitase; 4,isocitrate dehydrogenase; 5, α-ketoglutarate dehydrogenase; 6, succinate thiokinase; 7, succinate dehydrogenase; 8, fumarase; 9, malate dehydrogenase.

**Table 2.8** Typical rates of respiration for selected produce categorised by plant part.

| Unit of respiration ($CO_2$ in mg/kg/h) | Storage organ (dormant) | Fruit | Green vegetable | Flower |
|---|---|---|---|---|
| Very low (<5) | Grain, beans, nuts | Dates | | |
| Low (5–10) | Onion, potato | Apple, citrus, grapes, kiwifruit | | |
| Moderate (10–20) | Carrot | Apricot, banana, plum, fig, pepper, tomato, cherry, peach, nectarine, pear | Cabbage, lettuce | |
| High (20–40) | | Strawberry, blackberry, raspberry, avocado | Lima bean | Cauliflower |
| Very high (40–60) | | | Snap bean, Brussels sprouts, artichoke | Cut flowers |
| Extremely high (>60) | | | Asparagus, peas, spinach | Broccoli |

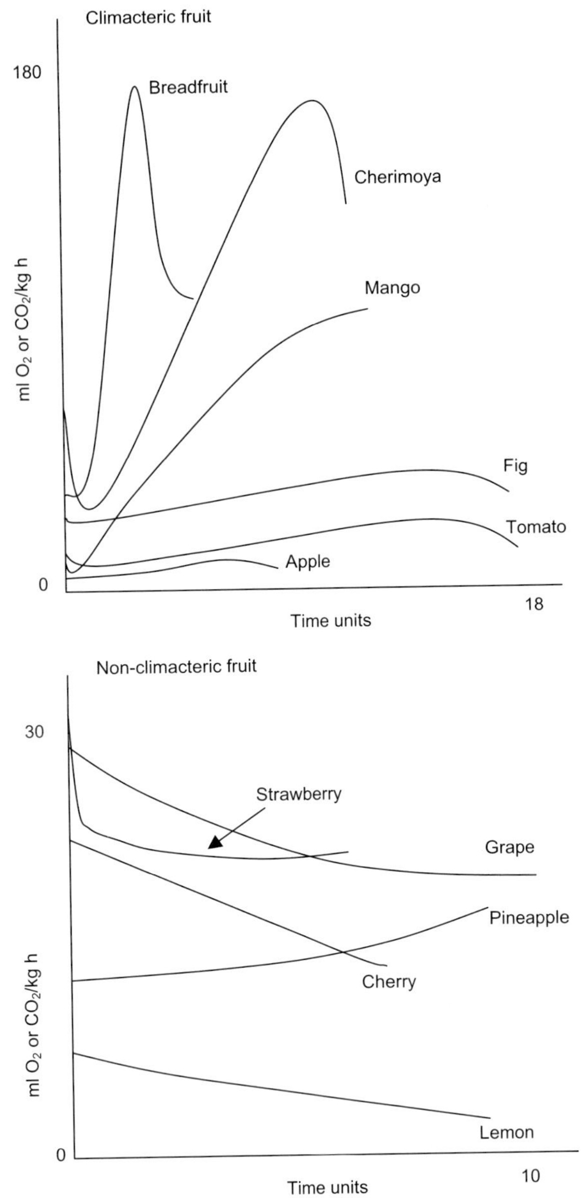


**Fig. 2.27** Respiration patterns during the ripening of climacteric and non-climacteric fruit. Source: Seymour *et al.* 1993.

The purpose of the climacteric rise in respiration is not completely clear. Respiration provides the energy for metabolic reactions and, as climacteric fruit tend to be those for which greater changes in composition occur during ripening, it was originally assumed that the rise in respiration had to occur to provide the energy for the changes associated with ripening to occur. However, it is now clear that in many cases the extent of respiration is greater than needed. It also appears that climacteric respiration is less efficient than normal in terms of ATP production. Within those fruit defined as climacteric, qualitative differences are evident in the intensity and duration of the climacteric respiratory peak. Examples of the different patterns of respiratory activity associated with some common climacteric fruits are shown in Fig. 2.27. Note that the timing of the climacteric with respect to the stage of optimum eating quality can also vary.

Another significant difference observed between climacteric and non-climacteric fruit stems from the natural levels of the gas, ethylene, found in fruit at the time of ripening. Climacteric commodities exhibit much higher internal concentrations of ethylene gas which is involved in the control of the ripening process.

### *Rates of respiration vary with temperature*

As a general rule, the rate of metabolism increases as the temperature is increased, and decreases as the temperature is decreased. For this reason, commodities are often stored at lower than ambient temperatures in order to slow down their rate of metabolic activity. However, this tendency is only true within a certain range of temperatures. Under normal conditions, living plant tissues will only function satisfactorily within a limited range of temperatures. This range is often referred to as the 'physiological range'.

It is a general rule that the physiological temperature range varies depending on the climatic conditions under which the plant originally evolved. Thus, for tropical produce the physiological range will tend to be at higher temperatures than for temperate produce.

At very low temperatures the rate of respiration slows and stops when the water in the tissue freezes. At high temperatures the structures of the enzymes which control metabolic reactions are disrupted. This phenomenon is described as denaturation of the enzymes. At this point a thermal death point is seen. However, at cool temperatures well above freezing and at warm temperatures below the thermal death point, cell metabolism is impaired. One reason for this is that the mobility of the membranes within the cells will change, becoming less at low temperatures and greater at high temperatures. This will affect the movement of substances across the membranes and the functioning of embedded complexes. Plants will adapt to some extent to their growth temperature by changing the composition of the membranes in terms of the extent of saturation of the component fatty acids. Thus, for example, a plant grown at low temperature will have more unsaturated fatty acids in the membranes to maintain greater mobility at low temperature.

Therefore, due to their source of origin, temperate crops can commonly withstand storage temperatures of 0–1°C whereas tropical crops are often damaged by temperatures below 8–10°C and many will even exhibit symptoms of chilling injury if held at temperatures below 13°C. Examples of commodities which are classified as chilling-sensitive or less chilling-sensitive are given in Table 2.9. Chilling sensitivity can also vary by cultivar, growing conditions or with ripeness in the case of fruit.

Chilling sensitivity is a very important phenomenon for perishables, especially those originating from tropical regions as it limits the extent to which low temperatures can be used to slow down metabolism, and therefore limits potential storage time.

### *Respiration produces heat*

The process of respiration is not completely efficient, so that not all the energy is retained as ATP and some is released as heat. Release of heat occurs particularly during the electron transport part of the reaction. The process is generally more efficient in growing tissues (10% lost as heat) than in non-growing harvested tissues where the major part of the energy is lost as heat. As an approximation the heat output of a commodity can be calculated from the rate of carbon dioxide release using the following equation:

$CO_2$ released at 1 mg/kg/h = 2.96 watts/tonne

**Table 2.9** Example of differential chill sensitivity of a range of commodities.

| Chilling-sensitive | Less chilling-sensitive |
|---|---|
| Avocado | Apples |
| Banana | Pears |
| Citrus | Asparagus |
| Guava | Broccoli |
| Mango | Brussels sprouts |
| Tomato | Cauliflower |
| Aubergine | Onion |
| Snap beans | Peas |

Estimates of heat production by commodities are especially important for assessing cold store specifications and operations.

A very important consequence of the heat produced by respiration is that, if produce is placed in an unventilated heap, the temperature at the centre of the pile will increase. A chain reaction is started (the heat of respiration warms the produce, the temperature of the produce increases and this increases its rate of respiration and hence the production of vital heat) which can lead to the overheating and rapid deterioration of the whole consignment.

Respiration → Heat → Faster respiration → More heat →

### *The rate of respiration is affected by levels of oxygen and carbon dioxide*

$$\underset{\text{glucose}}{C_6H_{12}O_6} + \underset{\text{oxygen}}{6O_2} \rightarrow \underset{\text{carbon dioxide}}{6CO_2} + \underset{\text{water}}{6H_2O}$$

*reactants* → *products*

The rate of a reaction tends to be reduced if the concentration of reactant(s) is reduced and/or if the concentration of the product(s) is increased. In the case of respiration the reactants are sugars and $O_2$, and the products are $CO_2$ and water. Hence, decreasing the $O_2$ concentration and/or increasing the $CO_2$ concentration of the air surrounding the produce will reduce respiration rates. This is the basis of controlled atmosphere (CA) storage where the effect of refrigeration is augmented by control of atmospheric composition. This technique is widely used to extend the post-harvest life of apples and it has potential for many other commodities.

When a large volume of a commodity is placed in a store, especially if the store is moderately gas-tight, the $O_2$ concentration will decrease as it is taken up for respiration and the $CO_2$ concentration will increase as it is released through respiration. In this way, the rate of respiration can again be decreased. This property is often used when produce is packaged (plastic films are available which have different permeabilities to respiratory gases) and is the basis of what is termed modified atmosphere (MA) storage.

Just as there is a physiologically accepted temperature range for each commodity, so care must be taken to keep $O_2$ and $CO_2$ levels within biologically acceptable limits. For example, $O_2$ levels must be kept sufficiently high to allow respiration to proceed. If there is not sufficient oxygen for electron transport to occur, the pyruvic acid formed by glycolysis is not taken into the mitochondria, but instead is converted into acetaldehyde and ethanol or lactic acid. This is known as anaerobic respiration. Anaerobic respiration produces much less energy per mole of glucose than aerobic pathways, but does allow some energy to be made available for keeping cells alive. However, high levels of ethanol or lactic acid are toxic and, as a consequence, prolonged anaerobic respiration can lead to cell death. Off-flavours are produced that can destroy product quality long before cell death occurs. The critical concentration of oxygen at which anaerobic respiration occurs differs with commodity. For example, whereas the critical concentration for most fruits is 1–3% oxygen, sweet potato roots will switch to anaerobic respiration at concentrations below 5–7%. Figure 2.28 shows the process of anaerobic respiration, while Tables 2.10 and 2.11 give the range of maximum acceptable carbon dioxide levels and minimum acceptable oxygen levels for a range of commodities.

### *Respiration results in the production of water*

As respiration results in the production of water, if there is insufficient ventilation then as a commodity respires there may be an increase in humidity as water is released. Condensation of water on the produce may lead to problems with rotting.

### *Damaged produce has increased rates of respiration*

When produce is physically damaged or is infected by rotting pathogens, the rate of respiration tends to increase. This is as a result of the tissues mobilising vari-

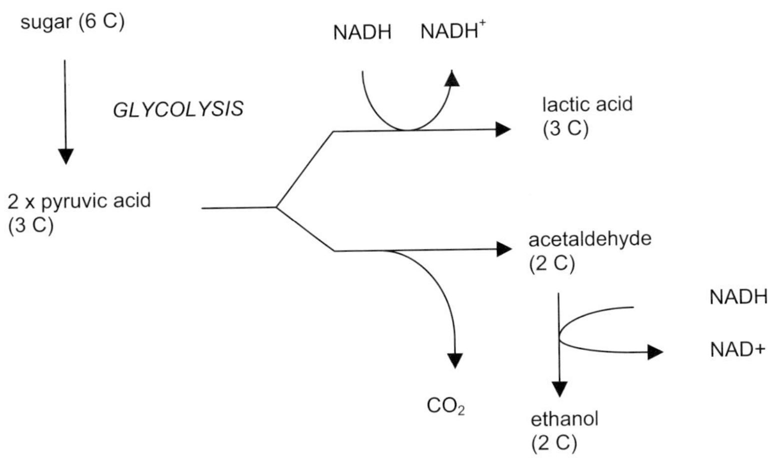


**Fig. 2.28** Anaerobic respiration.

**Table 2.10** Maximum acceptable carbon dioxide levels for some fruits.

| Maximum $CO_2$ | Fruit |
|---|---|
| 1% | Pears (Anjou, Bosc) |
| 2% | Apples (Delicious), apricots |
| 5% | Apples (McIntosh, Jonathan), pears (Bartlett), nectarines, peaches, avocados (Fuerte), bananas, mangoes, papaya |
| 10% | Cherries, olives |
| 14% | Avocados (Lula) |
| 20% | Strawberries, figs |

**Table 2.11** Minimum acceptable oxygen levels for some fruits.

| Minimum $O_2$ | Fruit |
|---|---|
| 5% | Citrus |
| 3% | Cherries, avocado |
| 2% | Apples, pears, pineapple, papaya, strawberries, nectarines, apricots, peaches, plums |

ous defence and repair mechanisms, all of which require extra energy. Increased respiration rates may have many effects for stored produce.

### Respiration and changes in oxygen and carbon dioxide levels

Since respiration results in the uptake of $O_2$ (through electron transport) and the release of $CO_2$, the measurement of these gases can be used to measure the rate of respiration. During the respiration of sugars the amount of $CO_2$ produced is the same as the amount of $O_2$ taken up. The ratio between $CO_2$ production and $O_2$ uptake, which is called the respiratory quotient, differs, however, for different substrates. This can be used in research to help determine which molecules or combination of molecules are being metabolised.

The respiratory quotient changes dramatically as commodities switch from aerobic to anaerobic respiration, as in the latter case carbon dioxide is produced but no oxygen is consumed. Experimentally, it is possible to determine at what point a commodity switches from aerobic to anaerobic respiration by measuring the ratio of $CO_2$ produced to $O_2$ consumed.

$$\text{Respiratory Quotient (RQ)} = \frac{CO_2 \text{ produced}}{O_2 \text{ consumed}}$$

Thus, for aerobic respiration of glucose RQ = 1.

The complete oxidation of malate (an organic acid) is described by:

$$C_4H_6O_5 + 3\,O_2 \rightarrow 4\,CO_2 + 3\,H_2O$$

Thus, the aerobic respiration of malate RQ = 4/3 = 1.333

For anaerobic respiration no oxygen is consumed and therefore RQ = infinity.

## The role of ethylene in controlling development

For many commodities it is important to be able to control (either slow down or speed up) rates of development. Many aspects of plant development are controlled by plant hormones. Plant hormones have the same role as hormones do in humans. Examples of plant hormones are auxin, gibberellin, cytokinin and abscisic acid. Ethylene is also a plant hormone and is particularly important in post-harvest systems.

It has been known for a long time that the behaviour of plants and fruits can be affected by gases. For example, for centuries the Chinese knew that if fruits were placed in a room with burning incense, then they would ripen more rapidly. Likewise, Puerto Rican pineapple growers and Filipino mango growers traditionally built bonfires near their crops as they found that this helped to initiate and synchronise flowering. It was also appreciated that the presence of ripe or rotten fruits could induce ripening in other fruits. All of these phenomena can be explained by the effects of ethylene.

### Ethylene as a plant hormone

The structure of ethylene is shown in Fig. 2.29. It is a two-carbon hydrocarbon with a double bond. It can be produced by almost all tissues of higher plants, and acts as a plant growth regulator, or hormone, controlling many processes such as ripening of fruits and senescence.

Ethylene is unusual among plant hormones in that it is a gas at normal temperatures and can therefore diffuse easily between plant tissues or from commodity to commodity. It has a very important role in the post-harvest behaviour of fruits, vegetables and root crops, both beneficial and detrimental. An understanding of the effects of this gas is, therefore, essential so that optimal storage and handling conditions can be maintained.

The concentration to which a plant tissue is sensitive to ethylene varies greatly between tissues and plant species, but can be as low as 0.1 ppm. At such low levels measurement was very difficult until gas chromatographic methods were developed in the 1960s. Since then the role of ethylene in plant development has gradually been elucidated.

### Developmental processes controlled by ethylene

#### *Ripening*

A very important process in post-harvest management

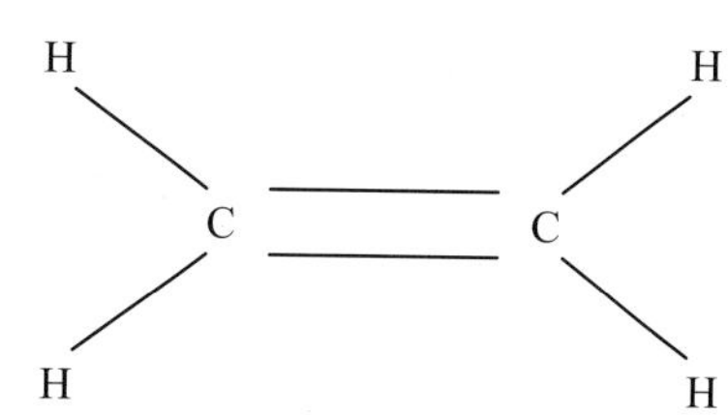


**Fig. 2.29** The structure of ethylene.

is ripening of fruit. Natural ripening is initiated and controlled by ethylene produced by the fruit itself, but can be induced artificially by the addition of exogenous ethylene. Climacteric and non-climacteric fruits differ in their synthesis of and response to ethylene.

#### *Flowering*

Ethylene stimulates flowering in some species. These include mango and pineapple. However, for other species this is not the case, and in many instances ethylene may even prevent bud opening in cut flowers. Some cut flowers can be extremely sensitive even to very low levels of ethylene.

#### *Sprouting*

Sprout development is stimulated by ethylene. Seed potatoes are an example.

#### *Mobilisation of defence mechanisms*

Ethylene is often produced under stress conditions, as a signal to the plant to mobilise defence mechanisms. For example, in certain produce following mechanical damage, ethylene production stimulates wound healing. On infection by pathogens, ethylene levels will also increase, and are involved in stimulating defence mechanisms such as production of antifungal toxins (phytoalexins). Note, however, that ethylene may also stimulate the growth of the pathogens themselves.

#### *Senescence and abscission*

Ethylene stimulates abscission and many other senescence processes. Thus, there are several changes that are induced by ethylene, associated with natural ageing or senescence processes that are detrimental to produce quality. These include:

- loss of green colour, as chlorophyll breaks down;
- stem browning of grapes;
- tissue maceration in water melons;
- abscission of calyxes, florets and leaves;
- stimulation of lignin synthesis and toughening of stems;
- bitter flavour in carrots and loss of flavour in sweet potato; and
- russet spotting of lettuce.

### Synthesis and mode of action of ethylene

It is clear that ethylene can have very diverse effects on plant tissues, depending on the tissue and stage of development. This raises the question of how one very simple chemical can have such a range of effects. The precise mechanism by which ethylene controls developmental processes is still not fully understood. Initially the plant tissue is stimulated to synthesise ethylene either by reaching a particular developmental stage, as for fruit that is mature for ripening, or by some external stimulus such as mechanical damage, or exposure to exogenous ethylene. As for other hormones, ethylene then brings about its effects by binding to receptors within the cell. This binding initiates a chain of events within the cell, which depend on the developmental stage of the tissue at that time, so that the final response likewise depends on the developmental stage.

#### *Synthesis of ethylene within plant tissue*

Ethylene is synthesised within plant tissue from the amino acid methionine. The pathway of the synthesis of ethylene from methionine is summarised in Fig. 2.30. Two key enzymes in the process are 1-amino-cyclopropane-1-cyclopropane (ACC) synthase and ACC oxidase. The regulation of these two enzymes is central to the control of ethylene synthesis. ACC synthase is the rate limiting step, so that any increase in ethylene production is accompanied by an increase in ACC levels within the plant tissue. Note that oxygen is necessary for the last step of ethylene synthesis. For this reason ethylene synthesis can be inhibited at low oxygen levels.

#### *Mode of action of ethylene during ripening*

Studies on the mode of action of ethylene have been facilitated by the use of chemicals that can mimic ethylene action, such as propylene, and chemicals that compete with ethylene for its binding site. It has been known for a long time that ethylene binding can be blocked by silver ions (Beyer 1976; Oetiker & Yang 1995). More recently, 1-methylcyclopropene (MCP) has been identified as a competitor (or antagonist) (Serek *et al.* 1995; Sisler *et al.* 1996a, b).

Fruit fall into two categories with respect to the characteristics of ripening, and the mode of action of ethylene:

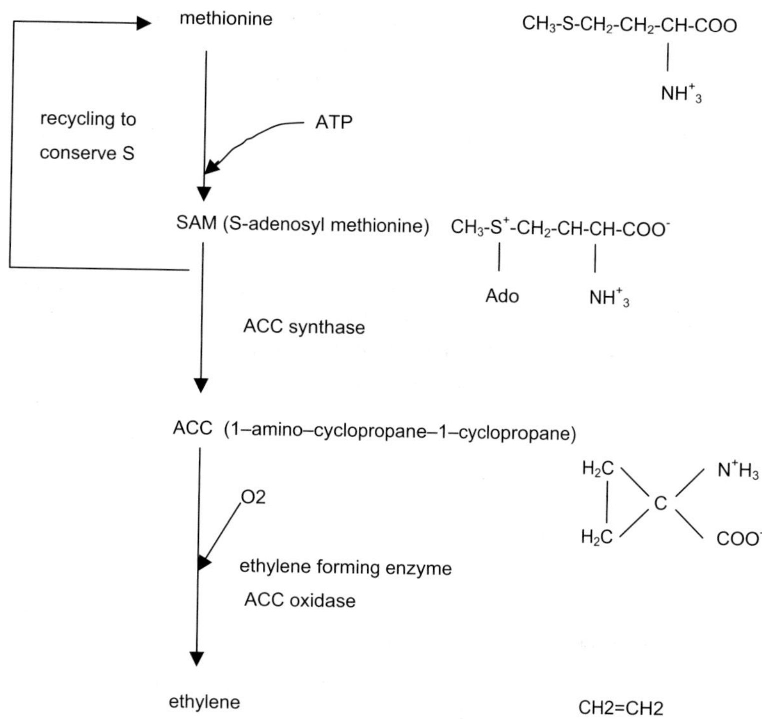


**Fig. 2.30** The synthesis of ethylene from methionine.

- non-climacteric fruit ripen at a steady rate with no great increase in metabolic rate – they produce low levels of ethylene which control ripening;
- climacteric fruit exhibit great changes during ripening, accompanied by an increase in metabolic rate (higher respiration rates) during ripening. They show a large increase in rates of ethylene production, which is autocatalytic (i.e. ethylene stimulates its own production). This can be demonstrated by the fact that addition of propylene (an ethylene mimic) can induce further ethylene production.

From the behaviour of fruit ripening, two systems of ethylene production and action have been postulated. These are termed system I and system II (McMurchie *et al.* 1972; Oetiker & Yang 1995). System I refers to the low rate of ethylene production found in climacteric fruit prior to ripening, non-climacteric fruit and vegetative tissues. Thus, ethylene controls its own synthesis. In this state ripening can be slowed down by antagonists of ethylene and speeded up by addition of ethylene. System II refers to the high levels of ethylene production during the climacteric phase for climacteric fruit. Ethylene stimulates its own synthesis and therefore there is a very fast rise in production. In this case ethylene synthesis can be triggered by addition of ethylene or an ethylene mimic, but cannot be stopped by the addition of an antagonist. The actions of system I and system II during ripening of a climacteric fruit are illustrated in Fig. 2.31.

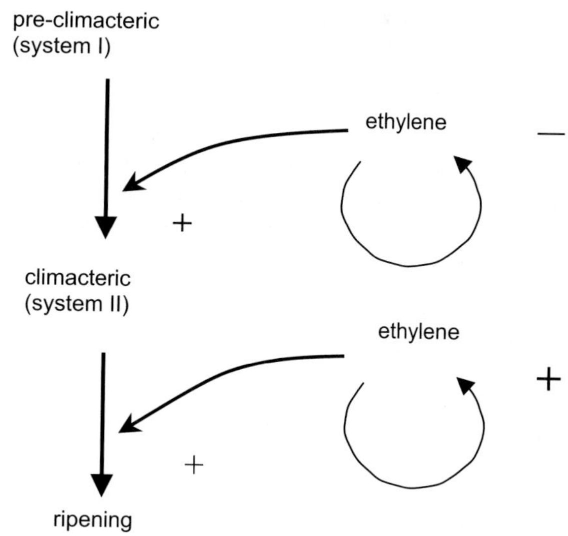


**Fig. 2.31** A simplified scheme showing system I and system II action during ripening of a climacteric fruit.

Figure 2.32 illustrates the effect of exogenous ethylene on climacteric and non-climacteric fruit. Non-climacteric fruit respond with an increased rate of respiration, and a corresponding increase in their rate of ripening. The effect is directly proportional to the level of exogenous ethylene and dependent on its continuous presence. At an early stage of ripening, climacteric fruit also respond to ethylene with an increase in respiration, but after a certain stage of maturity there will be an autocatalytic synthesis of ethylene as the fruit is now responding to system II. A certain threshold of exposure is required to initiate this autocatalytic ethylene, but once started further response is not proportional to exogenous ethylene and the process is essentially irreversible.

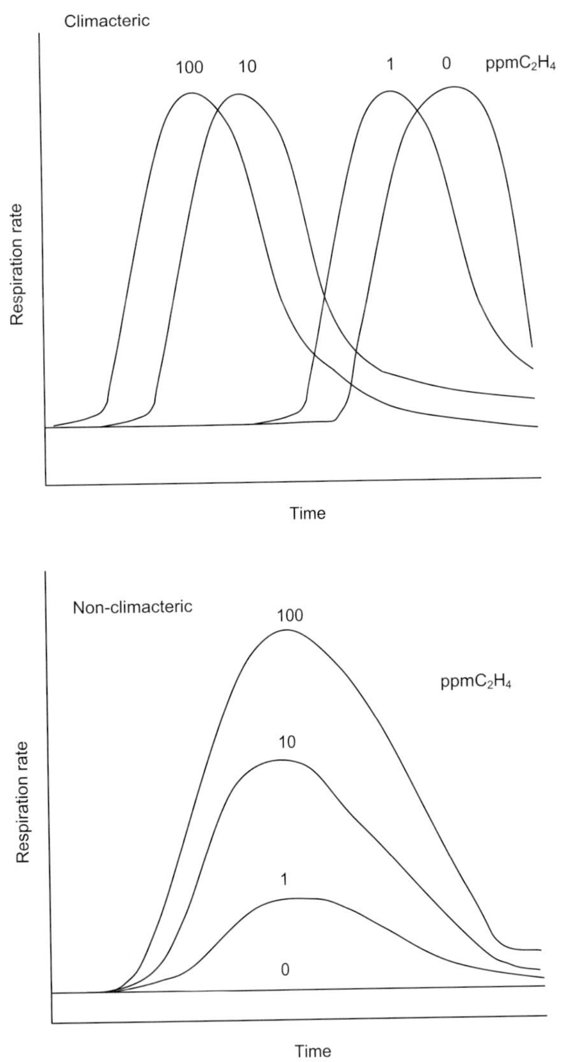


**Fig. 2.32** Effect of exogenous ethylene on fruit respiration. Source: Seymour *et al.* 1993.

### The importance of ethylene within the horticultural industry

It can be seen that ethylene plays a key role in plant development. This is of enormous importance for the post-harvest treatment of commodities, especially perishables. There are many cases where it is valuable to use the effects of ethylene to control the development of horticultural produce (Reid 1992).

#### *To promote ripening*

Ethylene can be used to promote faster and more uniform ripening in climacteric fruits. In many cases fruit is picked unripe, as it will transport and store better. Thus, ethylene may be added to stimulate ripening at the appropriate time.

#### *To promote degreening*

For some fruits customers expect the ripe fruit to be non-green, but the fruits themselves will not naturally lose their chlorophyll. This occurs for citrus fruits grown in warm climates, and also for some pineapples. In these cases in order to meet market expectations the fruit may be degreened by ethylene treatment.

#### *To promote fruit abscission*

Ethylene treatment can be used to promote fruit abscission and therefore facilitate harvesting. The treatments used involve spraying trees in plantations with ethylene-releasing chemicals.

#### *To promote sprouting*

Sprouting of seed potatoes may be induced by ethylene. However, there are many developmental processes that are undesirable for produce quality, such as leaf abscission, yellowing and russet spotting. Therefore, the ability to control ethylene levels, in some cases to treat produce with high levels and yet in other cases to exclude it, is very important.

### Levels of sensitivity to ethylene and levels produced by plant tissues

Both the sensitivity to ethylene and the levels of ethylene

produced by plant tissue vary depending on a range of factors, including commodity, variety, maturity stage, temperature, oxygen levels, carbon dioxide and stress. The low levels at which exogenous ethylene can have an effect are illustrated by Table 2.12, while Table 2.13 shows typical ethylene production rates of a range of horticultural commodities. Given that ethylene levels depend on so many factors, these figures should be treated with caution, and note that even for a single commodity, level of ethylene production can vary enormously between varieties. Table 2.14 illustrates the effects of temperature on ethylene production rates. However, one of the effects of altering atmospheric composition is to alter the synthesis of and reaction to ethylene. Low levels of oxygen inhibit ethylene synthesis and high levels of carbon dioxide inhibit the effect of ethylene by binding to ethylene receptors.

**Table 2.12** Threshold for ethylene action in various fruits. Source: Reid 1992.

| Fruit | Threshold concentration (ppm) |
|---|---|
| Avocado | 0.1 |
| Banana (var. Gros Michel) | 0.1–1 |
| (var. Lacatan) | 0.5 |
| (var. Silk fig) | 0.2–0.25 |
| Cantaloupe (var. P.M.R. No 45) | 0.1–1 |
| Honeydew melon | 0.3–1 |
| Lemon (var. Fort Meyers) | 0.1 |
| Mango (var. Kent) | 0.04–0.4 |
| Orange (var. Valencia) | 0.1 |
| Tomato (var. VC-243-20) | 0.5 |

**Table 2.13** Ethylene production rates. Source: Kader, undated.

| Class | Ethylene (μL/kg/h at 20°C) | Commodities |
|---|---|---|
| Very low | 0.01–0.1 | Cherry, citrus, grape, jujube, strawberry, pomegranate, leafy vegetables, root vegetables, potatoes, cut flowers* |
| Low | 0.1–1.0 | Blueberry, cucumber, okra, peppers, persimmon, pineapple, raspberry |
| Moderate | 1.0–10 | Banana, fig, honeydew melon, mango, tomato |
| High | 10–100 | Apple, apricot, avocado, cantaloupe, feijoa, kiwifruit, nectarine, papaya, peach, pear, plum |
| Very high | >100 | Cherimoya, passion fruit, sapote, mammee apple |

* Flowers produce high levels of ethylene just before they fade and wither, e.g. fading flowers of Vanda orchids have been recorded to produce 4300 μL/kg/h.

**Table 2.14** Effects of temperature on ethylene production by stone fruit.

| | Ethylene production (μL/kg/h) | | | |
|---|---|---|---|---|
| Fruit | 0°C | 5°C | 10°C | 20°C |
| Nectarine | 0.01–5 | 0.02–10 | 0.05–50 | 0.1–160 |
| Peach | 0.01–5 | 0.02–10 | 0.05–40 | 0.1–140 |
| Plum | 0.01–5 | 0.02–15 | 0.04–60 | 0.1–200 |

## Recommended further reading

### *Molecular components of living tissue*

MAF (1985)
Stryer (1995)
Whiteside (1991)

### *The structure and components of a cell; Plant tissues; Plant structures and life cycles; Cereal grains*

Fahn (1990)
Kays (1991)
Roberts (1976)
Salisbury & Ross (1992)
Taiz & Zeiger (1991)
Wills *et al.* (1998)

### *Processes occurring during ripening of fleshy fruit*

Oetiker & Yang (1995)
Seymour *et al.* (1993)

### *Metabolism and respiration*

Goodwin & Mercer (1983)
Lea & Leegood (1993)
Stryer (1995)

### *The role of ethylene in controlling development*

Kader (undated)
Kieber (1997)
Oetiker & Yang (1995)
Reid (1992)
Thompson (1996)

## References


Beyer, E. (1976) Silver ion: a potent anti-ethylene agent in cucumber and tomato. *HortScience,* **11**, 195–196.

Fahn, A. (1990) *Plant Anatomy*. Pergamon Press, New York NY, USA.

Goodwin, T.W. & Mercer, E.I. (1983) *Introduction to Plant Biochemistry,* Second Edition. Pergamon Press, Oxford, UK.

Grant, G.T., Morris, E.R., Rees, D.A., Smith, P.J.C. & Thom, D. (1973) Biological interactions between polysaccharides and divalent cations: the egg-box model. *FEBS Letters,* **32**, 195–198.

Jequier, E. (1999) Response to and range of acceptable fat intake in adults. *European Journal of Clinical Nutrition,* **53,** S84–S93.

Kader, A.A. (undated) Ethylene in postharvest biology and technology of horticultural crops. ASHS Slide Set **37**. American Society for Horticultural Science, Alexandria VA, USA.

Kays, S.J. (1991) *Postharvest Physiology of Perishable Plant Products*. Chapman & Hall, USA.

Kieber, J.J. (1997) The ethylene response pathway in Arabidopsis. *Annual Review of Plant Physiology & Plant Molecular Biology,* **48**, 277–296.

Lea, P.J. & Leegood, R.C. (1993) *Plant Biochemistry and Molecular Biology*. John Wiley & Sons, Chichester, UK.

McMurchie, E.J., McGlasson, W.B. & Eaks, I.L. (1972) Treatment of fruit with propylene gives information about the biogenesis of ethylene. *Nature*, **237**, 235–236.

Ministry of Agriculture and Food (1985) *Manual of Nutrition*. Reference Book 342 Tenth Edition. HMSO, London.

Oetiker, J.H. & Yang, S.F. (1995) The role of ethylene in fruit ripening. *Acta Horticulturae,* **395**, 167–178.

Reid, M.S. (1992) Ethylene in postharvest technology. In: A.A. Kader (Ed.) *Postharvest Technology of Horticultural Crops*. Second Edition. Publication **3311**. University of California, Davis CA, USA.

Roberts, M.B.V. (1976) *Biology. A Functional Approach*. Second Edition. Thomas Nelson and Sons Ltd. Sunbury-on-Thames, UK.

Salisbury, F.B. & Ross, C.W. (1992) *Plant Physiology*. Wadsworth, Belmont CA, USA.

Serek, M., Sisler, E.C. & Reid, M.S. (1995) Effects of 1-MCP on the vase life and ethylene response on cut flowers. *Plant Growth Regulation* **16**, 93–97.

Seymour, G.B., Taylor, J.E. & Tucker, G.A. (1993) *Biochemistry of Fruit Ripening*. Chapman & Hall, London.

Singer, S. & Nicolson, G.L. (1972) The fluid mosaic model of the structure of cell membranes. *Science,* **175**, 720–731.

Sisler, E.C., Dupille, E. & Serek, M. (1996a) Effect of 1-methylcyclopropene and methylenecyclopropane on ethylene binding and ethylene action on cut carnations. *Journal of Plant Growth Regulation,* **18**, 79–86.

Sisler, E.C., Serek, M. & Dupille, E. (1996b) Novel gaseous inhibitor of ethylene action 1-methylcyclopropene improves the postharvest life of bananas. In: Proceedings of the First International Conference on Bananas and Plantains for Africa, 13–18 October, Kampala, Uganda. *Acta Horticulturae.*

Stryer, L. (1995) *Biochemistry*. W.H. Freeman & Company, New York, USA.

Taiz, L. & Zeiger, E. (1991) *Plant Physiology*. The Benjamin Cummings Publishing Company, Inc., California, USA.

Thompson, A.K. (1996) *Postharvest Technology of Fruit and Vegetables.* Blackwell Science, Oxford, UK.

Whiteside, E. (1991) *Good Nutrition for Long Life. With Trends in Food Production.* Merlin Books Ltd, Braunton, Devon, UK.

Wills, R.B.H., McGlasson, W.B., Graham, D. & Joyce, D. (1998) *Post-harvest. An Introduction to the Physiology and Handling of Fruit, Vegetables and Ornamentals.* Fourth Edition. CABI International, Wallingford, UK.

Wolfe, S.L. (1972) *Biology of the Cell.* Fourth Edition. Wadsworth Publishing Company, Belmont CA, USA.

# Chapter 3
# Physical Factors in Post-Harvest Quality

A. D. Devereau, R. Myhara and C. Anderson

Stored food, both perishable and durable, may be considered to be an ecosystem, i.e. a system that includes a group of organisms and their environment (Multon 1988; Sinha 1995). The interactions between the physical, chemical and biological factors within the ecosystem lead to changes in the quality and nutritive value of the stored product (Jayas 1995). Knowledge of these factors is therefore essential if the quality and quantity of stored products are to be maintained. In this chapter we will look at the physical factors.

The physical factors of most importance during storage are temperature, moisture content of the crop, relative humidity of the atmosphere and concentration of atmospheric gases (oxygen, carbon dioxide). All living things within the store, including the stored product, insects, mites and micro-organisms, are affected by, and indeed interact with, these factors. Control of losses and maintenance of desirable quality traits depends to a great extent upon the measurement and control of these physical factors.

Physical factors do not exist in isolation; there are defined relationships between many of them, influenced by the presence and type of stored product. A change in one physical factor, caused perhaps by biochemical processes within the stored product, by changes in the ambient climate or by deliberate manipulation, may lead to changes in another, and may lead to conditions which either promote or prevent spoilage.

The properties of stored products affect the way in which they are influenced by physical factors, and so these are discussed first.

## Properties of stored products

The physical properties of grains affect some of the most basic choices made when planning storage and maintenance of grains – e.g. store size and shape, or choice and application of drying or aeration systems – in addition to influencing the effects of physical factors during storage. Hoseney and Faubion (1992) pointed out the importance of an understanding of cereal grain structure and presented a review of the physical properties of cereal grains.

Bulk solids have been described as a kind of fourth form of matter between solid and liquid. They consist of loose solid particles interspersed with voids, the latter usually filled with air (Fig. 3.1), and behave in some respects like a liquid, flowing to fill containers. Grain bulks are not homogeneous: properties inevitably vary, sometimes considerably, throughout the bulk (Christensen *et al.* 1982). In this section we review the density, void ratio, thermal properties and properties with respect to moisture of stored grains.

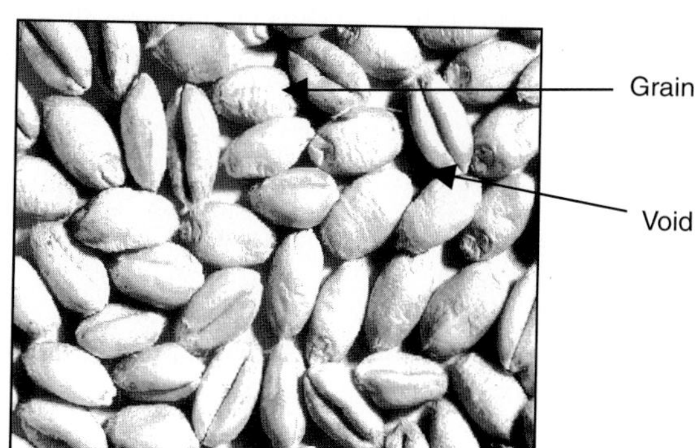


**Fig. 3.1** Wheat – a bulk solid – showing solid particles (grains) and voids filled with gas (air).

### Density

Density is the mass of a unit volume of a substance, i.e.

$$\text{Density} = \frac{\text{mass}}{\text{volume}}$$

The volume of bulk solids can be defined and measured in different ways and this leads to different definitions of their density (Fig. 3.2).

#### *Bulk density or test weight*

This is the most commonly used density measurement for stored products. A container of known volume is filled with the bulk solid and weighed. The volume thus includes the solid particles, air within any pores in the solids and air within the voids.

#### *Unit, apparent or envelope density*

The volume of the solids and any pores within them is measured and the sample weighed. The volume thus includes the solid particles and air in pores within them, but does not include air in the voids between the solid particles. This measure is not usually encountered in grain storage.

#### *Specific, absolute or kernel density*

The volume of the solid particles only is measured and the sample is weighed. The volume thus includes the solid particles only, and does not include air in voids between the solid particles or in pores within them.

The bulk density of grain is used as both a qualitative and quantitative measurement. It is used to indicate the breadmaking potential of wheat (Foster 1982) and has an effect on the performance of continuous flow grain driers (Brooker *et al.* 1992). Minimum test weights form part of US cereal grades, where the test weight is defined as the weight of 1 bushel of grain (0.0352 $m^3$ or 1.2445 $ft^3$) (Brooker *et al.* 1992). As a quantitative value it allows calculation of the capacity of existing storage facilities or the specification of storage capacity based on the mass of grain to be stored. Low test weights are found in grains that have been harvested before reaching physiological maturity or grains that have been invaded by field or storage fungi, resulting in excessive loss in dry matter (Hoseney & Faubion 1992), and thus give rise to lower crop value.

The bulk density of a product is lower than its specific density because air has a much lower density than the grain kernels. Some typical bulk densities are given in Table 3.1. The method of storage will affect the bulk density of the grain. For example, when products are stored in bags, the air spaces between the bags result in a further decrease in overall bulk density. This reflects a decrease in the overall quantity of grain per unit volume and not a decrease in the quality of the grain. Bulk density is also affected by the size and shape of the grains and their surface properties (Hosenay & Faubion 1992), the moisture content of the grains, the amount of impurities or foreign matter, degree of filling, kernel density and kernel packing. The latter is influenced by factors such as the method of filling and period of storage (Cenkowski

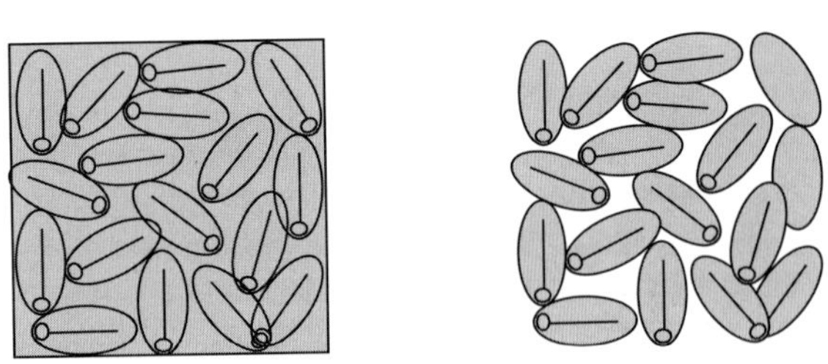

**Fig. 3.2** Bulk density (left); specific density (right). Shaded area = the volume included in the calculation of density.

**Table 3.1** Typical bulk densities of stored products.

| Products (bulk unless otherwise stated) | Bulk density (kg/$m^3$) |
|---|---|
| Wheat | 768–805 |
| Wheat (bagged) | 680 |
| Barley | 605–703 |
| Oats | 438–561 |
| Rye | 721 |
| Sorghum | 733 |
| Paddy | 576 |
| Paddy (bagged) | 526 |
| Oilseed rape (canola) | 689 |
| Linseed | 712 |
| Peas | 835 |
| Beans (field) | 859 |
| Rice | 579–864 |
| Rice (bagged) | 690 |
| Maize (shelled) | 718–745 |
| Maize (shelled, bagged) | 613 |
| Millet | 853 |
| Millet (bagged) | 743 |
| Groundnuts (in shells) | 352 |
| Groundnuts (in shells, bagged) | 322 |

& Zhang 1995). Bulk density increases during the drying process to a maximum around 14–16% (wet basis; w.b.): increases as high as 51.5 kg/m$^3$ during drying from 28% to 15.5% at 82°C are reported (Brooker *et al.* 1992). This can affect the capacity of storage bins – a decrease in test weight means that bins will hold less or must be increased in volume. Equations modelling the change in test weight with moisture content exist (Brooker *et al.* 1992).

Foster (1982) reported criticism of test weight as a quality measure of maize but stated that it would probably persist as a grading factor until more definitive measurements are devised. Brooker *et al.* (1992) stated that, although the test weight has formed a part of the US grain standards for some time, no consistent correlation has been established between high test weight and favourable end-user properties.

Standard apparatus exists (e.g. Parker *et al.* 1982; Hoseney & Faubion 1992) for measuring bulk density. In these methods a container of known volume is filled with grain and weighed, then the weight converted to a standard unit of bulk density using either a specially calibrated weighing scale or a standard weighing scale and conversion table. The grain is made to fall into the container from a fixed height and levelled without influencing its packaging to ensure, as far as possible, that the grain is packed in a repeatable way. The apparatus, known as a chondrometer, is illustrated in Fig. 3.3.

In some situations the specific volume (or stowage factor) is used. This is simply the reciprocal of the bulk density (i.e. 1/bulk density). For example, the bulk density of shelled maize is 720 kg/m$^3$, and the specific volume is 0.00139 m$^3$/kg or 1.39 m$^3$/tonne.

Specific density is measured using a pycnometer, a vessel which measures the displacement of a fluid such as mercury or helium from a vessel. Hoseney and Faubion (1992) refer to the use of air and toluene as the fluids in the pycnometer, and suggest an alcohol method if the volume of intact kernels is required. The specific density, like the bulk density, is moisture dependent and a mathematical model may be fitted to the data. Chung and Lee (1986) quote values of 1019 kg/m$^3$ to 1387 kg/m$^3$ for the specific density of rice and 1190 kg/m$^3$ to 1370 kg/m$^3$ for maize. Mohsenin (1980) describes the use of an air-comparison pycnometer for the measurement of specific density.

### Void ratio

The ratio of the volume of voids between grains to the total volume occupied by the bulk grain is called the void ratio or porosity. It is surprisingly high, ranging from 40–48% for wheat, maize and rice (Brooker *et al.* 1992). The porosity of grain depends upon the size and shape of the grains, their elasticity, surface state, dockage level, weight, compaction, storage period and distribution of moisture (Sinha & Muir 1973). Flour has a high void ratio – 55% – because the milling process produces highly irregularly shaped particles. The amount of fine material, degree of kernel filling and amount of foreign material also affect the porosity. The void ratio can be calculated from the ratio of bulk to specific density (Mohsenin 1970; Cenkowski & Zhang 1995). Mohsenin (1970) shows a simple apparatus for measuring void ratio.

The resistance of grain to airflow during drying or aeration is partly due to the porosity of the grains and their size (Brooker *et al.* 1992). The void volume of a grain store (i.e. the void ratio/store volume) may be needed when calculating the concentration of fumigant or the quantity of air needed to complete an air exchange during aeration (Hosenay & Faubion 1992).

### Thermal properties

Stored products tend to be good insulators with high

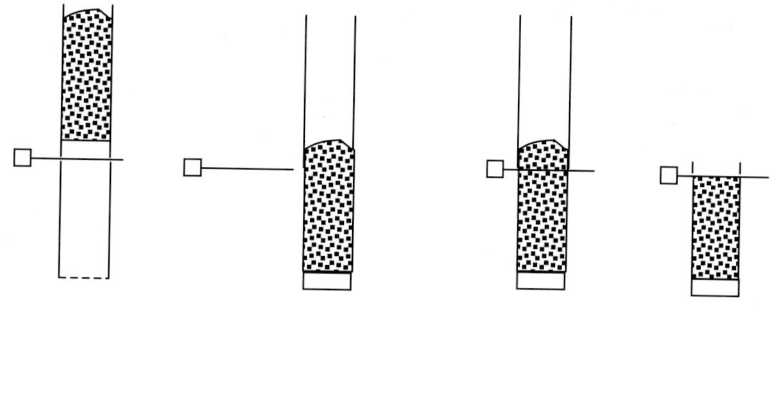

**Fig. 3.3** Standard US method for measuring grain bulk density.

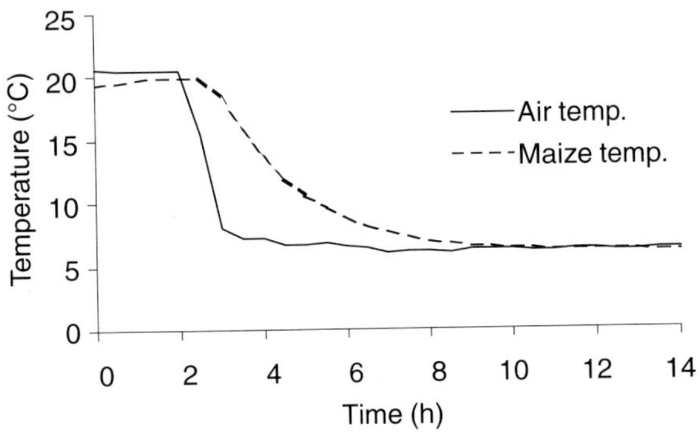


**Fig. 3.4** Thermal capacity of maize and air. The graph shows what happens when a jar full of air and a jar full of maize are taken from a room at 20°C and put into a refrigerator at around 5°C. The air cools very quickly, but the maize takes nearly 10 hours to cool to the same temperature. In a large silo it can take weeks or months for the grain to heat up or cool down.

thermal capacities. Sources of heat within stored grain bulks – usually caused by respiration of insects, mites, micro-organisms or the product itself – can therefore lead to significant temperature rises (Howe 1962). Sources of heat outside grain bulks (e.g. daily and seasonal ambient climate), or deliberate manipulation by cooling, aeration or drying, cause slow changes to the grain temperature or require a large amount of energy to achieve rapid changes. The most effective way of cooling or heating stored products is to force air through the voids between the grains. In a large silo it can take weeks or months for the grain to heat up or cool down (Fig. 3.4).

Heat transfer during product storage is complex, involving radiation, conduction and convection processes (Smith & Gough 1990). There are also heat transfers due to evaporation, condensation and adsorption. Thermal properties which are related to the transfer of heat in biological materials include dimensional characteristics, density, fluid viscosity, unit surface conductance (or heat transfer coefficient, unit film conductance or film coefficient), latent heat, specific heat, thermal conductivity, thermal diffusivity, thermal emissivity and coefficient of thermal expansion (Mohsenin 1980).

The thermal conductivity of grains has been measured in many ways (Mohsenin 1980). Many factors, including the form of the experiment, affect the results. Most determinations for grains have been carried out on bulks of grain held within spheres or cylinders and subjected to a steady-state temperature gradient. Many modes of heat transfer can occur in this arrangement, including conduction through the air and grain, convection currents, moisture migration (causing transfer of energy through the latent heat) and radiation. Table 3.2 shows typical figures for the thermal conductivity of some grains in comparison to other common materials.

Both moisture content and temperature have an effect on thermal conductivity, though the range of temperatures to which grains are exposed is limited in comparison to other materials (Mohsenin 1980). Data are often presented as a linear correlation with moisture content (Chung & Lee 1986). Factors such as dry matter, chemical composition, kernel size and shape, thickness of layer, grain variety, pack density, etc. have been cited as reasons for discrepancies between experimental results. The unit of thermal conductivity is W/m K. In some references the units used are Btu/ft hr °F (or Btu/ft hr R): this is converted to W/m K by multiplying by 1.731 (Rogers & Mayhew 1980). Moisture migration is often caused by temperature differences and is associated with transfer of latent heat which alters the temperatures and therefore the thermal conductivity. This needs to be considered when methods to determine the conductivity are chosen (Mohsenin 1980).

The specific heat of grain is dependent on the moisture content and has been expressed for wheat, paddy,

**Table 3.2** Thermal conductivities of some stored products.

| Product | Moisture content & temperature | Thermal conductivity (W/m K) | Reference |
|---|---|---|---|
| Wheat, maize, oats | | 0.13–0.18 | Oxley 1944 |
| Rice | 12–20% | 0.10–0.11 | Oxley 1944 |
| Sorghum | 1–22.5% | 0.10–0.13 | Oxley 1944 |
| Whole soybean | 11.2%, 10–66°C | 0.10–0.13 | Oxley 1944 |
| Rapeseed | 6.1%/4.4°C to 12.8%/31.7°C | 0.11–0.16 | Oxley 1944 |
| Fibreglass insulation | | 0.035 | CIBS A3, 1980 |
| Concrete | | 1.0–1.5 | Howatson *et al.* 1972 |
| Iron | | 50 | CIBS A3, 1980 |

finished rice, soybeans, sorghum and oats in terms of linear equations (Mohsenin 1980). Mohsenin (1980) and Brooker *et al.* (1992) give extensive tables of the thermal properties of food and non-food materials. The unit of specific heat is kJ/kg K. In some references the units used are Btu/lb °F (or Btu/lb R) or cal/g °C: these are equivalent and are converted to kJ/kg K by multiplying by 4.1868 (Rogers & Mayhew 1980).

Thermal diffusivity and specific heat have also been measured using the data from thermal conductivity experiments (Mohsenin 1980). Thermal diffusivity tends to fall slightly with increasing moisture content.

The heat of vaporisation or latent heat of water in grain is the amount of energy required to vaporise moisture from the grain. It can be calculated from the equilibrium moisture content curve. It falls as the moisture content increases until, at high moisture contents, it approaches the latent heat of free water. This means that more heat is required to vaporise water from grain at low moisture contents than from free water. For example, 1.175 times more heat is required to vaporise water from maize at 10% moisture content (wet basis) than from free water at 21.1°C, while at 15% moisture content only 1.019 times more heat is required (Chung & Lee 1986).

### Moisture diffusivity

Moisture movement within grain masses, like temperature movement, is a very slow process (Gough 1996). Diffusion of water within and between grains affects drying rates and the rate and pattern of moisture change or migration during storage. Many researchers studying drying have used models of diffusion within individual grains to model the drying processes. These models give rise to diffusion coefficients which are functions of temperature and moisture content. Cenkowski and Zhang (1995) presented a review of diffusion models. In general performance of models is variable, one reason being that their coefficients vary according to the way in which data are measured and fitted, and the type of diffusion equation used. Models tend to be fitted to the falling rate period of drying and do not perform as well if fitted to both the constant and falling rate periods. Taking account of different components of the grain structure, such as the pericarp, was found to be important (Cenkowski & Zhang 1995).

### Moisture capacity

Although durable stored products such as cereal grains are far drier than more perishable commodities they still contain significant amounts of water. Cereal grains typically contain 10% to 15% of water by weight during storage. Air also contains moisture and is the medium through which moisture changes in the grain often occur. The moisture-holding capacity of air is many times lower than grain: under typical conditions 1 m$^3$ of grain will contain 75–110 kg of water, while a similar volume of air can typically contain a maximum of 15–20 g of water, some 5000 times lower than the capacity of the grain (Fig. 3.5). Hence large volumes of air are required to alter the moisture content of stored products significantly.

**Fig. 3.5** Moisture capacity of maize and air. The picture shows a 25 kg sack of maize. It will typically contain 3 litres of water, shown on the right of the photograph. If the sack contained only air, the air would typically contain around 0.4 ml of water, shown in the tiny flask on the left of the photograph.

## The principal physical factors

### Temperature

Temperature affects the rate of all biochemical processes and is therefore of fundamental importance in any storage system. Together with moisture content, it largely determines the storage life of grain. The temperature of the stored product and that of the air around and within it are both important.

Insect, mite, fungal and mycotoxin development, germination loss and baking qualities are affected by temperature. At temperatures found in grain stores, biological activity of insects, mites, fungi and the grain itself doubles for every 10°C rise in temperature (HGCA 1999) (Table 3.3). At low temperatures insect breeding stops and less moisture is available for potential pests in cold grain. Grain should therefore be cooled immediately after drying and before it comes into store. Cooling during storage will also equalise temperature gradients and so prevent moisture translocation (HGCA 1999). Biochemical activity due to moulds, insects, etc. can also be a source of heating during storage, as can transfer of heat energy to or from the fabric of the store by radiation. Ventilation of the store and, in particular, movement of air directly through the stored product, is one of the main ways of modifying storage temperatures.

A temperature rise on the outside of a grain bulk causes a heat front to move through the grain at a rate of approximately two metres per month (Gough 1996). This typically occurs in tropical and subtropical climates at the onset of a season when the average ambient temperature rises. In a bag stack the heat front moves at approximately four metres per month.

### Moisture content

Many stored products are hygroscopic materials, which means that they can absorb and release water, rather like a sponge. They thus consist of an amount of dry matter and an amount of water.

The moisture content (m.c.) expresses the weight of water in a product as a proportion of its weight. The most common method of expression is known as the wet basis, shown in Equation 3.1, in which the weight of the product is taken as the weight of the dry matter plus water. In scientific work the dry basis, shown in Equation 3.2, may be preferred, in which the weight of the product is taken as that of the dry matter only.

$$\text{m.c. (wet basis)} = \frac{\text{weight of water in sample}}{\text{wet sample weight, i.e. weight of water + dry matter}} \times 100\% \qquad (3.1)$$

$$\text{m.c. (dry basis)} = \frac{\text{weight of water in sample}}{\text{dry sample weight, i.e. weight of dry matter}} \times 100\% \qquad (3.2)$$

Equations 3.3 and 3.4 can be used to convert between wet basis and dry basis moisture contents.

**Table 3.3** Effect of different temperatures on the risk of insect and mite infestation. Modified from HGCA 1999.

| Risk | °C | Effects on insects | Effects on mites |
|---|---|---|---|
| Least | 60 | Death in minutes | Death in minutes |
| | | Death in hours | Death in hours |
| | 50 | | |
| | | Development stops | Death in days |
| | 40 | Development slows | |
| | 30 | Maximum development rate | No increase |
| Greatest | | Development slows | Maximum development rate |
| | 20 | | |
| | | Development stops but all stages survive | Development slows |
| | 10 | | |
| | | Movement stops, death in months | |
| | 0 | Lowest development rate | |
| | | (Fungi can still grow in damp grain) | |
| | –10 | Death in weeks | Death in weeks |
| Least | | | |
| | –20 | Death in minutes, insects freeze | Death in minutes, mites freeze |

$$\text{m.c. (wet basis)} = \frac{100 \times \text{m.c. (dry basis)}}{100 + \text{m.c.(dry basis)}} \quad (3.3)$$

$$\text{m.c. (dry basis)} = \frac{100 \times \text{m.c. (wet basis)}}{100 - \text{m.c.(wet basis)}} \quad (3.4)$$

When presenting moisture contents the basis used should be stated, e.g. 15% (wet basis) or 15% (w.b.), though in situations where wet basis is always used it will often be omitted.

Moisture content is fundamentally important in establishing safe storage conditions. Changes in moisture content also cause a change in the overall weight of a commodity, and as products are often traded by weight this has obvious financial implications. The change in weight of a commodity when its moisture content changes can be calculated from Equation 3.5:

$$\text{Final weight} = \text{Initial weight}\left(\frac{100-a}{100-b}\right) \quad (3.5)$$

where $a$ = the initial moisture content, $b$ = final moisture content.

For example, if 1000 t of wheat is dried from 17% to 15% m.c. (w.b.) its final weight will be:

$$\text{Final weight} = 1000\left(\frac{100-17}{100-15}\right)$$

$$= 976.5 \text{ t}$$

**Relative humidity**

Air is a mixture of gases. One of the gases it contains is water vapour, that is, water in its gaseous form. Each constituent of a gas mixture exerts a pressure, called its partial pressure, in proportion to its volume. The sum of the partial pressures for each constituent of a gas mixture equals the total pressure exerted by the mixture. The amount of water vapour in the air can thus be expressed by its partial pressure in units of Pascals (Pa), millibars (mbar), millimetres of mercury or water (mmHg or $mmH_2O$) or pounds per square inch (psi).

The term humidity refers to the proportion of water vapour in the air. The absolute humidity is the mass of water in a certain volume of air, for example, the number of grams of water per cubic metre of air. The mixing ratio or moisture content of the air is the mass of water in a certain mass of air, e.g. grams of water per kilogram of air. Both these measurements are highly dependent on the pressure of the gases.

The ability of air to contain water vapour is limited: when the maximum limit is reached the air is said to be saturated, and can contain no more water. This saturation limit is highly dependent on the temperature of the air: as the temperature rises the air is able to contain an increasing amount of water vapour. The relative humidity is the amount of water vapour that is in the air as a proportion of the amount of water vapour required to saturate the air at the same temperature. The amount of water vapour is measured by its partial pressure, so the relative humidity can be written as shown in Equation 3.6.

$$\text{r.h.} = \left(\frac{p_w}{p_{ws}}\right) \times 100 \quad (3.6)$$

where $p_w$ = partial pressure of the water vapour, $p_{ws}$ = partial pressure of pure water at saturation.

Because the partial pressure of water at saturation is highly dependent on the temperature, the relative humidity is also highly dependent on temperature. As the temperature rises, the partial pressure at saturation rises so the relative humidity falls, provided the amount of water vapour in the air stays constant. This is illustrated in Fig. 3.6 which shows how the relative humidity within a jar of maize rises when the jar is placed into a refrigerator.

Relative humidity is more useful than absolute measures of humidity for determining the response of living things or biological materials to the air around them. For example, the ability of the air to dry water from crops or sweat from our skin depends on the degree to which it is saturated, i.e. on its relative humidity. Most importantly, in food storage the relative humidity of air around the product is one of the most important factors controlling spoilage by micro-organisms. Water vapour in the air around a stored product interacts with the water held in

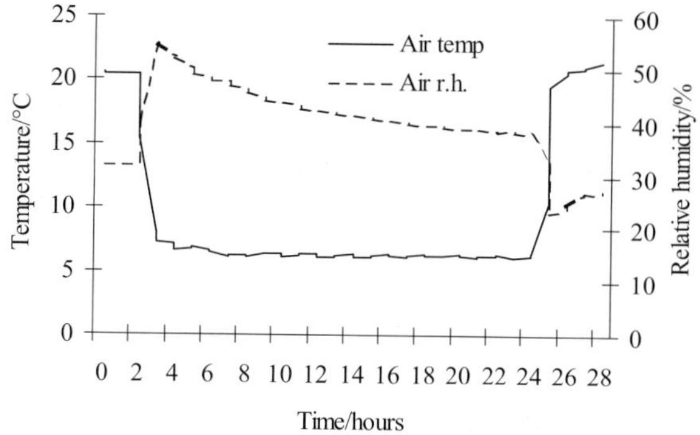


**Fig. 3.6** How relative humidity increases when temperature decreases.

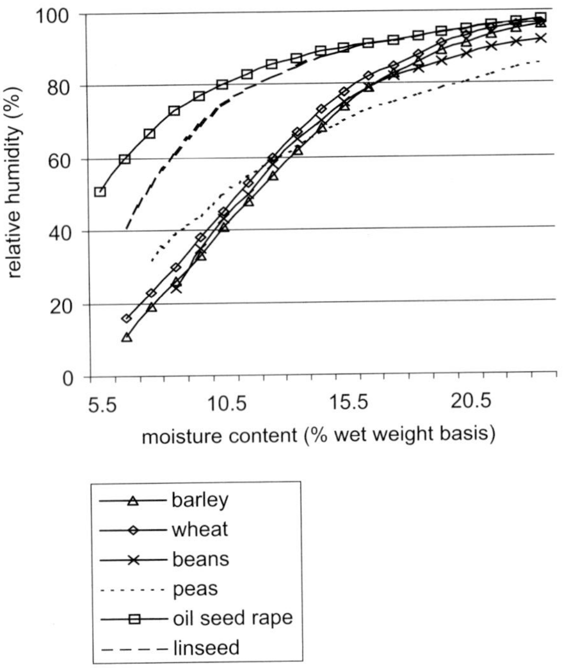


**Fig. 3.7** The relationship between relative humidity and moisture content in some stored commodities. Source: HGCA 1998.

the product, so when stored products are present the relative humidity depends not only on the partial pressure of the water and temperature but also on the moisture content of the stored product. The relationship between relative humidity and moisture content of some commodities is shown in Fig. 3.7.

Because air temperatures vary throughout the day and night the relative humidity of the air will also vary, often by a large amount, as shown in Fig. 3.8. This is important when ambient air is used to aerate or ventilate stored commodities. Hazards occur when the temperature falls at night and the relative humidity of the air rises. If the temperature falls by a large enough amount the relative humidity of the air can rise to 100% and liquid water will begin to condense out. It is clearly undesirable to bring air in this condition into contact with a stored commodity.

Problems can also occur within storage buildings or structures. The roof of a warehouse can get very cold at night while air rising from stored produce, for example, a bag stack of grain, may be relatively warm and moist. The air will cool and can reach saturation in contact with the cold roof. Condensation will then occur on the underside of the roof, causing water to drip on to the bag stack below and form a wet layer which will spoil due to mould growth (Fig. 3.9.) Another example occurs when commodities are transported within steel shipping containers from tropical to temperate climates (Gough 1996). The

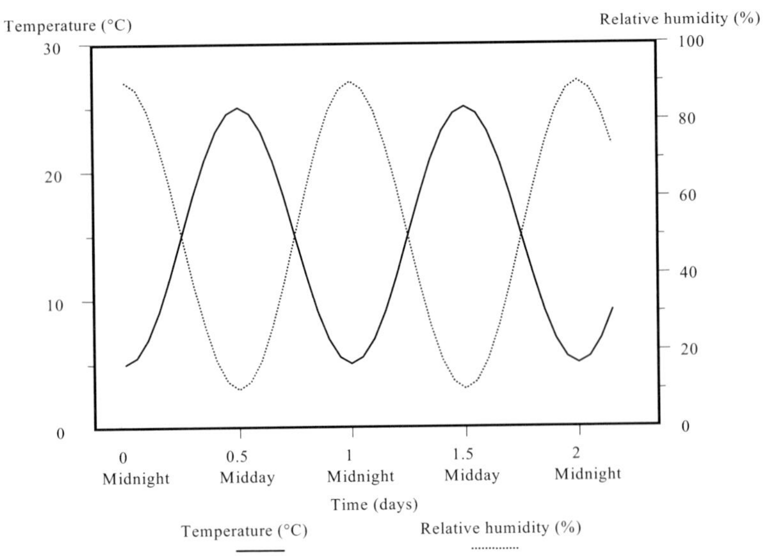


**Fig. 3.8** Fluctuations in temperature and relative humidity over 24 hours.

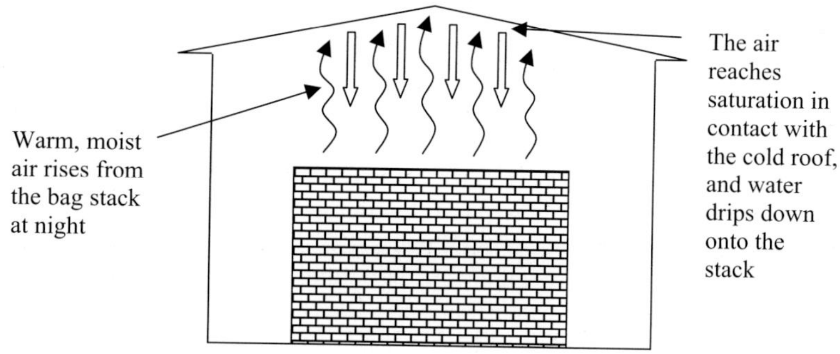


**Fig. 3.9** Night-time condensation, or raining, in a warehouse. The same phenomenon may occur inside shipping containers when they arrive in a cold climate from a hot one (internal raining).

containers remain warm when packed together in the hold of the ship, but their surfaces cool rapidly if they are unloaded in cold weather on arrival. Warm moist air from the commodity can reach saturation in contact with the sides and roof of the container, giving rise to the phenomenon of internal raining in which condensation rains down onto the commodity and leads to spoilage.

## Water

Liquid water is essential for life and so must be present in all living tissues. It is the major ingredient in most fresh foods, and contributes up to 92% of the total weight of some fruits and vegetables and up to 81% in some animal tissues (Table 3.4).

Water contributes to the bulk of many foods such as fresh fruits, vegetables and meats, but also contributes significantly to the structure, appearance and taste of foods.

Water also influences the susceptibility of foods to degradation and spoilage. Since water is essential for life, spoilage micro-organisms cannot grow in its absence. Because of this the water content of foods such as cereal grains, preserved by water reduction, is relatively low (Table 3.4). Additionally, most chemical reactions take place in water (as a solvent) or require water as a reactant. Consequently, the presence or absence of water (or its availability) determines to a great extent the chemical stability of foods.

**Table 3.4** Water content of some common foods.

| Food | Water content (%) |
|---|---|
| Meat | |
| Pork (trim, boneless) | 53–60 |
| Beef steak (boneless) | 56–59 |
| Fish (cod) | 81 |
| Fruit and vegetables | |
| Apple | 84 |
| Turnip | 92 |
| Grains | |
| Wheat | 12–13 |
| Maize | 14 |
| Rice | 12 |

Because most fresh foods contain large amounts of water, they are very susceptible to degradation. Preservation methods can be used either to remove the water, through drying, or by partitioning the water through freezing or chemical binding, thus denying water its solvative and/or reactive nature. Water influences the structure, appearance and taste of foods. Its removal and replacement causes changes to the food. These changes are often irreversible such that water replacement will not return food to its original state.

### Water–solute interactions

#### *Bound water*

The structure of water can be altered when various substances are dissolved in it. Indeed, the structure of the added substances, such as proteins, can themselves be changed. Hydrophilic substances interact with water through dipole–dipole, ionic–dipole or ionic–ionic interactions, depending upon the nature of the substance. The term 'bound water' or 'water of hydration' refers in general to the tendency of water to be attracted to hydrophilic substances.

All water can be described as free or bound. Water which is a constituent part of a molecule can be considered bound. This 'constitutional water' forms part of non-aqueous systems and gives structure and form to crystals (as water of crystallisation) or organic macromolecules (as interstitial water of proteins). The interactions involved in binding constitutional water may include ionic–dipole, dipole–dipole or ionic–ionic.

### *Charged, ionically bound groups*

At the molecular level, water, which is ionically bound to solutes (dipole–ionic), is some of the most strongly bound water found in foods. The degree to which various ions can impose order upon water surrounding the ions depends to a great extent upon the force of the electric field in the vicinity of the ion (the charge on the ion divided by the ionic radius). Thus multivalent ions, or ions with small radii ($Li^+$, $Na^+$, $H_3O^+$, $Ca^{2+}$, $Ba^{2+}$, $Mg^{2+}$, $Al^{3+}$, $F^-$ and $OH^-$) have strong electric fields and can impose considerable structure to the water in the vicinity of the ion. This tightly bound water forms a monolayer, immediately adjacent to the ion, and is referred to as 'monolayer' or 'vicinal' water. Conversely, monovalent ions or ions with large radii ($K^+$, $Rb^+$, $Cs^+$, $NH_4^+$, $Cl^-$, $Br^-$, $I^-$, $NO_3^-$, $BrO_3^-$, $IO_3^-$ and $ClO_4^-$) have weak electric fields and cannot impose significant order. In food systems, where binding of water is important, it is the former ionic species and not the latter that are used.

The influence of strong electric fields surrounding ions can impose structure upon water some considerable distance from the ionic surface. Of course the extent of the influence depends upon the strength of the electric field. The water well beyond that immediately surrounding the ion and adjacent to the monolayer water is referred to as multilayer water. The extent to which the ions can impose structure upon the multilayer water depends upon the strength of the electric field surrounding the ion and the distance the water is away from the surface of the ion. In reality, a range of structures can be found in multilayer water, again depending upon its distance from the ion.

The influence of the electric field is dependent upon the distance of the water from the ionic surface. Water not under the influence of the ionic charge is referred to as bulk water. Bulk water has properties similar to that of weak salt solutions.

### *States of water*

It is apparent that water can exist in several states, depending upon its degree of boundness. Previous classifications have described water as bound, partly bound or unbound. No distinction was made between constitutional and vicinal water. Current theory classifies water as either constitutional, vicinal, multilayer or bulk in nature. Each classification specifies unique properties for each group.

*Constitutional water:* The water in this category can simply be described as structural water. Water of crystallisation and water of hydration are examples of constitutional water. It is unfreezable at –40°C and possesses no potential solvating capability. In terms of bulk motion, constitutional water is practically immovable. The strong bonds place considerable constraints upon translational mobility of the molecules. Removal of constitutional water in foods requires virtual destruction of the material. Because of this high degree of binding, constitutional water contributes virtually no water vapour and, consequently, no water activity. High moisture foods, with water contents of 85–90% (dry weight), contain no more than 0.03% (dry weight) constitutional water.

*Monolayer or vicinal water:* Like constitutional water, monolayer water is not freezable at –40°C and cannot act as a solvent. Unlike constitutional water, however, vicinal water has some translational freedom. The enthalpy of vaporisation is low enough to contribute a small amount of water vapour and thus has some water activity. Conventional drying techniques are able to remove most of the vicinal water, but it is some of the most difficult water to remove, due to its high enthalpy.

High moisture foods, with water contents of 85–90% (dry weight), contain about 0.5% (dry weight) vicinal water. Some intermediate moisture foods, such as dried fruit, which contain relatively soluble sugar contents can contain as much as 18% (dry weight) vicinal water (Myhara *et al.* 1998).

*Multilayer water:* Water is attracted to the surface of charged and non-charged hydrophilic substances (also called humectants) until the monolayer moisture content is achieved. Additional water molecules are still attracted to the substances and are structured in layers adjacent to the monolayer. These molecules are referred to as multilayer water.

Multilayer water is mostly unfreezable at –40°C. The degree of freezability depends upon the degree of boundness of the water molecules. In practice a gradation of frozen and unfrozen water will exist simultaneously in the food matrix. In many cases, a glassy state will develop. The amount of multilayer water present in the food will greatly affect its glass transition temperature.

As the percentage of multilayer water increases, the partial pressure of water will decrease, with a concomitant decrease in water activity. High moisture foods will contain 3–5% (dry weight) multilayer water. As foods dry, however, the relative amount of multilayer water increases. Over half of the water in many intermediate moisture foods is present as multilayer water (Jayas & Mazza 1993). It is, therefore, obvious that as a food dries its water activity decreases.

*Bulk water:* Water molecules far from hydrophilic substances are not under their influence and behave like that of water in dilute salt solutions. This bulk water is found in large diameter capillary and void spaces. Bulk water can exist either free or entrapped in complexes of polymeric gels. Bulk water is freezable and obeys Raoult's law of freezing point depression. Bulk water has a large solvent capacity, virtually unrestricted translational mobility and an enthalpy of evaporation similar to that of pure water. In high moisture foods it constitutes 90% (dry weight) of the water present. Entrapped water behaves in an essentially identical way to free water, except that its gross movement is impaired. Bulk water is the easiest form of water to be removed by evaporation.

### *Drying and states of water*

The drying of high-moisture foods occurs in four distinct stages. Each stage represents a different type of drying, depending upon the form of water present (Toledo 1997). Typically, stage I exhibits a constant rate of drying. With an enthalpy of vaporisation identical to pure water, bulk water present on the product surface freely evaporates. All the heat put into the product evaporates water, so there is no increase in product temperature and the water is removed at the wet bulb temperature.

Stage II is the first of the falling rate stages. In this stage, all of the water present on the surface of the product has evaporated. As a consequence there is a linear decrease in drying rate, which corresponds to the diffusion of bulk water from the interior of the food product onto the surface. At the surface, the water evaporates with an enthalpy identical to that of pure water. A slight increase in product temperature can be measured at this point.

Stage III, the second of the falling rate stages, occurs when the equilibrium moisture content of the food product begins to fall below 100%. All the bulk water has been removed from the product and the first of the multilayer water is beginning to evaporate. More energy is required because both the heat of vaporisation of pure water and the heat of adsorption must be provided. Drying takes place in the interior of the food product. Therefore, the drying rate, which decreases in a linear fashion, depends upon the diffusion of water vapour from the interior of the food to the surface. At the surface air currents carry the water away. There is a moderate increase in product temperature at this point.

Finally, in stage IV, all of the multilayer water has been removed. The remaining vicinal water requires large amounts of energy for its removal. The temperature of the product begins to rise close to that of the drying air since the water activities of the product and the air are similar. Dehydration must be terminated at this point or severe thermal deterioration could occur.

## Water activity

### *Product stability*

It is apparent that a relationship exists between the amount of water in a food product and its stability. Microbes require water for their growth and reproduction, so a reduction in water content through drying will eventually lead to a shelf stable product, where no refrigeration is needed. The water content at which this microbial stability occurs differs from one food product to another. For example, maize is barely shelf stable at 10–15% moisture content, while dried dates achieve shelf stability with 20% or more water content.

Physical stability in many cases depends to a great extent upon the presence or absence of water. Compare the physical state of fresh tomatoes with that of tomatoes which have been sun-dried – high water content is necessary to maintain physical size and shape. Likewise, although chemical stability tends to increase with a decrease in moisture content, lipid oxidation can be accelerated at lower moisture contents. It seems therefore, that the situation is not straightforward. Complex interactions between water and various non-aqueous substances require a measure which is more descriptive in terms of the relationship between solvent and solute. The measure commonly used is that of water activity ($a_w$).

### *Definition of $a_w$*

Water in contact with an atmosphere will lose or gain water vapour, the rate of which depends upon the percentage relative humidity (r.h.) of the atmosphere. The point at which the number of water molecules leaving the liquid coincides with the number of molecules entering the liquid is known as the equilibrium relative humidity (ERH). For pure water contained in a small closed container, water vapour will accumulate until the headspace atmosphere is saturated (100% r.h.). The partial pressure of pure water ($P_0$) contained in the headspace will depend solely upon its temperature. Food products containing bound water will release fewer water molecules, at the same temperature, into the headspace since the enthalpy of evaporation of some of the molecules will be increased relative to pure water. Water activity ($a_w$) can be described as the point where a food product reaches equilibrium with the headspace percentage r.h. The ERH corresponds to specific equilibrium moisture content (EMC).

Water activity can also be described as the ratio of the partial pressure of water vapour in the headspace, relative to that of pure water:

$$a_w = \frac{P}{P_0} \tag{3.7}$$

where $P$ represents the partial pressure of water vapour in a headspace above the food product at the same temperature. Water activity can also be expressed as:

$$a_w = \frac{P}{P_0} = \frac{ERH}{100} \tag{3.8}$$

The relationship between the moisture content of the food product and that of its $a_w$ depends upon its temperature, physical structure and chemical composition. Therefore, its stability also depends upon these factors. It is important to note that $a_w$ is an intrinsic property of the food product, whereas the ERH is only a property of the headspace gas.

### *Water activity of high moisture foods*

High moisture foods contain large amounts of bulk water, which behaves as an ideal dilute salt solution. In this case $a_w$ equals $N$, where $N$ represents the mole fraction of water. The mole fraction can in turn be expressed in terms of the weight fraction of water (n) and the weight fraction of the solvent ($n_1$), thus:

$$a_w = \frac{n}{n + n_1} \tag{3.9}$$

Equation 3.9 is Raoult's law and can be used to calculate the $a_w$ of high moisture food products, where the molecular weight (or range of molecular weights) of the solute(s) are known. The measured $a_w$ of food products, however, deviates significantly from that predicted by Raoult's law, especially in intermediate and low moisture foods. Raoult's law deals with ideal systems, while most food products behave in non-ideal ways. Most importantly, water acts as a solvent where heat of crystallisation and heat of mixing are not accounted for.

### *Temperature dependence*

The water activity of a food material will increase with an increase in temperature. Intuitively this makes sense since individual molecules will have more energy at higher temperature. Additionally, the dipole–dipole bonds between water molecules weaken as the temperature increases, thus increasing molecular mobility. The relationship between $a_w$ and temperature can be described by the Clausius–Clapeyron equation:

$$\frac{d \ln a_w}{d(1/T)} = \frac{-\Delta H_{st}}{R} \tag{3.10}$$

where $\Delta H_{st}$ is the heat of vaporisation and $R$ is the universal gas constant. The linear form:

$$\ln\left(\frac{a_{w1}}{a_{w2}}\right) = \frac{-\Delta H_{st}}{R}\left(\frac{1}{T_1} - \frac{1}{T_2}\right) \tag{3.11}$$

can be used to calculate the heat of vaporisation given the ratio of the $a_{w1}$ at $T_1$ to the $a_{w2}$ at $T_2$. Adsorption of water vapour onto surfaces is an exothermic reaction, thus as water is adsorbed, the heat of vaporisation will come closer to that of pure water. This situation is not, however, always followed. Dried fruits, containing large amounts of crystalline sugar, will exhibit heats of vaporisation during rehydration, which are less than that of pure water (Saravacos *et al.* 1986). As water is adsorbed onto solid surfaces, the solvent effect of the water begins to dissolve the sugar crystals and the heat of crystallisation more than compensates for the heat of vaporisation.

It is important to note that the temperature must always accompany the measured water activity of a substance.

### *Moisture sorption isotherms*

The relationship between the EMC of a food and its corresponding $a_w$ at the same temperature, when plotted, is referred to as a moisture sorption isotherm. It is an isotherm because all the points on the line are at the same temperature. The data necessary to construct isotherms of a food material are obtained experimentally from direct observations of EMC and $a_w$. Traditionally, isopiestic methods are used to obtain these values. The isotherm shown in Fig. 3.10, shows the *s* or sigmoid shape characteristic of food materials.

Starting at low moisture contents (zone I), large increases in EMC result in small to moderate increases in $a_w$. As the plot moves into zone II, relatively small changes in EMC result in large changes in values of $a_w$. Finally, in zone III, large increases in EMC again result in low to moderate increases in $a_w$. The relationship between EMC and $a_w$, as represented in these three zones, reflects changes in the nature of the bonds between water molecules and solutes present in the food material.

In zone I, only constitutional and vicinal waters exist. They are the most tightly bound forms of water present in a food material, where ion–ion, ion–dipole and dipole–dipole interactions occur. Both constitutional and vicinal water have very high energies of vaporisation and therefore can contribute little to $a_w$ as the moisture content increases. As a result, the slope of the line is high.

Zone II represents that sector of the moisture sorption isotherm where all of the available sites on all the solute surfaces of the food material have been occupied. The EMC value corresponding to the transition point between zones I and II is referred to as the monolayer moisture content ($X_M$). Beyond this value, additional

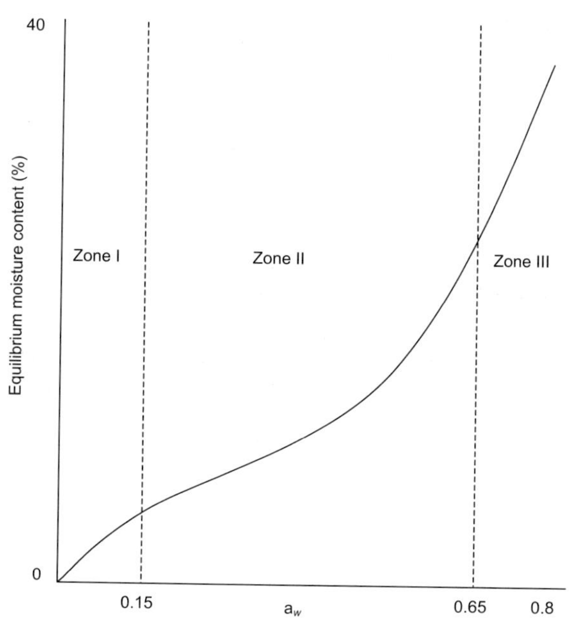


**Fig. 3.10** Idealised water sorption isotherm showing zones I, II and III.

water molecules must exist adjacent to vicinal sites bound by dipole–dipole interactions as multilayer water. Two properties of multilayer water – increased solvation and reduced heat of vaporisation – are crucial to the understanding of zone II. Firstly, because multilayer water can act as a solvent, new solute sites become available for hydration since the solvating water opens up the compact structure of the food material. Secondly, since the heat of vaporisation is decreased, more water can evaporate and contribute to $a_w$. Consequently the line in zone II is nearly horizontal where small changes in EMC correspond to large increases in $a_w$.

Zone III represents the sector of the moisture sorption isotherm where all non-solvated sites of the food material are hydrated. Additional water molecules are no longer constrained by dipole–dipole interactions and can exist as bulk water. This bulk water can fill voids and capillaries, and can be physically trapped in structures such as gels. In this zone, insoluble solids have a negligible effect upon $a_w$. Bulk water behaves in a similar fashion to dilute salt solutions, in that its contribution to $a_w$ depends upon the concentration of solutes as predicted by Raoult's law.

*Effect of temperature on isotherms:* Dipole–dipole and ionic–dipole interactions are strongest at low temperatures and decrease with an increase in temperature. Consequently, water molecules are less strongly bound to solutes at higher temperatures and isotherm plots shift to the right towards higher $a_w$ levels.

The degree to which the shift occurs depends upon the chemical composition of the food. Intermediate moisture foods (dried fruit, salt preserved foods), with large amounts of water-binding compounds (sugars, salts, various humectants), display the largest shifts (Myhara & Sablani 2000). This is due to the large amount of multilayer water present. Its degree of boundness depends mostly upon temperature. Food materials composed of large amounts of non-hydrated polymeric materials (cereal grains, dried pulses, dried tubers) such as cellulose, dietary fibre or starch contain little in the way of multilayer water. Consequently, these foods display isotherms which shift little as temperatures change (Mazza & Campbell 1985). It is important for food technologists to consider this shift, if the stability of the food is to be maintained. The water activities of those foods containing large amounts of multilayer water are very sensitive to changes in temperature, while those composed mostly of monolayer and/or bulk water are relatively insensitive.

Food materials containing large amounts of crystalline sugar will deviate from the normal trend, showing a reversal at higher $a_w$ levels where $a_w$ decreases at constant EMCs as temperatures increase. The result is a crossing of moisture sorption isotherms conducted at differing temperatures. This is due to the increased solubility of crystalline sugar as temperatures increase. As the crystalline sugar dissolves, an increasing number of binding sites become available, tending to reduce $a_w$ (Saravacos *et al.* 1986; Ayranci *et al.* 1990; Myhara *et al.* 1998).

*Hysteresis – adsorptive and desorptive isotherms:* A change in the EMC of a food can come about through either addition (adsorption) or removal (desorption) of water. Desorptive water sorption isotherms will always display lower $a_w$ values for all values of EMC, when compared to adsorptive water sorption isotherms. When desorptive and adsorptive water sorption isotherms of the same food substance are plotted together the two plots are not superimposable, but rather display hysteresis (Johnson & Johnson 2000).

It is thought that hysteresis results from the collapse of microcavities during drying. Sites previously accessible to water vapour molecules are permanently denied to additional molecules added during rehydration. The importance of hysteresis during food processing and formulation cannot be overemphasised. All foods preserved through moisture control will display enhanced stability when prepared by loss, rather than by addition of water.

### Models

Given the importance that $a_w$ plays in the determination of food stability, it is not surprising that many attempts have been made to predict the water sorption behaviour of food substances. Production specialists dealing with commodities such as maize, wheat or rice must be able to predict safe moisture contents for harvesting and storage. Food technologists must be able to predict the effect various ingredients may have upon the $a_w$ of many processed foods, especially those relying upon moisture control for stability.

*Theoretical:* Theoretical models based upon the adsorption of gases onto surfaces include the Brunauer Emmett Teller (B.E.T.) equation (Brunauer *et al.* 1938) and its modified version, the GAB equation, as described by Labuza *et al.* (1985). These models can accurately predict the $a_w$ of specific food materials given historical data on EMC, $a_w$ and product temperature.

*Empirical:* Several empirical models exist which attempt to more accurately predict water sorption behaviour of foods in a wider range of EMC, $a_w$ and temperature than that possible with theoretical modelling techniques. Empirical models include those by Henderson (1952), Halsey (1948), Chirife and Iglesias (1978) and the Oswin equation (Lim *et al.* 1995), to name but a few. The water sorption relationship of specific foods is a complex function of its chemical composition as well as its physical characteristics. Recently, a new technique (Myhara & Sablani 2000) predicts $a_w$ at a given EMC, temperature and chemical composition. The technique was shown to be more accurate than theoretical models.

### Chemical stability

Most degradation reactions take place in water (as a solvent) or require water as a substrate. Common chemical spoilage reactions, which occur in foods, include enzymic reactions, Maillard reaction, fat rancidity as well as vitamin and natural pigment degradation. Generally speaking, as $a_w$ decreases these chemical reactions take place at slower rates.

Enzymic as well as vitamin and natural pigment degradation reaction rates are highest in zone III of the water sorption isotherm, where bulk water is available as a participant. As the isotherm enters zone II, less water is available and reaction rates decrease as $a_w$ decreases. At the monolayer moisture content, corresponding to the boundary line between zone I and zone II of the water sorption isotherm, no bulk or multilayer water exists. Those reactions which require water as a solvent or reactant cannot occur. It is generally held that foods at their monolayer moisture content are chemically most stable.

As with all chemical reactions where water is a participant, fat oxidation is very low at the monolayer moisture content. However, in zone I, below the monolayer moisture content, water molecules interfere with fat oxidation by binding to hydroperoxides and hydrating metal catalysts, both necessary for the reaction to proceed. For these very dry foods, addition of water will actually decrease the rate of fat oxidation. As the moisture sorption isotherm enters zone II, the rate of fat oxidation again increases.

In some instances, such as the non-enzymic Maillard browning reaction, water is a product of the reaction. As a result, an increase in water content can cause end-product inhibition, resulting in a slowing of the reaction rate. Maillard reactions are most likely to occur at intermediate moisture levels in zone II, above the monolayer moisture content but below zone III.

It is clear that food materials are most stable, both microbiologically and chemically, at the monolayer moisture content. It is, therefore, important for food technologists to be able to predict at what moisture content that will occur. This is especially important for foods where shelf stability depends upon moisture control. Water sorption models, either theoretical or empirical in nature, are the only means by which such predictions can be made.

## Measurement of physical factors

### Temperature

Temperature is the easiest, cheapest and most reliable of the physical factors to measure. In tropical stored grain it is important that temperatures are related only to the small area in which they are measured, as the low thermal conductivity of grain means that, for example, temperatures at the surface of a grain bag stack do not necessarily correspond to temperatures within the stack. Thermometers should be capable of measuring up to 60°C in tropical stores with a resolution (the smallest discernible change in reading) of 1°C. Expensive and highly accurate temperature measurement instruments are available but the accuracy and reliability of liquid expansion thermometers is such that they can provide a sufficiently accurate standard for checking other thermometers used in grain storage.

Liquid expansion thermometers use mercury or coloured alcohol in a glass tube which expands and indicates the temperature when heated. They are accurate and versatile, but fragile, and as their scale must be visible it is difficult to use them to any great depth within grain. Shaded glass thermometers can be used to measure the air temperature in a warehouse, the shading preventing them being heated by radiation from the warehouse walls or through windows or doors, etc. Metal tubes may be used to help protect glass thermometers from damage.

Maximum and minimum thermometers work in a similar way to normal glass thermometers but allow the daily extremes of temperature to be recorded. They are filled with alcohol at each end with mercury in between. When the temperature rises the differing expansion of the two liquids causes the mercury to be pushed one way along the tube, pushing with it a small metal marker that cannot otherwise move. When the temperature reaches its maximum the marker is left in the tube as the mercury recedes, indicating the highest temperature that was reached. As the temperature falls a marker at the other end of the mercury is pushed in the opposite direction to indicate the lowest temperature reached. The maximum and minimum temperatures should be recorded every 24 h to give the daily maximum and minimum temperatures. The markers should then be returned into contact with the mercury using either a magnet or by pressing a button on the thermometer which allows them to fall by gravity.

Bimetallic thermometers use the principle that different types of metal expand at different rates when heated, so if two types of metal are joined together they will be forced to bend. This may be used to indicate the temperature. The bimetallic strip can be formed into a helix and fitted into a tube that can be placed into a grain mass. A bimetallic strip is used in the thermohygrograph: it is attached to a pen that marks a paper chart as it rotates over a one-day or one-week period, thus recording the temperature. This device also measures the relative humidity and is suitable for unattended monitoring and recording of warehouse conditions.

Electrical thermometers (Fig. 3.11) tend to be more accurate, easy to use, versatile and robust than other methods. They generally have a meter or display with one or more sensors either attached directly to it or on the end of a cable that can be extremely long. They work on a variety of principles:

- Platinum resistance thermometers measure the change in resistance of platinum metal with temperature.
- Thermistors are temperature-sensitive resistors based on oxides of manganese, nickel or silicon.
- Thermocouples have a pair of dissimilar metal wires joined to form a circuit. A small current is formed if the junctions between the two wires are at different temperatures. There are many types of thermocouple, the type depending on the pair of metals used, so care has to be taken to choose the right one for the measurement system being used. The different types are identified by a letter and colour code which is applied to the plug used to connect the sensor to the reading device, or to the thermocouple cable. Type K, coloured yellow, or type T, coloured blue, are both suitable for stored products work. The different types are suitable for different temperature ranges and have different accuracies: type T is more accurate than type K and is often cheaper, but the more common type K is sufficiently accurate. Other types are for more specialised use, e.g. at very high temperatures, so are likely to be more expensive.

Sensors using all three of these methods can be adapted into temperature probes suitable for inserting into stored

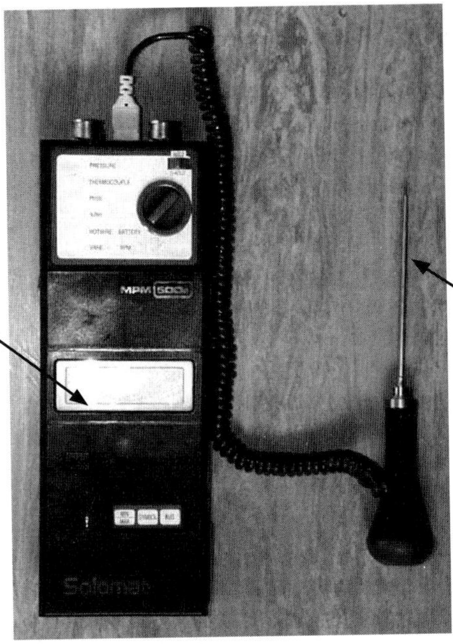


**Fig. 3.11** An electrical thermometer, with a thermocouple probe. Thermocouples can also be made as very long wires which can be installed in silos or bag stacks.

grain or for measuring air or surface temperatures. Cables of thermocouples or thermistors are commonly installed within grain silos to allow the temperatures during storage to be recorded.

The meters used to measure temperature range from small, hand-held devices using just one temperature probe, now very common in the food industry, to microprocessor-based systems which measure and record temperatures from many sensors automatically. Some moisture meters also include temperature sensors, usually thermistors, which can be inserted into the grain. Thermocouples can also be made as very long wires which can be installed in silos or bag stacks.

A wide range of probes and devices are available based on these principles, including simple hand-held thermometers for measurement of local air temperature, portable spear probes which can be used to measure the temperature up to 1 m below the surface of a grain bulk, and permanent probe arrays installed in grain silos or buildings which are measured and recorded using a datalogger or computer system and may be interfaced to record keeping software (HGCA 1999).

## Moisture content

Moisture content can be determined directly, by measurement of the amount of water present in the commodity, or indirectly by assessing parameters that correlate and change as the moisture content changes.

### *Standard (direct) methods*

The grain is dried in an oven. The moisture content is calculated as the loss in weight of the grain following drying under controlled conditions. The initial weight includes its water content.

$$\text{Moisture \%} = \frac{(\text{Initial weight} - \text{Dried weight}) \times 100}{\text{Initial weight}}$$

The measured value is affected by the method of grinding and by the temperature/time regime of the drying. It is therefore necessary to adopt a procedure as the accepted standard method. For cereals this is currently ISO 712: 1998 (ISO 1998). For material within the moisture range 7–17% (15% for oats and rice) a representative sample is ground to meet a particle size specification. Grinding conditions should not create significant heating as this causes moisture loss. Two sub-samples of known weight are then oven dried at 130±3°C for 2 h. The moisture content of the two sub-samples must not differ by a specified amount and the average is taken. If the initial moisture content is outside the specified range it must be preconditioned with measured weight change to this range before grinding.

**Table 3.5** Some standards for assessing grain moisture content using oven methods. Modified from HGCA 1999.

| Drying standard | Product | Temp. (°C) | Time (hours) |
|---|---|---|---|
| BS 4317 Part 2, 1987 = ISO 711, 1985 Reference method | Cereals and pulses | 45–50 | 100 |
| BS 4317 Part 3, 1987 = ISO 712, 1998 | Cereals | 130–133 | 2 |
| BS 4289 Part 3 = EN ISO 665, 2000 | Oilseeds | 101–105 | 3 + 1 (+1) |
| BS 4317 Part 15, 1981 = ISO 6540, 1980 | Maize | 130–133 | 4 |
| ASAE S 352.2, 1997 Whole grain method | Most seeds: | | |
| | – wheat | 130 | 19 |
| | – barley | 130 | 20 |
| | – oilseed rape | 130 | 4 |
| International Seed Testing Association (ISTA) 2001 | Cereals and legumes | 130–133 | 2 |
| Feeding stuffs (sampling and analysis) Regulations No. 1144, 1982, amended by S.I. 52, 1984 and S.I. 1119, 1985 | Feedstuffs | 99–101 | 3 +1 (+1) |

For oilseed rape, two sub-samples are oven dried at 103 ± 2°C to constant weight, currently ISO 665:2000 (ISO 2000). The difference must not exceed a specified amount and the average is taken.

Table 3.5 lists some standards for assessing grain moisture content using oven methods.

### *Secondary (indirect) methods – moisture meters*

As the standard methods are time-consuming, more rapid methods to estimate moisture content have been incorporated into moisture meters. A guideline on moisture meters has been published by the UK Home-Grown Cereals Authority (HGCA 2000).

The most common rapid methods use electrical properties of the moist grain that relate to and vary with moisture content. As they do not measure moisture content directly they are known as indirect or secondary methods.

The two properties most popularly used are electrical resistance and capacitance (or dielectric constant). A sample of grain is placed in a cell and, depending on the type of instrument, its resistance or its capacitance is measured. This is converted within the instrument to an estimate of the moisture content.

#### *Meter instrument types and factors affecting measurement*

*Electrical resistance:* The cell contains two electrodes that, with the circuitry in the instrument, enable the electrical resistance of the sample to be measured. There should be a good and reproducible contact between the sample and the electrodes. It is therefore usual for the grain to be milled before measurement. This means that the sample quantity is relatively small, typically around 10 ml (approximately 6–8 g of grain). Additionally, to further improve contact, the sample is compacted to constant pressure. The grinding and compression may be combined in one unit. Grinding mixes the grain components, improving measurement when the moisture distribution is not in equilibrium, such as immediately after grain drying or exposure to rain.

It is also possible to take measurements with whole grain. However, there is not such good contact between the material and the electrodes. Additionally, mixing of grain fractions has not taken place.

Electrical resistance is affected by temperature. The important values are the temperature of the grain that is being measured (not the room temperature at the time of testing) and the temperature at which the instrument was calibrated. At 15% moisture, a difference of +1°C between these two temperatures can require a correction of –0.15% moisture. This correction can be built into the instrument with temperature sensors as an integral part of the sample cell. Alternatively, manual correction may be made after the temperature of the grain has been taken.

*Capacitance (dielectric constant):* The ability of the grain to hold electrical charge, i.e. its capacitance, is measured by placing it between two conducting plates, forming a capacitor, and applying an alternating current. This capacitance value increases with increasing moisture content.

Measurements are made on whole grains using larger sample sizes than resistance measurement. Up to 350 ml (approximately 210–280 g of grain) are used for portable instruments and up to 1000 ml (approximately 700 g of grain) are used for tabletop (larger, mains operated) instruments.

The capacitance is affected by the packing density of the material. The instrument should be filled in the manner recommended by the manufacturer (using ancillary equipment if supplied): this may be to a specified height or weight. If the weight is not taken during the filling stage then some instruments will determine the weight in order to calculate density as part of the measurement process.

Grain temperature also influences the value. Automatic correction is usually a feature of the instrument via a built-in sensor.

Additionally, the distribution of water within the sample should be taken into account. The calibration selected should be that which most closely corresponds to the material being measured. For example, when testing freshly dried material, where the outer layers will be drier than the centre, a calibration that includes this type of sample rather than one derived from equilibrated grain should be used. A separate calibration should also be used when grain has been recently exposed to water (e.g. rain), where the outer layers will be wetter than the centre.

*General factors affecting measurement:* Different commodities have different electrical properties and each should have its own calibration. The variety and season-to-season variation will also affect the reading as will the presence of impurities.

### Instrument calibration and errors

The manufacturer calibrates a moisture meter by using a large number of samples of grain of known moisture content, i.e. determined using a standard oven method. The data used to establish the calibration may represent several hundred individual samples taken over several years to compensate for varietal and season-to-season variation.

A resistance or capacitance value is obtained for each sample (depending on the type of meter) and data analysis is carried out to build into the meter the relationship between these values and the oven moisture values.

### Accuracy

The accuracy is a measure of the agreement between the oven moisture value and the value predicted by the meter.

The calibration can be validated by testing how accurately it predicts oven moisture values. Samples of known oven moisture content are presented to the instrument and the predicted moisture values are recorded. This gives a number of pairs of oven and meter moisture values. Analysis produces a line-of-best-fit with the data points scattered about this line. The line describes the average relationship.

At any oven moisture content, the difference between the value predicted from the equation of the line-of-best-fit and the oven value is the offset.[2] The accuracy is made up of two components: (1) the offset and (2) the scatter of the data points. When available, information can be given individually for both components as, for example, offset (d) and standard deviation (s) or as a single figure combining the two as, for example, the root mean squared deviation (RMSD). For a large number of samples $d^2 + s^2 = \text{RMSD}^2$. The lower the RMSD, the more accurate is the instrument. Error values can be given at a specific critical moisture value, for example, 15% for wheat, or as representative values over a specified moisture range.

For secondary moisture meters, it is not possible to give a single number (or pair of numbers) that will describe the meter's overall accuracy. The accuracy of moisture meters decreases as the moisture content increases. Where average numbers are given for a moisture content range, they will probably understate the accuracy at the lower end of the range and overstate the accuracy at the higher end.

As shown in Fig. 3.12, after validation, calibrations may be:

*Correct:* The data does not show any significant difference from the meter-equals-oven line.

*Offset:* There is a constant difference between the line-of-best-fit and the meter-equals-oven line. Most meters allow adjustment to remove offset error. Making the correct offset (bias) adjustment will make the calibration correct. Any adjustment should be based on many, not few, measurements. The data scatter remains.

*Skewed:* There is misalignment. Although the line-of-best-fit is linear (a straight line) the difference between the line-of-best-fit and the meter-equals-oven line is not constant. Adjustment to correct this misalignment is not usually possible on portable meters: there is the need for the calibration to be updated by the manufacturer. Adjustment can be made at a critical point to remove the offset in order to increase the accuracy at this point.

*Curved:* The line-of-best-fit is significantly non-linear (not straight).

### Precision

The variation in the readings taken from different representative sub-samples from the same well-mixed sample over a short time period is the precision of the instrument. This figure is almost always lower than the RMSD and should not be confused with it.

This is random error due to small differences in procedure, as tests are not absolutely identical: for example, there will be small differences in the packing. Taking an average of several readings, preferably three or more, will tend to smooth this out.

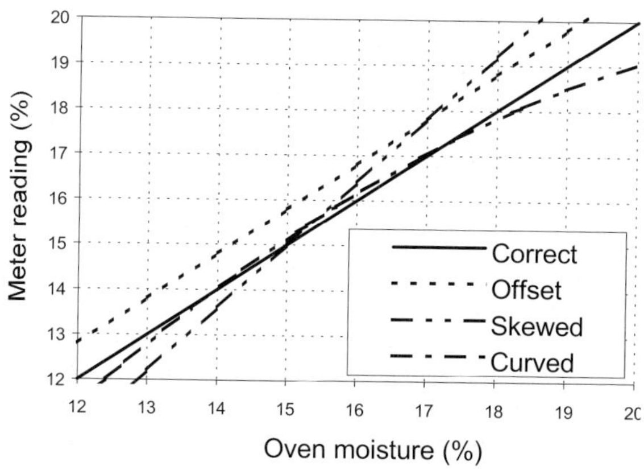


**Fig. 3.12** Moisture calibration classifications: offset can be negative or positive (illustrated); skew can be anticlockwise (illustrated) or clockwise; curvature can be concave (illustrated) or convex. Source: HGCA 2000.

Note that with secondary moisture measurement there will also often be systematic error due to the structural and compositional characteristics of the grain in the sample. Repeated measurement will not remove this error. If the inherent accuracy of the instrument is to estimate the moisture content of a particular sample with an error of, for example, 0.2%, then repeated measurement will reduce the random error but will not reduce this systematic error.

*Accuracy in use*

In practice, the accuracy of an individual moisture meter depends on a number of factors including how well it has been maintained, whether the calibration is up-to-date and whether any individual bias corrections have been made. One major consideration is that at harvest time meters are being used for new season material with previous season(s) calibrations. Previous accuracy data can only be a guide until information specific to the new season becomes available. The best course of action is for a user to check the meter against a proportion of the samples that are of known oven moisture content and make any bias adjustment that is considered relevant. Updated calibrations from the manufacturer may only be available after the harvest. There are some situations in which updates can be downloaded from the manufacturer via a modem.

Manufacturers sometimes recommend that users should make an allowance for meter error when using the instrument. This begs the question 'What is the error?' As illustrated in the preceding text it is virtually impossible for the manufacturer to give a definitive figure because of all the factors involved in the instrument's use. They may provide information for a correctly set-up instrument.

The following is a guide to the allowance that should be made if the meter does not have an offset. If batches of grain are offered for sale at a meter moisture reading equal to the trading value, on average, around half will be rejected. This will fall to around 1 in 6 rejected if the submitted material's meter moisture value is one RMSD value below the trading level and to around 3 in 100 rejected if the allowance is twice the RMSD value below the trading level. The error values given in the HGCA guideline (1999 harvest data) show varying accuracy depending on the instrument (HGCA 2000).

***Using moisture meters***

The following factors are important when measuring moisture using a secondary moisture meter.

*Sampling*

The material being measured must be a representative sample, with removal of impurities such as green grains, chaff, etc. A single sample from the top of a heap is not sufficient to determine the average moisture content of a large bulk of material. Several samples should be taken using the recognised standard sampling method and they should be thoroughly mixed before moisture measurement. Individual well-mixed sub-samples may be used if the variation in the moisture within the bulk is required. At least duplicate measurements on different material from each sub-sample should be made in order to reduce the random error in the measurement.

*Equipment*

The instrument should be in good working order. Check that the battery condition is good, as a low battery will give erroneous results. The instrument should be serviced annually so that it is functioning correctly and so that updated calibrations, if available, can be installed. Follow the manufacturer's operating instructions. Before making any measurements, carry out any preliminary instrument checks as specified. Where an instrument needs other specific checks, such as weight, temperature, capacitance or resistance measurement, make sure these are carried out at regular intervals. Regularly check the instrument with several samples of known moisture content to demonstrate that the readings are within the acceptable instrument measurement errors; in a laboratory the use of control charts is recommended. If the errors are not acceptable then take in-house remedial action if possible, otherwise the instrument manufacturer should be contacted for advice and the instrument should be repaired if necessary. Ring tests between sites to check meters without oven moisture values on the samples are not acceptable; meters could agree about the predicted value but be incorrectly calibrated or adjusted. Use the calibration most appropriate to the material under test. The standard method to which the moisture meter is calibrated should be the one currently in place. If in doubt, check this with the manufacturer.

*Temperature*

The grain and instrument should preferably be at the same temperature, with the temperature being that at which the instrument was calibrated. Make sure that temperature compensation is applied, either automatically or manually. The grain temperature should be stable when the moisture reading is taken and should be within the temperature compensation range of the instrument, if applicable. If not, the grain temperature should be allowed to adjust to within this range. There should not be any surface water in the instrument cell or on the grain. When a cold instrument or cold grain is brought into a warm room, condensation will arise on the surface, producing anomalous results. Cold grain should be sealed in a closed container and allowed to come to room temperature before testing.

#### *Secondary (indirect) methods – near infrared spectroscopy*

The technique of near infrared spectroscopy (NIR) is widely used to measure moisture content and water activity. NIR has found favour with some parts of the food industry because of rapid sample turnaround. For example, each NIR analysis takes only 2 min to determine grain moisture content in malting barley, rather than 2–3 h using the standard oven method.[1] Water in perishable produce can also be measured with NIR. Raisins and most other dried fruit are initially dried to low moisture levels for preservation in bulk. Just before packaging, water is added to make the fruit soft and palatable, but the amount of water must be carefully controlled so that the produce does not become too wet and spoil. NIR can be used to predict water activity accurately and precisely in this situation even though the fruit has not reached moisture equilibrium (Huxsoll 1998).

### Relative humidity

#### *Measurement of relative humidity of intergranular and ambient air*

Instruments that measure relative humidity are called hygrometers. Relative humidity is a difficult quantity to measure: compared to other physical parameters the accuracy achievable with even the most elaborate methods is relatively low. As with moisture content there are direct and indirect methods, the latter being the most suitable for portable measuring devices.

*Direct methods*

Gravimetric hygrometers detect the change in weight of a hygroscopic substance when a measured mass of air is passed through them. It is an elaborate laboratory method used for reference purposes.

Diffusion hygrometers are also laboratory methods, which detect the differences in pressure across chambers containing porous plates, one with a drying agent within the chamber and one with water within the chamber. The relative humidity can be deduced from these measurements.

Dew-point hygrometers can be used in the laboratory and also to some extent as portable instruments. A mirror in contact with the sample air is cooled by passing electricity through a Peltier pump attached to it until the air reaches saturation, i.e. 100% relative humidity. This is detected by a beam of light that is reflected from the mirror onto a sensor: when saturation is reached condensation forms on the mirror and interrupts the beam of light. The temperature at this point is called the dew-point temperature, and it is directly related to the relative humidity of the air. It has been suggested that automatic dew-point hygrometers are not suitable for making measurements from grain because dust in the air from the grain contaminates the mirror, which must remain clean to ensure accurate readings. They can be used in the laboratory for measuring the ERH of grain samples when constructing isotherms, and provide a robust and affordable standard method for the laboratory.

*Indirect methods*

Wet and dry bulb hygrometers use the different cooling effects of air upon wet and dry thermometer bulbs to measure the relative humidity of the air. When air passes over a wet object it causes the water to evaporate and this results in cooling. The extent of evaporation, and therefore of cooling, depends on the relative humidity and temperature of the air. The dry thermometer bulb measures the air temperature, thus allowing the relative humidity to be calculated.

In the whirling hygrometer (Fig. 3.13) the thermometers and water reservoir are mounted on a handle that allows them to be spun in the air by hand. This device can measure the relative humidity to an accuracy of ± 3%. It cannot be used for intergranular air, and is limited in its accuracy because the bulbs need to be spun at 3 m per second or greater, which is difficult to achieve by hand for any period. The motorised or Assman hygrometer uses a clockwork or electric motor to produce air movement and improves accuracy to about ± 2%. It is possible to devise similar instruments to pump air over electrical thermometers and thus make more automated devices that can withdraw air from inside a grain stack, for example.

Hair hygrometers use the properties of certain animal or synthetic hairs which change in length depending on the relative humidity. The thermohygrograph uses a bundle of hairs under tension linked to a recording pen to allow relative humidity in a warehouse to be recorded. The accuracy of these methods can be ± 3% over a short period but they can lose calibration, especially if the relative humidity rises above 85% or falls below 20%, resulting in serious errors.

Electrical hygrometers are, where available, the most suitable devices for portable use; they are relatively cheap, robust, accurate, respond rapidly and can be calibrated relatively easily.

Many different electrical methods have been devised but in recent years the capacitance sensor has become established as the most common method. It measures the change in electrical capacitance of a small square of a hygroscopic polymer material (Fig. 3.14) as it adsorbs and desorbs moisture from the air. The sensor

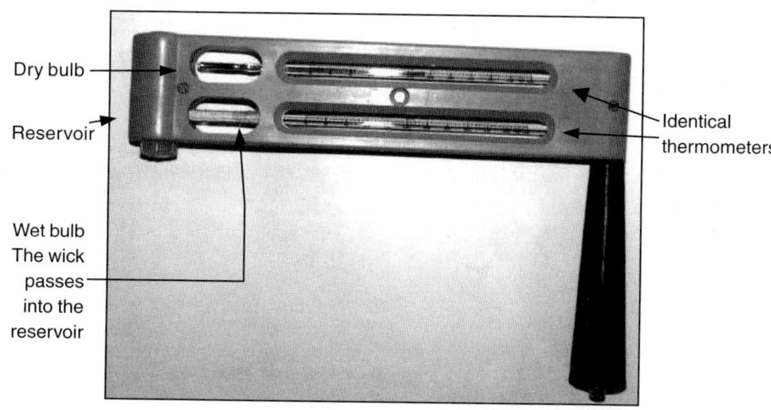


**Fig. 3.13** A whirling (wet and dry bulb) hygrometer.

**Fig. 3.14** An electrical relative humidity sensor. The sensor is a very thin square of hygroscopic polymer with flat electrodes joined to it. The temperature is measured very close to it.

is very small and can be placed within probes suitable for inserting into grains, between sheets of paper (another hygroscopic material) or for use in open air. Figure 3.15 shows a hand-held electrical hygrometer for use in the air, Figure 3.16 shows a probe that may be placed within the air or inserted into a material. It is plugged into a meter which can be some distance away, if necessary.

As temperature has a very significant effect on relative humidity it is essential that the temperature of the air close to the sensor is known. Probes almost always have a temperature sensor placed next to the relative humidity sensor and relative humidity meters display temperature as well as relative humidity.

Measurement of relative humidity is very common in building services and industry for the control and monitoring of air conditioning and process control. Probes for temporary or permanent installation in buildings, ducts or process vessels are available based on either dew point or capacitance sensors. A permanently mounted capacitance sensor is shown in Fig. 3.17.

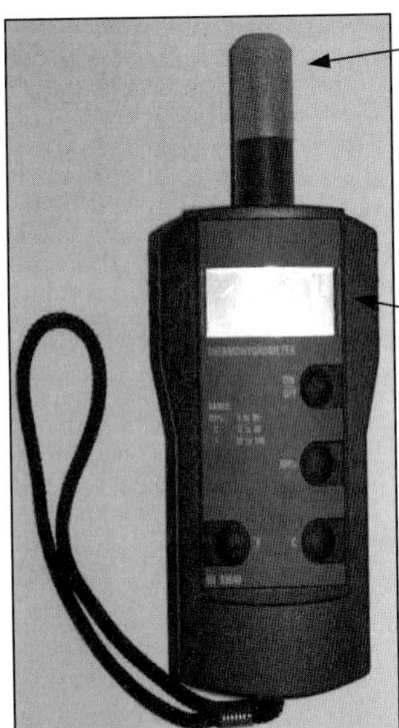


**Fig. 3.15** A hand-held electrical hygrometer.

***Measurement of equilibrium relative humidity (ERH)/water activity ($a_w$)***

ERH can be measured using an r.h. sensor mounted in a sealed chamber in which the stored product is able to reach equilibrium with the air. The AquaLab™ (Decagon Services, Washington, USA) series of water activity meters use a dew-point hygrometer; the sample is placed into a small plastic pot which is sealed within a chamber containing the chilled mirror, a temperature sensor and fan to circulate the air. Other devices such as those made by Novasina and Rotronic use capacitance r.h. sensors. In all cases the time required for the sample to reach equilibrium should be borne in mind. Whole grains take some hours to reach equilibrium in small volumes so should not be taken from a situation in which they are not in equilibrium with their surroundings then placed into a water activity meter. Instead they should be stored in sealed jars prior to making readings – the AquaLab™ meters allow samples to be sealed into sample pots prior to taking measurements. Samples of whole grains should not be ground to allow them to reach equilibrium more quickly as this will alter their isotherm relationship – in general milled products have a higher ERH than unmilled products.

## Moisture migration during storage

Movement or migration of moisture during storage and transport of grain can lead to problems with the quality and quantity of commodities. Although the movement of moisture is generally very slow, the proportion of moisture by weight in a commodity is relatively high so changes can be significant and easily lead to areas of the commodity becoming unsafe for storage.

Moisture migration can occur when the relative humidity of the air around stored grains and their interstitial equilibrium relative humidity differs but the temperatures are equal – this is called isothermal diffusion.

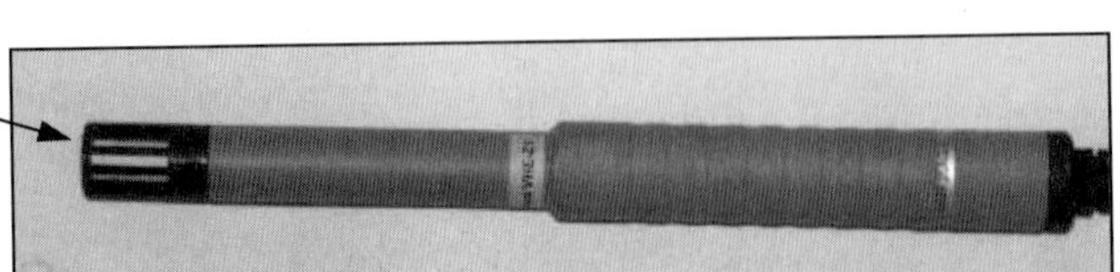


**Fig. 3.16** A relative humidity probe which can be inserted into grain or used in the air. It is plugged into a meter which can be some distance away if necessary.

**Fig. 3.17** A wall mounted relative humidity probe. The unit contains a capacitance sensor and a temperature sensor.

It is a very slow process: grain at 11% moisture content exposed to 75% relative humidity will be changed by 1% or more to a depth of only 0.3 metres after four months, though the changes can be significant in terms of the safety of the grain and its weight (Gough 1996). Thorpe (1981) assumed that diffusion was controlled solely by water vapour pressure gradients within the grain bulk, and presented a theory to allow the effects of moisture diffusion at different temperatures to be predicted.

Temperature gradients across grain bulks also cause moisture movement, called non-isothermal moisture migration. Moisture moves from warm to cold grain. This can lead to the build-up of moisture at the cold sides of silos or ships and is the cause of caking on the silo walls or ships' sides. Thorpe (1982) again postulated that the moisture migrates through vapour diffusion. Temperature gradients also give rise to convection currents within stored grains which can lead in turn to moisture migration. Moisture is carried in the vapour state from warm to cold regions, for example, from the interior of a silo into its headspace, where is may be absorbed by the upper layers of grain or condense onto the underside of the roof and fall back onto the surface layers of grain (Gough *et al.* 1987). It can be a major problem, and attempts have been made to model the process (Nguyen 1986) and measure the magnitude of convection currents through grains (Gough *et al.* 1987).

Mathematical models have been used to explain and predict the related, and simultaneous, phenomena of heat, moisture and gas transfer in grain storage. Models of heat transfer alone have given good results but there have been few models and limited validation of those that include moisture migration (Jayas 1995).

## Notes

1 http://www.nmpuk.co.uk/publications/case_studies/crisp_malting.html.
2 The more scientific term is bias. However, when correcting instruments the usual instruction is to enter a bias value. This has the same magnitude but has the opposite sign to the offset described.

## Suggested reading

Kress-Rogers, E. & Brimelow, C.J.B. (Eds) (2001) *Instrumentation and Sensors for the Food Industry.* Second Edition. CRC Press, USA.

## References

Ayranci, E., Ayranci, G. & Zafer, D. (1990) Moisture sorption isotherms of dried apricot, fig and raisin at 20°C and 36°C. *Journal of Food Science,* **556**, 1591–1593.

Brooker, D.B., Bakker-Arkema, F.W. & Hall, C.W. (1992) *Drying Cereal Grains.* AVI Publishing, Westport CT, USA.

Brunauer, S., Emmett, P.H. & Teller, E. (1938) Adsorption of gases in multimolecular layers. *Journal of the American Chemical Society,* **60**, 309–319.

Cenkowski, S. & Zhang, Q. (1995) Engineering properties of grain and oilseeds. In: D.S. Jayas, N.D.G. White & W.E. Muir (Eds) *Stored-grain Ecosystems.* Marcel Dekker, New York NY, USA.

Chirife, J. & Iglesias, H.A. (1978) Equations for fitting water sorption isotherms of foods. *Journal of Food Technology,* **13**, 159.

Christensen, C.M., Miller, B.S. & Johnston, J.A. (1982) Moisture and its measurement. In: C.M. Christensen (Ed.) *Storage of Cereal Grains and Their Products*. American Association of Cereal Chemists, Minnesota, USA.

Chung, D.S. & Lee, C. (1986) Physical and thermal properties of grains. In: B.R. Champ & E. Highley (Eds) *Preserving grain quality by aeration and in-store drying*. Proceedings of an international seminar. 9–11 October 1985, Kuala Lumpur, Malaysia. ACIAR, Canberra, Australia.

CIBS (1980) *Thermal Properties of Building Structures.* CIBS Guide A3. Chartered Institute of Building Services, London, UK.

Foster, G.H. (1982) Drying cereal grains. In: C.M. Christensen

(Ed.) *Storage of Cereal Grains and Their Products*. American Association of Cereal Chemists, Minnesota, USA.

Gough, M.C. (1996) Marine transport of some tropical food cargoes to cold climates: maintenance of quality. *NRI Bulletin* **69**. Natural Resources Institute, University of Greenwich, Chatham, UK.

Gough, M.C., Uiso, C.B.S. & Stigter, C.J. (1987) Convection currents in bulk grain. *Tropical Science*, **27**, 29–37.

Halsey, G. (1948) Physical adsorption on non-uniform surfaces. *Journal of Chemistry and Physics,* **16**, 931–945.

Henderson, S.M. (1952) A basic concept of equilibrium moisture. *Agricultural Engineering,* **33**, 29–32.

HGCA (1998) *Bulk storage drying of grain and oilseeds.* Topic Sheet **16**. Home-Grown Cereals Authority, London, UK.

HGCA (1999) *The Grain Storage Guide*. Home-Grown Cereals Authority, London, UK.

HGCA (2000) *Moisture meter guidelines*. Home-Grown Cereals Authority, London, UK.

Hoseney, R.C, & Faubion, J.M. (1992) Physical properties of cereal grains. In: D.B. Sauer (Ed.) *Storage of Cereal Grains and Their Products*. American Association of Cereal Chemists, Minnesota, USA.

Howatson, A.M., Lund, P.G & Todd, J.D. (1972) *Engineering Tables and Data.* Chapman & Hall, London, UK.

Howe, R.W. (1962) A study of the heating of stored grain caused by insects. *Annals of Applied Biology,* **50**, 137–158.

Huxsoll, C.C. (1998) *Assessment of near infrared (NIR) diffuse reflectance analysis for measuring moisture and water activity in raisins.* Agricultural Research Service. United States Department of Agriculture. http://http://www.nal.usda.gov/ttic/tektran/data/000009/57/0000095710.html

ISO (1998) *Cereals and cereal products – determination of moisture content-routine reference method.* ISO 712:1998. International Organization for Standardization, Geneva, Switzerland.

ISO (2000) *Oilseeds – determination of moisture and volatile matter content*. ISO 665:2000. International Organization for Standardization, Geneva, Switzerland.

Jayas, D.S. (1995) Mathematical modelling of heat, moisture, and gas transfer in stored-grain ecosystems. In: D.S. Jayas, N.D.G. White & W.E. Muir (Eds) *Stored-grain Ecosystems.* Marcel Dekker, New York NY, USA.

Jayas, D.S. & Mazza, G. (1993) Comparison of five, three-parameter equations for the description of adsorption data of oats. *Transaction of the American Society of Agricultural Engineers,* **36**, 119–125.

Johnson, P.-N.T. & Johnson, J.G. (2000) Moisture sorption isotherm characteristics of plantain (Musa, AAB). *Journal of Food Engineering,* **44**, 79–84.

Labuza, T.P., Kaanane, A. & Chen, J.Y. (1985) Effect of temperature on the moisture sorption isotherms and water activity shift of two dehydrated foods. *Journal of Food Science*, **50**, 385–391.

Lim, L.T., Tang, J. & He, J. (1995) Moisture sorption characteristics of freeze dried blueberries. *Journal of Food Science,* **60**, 810–814.

Mazza, G. & Campbell, C.G. (1985) Influence of water activity and temperature on dehulling of buckwheat. *Cereal Chemistry,* **62**, 31–34.

Mohsenin, N.N. (1970) *Physical Properties of Plant and Animal Materials.* Gordon & Breach, New York NY, USA.

Mohsenin, N.N. (1980) *Thermal Properties of Foods and Agricultural Materials*. Gordon & Breach, New York NY, USA.

Multon, J.L. (1988) Spoilage mechanisms of grains and seeds in the post-harvest ecosystem, the resulting losses and strategies for the defense of stocks. In: J.L. Multon (Ed.) *Preservation and Storage of Grains, Seeds and Their By-products.* Lavoisier, New York NY, USA.

Myhara, R.M. & Sablani, S.S. (2000) Unification of fruit moisture sorption isotherms using artificial neural networks. In: *Proceedings of the 12th International Drying Symposium.* Elsevier, Amsterdam, Netherlands.

Myhara, R.M., Taylor, M., Slominski, B. & Al-Balushi, I. (1998) Moisture sorption isotherms and chemical composition of Omani dates. *Journal of Food Engineering,* **37**, 471–479.

Nguyen, T.V. (1986) Modelling temperature and moisture changes resulting from natural convection in grain stores. In: B.R. Champ & E. Highley (Eds) *Preserving grain quality by aeration and in-store drying.* Proceedings of an international seminar. 9–11 October 1985, Kuala Lumpur, Malaysia. ACIAR, Canberra, Australia.

Oxley, T.A. (1944) The properties of grain in bulk. III. The thermal conductivity of wheat, maize and oats. *Journal of the Society of Chemical Industry,* **63**, 53.

Parker, P.E., Bauwin, G.R. & Ryan, H.L. (1982) Sampling, inspection and grading of grain. In: C.M. Christensen (Ed.) *Storage of Cereal Grains and Their Products*. American Association of Cereal Chemists, Minnesota, USA.

Rogers, G.F.C. & Mayhew, Y.R. (1980) *Thermodynamic and Transport Properties of Fluids*. Third Edition. Basil Blackwell, London, UK.

Saravacos, G.D., Tsiourvas, D.A. & Tsami, E. (1986) Effect of temperature on the water adsorption isotherms of Sultana raisins. *Journal of Food Science,* **51**, 381–383, 387.

Sinha, R.N. (1995) The stored-grain ecosystem. In: D.S. Jayas, N.D.G. White & W.E. Muir (Eds) *Stored-grain Ecosystems.* Marcel Dekker, New York NY, USA.

Sinha, R.N. & Muir, W.E. (1973) *Grain Storage: Part of a System.* AVI Publishing, Westport, CT, USA.

Smith, C.V. & Gough, M.C. (1990). *Meteorology and grain storage.* Technical Note **101**. WMO – No. 243.TP.133 (Rev.), World Meteorological Organization, Geneva, Switzerland.

Thorpe, G.R. (1981) Moisture diffusion through bulk grain. *Journal of Stored Products Research,* **17**, 39–42.

Thorpe, G.R. (1982) Moisture diffusion through bulk grain subjected to a temperature gradient. *Journal of Stored Products Research,* **18**, 9–12.

Toledo, R.T. (1997) *Fundamentals of Food Process Engineering*. Third Edition. Chapman & Hall, New York NY, USA.

# Chapter 4
# Biological Factors in Post-Harvest Quality

G. Farrell, R. J. Hodges, P. W. Wareing, A. N. Meyer and S. R. Belmain

## Post-harvest pests and the damage they cause

Stored produce is at risk from problems not faced by crops in the field. This arises from the fact that seeds, fruits and tubers are essentially dormant structures; their cells are physiologically different to those of the growing plant. In addition, bulking of produce, in transit or in store, gives rise to conditions very different from those in the field (Wheeler 1969).

A wide range of biological factors influence stored products. Any organic product that is not kept in a sterile manner is liable to be degraded by some biological agent, if it is kept for long enough. Losses may be qualitative or quantitative, or a combination of both, and result from the inability of the host to limit biological damage. In addition, produce may be predisposed to further attack because points of entry for secondary insect pests and saprobic fungi have been created by the existing damage.

Produce with surface blemishes, such as discoloration, blotches or signs of insect activity, may be rejected by graders or consumers and hence give rise to qualitative losses. Qualitative losses are particularly important in the international trade in fruit and vegetables, where emphasis is placed on visual quality, and where even a small cosmetic defect may render the produce unsaleable (Waller 2001). However, organic growers have persuaded buyers to accept produce that is less than perfect in appearance, emphasising that surface imperfections do not necessarily imply that the underlying tissue has also deteriorated.

Quantitative losses arise when stored produce is directly consumed by primary insect pests and rodents, or from the rapid and extensive decay caused by the action of micro-organisms. Attack usually begins with one or a few species, followed by invasion of a broad range of non-specific micro-organisms or secondary insect pests. The feeding habits of primary pests may lead to quality losses, since some insects exhibit a preference for feeding on the germ region of the seed, leading to a loss in nutritive value in cereal grains and viability in seeds.

A stored product pest can be defined as 'any organism injurious to stored foodstuffs of all types (especially grains, pulses and fruits), seeds, and diverse types of plant and animal materials' (Hill 1990). These pests can be categorised in various ways, for example, on the basis of systematics, dependence on stored products, feeding habits and damage caused, commodity infested, ecological niches, time taken to reach economically important levels, the value losses they cause or the type of handling or manufacturing industry in which they are found. For the purposes of this chapter, the degradation of organic material caused by any biotic agency, after the material has crossed the harvest divide, will be considered. However, many post-harvest problems originate before the commodity has been gathered from the field, and these will also be discussed as necessary.

Post-harvest pathogens may initiate their attack in the field or after harvest. Before harvesting, disease symptoms may follow soon after infection begins, for example, with bacterial agents and fungal late blight of potatoes. With other pathogens, infection may remain dormant and hidden and not be revealed until later, as occurs with anthracnose disease of bananas. Infection that takes place at or after harvesting often occurs at the site of wounds resulting from insect feeding or mechanical injury during harvest and post-harvest handling. Another route of entry is provided by points of attachment where the fruit or vegetable joined the parent plant. However, some pathogens are able to penetrate the host through natural openings such as stomata and lenticels. Others can enter directly through undamaged epidermis (Waller 2001).

The term 'pest' includes representatives from all types of living organism, but most are insects, fungi and bacteria. The main storage fungi include species of *Fusarium, Phytophthora, Alternaria, Rhizopus, Pythium, Sclerotinia, Botrytis, Penicillium, Cladosporium, Phoma, Mucor* and *Rhizoctonia*. Important post-harvest bacteria are in the genera *Pseudomonas, Xanthomonas* and *Erwinia*. There are no true plant pests on stored products. The major orders of stored product insect pests include species from the Coleoptera, Lepidoptera, Diptera, Psocoptera and Dictyoptera (Hill 1990; Haines 1991).

Fungal growth may lead to unsightly discoloration, distortion, unpleasant odours and off-flavours, and loss of seed viability. If infection is severe then the product may be completely destroyed. In addition to the physical injury caused, important additional chemical by-products of their activity can be produced by the fungus to aid attack or by the host to limit the damage inflicted by the invader. Examples of the former are mycotoxins, such as carcinogenic aflatoxin produced by *Aspergillus flavus* on groundnuts. The stress metabolite psoralen, produced by celery in response to attack by *Sclerotinia sclerotiorum*, is an example of the latter.

The wide range of non-specific fungi, such as *Botrytis, Fusarium, Rhizopus* and *Mucor*, are weak pathogens or saprobes on the dead or moribund tissue remaining from the primary infection. Notwithstanding their status as secondary invaders, such fungi may still be very aggressive and often exacerbate the damage resulting from attack by primary pathogens. The spoilage of root and fruit crops after harvest mostly involves fungi; 20–25 different species have been associated with decayed tomatoes and yams and 25 fungi have been isolated from the sites of banana crown rot (Waller 2001).

Bacteria are also important agents in the spoilage of fruit and vegetables, particularly those that cause soft rots, of which *Erwinia* is the most significant. Bacteria are not able to penetrate healthy epidermal tissue in the way that some fungi can, and so they gain entry through natural openings or through wounds.

As well as causing direct losses to the produce, some bacterial infections of stored foods can be serious when the raw or processed food is eaten and poisoning results, particularly for the elderly, the very young and the sick. Most cases of food poisoning are caused by *Staphylococcus aureus, Salmonella* spp, *Clostridium perfringens, Campylobacter, Listeria monocytogenes, Vibrio parahaemolyticus, Bacillus cereus,* and enteropathogenic *Escherichia coli.*

Dried produce, whether perishable or durable, can reabsorb moisture from the air, unless it is stored in sealed containers. Thus, losses from fungi and bacteria are usually more frequent in the humid tropics. Higher ambient temperatures in tropical climates enhance the speed of decay.

Nutritional disorders originate before harvest, and may result from imbalances of minerals within the plant. The mineral may be taken up by the plant but not be adequately distributed. Calcium deficiency is an example, causing bitter pit in apples and pears, blossom end rot of tomato and tip burn in lettuce. Lack of boron can lead to internal black spot of beet and internal gumming of citrus, and potassium deficiency is associated with blotchy ripening of tomatoes. Nutritional factors also influence senescent breakdown in stored apples (Snowdon 1990).

Respiration may be upset if produce is stored in an unsuitable atmosphere. Lack of oxygen results in anaerobic conditions within the produce, and causes black heart of potatoes and invasive alcohol poisoning in apples. Too much carbon dioxide can also be harmful. Some apple cultivars exhibit external injury, and brown heart can arise in apples and pears. Non-green tissue in vegetables is particularly at risk from respiratory damage, for example pale midribs of lettuce leaves and cauliflower curd (Snowdon 1991).

## Pests of durable crops – insects and arachnids

Nearly all of the important insect pests of stored products are either beetles or moths (members of the orders Coleoptera and Lepidoptera) and most of the following account will concern them. Rather less will be said about arachnids (important in stores only as mites) as these are less frequently encountered storage pests but are important when present in large numbers. Although about 100 species of insects and mites are commonly associated with stored products, relatively few species account for a large proportion of commodity damage.

Not all insects and mites found in stores are pests: some may have just found their way there by accident and are referred to as 'incidentals', while others may be predators or parasites of pests and hence are 'beneficials'. The following account describes the biology, behaviour and recognition features of some representative pest and non-pest species. For more details and a wider coverage of insects, consult the two volumes of *Insect Travellers* (Aitken 1975, 1984), which deal with the Coleoptera and Lepidoptera imported into the UK on grain, and the Natural Resources Institute's training manual (Haines 1991) that focuses on tropical species. Identification keys to storage insects can be found in Haines (1991), Mound (1989) and Gorham (1987). Hughes (1976) has published an excellent monograph on mites of house and store.

### Grain pests

It is convenient to describe insects and mites that attack stored cereals, grain legumes (pulses) or oilseeds as either primary or secondary pests. Primary pests are those species that are capable of invading undamaged grain and establishing an infestation. Secondary pests are restricted to commodities that have already been damaged. The damage may have been caused by man in the process of milling or grinding the product, or by the activity of a primary pest. Primary and secondary pests are defined as follows:

> *Primary pests* are capable of successfully attacking and breeding in previously undamaged solid grains, for example, whole cereal and pulse grains. That is, they are capable of penetrating an undamaged seed-coat and sometimes also a pod or sheath in order to feed on the embryo, endosperm or cotyledons of the seed. Such pests are also sometimes capable of feeding on other solid but non-granular commodities, for example, dried cassava, but they are rarely successful on milled or ground commodities.
>
> *Secondary pests* are not capable of successfully attacking previously undamaged solid commodities. They can only attack and breed in commodities that have been previously damaged by some other agency: (a) other pests, especially primary pests, (b) bad threshing, drying, handling, etc. or (c) intentional processing of the commodity.

The main characteristics of primary and secondary pests are given in Table 4.1. Some pests do not fall easily into either category: specifically, those pests that are just capable of attacking undamaged commodities but develop much more rapidly if some previous damage is present, for example, *Trogoderma granarium*. In such cases, it is best to classify these species as secondary pests, partly because they do not develop successfully on undamaged grains but also because they usually exhibit other secondary pest characteristics such as a wide range of food preferences. It is useful to distinguish between primary and secondary pests of cereal grains but it is inappropriate to refer to a species as a secondary pest of flour because flour by definition cannot have primary pests. Similarly, it is inappropriate to use the categories of primary and secondary pests with reference to non-granular commodities lacking a surface layer that presents a physical barrier, for example, dried fish and dried cassava chips.

On whole undamaged grains, primary pests usually have to attack the commodity first before there can be any significant secondary pest attack. On any batch of stored whole grain there is therefore usually a succession of pest attack. The first attack, which may begin before harvest, is usually by primary pests. This is followed by an invasion by secondary pests. The word 'primary' does not mean 'more important', it refers to the order of succession in the infestation of whole grains. In different situations, primary and secondary pests differ in their relative importance.

**Table 4.1** The main characteristics of primary and secondary pests of cereals and pulses.

| | Primary pest | Secondary pest |
|---|---|---|
| Diet | Narrow range of foods attacked, whole grain cereals or pulses. | Wide range of hosts including damaged whole grains, flours and other processed foods. |
| Initial attack | Infestation is frequently pre-harvest in the mature crop in the field. Infestation passes from field into store, sometimes vice versa. | Infestations are normally confined to stores and rarely start before harvest. |
| Life cycle | Life cycle involves development within single grains so that immature stages are hidden from view and hence difficult to detect. Eggs usually carefully placed in or on a grain. | Larvae move freely in the stored food and so are clearly visible. Eggs laid close to food but generally scattered (placed less precisely than by primary pests). |
| Damage | Damage to whole grain is often very distinctive in form, so that the pest may be recognised from samples of damaged grain. | Damage caused by surface feeding is indistinct and cannot be used for pest recognition. |

### *Primary pests of cereals*

Sitophilus *spp (Coleoptera: Curculionidae)*

Beetles of the genus *Sitophilus* are important primary pests of whole cereal grains. They are members of the 'weevil' family, the Curculionidae. Three species are well known as pests of stored grain – *Sitophilus zeamais* known as the 'maize weevil', *Sitophilus oryzae*, often called the 'rice weevil', and *Sitophilus granarius*, referred to as the 'granary weevil'. Although the common names are frequently used, they are misleading and should not be used in scientific communications. The biology of all three of these species has been reviewed in detail (Longstaff 1981).

The adults of all three species are small (about 3 mm long) active animals which are characterised by a narrow snout-like forward extension of the head which carries the mouthparts. The antennae are 'elbowed' in form, and are held in a right-angled position when the insect is at rest. The body colour can range from light to dark brown, and both *S. zeamais* and *S. oryzae* often have four large reddish-orange spots on the elytra.

Adult *S. granarius* can be distinguished from *S. zeamais* and *S. oryzae* by the pattern of punctures on the dorsal surface of the prothorax (compare Figs 4.1 and 4.2), and by the very reduced vestigial flight-wings of *S. granarius*. It should be noted that it is not possible to separate *S. zeamais* from *S. oryzae* by external morphological characteristics, despite contrary statements in some publications; examination of the genitalia is required (Haines 1991).

The larvae of *Sitophilus* spp are whitish, legless grubs that spend all of their pre-adult life tunnelling in the endosperm of a cereal grain. The adult female weevils lay eggs singly in tiny holes that they gnaw in the seed using the mouthparts that are located at the end of the snout. Each egg is protected by a gelatinous, mucopolysaccharide 'egg-plug' that is secreted by the egg-laying female (Fig. 4.1). Upon hatching from the egg, the larva begins to feed in the endosperm, producing a cavity in the grain as it increases in size. Eventually the fully grown larva pupates within the grain, and the adult that emerges bites its way out of the grain, leaving a characteristic, large, somewhat irregular emergence hole. The immature stages of *Sitophilus* spp spend all of their pre-adult life hidden inside a cereal grain, so that they are extremely difficult to detect. Adult *Sitophilus* feed especially on grain endosperm that has been exposed by breakage, or by entering emergence holes.

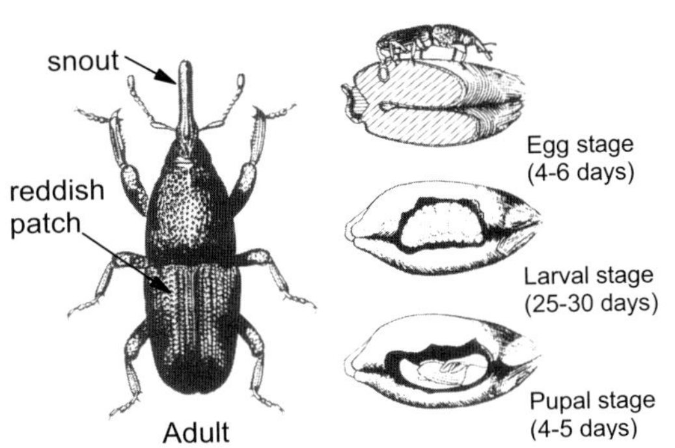


**Fig. 4.1** *Sitophilus zeamais* (adult life-size 2.5–4.5 mm) showing its life cycle in a wheat grain; note female laying egg in hole in the grain (courtesy of Central Science Laboratory, UK).

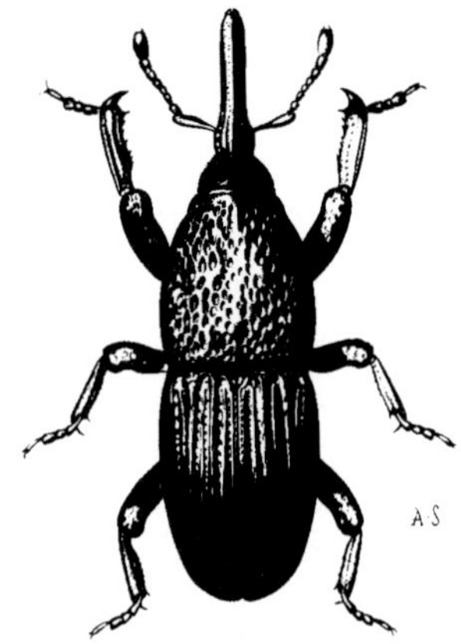


**Fig. 4.2** *Sitophilus granarius* (life-size 2.5–4.5 mm) (courtesy of Central Science Laboratory, UK).

*Sitophilus granarius* is essentially a temperate pest and is not found in tropical countries except occasionally in cooler, upland areas. *Sitophilus zeamais* and *S. oryzae* are commonly found throughout the world in tropical and subtropical regions, especially where ambient humidities are fairly high. *Sitophilus zeamais* and *S. oryzae* thrive best at warm temperatures (around 27°C) and in grain in which the moisture content is not much less than 13%. Under favourable conditions, such as 27°C and 70% r.h., development from egg to adult in all three species is completed in about 35 days. In *S. zeamais* and *S. oryzae*, development periods are very protracted at temperatures below 18°C, whereas *S. granarius* can develop, albeit slowly, at 15°C when at 70% r.h. Development can be completed in about 140 days.

Both species can infest any cereal grain that is large enough to support their development, as well as other starchy foods such as dried cassava roots and, even, spaghetti. *Sitophilus zeamais* is particularly associated with maize and milled rice, but is often found on wheat and other small grains. *Sitophilus oryzae* is particularly associated with small grains, but is often found infesting maize and paddy rice. Certain rare strains of this species have been found breeding on lentils and split peas (Coombs *et al.* 1977).

*Sitophilus zeamais* flies strongly, and often infests commodities in the field before harvest (Giles 1969). Most strains of *S. oryzae* are infrequent fliers, and are particularly common as pests of commerce in warehouses, transport vehicles and silos where they have been introduced by man in transported grain.

### Rhyzopertha dominica *and* Prostephanus truncatus *(Coleoptera: Bostrichidae)*

Most members of the Bostrichidae are wood-boring insects. Two species, *Rhyzopertha dominica* and *Prostephanus truncatus,* are able to thrive in stored products such as cereals and dried cassava roots and are important primary pests. However, *P. truncatus* is not confined to stores and large populations live in the natural environment (Nang'ayo *et al.* 1993).

Adult *R. dominica* are small (about 2–3 mm) and *P. truncatus* somewhat larger (about 3–4.5 mm); both are cylindrical brown beetles (Fig. 4.3). In *R. dominica* the apex of the wing cases is rounded and gently sloping downwards, whereas in *P. truncatus* it is flattened and sheer, hence the name '*truncatus*' referring to the truncated or 'cut-off' back end of the beetle. In both species, the head is held beneath the prothorax so that it is obscured when the insects are viewed from above. The antennae have a three-segmented club.

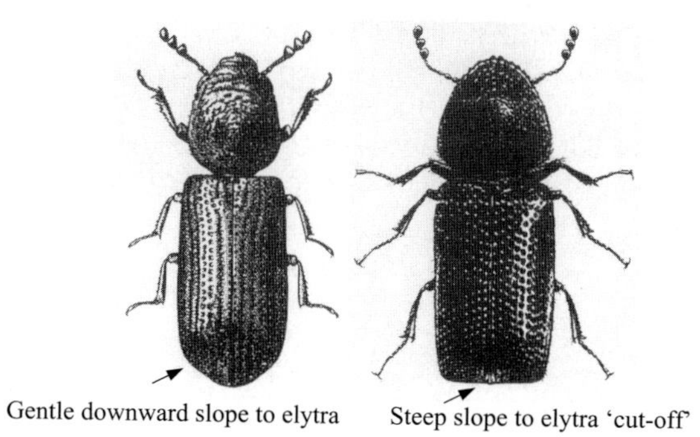


**Fig. 4.3** (left) *Rhyzopertha dominica* (life-size 2–3 mm) (courtesy of Central Science Laboratory, UK); (right) *Prostephanus truncatus* (life-size 3–4.5 mm) (courtesy of Syngenta, UK).

*Rhyzopertha dominica* is widespread throughout the tropics and subtropics and is most important as a pest of wheat and paddy rice, but also occurs on other cereals and roots such as dried cassava (Potter 1935). *Prostephanus truncatus* is a sporadic but locally serious pest of maize stored on the cob in Central America. In the late 1970s, it became established in western Tanzania, where it became an extremely serious pest of farm-stored maize and dried cassava, roughly doubling average farm store losses from 5% to 10% (Hodges *et al.* 1983). In individual cases, farmers might lose as much as 30%. Subsequently, it has spread to many countries in both East and West Africa. It is a serious pest in a wide range of environments but is particularly favoured by hot drier habitats. *Prostephanus truncatus* is now a quarantine pest in many countries.

Like *Sitophilus* spp, the pre-adult stages of *R. dominica* and *P. truncatus* develop within cereal grains (Fig. 4.4). Adult females lay eggs at the end of tunnels excavated in the grain. Subsequent development usually takes place within the grain, but unlike *Sitophilus*, larvae may bore out of one grain and into another. After pupation the newly developed adult escapes from the grain by chewing its way out, then continues to bore through the food.

Adult *R. dominica* and *P. truncatus* feed throughout their lives, producing large quantities of dust and frass containing a high proportion of undigested fragments, which can support the development of larvae if sufficiently compacted. Similarly, ground produce or very small grains can be infested to a limited extent if compacted and stabilised.

*Rhyzopertha dominica* and *P. truncatus* are adapted to rather higher temperatures and lower moisture contents than *Sitophilus* spp and they are therefore the dominant pest in hot, drier areas. Sorghum is often grown in such areas, so *R. dominica* is frequently associated with this crop. *Prostephanus truncatus* is very rarely associated with sorghum as the grain size is usually too small to support its development. Instead it is almost exclusively

**Fig. 4.4** Larva and pupa of *Rhyzopertha dominica* in grain (courtesy of Central Science Laboratory, UK).

found in maize and dried cassava chips. Under favourable conditions (27°C and 70% r.h.), full development of *P. truncatus* can be completed in about 32 days, with an estimated intrinsic rate of increase for the population of 0.73/week. This is very similar to 0.72/week estimated for the well-known storage pest, *Tribolium castaneum*. When the humidity is lowered to 40% r.h., development is still possible and completed in 52 days and the rate of increase is reduced to 0.46/week. Detailed reviews have been published on the biology of *P. truncatus* (Hodges 1986; Markham *et al.* 1991) and *R. dominica* (Potter 1935).

### Sitotroga cerealella *(Lepidoptera: Gelechiidae)*

*Sitotroga cerealella* is an important primary pest of cereals and can infest grain in the field before harvest, especially maize and sorghum. In *S. cerealella*, the fore-wings of newly emerged adults are covered with yellowish-golden scales, but in older adults that have lost most of their scales, the body is entirely grey. The hind-wings carry a fringe of very long hairs (Fig. 4.5).

Female *S. cerealella* lays eggs in masses on the commodity, and, upon hatching, the larvae bore into the grain. Subsequent development takes place within the grain. Pupation takes place inside the grain, or sometimes just outside. If pupation takes place inside the grain, then before pupation the larva prepares its emergence point by chewing its way to just beneath the surface of the grain. It leaves only a thin area of grain coat, known as a window, separating the feeding chamber from the exterior. After pupation, the relatively feeble adult is able to push its way out through the window, leaving a characteristic hole behind. A partial covering remains at the edge of the hole in the form of a 'trap-door' (Fig. 4.5). The adult is rather short-lived (typically 7–14 days) and is an active flier. Egg to adult development is completed in as little as 25 to 28 days at 30°C and 80% r.h.

*Sitotroga cerealella* attacks any cereal with grains large enough to support larval development. This moth is widespread over tropical and subtropical parts of the world, sometimes entering warmer temperate areas. The adults are good fliers and cross-infestation occurs easily. They are delicate and cannot penetrate far into densely packed grain; as the larvae also stay within the first seed they penetrate, infestations in bulk grain are generally confined to the outer, most exposed layers. However, quite serious infestations can develop in cereals stored in bag stacks, especially if the pre-harvest infestation has been heavy. Infestations of the pest are most frequently encountered in farm storage. Because the larvae compete with those of *Sitophilus* spp, *S. cerealella* is relatively more important in dry conditions that are less favourable to *Sitophilus* spp.

### *Primary pests of legume seeds*

Pulses, (peas, beans, grams, etc.) are the edible seeds of legumes. They are fairly resistant to attack by most storage pests, but one family of beetles, the Bruchidae, are adapted to attack them. All major pests of pulses are Bruchidae.

The Bruchidae are very closely associated with legumes, all are seed eaters and most attack the seeds of legumes. Many legumes contain chemical substances that make them resistant to invasion by most insects, but bruchid species have developed tolerance to many of these chemicals. This, in conjunction with the evolution of different behavioural strategies, has ensured that different species of Bruchidae are specialised to feed on specific legume species. Only a few species regularly attack stored products and details of these can be found in Southgate (1978), Gorham (1987) and Haines (1991).

The plant family Leguminosae contains many species of food plants which form part of basic human diets all over the world. The widespread use of legumes has resulted in a proliferation of local names for different pulses, which makes it rather difficult to communicate information about them using common names. Unfortunately, the taxonomy of the Leguminosae is extremely

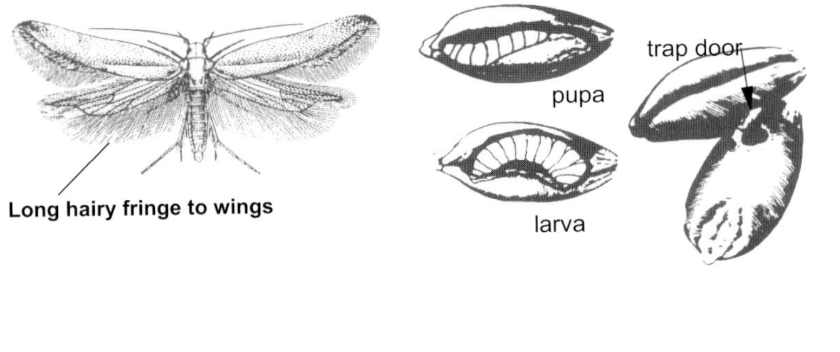


**Fig. 4.5** *Sitotroga cerealella* adult (wing span 10–18 mm), pupa, larva and grain with emergence window (courtesy of Central Science Laboratory, UK).

**Table 4.2** Common food legumes.

| Common name | Latin name |
|---|---|
| Groundnut (= peanut) | *Arachis hypogaea* |
| Pigeon pea | *Cajanus cajan* |
| Chickpea | *Cicer arietinum* |
| Soybean | *Glycine max* |
| Lentil | *Lens esculenta* |
| Adzuki bean | *Phaseolus angularis* |
| Butter bean (= lima bean) | *Phaseolus lunatus* |
| Kidney bean (= common bean) | *Phaseolus vulgaris* |
| Pea | *Pisum sativum* |
| Broad bean (= fava bean) | *Vicia faba* |
| Green gram | *Vigna radiata* |
| Black gram | *Vigna mungo* |
| Bambara groundnut | *Vigna* (= *Voandzeia*) *subterranea* |

complex and over recent years frequent revisions have resulted in name changes for many of the common food legumes. Most of these revisions have involved transferring species from the genus *Vigna* to the genus *Phaseolus* and vice versa. Some species have recently oscillated between these genera at a confusing rate (for example, green gram is at the time of writing classified as *Vigna radiata* but at various times has been classified as *Phaseolus radiatus*). The partial list of species in Table 4.2 gives the most common English names of a few species of Leguminosae and their current scientific names.

### Acanthoscelides obtectus *(Coleoptera: Bruchidae)*

*Acanthoscelides obtectus* (Say) is a common pest of *Phaseolus* beans. It sometimes attacks other legumes, but on these is seldom a serious pest.

The adult is a robust, active beetle, the body colours of which are greys, browns and reddish-browns, forming vague and indistinct patterns (Fig. 4.6). The antennae are dark grey with reddish apical segments. Each femur of the hind legs bears a large spine followed by two or three smaller, sharp spines. The elytra do not completely cover the abdomen, leaving the upper surface of the last abdominal segment, the pygidium, visible from above.

The adult beetles are able to infest beans before or after harvest. Eggs are laid loosely in or around pods or beans, often under cracks in the testa. After hatching, the larvae bore into the beans and spend their larval life feeding on the cotyledon, excavating a feeding chamber as they grow. The larvae pupate within the bean, but prepare the site of eventual escape by chewing their way to the outside, leaving only the testa of the seed separating the pupation chamber from the exterior.

The adult, which has relatively feeble mouthparts, is able to penetrate the testa and escape. The area of undermined testa is easily seen before adult emergence and is known as a window. The window itself is usually completely removed on emergence, leaving a neat round hole in the bean.

Adult beetles are short-lived (typically 7–14 days) and do not feed in store. In the field, however, they may feed on the pollen of many species of plant. *Acanthoscelides obtectus* develops most rapidly at around 30°C and 70% r.h. when egg to adult development can be completed in about 25 days. The species is capable of tolerating quite low temperatures, which has resulted in it being able to spread to cool highland regions of the world and into some temperate areas. It is less common in those parts of south and southeast Asia where grams, peas and lentils *(Vigna, Lens,* etc.) are more commonly grown than *Phaseolus* beans.

### Callosobruchus *spp (Coleoptera: Bruchidae)*

Species of *Callosobruchus* are important primary pests

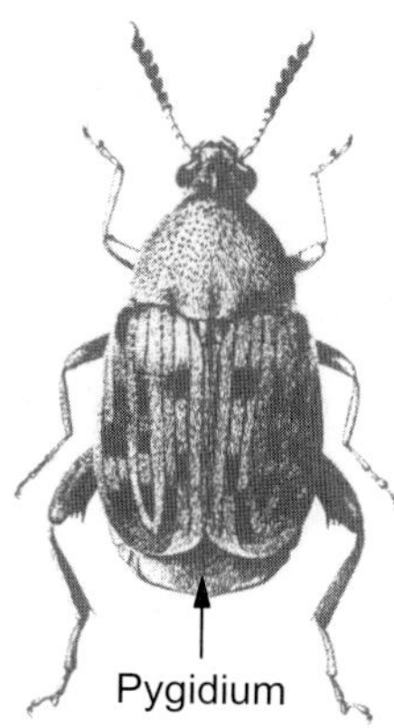


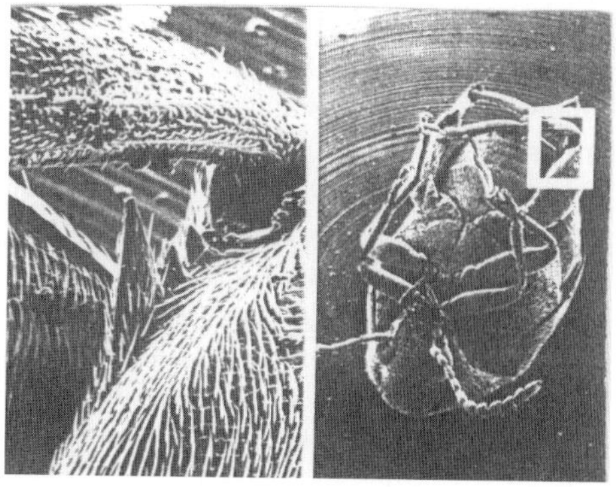

**Fig. 4.6** *Acanthoscelides obtectus* (life-size 3.0–4.5 mm), showing spines on femur of hind leg (drawing courtesy of Central Science Laboratory, UK).

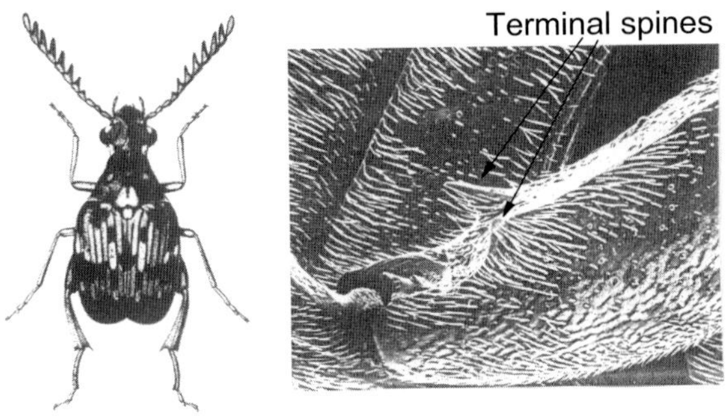


**Fig. 4.7** *Callosobruchus chinensis* (life-size 2.0–3.5 mm), male with typical comb-like antennae and hind femur showing terminal spines typical of the species (drawing courtesy of Central Science Laboratory, UK).

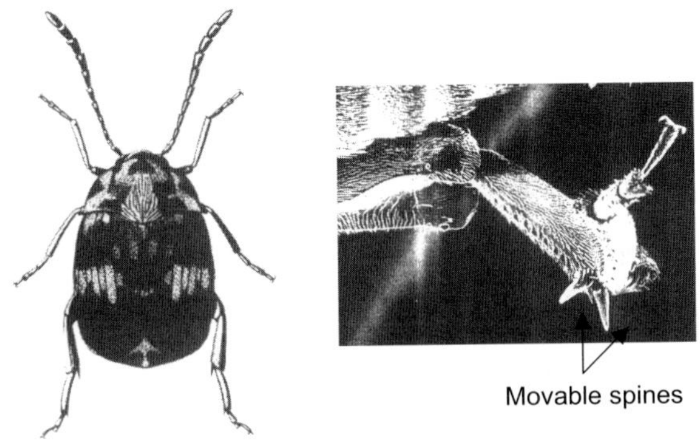


**Fig. 4.8** *Zabrotes subfasciatus* (life-size 2.0–2.5 mm), showing movable spines on the tibia of the hind leg (drawing courtesy of Central Science Laboratory, UK).

of a number of legumes including cowpeas, pigeon peas, bambara groundnuts, chickpeas, adzuki beans, peas, grams and (occasionally) soybeans. They do not usually attack kidney beans or butter beans (*Phaseolus* spp).

The adults are of the same general form as *A. obtectus* (Fig. 4.6), but are usually somewhat smaller. The elytra of some species are distinctly patterned, especially in the females, also in the female the elytra do not cover the last abdominal segment (pygidium). Each hind femur bears two longitudinal ridges, each of which bears a terminal spine (Fig. 4.7).

The life cycle of *Callosobruchus* spp is similar to that of *A. obtectus*, except that the eggs are stuck firmly to the testa of the host seed or to the wall of a pod. Upon hatching, the larva bores through the floor of the egg, directly into the seed or the pod. There are several species including *Callosobruchus chinensis*, a common Asian pest which is also found throughout the tropics and subtropics. The antennae of the male are comb-like in form (Fig. 4.7). *Callosobruchus maculatus* originated in Africa, but is now widespread. *Callosobruchus analis*, a common species in Asia, has frequently been confused with *C. maculatus* in the past. *Callosobruchus rhodesianus* is common to southern Africa and of economic importance.

### Zabrotes subfasciatus *(Coleoptera: Bruchidae)*

*Zabrotes subfasciatus* (Boheman) is a common pest of kidney beans and butter beans, and seldom attacks other legumes. It is similar in size to *Callosobruchus* spp, the elytra are rather square and broad and are strongly marked with white markings on a dark (almost black) background (Fig. 4.8). Unlike other bruchids infesting stored products, it has two movable spines on each hind tibia.

The life cycle of *Z. subfasciatus* is similar to that of *Callosobruchus* spp, with eggs glued onto the testa of the pulse. *Zabrotes* originated in tropical America, but is now common in many tropical and subtropical regions, especially Central and East Africa, Madagascar, the Mediterranean and India. *Zabrotes* and the other bruchid pest of *Phaseolus* beans, *Acanthoscelides obtectus,* can be distinguished in the pupal stage prior to emergence. The difference lies in the appearance of the pupal 'window', at least in beans with a white coloured testa. In *Z. subfasciatus,* the peripheral ring marking the outer edge of the window is more heavily eroded inside the bean, so when viewed from the outside it is much darker and more prominent (McFarlane & Wearing 1967).

### Caryedon serratus *(Coleoptera: Bruchidae)*

*Caryedon serratus* is a major pest of groundnuts, and also attacks tamarind seeds (*Tamarindus indica*). It is a large, robust beetle (Fig 4.9). The adult is reddish-brown in colour with dark, irregular markings on the elytra. Each hind femur bears a conspicuous comb of spines.

Adults glue their eggs to groundnut shells or on to the kernel testa after decortication. The larvae bore into the seeds, but may leave one seed and attack another. Pupation takes place outside the seed in a paper-like cocoon that is spun by the larva.

Under warm humid conditions (30–33°C, 70–90% r.h.) damage to groundnuts can be severe, although serious injury is usually restricted to the outer layers of bulks. The pest is of greatest significance in high-value

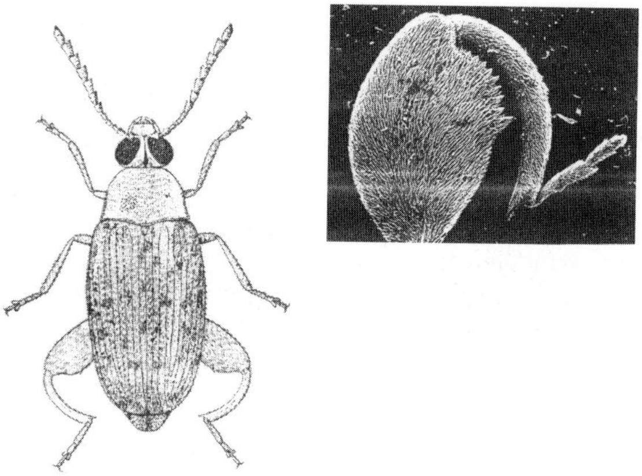

**Fig. 4.9** *Caryedon serratus* (life-size 3.5–6.8 mm), showing characteristic spines on hind femur (drawing courtesy of Central Science Laboratory, UK).

confectionery grade groundnuts that may be downgraded as a result of moderate infestation. It is now established that residual infestation in store is rare, while infestation of the drying crop occurs by cross-infestation from other hosts in the field (Conway 1983). This shifted pest control measures from residual pesticide treatments on groundnuts in store to the fumigation of stocks before the commencement of storage. Groundnuts have a high oil content and so are not suitable for fumigation with methyl bromide. Instead, phosphine gas should be used. When treating groundnuts in shell the phosphine dosage needs to be higher than normal, due to excessive absorption.

*Caryedon serratus* is found mainly in parts of Asia, north-eastern, western and south-western Africa, the West Indies and parts of Central and South America as far north as Mexico.

### Secondary pests

A large number of unrelated pests can be conveniently classified as secondary pests. They are predominantly associated with commodities that have suffered previous physical damage caused by a primary infestation or a milling process. Many are pests of cereal products, but others are associated with oilseeds, spices and other commodities.

#### Trogoderma granarium *(Coleoptera: Dermestidae)*

*Trogoderma granarium* is a very serious pest of cereal grains and oilseeds and many countries have specific quarantine regulations against possible importation. Massive populations may develop and grain stocks can be almost completely destroyed. Attack occurs in large-scale stores; it appears not to have been reported from farm stores. The presence of *T. granarium* on grain exported to some countries will result in an order to carry out expensive pest control measures or a rejection of the shipment.

*Trogoderma granarium* adults are small (2–3 mm) oval beetles (Fig. 4.10). The females are larger than the males. The elytra are lightly clothed with fine hairs and are mid-brown in colour or irregularly mottled. Although the adults have wings they are not known to fly and appear to rely on transport in old bags, etc., to get from one store to another. The larvae are extremely hairy (see Fig. 4.10) and their cast skins may cover the surface of infested grain. Hairs from the skins are allergenic, presenting a health hazard to storage workers and consumers.

*Trogoderma granarium* is very tolerant of high temperatures (up to 40°C) and low humidities (down to 2% r.h.). It is therefore a pest in hot, dry regions where other storage pests cannot survive (Banks 1977). In addition, the larvae are able to enter diapause (a resting stage) when physical conditions are unfavourable (Burges 1963). When in diapause the larvae move very little, or not at all, and their metabolic rate is lowered. In this state they can survive several years of adverse conditions. In diapause, larvae usually hide in cracks or crevices in the store, and are thus protected against contact insecticides. Their low metabolic activity also helps to reduce

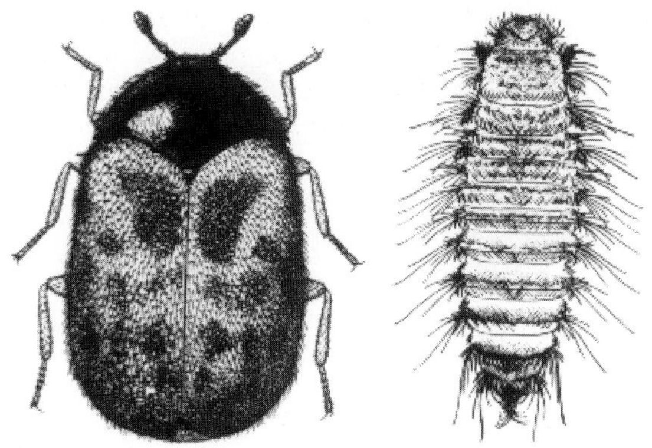

**Fig. 4.10** *Trogoderma granarium* adult (life-size 2.0–3.0 mm) and larva (courtesy of Central Science Laboratory, UK).

the rate of pesticide uptake and translocation. They are therefore very difficult to kill with residual insecticides or fumigants, although otherwise, out of diapause, they would be susceptible to the usual storage insecticides and fumigants.

*Trogoderma granarium* is widespread in the Indian subcontinent and adjacent areas and in many hot dry regions around the world. It is usually not found in humid regions.

### Tribolium castaneum *(Coleoptera: Tenebrionidae)*

*Tribolium castaneum* feeds on a range of commodities, especially cereals, but also groundnuts, nuts, spices, coffee, cocoa, dried fruit and occasionally pulses. They will also feed on animal tissues, including the bodies of dead insects, and will attack and eat small or immobile stages of living insects, especially eggs and pupae. Under conditions of overcrowding there is considerable cannibalism.

Adult *T. castaneum* are brown, medium-sized (2.5–4.5 mm), parallel-sided beetles that are partially dorso-ventrally flattened (Fig. 4.11). The larvae are cream or pale brown, have little hair and are very active.

Under optimum conditions (33–35°C at about 70% r.h.) adults live for many months. Throughout their lives females lay eggs loosely among their food and the larvae feed and complete their life cycle without necessarily leaving the food commodity. Development can be very quick (about 30 days) and population growth is very rapid.

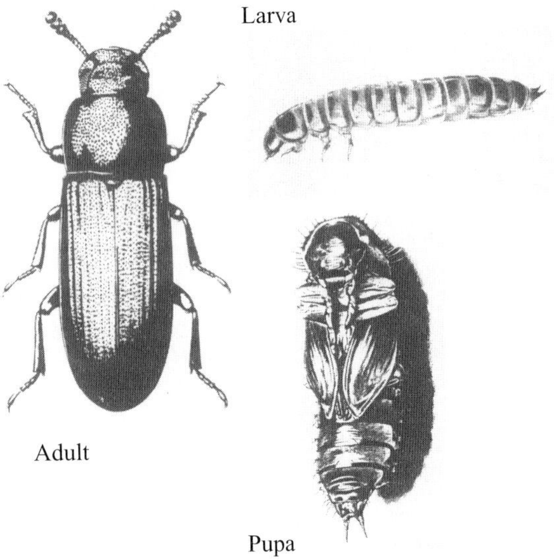


**Fig. 4.11** *Tribolium castaneum*, adult (life-size 2.5–4.5 mm), larva and pupa (courtesy of Central Science Laboratory, UK).

Heavy infestations by *T. castaneum* and other tenebrionids can produce disagreeable odours and flavours in commodities due to the production of quinones from the abdominal and thoracic defence glands of the adults (Leconti & Roth 1953). Flour exposed to *T. confusum*, at 100 adults/kg for three weeks, showed a distinct change in viscosity and extensibility when made into dough (Payne 1925). El-Mofty *et al.* (1992) observed tumours in mice that had been fed flour on which an initial population of *T. castaneum* at 20 adults/kg had been allowed to develop for one year. In contrast, Hodges *et al.* (1996) found that quinones did not accumulate on milled rice. It was concluded that flour absorbed quinones, probably due to its finely divided nature, while solid semi-crystalline grains did not.

### *Other tenebrionids*

Many species of Tenebrionidae, at least ten of which are very similar in appearance to *T. castaneum*, are found in farm and central stores. As *T. castaneum* is a very well-known insect, these other species are often mistaken for it.

*Tribolium confusum* closely resembles *T. castaneum* but can be differentiated by examining details of eye and antennal structure. The two species are biologically quite distinct: the optimum temperature for development for *T. confusum* is 2–3°C lower than for *T. castaneum*, and *T. confusum* shows a greater preference for food with a high oil content. *Tribolium castaneum* is found more commonly than *T. confusum* in the tropics although mixed populations of the two species do occur.

*Latheticus oryzae* is well known as a pest in cereals and oilseeds although it is particularly tolerant of low oil diets and can develop considerably better on these diets than *T. castaneum*. It is also unusual in being unable to develop at temperatures below 26°C. Several species of *Palorus* are locally common pests of cereals. The males of *Gnatocerus* spp have very distinctive large horn-like mandibles, but the females lack this feature. They attack mostly cereals, but also oilseeds, groundnuts, etc. *Alphitobius* spp attack damp and mouldy produce.

### Cryptolestes *spp. (Coleoptera: Cucujidae)*

Several species of *Cryptolestes* are common in mills and

stores where they are secondary pests of cereals, nuts, oilcakes, dried fruit and other commodities. The adults are small (2–2.5 mm), elongate, very flat, light-coloured beetles. They have long thin antennae. The head and prothorax are large and account for half of the length of the body (Fig. 4.12).

The small larvae of *Cryptolestes* spp may enter cereal grains at points of minor damage, especially in wheat, where the embryo is often exposed. The embryo of cereals is often attacked preferentially. *Cryptolestes* spp prefer high moisture content food and the presence of large numbers may indicate moisture problems.

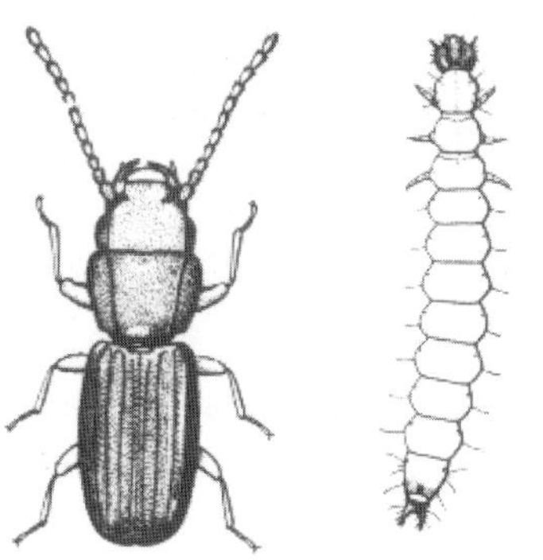

**Fig. 4.12** *Cryptolestes* sp., adult (life-size 2–2.5 mm) and larva (courtesy of Central Science Laboratory, UK).

### Oryzaephilus *spp. (Coleoptera: Silvanidae)*

*Oryzaephilus* spp are moderately small (2.5–3.5 mm) rather flat, parallel-sided beetles, which are distinguished by six large tooth-like projections on each side of the prothorax (Fig. 4.13).

There are two common species, *Oryzaephilus surinamensis* and *Oryzaephilus mercator*, which are similar in appearance but differ biologically. *Oryzaephilus surinamensis* develops more quickly than *O. mercator* at high temperatures and humidities (35°C, 90% r.h.) and is more tolerant than *O. mercator* of extremely high and low temperatures and humidities. Both species attack cereals, cereal products, oilseeds, copra, spices, nuts and dried fruit. However, *O. surinamensis* is most successful on starchy, cereal diets, while *O. mercator* prefers diets with a high oil content.

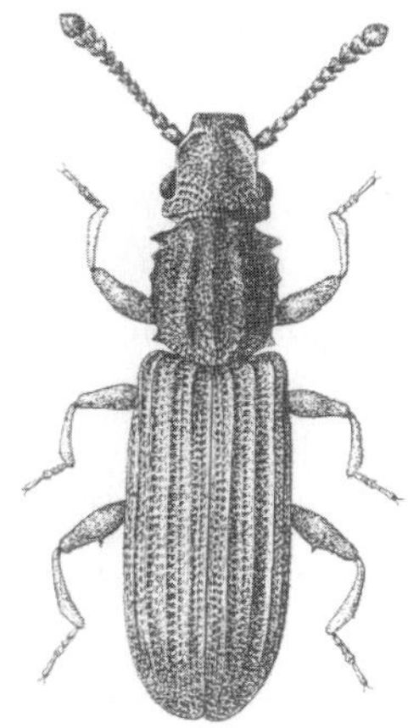

**Fig. 4.13** *Oryzaephilus surinamensis* (life-size 2.5–3.5 mm) (courtesy of Central Science Laboratory, UK).

### Lasioderma serricorne *(Coleoptera: Anobiidae)*

*Lasioderma serricorne* is a rounded, light-brown beetle with smooth elytra. The head is carried beneath the prothorax and so cannot be seen from above. The antennae are long with saw-like apical segments (Fig. 4.14). This beetle is a well-known pest of dried tobacco and manufactured tobacco products. However, it will attack a great range of commodities, including cocoa, cereals, spices, dried fruit and processed foodstuffs.

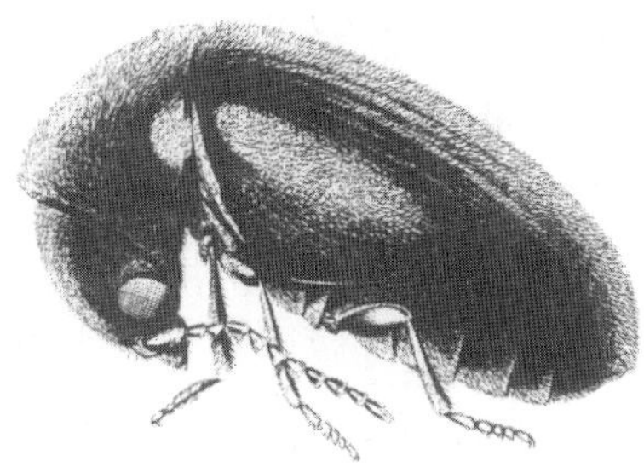

**Fig. 4.14** *Lasioderma serricorne* (life-size 2.0–2.5 mm) (courtesy of Central Science Laboratory, UK).

A closely-related species, *Stegobium paniceum* (Fig. 4.15), attacks a similar range of commodities, but is seldom found on tobacco. It can be distinguished from *L. serricorne* by the shape of its antennae.

### Cadra cautella *(Lepidoptera: Pryalidae)*

*Cadra cautella* is a common and important secondary pest of cereals, cereal products, cocoa, dried fruit, nuts and many other commodities. In newly emerged adults the fore-wings are greyish-brown in colour, with an indistinct pattern. Older specimens which have lost most of their scales are dull grey in colour (Fig. 4.16).

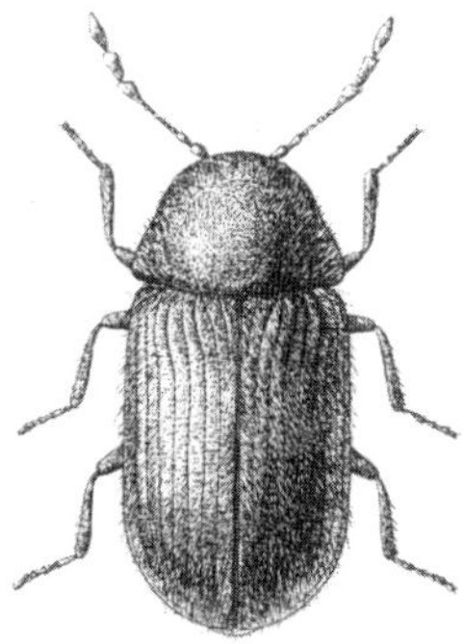

**Fig. 4.15** *Stegobium paniceum* (life-size 2.0–3.0 mm) (courtesy of Central Science Laboratory, UK).

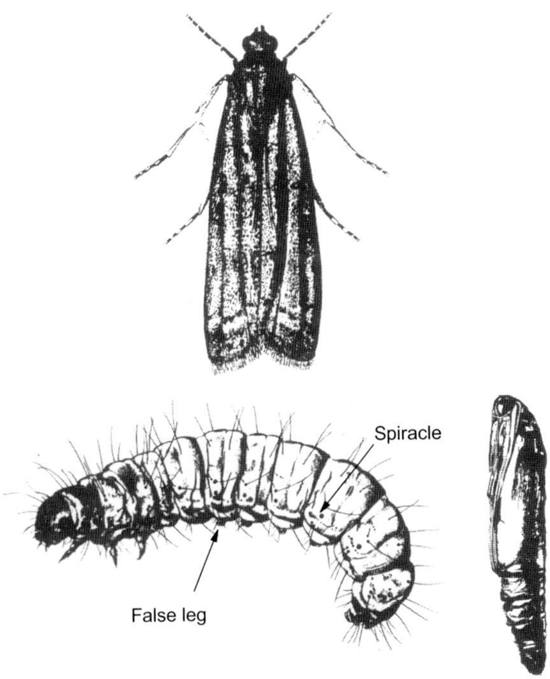


**Fig. 4.16** *Cadra cautella,* adult (wing span 11–28 mm), larva and pupa (courtesy of Central Science Laboratory, UK).

Adult *Cadra cautella* are fairly short-lived (usually 7–14 days) and do not feed. The females lay their eggs loosely on the surface of the commodity. The larvae move extensively through the produce as they feed and, as they move about, they spin copious quantities of silk, called webbing. The webbing from heavy infestations can mat together the commodity and render it unfit for consumption. Last instar larvae move out of the produce and wander about freely until they find a suitable site for pupation. Pupation sites are usually cracks, crevices and frequently the gaps between grain bags.

Newly-emerged adults can mate within a few hours of emergence, and eggs are laid soon afterwards (usually within 24 hours). Adult *C. cautella* usually remain at rest during daylight. The peak periods of flight are around dawn and dusk. Egg-laying behaviour follows the same rhythm.

#### *Other moth secondary pests*

Two other moth species are common secondary pests: *Corcyra cephalonica* and *Plodia interpunctella.* The life history of both is very like that of *C. cautella* and both also attack a very wide range of stored commodities. However, *C. cephalonica* is particularly associated with rice while *P. interpunctella,* which is very common in tropical America, often behaves rather like a primary pest, infesting undamaged maize by attacking the embryo.

*Corcyra cephalonica* are robust, medium sized (5–10 mm) moths. The wings are greyish-brown in newly emerged specimens. The eggs are rather sticky and the webbing produced by the larvae is very tough. Newly emerged specimens of *P. interpunctella* are easily identified by their fore-wings, the inner parts of which are pale yellow, while the basal parts are dark reddish brown (Fig. 4.17).

**Fig. 4.17** *Plodia interpunctella,* adult (wing span 11–28 mm) (courtesy of Central Science Laboratory, UK).

### Liposcelis *spp (Psocoptera: Liposcelididae)*

Psocids, or booklice as they are sometimes called, are small insects only about 1 mm long and for that reason are often confused with mites. However, the presence of six legs and very long antennae (mites have no antennae) should make them easy to distinguish (Fig. 4.18).

The most common species of psocid found in stores belong to the genus *Liposcelis*. Large numbers of psocids sometimes occur in stores, especially in the humid tropics. Psocid problems are well known in Southeast Asia where the numbers may be so great that they form a thick carpet on the floor of dead, dying and living individuals. Psocids swarm over store and bag stack surfaces and are thus a serious contaminant. For many years they were considered to be only scavengers, feeding on dust, debris and moulds found in stores. However, they eat the surface of milled rice, especially when the rice itself has not been fully processed, when a considerable amount of the nutritious brown layer remains after polishing (Pike 1994a). Psocids are attracted to flour, meal and other cereal products. Besides their effect on stored food and hygiene, they are also a nuisance to storage workers as contact with skin causes some irritation.

In recent years it has become clear why psocids reach very high numbers in stores, especially those subject to regular pest control operations. The eggs of psocids are particularly tolerant of the fumigant phosphine, so that unless fumigation is done to a very high standard they may not be killed (Pike 1994b). However, *Tribolium castaneum* and *Cryptolestes* spp, which are known to be psocid predators, will not survive. The psocids that emerge after fumigation are thus free to increase without the constraint of competition or predation from other storage pests (Santoso *et al.* 1996).

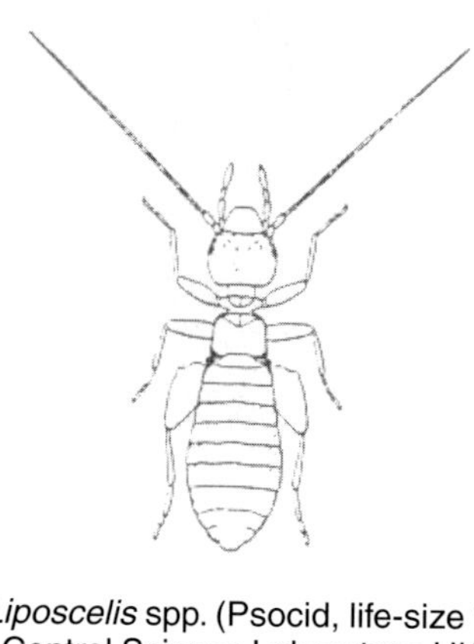

**Fig. 4.18** *Liposcelis* spp. (Psocid, life-size 1.0–1.5 mm) (courtesy of Central Science Laboratory, UK).

Some species of psocid can undergo parthenogenesis (the eggs develop without fertilisation) and all resulting individuals are female which results in a very rapid population increase. So, how can psocids be controlled? They are very susceptible to the fumigant methyl bromide, but this is being phased out because it is an ozone depletor and should not be used. The suggested practice in cases where psocids are a problem is either to ensure that enough gas is retained during the 5-day treatment (at 27°C), so that 1.7 mg/L (1275 ppm) is still present on the last day, or that the treatment is extended to 8 days and not less than 0.05 mg/L (38 ppm) is retained on the last day. If sufficient gas cannot be retained over an 8-day period then a strategy first suggested for the control of mites needs to be used. This requires two fumigations to be undertaken in succession. The second fumigation follows about 10 days after the first, when any surviving eggs will have developed into nymphs.

### Tyrophagus putricentiae *(Astigmata: Acaridae)*

Astigmatan mites are small or very small. They are whitish, cream or milky translucent in colour (Fig. 4.19). The name of the order indicates the absence of any sigmata (breathing holes); respiration is by diffusion over the whole body surface. They are important pests of produce, either feeding directly on the commodity or by feeding on, and thus spreading, fungi.

*Tyrophagus putricentiae* is a cosmopolitan and common mite pest in food stores, recorded from a wide range of commodities. It is known to feed on storage fungi, and

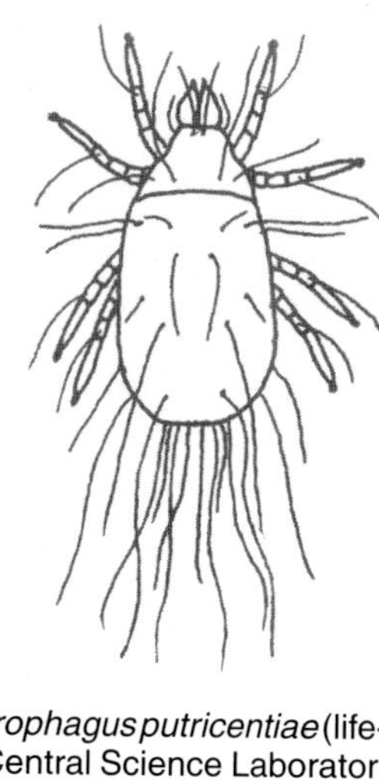

**Fig.4.19** *Tyrophagus putricentiae* (life-size 0.3–0.8 mm) (courtesy of Central Science Laboratory, UK).

to develop rapidly when preferred fungi are present. It can develop successfully in the absence of fungi, for example on sterile wheatgerm in laboratory tests, and thus can cause direct damage to the produce. Like all acarids, it develops more rapidly in higher humidities but development is possible at all humidities over 60%. Its upper temperature limit is about 35–37°C, with an optimum about 32°C.

There is little information on the economic losses caused by stored product mites. Some studies from the 1950s (reported in Hughes 1976) suggested that feeding various animals on diets contaminated with the mite *Acarus siro* had no effect on weight or condition. However, other studies in chickens, rats and rabbits fed on mite-infested food showed organ degeneration. This may be due, at least in part, to a reduction in vitamin A and D caused by mite infestation. An unpublished study found that the nutritive value of pig feed, infested by 2–3 million mites/kg, after 10 weeks was seriously reduced; this resulted in a 16% reduction in the growth rate of the pigs fed this diet. Mites that feed directly on the commodity, whether or not they also consume fungi, obviously cause quantitative losses, though these may only become significant when the number of mites is very high. Such infestation may taint the stored commodity with off-odours or off-flavours. Mites can also cause allergenic responses in animals eating the infested food and in the storage workers handling it.

### Pests of dried fish and other animal products

Dried fish, hides and skins and other dried animal products create an environment which can be exploited by a number of beetles, particularly of the genus *Dermestes*. During drying, fish may be heavily damaged by the larvae of various species of Diptera (flies).

#### Dermestes *spp (Coleoptera: Dermestidae)*

*Dermestes* spp are the most important pests of dried fish, and also may attack dried meat, hides and skins. Salted products are also attacked if salting is insufficient. The adult beetles are large and robust (Fig. 4.20) and usually dark in colour with areas of contrasting light-coloured hairs. The antennae are clubbed. The larvae are conspicuous and very hairy.

*Dermestes maculatus* is a common pest of dried fish. It has a low tolerance to salt and is therefore usually associated with unsalted freshwater fish rather than with marine species. In contrast, *Dermestes frischii* has frequently been reported attacking dried marine fish.

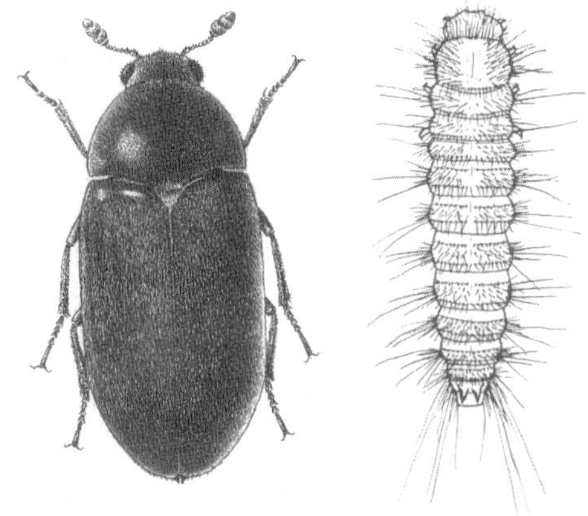

**Fig. 4.20** *Dermestes maculatus*, adult (life-size 6–10 mm) and larva, pests of animal products (courtesy of United States Department of Agriculture).

*Dermestes* spp are scavengers, and may feed on a variety of foods of animal origin. They may scavenge the remains of dead insects, feed on the remains of other animals or infest farm buildings where there is intensive animal production. The late instar larvae bore into solid materials, and may severely weaken wooden structures, or damage food commodities or other materials.

#### Necrobia rufipes *(Coleoptera: Cleridae)*

*Necrobia rufipes* is a shiny green or bluish-green beetle with the base of the antennae and legs red (Fig. 4.21). It is found throughout the tropics and subtropics where

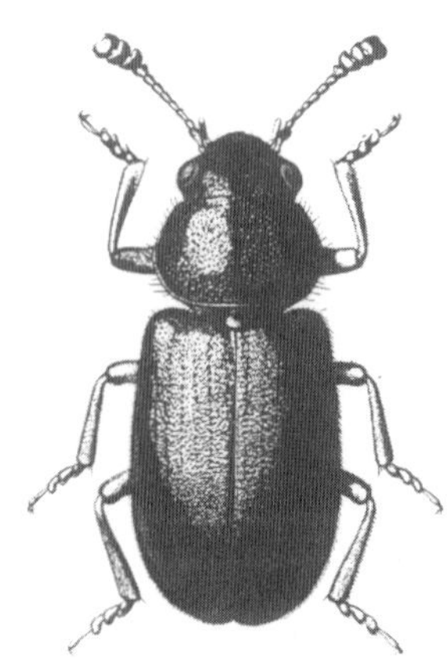

**Fig. 4.21** *Necrobia rufipes* (life-size 4.0–5.0 mm), a pest of animal products (courtesy of Central Science Laboratory, UK).

it will attack animal products such as ham and cheese, dried fish and fishmeal. It is also a predator on other insects and its infestations are often found in association with *Dermestes* spp. It may also feed from plant materials with a high fat content such as copra, palm kernels and certain oilseeds.

*Necrobia rufipes* thrives under warm (30–34°C) damp conditions. Adults are very active fliers and, when present in large numbers, may cause annoyance to people handling infested commodities. The presence of these species on high-value commodities such as processed meats and cheese, even in small numbers, is often unacceptable and can lead to rejection by merchants and consumers.

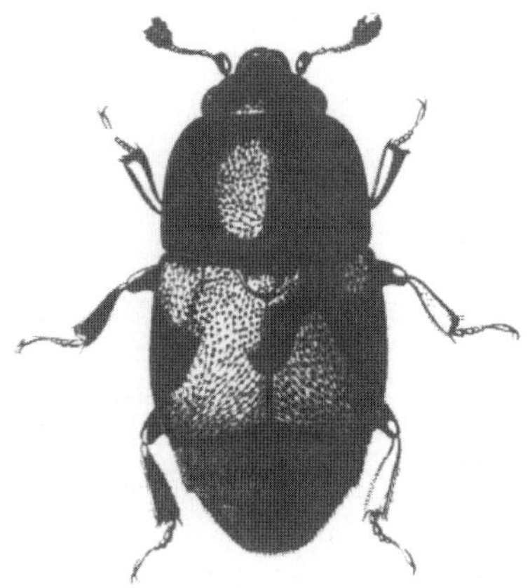

**Fig. 4.22** *Carpophilus hemipterus* (life-size 2.0–4.0 mm), feeder on damp produce (courtesy of Central Science Laboratory, UK).

### Pests associated with damp and mouldy produce

A number of pests are particularly associated with commodities that have a high moisture content or that have become mouldy. They are found on freshly harvested commodities, commodities which have been stored at an excessively high moisture content or commodities that have become damp and mouldy as a result of infestation by other species of insect or by water ingress. Neglected grain residues in stores are often very damp as a result of insect activity and therefore often contain many damp-associated species which are unlikely to attack the main bulk of stored commodity. However, if damp-associated species are found in large numbers roaming freely in a store or moving about the stored commodity, then they may indicate the presence of seriously wet produce.

#### Carpophilus *spp (Coleoptera: Nitidulidae)*

*Carpophilus* spp are common pests of mature maize before harvest and are often transported to stores after harvest (Fig. 4.22). Some species, such as *Carpophilus dimidiatus,* may persist for some time during storage, especially in maize cobs in farm stores where moisture contents remain high. Other species, such as *Carpophilus pilosellus*, are common in milled rice stores in Southeast Asia where the ambient humidity is at least 70%. *Carpophilus* spp are also associated with other cereals at harvest, with cocoa, and will attack dried fruit.

*Cathartus quadricollis* is another insect that is common on newly harvested cereals, cocoa and other crops, and may persist for a time in store. It is particularly common in the humid parts of West Africa.

#### Typhaea stercorea *(Coleoptera: Mycetophagidae)*

Many species, including the beetle *Typhaea stercorea* (Fig. 4.23), feed on damp produce. Some are favoured by the high moisture content of the produce, some supplement their diet with moulds and some feed exclusively on mould. The presence of any such species in a store in large numbers may serve as a warning that some of the commodity is in a dangerously damp condition.

*Ahasverus advena* attacks many products, and is favoured by the presence of mould. *Alphitobius diaperinus* and *Alphitobius laevigatus* both attack damp and mouldy produce.

Many mite species are also specifically mould feeders, these include *Glycyphagus* spp (Fig. 4.24) which are common in tropical storage.

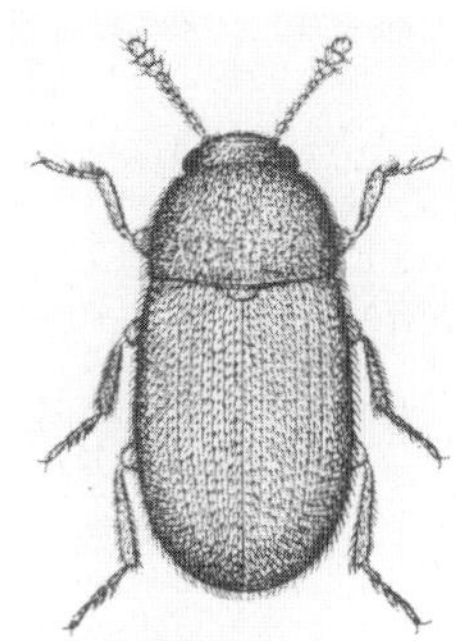

**Fig. 4.23** *Typhaea stercorea* (life-size 2.0–3.0 mm), the hairy fungus beetle (courtesy of Central Science Laboratory, UK).

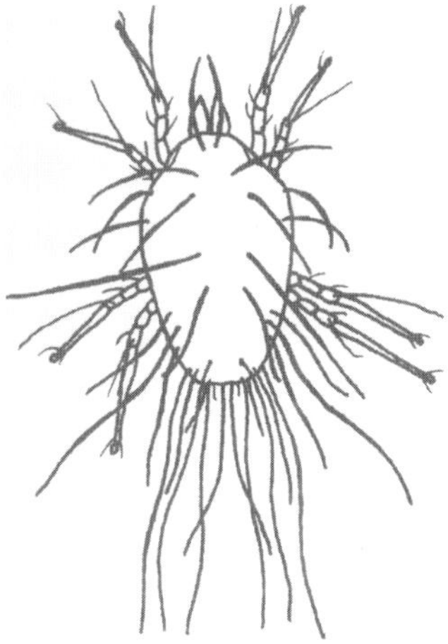

**Fig. 4.24** *Glycyphagus* spp. (life-size 0.3–0.8 mm), fungus mite (courtesy of Central Science Laboratory, UK).

### Miscellaneous pests of specific commodities

Many stored product pests are fairly general feeders attacking a range of commodities (for example, *Lasioderma serricorne*: see Secondary pests). Others are restricted to a narrower range of commodities (for example, *Sitophilus* spp attack large-grained cereals: see Primary pests of cereals). Most commodities can be attacked by a number of pest species. However, a few commodities are particularly associated with certain pest species.

*Araecerus fasciculatus* is the most important pest of stored coffee (Fig. 4.25). It attacks coffee beans, especially those that have been in store for a considerable period. The larvae bore into the beans and once damaged they may then be attacked by secondary pests, especially *Cadra cautella*. *Araecerus fasciculatus* also attacks cocoa beans, nutmegs and, occasionally, maize, cassava and sweet potato.

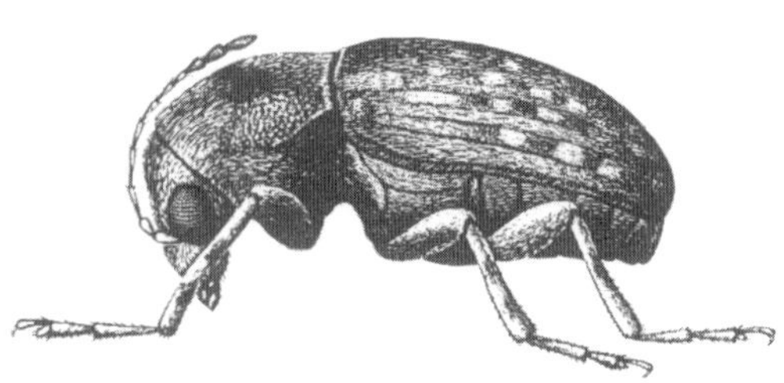

**Fig. 4.25** *Araecerus fasciculatus* (life-size 3.0–5.0 mm) (courtesy of Central Science Laboratory, UK).

*Lasioderma serricorne* is a very general feeder (Fig. 4.14). However, it is most important as a pest of cured tobacco, which it can damage severely and cause extensive economic problems, requiring the closure of tobacco auctions. Few other pests ever attack tobacco, except the moth *Ephestia elutella*. Copra is commonly attacked by *Necrobia* spp (although this species is also a predator).

### Scavengers

A number of the insects that occur in stores do not attack the grain directly but feed on dust and residues or on the dead bodies of insects and other animals. These insects are referred to as scavengers and their numbers are high when conditions in store are unhygienic. Typical scavengers are cockroaches (Dictyoptera), earwigs (Dermaptera) and silverfish (Thysanura) (Fig. 4.26).

### Structural pests

A number of species may attack the wooden structure of stores such as *Lyctus brunneus* (Fig. 4.27), some pest species such as *P. truncatus* (Fig. 4.3) and *Dinoderus* spp, and the last instar of *Dermestes* larvae (Fig. 4.20).

Termites are a common problem in the tropics, causing considerable damage to wooden storage structures and sometimes to the commodity contained within them. Termite damage to the crop in the store is aggravated by the damp soil carried in by the termites.

There are 3000 species of termite worldwide but only 300 species are considered to be pests. Termites are often referred to as white ants but they are more closely related to cockroaches. The main visible differences between

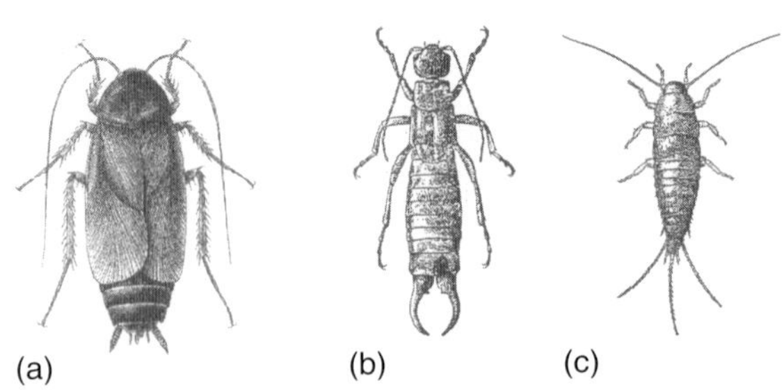


**Fig. 4.26** Typical scavengers: (a) cockroach (courtesy of United States Department of Agriculture); (b) earwig and (c) silverfish (both courtesy of Central Science Laboratory, UK).

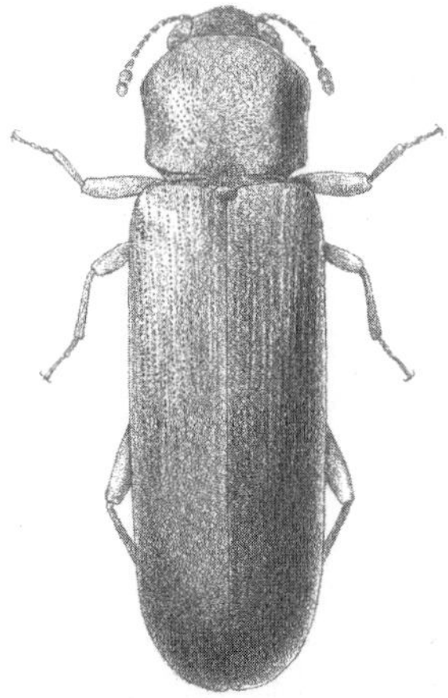

Fig. 4.27 *Lyctus brunneus* (life-size 3.0–7.0 mm) (courtesy of United States Department of Agriculture).

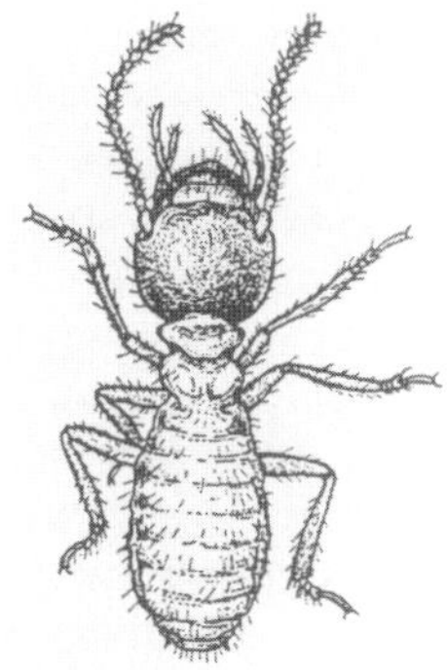

Fig. 4.28 Worker termite (*Ancistrotermes latinotus*) (courtesy of Rentokil Initial, UK).

termites and ants are that termites have no waist and their antennae are straight, not angled. Both ants and termites are social insects living in colonies. Ants and many species of termite forage out from a central nest to collect food.

The vast majority of termite species are not pests. They do not damage crops, plantations, forests or buildings. Instead they play a vital part in the ecosystem by recycling dead wood and plant litter, breaking it down to material which increases the fertility of the soil. Their tunnelling, mound building and soil moving activities carry subsoil rich in minerals to the surface. They also open up the soil, which increases aeration, percolation of water and root growth. Termites are a food source for a wide range of animals including ants, spiders, frogs, lizards, birds, mammals and man.

Termites belong to the order Isoptera that is divided into seven families, of which the Kalotermitidae (dry wood termites), Termopsidae (damp wood termites), Rhinotermitidae (subterranean termites) and Termitidae (higher termites with the most elaborate caste systems, containing 70% of known termites) are all known to attack buildings.

### *Termite castes*

Each termite occurs in a number of different forms or castes: reproductives, workers (Fig. 4.28), soldiers (Fig. 4.29) and alates (Fig. 4.30).

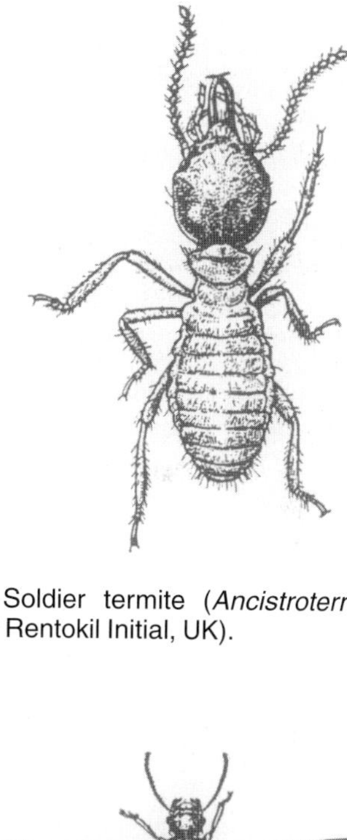

Fig. 4.29 Soldier termite (*Ancistrotermes latinotus*) (courtesy of Rentokil Initial, UK).

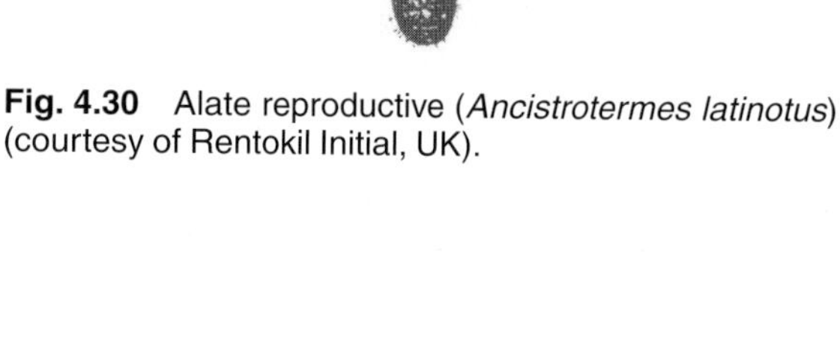

Fig. 4.30 Alate reproductive (*Ancistrotermes latinotus*) (courtesy of Rentokil Initial, UK).

- *Reproductives*: Only the king and queen termites reproduce. The king remains the same size throughout his life but the queen in some species grows to many times her original size as her ovaries develop. She becomes an egg-laying machine. The reproductives have eyes; the other castes are usually blind.
- *Workers*: Workers collect food and feed the reproductives and soldiers, which cannot feed themselves. They also tend the young and, in subterranean and mound building species, construct tunnels.
- *Soldiers*: Soldiers guard both the nest and foraging parties of workers. Soldiers often have specially adapted mandibles for attacking predators (including man) and salivary glands which produce a repellent, irritating or sticky material.
- *Alates*: Alates are young, winged virgin reproductives that fly from the colony to start new colonies. Alates can occur in vast numbers after rains. The alates fly out of the nest, generally for a short distance only, then settle on the ground or wood (depending on the species), shed their wings and start to reproduce.

### *Characteristics of termite damage*

#### *Dry wood termites (Kalotermitidae)*

The dry wood termites can set up colonies in wood in any part of a building and can attack wood directly. The flying alates tunnel into the wood and establish colonies. Damage occurs within the wood and there may be no evidence of the damage until the beam is completely eaten out. Generally the only sign of dry wood termite damage is the presence of piles of small dry faecal pellets which are ejected from the colony.

#### *Subterranean termites (Rhinotermitidae and some other families)*

Subterranean termites live underground or in mounds and attack buildings by tunnelling towards them and either enter wood in contact with the soil or build soil runways over the concrete foundations. Such runways can travel within or over the walls to reach rafters several floors above the ground. Storage structures should be inspected for such runways, which can easily be brushed off, thereby destroying the termite route into the structure before they reach any wood.

## Beneficial feeders (predators and parasites)

### *Predators*

When produce is heavily infested, a number of specialised predators may become established. Most of these are members of the order Hemiptera (Fig. 4.31), which attack insect larvae, kill them and feed by sucking out the insect's body fluids through specialised piercing mouthparts.

A few other specialised predators, including beetles of the families Histeridae, Staphylinidae and Carabidae, also sometimes attack stored products pests. These may either feed on a wide range of different prey or on just a few species. A good example of a predatory beetle feeding on a narrow range of prey items is *Teretrius nigrescens* (Fig. 4.32), formerly called *Teretriosma nigrescens*, which feeds largely on the developing stages of *Prostephanus truncatus*.

Some species of mites are predators, including *Blattisocius* spp (Fig. 4.33), that feed on insect eggs, especially those of moths, and *Cheyletus* spp which eat the eggs and immature stages of insects.

In addition to the specialised predators, some stored products' beetles can act as predators some of the time. Their predatory activity is likely to occur most when produce is heavily infested and prey is easily available. Adult insects are seldom attacked: eggs, larvae and pupae are the usual prey, especially the immobile eggs and pupae.

*Tribolium* spp (see Secondary pests) are capable of preying on eggs and pupae of their own and other species. *Dermestes ater* is known to prey on fly pupae, and *Necrobia* spp are general predators on many species

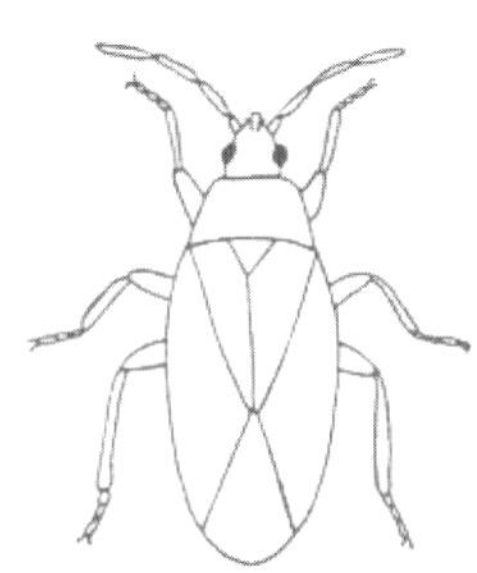

**Fig. 4.31** Hemiptera; some species are predators (life-size 1.5–15.0 mm) (courtesy of Natural Resources Institute).

**Fig. 4.32** *Teretrius nigrescens* (life-size 2.0–3.0 mm), predator of the pest *Prostephanus truncatus* (courtesy of Ms. J. Hodges).

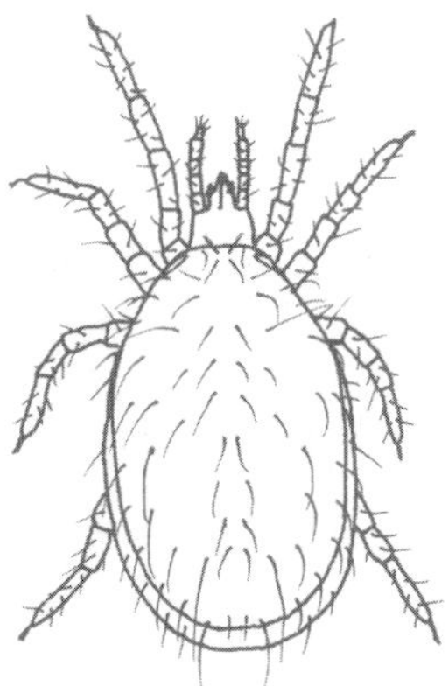

**Fig. 4.33** *Blattisocius* spp. (life-size 0.2–1.0 mm), mite predator of moth eggs (courtesy of Natural Resources Institute).

(see Pests of dried fish and other animal products). *Tenebroides mauritanicus* (Fig. 4.34) is partially predatory on other insects, but also feeds on cereals.

### *Parasites*

Parasites that may become established are mostly small wasps (Hymenoptera). Female wasps possess a longish egg-laying tube (ovipositor), which they use to pierce the eggs or larvae of storage pests and so deposit their eggs within the bodies of their hosts. The wasps develop within the host and subsequently emerge as adults; the host usually dies as a result of this. Good examples of such parasites are *Uscana lariophaga* that parasitises

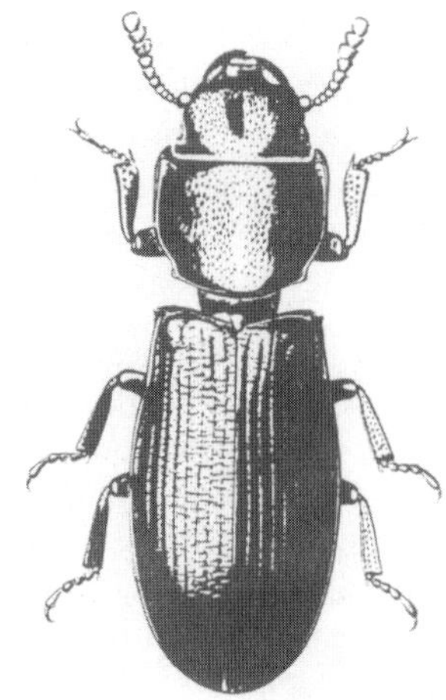

**Fig. 4.34** *Tenebroides mauritanicus* (life-size 5.0–11.0 mm), predator and pest of stored grain (courtesy of Central Science Laboratory, UK).

the eggs of bruchids, *Anisopteromalus calandrae* that is a parasite of beetle larvae, especially *Sitophilus* spp, and *Bracon hebetor* (Fig. 4.35) that is a parasite of moth larvae.

Stored product insects may also be parasitised by mites. The best-known parasitic mite species are *Pyemotes* spp and *Acarophenax tribolii*. They live on the surface of the insects, attaching themselves to soft areas of cuticle, which they pierce; the body fluids are then sucked from the insect. Although these parasites may exert considerable control on insect populations, a few can also burrow into human skin and cause great irritation. Consequently, they would not be of practical use in biological control. These mites can destroy experimental insect cultures and are an important reason why cultures prepared from newly caught storage pests should be kept

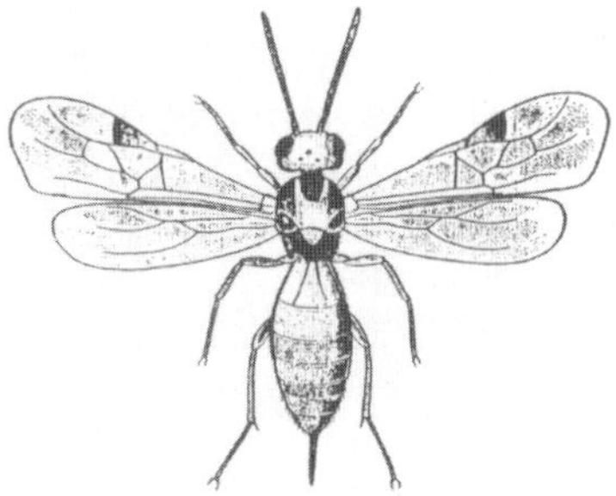

**Fig. 4.35** *Bracon hebetor* (life-size 4.0–7.0 mm), parasitic wasp (courtesy of Central Science Laboratory, UK).

in quarantine, i.e. away from longer established cultures, until it is certain that they are free of such mites.

## Pests of durable crops – vertebrates

### Rodents

Rodents comprise nearly 40% of all species of mammals with some 1700 species roughly divided into 390 genera within 35 families. The range of species within the Rodentia covers beavers, chinchillas, squirrels, gophers, porcupines, voles, hamsters, gerbils, mice and rats. The principal identifying feature of a rodent is that one pair of incisors in the upper and lower jaw is greatly enlarged and used for gnawing. In fact, the taxonomic name of the order is derived from the Latin *roderus*, meaning to gnaw or chew. The ability of rodents to gnaw extensively is derived from the design of their incisors. The incisors grow continuously so that rodents must gnaw to keep the length of the teeth in check. Only the front portion of the tooth is covered in a hard enamel, so as the rodent chews the softer dentine wears away leaving the enamel front of the tooth longer; hence gnawing ensures the incisors are always kept sharp. The other major defining feature of a rodent is the presence of a gap, the diastema, between the incisors and the rest of the teeth. The diastema allows a rodent to close off its incisors from the rest of its mouth by sucking in its cheeks, which permits gnawing non-food items with no danger of ingestion (Plate 1). Further information on the taxonomy of rodents can be found on the websites of BIOSIS UK, George Washington University and the University of Michigan.

Rodent species vary greatly in general morphology and in their ability to exploit different ecological niches. They are normally known for being prolific breeders, but this is not always the case. For example, the African spring hare, *Pedetes capensis* (Forster), produces only a single offspring each year (Hanney 1975). However, most rodents, especially those considered to be pests, follow *r*-selection breeding patterns with explosive reproduction rates but often poor individual survival (Macdonald & Fenn 1994). In *r*-selected communities population growth is potentially exponential, i.e. the rate of addition of individuals to the population is proportional to the number of individuals already present. Animals that have an *r* strategy produce large numbers of offspring, reproduce early in life, have a short development period, small body size and a high intrinsic rate of increase (Haines 1991).

Rodents generally have very acute senses, with excellent ability in sight, smell, hearing, touch and taste. Odour is very important to social interactions, affecting their behaviour and physiology. Through scent-marking of territory, rodents can negotiate their way in total darkness. Many produce ultrasounds which are used in courtship, alarm signals and as signs of aggression. Agility is another trait they possess, and many species are extremely good climbers or swimmers.

Out of all the species of rodents, only a handful are considered to be pests. Nearly all pest species are commensal rats and mice, with a long history of association with man and human habitation.

#### *Rodents as pests of stored products*

Generally, rodent species are best suited to living in the natural environments in which they have evolved; they are not well adapted to living in the very specialist and evolutionarily new storage environments created by man.

The most comprehensive worldwide survey of rodent occurrence in storage environments was undertaken in the early 1970s (Hopf *et al.* 1976). Respondents to the survey indicated that, whereas there were some 200 species recorded as damaging field crops, less than 40 species caused problems in food stores (Table 4.3). The majority of these show strict geographical limitations in their ranges and only the three main commensal species, *Rattus rattus* (ship rat), *Rattus norvegicus* (Norway rat) and *Mus musculus* (house mouse), are found on all the major continents in both tropical and temperate climates.

By their very nature, commensal rodent species are behaviourally and physically well adapted to take advantage of the wide variety of storage environments that are found within environments created by man. Other species are best seen as storage opportunists, perhaps better adapted at living outside the store. These may be able to take advantage of the store within their normal environment, but are not well suited to living in stores which are outside their usual home range.

While there may be some variation in the species inhabiting and infesting commodities within a store, the control strategies applied are broadly similar and are based upon a programme that takes account of all the variables which may be identified both within and outside the store. The need to identify clearly the species that are causing the problem must be the first requirement.

**Table 4.3** Rodent species associated with storage environments. Source: Hopf *et al.* 1976.

| Family | Species | Region |
|---|---|---|
| Cricetidae | *Akodon azarae* (Fischer) Tate | South America |
| | *Calomys callosus* (Rengger) | South America |
| | *C. laucha* (Olfers) Cabrera | South America |
| | *C. musculinus* (Thomas) (syn. *C. laucha musculinus*) | South America |
| | *Microtus arvalis* (Pallas) | Europe |
| | *M. pennsylvanicus* (Ord) | North America |
| | *Oryzomys flavescens* (Waterhouse) | South America |
| | *Tatera indica* (Hardwicke) | Asia |
| | *T. leucogaster* (Peters) | Africa |
| Muridae | *Acomys cahirinus* (Desmarest) | Africa |
| | *Aethomys namaquensis* (Smith) | Africa |
| | *Apodemus sylvaticus* L. | Europe |
| | *Arvicanthis niloticus* (Desmarest) | Africa |
| | *Bandicota bengalensis* (Gray & Hardwicke) | Asia |
| | *B. indica* (Bechstein) | Asia |
| | *B. savilei* (syn. *B. bengalensis*) | Asia |
| | *Gunomys gracillis* (syn. *B. bengalensis*) | Asia |
| | *Mastomys natalensis* Smith | Africa |
| | *Mus cantaneus* (syn. *M. musculus*) | Asia |
| | *M. domesticus* ( syn. *M. musculus*) | Global |
| | *M. molissimus* (syn. *M. musculus azoricus*) | Asia |
| | *M. musculus* L. | Europe |
| | *Praomys natalensis* (syn. *Mastomys natalensis*) | Africa |
| | *Rattus argentiventer* (Robinson & Kloss) | Asia |
| | *R. exulans* (Peale) | Asia, Oceania |
| | *R. norvegicus* (Berkenhout) | Global |
| | *R. rattus* L. | Global |
| | *R. r. diardii* (syn. *R. tanezumi*) | Asia |
| | *R. r. flavipectus* (syn. *R. tanezumi*) | Asia |
| | *R. r. mindanensis* Mearns | Asia |
| | *R. tanezumi* Temminck (syn. *Rattus griseiventer annandelei*) | Asia |
| | *R. tiomanicus* (Miller) | Asia |
| | *Rhabdomys pumilio* (Sparman) | Africa |
| Sciuridae | *Funambulus palmarum* L. | Asia |

### *What makes the post-harvest environment so suitable for rodents?*

Rodents generally require three main things for survival: food, water and shelter. However, because they are highly adaptable, rodents are usually able to find sufficient shelter, digging holes in the ground, nesting in trees or chewing their way into relatively inaccessible and marginal areas. Their ability to exploit both the horizontal and vertical components of their environment means that finding shelter is usually not a particular constraint to their survival. Rodent species have differing requirements for water intake; many species can find sufficient water from their food and are capable of living in semiarid environments. Food availability is the principal constraint to rodents, and areas of high food availability, such as a food store, can allow them to fully exploit their high reproductive capabilities. In fact, most food stores not only provide a constant and virtually unlimited food supply, often with water nearby, but also provide excellent shelter away from any potential natural enemies and adverse weather.

Rodents can cause a variety of problems in a food store, and these are generally applicable to all food store situations, from small-scale, on-farm storage to large-scale silos or warehouses. Direct food consumption by rodents in a store can increase considerably. On average, they consume about 10% of their body weight per day. There have been various attempts to measure food lost to rodents under real conditions (LaVoie *et al.* 1987; Ahmad *et al.* 1995), but this has been difficult to accomplish. Measuring the actual consumption of a rodent population in a food store is difficult. Measuring food loss is arduous and prone to large error because rodents can actually physically remove entire grains, and loss assessment techniques developed for measuring insect damage are inappropriate. However, the impact for the small-scale farmer is likely to be most significant as a greater proportion of food will be lost.

Rodents damage and contaminate much greater quantities of food than they actually consume, and these are more important problems than the direct loss of food. The injury is caused mainly by urine and droppings which can spread disease, and by rodent hairs. As rodents are known carriers of many gastroenteric diseases such as leptospirosis and toxoplasmosis (Gratz 1988), grain contamination by rodents is a particular concern (Gorham 1989). Contamination can make food unfit for human consumption or lead to it being downgraded in quality. Methods for measuring mouse contamination are described by LaVoie *et al.* (1987).

Extensive damage to the store structure, requiring an increase in maintenance and expense, can result from rodent action. The overall impact of these pests upon the storage environment is particularly severe in financial terms, in that farmers have already heavily invested in seed, fertilisers, pesticides, equipment and labour to get the crop to the point of harvest. The harvested crop is, therefore, the most valuable stage of the agricultural production system.

### *Biology and behaviour*

Understanding rodent biology and behaviour is crucial to achieving effective control. While a great deal is known about the physiology, toxicology and biochemistry of the Norway rat (*Rattus norvegicus*), the majority of this work has been undertaken on the domestic and laboratory forms of this species and comparatively little on the wild forms (Berdoy & Macdonald 1991). Even less is known about the house mouse (*Mus musculus*), and less still the ship rat (*Rattus rattus*) and the other species of rodent and their biology and control in relation to the storage environment.

Nonetheless, it is possible, using what is known of the three main commensal species, to extrapolate and construct a relatively sound management and control strategy for use in food stores for all commensal rodents.

The success of rodents, particularly the commensal rodents as pests, depends largely upon three main characteristics: physical capability, reproductive potential and behaviour. It is necessary to have as detailed an understanding as possible of these characteristics if an effective management strategy is to be developed.

#### *Physical capability*

It is essential to appreciate the extraordinary ability that rodents have, particularly the commensal species, to colonise and exploit new environments as a consequence of their physical capabilities. Stores are particularly susceptible, not only because they provide often limitless food and harbourage, but also because they are usually situated within or adjacent to areas where human habitation is present and, consequently, the commensal rodents.

#### *Movement*

On farmland in England, *R. norvegicus* have been shown to travel up to 3.5 km in a single night with average distances travelled by males of some 700–800 m and females some 400–500 m (Taylor & Quy 1978). Much of this activity was within the home ranges of the animals concerned, but nevertheless gives some idea as to the vulnerability to invasion of buildings within foraging distances of neighbouring infestations. *Rattus rattus* and *M. musculus* can move some hundreds of metres in a single day if conditions are suitable (Meehan 1984).

It is also important to remember that a major source of infestation of storage facilities comes from introductions with imported commodities. The smaller and less conspicuous the rodent, the more likely it is to be imported. In this case the rodents are passively introduced into the store and do not have to find their own way in.

#### *Climbing*

Almost all rodent species are extremely agile and are capable of climbing relatively smooth surfaces. Walls and pipes can be traversed with some ease. *Rattus rattus* is by far the most successful of the main commensals as far as its climbing ability is concerned, but locally abundant species such as *R. argentiventer* and *R. exulans* are also specialist climbing species. Not only does this climbing ability enable the rodents to gain access to storage facilities, but also means they will frequently colonise the upper areas of the store in preference to the lower areas. *Rattus rattus*, for instance, will spend most of its time in the higher areas of a store and very little time at ground level (Meehan 1984). This is frequently overlooked when applying control measures, which are often and most easily applied to the ground floor areas.

Rodent agility also enables them to walk along thin wires and ropes. This route is frequently used as a means of travelling from one building to another along telephone or electricity cables or to and from ships along mooring ropes.

*Digging*

All the rodent species are capable of digging and tunnelling. Some, like *R. norvegicus*, are largely terrestrial and live in underground burrow systems. Their digging ability is particularly good, but the other species are also capable of such behaviour when the situation demands. *Mus musculus*, principally seen as an indoor animal in wetter and colder climates, is capable of producing extensive burrow systems out of doors, in warmer and drier regions. *Rattus rattus* is capable of similar behaviour although it prefers to live off the ground if a three-dimensional habitat is available. *Rattus norvegicus* is also able to live in sewage and drainage systems and will often find its way into buildings and stores through such a route. It can dig its way out of drainage systems when faults occur in the system.

*Gnawing*

The highly specialised incisors possessed by all rodents are capable of gnawing through a very wide range of materials that may seem relatively resistant to their attack. All materials that are softer than the enamel on the outside of rodent teeth (5.5 on the Mohr hardness scale), lead, aluminium, tin and many metals, are all susceptible to rodent gnawing. The availability of edges or other irregularities against which the rodent can obtain a purchase for gnawing should be avoided when proofing against rodent access.

*Jumping*

Rodents are able to jump significant distances both vertically and horizontally. It is generally accepted that *R. norvegicus* can jump up to 75 cm vertically and up to 120 cm horizontally. Adult *M. musculus* can leap 24 cm vertically and *R. rattus* can jump 150 cm diagonally upwards (Meehan 1984).

*Squeezing*

While some rodents may look fairly robust, they are in fact capable of much body manipulation and are thus capable of squeezing their bodies through what look like impossibly small gaps. Any gap larger than 5 mm can allow access to at least very young rodents. It is also important to remember that if a rodent finds a gap of less than 5 mm, but wishes to gain access through it, it can relatively easily increase the size of the gap through gnawing.

*Swimming*

Water is not necessarily a barrier to the movement of rodents. They can be good swimmers and infestations of *R. norvegicus* are frequently found beside water, such as river banks, canal sides and sewers (Meehan 1984).

*Reproductive capability*

Rodents have very high mortality rates, but reproduce so rapidly, when conditions are favourable, that there are always sufficient individuals to maintain a continuing population. In addition, the production of large numbers of offspring encourages emigration and the colonisation of new habitats. This survival strategy is a key factor in the success of rodents. In pest management terms, rodents have evolved to cope with high mortality rates and efforts at control are simply an additional mortality factor.

It is noteworthy how many young are produced per litter and the short gestation and weaning times (see Table 4.4). Furthermore, unlike many other larger mammals, a female, having had one litter, can come into breeding condition again (oestrus) almost immediately,

**Table 4.4** Some reproductive characteristics of the commensal rodents.

| | *Rattus norvegicus* | *Rattus rattus* | *Mus musculus* |
|---|---|---|---|
| Life expectancy (months) | 10–12 | 10–12 | 10 |
| Gestation period (days) | 22 | 20–22 | 19–21 |
| Young per litter | 8–12 | 4–8 | 4–7 |
| Litters per year | 4–7 | 4–6 | 6–8 |
| Weaning time (days) | 19–22 | 19–22 | 18–21 |
| Average age at maturity (weeks) | 8–12 | 8–12 | 7–10 |

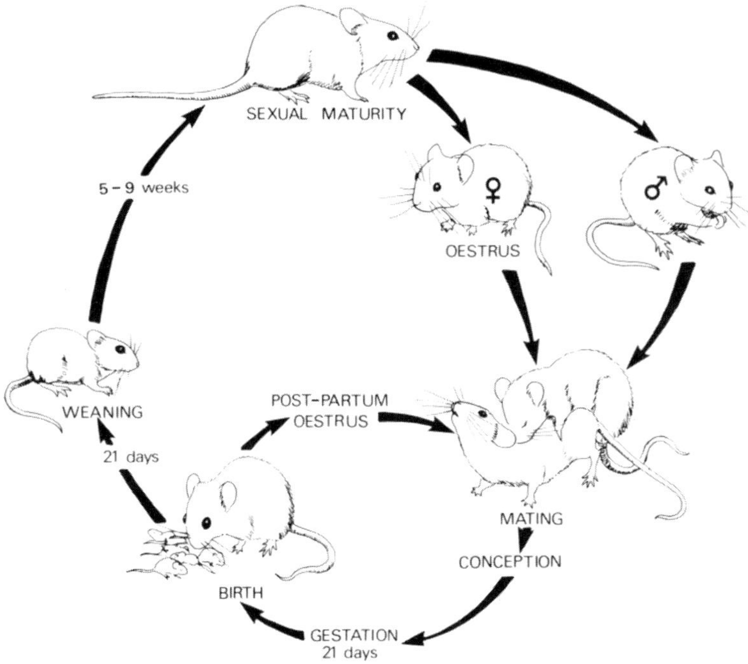


**Fig. 4.36** Typical life cycle of rats and mice. Source: Meehan 1984 (courtesy of Rentokil plc).

often within 24 h (see Fig. 4.36). Thus the potential exists for litters to be produced every 20–25 days. In storage environments, where there may be unlimited access to food, a normally restricting commodity, this reproductive potential is most likely to be met. Combine this with the rapid development of mature and breeding animals and it is clear that there is significant potential for the rapid development of sizeable rodent populations.

The average lifespan for many of these commensal rodents is estimated to be about one year (Meehan 1984). In such a case a female, maturing 10 weeks after birth and with a gestation period of 23 days, can have as many as ten litters a year under ideal conditions. With an average litter size of eight young she alone is capable of producing 80 direct offspring during her short lifetime, and her own offspring are capable of breeding after 10 weeks. These high reproductive rates set the objectives for control operations. It is clear that complete eradication of infestations, with 100% mortality, must be the objective if cost-effective control is to be achieved. Failure to do this will result in rapid return of the population to pre-control numbers. For example, the achievement of 80% mortality in a food store following extensive control operations may seem highly successful. However, 20% of the population will survive. With a sex ratio of 1:1 this leaves 10% females. Each female is capable of coming into breeding condition almost immediately and of having a litter of about eight young about 3 weeks later. Thus, within about 21 days of achieving 80% mortality, an original population of 100 rodents may have returned to its original size.

Pest management objectives are thus set by this high reproductive rate, to obtain 100% mortality and to combine the control with effective environmental management, which reduces the suitability of the environment for the rodents. This will reduce the breeding potential of the population and the carrying capacity of the environment. The population cannot then return to its original size. These additional elements will involve time and expense but provide the best opportunity to achieve lasting and cost-effective control.

### *Behaviour related to control operations*

An understanding of those aspects of behaviour that relate directly to control and management is essential if cost-effective control is to be achieved.

### *Nocturnal foraging*

All the commensal rodent species are largely active during the night, although some diurnal activity is present in most populations. This is relevant to control only in so far as it means that not seeing rodent activity during the day does not mean that there is not a significant rodent population present. The assessment of the size and activity patterns of rodent infestations is not dependent on seeing the rodents, but on developing the skills to assess the infestation from the tracks and traces that they leave behind. Thus it is essential to develop surveying skills as an essential part of a control strategy.

### *Neophobia*

Neophobia is the avoidance or fear of new objects. In commensal rodent populations the full range of neophobic responses can be identified, both within species and between them, the most obvious being in *R. norvegicus*. In the late 1940s neophobia was identified as a major constraint to achieving effective control (Chitty & Southern 1954) and the technique of prebaiting was identified as one method of overcoming this problem. It has since become clear that even within *R. norvegicus* the responses to new objects varies. Seemingly this is related to a number of variables, the most important being the stability of the rats' environment (Shorten 1954; Cowan & Barnett 1975; Quy *et al.* 1992). For example, *R. norvegicus* living on a rubbish tip, where the environment may change on a daily basis, are likely to exhibit a much lower (if any) level of neophobia than those living around a grain store where activity and change may be minimal.

This neophobic response has been used to explain differing levels of efficacy of anticoagulant baits, both in terms of rat reluctance to feed on the baits themselves, and on their reluctance to enter bait containers with which they are not familiar (Taylor & Quy 1978). More recent research indicates that the *R. norvegicus* may also be very unwilling to spend any length of time within an unfamiliar container, again affecting the amount of rodenticide consumed (R.J. Quy, pers. comm.). Thus these responses in *R. norvegicus* affect the consumption and the efficacy of rodenticide baits. In addition, research on *R. norvegicus* feeding behaviour suggests that they tend to feed from a relatively few (three or four) feeding points during the night (Meehan 1984). Hence the establishment of familiar feeding points at which the rats are able to consume relatively large amounts of any rodenticide bait is more likely to achieve effective control.

*Mus musculus*, in contrast, shows low levels of neophobia and investigates new objects in its environment. However, mice appear to become bored very quickly at any one feeding point and move on to another, thus consuming very little from any one site. Typically, a mouse may feed at 30 different points in a night, consuming a small quantity at each (Meehan 1984). More recent work, however, indicates that, as with *R. norvegicus*, there may be more variation from this norm than is generally assumed and that some wild *M. musculus* infestations exhibit strong neophobia. *Rattus rattus* is far less well researched and again its response is likely to vary, but to lie between *R. norvegicus* and *M. musculus* in its neophobic response.

In stores, which are generally very stable environments, it is essential when baiting for rodents to take into account both the concept of neophobia and the variability between and within species. The setting of large numbers (possibly every 2 m) of bait points for *M. musculus* is recommended, with far fewer but more strategically placed bait points for rats. The most effective strategy for rat control might be to place baiting containers prior to the development of infestation, so that they are there when the rats arrive and do not initiate a neophobic response, although the subsequent placement of bait may do so.

### *Hoarding and food removal*

A characteristic shared by almost all rodent species, and certainly the commensals, is their habit of removing food from where they find it and taking it to somewhere where they can feed on it more safely. This behaviour is often confused with hoarding though it may be a part of hoarding behaviour. In many instances, however, it simply reflects the determination of a rodent to protect the food from being eaten or taken by other rats. Thus *R. norvegicus* individuals under mild attack by other rats have been observed hoarding food in burrows (Calhoun 1949). *Bandicota bengalensis* in paddy fields in West Bengal has been reported hoarding at least 5.7% of the total rice production (Roy 1974). Hoarding behaviour has a significance for control operations, in that removal of rodenticides by rodents and their subsequent placement in another area has the potential both to contaminate food in the store and to provide access to the rodenticide to non-target species.

### *Territoriality*

*Rattus norvegicus, R. rattus* and *M. musculus* (although far less information is available for *R. rattus*) live in small social or family groups (Meehan 1984). *Rattus norvegicus* live in groups of about 8–14, in which age is the primary determinant of dominance (Macdonald

& Fenn 1994). *Mus musculus* live in slightly smaller groups of about 5–8 in which a dominant male attempts to control the group. Each group will try to control the territory in which it resides and will exclude intruders from outside the group. However, dominant male *M. musculus* will attempt to encourage stray females into their own territory. Individuals, usually males, who do not conform or who attempt to compete with these dominants are usually excluded from the group. These territories are maintained by regular patrols and by scent-marking with urine, faeces and secretions from scent glands. Territorial boundaries are probably clear to other members of the same species, but not to those who undertake pest control. The only outward sign of territory marking is found in *M. musculus*, who may place their urine in exactly the same spot every time, forming what are known as 'urination pillars'. These pillars may have very significant functions in the maintenance of territories. With this in mind it is clearly essential to identify all areas where rodents may be active in and around a storage facility before applying baits and other control techniques. In this way, a thorough application of control measures should ensure that all territories are adequately covered. Thorough mapping of activity assists in this assessment.

#### *Communication*

Communication within and between rodents is largely mediated through smell and ultrasonic sound. Individual animals recognise each other by their smell, which communicates information relating to sex, dominance, age and breeding status. Pheromones are also active in the broadcast of alarm signals, defence and social organisation (Stoddart 1988). They also communicate through ultrasound. While responding to frequencies within the human hearing range (below 20 kHz), rodents also communicate at frequencies up to 60–70 kHz (Smith 1994). There may not be a great deal of sophistication associated with these sounds, but basic signals denoting fear and wellbeing are transmitted. The ability to communicate ultrasonically is reinforced by very acute hearing. Rodents have a sensitivity of hearing which is some 3–4 times greater than that of humans (Meehan 1984).

#### *Taste and smell*

Rodents have developed very acute senses of taste and smell, reinforcing their ability to recognise objects in their territory. They are also able to relate such tastes and smells with their own experiences. For instance, if a food has caused pain or discomfort soon after its consumption, the same taste or smell will be avoided in the future. This is called induced avoidance or bait shyness and can lead to shyness to foods, baits, poisons and other control techniques or experiences which may have caused discomfort, pain, shock or fright (Macdonald & Fenn 1994). The ability of rodents to avoid experiences which they have not enjoyed but have nevertheless survived strengthens their learning ability; this leads to a learned avoidance of many active techniques that are used in rodent control, particularly the acute rodenticides and killing or break-back traps.

#### *Tracks and traces*

While rodents infest a particular area they will inevitably leave behind traces of their activity. These can and should be used during a survey, in determining whether or not an area is infested and the extent and severity of infestation.

The types of evidence that rodents leave behind include:

- smears – greasy smears left behind by the rodent fur as it rubs against surfaces;
- runs – compacted earthen runways formed by continual rodent activity along a particular route;
- faecal droppings;
- damaged goods and structures – materials damaged or partially eaten by rodents;
- urination pillars – caused by *M. musculus* urinating regularly on a particular spot;
- smell – different rodent species generate their own particular smell when infesting closed areas.

Many of these signs and traces are species-specific in their characteristics and enable identification of the species involved. A thorough and practical understanding of rodent biology and behaviour will lead to the more effective application of control techniques.

## Birds

### *Which birds cause problems?*

There are relatively few cosmopolitan birds that regularly attack stored grain (Smith 1995). However, virtually any granivorous bird has the potential to become a stored grain pest. Depending on the store conditions and environment, it is possible to find dozens of bird species exploiting a stockpile of grain. Generally, the more open the access to the stored grain, the greater the bird problems and number of bird species that will be found. Bird pests would normally be described as store invaders as they may only reside for very short periods of time and/or migrate to and from the store. Some species of granivorous bird are capable of exploiting grain stores; their biological characteristics are described below. Bird species that are commonly considered to be potential post-harvest pests are usually those that can roost or perch in the store structure and belong to the groups containing pigeons and sparrows.

The Order Passiformes is the largest order of birds, comprising more than 5000 species. Also known as the perching birds, the order contains many important grain pest species including crows (*Corvus* spp), quelea (*Quelea* spp), starlings (Sturnidae), finches (Ploceidae) and sparrows (*Passer* spp). Of these, the sparrows, particularly the house sparrow (*Passer domesticus*), is often cited as a pest of stored grain. The relatively small size of *P. domesticus* means that it can gain access to the store through small holes in the building. *Passer domesticus* can breed rapidly, producing 3–5 young 2–3 times per year. This species tends to live and feed in large flocks and is found throughout the world. Like the pigeon, the house sparrow is often associated with urban areas and will nest in man-made structures. The starlings, such as *Sturnus* spp and *Lamprotornis* spp, can also be serious post-harvest pests in both temperate and tropical areas. Other granivorous bird pests of Africa are described by Allan (1996) and of Asia by Brooks and Ahmad (1990). However, many of these species do not have a reputation for attacking food in store.

The Order Columbiformes comprises the pigeons and doves. The major pigeon species, *Columba livia* (the feral or rock pigeon), is found throughout temperate and tropical areas. It is has been cited as one of the most common and widespread bird pests (Shuyler 1970; Meyer 1989). *Columba livia* has been domesticated for food (China and parts of Africa) and sport (racing pigeons in the UK). Domesticated stock may become feral, leading to pest populations that have little fear of humans. Along with *Columba* species, various doves (*Streptopelia* spp) can also damage stored grain, particularly in the tropics. Both pigeons and doves are often associated with human habitation and urban areas. They can live and feed in flocks or as individuals, foraging over distances of up to 8 km.

#### *What post-harvest problems do birds cause?*

Large pests such as birds are often perceived to be more of problem than they are. Birds, however, can cause a range of problems, similar to those described for insect and rodent pests. Birds are particularly effective in causing direct losses to grain through spillage and consumption. When commodity is stored in bags, pecking birds can rip sacks open, causing major spillage. Birds are, therefore, less of a problem for bulk-stored commodities. Adult pigeons consume about 35 g of grain per day, and 20 pigeons eat about as much as a single adult human. A large store that supports 1000 pigeons may lose enough grain through direct consumption to feed 50 people. As with rodent pests in store, loss estimates for damage caused by birds are difficult to gauge, and methodologies for estimating losses caused by birds are lacking.

Economically, contamination is likely to be a more serious problem than direct grain loss. Birds contaminate stored food and spread pathogens. Because the main stored-grain bird pests such as pigeons and sparrows often live in flocks, diseases can easily spread among individuals. The defecation of birds within the store not only produces visible contamination but can lead to high levels of zoonoses such as salmonella infection. It is not known to what extent fungi, insects or mites are carried on the feet or feathers of birds. However, the mobility of birds makes it likely that cross-contamination among stores is a possibility. Birds nesting within the store are likely to increase the risk of store contamination.

#### *Biology and behaviour*

Birds can be successful pests of the storage environment. This is not because they have evolved in such environments, but because they happen to be present in the area in which the storage facility has been constructed. They are then able to take advantage of the food and shelter that the store provides. The availability of these resources, often in relatively unlimited quantity, leads to increased reproductive success of those species that can utilise the facility, and/or the attraction of birds from some distance away on a regular basis. Both scenarios can lead to increased exploitation of the resource and hence damage, contamination and loss.

The birds that are most likely to cause such problems are either granivorous or omnivorous. It is important, however, that consideration of any bird problem starts with identification of the species. Specific knowledge of the ecology of the bird will assist in the development of management programmes. In general, those species that are most likely to cause serious problems have a number of common characteristics.

*Mobility*

Perhaps the single greatest characteristic that separates birds from the other vertebrate pests of the post-harvest environment is their mobility. Most other pests are relatively immobile and so the problems they cause can usually be solved in and around the store since they do not migrate large distances as a part of their daily activity. In contrast, birds may establish themselves around the store, but are also capable of migrating many miles to and from the store, as and when it suits them. This is usually part of a learned exploitation behaviour.

In addition, birds' mobility allows them to exploit available niches more rapidly as vacancies become available. Thus, the killing of birds as a part of a control programme leaves niches available that are no longer occupied. Their high mobility means that these vacant niches can become reoccupied more rapidly than would be the case with less mobile species. Rapid immigration is often a reason for the failure of culling strategies. The mobility of birds, whose roosting and breeding sites might be some distance from the store, reduces access to them for management purposes.

The use of any toxicant is also complicated by this mobility. The dangers of feeding a poisonous or toxic substance to birds, which then fly away before the toxicant has had its effect, means that birds may die at sites well away from the storage facility. The hazards associated with secondary poisoning are clear in these circumstances and must be avoided.

The ability to fly not only provides birds access to almost all areas of a food store, such as the tops of stacks for feeding and ledges and beams for roosting, but also enables them to get into the store through holes, gaps and spaces in the upper structure. Any exclusion strategy must take account of this three-dimensional behaviour.

*Reproduction*

Those bird pests which most often cause problems at stores are also, like other pests of food stores, effective at reproducing rapidly and can, more importantly, increase their breeding rates in response to the availability of normally limiting factors such as food availability. For example, *C. livia* are able to increase the numbers of broods they have per year in response to increased food availability. If this were not the case they would be less likely to build up sizeable populations so rapidly and cause problems.

*Learning ability*

Birds learn very quickly. If a new resource becomes available in an area, birds will find it rapidly because of their mobility and will then remember where it is and return to it. The more reliable and predictable the availability of a resource, such as food, the more the birds will return to it.

*Additional aspects*

There are additional characteristics that make some bird species good pests of food stores. They are often small. *Passer domesticus* (house sparrows) and *Sturnus vulgaris* (starlings) can pass through very small holes and are therefore difficult to exclude from buildings.

Natural predators, such as birds of prey, do not frequent food stores because the stores are in artificial, man-made environments, which they find unsuitable. Predators are in any case unlikely to be able to control birds to any degree in these artificial environments. Stores are often situated in or around towns and other urban areas. This provides a ready source of bird pests who simply migrate from the surrounding area to take advantage of the food available.

Birds are diurnal and thus active during the day, at the same time as the food store is open and working. They are thus able to take immediate advantage of any spillages that occur and enter the building through open doorways.

The size of the bird population is directly related to the food available. The size of the food source is the most important factor controlling populations. Thus control of this food is the primary concern in managing a bird infestation.

## Pests of durable crops – moulds

Fungi are plant-like organisms that lack chlorophyll and are therefore unable to produce their own food by photosynthesis. They grow by breaking down complex substrates such as carbohydrates, proteins and lipids by enzymatic hydrolysis, absorbing the simpler compounds through their cell walls (Gravesen *et al.* 1995).

They may be single-celled organisms (for example, yeasts) or possess tubular filaments called hyphae, which form the main body of the fungus in the so-called filamentous fungi. They reproduce by forming spores, often in enormous numbers. Since fungi lack chlorophyll they are referred to as saprobes, living on dead or decaying tissue, or parasites of plants and animals, actively invading living tissue and causing disease. They are extremely important as agents of plant disease and in the decay of food, fabrics and timber. Fungi are a very large group with over 70 000 known species worldwide (Gravesen *et al.* 1995), though this number may represent only a third of the total species (Hawksworth & Kirksop 1987).

Filamentous fungi can take a range of forms. The larger fungi (macro-fungi) can be identified with the naked eye by their fruiting bodies – mushrooms or toadstools. However, the fungi that are of concern in crop storage are the moulds that do not develop large fruiting bodies and often require a microscope for identification.

Some fungi are beneficial to humans and some harmful, depending on the result of their growth on food. Beneficial activities include many of the food fermentations (for example, bread, beer, wine, involving yeasts) and fermentations of cassava and other roots, of rice, soy, grams, maize and fish, involving yeasts and/or filamentous fungi. The first antibiotics were isolated from fungi, for example, penicillin from *Penicillium chrysogenum* Thom, discovered by Alexander Fleming in 1929.

Harmful activities include pathogenic attack on plants or animals, causing disease, or food spoilage, including the loss of germination, discoloration, caking, mouldy smells, and the production of toxins. Saprobic fungi can grow on or in decaying tissue. Their role in putrefaction, fermentation and as agents of disease was largely established during the nineteenth century. Some fungi, the parasites, can only grow on other living organisms.

The lifespan of different fungi is very variable; in the wild, some fungal colonies have been growing for several hundred years (Cooke 1979). Some fungal spores can survive over 30 years in herbarium collections, over 15 years in grains, and 10 years in field plots (Sussmann 1968). Viable sclerotia of *Sclerotium cepivorum*, the pathogen causing white rot of onions, have been recovered from field plots 20 years after infection (Coley-Smith *et al.* 1990). Anecdotal evidence (J.R. Coley-Smith, pers. comm.) suggests that sclerotia of this pathogen could survive for up to 50 years, in the absence of the host plant. Wilhelm (1955) noted that microsclerotia of *Verticillium alboatrum* were still viable 13 years after burial in field plots. Mycelium may survive in infected crop debris for 10–60 years, particularly in root debris (Gaumann 1950; Chang 1996). These long survival times have important implications in the management of fungal damage.

Moulds are more important spoilage agents than bacteria within stored durables, because they are adapted to grow at lower water activities and are better able to penetrate and grow within the grain.

Many fungi can grow over a wide range of environmental conditions. For example, some can grow below –10°C, others to at least 60°C. Many of the fungi that grow on stored grain flourish most rapidly between 25 and 35°C. This can mean that fungal growth is particularly fast in the tropics. As fungi grow, they respire, producing heat and free water, often raising the moisture content and temperature of the grain, accelerating spoilage. Others can grow in fresh water, in saturated salt solutions, and still others in atmospheres of greater than 60% carbon dioxide and less than 0.5% oxygen. Many important storage fungi are able to grow in partially dried crops (and so are termed xerophilic). Some fungi are able to survive periods of freezing or drying. Fungi can be isolated on crops before and after harvest, from soil, water, and from the air. Many fungi have a worldwide distribution: *Aspergillus flavus,* for example, can be found throughout Africa, Asia, North and South America, and Western and Eastern Europe; others are restricted to localised areas.

### Classification of fungi

For practical purposes it is useful to classify fungi associated with food plants into those that cause spoilage before harvest (field fungi) and those that grow on plants after harvest, the storage fungi. Some fungi can grow both before and after harvest. So-called storage fungi often contaminate crops in the field at very low levels. They only start to grow and reach significant levels in store when the moisture level drops. Table 4.5 gives examples of common field and storage fungi.

#### *Field fungi*

These invade the grains before harvest and require water activity values ($a_w$) for growth greater than 0.93 (equivalent to a moisture content of 22–25% wet weight for most cereals). They cause blemishes, blights and discolorations. Some can also cause diseases in plants that are grown from infected seeds. The dormant mycelium of

**Table 4.5** Selected genera of important field and storage fungi.

| Field fungi | Storage fungi |
|---|---|
| *Fusarium* | *Eurotium* |
| *Alternaria* | *Aspergillus* |
| *Cladosporium* | *Penicillium* |
| *Phoma* | *Emericella* |
| *Colletotrichum* | *Paecilomyces* |
| *Drechslera* | *Wallemia* |
| *Curvularia* | *Xeromyces* |
| *Aspergillus* (some species) | *Rhizopus* |
| *Bipolaris* | *Mucor* |
| *Penicillium* (some species) | |

field fungi may survive for years in dry grain but most die rapidly in seeds held at moisture contents in equilibrium with relative humidities of 75–90% (m.c. 15–19%).

### *Storage fungi*

These fungi invade the grains after harvest during drying, and in storage. Generally they do not invade seeds to any extent before harvest although their spores may be found on the outsides of the grain. One exception is the growth of the storage mould *A. flavus* in maize cobs in the field. This is of particular importance as this mould is capable of producing the mycotoxin, aflatoxin, under these conditions. These fungi are able to grow under conditions of reduced $a_w$, some down to an $a_w$ of 0.70 (m.c. 14%).

Fungi belong to their own kingdom, that is, they are neither plants nor animals. This kingdom has five subkingdoms of which three contain important food spoilage fungi. An example of the classification of a fungus in each of the three subkingdoms is shown in Table 4.6. Some of the most important fungi in affecting food storage are listed in Table 4.7 along with the most important field recognition features for these fungi.

### Pre-harvest disease and rots

Pre-harvest disease of durable crops is often initiated after in-field damage has occurred, for example by insects, birds, rodents, cultivation practices or adverse weather. Thus, infection of maize by *Fusarium moniliforme*, *Fusarium graminearum* and *Stenocarpella* (*Diplodia*) *maydis* is often associated with attack by the European corn borer (Michaelson 1957). Cultivar differences exist for many of the pre-harvest factors, including pest resistance, ability to withstand drought, stalk strength and sheath length.

Pre-harvest disease often has post-harvest consequences. Fungal inocula can continue to grow if harvesting and storage operations do not reduce the moisture content sufficiently rapidly to prevent further growth. Damage caused by pre-harvest disease may allow ingress of storage fungi. Mycotoxins produced pre-harvest can survive in the grain in-store, contributing to a loss of food safety and quality (see below). *Fusarium* toxins are generally produced pre-harvest, with limited production post-harvest if the crop remains very moist, for example in the case of crib-dried North American maize where the crop is being stored above 22–25% moisture content. The crop dies slowly in the crib over the autumn period. *Aspergillus flavus* can produce aflatoxin in the field, both in maize and in groundnuts.

Various types of ear rots, blemishes, blights and discolorations occur on cereals and other durables. Some of these pre-harvest blights can also cause reduced germination and diseases in plants that are grown from infected seeds; for example *Fusarium culmorum*, *A. flavus*, *A. niger*, *Penicillium* spp, *Lasiodiplodia* (*Botryodiplodia*) spp, *Tilletia* spp, *Ustilago* spp and *Bipolaris sorokiniana* (= *Cochliobolus sativus*) (Warham *et al.* 2000).

Cereal crops grown predominantly in temperate zones, such as wheat, barley and oats, tend to be infected with *Alternaria alternata* and *Alternaria* (*Pleospora*)

**Table 4.6** Classification of three important spoilage fungi.

| Kingdom | Fungus | Fungus | Fungus |
|---|---|---|---|
| Subkingdom | Zygomycotina | Ascomycotina | Deuteromycotina |
| Class | Zygomycetes | Plectomycetes | Hyphomycetes |
| Order | Mucorales | Eurotiales | Hyphomycetales |
| Family | Mucoraceae | Trichocomaceae | Anamorphic Trichocomaceae |
| Genus | *Rhizopus* | *Eurotium* | *Aspergillus* |
| Species | *oryzae* | *amstelodami* | *flavus* |

**Table 4.7** Features of the three important subkingdoms of fungi associated with food storage.

| Fungus | Growth rate on artificial media | Mycelium | Asexual reproduction | Sexual reproduction | Water tolerance | Production of mycotoxins |
|---|---|---|---|---|---|---|
| Zygomycotina *Mucor plumbeus* Bonord, *Rhizopus stolonifer* (Ehrenb.) Vuill., *Rhizopus oryzae* Went & Prins. Geerl, *Absidia corymbifera* (Cohn) Sacc. & Trotter | Very fast (up to 45 mm a day) | Mycelia at the growing edge of the colony lack cross-walls (non-septate) and are relatively wide which means cellular contents can move quickly from one part of the mycelium to the other | By sporangiospores produced in a sporangium found on the end of fast-growing, long hyphae. Sporangiospores are genetically identical to the mother cells | Sexual spores rarely produced by Zygomycotina species associated with food | Not tolerant of dry conditions (not xerophilic) | None known |
| Ascomycotina *Eupenicillium, Byssochlamys, Eurotium* | Fast (up to 20 mm a day) | Septate mycelium | Asexual spores (conidia) may be produced | Sexual spores (ascospores) produced within an ascus, themselves within a larger, more visible body, the ascocarp. Often highly resistant to heat and chemicals | May be xerophilic | Patulin from *Byssochlamys*; possible feed refusal factors from *Eurotium* |
| Deuteromycotina *Penicillium, Aspergillus flavus, A. niger* Lal & Ram Chandra, *A. ochraceus* G. Wilh. | Fast (up to 30 mm a day) | Septate mycelium | Asexual spores (conidia) only produced | Not produced | May be xerophilic | Numerous toxins from *Aspergillus* and *Penicillium* spp. |

*infectoria* causing leaf blights and black point of wheat, *Cladosporium herbarum* and *C. macrocarpum* causing sooty head mould of wheat, and *B. sorokiniana* causing black point, seedling blight and spot blotch of cereals. *Alternaria* and *Cladosporium* are commonly found in more than 85% of wheat grains, pre-harvest, and can be considered as an indicator of freshly harvested or well-stored grain (Flannigan 1980). *Fusarium* species cause foot rot (*F. culmorum*), pre-emergence blight (*F. graminearum*), root rots (*F. avenaceum*) and head blight of winter wheat (*F. dimerum* (= *Microdochium dimerum*) and *M. nivale*). On tropical and subtropical crops, for example, maize, rice, sorghum and millet, *Fusarium* species are again important, but with *F. moniliforme* and *F. pallidoroseum* predominating on cereals. Higher temperatures and humidities in the tropics mean that *Penicillium* and *Aspergillus* species can occur on crops pre-harvest, sometimes causing significant damage. Examples include *Aspergillus flavus* and *A. niger* ear rots of maize, with the danger of pre-harvest mycotoxin production from *A. flavus*. Ear rot infection starting in the field can carry over into store. Other major fungal pathogens of maize are *Stenocarpella* (*Diplodia*) *maydis* and *S. macrospora, Fusarium moniliforme*, *F. graminearum*, *Penicillium oxalicum* and *Penicillium chrysogenum*.

Sorghum can be infected with a wide range of fungi in the field, such as *F. pallidoroseum, F. moniliforme, Aspergillus* spp. (*A. flavus*), *Curvularia* spp. (*C. lunata* and *C. pallescens*), *Bipolaris* (*Drechslera*) spp., *Cladosporium* spp., *Nigrospora oryzae, Epicoccum* spp., *Phoma* spp. (*P. sorghina*), *Alternaria alternata* and *Alternaria longissima*, *Lasiodiplodia theobromae* and *Pleospora* spp. (Pitt & Hocking 1997). Most of these are termed grain or head moulds and can cause some yield and quality losses in the harvested crop (Tripathi 1974).

Important pre-harvest pathogens of rice include the stem pathogens *Trichoconiella* (*Alternaria*) *padwickii* (associated with stackburn), *Bipolaris maydis*, *Bipolaris oryzae* (brown spot), *Pyricularia oryzae* (blast), *Rhizoctonia solani* (sheath blight) and sheath rot (*Sarocladium oryzae*). Head moulds include *A. flavus*, *Tilletia barclayana* (kernel smut), *F. graminearum* (scab), and *N. oryzae*, *F. chlamydosporum* and *Curvularia* spp.

The major pre-harvest problems with groundnuts include infections with *A. flavus*, particularly if the crop is attacked by insects or suffers drought stress close to harvest.

Tree nuts are virtually sterile before harvest, unless insect damage occurs (Wareing *et al.* 2000). Most fungal problems occur if the crop lies on the ground for an extended period after maturity and nut fall. Brazil nuts, pecans and pistachios are particularly susceptible to infection by *A. flavus* and the production of aflatoxins.

### Post-harvest disease

Post-harvest handling practices can have direct effects on the infection of grains and growth by storage fungi. For example, grain breakage, insect damage and drying at too high a temperature may lead to cracking of the grain. As noted above, pre-harvest fungal attack can lead to the production of mycotoxins that survive in the stored crop for a considerable period of time, even after the associated fungi are no longer viable.

Slow drying or rewetting can allow fungi infecting in the field to continue to grow in store, unless or until the commodity becomes too dry to support further development. It should be noted that fungi can grow in a commodity at lower water activities than those required for germination of spores or initial infection.

If the commodity is allowed to become wet after drying, or is improperly dried, then a succession of fungi will invade the grain, dependent upon the moisture content, or more properly, the water activity (see below). Species of *Eurotium* (*E. amstelodami chevalieri*, *E. rubrum* and *E. repens*) are the most common early invaders of improperly dried grain, leading to a loss of germination and other effects noted below. A succession of *Aspergillus* and *Penicillium* species then invade, if the water activity is high enough. Species of fungi from these three genera are the most frequently isolated organisms from most stored durable crops.

The most common species from cereals are *Penicillium aurantiogriseum, Penicillium viridicatum, Penicillium verrucosum, Penicillium hordei, Aspergillus flavus, Aspergillus candidus, Aspergillus restrictus* and *Aspergillus fumigatus* (Samson *et al.* 1996). *Aspergillus flavus* is more common on subtropical and tropical grain post-harvest, though it has been recorded on temperate cereals, and occasional reports of aflatoxin contamination of wheat have been recorded (Shotwell *et al.* 1976).

Slow drying of grains and oilseeds after harvest can allow the continued growth of *A. flavus* or *A. parasiticus* and a rapid rise in aflatoxin concentration. For example, in Thailand, there can be considerable delays after harvest before maize is dried to a safe moisture content (Cutler 1991). Maize is commonly shelled above 23% moisture, resulting in grain breakage, increasing the risk of toxin formation.

In addition to the species noted above, *Penicillium islandicum*, *Penicillium citrinum* and *Penicillium citreonigrum* are important on rice post-harvest, and can produce the mycotoxins islanditoxin, citrinin and citreoviridin (Samson *et al.* 1996).

Cereals preserved by acidification (using lactic acid, propionic acid or an inorganic acid) have a different flora, with *Penicillium glandicola*, *Penicillium roqueforti*, *Aspergillus flavus*, *Aspergillus candidus* and *Aspergillus terreus* predominating.

Fungi that grow under conditions of low oxygen tension are adapted to growing in cereals stored under an airtight regime. These include *Paecilomyces variotii*, *Scopulariopsis candida*, *P. roqueforti*, *Candida* spp, *Byssochlamys fulva* and *B. nivea*. Patulin is the predominant mycotoxin expected in this case, produced by *P. variotii* and *Byssochlamys* spp. (Samson *et al.* 1996).

*Aspergillus parasiticus* is the most important post-harvest fungus on groundnuts, with about 70% of isolates taken from groundnuts able to produce aflatoxins. Copra is produced by drying the flesh of coconuts after splitting the nut in half. Sun drying or artificial drying is employed. During the drying process, infection by *A. flavus* and the production of aflatoxins is common, until the crop reaches a safe water activity (below 0.8 for aflatoxin production, below 0.7 for safe storage).

Tree nuts are invaded by an entirely different fungal flora compared with cereals; these fungi are adapted to growing in crops with a high lipid content. Fungi include *Penicillium commune*, *P. crustosum*, *P. solitum*, *P. funiculosum*, *P. oxalicum*, *P. citrinum*, *A. flavus*, *A. wentii* and *A. versicolor*. Mycotoxins include cyclopiazonic acid, citrinin, penitrem A, aflatoxin and sterigmatocystin.

## Factors affecting the growth of fungi

The main requirements for growth of storage fungi are nutrients, moisture, favourable temperature, stable atmosphere and degree of acidity (pH).

### *Nutrients*

Fungi grow on most food and plant materials. Species of *Aspergillus* and *Penicillium*, for example, can subsist on thousands of different kinds of plants, on leather, paper, wood, animal dung, ink, syrup, seeds of all kinds, manufactured cereal products and even the boxes in which they are packed. They also grow on stored grain and grain products, fruits and vegetables. Dermatophytes are an example of a group of fungi that are specialised to grow on a narrow range of food, in this case animal skin tissues (e.g. the fungus causing athlete's foot, *Trichophyton mentagrophytes*).

### *Moisture – $a_w$*

Water activity ($a_w$) and moisture content (m.c.) are two terms used to describe how wet a commodity is. The moisture content of a commodity is the percentage of the weight of that commodity that is water. The water activity is a measure of the water available for biological processes such as fungal growth and depends on the type of the commodity and its m.c. and temperature. Water activity rises by approximately 0.03 for a 10°C rise in temperature.

The water activity of a food greatly affects its susceptibility to attack by fungi and it is principally by reducing $a_w$ that drying techniques help in the prevention of spoilage of foodstuffs. Different foods in equilibrium with the same relative humidity of the atmosphere have different moisture contents because of differences in composition (see Table 4.8). Some moulds are capable of survival in products with an $a_w$ as low as 0.60, although 0.70 is the minimum $a_w$ which will sustain the growth of storage moulds. Moulds which grow best at low $a_w$ are called xerophiles, which are defined as those that can grow below $a_w$ 0.85 at a particular temperature. The more common storage moulds and the minimum $a_w$ required for their growth are given in Table 4.9.

**Table 4.8** Moisture contents of various grains and seeds in equilibrium with different relative humidities at 25–30°C.

| Relative humidity (%) | Wheat, maize, sorghum | Soybeans | Sunflower | Groundnuts |
|---|---|---|---|---|
| 65 | 12.5–13.5 | 12.5 | 8.5 | 6.2 |
| 70 | 13.5–14.5 | 13.0 | 9.5 | 7.0 |
| 75 | 14.5–15.5 | 14.0 | 10.5 | 7.9 |
| 80 | 15.5–16.5 | 16.0 | 11.5 | 9.0 |
| 85 | 18.0–18.5 | 18.0 | 13.5 | 10.5 |

| Fungal species | Minimum ERH % | Minimum $a_w$ |
|---|---|---|
| *Aspergillus halophilicus* | 68 | 0.68 |
| *Aspergillus penicillioides, Wallemia sebi* | 70 | 0.70 |
| *Eurotium* spp. | 73 | 0.73 |
| *Aspergillus candidus, A. ochraceus, A. flavus* | 80 | 0.80 |
| *Penicillium* (dependent on species) | 80–90 | 0.80–0.90 |

**Table 4.9** Minimum equilibrium relative humidity (ERH) and water activity ($a_w$) for the growth of common storage fungi at their optimum temperatures (26–30°C).

### *Temperature*

All mould growth is temperature sensitive; the optimum temperature for several species is about 30°C, a common ambient temperature in tropical regions (Table 4.10). Moulds can be classified into those able or preferring to grow at low temperatures (psychrophiles), those growing at medium temperatures (mesophiles), and those only able to grow at elevated temperatures (thermophiles). Most moulds are mesophiles. Some moulds are able to tolerate high temperatures, but are able to grow at mesophilic temperatures (thermotolerant).

A general definition of a thermophile is an organism unable to grow below 20°C, and having its maximum above 50°C. Examples include *Thermomyces lanuginosus*, which does not grow below 30°C, and can grow up to 60°C, *Thermoascus aurantiacus*, with similar growth criteria, *Thermoascus crustaceus*, unable to grow below 25°C, and able to grow up to 55°C. *Aspergillus fumigatus* is an example of a thermotolerant mould, able to grow to 55°C, but also down to 15°C. These types of moulds are able to cause heating of grain and gross spoilage of commodities.

Most mesophilic moulds have restricted growth at high temperatures but species such as *A. candidus* and *A. flavus* grow up to 45–50°C, and can increase the temperature of infected grain to 50°C and hold it there for weeks, eventually causing germ damage, mustiness, heating, caking, and burned grain.

Species of *Penicillium*, for example, *P. aurantiogriseum* and *P. expansum*, can grow slowly at temperatures down to –2°C, but require a high moisture content.

### *Atmosphere*

The gaseous environment greatly affects the growth of moulds, and the assumption, in the main, is that they are aerobic (i.e. they need free oxygen for growth). However, many of the fungi that cause the deterioration of stored grains can grow in an atmosphere containing only 0.1–0.2% oxygen or in an atmosphere containing more than 80% carbon dioxide, and a few at least are anaerobes. Thus, although airtight storage may be used to limit fungal spoilage, if the moisture content–temperature–time combination permits the fungi to grow, the grain may develop undesirable off-flavours. One of the most xerophilic fungi, *Monascus bisporus*, which has been isolated from dried prunes, currants and chocolate sauce, has been reported to tolerate extremely low levels of oxygen, be able to survive 95% carbon dioxide and grow at water activities as low as 0.65.

### *pH*

At high $a_w$ moulds can grow over a wide range of pH but in general most food-invading moulds prefer an acid pH (below pH 7) for active growth. The optimum is usually between 4.0 and 6.5, but the pH range for the actual initiation of growth, 2.0–8.5, is quite wide.

## Effects of moulds on stored commodities

### *Loss in germination*

Storage fungi preferentially invade the embryo (germ)

| Fungus | Minimum °C | Optimum °C | Maximum °C |
|---|---|---|---|
| *Aspergillus penicillioides* | 5–10 | 30–35 | 40–45 |
| *Eurotium* spp. | 0–5 | 30–35 | 40–45 |
| *A. candidus* | 10–15 | 45–50 | 50–55 |
| *A. flavus* | 10–15 | 30–35 | 45–50 |
| *Penicillium* spp. | –5–0 | 20–25 | 35–40 |

**Table 4.10** Approximate minimum, optimum and maximum temperatures for the growth of common storage fungi on grain.

**Table 4.11** The effect of *Eurotium* spp. on seed germination.

| Storage time (months) | Percentage infection with *Eurotium* spp. | Germination % |
|---|---|---|
| 0 | 40 | 50 |
| 2 | 80 | 12 |
| 4 | 100 | 0 |

of grain, consequently the first effect of fungal spoilage is a loss of seed germination. Table 4.11 shows the effects of a storage fungus on germination of grain that is colonised internally by *Eurotium* spp.

### *Deterioration in nutritional value and quality*

Preferential attack of the embryo reduces the nutritional value of the grain. The level of certain amino acids may be reduced, starches broken down to free sugars and then utilised by the fungus, and fats may be metabolised. A loss of the characteristic grain odour and flavour, and replacement with musty odours and flavours, may occur, as well as changes in the texture of the grain. Other associated changes include souring or bitterness and high fat acidity from an increase in free fatty acids.

Discoloration of the grain can occur, both from pre-harvest fungi, and storage fungi. For example, *Fusarium* can cause red streaks or an overall pink discoloration of grain, *Alternaria* and other dark fungi produce black streaks and spots on the grain surface and under the pericarp. Surface fungal growth often makes the grain appear dusty or powdery. These discolorations may result in a loss of value due to trader downgrading in the marketplace.

### *Food loss*

Continued growth of fungi within the grain can lead to eventual food loss. Grain and grain products are sometimes invaded by storage fungi to such an extent as to render them unfit for human or even animal consumption. This can be accompanied by caking of grains, so that they become matted together in large lumps. It has been estimated that between 10% and 30% of food production in the tropics is lost due to fungal damage. Examples of losses during the transit and storage of food aid grain are detailed in Table 4.12.

**Table 4.12** Food aid losses due to fungal damage.

| Commodity | Transit losses |
|---|---|
| Maize | 602 t shipped to Cameroon, 30% lost; 1982 |
| Maize | 10 000 t to Burkina Faso and Côte D'Ivoire, 35% lost; 1981 |
| Maize | 3005 t to Togo, 46% lost; 1982 |
| Wheat | 132 000 t to Ethiopia, 11% lost; 1985 |
| Wheat | 6000 t to Senegal, 4% lost; 1982 |
| Rice | 730 t to Honduras, 83 t lost, 255 t downgraded for animal feed |

### *Production of mycotoxins*

Several species of *Aspergillus* and *Penicillium* are associated with naturally occurring mycotoxins such as aflatoxins in groundnuts, Brazil nuts, sorghum and rice, and ochratoxin A in corn, oats, barley and Brazil nuts. Mycotoxins represent a hazard to human and animal health and are discussed more fully below.

### *Fungal heating*

Fungi produce heat as a by-product of metabolising a food source. Typically, the faster a fungus grows, the higher its metabolic rate, and, therefore, the more heat energy it releases into its surroundings. All fungi have minimum, maximum and optimum temperatures for growth; at the optimum, the fungus is likely to be generating the greatest amount of waste heat. The effects of fungal heating of grains and other commodities include discoloration, caking, production of odours and taints, loss of nutritional value, leading to total loss of the grain, and in some cases, the production of mycotoxins.

The two groups associated with the start of deterioration of stored grain are the *Aspergillus restrictus* group and *Eurotium* species. In grain and other dried foods with an $a_w$ less than 0.78–0.8, these are the main species that can grow. However, in dried foods with an $a_w$ above 0.8, these two species may be followed by *A. candidus*, *A. ochraceus*, *Aspergillus versicolor*, *A. flavus* and *Penicillium* species. There is a regular succession during the invasion of dried foods by fungi and all these species above have their own sharply defined lower limit of $a_w$ for growth.

Grain heating is also a continuum, often starting with the grain being slightly too wet, at an $a_w$ of 0.70–0.75, allowing the growth of extreme xerophilic fungi, which grow slowly and produce little heat, but causing a localised increase in moisture. Other, less xerophilic fungi then grow, producing more metabolic water, more heat, until the thermotolerant and thermophilic fungi are able to grow. Under these circumstances, grain heating may take many months to be noted.

If grain is stored or shipped at less suitable moisture contents, for example, 16% and above, equivalent to a water activity of 0.80 or above, then spoilage is more rapid, quickly progressing to the point where thermophiles can grow.

The eventual effects of excessive fungal growth can be spontaneous combustion of the commodity. This usually only occurs with oil-rich commodities, for example, soybeans, cottonseeds, groundnuts and sometimes maize.

### *Human disease and allergies due to handling mouldy commodities*

Handling of commodities contaminated by fungi can have health implications. The inhalation of mould spores can cause allergenic responses in susceptible individuals. These are categorised into type 1 and type 3 reactions. In the former, the immune system overreacts upon repeated and excessive exposure to the allergens, often over a period of years. Once the response is triggered, even small amounts of the antigen will elicit a response. This type of reaction is found in hay fever, asthma and various types of dermatitis. In type 3 reactions (farmers' lung, maltsters' lung, wood trimmers' disease and cheese washers' lung), an extrinsic allergic alveolitis occurs, presenting as a flu-like illness with raised temperature and aching joints within 6–8 hours of exposure to the allergen. Irreversible fibrosis of lung tissue may result (Gravesen *et al.* 1995).

### *Mycotoxins*

The first account of a disease caused by a fungal metabolite was in Europe in the tenth century. The disease arose from the consumption of rye grain contaminated with alkaloids produced by the mould *Claviceps purpurea*, within a sclerotium known as 'ergot'.

Toxic secondary metabolites, for example ergot alkaloids, which are produced by certain moulds growing on foods and feeds, are described as mycotoxins; and the diseases caused by these toxins are termed mycotoxicoses. The moulds which are currently considered to be the main producers of mycotoxins belong to the genera *Aspergillus*, *Fusarium* and *Penicillium*. A selection of the more important moulds which produce mycotoxins are shown in Table 4.13.

**Table 4.13** Important moulds and mycotoxins.

| Mould species | Mycotoxins produced |
|---|---|
| *Aspergillus parasiticus* | Aflatoxins B1, B2, G1, G2 |
| *A. flavus* | Aflatoxins B1, B2 |
| *Fusarium sporotrichioides* | T-2 toxin |
| *F. graminearum* | Deoxynivalenol, zearalenone |
| *F. moniliforme* | Fumonisins |
| *Penicillium verrucosum* | Ochratoxin A |
| *A. ochraceus* | Ochratoxin A |

Mycotoxin production depends on mould growth, and can therefore occur at any stage during processing or storage of crops provided conditions are conducive to mould growth. However, growth and toxin production are not necessarily synonymous; many factors interact. Mycotoxins may be present in a food long after the moulds responsible have died.

Mycotoxins are a diverse group of compounds. The structure of aflatoxin B1, the most important toxic metabolite, is shown in Fig. 4.37.

Mycotoxins occur in many foods and feeds and have been implicated in a range of human and animal diseases. Exposure to mycotoxins can produce both acute and chronic effects upon the central nervous, cardiovascular and pulmonary systems, and upon the alimentary tract. In serious cases death may result. Mycotoxins may be carcinogenic, mutagenic, teratogenic and immunosuppressive.

Significant economic losses are associated with the impact of mycotoxins on human health, animal productivity and both domestic and international trade. It has been estimated, for example, that annual losses in the US and Canada from mycotoxins are of the order of \$5 bn. Approximately 50 countries now regulate against the presence of mycotoxins (especially the aflatoxins) in foods and feeds.

#### *Aflatoxin*

The term aflatoxin was first used in early 1961 in the UK when the death of thousands of turkeys (Turkey X disease), ducklings and other domestic animals was attributed to the presence of toxins from *A. flavus* and *A. parasiticus* in groundnut meal imported from South America. Aflatoxins may be produced, both before

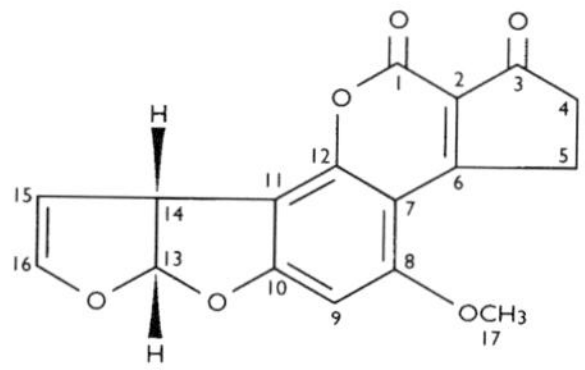

Aflatoxin B1

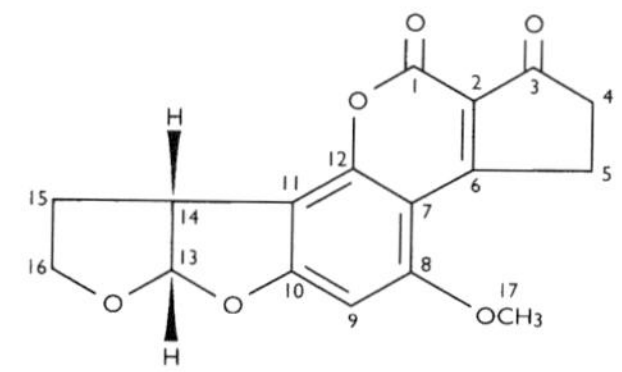

Aflatoxin B2

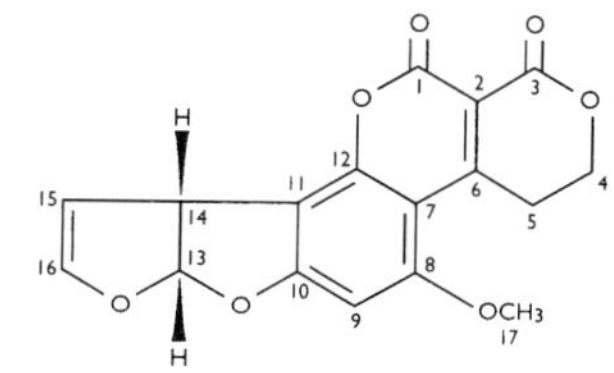

Aflatoxin G1

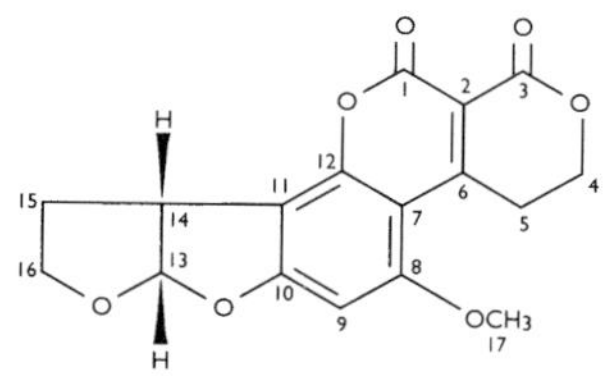

Aflatoxin G2

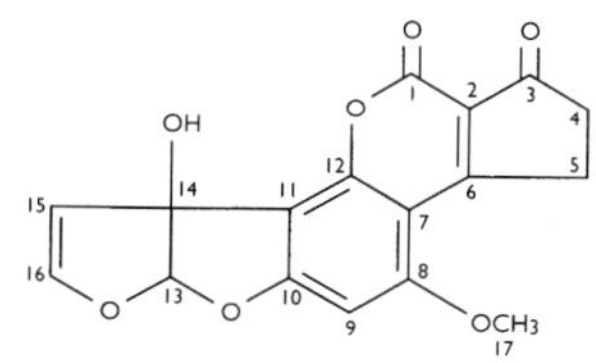

Aflatoxin M1

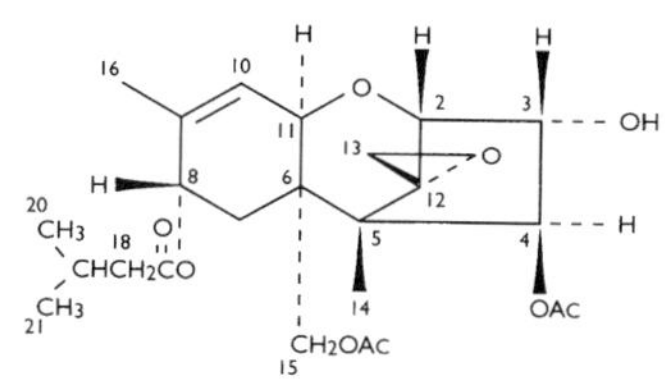

T-2 toxin

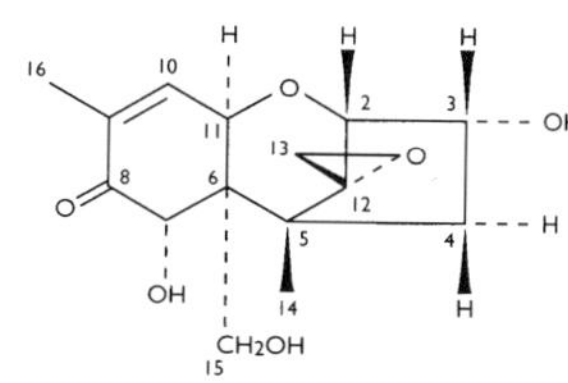

Deoxynivalenol

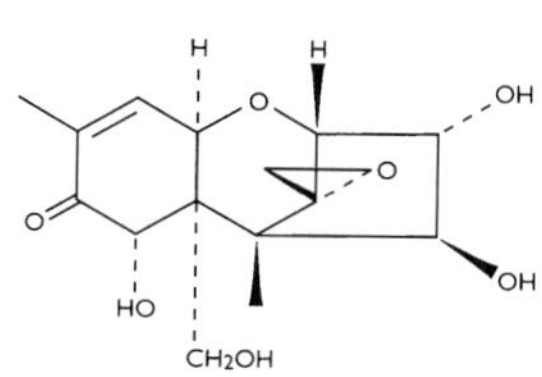

Nivalenol

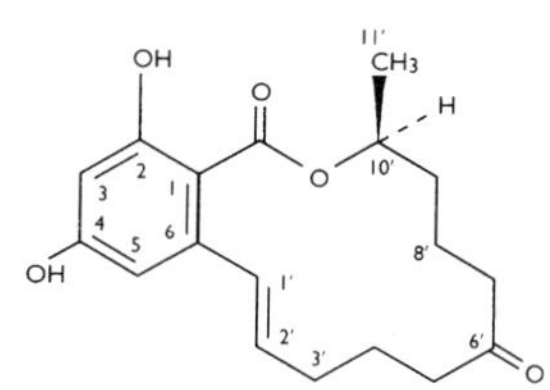

Zearalenone

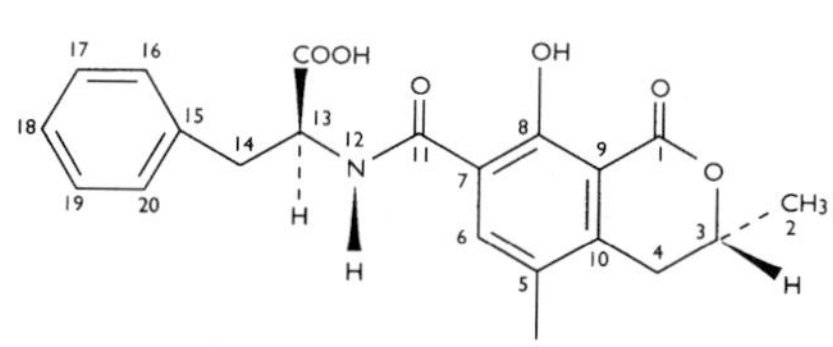

Ochratoxin A

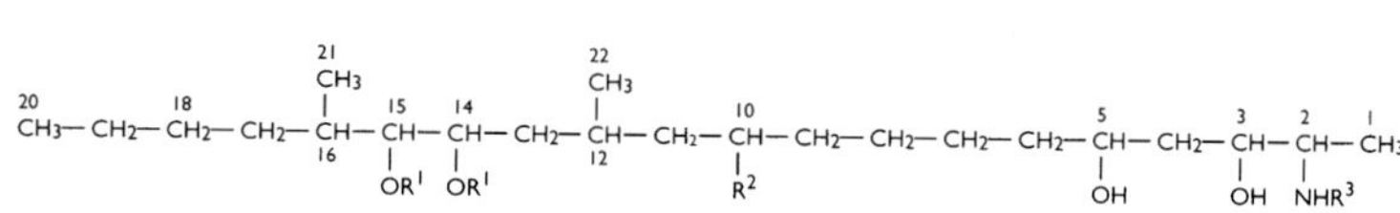

Fumonisins

A1: $R^1 = COCH_2CH(CO_2H)CH_2CO_2H; R^2 = OH; R^3 = C^{23}OC^{24}H_3$
A2: $R^1 = COCH_2CH(CO_2H)CH_2CO_2H; R^2 = H; R^3 = COCH_3$
B1: $R^1 = COCH_2CH(CO_2H)CH_2CO_2H; R^2 = OH; R^3 = H$
B2: $R^1 = COCH_2CH(CO_2H)CH_2CO_2H; R^2 = R^3 = H$

**Fig. 4.37** Structures of some mycotoxins. Source: Nicolaides *et al.* 1998.

and after harvest, on many foods and feeds, especially oilseeds, edible nuts and cereals.

Later evidence suggests that cyclopiazonic acid, another mycotoxin produced by *A. flavus*, is implicated in the development of Turkey X disease. The chronic effects of low dietary levels (parts per billion) of aflatoxin on livestock are also well documented and include decreased productivity and increased susceptibility to disease (IARC 1993).

Aflatoxin B1 is a human carcinogen and is one of the most potent agents causing cancer of the liver. Human fatalities have also occurred from acute aflatoxin poisoning when unseasonal rains and a scarcity of food have prompted the consumption of heavily contaminated maize (Krishnamachari *et al.* 1975). Aflatoxins have immunosuppressive effects in livestock; if they can also affect humans, it is possible that the aflatoxins (and other mycotoxins) could play a significant role in the development of human disease in some developing countries, where a high exposure to these toxins has been reported.

### *Trichothecenes*

Several species of *Fusarium*, including *F. chlamydosporum*, *F. graminearum* and *F. equiseti*, can produce trichothecenes. T-2 toxin is produced on cereals in many parts of the world and is particularly associated with prolonged wet weather at harvest. It is the probable cause of alimentary toxic aleukia (ATA), a disease which affected thousands of people in Siberia during World War II. T-2 toxin is responsible for haemorrhagic disease in animals and with the formation of oral lesions and neurotoxic effects in poultry. The most significant effect of T-2 toxin and other trichothecenes is immunosuppressive activity (Schiefer *et al.* 1986; Bhavanishankar *et al.* 1988).

Deoxynivalenol (DON) is probably the most widely occurring *Fusarium* mycotoxin, contaminating a variety of cereals, especially maize and wheat. Low concentrations of DON commonly occur in grains in North America, Japan and Europe, whereas higher levels may occur, intermittently, in some developing countries. Deoxynivalenol is also known as vomitoxin because of its emetic and feed refusal effects upon livestock.

The ingestion of DON has caused outbreaks of acute human mycotoxicoses in India, China and rural Japan. The Chinese outbreak, in 1984–85, was caused by mouldy maize and wheat; symptoms which occurred within 5 to 30 minutes included nausea, vomiting, abdominal pain, diarrhoea, dizziness and headache.

### *Zearalenone*

Fungi associated with zearalenone production are species of *Fusarium*, particularly *F. graminearum* and *F. equiseti*. The occurrence of zearalenone is related to low temperatures and wet weather. Maize is invaded at the silking stage; the fungus continues to grow until grains are sufficiently dried.

Zearalenone is a widely distributed oestrogenic mycotoxin occurring mainly in maize, in low concentrations, in North America, Japan and Europe. However, high concentrations can occur in developing countries, especially when maize is grown under more temperate conditions in, for example, highland regions.

Zearalenone is co-produced with deoxynivalenol by *F. graminearum* and has been implicated, with DON, in outbreaks of acute human mycotoxicoses. Exposure to zearalenone-contaminated maize has caused hyperoestrogenism in livestock, especially pigs, characterised by vulvar and mammary swelling and infertility.

### *Fumonisins*

The fumonisins are a group of recently characterised mycotoxins produced by *F. moniliforme*, a mould that occurs worldwide and is frequently found in maize. Exposure to fumonisin $B_1$ ($FB_1$) in maize causes leucoencephalomalacia (LEM) in horses and pulmonary oedema in pigs. Leucoencephalomalacia has been reported in many countries including the US, Argentina, Brazil, Egypt, South Africa and China. $FB_1$ is also toxic to the central nervous system, liver, pancreas, kidney and lung in a number of animal species. Fumonisin-containing cultures of *F. moniliforme* may also be carcinogenic, whereas there is limited evidence, in experimental animals, for the carcinogenicity of $FB_1$. The presence of the fumonisins in maize has been linked with the occurrence of human oesophageal cancer in the Transkei, southern Africa and China.

### *Ochratoxin A*

Ochratoxin A is a highly toxic metabolite that produces both liver and kidney damage as chronic toxicity studies with both poultry and swine have demonstrated. Fungi associated with ochratoxin A production are *A. ochraceus* and *P. verrucosum.*

Exposure to ochratoxin A (OA) appears to occur mainly in wheat- and barley-growing areas in temperate zones of the northern hemisphere, although it was first isolated from maize in South Africa on which *A. ochraceus* had grown. Ochratoxin A has been found in human blood samples in several European countries. It also occurs in maize, rice, oats, peas, groundnuts, beans, cowpeas and coffee; developing country origins of OA include Brazil, Chile, Egypt, Senegal, Tunisia, India and Indonesia. Ochratoxin A has been found in pig meat as a result of its transfer from animal feeds.

Ochratoxin A has been linked with the human disease, Balkan endemic nephropathy, a fatal, chronic renal disease occurring in some areas of Eastern Europe. Ochratoxin A also has immunosuppressive effects in several animal species and has been shown to be carcinogenic in experimental animals.

#### *The co-occurrence of mycotoxins*

The complex ecology of mould growth and mycotoxin production can produce mixtures of mycotoxins in foods and feeds, especially in cereals. The co-occurrence of mycotoxins can affect both the level of mycotoxin production and the toxicity of the contaminated material. The production of the aflatoxins in stored grains, for example, may be enhanced by the presence of trichothecenes, whereas naturally occurring combinations of some *Fusarium* toxins are far more toxic than the toxins presented separately to animals. Production of other toxins may be suppressed in the presence of other fungi.

#### *Monitoring for mycotoxins*

In order to reduce the occurrence of mycotoxins in oilseeds, and other foods and feeds it is essential that effective monitoring, regulatory and quality control procedures are made available. An integrated package of sampling, sample preparation, analytical and detoxification methods is required to meet this need (Coker 1998; Coker *et al.* 2000).

Worldwide, 5 ppb (5 μg/kg) is the most common maximum level of total aflatoxins accepted in foodstuffs. Consequently, it is essential that the analytical methods used for monitoring, regulatory and quality control purposes are accurate and precise at these extremely low levels of contamination (Van Egmond 1989).

Analysis of the mycotoxin content of food products may be described in terms of extraction, clean-up, quantification and confirmation steps. Quantification methods include high performance liquid chromatography (HPLC), thin layer chromatography (TLC), high performance TLC (HPTLC) and enzyme-linked immunosorbent assay (ELISA) methods. Rapid methods should be simple, robust, and cost-effective, for example, those using mini-column and immunodiagnostic procedures.

Surveillance programmes are costly in terms of time and money. Exposure data can be obtained much more conveniently by the collection and analysis of urine or blood samples for biomarkers, which reflect the levels of recent dietary exposure to mycotoxins. A recent study of a blood biomarker has demonstrated the significantly higher exposure to the aflatoxins that occurs in Gambia, Kenya and parts of China, compared with Thailand and Europe.

#### *Prevention and cure*

Prevention of mycotoxins is best achieved by controlling the fungi that produce the toxins, as already described. For example, pre-harvest contamination of groundnuts with aflatoxins is enhanced by both insufficient and excessive rainfall during critical phases of crop development. Although the climate cannot be controlled, effective drying and storage procedures, after harvest, can help prevent the production of aflatoxins in crops.

Consignments of contaminated commodity may be segregated from wholesome material by the implementation of acceptance and rejection criteria for levels of mycotoxins on batches, where acceptable levels of contamination will be specified by regulators, commercial agreements and individual customers. Segregation may also be undertaken during processing. Groundnuts, for example, may be subjected to a combination of automatic and manual sorting in order to segregate kernels which are either damaged or of abnormal appearance.

## Diseases and pests of perishable crops

For the purpose of this discussion, perishable produce refers to commodities that are liable to speedy decay. Perishable crops in their dried forms, after processing, are considered as durable products. Because of the high water content of fresh fruits and vegetables (usually more than 50%) they are more likely to be attacked by micro-organisms than by insects.

### Fungi and bacteria

Fungi and bacteria are responsible for most storage problems in perishables, either alone or in combination. Based on the number of plant species attacked, the major fungi of stored perishable produce are species of *Fusarium* (teleomorphs *Gibberella* and *Nectria*), *Phytophthora, Alternaria* (teleomorph *Lewia*), *Rhizopus, Pythium, Sclerotinia, Botrytis* (teleomorph *Botryotinia*), *Penicillium* (teleomorphs *Eupenicillium* and *Talaromyces*), *Cladosporium* (teleomorphs *Mycosphaerella* and *Venturia*), *Phoma* (teleomorph *Pleospora*) and *Rhizoctonia* (teleomorphs *Thanatephora, Helicobasidium*) (Snowdon 1990, 1991). Wills *et al.* (1998) include species of *Botryosphaeria* (anamorph *Diplodia*), *Monilinia* and *Phomopsis* (teleomorph *Diaporthe*). Important post-harvest bacteria are in the genera *Pseudomonas, Xanthomonas* and *Erwinia* (though some species of *Erwinia* have recently been moved to the genus *Pectobacterium*). However, some diseases result from a complex interaction of several fungi and bacteria. Crown rot in bananas is one such example. This disease causes shrinking, blackening and rotting of the crown area (Dadzie 1999). In crown rot, different organisms predominate in different locations and times of year. *Colletotrichum musae, Fusarium pallidoroseum* and *Verticillium theobromae* are historically the major causes of crown rot in Central America and the Caribbean (Wallbridge & Pinegar 1975), whereas *Acremonium* species became more prominent in the 1990s (Snowdon 1990). In South America, *Ceratocystis paradoxa* is one of the main agents of crown rot, while in Africa and Asia *Botryodiplodia theobromae* and *C. paradoxa* are found instead of or as well as some of the fungi already mentioned (Snowdon 1990). These fungi are present throughout the year on decaying leaf litter in banana plantations. Most are weak pathogens that are unable to initiate infection unless the fruits are damaged (Dadzie 1999).

Bottom rot of lettuce also comprises a disease complex, with *Rhizoctonia solani, Botrytis cinerea, Pythium* spp, *Sclerotinia minor* and *S. sclerotiorum* and bacteria variously implicated. Blossom rot has been recorded on lettuce in the field and glasshouse (Snowdon 1991).

### *Viruses*

Viruses are not normally of post-harvest significance but may detract from the market value of the produce. They enter plants through wounds, from sap-sucking insects, mechanical abrasion or harvesting activity. Viruses are mainly disease agents of field crops, and the damage they cause is usually obvious so that diseased plants are removed from the food chain during harvest or sorting. Thus, viruses are not often encountered in storage, though there are a few examples, such as spraing (brown lines in the flesh of infected potato tubers). Spraing is caused by tobacco rattle virus or potato mop-top virus. Infection by cauliflower mosaic virus and turnip yellows virus leads to mosaic and yellowing of Cruciferae, and cucumber mosaic virus is sometimes found on cucumber fruits, varying in degree from a slight pale mottling to the development of white distorted areas. Other viruses of post-harvest interest are internal cork in sweet potato (caused by sweet potato feathery mottle virus), internal brown spot of yams (caused by dioscorea badnavirus), apple ringspot (from infection by apple chlorotic leafspot virus and apple stem pitting virus) and stony pit of pears (also caused by apple stem pitting virus) (Brunt *et al.* 1996).

### *Arthropods*

Insects and mites may cause extensive damage to stored durable produce, and to all perishable and durable crops in the field. However, they are of much less importance in stored perishable produce. Some arthropod species are noteworthy, such as the tuber moths *Phthorimaea operculella* and *Scrobipalpopsis solanivora* on stored Irish potatoes. *Phthorimaea operculella* is cosmopolitan, whereas *S. solanivora* appears to be limited to Central America (Povolny 1973). The larvae of these moths bore into the leaves, shoots and stems of the plants. However, post-harvest damage results from larvae mining in the tubers. Small larvae usually gain entry through the eyes, and deposits of frass in webbing around an eye are apparent where a larva has begun to tunnel. Frequently, the larvae feed just below the surface of the potato but they sometimes bore deep into the tuber. In either case, the tunnel is filled with frass. Tubers in the field that are exposed due to shallow setting or cracks in the soil are most frequently infested. Once in store the moths can undergo a limited number of generations before they have to transfer to the aerial parts of their hosts (Haines 1977).

Important beetle pests of stored sweet potato are the weevils *Cylas formicarius* (pan-tropical), and *C. puncticollis* and *C. brunneus* (in Africa). These species can cause serious damage in the field but continue to develop and breed in stored tubers. The smaller *C. formicarius* is more common and the most damaging. Eggs are laid on the stems or inserted directly into the tubers, in holes excavated by the adult female. The larvae bore through the tubers leaving dark-stained tunnels (Hill & Waller 1988). Low levels of infestation are sufficient to markedly reduce tuber quality and acceptability by consumers because the plants produce unpalatable terpenoids in response to weevil feeding (Uritani *et al.* 1975).

The turnip gall weevil, *Ceutorhynchus pleurostigma*, causes galls on the roots of cultivated brassicas. Galls are marble-like swellings on the roots or stem bases formed by the proliferation of host tissue. Larvae feed inside the gall. Damage is largely cosmetic and not of major economic importance (Jones & Jones 1984). A scolytid beetle, *Pagiocerus frontalis*, breeds on avocado seeds in South America and *Curimosphena villosa* (= *Himatismus villosus*) has been found infesting chilli peppers in Ethiopia (Haines 1991).

Mites such as *Rhizoglyphus echinopus* and *R. robini* are associated with onions, flower bulbs and fresh root crops. Both mite species are distributed virtually worldwide, because they can survive on numerous food sources and are often shipped long distances on bulbs, corms and tubers. The most recent review of *Rhizoglyphus* is by Díaz *et al.* (2000).

### *Yeasts*

Yeasts are unicellular, budding fungi, some of which have been reported on stored produce. They gain entry through wounds and may produce ethanol as they grow, so that infected produce gives off an odour of fermentation. Some examples are *Saccharomyces* on strawberries, *Candida* on carrots and yams, *C. krusei* on citrus, *Nematospora* on beans and *Kluyveromyces marxianus* on onion. Pineapples and litchis are also attacked. Yeasts are sometimes found in association with filamentous fungi (Snowdon 1990, 1991).

### *Nematodes*

Nematodes are not major pests of stored produce, but can cause some problems on sweet potatoes and yams if they are brought into the store on infested tubers. The sweet potato nematode is *Meloidogyne incognita*, which causes cracking, shrinkage necrosis and occasional hyperplasia in storage (Lawrence *et al.* 1986). Stored yams are attacked by *Pratylenchus coffeae*, which causes cracking of the skin and gives the tuber a corky appearance (Thompson *et al.* 1973).

## Types of disease

### *Anthracnose*

Fungi that cause anthracnose are generally acervular coelomycetes. The term anthracnose describes the black, coal-like lesions on the surface of infected produce, usually with pink spore masses covering the lesion. Conidia are hyaline, single-celled and ovoid to oblong, and are borne in cushion-like masses of conidiophores housed in erupting acervuli (Wheeler 1969). The spores are produced in large numbers in sticky masses that are dispersed by rain splash. Damage to the host is characterised by limited lesions, necrosis and hypoplasia (Kirk *et al.* 2001). However, the gross injuries and symptoms of anthracnose are not exclusive to the group, being similar to the damage caused by some leaf spots and blights. For example, the fungus *Marssonina panattoniana* causes a leaf spot (shot hole spot) of lettuce, but an anthracnose disease on walnut and cottonwood trees. Anthracnose diseases comprise a small group in which the fungi are similar in morphology and behaviour. The causal agents require high humidities for infection and are most destructive when the host tissue is water-soaked (Wheeler 1969).

Anthracnose fungi have a cosmopolitan distribution. Species of particular concern on post-harvest crops include *Colletotrichum musae* on bananas, *C. pisi, C. truncatum* and *C. lindemuthianum* on peas and beans, *Glomerella cingulata* on mango, *C. dematium, C. capsici* and *Glomerella cingulata* attacking tomatoes, peppers and eggplants, and *C. acutatum, C. dematium* and *C. fragariae* on strawberry (Snowdon 1990, 1991).

### *Blight*

Blight generally indicates a sudden and significant decline in plant health, with blighted areas showing extensive tissue death. Many pathogens can cause blight but there are a select few whose effects are so devastating that they fully merit the term (Wheeler 1969). Post-harvest blight pathogens are a disparate group, with genera from different families of fungi and bacteria. The major examples are alternaria, bacterial and mycosphaerella blight of peas and beans (caused by *Alternaria alternata, Pseudomonas* and *Xanthomonas,* and *Mycosphaerella pinodes),* and early and late blight of potato (caused by *Alternaria alternata, A. solani* and *Phytophthora infestans*). Alternaria blight on potato tubers is characterised by many dark, shrunken lesions that are sharply delineated from surrounding healthy tissue. The external lesions of potato late blight are similarly dark and shrunken, but internal infection is revealed by reddish-brown marbling of the flesh (Snowdon 1991). The historical importance of *P. infestans* is well documented (see Large 1940; Woodham-Smith 1963). However, it remains a serious problem in stored potatoes because tubers may carry internal infection but appear healthy at harvest, thus rotting the bulk during storage.

#### *Blotch*

Blotch denotes a lesion of irregular and indefinite shape that often remains reasonably superficial. The lesions may become blanched and brittle. Examples of this disease include sooty blotch in apples and pears (caused by *Gloeodes pomigena*) in which the symptoms are dark, granular blemishes on the fruits. Brown blotch in avocado is caused by *Pseudocercospora purpurea,* purple blotch of onions by *Alternaria porri*, and *Septoria pisi* causes septoria blotch of peas (Snowdon 1990, 1991).

#### *Brown rot*

Stone and pome fruits are particularly at risk from brown rots. Early symptoms are small, circular lesions, light brown in colour, but the flesh remains firm. The lesions can spread very quickly, in as little as 24 h, if the temperature is suitable (15–17°C) (Wheeler 1969). In the presence of light, copious amounts of grey-brown conidia may be produced, often in concentric rings around the site of original infection. In store, in the absence of light, fruits take on a shiny black appearance. Infected fruit tend to remain firm and dry. Brown rots are caused by three species of *Monilinia*, in different parts of the world. *Monilinia fructigena* attacks apples and pears in Europe, Asia and South America, *Monilinia fructicola* is found on the same crops in North and South America, South Africa, Japan, Australia and New Zealand, whereas *M. laxa* is of occasional importance in Europe, Asia and the Pacific coast of North America (Snowdon 1990).

#### *Canker and scab*

Disease tissue in cankers and scabs develops in an orderly way such that areas of necrosis are surrounded by successive layers of callus. The host responds to attack by forming layers of cork within the cortex, near the necrotic tissue. This results in conspicuous, raised, scab-like pustules. Cankers are not of major importance in stored produce, except for *Itersonilia pastinacae* on parsnip, and bacterial cankers of citrus and tomato caused by *Xanthomonas campestris* and *Clavibacter* (= *Corynebacterium*) *michiganensis* ssp. *michiganensis,* respectively. Infection by *Pestalotiopsis psidii* causes a minor canker of guava. Fungal scabs are found on cucurbits as a result of infection by *Cladosporium* species and on apples and pears after attack by *Venturia*. In cool humid climates scab is one of the most important diseases of apples and pears. Although primarily a problem in orchards, infection late in the season can lead to problems in store. *Venturia carpophila*, and its conidial state *Fusicladium carpophilum*, may cause problems on apricots, nectarines, peaches and plums. Scab pathogens of roots and tubers are mainly soil-dwelling actinomycetes (filamentous, Gram positive bacteria); most examples belong to the genus *Streptomyces*.

#### *Dry rot*

These rots are caused by soil-borne fungi that infect roots and tubers. The pathogens grow internally, thereby dehydrating the host tissue. Potato dry rot is first noticed as small, sunken dry patches on the skin, which wrinkle as infection proceeds. Later, white tufts of mycelium emerge, together with pink spores if the tuber is exposed to light. Cavities containing mycelium form within the tuber. Tubers lose water and fully rotted examples are reduced to a hard, dry mass (Wheeler 1969). The causal agents are species of *Fusarium,* such as *F. solani* var. *coeruleum* in Europe and the US, *F. sulphureum* in UK, US and Cyprus, and *F. trichothecioides* in Australia, US and western Canada. Dry rot lesions in carrots are leathery in appearance and more limited in extent. *Fusarium avenaceum* is the disease agent in carrots. Another dry rot fungus is *Leptosphaeria maculans*, found on swedes and turnips. This pathogen is harboured by cruciferous weeds and infected brassica crop debris, which provide inocula for new plantings. In the US, dry rot of sweet potato is widely distributed. It is caused by *Diaporthe phaseolorum.*

#### *Hyperplasia and hypertrophy*

In response to infection some host tissues become overdeveloped, resulting in an increase in size (hypertrophy) and number (hyperplasia) of cells within the tissue (Wheeler 1969). This results in galls, warts, witches' brooms or other swellings. Of post-harvest importance in this regard are at least three fungal infections: black wart disease of potato, caused by *Synchytrium endobioticum*, club root of brassicas resulting from attack by *Plasmodiophora brassicae*, and *Thecaphora solani* (causing a smut disease of potato). *Synchytrium* and *Plasmodiophora* are primitive, soil borne fungi in which zoospores infect the hosts. Hyphae are not produced. Inside potato tubers, *Synchytrium* exists as a sporangium that eventually releases zoospores, whereas an amoeboid thallus, or plasmodium, is produced by *Plasmodiophora* (Jones 1987). The resting spores are very long-lived, up to 30 years in the case of *Synchytrium*. In wart disease the tubers are converted into shapeless, black warty masses or bear one or more warty outgrowths erupting from the eyes. After a dry season, the warts may be inconspicuous when the crop is harvested, but the disease will continue to develop in store. Swede and turnip with club root exhibit club-like swellings and galling. Sectioning of the swollen tissue reveals a marbled or mottled appearance. Both diseases are favoured by wet conditions which aid the motility of the zoospores.

The smut pathogen *Thecaphora solani* attacks tubers, stem bases and stolons. As a result of infection the host tissue proliferates to produce outgrowths and tumours.

### Leaf spot

Leaf spots are small, well delineated areas of discoloration that may coalesce into larger areas of necrosis as seen in blight or blotch diseases. Such spots can result from unfavourable weather or water, or virus attack, but diseases in which leaf spots are the principal symptom are most commonly caused by bacteria and fungi (Wheeler 1969). In post-harvest terms, leaf spots are only important when leaves are the product that is brought to market. Cruciferae, for example, suffer light leaf spot as a result of infection by *Pyrenopeziza brassicas* in temperate parts of the world. The bacterium *Pseudomonas* also causes leaf spots on brassicas, and *P. cichorii* infection of lettuce leads to varnish spot on the leaves.

### Mildew

A mildew is a plant disease in which the pathogen can be seen on the surface of the host. The term has been used loosely in the past to describe any fungal growth, but in mycology is used for the true, powdery mildews (Erysiphaceae) and false, downy mildews (Peronosporaceae) (Kirk *et al.* 2001). Mildews are obligate parasites and biotrophic. Hyphae do not penetrate deeply into the host tissue, but produce haustoria that absorb nutrients from the surface layers. Mildews are essentially field diseases, but late season infections can be carried into, and develop further, in store. Examples are *Plasmopara viticola* and *Uncinula necator* on grape, *Podosphaera* and *Sphaerotheca* on apples, apricots, peaches, pears, nectarines and plums, and *Peronospora parasitica* and *Erysiphe cruciferarum* on Cruciferae. Mildews are found wherever these crops are grown.

### Mould

Mould has been used to describe any fungal growth that is obvious to the naked eye. Humid conditions which lead to the massive production of coloured spores can make visible infestations that were previously hidden. Mould fungi are cosmopolitan in distribution, come from different fungal families and have a wide host range. Their common names often relate to the colour of the spores, hence, for example, black mould (*Aspergillus niger* on apples and onions), blue mould (*Penicillium expansum* on apples and pears and *P. italicum* on citrus), green mould (*P. digitatum* on citrus), grey mould (*Botrytis cinerea* on grape and Cruciferae), pink mould (*Trichothecium roseum* on apples, pears, peas and beans) and yellow mould (*A. alliaceus* on onion).

### Mosaic

Changes in leaf colour occur in many diseases, but they are particularly common in virus diseases where they may be valuable in diagnosis. Mosaic spotting of leaves is particularly characteristic of virus infections. Most virus diseases are problems in the field, but cauliflower mosaic virus (CaMV) and turnip mosaic virus (TMV) can develop after harvest in the store. Cauliflower mosaic virus and TMV cause a characteristic mosaic on the leaves of cruciferous plants. These two viruses frequently occur together and are transmitted by aphids (Snowdon 1991).

### Rust

Rusts are fungal diseases that reveal themselves as orange or brown spore masses on plant surfaces. Infections are superficial and generally localised in small, circular lesions, usually sunken. Most rusts are obligate parasites. Like the downy mildews, they colonise the host by intercellular mycelia and intracellular haustoria. The life cycle of rusts is complex, and may involve up to five different spore forms, sometimes on two different host species. Rust fungi are mainly important as field pathogens, but a few species can be important in store. For example, rust of stone fruits, caused by *Tranzschelia pruni-spinosae* var. *discolor*, does not develop after harvest but the presence of field lesions leads to a loss in market value. Infected tissue is tough and the flesh beneath becomes leathery, which may cause rejection of the crop by processors (Snowdon 1990). This fungus is cosmopolitan in distribution, as is *Uromyces* on peas, beans and fava beans. Rust disease on peas and beans predisposes infected plants to attack by secondary invasion by pathogens such as *Ascochyta*. *Puccinia allii* is also widespread, on onions, leeks and chives. Onions are particularly at risk from rust disease when they are stressed or overcrowded (Dixon 1976).

#### *Smut*

The smut fungi are characterised by the production of black or brown sooty spores in very large numbers, in discrete areas of the host. The spores are produced in a sorus (an agglomeration of fungal and host tissue), and the infection may be superficial or systemic, on leaf surfaces or within tubers. Spores may survive for many years in the soil. Post-harvest smut problems arise from field infections, and are of note in onions, leeks and chives, and potatoes. The causal agent in the former is *Urocystis cepulae*, which results in shrinkage of the bulbs and increased susceptibility to secondary infections. Smut disease of onions and leeks can cause heavy losses in temperate growing areas. *Thecaphora solani* inflicts serious losses in potatoes in Central and South America. This smut disease can continue to develop in store, such that all of the internal tissues are replaced by a crumbling mass of sori and spore balls (Snowdon 1991).

#### *Vascular disease*

Infection of vascular tissue leads to wilting and collapse as the disease agent prevents the transport of water in the host, either by blocking the xylem or by damaging the root system. Vascular interference in field crops can be caused by fungi, bacteria, insects and parasitic plants, and, to a lesser extent, by viruses and nematodes. In post-harvest situations, bacteria and fungi are more important than the other agents. For example, the bacterium *Ralstonia* (= *Pseudomonas*) *solanacearum* causes a widespread and periodically serious brown rot disease on potato crops in tropical and warm temperate regions. In an infected tuber cut across the heel, the vascular tissue shows as a brown ring which exudes bacterial ooze after a few minutes. Infected haulms may not show disease symptoms but can give rise to infected tubers that rot in store. Brown rot can be distinguished from bacterial ring rot (caused by *Clavibacter michiganensis* ssp. *sepedonicus*), in which the exudate only becomes apparent if the tuber is squeezed (Snowdon 1991). Ring rot is a very serious disease of potatoes in temperate regions. Vascular browning is also found in the tubers of potatoes systemically infected by the fungus *Verticillium alboatrum*.

#### *Wet and soft rot*

In warm, humid conditions stored produce is susceptible to wet or soft rots. These are characterised by the rapid speed of decay, often accompanied by a foul smell from the action of secondary invaders. Wet rots are caused by some fungi, such as *Rhizopus*, *Sclerotium* and *Sclerotinia*, but mainly result from infection by bacteria. The bacteria release enzymes that degrade the pectins of the middle lamellae and cellulose in the cell walls, so that the host cells separate and rupture. Wet rots develop in poorly ventilated stored produce, and in advanced cases the whole of the internal tissue becomes rotted, held together only by the skin (Wheeler 1969). Some examples of wet rot bacteria are *Erwinia carotovora* and *Burkholderia* (= *Pseudomonas*) *cepacia* on onions, shallots and garlic, and *E. carotovora* ssp. *atroseptica* and *E. carotovora* ssp. *carotovora* on potatoes and crucifers; in potatoes, the associated field disease is known as black leg. If the lesions are centred on the lenticils they are typically circular, whereas those associated with injuries are irregular in shape (Snowdon 1991). Other bacterial soft rots are caused by *Paenibacillus* (= *Bacillus*) *polymyxa* (on pepper and eggplants), *Pseudomonas aeruginosa* and *Xanthomonas campestris* (on tomato) and *Cytophaga* (on bell peppers).

### Toxins in perishable produce

In addition to the rots and decay mentioned above, many fungi and/or host plants produce toxins as a consequence of infection. These mycotoxins are toxic to the host and aid the disease process, but they may also pose a health risk to consumers. Some toxins are stress metabolites produced by the host in response to attack, such as psoralen by celery infected with *Sclerotinia sclerotiorum*. This fungus causes pink rot disease. Psoralen is a photoreactive furanocoumarin, which can lead to serious skin burns in pickers if the skin is later exposed to sunlight (Pathak 1986). Other stress metabolites include 6-methoxymellein, produced in carrots and other crops as a result of infection by *Ceratocystis fimbriata*, and pisatin, produced by peas attacked by *Monilia fructicola* and *Rhizoctonia solani*. Many metabolites derived from phenolics, terpenes and coumarins are produced in infected sweet potatoes and Irish potatoes. Alkaloids are also found in Irish potato damaged by *Phytophthora infestans*. In sweet potatoes, ipomeanaronone, ipomeanine, batatic acid and furan-β-carboxylic acid are formed in reaction to infection by *C. fimbriata, Fusarium solani, F. oxysporum* and *Thielaviopsis basicola*.

Patulin is found in juice from fruits infected with a variety of moulds (*Penicillium expansum*, *Aspergillus clavatus*, *A. terreus* and *Byssochlamys nivea*). Cider from organic apples has much higher patulin levels than that from conventionally treated fruit; one diseased apple can contaminate a whole production run of juice. This is particularly a problem with 'cloudy' apple juice; specifically on apples stored for prolonged periods and crushed on a periodic basis, rather than in a continuous process straight from the harvest. Patulin seems to be associated with particulate matter, which accounts for its greater concentrations in cloudy, unfiltered juice.

Aflatoxin B1 is found in figs infected by *Aspergillus flavus*. Fresh figs are sorted under ultraviolet light to reveal the presence of the fluorescent aflatoxin. Aflatoxin has also been detected in tomatoes and bananas. Diseased tomatoes and apples may also contain tenuazonic acid from *Alternaria alternata*. The trichothecenes T-2 toxin, diacetoxyscirpenol and trichothecolone occur in bananas infected by *Fusarium moniliforme, F. oxysporum* and *F. pallidoroseum*, and ochratoxin A, produced by *A. ochraceus* and several *Penicillium* species, has been isolated from citrus fruits.

Some toxins are deleterious to the plant host but have not been implicated in animal pathogenicity, such as AK-toxin produced by the pear black spot fungus *Alternaria kikuchiana* (the pear pathotype of *Alternaria alternata*).

## Literature

The most recent and comprehensive practical guides to the identification of post-harvest problems in fruit and vegetables are the atlases of Snowdon (1990, 1991), though these do not include insect pests. Plant physiological disorders are covered by Anon. (1985) and microbial deterioration by Coursey and Booth (1972), Hall and Scott (1977), Vock (1982) and Dennis (1983). Post-harvest technology is extensively described by Thompson (1996) and Wills *et al.* (1998). These latter volumes concentrate on the physiology, handling and protection of stored fruit and vegetables, and do not treat pathogens or insect pests in any depth. Mycotoxins of perishable and durable stored products are covered by Dupaigne (1978) and Champ *et al.* (1991). Chelkowski (1991) considers the quality and mycotoxin contamination issues of grain storage on farms and in industry. The biology, ecology and management of *Rhizoglyphus*, the main mite pest of stored perishables, is covered comprehensively by Díaz *et al.* (2000).

## References


Ahmad, E., Hussain, I. & Brooks, J.E. (1995) Losses of stored foods due to rats at grain markets in Pakistan. *International Biodeterioration and Biodegradation,* **36**, 125–133.

Aitken, A.D. (1975) *Insect Travellers. Volume I – Coleoptera.* Ministry of Agriculture Fisheries and Food, Technical Bulletin 31, HMSO (UK).

Aitken, A.D. (1984) *Insect Travellers. Volume II.* Ministry of Agriculture Fisheries and Food, Reference Book 437, HMSO (UK).

Allan, R. (1996) *The grain-eating birds of sub-Saharan Africa: identification, biology and management.* Natural Resources Institute, University of Greenwich, Chatham, UK.

Anon. (1985) *Plant Physiological Disorders*. Reference Book No. 223. Ministry of Agriculture, Fisheries and Food, UK.

Banks, H.J. (1977) Distribution and establishment of *Trogoderma granarium* Everts (Coleoptera: Dermestidae): climatic and other influences. *Journal of Stored Products Research,* **13**, 183–202.

Berdoy, M. & Macdonald, D.W. (1991) Factors affecting feeding in wild rats. *Oecologia*, **12**, 261–279.

Bhavanishankar, T.N., Ramesh, H.P. & Shantha, T. (1988) Dermal toxicity of Fusarium toxins in combinations. *Archives of Toxicology*, **61**, 241–244.

Biosis http://www.biosis.org.uk/htmls/free_resources/classifn/classifn.html

Brooks, J.E. & Ahmad, E. (1990) Pest birds of Pakistan: identification and distribution. In: J.E. Brooks, E. Ahmad, I. Hussain, S. Munir, & A.A. Khan (Eds) *Vertebrate Pest Management*. GOP/USAID/DWRC, Islamabad, Pakistan. pp. 173–180.

Brunt, A.A., Crabtree, K., Dallwitz, M.J., Gibbs, A.J., Watson, L. & Zurcher, E.J. (Eds) (1996 onwards) Plant Viruses Online: Descriptions and Lists from the VIDE Database. Version: 20 August 1996, http://biology.anu.edu.au/Groups/MES/vide/.

Burges, H.D. (1963) Studies on the dermestid beetle *Trogoderma granarium* Everts. VI. Factors inducing diapause. *Bulletin of Entomological Research*, **54**, 571–587.

Calhoun, J.B. (1949) A method for self-control of population growth among mammals living in the wild. *Science*, **109**, 333.

Champ, B.R., Highley, E., Hocking, A.D. & Pitt, J.I. (Eds) (1991) *Fungi and Mycotoxins of Stored Products*: Proceedings of an International Conference, 23–26 April 1991, Bangkok, Thailand. ACIAR Proceedings No. 36.

Chang, T.T. (1996) Survival of *Phellinus noxius* in soil and in the roots of dead host plants. *Phytopathology*, **86**, 272–276.

Chelkowski, J. (Ed.) (1991) *Cereal Grain Mycotoxins, Fungi and Quality in Drying and Storage*. Elsevier, The Netherlands.

Chitty, D. & Southern, H.N. (Eds) (1954) *The Control of Rats and Mice*. 3 Vols, Clarendon Press, Oxford, UK.

Coker, R.D. (1998) Design of sampling plans. In: K.K. Singh & D. Bhatnagar (Eds) *Mycotoxins in Agriculture and Food Safety*. Marcel Dekker, NY, USA.

Coker, R.D., Nagler, M.J., Defize, P.R., *et al*. (2000) Sampling plans for the determination of aflatoxin B1 in large shipments of animal feedstuffs. *Journal of the AOAC International*, **83**, 1252–1258.

Coley-Smith, J.R., Mitchell, C.M. & Sansford, C.E. (1990) Long-term survival of sclerotia of *Sclerotium cepivorum* and *Stromatinia gladioli*. *Plant Pathology*, **39**, 58–69.

Conway, J.A. (1983) The biology and ecology of the groundnut seed beetle (*Caryedon serratus*) (Ol.) under field conditions in Senegambia. *Tropical Stored Products Information*, **45**, 11–13.

Cooke, W.B. (1979) *The Ecology of Fungi*. CRC Press, Boca Raton, FL, USA.

Coombs, C.W., Billings, C.J. & Porter, J.E. (1977) The effect of yellow split-peas (*Pisum sativum* L.) and other pulses on the productivity of certain strains of *Sitophilus oryzae* (L.) (Col., Curculionidae) and the ability of other strains to breed thereon. *Journal of Stored Products Research*, **13**, 53–58.

Coursey, D.G. & Booth, R.H. (1972) The post-harvest phytopathology of perishable tropical produce. *Review of Plant Pathology*, **51**, 751–765.

Cowan, P.E. & Barnett, S.A. (1975) The new object and new place reactions of *Rattus rattus* L. *Zoological Journal of the Linnean Society*, **56**, 219–234.

Cutler, M. (1991) Strategies for managing spoilage fungi and mycotoxins: a case study in Thailand. In: B.R. Champ, E. Highley, A.D. Hocking & J.I. Pitt (Eds) *Fungi and Mycotoxins in Stored Products*: Proceedings of an International Conference, 23–26 April, 1991, Bangkok, Thailand. ACIAR Proceedings No. 36, 168–176.

Dadzie, B.K. (1999) *Banana Fruit Quality*. Natural Resources Institute, University of Greenwich, Chatham, UK.

Dennis, C. (Ed.) (1983) *Post-harvest Pathology of Fruit and Vegetables*. Academic Press, London, UK.

Díaz, A., Okabe, K., Eckenrode, C.J., Villani, M.G. & O'Connor, B.M. (2000) Biology, ecology, and management of the bulb mites of the genus *Rhizoglyphus* (Acari: Acaridae). *Experimental and Applied Acarology*, **24**, 85–113.

Dixon, G.R. (1976) Observations on the incidence of leek rust in NIAB trials. *Journal of the National Institute of Agricultural Botany*, **14**, 100–104.

Dupaigne, P. (1978) Les mycotoxines et les fruits. *Fruits*, **33**, 97–122.

El-Mofty, M.M., Khudoley, V.V., Sakr, S.A. & Fatala, N.G. (1992) Flour infested with *Tribolium castaneum*, biscuits made of this flour, and 1,4-benzoquinone induce neoplastic lesions in Swiss albino mice. *Nutrition and Cancer*, **17**, 97–104.

Flannigan, B. (1980) Comparison of seed-borne mycofloras of barley, oats and wheat. *Transactions of the British Mycological Society*, **55**, 267–276.

Gaumann, E.A. (1950) In: W.B. Brierley (Ed.) *Principles of Plant Infection*. Lockwood, London, UK.

George Washington University http://gwis2.circ.gwu.edu/~atkins/newwebpages/Mammalia/rodentia/rodentia.html

Giles, P. H. (1969) Observations in Kenya on the flight activity of stored products insects, particularly *Sitophilus zeamais* Motsch. *Journal of Stored Products Research*, **4**, 317–329.

Gorham, J.R. (1987) *Insect and Mite Pests in Food: an Illustrated Key*. Vols I & II. US Dept. of Agriculture, Agriculture Handbook Number 655.

Gorham, J.R. (1989) HACCP and filth in food: the detection and elimination of pest infestation. *Journal of Environmental Health*, **52**, 84–86.

Gratz, N.G. (1988) Rodents and human disease: a global appreciation. In: I. Prakash (Ed.) *Rodent Pest Management*. CRC Press, Boca Raton, FL, USA.

Gravesen, S., Samson, R.A. & Frisvad, J.C. (1995) *Microfungi in Daily Life*. Cardiff Academic Press, Cardiff, UK.

Haines, C.P. (1977) The potato tuber moth, *Phthorimaea operculella* (Zeller): a bibliography of recent literature and a review of its biology and control on potatoes in the field and store. Tropical Products Institute G112, Slough, UK.

Haines, C.P. (Ed.) (1991) *Insects and Arachnids of Tropical Stored Products: Their Biology and Identification (A Training Manual)*. Second Edition. Natural Resources Institute, Chatham, UK.

Hall, E.G. & Scott, K.J. (1977) *Storage and Market Diseases of Fruit*. CSIRO, Melbourne, Australia.

Hanney, P.W. (1975) *Rodents – Their Lives and Habits*. David & Charles, London, UK.

Hawksworth, D.L. & Kirskop, B.E. (1987) *Filamentous Fungi. Living Resources for Biotechnology*. Cambridge University Press, Cambridge, UK.

Hill, D.S. (1990) *Pests of Stored Products and their Control*. Belhaven Press, London, UK.

Hill, D.S. & Waller, J.M. (1988) *Pests and Diseases of Tropical Crops*. Longman, Harlow, UK.

Hodges, R.J. (1986) The biology and control of *Prostephanus truncatus* (Horn) (Coleoptera: Bostrichidae) – a destructive storage pest with an increasing range. *Journal of Stored Products Research*, **22**, 1–14.

Hodges, R. J., Dunstan, W. R., Magazini, I. & Golob, P. (1983) An outbreak of *Prostephanus truncatus* (Horn) (Coleoptera: Bostrichidae) in East Africa. *Protection Ecology*, **5**, 183–194.

Hodges, R.J., Robinson, R. & Hall, D.R. (1996) Quinone contamination of dehusked rice by *Tribolium castaneum* (Herbst) (Coleoptera: Bostrichidae). *Journal of Stored Products Research*, **32**, 31–37.

Hopf, H.S., Morley, G.E.J. & Humphries, J.R.O. (Eds) (1976) *Rodent Damage to Growing Crops and to Farm and Village Storage in Tropical and Sub-Tropical Regions*. Centre for Overseas Pest Research and Tropical Products Institute, Slough, UK.

Hughes, A.M. (1976) *The Mites of Stored Food and Houses*. Technical Bulletin 9. Ministry of Agriculture Fisheries and Food, UK.

IARC (1993) Aflatoxins. *IARC Monographs on the Evaluation of Carcinogenic Risks to Humans*, **56,** 245–395.

Jones, D.G. (1987) *Plant Pathology*. Open University Press, Milton Keynes, UK.

Jones, F.G.W. & Jones, M.G. (1984) *Pests of Field Crops*. Third Edition. Edward Arnold, London, UK.

Kirk, P.M., Cannon, P.F., David, J.C. & Stalpers, J.A. (2001) *Ainsworth and Bisby's Dictionary of the Fungi*. Ninth Edition. CAB International, Wallingford, UK.

Krishnamachari, K.A.V., Bhat, R.V., Nagarajan, V. & Tilak, T.B.G. (1975) Hepatitis due to aflatoxicosis. An outbreak in western India. *Lancet*, **1**, 1061–1063.

Large, E.C. (1940) *The Advance of the Fungi*. Jonathan Cape, London, UK.

LaVoie, G.K., Michell, G.C. & Bornstein, M.S. (1987) Contamination and consumption of stored wheat by a closed population of *Mus musculus*. International Congress of Plant Protection, 5–9 October 1987, Philippines, Manila.

Lawrence, G.W., Clark, C.A. & Wright, V.L. (1986) Influence of *Meloidogyne incognita* on resistant and susceptible sweet potato cultivars. *Journal of Nematology*, **18**, 59–65.

Leconti, J.D. & Roth L.M. (1953) Composition of the odorous secretion of *Tribolium castaneum*. *Annals of the Entomological Society of America*, **4,** 281–289.

Longstaff, B.C. (1981) Biology of grain pest species of the genus *Sitophilus* (Coleoptera: Curculionidae). A critical review. *Protection Ecology*, **2**, 83–130.

Macdonald, D.W. & Fenn, M.G.P. (1994) The natural history of rodents: preadaptations to pestilence. In: A.P. Buckle & R.H. Smith (Eds) *Rodent Pests and their Control*. CAB International, Wallingford, UK.

McFarlane, J.A. & Wearing, A.J.S. (1967) A means of differentiating between *Acanthoscelides obtectus* (Say) and *Zabrotes subfasciatus* (Boh.) (Coloeptera: Bruchidae) in white haricot beans at the pupal stage. *Journal of Stored Products Research*, **3**, 261–262.

Markham, R.H., Wright, V.F. & Rios Ibarra, R.M. (1991) A selective review of research on *Prostephanus truncatus* (Col.: Bostrichidae) with an annotated and updated bibliography. *Ceiba*, **32**, 1–90.

Meehan, A.P. (1984) *Rats and Mice*. Rentokil, East Grinstead, UK.

Meyer, A.N. (1989) Vertebrate pest control report on a visit to the food department, Sri Lanka. Report R1581. Natural Resources Institute, Chatham, UK.

Michaelson, M.E. (1957) Factors affecting development of stalk rot of corn caused by *Diplodia zeae* and *Gibberella zeae*. *Phytopathology*, **47,** 499–503.

Mound, L. (Ed.) (1989) *Common Insect Pests of Stored Food Products*. Economic Series No.15 (7th Edn). British Museum (Natural History), London.

Nang'ayo, F.L.O., Hill, M.G., Chandi, E.A., Chiro, C.T., Nzeve, D.N. & Obiero, J.W. (1993) The natural environment as a reservoir for the *Prostephanus truncatus* (Horn) (Coleoptera: Bostrichidae) in Kenya. *African Crop Science Journal*, **1**, 39–47.

Nicolaides, L., Coker, R., Wareing, P. & Nagler, M. (1998) *Food Safety and the Developing World*. Natural Resources Institute, University of Greenwich, UK.

Pathak, M.A. (1986) Phytophotodermatitis. *Clinical Dermatology*, **4**, 102–121.

Payne, N. (1925) Some effects of *Tribolium* on flour. *Journal of Economic Entomology*, **8,** 737–743.

Pike, V. (1994a) Impact of milling degree on *Liposcelis paeta* population growth rate and assessment of milled rice weight loss due to infestation. *Crop Protection*, **13**(6), 425–428.

Pike, V. (1994b) Laboratory assessment of the efficacy of phosphine and methyl bromide fumigation against all life stages of *Liposcelis entomophila* (Enderlein). *Crop Protection*, **13**(2), 141–146.

Pitt, J.I. & Hocking, A.D. (1997) *Fungi and Food Spoilage*. Second Edition. Blackie Academic & Professional, London.

Potter, C. (1935) The biology and distribution of *Rhyzopertha dominica* (F.). *Transactions of the Royal Society of London*, **83,** 449–482.

Povolny, D. (1973) *Scrobipalpopsis solanivora* sp. n. – a new pest of potato. *Acta Universitas Agriculturae*, **21**(1), 133–146.

Quy, R.J., Shepherd, D.S. & Inglis, I.R. (1992) Bait avoidance and effectiveness of anticoagulant rodenticides against warfarin- and difenacoum-resistant populations of Norway rats (*Rattus norvegicus*). *Crop Protection*, **11**, 14–20.

Roy, S.K. (1974) Pre-harvest loss of rice due to field rodents. *Economic and Political Weekly*, **9**, 66.

Samson, R.A., Hoekstra, E.S., Frisvad, J.C. & Filtenborg, O. (1996) *Introduction to Food-borne Fungi*. Centraalbureau voor Schimmelcultures, Baarn, The Netherlands.

Santoso, T., Sunjaya, Dharmaputra, O.S., Halid, H. & Hodges, R.J. (1996) Pest management of psocids in milled rice stores in the humid tropics. *International Journal of Pest Management*, **42**, 189–197.

Schiefer, H. B., Hancock, D. S. & Bhatti, A. R. (1986) Systemic effects of topically applied trichothecenes. I. Comparative study of various trichothecenes in mice. *Journal of Veterinary Medicine*, **33A**, 373–383.

Shorten, M. (1954) The reaction of the brown rat towards changes in its environment. In: D. Chitty (Ed.) *Control of Rats and Mice*. Vol. II. Clarendon Press, Oxford, UK.

Shotwell, O.L., Goulden, M.L. & Hesseltine, C.W. (1976) Survey of US wheat for ochratoxin and aflatoxin. *Journal of the Association of Official Analytical Chemists*, **59**,122–124.

Shuyler, H.R. (1970) *Control of rodents and grain-eating birds: report to the government of the United Arab Republic*. FAO Report RP7. Food and Agriculture Organization of the United Nations, Rome, Italy.

Smith, R.H. (1994) Rodent control methods: non-chemical and non-lethal chemical. In: A.P. Buckle & R.H. Smith (Eds) *Rodent Pests and their Control*. CAB International, Wallingford, UK.

Smith, R.H. (1995) Rodents and birds as invaders of stored-grain ecosystems. In: D.S Jayas, N.D.G. White & W.E. Muir (Eds) *Stored Grain Ecosystems*. Marcel Dekker, NY, USA.

Snowdon, A.L. (1990) *A Colour Atlas of Post-Harvest Diseases & Disorders of Fruits & Vegetables*. Volume 1. Wolfe Scientific, London, UK.

Snowdon, A.L. (1991) *A Colour Atlas of Post-Harvest Diseases & Disorders of Fruits & Vegetables*. Volume 2. Wolfe Scientific, London, UK.

Southgate, B.J. (1978) The importance of Bruchidae as pests of grain legumes, their distribution and control. In: S.R. Singh, H.F. van Eden & T.A. Taylor (Eds) *Pests of Grain Legumes: Ecology and Control*. Academic Press, London, UK.

Stoddart, D.M. (1988) The potential for pheromonal involvement in rodent control programs. In: I. Prakash (Ed.) *Rodent*

*Pest Management.* CRC Press, Boca Raton, FL, USA.

Sussmann, A.S. (1968) Longevity and survival of fungi. In: G.C. Ainsworth & A.S. Sussmann (Eds) *The Fungi. An Advanced Treatise*. Academic Press, NY, USA.

Taylor, K.D. & Quy, R.J. (1978) Long distance movements of a common rat (*Rattus norvegicus*) revealed by radio tracking. *Mammalia*, **42**, 63–71.

Thompson, A.K. (1996) *Post-harvest Technology of Fruit and Vegetables.* Blackwell Science, London, UK.

Thompson, A.K., Been, B.O. & Perkins, C. (1973) Nematodes in stored yams. *Experimental Agriculture*, **9**, 281–286.

Tripathi, R.K. (1974) Head fungi of sorghum, phytotoxins and their effect on seed germination. *Indian Journal of Experimental Botany*, **11**, 361–362.

University of Michigan http://animaldiversity.ummz.umich.edu/chordata/mammalia/rodentia.html.

Uritani, I., Saito, T., Honda, H. & Kim, W.K. (1975) Introduction of furanoterpenoids in sweet potato roots by the larval components of the sweet potato weevils. *Agricultural Biology and Chemistry*, **37**, 1857–1862.

Van Egmond, H.P. (1989) Current situation on regulations for mycotoxins, overview of tolerances and status of standard methods of sampling and analysis. *Food Additives and Contaminants*, **6,** 139–188.

Vock, N.T. (1982) *A Handbook of Plant Diseases in Colour.* Volume 1. Fruit and vegetables. Second Edition. Queensland Department of Primary Industries, Brisbane, Queensland, Australia.

Wallbridge, A. & Pinegar, J.A. (1975) Fungi associated with crown-rot disease of bananas from St Lucia in the Windward Islands. *Transactions of the British Mycological Society*, **64**, 247–254.

Waller, J.M. (2001) Postharvest diseases. In: J.M. Waller, J.M. Lenné & S.J. Waller (Eds) *Plant Pathologist's Pocketbook.* Third Edition. CABI Publishing, Wallingford, UK.

Wareing, P.W., Nicolaides, L. & Twiddy, D.R. (2000) Nuts and nut products. In: B. Lund, A.C. Baird-Parker & G.W. Gould (Eds) *The Microbiological Safety of Foods.* Aspen Publishers, Gaithersburg, Maryland, USA.

Warham, E.J., Butler, L.D. & Sutton, B.C. (2000) *Seed Testing of Maize and Wheat. A Laboratory Guide.* CIMMYT, Mexico; CAB International, Wallingford, UK.

Wheeler, B.E.J. (1969) *An Introduction to Plant Diseases.* John Wiley, London, UK.

Wilhelm, S. (1955) Longevity of the *Verticillium* wilt fungus in the laboratory and the field. *Phytopathology*, **45,** 180–181.

Wills, R., McGlasson B., Graham, D. & Joyce, D. (1998) *Postharvest.* Fourth Edition. CAB International, Wallingford, UK.

Woodham-Smith, G. (1963) *The Great Hunger, Ireland, 1845–9.* Hamish Hamilton, London, UK.

# Chapter 5
# Technology and Management of Storage

**R. A. Boxall, J. R. Brice, S. J. Taylor and R. D. Bancroft**

Farmers produce a wide range of crops. Depending on the nature of the commodity and its source location, many of these will need to be recovered from the field at particular seasons during the year. In contrast to this periodic availability, the need for a product often continues year round and, as a consequence, some form of storage is required to more readily match produce availability with demand.

The storage requirements of the range of crops produced are extremely varied. For durables, such as cereal grains, the requirements are comparatively simple, while the physiological characteristics of fresh fruit, vegetables and root crops demand a broader range of technical interactions. As a general rule, the more exacting the quality standards demanded by the market for a particular product and the more sophisticated the storage structures and regimes required, the more expensive the storage protocols are likely to be. To some extent, the difficulties inherent in the storage of fresh, perishable produce may be reduced by lengthening the period of active production and harvesting to obviate the need for storage or, alternatively, by partially or completely processing the commodity into a more stable product with a greater potential shelf-life. All such factors need to be considered with great care when planning storage strategies appropriate to any given commodity.

In many societies, especially in the developing world, durable commodities, notably the grains and legumes, have been the most important foodstuffs in terms of quantities produced. Once harvested, these are predominantly stored in compliance with local traditions. The efficiency or otherwise of these processes may well determine the security and survival of the local communities.

Generally, it is easier to protect dormant dried grains and legumes from external attack by pests such as insect or rodents during storage than it is to prevent the post-harvest physiological deterioration, pest infestation and microbial decay of perishables. As a consequence, where storage technologies are not readily accessible or affordable, the successful marketing of perishable crops often depends on the serial harvesting of produce and its prompt delivery to the consumer. This is particularly true of fruits and vegetables. In the case of the more robust root and tuber crops, however, harvesting windows may be extended and, depending on the pressure to re-utilise the land, produce may be retained in the ground for reasonable periods until required for marketing, so obviating the need for interim storage. In very general terms the crop product must be stored so that:

- the quality does not deteriorate during the storage period;
- the quantity in storage is not unintentionally reduced;
- it is secure against pests, diseases and physical loss; and
- it is accessible at the time and in the quantity required.

The main crop products which may require storage facilities are certain perishables (fruits and vegetables), the semi-durables (roots and tubers) and the durables (cereal grains).

Durable crop products have favourable natural properties that make them suitable for storage over an extended period, but the fresh, non-processed forms of perishable crops are less conducive to storage and, without specialist storage environments, their conservation is limited to days or possibly weeks. Semi-durable root crops may have an intermediate storage potential of several months if handled correctly. The factors determining the difference between durable commodities and those that are more perishable are illustrated in Table 5.1.

**Table 5.1** Properties of durable and perishable crops.

| Durable crops | Perishable crops |
|---|---|
| Pronounced seasonal harvest. Long-term storage necessary | Continual or semi-continual harvesting is often possible; long-term storage may therefore be avoided |
| Harvest time moderately critical over 1–2 weeks | Harvest windows are very variable. Critical periods may be 1–3 days; for semi-durable crops the harvest period may be over many weeks |
| Processing (other than threshing) to prepare the produce is rarely necessary | Processing to physiologically stable products often used as an alternative to storing fresh produce |
| Often symmetrical in shape – facilitates handling | Many shapes, often asymmetrical and therefore difficult to handle without damage |
| Small units mostly weighing less than 1 g | Large units, mostly weighing between 5 g and 5 kg |
| Low moisture content, usually in the range 10–15% | High moisture content, commonly 50–90% |
| Slow respiratory activity of stored crops and thus a low rate of development of heat in stored produce leads to the generation of considerable heat and moisture, particularly in the absence of refrigeration | Moderate to high respiratory activity |
| Hard texture, good protection from injury | Soft textured tissues prone to injury and predation |
| Physiologically stable, inherent storage life of several years | Physiologically perishable, inherent storage life of few days or weeks, although longer for semi-durables, some of which may be stored for several months |
| Storage losses mostly caused by external factors (moulds, insects and vertebrate pests) | Storage depreciation caused by both inherent processes (respiration, sprouting, and compositional changes – especially ripening) and external factors (water loss, temperature stress, moulds, bacteria, insects and vertebrate pests) |

## Deterioration in storage

Loss and deterioration of produce during storage is likely to occur unless adequate precautions are taken. The losses may be quantitative (loss of units or weight) and/or qualitative (deterioration leading to loss of acceptability), resulting in serious financial discounting of part or all of the stored consignment. Once depreciation has occurred, corrective treatments cannot restore the quality or replace the lost produce; thus the major objective of storage management is to ensure that both biological and economic losses are kept to a minimum. There are many causes of loss in storage and they may be categorised in a number of ways. One method identifies primary and secondary causes of loss.

### Primary and secondary causes of loss

Primary causes of loss are those that directly affect the stored commodity and include the following.

*Biological* – for example, infestation by rodents, birds and insects, which (a) result in removal of food from the system and (b) render it unpalatable or spoiled because of contamination.

*Microbial (or microbiological)* – for example, damage by fungi and bacteria. Growth of fungi and bacteria will be at the expense of the commodity on which they are growing and so there will be not only weight loss but damage to the food crop to the extent that it becomes unacceptable because of heating, rotting and other defects. The growth may also result in the production of mycotoxins. Microbial damage to fruit and vegetables leads to a loss in appearance resulting in produce being outgraded or sold as a lower class. (Microbial loss may also be considered as a biological cause of loss.)

*Chemical* – this will include contamination by pesticides and the effect of chemical reactions such as the loss of colour, flavour, texture and nutritional value. An example of a chemical reaction is the Maillard reaction that causes browning and discoloration in fruit.

*Biochemical* – some enzyme-activated reactions can occur in stored food commodities which give rise to off-flavours, discoloration and softening.

*Mechanical* – for example, breakage, bruising and other damage or injury during handling and processing, including harvesting.

*Physical* – usually related to climatic conditions, whereby high or low temperatures, high humidity and improper storage atmospheres in confined storage lead to deterioration. This is a factor that influences chemical and biochemical causes of loss.

*Physiological* – natural respiratory losses which occur in all living produce may account for significant weight losses and, as the process generates heat, may result in conditions conducive to development of biological agents of deterioration. Changes occurring during ripening, senescence, wilting and sprouting often increase the susceptibility of the commodity to damage and infection by pathogens and may render the commodity unsuitable for normal processing.

Of these primary causes of loss, microbiological, mechanical and physiological factors are responsible for most of the losses in perishable commodities, whereas biological and microbiological factors are important in the case of grains.

The secondary causes of loss are those that lead to conditions that encourage primary causes of loss and are usually a result of inadequate or non-existent handling equipment, technology and control. They include:

- poor harvesting facilities (including a lack of equipment and lack of skill);
- lack of adequate containers and packaging for transport and handling;
- inadequate transport to move produce quickly from the field;
- inadequate drying facilities or equipment;
- inadequate storage facilities;
- traditional processing methods giving rise to considerable injury to perishables, breakage in grain or excessive removal of nutrients; and
- inadequate management.

## Storage losses

It is often reported that, globally, a minimum of 10% of cereals and legumes and 20% of perishable commodities are lost after harvest. However, these figures are often cited for planning purposes or to draw attention to a priority area for development, and there is often a temptation to cite 'worst case' figures to dramatise the problem. In reality the total magnitude of storage losses can never be reliably known. Losses vary tremendously and are a function of crop variety, pests and pest combinations, climate, the methods of harvesting, handling, storage and distribution, and the social and cultural settings in which they occur. The importance of losses in particular locations varies according to the availability of food and the purchasing power of the various sectors of the community.

Nevertheless, it is widely agreed that food losses after harvest can be substantial and are important, especially in the developing world, in terms not only of quantity but also of quality and nutritional and economic value. Evidence for this in Pakistan is provided by Chaudhry (1980), in Kenya by De Lima (1979), in Malawi by Golob (1981), on rice in Bangladesh by Huq (1980) and Haque *et al.* (1985), and on millet in Mali by Rowley (1984). A general review on grain loss in store is provided by Tyler (1981).

In the early 1970s a great deal of attention focused upon the opportunity for increasing food availability by reducing post-harvest losses, and storage losses of cereals and pulses in particular. It was believed that a 50% reduction in post-harvest losses was possible and, in 1975, the United Nations General Assembly committed member nations to work towards this goal. Unfortunately, there were no reliable figures from which such a reduction could be measured and so a major international effort was launched to improve the reliability of loss assessment methods and to provide the baseline figures for loss reduction activities. In countries where farming is predominantly at or near subsistence level, it is estimated that 70% of food grain production is retained on the farm. Consequently, much effort was devoted to this sector in the belief that the greatest returns might be achieved here. A methodology was published to provide the means whereby post-harvest losses in grains might be estimated in a standardised and meaningful way so that effective loss reduction programmes might be undertaken (Harris & Lindblad 1978).

At about the same time the Food and Agriculture Organization of the United Nations (FAO) established a 'Prevention of Food Losses Programme' and this prompted similar national and international loss assessment and loss reduction programmes. Initially, in order to achieve significant impact within a limited time, these programmes concentrated on the reduction of losses of staple foods (principally, though not exclusively, food grains). Experiences in applying the loss assessment methodology for assessing grain losses after harvest were later summarised by Boxall (1986).

One of the conclusions from the decade of activity in developing methods for assessing grain storage losses was that there was no simple technique or procedure that could be universally applied. The handling and storage of produce is so varied and the irregular movement and mixing of batches of produce makes it difficult to apply generalisations about analysis and sampling procedures. However, it was clear that loss assessment procedures should be designed so that the methodology is meaningful, economic and acceptable to all those involved. Moreover, loss assessment should be undertaken with a positive aim in view, i.e. that of loss reduction.

Less attention has been paid to the development of a methodology for assessing losses of perishable commodities and there are no generally accepted methods. This may simply reflect the importance given to grains as staple foods. However, the relative uniformity of grains and the broadly similar methods of storage are also important considerations. The development of standard techniques for assessing losses of perishables is complicated by a number of factors that are related to the nature of the commodities themselves.

The rate of deterioration of perishable commodities is higher than in grains and more frequent observation or measurement may be needed compared to studies of grain losses. Weight losses of grains are usually assessed by comparing damaged and undamaged grains at a standard moisture content or on a dry matter basis. This approach is virtually impossible with perishable commodities because they characteristically have a much higher moisture content and individual items in a consignment may often exhibit considerable variability in moisture content. Moreover, there may be a lack of uniformity in weight, size and shape of individual items of perishable commodities compared to individual grains, making estimation of weight loss by comparison of damaged and undamaged items impossible. Damaged grains are not generally acceptable as food and will form part of the overall weight loss, but perishable commodities may suffer only a partial loss. Because of their size, it may be possible to salvage some acceptable or edible parts of damaged fruits and vegetables by cutting away the defective portion. Finally, losses of perishable produce may be more difficult to value. Not only do the individual items have a higher value than individual grains, but there may also be differences in the relative value of each individual item.

### Loss and damage

The term 'loss' when applied to food commodities has been defined in many different ways and confusion has sometimes arisen when 'loss' has been used synonymously with the term 'damage'. In the context of storage losses it is generally agreed that 'loss' means a measurable decrease of the foodstuff, which may be qualitative or quantitative. 'Damage', however, generally refers to the superficial evidence of deterioration, for example, broken grains, bruised fruits or physical spoilage, which may later result in loss.

The importance of damage to consumers will depend upon their economic status; for example, subsistence farmers may have no alternative but to consume a certain amount of damaged produce. In such cases the loss, expressed as a simple weight loss, will be quite small. On the other hand, more affluent consumers may be in a position to be more selective of the produce they consume, in which case the loss of food, i.e. that which is rejected, may be quite high.

### Categories of storage loss

It is difficult to draw up precise categories of loss; nevertheless, loss may generally be considered in terms of quantity or quality, each of which will have economic or financial implications.

*Quantitative loss* is a physical loss of produce that can be measured as a reduction in weight or volume and, therefore, can be measured and valued most readily.

*Qualitative loss* is more difficult to assess since it is frequently based upon subjective judgements, and is perhaps best identified through comparison with locally accepted quality standards. It may include the presence of contaminants, and changes in appearance, taste and texture that may cause the produce to be rejected by consumers. Loss of nutritional value may be considered as an aspect of quality loss.

The following categories of storage loss, which relate primarily to losses of grains, are listed for convenience to demonstrate that loss may be expressed in terms other than weight loss, and should not be regarded as exclusive. In any assessment of losses, the investigator will need to draw up definitions relative to the specific situation under study.

#### *Weight loss (loss of quantity)*

A reduction in weight is easily detected but it may not necessarily indicate a loss of food material. In the case of grains, for example, it may be due to reduced moisture content. This is recognised and allowed for in commercial transactions by a 'shrinkage factor'. Moisture loss may be an economic loss if it is not taken into account by grading for price control, but it is an artificial loss. True weight loss may result from feeding by insects, rodents and birds or from growth of micro-organisms.

Moisture changes may lead to an increase in weight and, in some circumstances, may partly offset the weight loss, for example, production of water by an insect infestation in grain. If we consider the example of the comparison of the weight of a sack of grain before and after storage, any reduction in weight may be described as an apparent loss. However, during storage, the moisture content of the grain may have increased and insects may have consumed some of it, producing non-edible dust. Since the additional moisture and non-edible material do not constitute food, they form part of the true loss.

Weight losses caused by insects feeding in grain may go undetected at a village market if the trader sells grain by volume. A useful way of indicating a loss in these circumstances would be to take equal volumes of sound and infested grain (cleaned to remove the non-edible materials, dust, etc.) and to grind them into flour. The yield of flour from the infested sample will be much less than from the good, wholesome sample.

In commercial grain storage, produce is invariably sold by weight and one must be aware of malpractice such as adulteration with water, stones, earth or sand to make up a deficiency due to insect attack. In assessing loss it is important, therefore, to take account of changes not only of moisture content, but also changes in the amount of foreign matter present.

### *Loss of quality*

Quality is difficult to define and measure objectively unless it can be related to economic value. Quality of produce is assessed in different ways according to the circumstances considered important by local traders and consumers. Generally, quality is assessed and products graded on the basis of appearance, shape, size, etc., but biochemical indices (e.g. sugars, acidity, smell and flavour) are sometimes important quality parameters, especially for some perishable commodities. Foreign matter content and contaminants are factors in loss of quality in grains. Foreign matter may include insect fragments, frass, rodent hairs and excreta, weed seeds, parts of plants, earth, stones and glass. Contaminants, which cannot be readily removed, include soluble excretions of pests, oils, pesticides, pathogenic organisms spread by rodents, and toxins arising from fungal infections. The higher the standard set by the consumer the greater will be the potential for loss.

### *Nutritional loss*

This, in a sense, is the product of the quantitative and qualitative losses; but more specifically, it is the loss in terms of nutritional value to the human population which, in turn, will depend on the nutritional status of that population.

Weight loss of grains during storage is a measure of food loss, but the nutritional loss may be proportionately larger owing to selective feeding by pests. Rodents and some insect larvae, for example *Ephestia* and *Plodia,* feed preferentially on the germ of the grain, removing a large percentage of the protein and vitamin content. Weevils feed mainly on the endosperm and reduce the carbohydrate content. Many pests eat the bran of cereals, thereby reducing the vitamin content; *Liposcelis* spp (Psocidae) feed selectively on the germ and bran of rice (Pike 1994). High moisture content and the associated growth of micro-organisms also lead to changes in vitamin content of grain. Bruchids feed on the cotyledons of pulses (Haines 1991) and loss of protein due to such infestation may be serious as up to 25% of the dry bean matter may be crude protein.

### *Loss of seed viability*

This relates to loss in seed germination – important because of its effect on future food supplies. More care is usually exercised in the storage of seed grain owing to its greater potential value. Loss may be caused by changes of temperature, moisture content, excessive respiration, light, insect infestation and, perhaps, the methods used to control infestation. Insects that selectively attack the germ will cause a greater loss in germination than others. Loss of seed can be determined using standard germination tests (ISTA 1966).

### *Commercial losses*

Commercial losses may occur as a direct consequence of any of the foregoing factors, or indirectly as the cost of preventive or remedial actions required, including the costs of the necessary equipment. Commercial losses may be expressed in terms of monetary loss, a loss of goodwill and loss due to legal action. Commercial losses may affect inter-country trade: for example, after an outbreak in Tanzania of the destructive maize pest *Prostephanus truncatus*, Malawi and Somalia refused to accept Tanzanian maize because of the risk of the insect spreading to these countries (Tyler *et al.* 1990).

All of these losses can be reduced. In most cases knowledge and experience can be major factors and improvement may be relatively rapid. However, in cases where attitudes or beliefs are involved, such as consumer preference, much slower progress must be expected.

### Losses of perishables in storage

Fruit and vegetables, roots and tubers (perishable crops) are quite different in nature from cereal grains and identification of storage losses *per se* may be more difficult. When recording losses of perishable commodities it is essential to provide as much information as possible about what is being measured. For example, when recording weights of roots and tubers it should be noted whether observations are made on fresh, cured, or aged material, whether with or without skins and whether vegetative reproductive parts have or have not been removed. For fruits and vegetables, weights should specify whether observations are made on fresh, whole material and whether skins, peels, cores, etc. have been removed.

Post-harvest deterioration of perishable commodities can arise from a variety of causes, not all of which are directly related to storage. The most common causes of loss are mechanical injury, injury from temperature effects and pests and diseases. The causes and degree of deterioration, and the rate at which it occurs, are substantially different than for the cereals, as are the actions needed to reduce it. Because of these differences it is usually necessary to design a different set of intervention programmes to reduce post-harvest losses in perishable products. Whereas it may be possible to reduce storage losses of cereals by intervention at one point in the post-harvest system, a much broader approach will be necessary for perishable commodities. Here emphasis will be on good management of the supply chain (which may or may not involve storage) with the objective of delivering a high quality product from field to market.

#### *Mechanical injury*

The soft texture and high moisture content of perishable crops render them susceptible to mechanical injury, which can occur at any stage from the field to market and not during storage alone. Injury may arise because of poor harvesting practices, the use of unsuitable containers to transport the crop from field or to market, improper packing (over- or under-packing) of containers, and careless handling of the produce or the containers in which it is packed.

The injuries, which may result in immediate loss or lead to further deterioration, can take many forms. Produce that is dropped may split on impact, or may suffer internal bruising (which may not be visible externally). Superficial grazing of the skin of the produce may result from poor handling, and soft produce, especially leafy vegetables, will be susceptible to crushing. When the outer skin of produce is damaged, fungi and bacteria can enter, leading to rapid decay, and increased water loss from the damaged area can occur. The rate of respiration may also increase, resulting in a rise in temperature within the consignment.

#### *Injury from effects of temperature*

Perishable crops show a wide range of temperature tolerance but are often particularly vulnerable to injury when exposed to extremes of temperature. It is generally accepted that perishable commodities should be kept cool to delay the onset of deterioration as long as possible. Fresh produce exposed to high temperatures caused by solar radiation will deteriorate rapidly. Long exposure to tropical sun will cause severe water loss, especially from thin-skinned produce and leafy vegetables. It is not unusual for produce left in the sun after harvesting in the tropics to reach temperatures as high as 50°C. Respiration increases with increased temperature, and produce that is packed or transported without cooling or adequate ventilation will quickly become unusable.

Tolerance of low temperatures is important in relation to cool storage. Injury from freezing is likely at temperatures between 0 and –2°C. Although some commodities may be tolerant of slight freezing their storage life will be reduced since produce recovering from freezing will be very susceptible to decay. Some commodities, particularly those of tropical or subtropical origin, are susceptible to chilling injury when exposed to low but non-freezing temperatures (which can be as high as 12–14°C). The effects of chilling injury (which may not become apparent until the produce is removed from the chilled environment) include discoloration, skin pitting, abnormal or uneven ripening and susceptibility to rapid decay.

#### *Pests and diseases*

Insect pests are an important cause of loss during storage of grains and legumes but they are rarely a cause for major concern in perishable crops. However, when perishable crops are attacked, damage can be serious. Insect infestation usually occurs in the field before harvest and damage is caused by larvae burrowing through the produce (e.g. fruit fly, sweet potato weevil, potato tuber moth). Further development of the infestation can be a problem if produce is stored for long periods. As in the case of stored cereals and pulses, rodents and birds sometimes cause damage and loss during storage.

More serious loss and deterioration arise from diseases caused by fungi and bacteria, often the result of infection of the crop in the field. Loss in quantity occurs where deep penetration of decay makes the infected produce unusable and loss of quality occurs when the disease affects only the surface of the produce. It is sometimes possible to remove the affected areas and retain the undamaged portion for consumption.

### Loss assessment of perishables

There are few accurate figures available for losses of perishable commodities measured by a described methodology since there are no generally accepted methods for assessing post-harvest losses of fresh produce comparable to those for grains. It is possible to find individual cases with losses ranging from 0% to 100%. Even when figures have been obtained by direct measurement, they may be of limited value because they refer to loss for one specific commodity, in one location and for one specific set of conditions. Moreover, the extent of loss can vary tremendously within a short time. The figures generally indicate a total weight loss and do not normally distinguish between loss of food *per se* and loss of moisture during storage or due to the metabolic processes that continue after harvest.

It would be useful to have a standard method for assessing losses for each type of commodity but this is a difficult task due to crop diversity, inherent perishability and the complexity of the marketing and distribution channels. However, the international standard for sampling of fresh fruits and vegetables (ISO 1980) can provide a starting point for assessing losses. The standard is recognised as good practice and specifies a method of sampling for commodities in cartons and in bulk and the size of samples necessary for assessing quality and other characteristics of produce.

Assessment of losses of perishable commodities will usually involve an assessment of various external and internal quality characteristics such as:

- unit size and weight;
- general colour/appearance;
- colour and firmness of flesh;
- external and internal blemishes/bruising (arising from physiological and mechanical damage);
- damage due to insects or disease; and
- taste and smell.

Special equipment, calibrated to international standards, may be used for the assessment of some of these characteristics.

### Estimation of grain losses in storage

Of the three principal agents causing loss of grain during storage – insects, moulds and rodents – insects are generally regarded as being the most important. It is not surprising, therefore, that the assessment of losses due to insects has received the most attention. Moreover, although the action of insects boring into and feeding on grains will result in both quantitative and qualitative loss, emphasis has been given to measuring weight losses.

#### *Weight losses caused by insects*

The assessment of grain storage losses can involve the observation of the samples of grain collected from stores at specific points in time or over a long period. Ideally, samples should be collected at intervals throughout the storage period. If this can be arranged, then a sample should be collected when the grain is first stored to provide baseline information with which subsequent samples can be compared. Sampling at intervals until a store is empty, coupled with records of the quantity of grain stored and the quantities removed on each occasion, will enable an accurate estimate of storage loss to be made. Samples should be collected from each batch of grain removed and the losses recorded in samples applied to these quantities. Sampling from all parts of the store should be avoided since, in bulk stores, this may cause grain to move and disturb the pattern of infestation within the store. When such regular sampling is impossible, it is recommended that at least three samples should be collected: the first at the time the grain is stored, the next approximately halfway through the storage period and finally during the last few weeks of storage. The pattern of grain use and quantities removed should be noted (Boxall 1986).

Two methods for measuring weight losses caused by insects are appropriate when sampling throughout the storage period is possible, namely, the volumetric method and the thousand grain mass (TGM) method. However, there are occasions when regular sampling is impossible or when an estimate of loss at a certain point in time is required. Under these circumstances the choice of method is between the so-called 'count and weigh' or 'converted percentage damage' methods.

*The volumetric method*

This method is also known as the bulk density or the standard volume weight (SVW) method. It is based upon the use of equipment for measuring the bulk density of a clean, sieved sample of grain. At the beginning of a storage period a baseline SVW is determined from a representative sample of the grain put into store. Losses are recorded by following changes in the SVW on subsequent occasions throughout the storage period. Although the method strictly records changes in bulk density, the change in weight over time is taken to reflect the weight loss due to the damage caused by grain-boring insects.

The difference in moisture content in grain samples collected at different times will affect the weight of grain in the standard volume container. This effect can be excluded by expressing all weight measurements in terms of a constant moisture content – usually the dry weight. However, changes in moisture content also affect the volume and frictional properties of grain. Generally speaking, an increase in moisture content will increase the volume of the grain and cause it to pack more loosely, leading to a decrease in the dry weight of a given volume. To allow for the effect of moisture on the volume of the grain it is necessary to calculate, by experiment, the dry weight of a standard volume of a reference sample of grain at different levels of moisture content. The dry weight of grain filling the standard volume container for subsequent samples, taken at the prevailing moisture content, can then be related to the dry weight of the reference sample at the same moisture content by reference to a specially prepared graph or chart. The procedure requires a great deal of care and time and an adequately equipped laboratory (Adams & Schulten 1978).

Another factor affecting the weight of a standard volume of grain is the addition of insecticide dust. The dust adheres to the surface of the grains, causing an increase in the volume and a change in the frictional properties. Sieving the grain is unlikely to remove all the dust; where insecticides have been applied, therefore, the volumetric method is less useful since it will tend to lead to overestimates of loss.

*The thousand grain mass method*

This method is similar to the standard volume weight method but instead of comparing weights of a fixed volume of grain, the weights of a fixed number of grains are compared. The thousand grain mass (TGM) is the mean grain weight multiplied by 1000 and corrected to a dry weight, and is calculated by counting and weighing the number of grains in a given sample.

A baseline TGM is determined from a sample of grain collected in a representative manner as the grain is put into store. Subsequent measurements of the TGM made throughout the season are compared to the baseline value.

*The count and weigh method*

The count and weigh method (sometimes called the gravimetric method) provides an estimate of loss where a baseline sample cannot be obtained at the beginning of the storage season (Anon. 1969). It uses only minimal equipment and a sample of about 1000 grains. The method, which is applied to a single sample, involves separating the damaged and undamaged grains and then counting and weighing each fraction. The data are then substituted into the following equation:

$$\frac{(U \times N_d) - (D \times N_u) \times 100}{U\,(N_d + N_u)} = \%\text{ weight loss}$$

where U = weight of undamaged grain
D = weight of damaged grain
$N_u$ = Number of undamaged grains
$N_d$ = Number of damaged grains

The method uses a single sample and it is considered unnecessary to determine the moisture content of the separate fractions, on the assumption that the differences are likely to be small.

This method assumes that insects choose grains at random, which may not be true. It also does not account for hidden infestation, because grains containing such an infestation are classed as undamaged (Adams & Harman 1977). Both of these factors may cause misleading, or even negative, results at very low levels of infestation. At very high levels of infestation, misleading results occur because of multiple infestation in large grains such as maize, beans, and in some sorghum varieties. However, it is a useful, quick field method if allowance is made for the problems occurring at the extremes.

Several refinements to the technique have been suggested to address potential biases arising from, for example, preferential attack of large or small grains, differences in moisture content between damaged and undamaged grains, and the presence of hidden infestation. The refinements include separating grains into size categories before counting and weighing (Boxall 1986), separating superficially from severely attacked grains (Ratnadass *et al.* 1994) and making a second assessment after the emergence of hidden infestation (Ratnadass & Fleurat-Lessard 1991).

Attempts have been made to overcome the problem of hidden infestation by dissecting grains to establish the undamaged fraction, but this is tedious and all calculations will then need to be done on dry weights as moisture contents change during the procedure.

### *A modification of the count and weigh method*

A modification of the count and weigh method has been developed for use in the situation where insects completely destroy grains on maize cobs. If such missing grains are not taken into account when assessing losses of cob maize, the count and weigh method is likely to underestimate the loss (Compton *et al.* 1998). The development of the method was prompted by the arrival in Africa of the larger grain borer, *Prostephanus truncatus*, a new pest of cob-stored maize which reduces many grains to powder.

It is recommended that a sample of about 30 maize cobs is used and that each cob is shelled separately. The number of destroyed or missing grains are counted for each cob and then summed over all cobs in the sample to give the total number of destroyed and missing grains (TND). The shelled grains from all cobs are pooled and weighed and the weight recorded (FW).

Two sub-samples, each of about 500 grains, are extracted and the grains in each are then sorted into damaged and undamaged groups and counted and weighed as in the conventional method.

U = weight of undamaged grain
D = weight of damaged grain
$N_u$ = Number of undamaged grains
$N_d$ = Number of damaged grains

The percentage weight loss is then derived from the formula:

$$\frac{[TND(D + U)U] + FW(N_d U - N_u D)}{[TND(D + U)U] + [FW(N_d + N_u)U]} \times 100$$

The weight loss is calculated separately for the two sub-samples and the average taken as the weight loss in the cob sample.

### *Derivation of the equation*

The percentage weight loss in the sample is defined as:

$$\frac{\text{Undamaged weight (UW)} - \text{final weight (FW)} \times 100}{\text{Undamaged weight (UW)}}$$

The final weight (FW) is explained above. The undamaged weight (UW) is the estimated sample weight in the absence of destroyed and damaged grains and is estimated by applying the same assumptions used in the conventional count and weigh method (damage is equally distributed over large and small grains in the sample). On this basis the average weight of undamaged grains in the original cob sample will be equal to the average unit weight of the remaining undamaged grains in the grain sub-sample. The undamaged weight of the whole original cob sample is calculated as the product of the total number of grains estimated to be in the original sample and the unit weight of undamaged grain in the sub-sample. The total number of grains is the sum of the destroyed and missing grains and the number of grains (damaged and undamaged) in the pooled sample of shelled maize. This last parameter has to be estimated and is obtained from the final sample weight divided by the average unit weight of all grain.

The modified count and weigh method suffers less from systematic bias associated with destroyed grains than the conventional count and weigh method, which seriously underestimates true weight loss when many grains are destroyed by insects. Hence, the modified count and weigh method is recommended for studies where destroyed grains are likely to be significant (Compton *et al.* 1998).

### *The converted percentage damage method*

This method is suitable where a quick assessment of loss caused by grain-boring insects is required, without the need for equipment, for example, during a rapid field appraisal. Weight losses in samples of grain may be estimated by reference to the percentage of damaged grains in a sample. A laboratory study must be undertaken first to establish the relationship between damage and weight loss. A conversion factor can then be calculated and subsequently used to determine weight losses in other samples of the same type of grain.

It is usual to determine the conversion factor from the results of the count and weigh method and so this technique will be subject to the same sources of error.

The conversion factor is calculated from the following formula:

$$\frac{\%\ \text{damaged grain}}{\%\ \text{weight loss}} = \text{conversion factor}$$

In order to avoid some of the sources of error arising from the use of the count and weigh technique to derive a conversion factor, it is recommended that a sample of grain with 10% or more damaged grains be used in the first step. This is because the count and weigh method tends to underestimate loss at low levels of infestation. The sample size should never be less than 500 grains.

When a subsequent sample of grain is collected, the number of insect-damaged grains in a sub-sample (of not less than 500 grains) is counted and expressed as a percentage of the total grains present. This figure is converted to a weight loss using the predetermined conversion factor.

Some approximate conversion factors have been established. They all relate to cases where larval stages of insects develop within grains, for example *Sitophilus* spp and *Sitotroga cerealella* infestations (Table 5.2). They are only approximate and should be regarded as rough guides; it is preferable to determine conversion factors for the particular grains being studied.

### *Rapid loss assessment technique based on grain damage/weight loss relationships*

A rapid field assessment technique for predicting weight loss in grains has been developed, based on the relationship of grain damage to weight loss. Although specifically developed for cowpea and bambara groundnut, the technique could be used for any grain, especially larger types such as maize. The technique requires the use of standard graphs relating percentage damage and weight loss for the commodities under study.

When preparing the reference graphs, at least ten working samples consisting of around 500 grains are required for the preliminary laboratory work. The weight loss of each sample is calculated using the count and weigh method, and the percentage of damaged grains is determined. From these results, the percentage weight loss is plotted against the percentage damage and a graph of the best fit is produced.

When using the graphs in the field, a clean sample of approximately 200 grains is collected and the proportion of damaged grains is determined and the corresponding loss figure is read off by reference to the graph. Experience has shown that, using a sample of approximately 200 grains for cowpeas or bamabara groundnut, the technique has an accuracy of ± 7%. Although this is not as precise as some other methods, it is better than the converted percentage damage method. However, the ability to take many individual samples in a short period of time at the farm allows a good working estimate of average losses to be determined.

**Table 5.2** Conversion factors for a selection of crops.

| Grain | Conversion factor |
|---|---|
| Maize (stored as shelled maize or as cobs without husk) | % bored grain ÷ 8 |
| Maize (stored as cobs in husk) | % bored grain ÷ 4.5 |
| Wheat | % bored grain ÷ 2 |
| Sorghum | % bored grain ÷ 4 |
| Paddy rice | % bored grain ÷ 2 |

### *Rapid loss assessment using visual scales*

All of the above loss assessment techniques (with the exception of the last described) are, to a greater or lesser extent, time-consuming, require well-trained personnel and appropriate equipment. These shortcomings can be overcome by using visual scales. Visual scales are used routinely for assessing damage in field crops and are ideal for field use, rapid to use, require nothing beyond reference scales (for example, photographs), and have low levels of operator bias.

A technique has been developed for cob maize and dried cassava pieces in which sampling and scoring can take less than 15 minutes. This enables increased sampling and wider coverage or reduced sample error. A loss estimate is obtained on the spot, leading to a reduced risk of spoilage, and anomalous results can be double-checked before leaving the site. After analysis the cobs or cassava pieces can be handed back to the owner intact, avoiding the common problem of how to compensate farmers for any samples removed (Compton & Sherington 1999).

Visual scales have to be calibrated using a laboratory technique and can be calibrated according to the objectives of the study, i.e. the scale can be calibrated according to weight loss, end use or farmer perception of value.

### *Preparing and using visual scales for maize cobs*

A supply of at least 100 cobs showing a wide spectrum of damage is required. The dehusked cobs are sorted into five or six 'damage classes' ranging from Class 1 (no damage) to Class 5 or 6 (severest damage) (see Plates

2a–d). Photographs are prepared for each damage class using representative cobs. The visual scales are calibrated for weight loss using a large number of cobs from each class and the modified count and weigh method.

When using the scales to estimate a weight loss, a sample of at least 30 cobs must be used. Each cob is compared with the visual scales photographs and assigned to a damage class. The average percentage weight loss is then estimated as in Box 5.1.

**Box 5.1 Example calculation of average percentage weight loss using visual scales.**

A sample of 30 cobs was collected from store: 9 were assigned to damage class 2, 6 to damage class 3, 5 to damage class 4 and 1 to damage class 5. The remainder were undamaged.

The predetermined weight loss (by modified count and weigh method) for each class was as follows:

| Damage class | 1 | 2 | 3 | 4 | 5 |
|---|---|---|---|---|---|
| % Weight loss | 0 | 8.9 | 12.6 | 26.3 | 57.9 |

Thus the average weight loss for the sample =

$$\frac{(9\times 8.9)+(6\times 12.6)+(5\times 26.3)+(1\times 57.9)}{30}=11.5\%$$

### *Losses caused by vertebrate pests*

Vertebrate pests such as rodents and birds frequently remove whole grains from store and so it is impossible to estimate the loss from an examination of grain samples (Elias & Fall 1988). An exception is perhaps when maize is stored on the cob. In this case the loss of grain can be estimated by reference to the percentage of grain removed and average grain weights (Boxall & Gillet 1982).

Losses due to vertebrate pests, especially rodents, may be calculated from population studies and feeding trials, but often with limited accuracy in relation to the effort expended (Hernandez & Drummond 1984). These pests utilise stored grain as part of their diet only, so feeding trials may overestimate the consumption of stored grain. Mian *et al.* (1987) used a system of tracking tiles and traps to assess populations of rodents in rice stores in Bangladesh, and from food consumption data estimated that farm families lost around 53 kg of rice per year.

The need for figures for loss caused by rodents might, however, be questionable. If it can be established from a general survey of rodent activity that a rodent problem exists and is rated important by the community, then this may be enough to justify the introduction of a control programme. Losses of food due to rodents may be relatively insignificant when compared to the loss and damage to personal property, storage containers and buildings, etc., and the potential health risks.

### *Weight losses caused by moulds*

Grain which is attacked by mould will lose weight, and the methods of assessing weight losses due to insects might also be used to assess loss due to moulds. However, mouldy grains may increase in weight due to moisture absorption, and so any loss due to the growth of moulds will be masked. Moreover, the methods may not give an indication of the real loss since some internally infested grains showing no outward sign of mould damage may be counted as undamaged. More importantly, the consumption of mouldy grain is undesirable and in many cases such grain will be rejected. The impact of mould damage on storage loss can be estimated by including a separation of mouldy grains from other types of damage during the analysis. In this case the weight of mould damaged grains is equal to the loss.

## Estimation of total loss in a season

The methods of estimating losses described above will indicate the loss at a given point in time. These figures if used in isolation may be misleading and so they must be considered in relation to the pattern of grain use. If grain remains undisturbed throughout the storage season and at the time the store is emptied a sample shows a loss of 10%, then this is the total loss caused by insects. More commonly grain is removed at intervals throughout the season and so each quantity removed will have suffered a different degree of loss since it will have been exposed to insect infestation for different lengths of time. During the first few months of storage, insect infestation and the percentage loss recorded in samples may be low, but the number of insects typically increases with time and so at the end of the season there will be a very high percentage loss in the samples. However, this high loss is only applicable to the small quantity of grain remaining in store.

The total quantity of grain lost in a season can be calculated from the difference between the weights of grain put into and taken out of store, after allowing for changes in grain moisture content. An indication of the loss due to causes other than insects can be obtained by subtracting the loss to insects.

As a result of this approach, overall losses during storage have been shown to be considerably lower than the early (often extreme) figures suggested. Tyler and Boxall (1984) in their review of a decade of loss assessment activities presented selected examples of results of studies at farm level and in the commercial sector (Tables 5.3 and 5.4). The criteria for selecting these examples were that an acceptable methodology was adopted, it was fully described and the results represent the best estimate of loss.

There were few figures for losses in commercial operations and virtually none for small trader or co-operative level storage, and the situation has not changed significantly. This may reflect the fact that many grain traders in the developing world frequently buy and sell grain over relatively short periods of time. There has been increased involvement of the private sector since market liberalisation and the decline of parastatal marketing boards, but few figures for storage loss are available. Certainly some entrepreneurs are storing larger quantities of grain for longer periods, but market liberalisation has, in some areas, resulted in increased levels of on-farm storage. Wright (1995) suggested that investigation of the nature and extent of losses at trader/intermediary level combined with an appraisal of trader attitudes to the losses incurred was an area that need further work.

The results for the farm level showed losses to be fairly well contained about or below 5% over the storage season. The one exception (Hodges *et al.* 1983) was for losses in stored maize, caused primarily by the then newly established pest, *Prostephanus truncatus*, to which farmers were unaccustomed and for which

**Table 5.3** Examples of comprehensive studies to measure storage losses in farm-level storage.

| Country | Crop | Period of storage (months) | Cause of loss | Percentage loss | Reference |
|---|---|---|---|---|---|
| Bangladesh | Paddy | 3–4 | Insects, rodents | 2.4 | Huq 1980 |
| Honduras | Maize | 7 | Insects | 5.5 | De Breve *et al.* 1982 |
| India | Paddy | 7 | Insects, rodents, moulds | 4.3 ± 1.3 | Boxall *et al.* 1978 |
| Kenya | Maize | up to 9 | Insects, rodents | 3.5 ± 0.3 | De Lima 1979 |
| Malawi | Maize | up to 9 | Insects | 3.2 ± 1.8 | Golob 1981 |
| Malawi | Sorghum | up to 9 | Insects | 1.7 ± 0.5 | Golob 1981 |
| Nepal | Maize | 6 | Insects, rodents | 5.7 ± 3.2 | Boxall & Gillet 1982 |
| Nepal | Wheat | 3 | Moulds, insects | 2.4 ± 1.9 | Boxall & Gillet 1982 |
| Nepal | Paddy | 8 | Insects | 3.4 ± 2.2 | Boxall & Gillet 1982 |
| Swaziland | Maize | unspecified | Insects , moulds, rodents | 4.35 | De Lima 1983 |
| Tanzania | Maize | 3–6.5 | Insects | 8.7 | Hodges *et al.* 1983 |
| Turkey | Wheat | 8 | Insects, moulds | 3.7 ± 1.9 | Boxall (pers. comm.) |
| Zambia | Maize | 7 | Insects | 1.7–5.6 | Adams & Harman 1977 |

**Table 5.4** Examples of comprehensive studies to measure storage losses in commercial storage.

| Country | Commodity | Sector and storage period (months) | Cause of loss | Loss % | Reference |
|---|---|---|---|---|---|
| Cuba | Various processed foods | Warehouse (3) | Rodents | 1.0 | Hernandez & Drummond 1984 |
| Cyprus | Barley and wheat | Storage (various) (3.5) | Insects | 3.54 | Tyler 1981 |
| Mali | Millet | Public sector warehousing (8) | Insects | 1.0 | Rowley 1984 |
| Dominican Republic | Rice | Public sector warehousing | Spillage | 0.35 | La Gra *et al.* 1982 |
| Pakistan | Paddy | Market storage | Various | 1.8 | Chaudhry 1980 |
| | | Wholesale storage | Various | 3.2 | |
| | | Public sector warehousing | Various | 3.0–5.0 | |

locally traditional storage methods provided an ideal environment.

There is a danger in accepting overall figures for storage loss since the importance of high seasonal losses may be overlooked, particularly if they occur at the end of a storage period and relate only to a small proportion of the total amount stored. Some farm households may have no option in the hungry season but to consume this grain, which may be heavily damaged by insects. Clearly there is a need for intervention although the loss figures may not seem to justify it. This situation arises because of the emphasis placed on weight loss as the simplest factor to measure. Hence more attention should perhaps be paid to other criteria for loss, such as grain quality. A measure of storage loss in terms of the degree of qualitative deterioration from the time of storing is likely to result in substantially higher figures.

The early farm-level studies undertaken in the 1970s tended to concentrate upon estimates of loss for the unimproved, traditional system, with little regard to loss reduction. However, it was soon realised that the approach took too long and provided insufficient information on which to develop a loss reduction strategy. It is usually possible to identify, at an early stage, improvements in store design or storage practices that could be evaluated as part of a loss assessment study. Surveys providing information on loss under existing conditions, an indication of the extent to which the loss might be reduced, and an evaluation of the acceptability of new measures, will be of greater benefit than those which simply describe loss in the traditional system.

Early studies of storage losses tended to be undertaken in isolation rather than in the context of the post-harvest system or even the total production system. Many storage problems can be attributed to events occurring prior to storage, and post-harvest problems generally may arise because of deficiencies in the production system. This systems approach is necessary if the results of loss assessment studies are to be a meaningful step towards loss reduction. Participation by farmers in loss assessment studies is essential; they are an important source of detailed information about the farming system, the problems that occur and the importance of losses. A failing of some studies has been a reluctance to consult farmers on their views on losses and even the need for the work. There is not much point in measuring losses to which the farmer attaches little importance. In Bangladesh, for example, farmers are aware that losses occur, but most regard this as unavoidable and accept them (Haque *et al.* 1988).

Since the late 1980s, participatory methods developed primarily for socio-economic assessment of rural households have been applied in studies of post-harvest losses with the result that problems are approached more from the farmers' viewpoint. As a consequence, detailed studies of losses are not always necessary. If researchers and farmers agree that there is a storage problem, the aim should be to find an acceptable solution rather than simply deriving figures for loss. Some measurement may be required to demonstrate the value of any intervention.

Wright (1995) commented that, over a period of 20 years, much time, money and effort had been devoted to measuring storage losses, particularly at the farm level, and that there was a danger that ongoing projects might exhibit some of the failures of earlier studies. He went on to make or reiterate a number of suggestions to improve future loss assessment studies. An important prerequisite is a knowledge of the systems in place, their limitations and constraints, and an understanding of farmers' needs as perceived by farmers themselves. This requires strong socio-economic inputs to complement the technical components of loss assessment.

Moreover, studies of storage losses should not be undertaken in isolation but as part of an analysis of the post-harvest sector as a whole. Planners or donor agencies may call for precise figures for loss to justify projects, but emphasis should be away from precise measurement of figures. Loss assessment should be undertaken with a clear idea of the use to which figures will be used. Social surveys will help to identify perceived farmer problems, will indicate the need for an assessment of loss and will allow selection of the most appropriate measurement techniques.

## The influence of pre-storage factors on safe storage of produce

### Pre-harvest factors

As with many other characteristics, the storage potential of different products may be influenced by their genetic composition and the quality of the environment in which they were nurtured. As a consequence the behaviour of many commodities during storage can be influenced by the type of cultivar selected, the conditions under which produce is grown, and possibly the nature and efficacy of any chemical treatments that may have been applied during the development of the crop.

### *Cultivars*

The sheer diversity in the physical forms of produce presently committed to stores suggests that different genera and species of produce will naturally have contrasting storage potentials and require different storage conditions. It is perhaps not quite so obvious that the storability of a particular commodity may also vary quite markedly between different varieties and cultivars. Much of our knowledge regarding these storage conditions has been acquired empirically over time. On occasion this has led to the classification of horticultural commodities according to their storage characteristics in, for example, seed catalogues. The extent of variation in optimal storage periods of a commonly traded commodity such as citrus helps to illustrate the point. For instance, under appropriate regimes of temperature and humidity, tangerines may be stored for 3–5 weeks, mandarins for 3–8 weeks and navel oranges sourced from Australia for 1–2 months. In contrast, navel oranges from Spain may be conserved in refrigerated stores for periods of 3–4 months.

Considerable effort has been devoted to increasing the production of food grains and as a result new cereal and legume varieties have been developed which show substantial yields over traditional varieties. Some of the new varieties can be grown outside the normal growing season, thus allowing two or more crops per year. These additional crops invariably mature in the rainy season, creating problems in drying and storage of wet grain. The inherent qualities of traditional grain varieties (namely, hard endosperm and, in the case of maize, good husk cover) help protect the grain against insect attack but the high-yielding varieties, although possessing improved nutritional value, have characteristics that render them liable to spoilage in store.

In the past, the successful exploitation of varieties and cultivars on account of their post-harvest characteristics has been the province of either the producers themselves or commercial breeders responding to the demands of the market. The fast-developing potential to genetically engineer chosen characteristics into a germline opens up many possibilities in the future whereby the fundamental attributes of a commodity may be specifically designed to meet a particular market need.

### *Soil type, fertiliser and irrigation treatments*

The soil type or, more exactly, the soil fertility in which the commodity has grown can influence the behaviour of the commodity within the store. Lack of plant nutrients in the soil will affect the quality of the produce at harvest and subsequent keeping qualities. The application of nitrogen during growth may be important for growth, but excessive nitrogen applied towards harvest time can lead to increased losses during storage. For example, nitrogen in excess of 1.8 kg per tree can produce unfavourable effects on external colour and flesh firmness of mangoes (Young & Koo 1974) and when leaf nitrogen levels are greater than 1.5% there is an increased incidence of soft nose and soft brown rots (Nakasone & Paull 1998). Similarly, while irrigation may be required to obtain the size and yield required, excessive or late irrigation can lead to higher levels of rotting during storage.

### *Chemical treatments*

Chemical treatments of the commodity, for example, treatment of onion seed with a benomyl fungicide prior to planting, can dramatically reduce storage losses from neck rot (*Botrytis allii*). An appropriate pre-harvest spray programme, for example, using carbendazim and mancozeb fungicide, can significantly reduce storage losses of mangoes from anthracnose (*Colletotrichum gloesporioides*).

### *Pre-harvest pests and diseases*

Grain crops in the tropics are commonly attacked in the field by storage insects and the hidden, developing infestations within grains may be carried into store where they will later emerge if conditions are favourable. Some perishable crops may also be infested in the field.

Fungal and bacterial diseases are spread by microscopic spores, which are widely distributed in the air and soil and on dead and decaying plant material. Both durable and perishable crops are therefore vulnerable to infection in the field. Grains may be affected during wet harvesting conditions. Field infections in perishable commodities may not become visible until after harvest; for example, decay of root crops caused by soil moulds will usually develop during storage. Similarly, tropical fruits infected at any time during their development may show decay only during ripening.

Post-harvest diseases may be spread in the field before harvest by the use of infected seed or other planting material. Many diseases can survive by using weed plants or other crops as alternate or alternative hosts. They are also spread by means of infected soil carried on farm implements, vehicles, boots, etc. and from crop residues or rejected produce left decaying in the field. Infection may also occur if the containers used to transport commodities from the field are contaminated by soil or decaying produce.

Rodents and birds frequently attack crops in the field, leaving damaged produce that will be susceptible to invasion by micro-organisms. Damaged grains in particular will be susceptible to insect attack in store.

## Harvesting and field handling

### *Harvesting*

Harvesting is a major operation and one that is often regarded as being the last step in crop production. However, it is also the first step in the post-production (post-harvest) system since the condition of the crop at harvest and the methods of harvesting will influence subsequent handling, processing and storage of the crop.

Timing of harvest is very important but a compromise often needs to be reached. Premature harvesting of grain can result in a high proportion of immature grains that, because of their high moisture content, will deteriorate rapidly in store. If harvesting is delayed too long then mature grain may suffer attack from insects, micro-organisms and vertebrate pests, and may be physically damaged by cracking through repeated wetting and drying.

Delaying the harvest of horticultural produce will, within reason, allow the yields to be maximised. However, the optimum time for maximum quality will invariably be earlier, while the optimum time of harvest for the storability of the commodity may be earlier still. Some commodities are best stored in an immature stage, for example, mature green tomatoes will store for up to 3 weeks compared with up to 1 week when ripe. Ideally, horticultural produce should be stored or transported at similar stages of maturity or ripeness. Fruit at a more advanced stage of maturity or ripeness, for example, mangoes and papaya, may stimulate the ripening of other fruit in the consignment. Newly harvested perishable produce left in the sun deteriorates rapidly as a result of increased respiration and water loss and so harvesting is best done during the coolest part of the day.

Harvesting of produce that is wet from dew or rain should be avoided. Wet produce will overheat if not well ventilated and will be more susceptible to decay. Some perishable commodities are especially susceptible to damage when wet; for example, some citrus fruits will be subject to oil spotting and rind breakdown. Harvesting of damp grain may be possible as long as grain can be threshed quickly and adequate drying facilities are available.

Harvesting methods differ according to the part of the plant to be used. Harvesting of forage crops involves cutting of the whole plant whereas with cereals the crop is first cut, either as a whole or partially (cutting of ears heads and cobs), and then threshed and cleaned (winnowed) to separate the grain from the chaff and straw.

Grain harvest may be carried out as a separate exercise to be followed by threshing and cleaning or in combination with threshing – using combine harvesters. In developing countries, harvesting and threshing are commonly carried out by hand, and although the act of cutting is unlikely to result in damage that could lead to deterioration of the crop in store, there may be risk of damage with some methods of threshing. Large-scale commercial producers may have some mechanical harvesting equipment but the use of more sophisticated machinery will generally be limited to agroindustrial production of cash crops. Small-scale producers of cereals may have access to mechanised harvesting and threshing equipment through community groups and co-operatives. Generally, hand harvesting results in lower levels of damage to produce, thereby reducing the risk of deterioration during subsequent handling and storage.

Traditionally, grains may be threshed by beating bundles of the crop against a hard surface (e.g. a log of wood, a wooden bar, a stone, or a wooden or metal tub). Alternatively grain heads, or ears (of wheat, sorghum and millet, for example) may be beaten with sticks. Both methods can result in damage to the seed coat or cracking of the grain, rendering it more susceptible to pest attack or mould infection during storage. Sometimes, grain crops other than maize may be threshed by being trodden underfoot (by humans or more usually by animals) and this is considered to be less damaging to the grain. Small quantities of maize cobs are often shelled by hand but larger quantities may be shelled by beating with sticks – which obviously results in high levels of damaged and broken grains. Small disc shellers, either hand operated or mechanised, speed up the process of shelling but again tend to lead to increased numbers of damaged grains.

Mechanised threshers invariably result in damage to the grain although larger more sophisticated models may be adjusted according to the type of grain to minimise damage. The amount of damage to grain harvested by combine harvester will depend on the skill of the operator in adjusting the equipment to suit the crop and conditions of harvesting.

Methods of harvesting vegetables include hand harvesting of parts of the plant such as the leaves (spinach), lateral buds (Brussels sprouts) and immature flower heads (broccoli), cutting of the above-ground parts of the plant (cabbage) and digging, pulling or lifting of root crops (carrots), tubers (potatoes) and bulbs (onions) from the ground and removal of the soil sticking to the produce.

The risk of causing damage to vegetables that may result in deterioration in store will be minimal when produce is harvested by hand. When harvesting involves cutting, there may be no damage to produce but there is a risk of disease and infection if knives are not kept clean and sharp. Root and tuber crops may suffer some mechanical injury from digging or lifting tools, but this can be minimised if the crops are grown on raised beds so that tools can be pushed into the soil well below the roots or tubers. The risk of damage to vegetables increases where mechanical harvesting is possible, for example, in the case of potatoes, onions and some other root crops.

Many fruits have a natural break point on the fruit stalk that can be easily broken by hand at harvest. However mature, green produce with woody stalks which break at the junction of the fruit and stalk, and immature fruits with fleshy stems, are best harvested by cutting. Care is needed in such cases to ensure that produce is separated cleanly and without damage that may lead to decay not only of the produce but also in the plant itself.

Excess vegetation such as binds on potatoes, tops of onions, and leaves on carrots should be carefully removed so as to minimise damage to the commodity. Damage during topping can result in increased respiration rates and increased skin loss and may allow the entry of pathogenic organisms, leading to rotting. It may be necessary, either for the subsequent storability or the marketability of the produce, to leave a proportion of the foliage on the commodity. For example, it is recommended that tops of 25–35 mm be left on onions to help reduce moisture loss and fungal infection as well as to maintain skin quality during storage. The method of storage may sometimes dictate that the vegetation is left on the commodity. Onion leaves must be retained if the bulbs are to be hung by their tops during storage. However, if onions are to be stored in bulk or containers the presence of leaves is undesirable as they affect the flow of the material into the store, they add to the cost of drying, take up storage space and create more wastage during grading.

### *Handling and packaging*

Care is needed in handling perishable crops during all post-harvest operations from the field to store. Their soft texture and high moisture content render them susceptible to mechanical injury. Minimum levels of drop are often recommended, for example, potatoes should not be allowed to fall more than 230 mm onto a hard surface or 600 mm onto other potatoes at any stage during harvest. Careless handling of produce, for example, dropping fruit from a height during harvesting, can result in bruising which may only be apparent when the fruit starts to ripen. Tomatoes are sensitive to a physiological disorder known as internal bruising which only becomes apparent after the tomato reaches the full-red ripeness stage. It develops when a tomato receives an impact above the locule during harvest or handling. It is usually advisable for perishable produce, especially fruit, to be transported from the field to packhouse or store in suitable containers with some form of cushioning to ensure that damage during transit is minimised.

Lack of attention to packaging can have a significant impact on the storage life of fruits and vegetables. Good-quality packaging should be constructed from material that will protect the commodity from a wet environment and allow it to be stacked for a period of several weeks. Cartons made from fibreboard have a poor wet strength and there is a limit to the height to which these cartons can be stored without the load gradually compressing. Carton collapse can lead to mechanical damage of the contents, and reduced air (and gas) circulation during transit and storage, thus creating temperature and gas gradients. Packages need to be perforated to facilitate ventilation of the product. This is important in terms of the efficient removal of field heat during cooling and for maintenance of product temperature during handling and storage.

### *Transport*

The transport of durable commodities from the field presents few problems with respect to deterioration that may subsequently occur during storage. The main concern will be to ensure that the commodities are kept dry. There is a small risk that grain may become infested if the transport or containers in which it is carried have a residual insect infestation. Perishable commodities, on the other hand, are more susceptible to overheating and mechanical damage or injury during transportation whether it is from field to store or from store to market. If damage, which may subsequently result in deterioration or loss of quality, is to be avoided, the produce must be kept dry, as cool as possible, and moved quickly.

Damage and loss incurred during non-refrigerated transport arise primarily from mechanical damage and overheating. Mechanical damage can arise from: careless handling; use of unsuitable containers; insecure stacking of containers; and poor driving, a poor condition of the vehicle and poor road conditions, all of which may lead to excessive movement of produce.

Overheating of produce may occur from exposure to external sources of heat (the sun or heat from vehicle engines) or as a result of heat generated by the produce within its packaging. This overheating increases moisture loss and the processes of the natural breakdown and decay.

## Post-harvest factors

### *Pre-storage treatment*

It is generally advisable to place fruits and vegetables in store as quickly as possible after harvest. However, in certain cases it has been shown that pre-storage treatments can be beneficial. Perishable crops which are to be stored for long periods may have to be treated with chemicals to prevent attack by micro-organisms. The use of such chemicals is controlled by legislation, and maximum residue levels are stipulated for the edible portions of fruits and vegetables. Other chemicals may be applied to prevent the development of physiological disorders and sprouting of root crops during storage.

Care needs to be exercised in the application of chemical treatments since they may have a detrimental effect. For example, hot-water fungicide (benomyl or procloraz) treatment of mangoes to control diseases (anthracnose), if not properly carried out, can lead to spoilage of fruit during storage. Temperature and exposure times are important and often variety-specific, and fruit must be allowed to cool before packaging and storage.

The storage period of commodities, such as potatoes, yam, sweet potato, garlic and onion, and to some extent citrus fruits, may be enhanced by such processes as curing and drying. These practices are standard treatments on some crops and applied as routine whatever type of subsequent storage is carried out. For example, holding potatoes at approximately 15°C for up to 2 weeks before storage will help the sealing of damaged areas (through the formation of a corky 'skin'), thereby reducing subsequent moisture loss and infection. Drying and curing of the outer skins of onions also help to reduce moisture loss and infection during storage.

For other crops, pre-storage exposure to high temperature is less well established and still in an experimental stage. For example, Granny Smith apples kept at 46°C for 12 h, or 42°C for 24 h, or 38°C for 72 and 96 h before storage at 0°C for 8 months in 2–3% oxygen with 5% carbon dioxide were firmer at the end of storage. They also had a higher soluble solid to acid ratio and a lower incidence of superficial scald than non-treated fruit. Pre-storage regimes with longer exposures to high temperature of 46°C for 24 h or 42°C for 48 h resulted in fruit damage after storage (Klein & Lurie 1992).

### *Drying*

Cereal grains and legumes attain physiological maturity at moisture contents ranging between 35% and 45% depending on the crop. However, appropriate moisture levels for safe storage vary between 10% and 14%, depending on the grain and factors such as the storage temperature. Timely harvesting and drying at crop maturity are important for achieving a high-quality product. Under most circumstances, stored grain deteriorates because it is biologically active and contaminated by fungi, but it can also be damaged by insects and other pests. The primary objective of drying is to reduce grain respiration through removal of excess moisture, but drying will also prevent the qualitative deterioration which may arise from growth of fungi and insect pests.

The process of drying may itself adversely affect grain quality unless the method of drying is operated with care. Whatever method is used, when grain is dried there is likely to be some change in quality. Generally, the adverse effects on grain quality will increase with the severity of the drying conditions. Damage to grain may arise if drying is carried out at too high a temperature or too quickly.

Whenever possible, most grain crops will be harvested during a dry season and simple drying methods such as exposure to sun and wind may be adequate. However, crop maturity does not always coincide with a suitably dry period. The introduction of high-yielding varieties, irrigation and improved farming practices have all led to the need for alternative drying practices to cope with increased production or with grain harvested in a wet season as a result of multi-cropping.

Drying may be used to preserve perishable commodities, some of which may then be stored for long periods. However, extra care is needed when drying perishables since their nutritional quality and palatability may be impaired. For example, roots and vegetables are a source of the essential vitamin C, and this may be destroyed if the produce is exposed to high temperatures.

### *Sun drying*

Sun drying of grain and some perishable commodities is still a common drying method in tropical developing countries. It may be first employed when the crop is standing in the field prior to harvest; for example, maize cobs may be left on the plant for several weeks after attaining maturity, to allow them to dry. Alternatively, drying in the field may involve stacking the harvested plants, cobs or panicles, sometimes on frames or platforms, where they are exposed to the sun. During these periods of drying, crops are vulnerable to insect infestation, mould damage and possible attack by rodents or birds.

Drying of threshed grain and fresh produce on flat exposed surfaces is also common but, unless the commodity is spread on trays or sheets, there is a risk of contamination by soil or stones. Large quantities of grain (e.g. paddy at rice mills) may be dried on specially constructed drying floors that are easy to maintain and will allow rapid run-off of rainwater. The grain is spread in thin layers and raked or turned at regular intervals during the day to facilitate drying, but must be gathered up and covered with sheets at night.

Sun drying has a number of disadvantages:

- the temperature cannot be controlled (important when drying perishable crops and certain grains, e.g. paddy rice dried at high temperature may result in stress cracks leading to high levels of broken grains during milling);
- produce is exposed to dust and atmospheric contamination;
- produce may be disturbed or removed by humans or animals; and
- produce is exposed to a risk of insect infestation.

### *Solar drying*

Solar drying is a modification of sun drying in which the sun's rays are collected inside a specially designed unit with adequate ventilation for removal of moist air. The temperature in the unit is usually 20 to 30 degrees higher than in open sunlight, which results in a shorter drying time. Solar driers of similar designs may be used for both grains and perishable commodities. Solar drying of perishables in driers that allow some degree of temperature and airflow control is preferable to natural drying.

In solar driers, air is heated in a solar collector and then passes through a bed of grain. There are essentially two basic designs: natural convection driers where air flow is induced by thermal gradients, and forced convection driers where a fan is used to force air through a solar collector and the layer of grain.

Natural convection driers are often suitable for on-farm use. An early design from the Asian Institute of Technology in Bangkok (Boothumjinda *et al.* 1983) has been used as a blueprint for many such natural convection driers and consists of a solar collector, a drying bin and a solar chimney. The base of the solar collector consists of a layer of black polythene sheet or layer of burnt paddy husk and is covered with clear polythene sheeting. The drying bin consists of a perforated platform. The solar chimney provides a column of warm air that increases the thermal draught of air through the drier. Disadvantages of the drier are the high structural profile, which poses stability problems in windy conditions, and the need to replace the polythene sheet regularly.

### *Mechanical driers*

The forced convection solar drier can be regarded as a conventional mechanical drier in which air is forced through a bed of grain, but the air is heated by a flat plate solar collector rather than by more conventional means.

In modern, fully mechanised storage systems, drying may take place at one of two points within the system – prior to loading grain into store using 'free-standing driers' (pre-storage driers) or after the grain has been loaded into the final storage receptacle (in-store drying). Pre-storage driers include batch or continuous flow driers in which ambient air is used, occasionally being heated using a thermostatically controlled furnace powered by gas, diesel or electricity. Heat may be delivered either directly or indirectly. The latter is preferred since the products of combustion are vented to the atmosphere and not through the grain. With batch dryiers, the grain is dried in clearly defined batches, whereas, in a continuous flow drier, grain is fed into the system and then moves slowly through the drier until it emerges at the desired moisture content.

*Tray driers,* sometimes called flat bed driers, are the simplest of the batch driers. Grain is loaded onto mesh trays within the drier to a depth of 600–700 mm and warm, dry air is blown through the trays until the grain has been dried sufficiently.

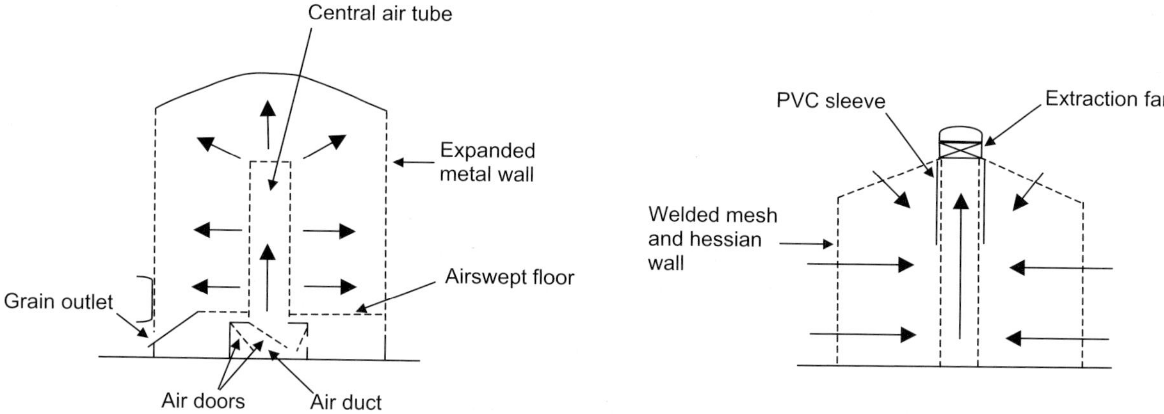


**Fig. 5.1** Principle of operation of simple radial drying bins.

*Radial drying bin*: The basic design of the radial drying bin consists of two vertical metal mesh cylinders, one inside the other (Fig. 5.1). Grain is loaded into the space between the two cylinders and air is blown into the inner cylinder; from there, it passes through the mesh wall, and the bed of grain and out of the outer mesh wall. Air can be sucked out of the central cylinder, reversing the airflow through the grain (Fig. 5.1). These driers are popular, but there is a risk that grain nearest to the inner cylinder can be overdried as this is where hot, dry air first makes contact with the grain. Grain towards the outside where the air is leaving the grain is cooler and wetter.

*Continuous flow driers* are designed to remove moisture from the grain rapidly by blowing or sucking hot air through the grain as it passes through the system from top to bottom. The simpler designs of continuous flow driers have a holding bin on top of the tall drying section. A final cooling section, where air at ambient temperature is blown through the grain, is usually incorporated at the base of the drier to avoid conveying hot grain into the rest of the system. The bed of grain may be vertical, horizontal or inclined. Grain may be moved by vibration, scrapers, conveyors, or by gravity. The degree of drying is determined by the rate of grain flow through the drier, and is usually controlled by varying the speed of outlet conveyors or the size of the outlet itself.

There are three categories of continuous flow driers based on the relative directions of the grain flow and the air stream:

*Cross-flow*, in which grain moves downward in a column between two perforated metal sheets while air is forced through the grain horizontally. These driers are relatively simple but as moisture gradients can be set up across the bed of grain mixing of the grain at some stage is advantageous.

*Counter-flow*, using a round bin, unloaded at the base and with an upward airflow. In these driers the hottest air meets the driest grain, causing little evaporative cooling, and so air temperatures must be limited, reducing efficiency. However, grain drying is uniform throughout the drier.

*Concurrent-flow*, in which the air moves down through the bed of grain so the hottest air meets the wettest grain. Moisture is evaporated thereby cooling the grain. High temperatures can be used for drying, thus improving efficiency.

Cross-flow type driers have been used widely but in recent years mixed-flow driers, incorporating a mixture of cross, concurrent and counter-flows, have become more popular, largely because of their high fuel efficiency. However, wet grain and high levels of trash can result in an uneven flow of grain through the drier, which in turn leads to uneven grain drying and reduced throughputs.

*Tower (mixed-flow) driers* resemble tall rectangular storage bins, fitted with a large number of horizontal, triangular-shaped ducts across the width of the drier (Fig. 5.2). Approximately half of the ducts introduce warm dry air into the grain while the others remove the cooled, damp air. Grain is fed in at the top and passes down, around the ducts, by gravity. With the drier full of grain, the grain moves slowly down through the drier, first through the drying section and then a final cooling section using ambient air.

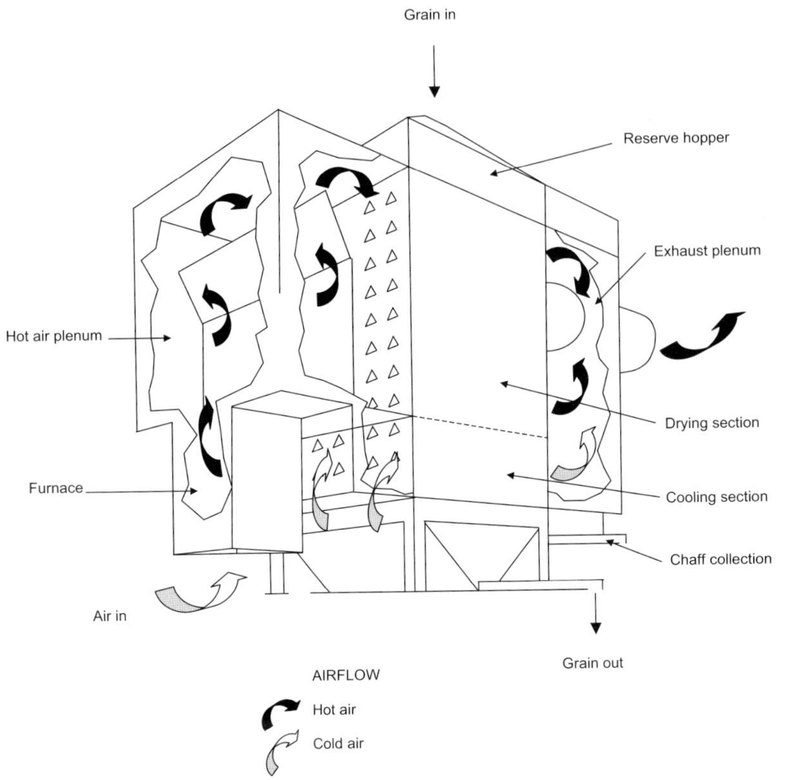


**Fig. 5.2** Tower (mixed-flow) drier.

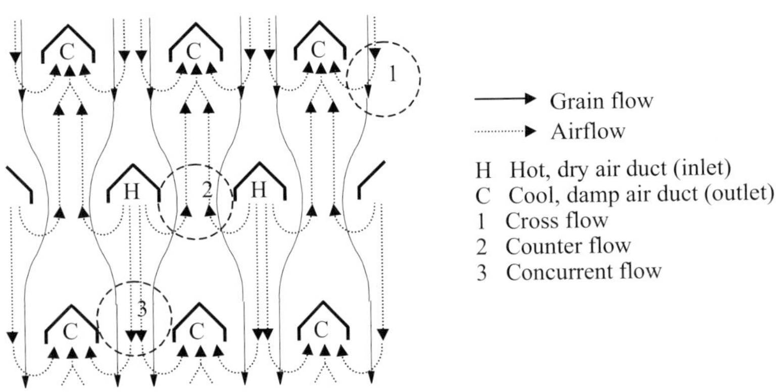


**Fig. 5.3** Tower (mixed-flow) drier – principles of operation.

A cross-section through part of the drier (Fig. 5.3), shows that the relative directions of grain versus airflow varies between cross-, concurrent- and counter-flow, i.e. it is a true multi-flow dryer.

*Louvered-bed drier*: This drier is similar in design to the tray batch drier except that the grain is dried as it passes over louvered-beds through which heated air is blown rather than being dried *in situ*. These are generally cross-flow driers. The depth and speed of the moving bed of grain is controlled to determine the degree of drying. There are two basic designs, the cascade drier and conveyor drier.

The cascade drier is a gravity-fed cross-flow drier with grain depth controlled by a series of roller dams

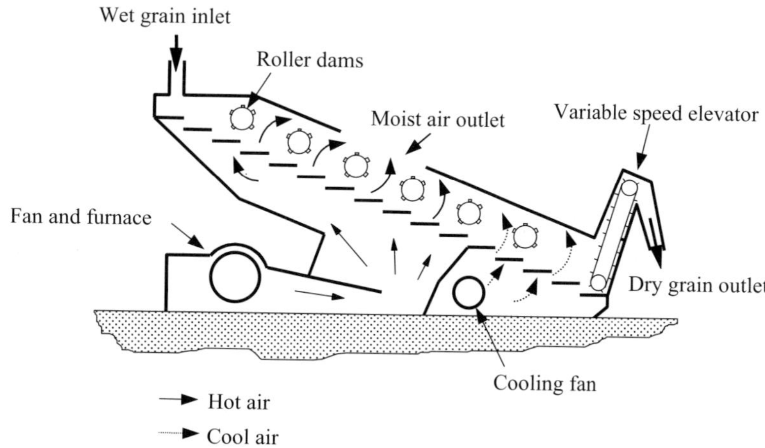


**Fig. 5.4** Single-directional cascade drier.

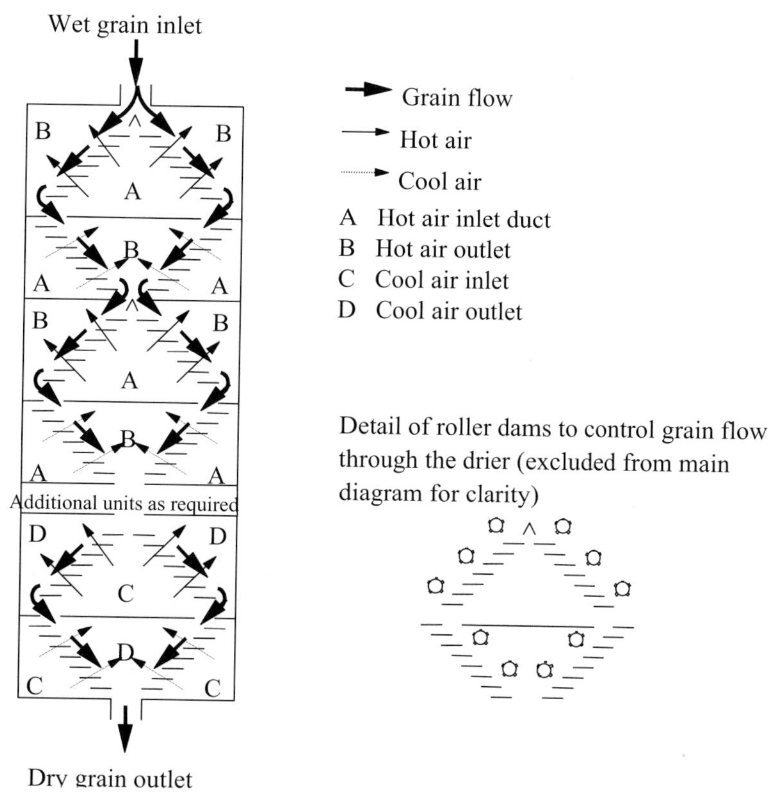


**Fig. 5.5** Multi-directional cascade drier.

along the length of the louvered-bed (Fig. 5.4). The degree of drying is controlled by the speed of the output elevator. To avoid excessively long lengths, larger driers may involve several changes in direction of grain flow (Fig. 5.5).

In conveyor driers (Figs 5.6 and 5.7) grain is also supported on an inclined louvered-bed through which air is blown, but grain flow through the drier is by a variable speed, heavy duty, roller chain conveyor. These driers may be single-directional, two-directional or multi-directional.

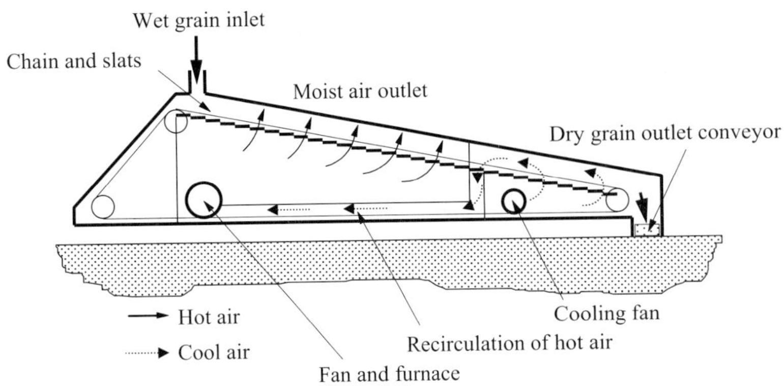


**Fig. 5.6** Single-directional conveyor drier.

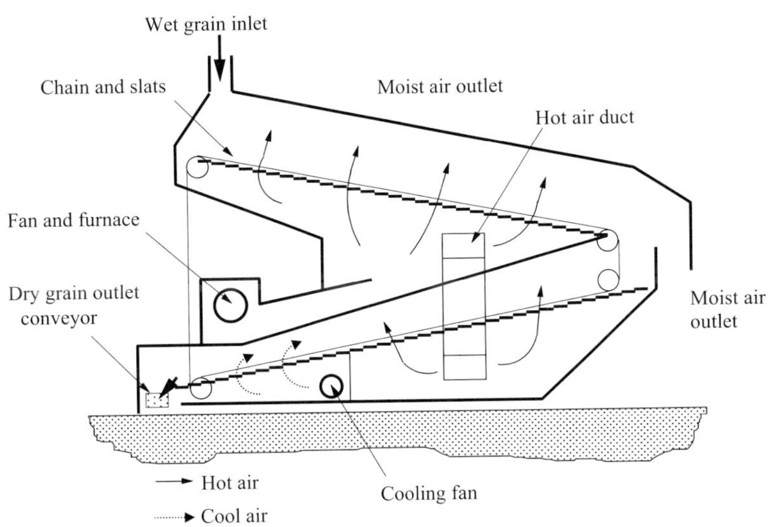


**Fig. 5.7** Double-directional conveyor drier.

The changes in direction of grain flow mixes the grain, allows some of the dried waste material to be blown from the grain and reduces the overall size of the machine.

### *In-store drying*

As an alternative to drying the commodity before storage, grain may be loaded into bulk floor storage, and bins and then dried in-store.

*Bulk on-floor storage*: The store usually consists of a simple, clear span, concrete floored building with specially strengthened retaining walls to withstand the weight of grain when loaded against them to produce a bed of grain of uniform depth. The aeration system consists of a fan located at one end of the building connected to a main duct or plenum chamber running either along one wall or along the centre of the store and with smaller perforated lateral ducts running above or below floor level, under the grain bulk (Figs 5.8 and 5.9).

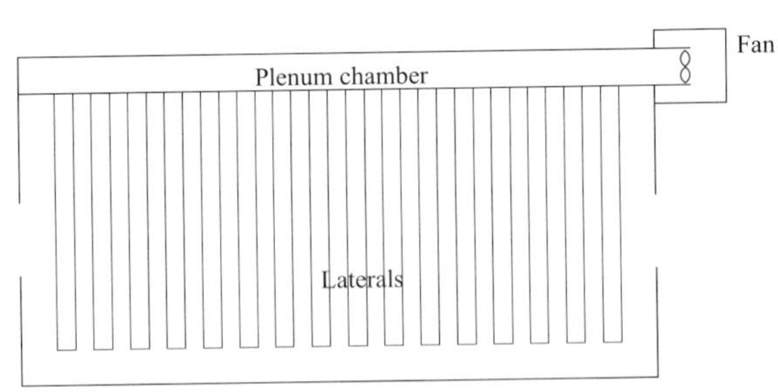


**Fig. 5.8** Overhead view of the floor of a ventilated store with the plenum chamber running along one side of the store.

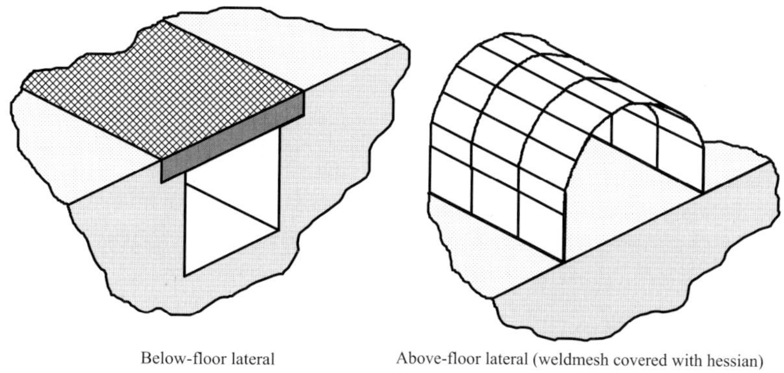


**Fig. 5.9** Typical above-floor and below-floor laterals.

*In-bin drying*: Bin storage systems may consist of one or more bins with drying floors, with the remaining bins used for storage. After drying, grain can be transferred to the storage bins. Alternatively all bins may have drying floors, thus avoiding the need to transfer the dried grain. This has the advantage of reducing handling which in turn reduces the risk of physical damage. Also, high moisture content grain can be spread between several bins in relatively shallow layers, making drying quicker and safer. Once the grain has been partially dried batches can be combined, freeing space for more incoming grain. Bins may be ventilated using a system of laterals (Fig. 5.10), or a fully ventilated floor usually consisting of expanded metal mesh, supported approximately 0.5 m above the base.

*Bag driers*: Drying of grain in bags is difficult – primarily because of the need to ensure that the drying air is forced through the bags and therefore the grain and not through the air spaces between the bags, the path of least resistance.

Sack platform driers consist of a specially constructed floor supported over an air duct or plenum chamber while a fan blows heated air under the floor and through apertures over which bags can be placed.

Moisture extraction units can be used with larger quantities of bags. Bags are stacked specially for drying by incorporating a tunnel in the centre of the stack. A large fan blows heated air via ducting into this tunnel (Fig. 5.11). Care is needed in selecting the correct dimensions of the stack to ensure even drying of the grain.

Drying by ventilation of conventional bag stacks is difficult since air will often short-circuit through certain areas of the stack, leaving others unventilated. Trials in Indonesia using small stacks ventilated with low volume fans were reasonably successful. Shallow pallets were laid on top of the stack and the stack covered in polythene sheeting, leaving a gap around the dunnage at the base. Air was extracted with a small fan through a hole in the sheeting at the top of the stack. During operation, the sheeting was

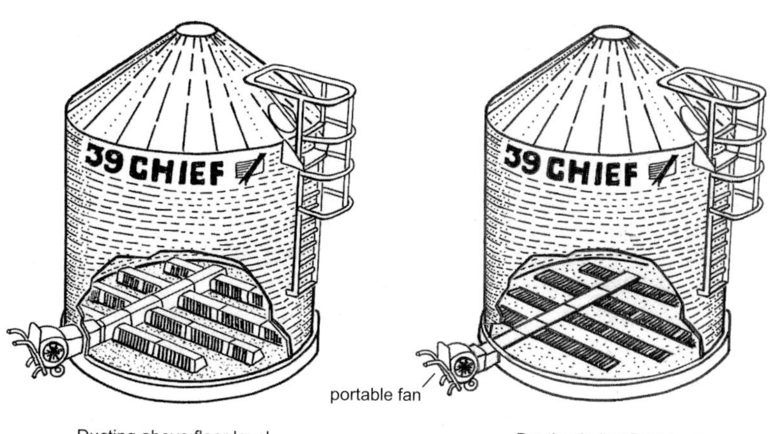


**Fig. 5.10** Ventilation of round bins using laterals.

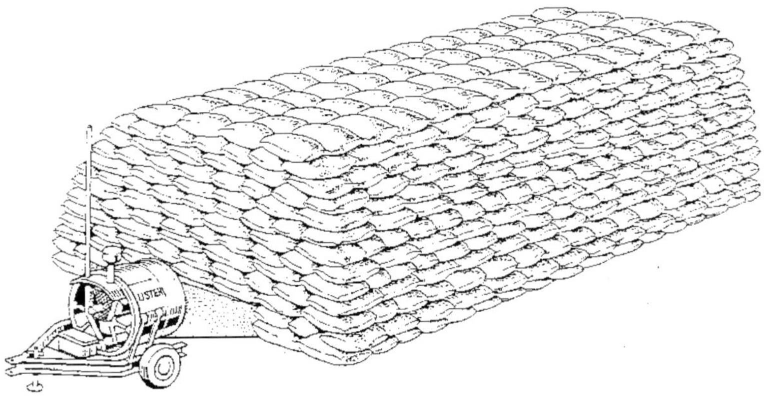

**Fig. 5.11** Moisture extraction unit.

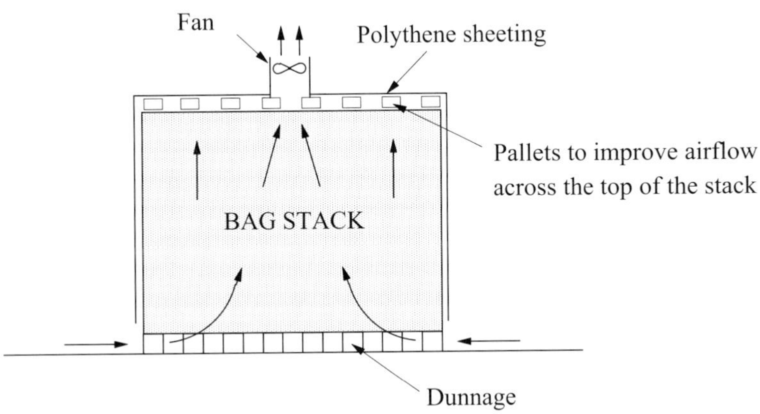


**Fig. 5.12** Ventilation of conventional bag stacks.

sucked against the stack and air was drawn into the base of the stack from under the dunnage (Fig. 5.12). Although effective in ventilating the stack, there is no modification of the incoming air through heating or dehydration and so the degree of drying is solely dependant on the conditions of the air within the warehouse.

### *Cooling of perishable crops*

Throughout the period between harvest and consumption, temperature control has been found to be the most important factor in maintaining the quality of perishable crops. The temperature of the product is the greatest determinant of the rate of deterioration by decay and senescence and consequently of the potential market life; each 10°C reduction in product temperature reduces the respiratory activity by a factor of 2–4. Lowering the temperature also decreases the product's sensitivity to ethylene and reduces moisture loss. By reducing the temperature of the product to the lowest safe temperature (0°C or 32°F for temperate crops or 10–12°C or 50–54°F for chilling-sensitive crops) the potential storage life of the product can be maximised. Any delays in cooling or suboptimal cooling, and storing or shipping above or below optimal temperature, will reduce the quality of produce and its storage life.

At harvest, the temperature of perishable commodities often exceeds the optimum temperature for maintaining quality and the rapid removal of this field heat is of critical importance to overall temperature management of the crop. Pre-cooling is the rapid removal of field heat to bring the product temperature down to or near to its subsequent storage or shipping temperature. This is particularly important for products that produce a lot of heat, for example, artichokes, asparagus, snap beans, blackberries, broccoli, Brussels sprouts, carrots, sweetcorn, lettuce, mushrooms, green onions, okra, parsley, peas, raspberries, spinach, strawberries and watercress. The benefit of reducing the temperature of produce rap-

**Table 5.5** Effects of delaying cooling on potential storage life (days) of lettuce. Source: Longmore 1974.

| Delay between harvesting and cooling (h) | Harvested at | | |
|---|---|---|---|
| | 16°C | 20°C | 24°C |
| 2 | 9.5 | 7.5 | 5.5 |
| 4 | 8.5 | 6.5 | 5.0 |
| 6 | 7.0 | 6.0 | 4.0 |
| 8 | 6.0 | 5.0 | 3.5 |
| 10 | 5.0 | 4.0 | 2.5 |

idly after harvest by extending the storage life and in maintaining market quality is illustrated in Table 5.5.

### *Pre-cooling factors*

The success of pre-cooling is dependent on a number of factors:

- the duration between harvest and pre-cooling;
- type of shipping container;
- if the product is packed beforehand;
- the initial product temperature;
- velocity or amount of cold air, water or ice provided;
- the final product temperature;
- sanitation of the pre-cooling air or water to reduce decay organisms; and
- maintenance of the recommended temperature after pre-cooling.

Pre-cooling should occur as soon as possible after harvest. Harvesting should be done in the early morning to minimise field heat and the refrigeration load on pre-cooling equipment. Harvested products should be protected from the sun with a covering or stored under the shade of a tree until they are transported to the pre-cooling facility.

The cool chain should be maintained between the field and the pre-cooling facility – where possible refrigerated vans should be used. When these are not available, the produce should be stowed in the transport vehicle in such a way as to ensure the produce is not exposed to direct sunlight (e.g. through the use of protective coverings), and is well ventilated. Harvesting and packaging of most products should be closely co-ordinated with transportation to minimise time in transit and storage, and maximise product freshness in the hands of consumers.

Many products are field or shed packed and then pre-cooled. Pre-cooling of products packed in boxes and stacked in unitised pallet loads is especially important, as air circulation around and through the packaging may be limited during transportation and storage. After pre-cooling, the products must be properly loaded and transported at or near the recommended storage temperature and relative humidity to maintain quality.

### *Pre-cooling methods*

Cooling is a term that is often used quite loosely. In order to be effective and significantly benefit the shipping and storage life of the product, an appropriate definition of commercial cooling for perishable crops is: 'the rapid removal of at least 7/8 of the field heat from the crop by a compatible cooling method'. The time required to remove 7/8 of the field heat is known as the *cooling time*. Removal of the remaining 1/8 of the field heat will occur during subsequent refrigerated storage and handling with little detriment to the product.

The rate of heat transfer, or the cooling rate, is critical for efficient removal of field heat in order to achieve cooling. During cooling, the field heat from the product is transferred to the cooling medium. The efficiency of cooling is dependent on time, temperature and contact. In order to achieve maximum cooling, the product must remain in the pre-cooler for sufficient time to remove heat. The cooling medium (air, water, crushed ice) must be maintained at constant temperature through the cooling period. The cooling medium also must have continuous, intimate contact with the surfaces of the individual fruits or vegetables. Inappropriately designed containers with insufficient vent or drain opening or incorrectly stacked pallets can markedly restrict the flow of the cooling medium, increasing cooling time.

The choice of pre-cooling method depends on the nature, value and quantity of the product, as well as the cost of labour, equipment and materials. Pre-cooling methods include evaporative cooling, positive ventilation and an ice bank cooler, ice cooling, room cooling, forced air-cooling, hydro-cooling and vacuum-cooling. Table 5.6 summarises the various cooling methods and their suitability for various products.

Hydro-cooling and vacuum cooling are the fastest cooling methods, with cooling times of half an hour possible. However, the produce and packaging must be able to withstand direct water contact in hydro-cooling and, for vacuum cooling, the produce should have a large surface area, low density, and high moisture content.

**Table 5.6** Cooling methods for fruits and vegetables.

| Cooling method | Products cooled | Comments |
|---|---|---|
| Evaporative cooling | Non chill-sensitive products | Low cost equipment<br>Mobile equipment<br>Need for temperature differential between wet and dry bulb |
| Positive ventilation with ice bank cooler | Non chill-sensitive products | Low operating costs<br>Moderately effective<br>Maintains high relative humidity |
| Ice cooling | Leafy vegetables, some roots, e.g. carrots, stem vegetables, etc. | Rapid<br>Limited to commodities which can tolerate contact with ice and water |
| Room cooling | All products | Relatively slow<br>Uneven cooling |
| Forced air cooling | Flowers, soft fruits, fruits, fruit type vegetables | Versatile<br>Efficient<br>Uniform |
| Hydro-cooling | Roots, tubers, some fruits, e.g. melons, leafy and stem vegetables | Rapid<br>Produce must tolerate contact with water<br>Water-resistant containers needed |
| Vacuum cooling | Leafy vegetables, e.g. lettuce, some fruit and flower vegetables | Rapid |

Forced-air cooling can take 1 or 2 hours, depending on the amount of packaging, while room cooling may take 24–72 hours. In both of these methods it is important that the packaging allows ventilation of heat.

Portable ice plants, hydro-coolers, vacuum coolers, forced-air coolers and package-icing machines are available for use in the fields and thus useful for remote or small-scale operations that cannot justify investment in a fixed pre-cooling facility. When mounted on skids, dollies or tractor-trailers, the equipment can follow the harvest from field to field and can be shared by many growers.

*Evaporative cooling* is an effective means of providing reduced air temperature and high relative humidity for cooling produce. However, a well-designed system will only provide air at approximately 1°C above the wet bulb temperature of the outside, ambient air. The system is unsuitable for some regions with low wet bulb temperatures where there is a need to cool chill-sensitive crops. Designs for small commercial units with a cooling capacity of approximately 300 kg in 2 hours are available (Thompson & Kasmire 1981). The relatively small investment cost of an evaporative cooling system and the low energy requirement make it a useful commercial tool for some regions.

Evaporative coolers can be constructed from simple materials, such as burlap and bamboo. Wetting the walls and roof first thing in the morning creates conditions for evaporative cooling of a packinghouse that is made of straw. This dripper cooler operates solely through the process of evaporation, without the use of a fan. Cooling will be enhanced if the unit is kept shaded and used in a well-ventilated area (Kitnoja & Kader 1995).

The low-cost cooling chamber is constructed from bricks. The cavity between the walls is filled with sand and the bricks and sand are kept saturated with water. Fruits and vegetables are loaded inside, and the entire chamber is covered with a rush mat, which is also kept moist. The relatively high cost of construction materials may limit the use of this cold storage chamber to the handling of high-value products. During the hot summer months in India, this chamber is reported to maintain an inside temperature between 15 and 18°C and a relative humidity of about 95% (Kitnoja & Kader 1995).

Evaporative coolers can be constructed to cool the air in an entire storage structure or just a few containers of produce. These coolers are best suited to lower humidity regions, since the degree of cooling is limited to 1–2°C above the wet bulb temperature. A cooling pad of wood fibre or straw is moistened and air is pulled through the pad using a small fan. Water is dripped, at around 2 litres per minute, onto a pad of around 1 $m^2$, providing enough moist air to cool up to 18 crates of produce in 1 to 2 h. Water is collected in a tray at the base of the unit and recirculated (Kitnoja & Kader 1995).

An evaporative cooler can be combined with a forced air cooler for small lots of produce. Air is cooled by passing through the wet pad before it passes through the packages and around the produce. The air can be cooled to within a few degrees of the wet bulb temperature of ambient air.

*Positive ventilation and an ice bank cooler*: A large percentage of the total refrigeration capacity of a conventional refrigerated store is needed only to remove the field heat at the time of pre-cooling. The principle of the ice bank cooler is that the initial demand for refrigeration for heat removal is achieved by melting ice, which has been built up over a number of days by a small refrigeration unit. The heat is removed from the air in the store by passing it through sprays of ice-cold water in a cooling tower. Positive ventilation is achieved by directing the air through the produce in the store.

*Ice cooling*: Ice may be used in a number of ways to reduce product temperature. Contact icing has been used for both cooling and temperature maintenance during shipping. Heat from the product is absorbed by the ice, causing it to melt and maintain a high relative humidity. There are two types of contact icing:

*Top icing*: usually applied on loads in rows rather than a solid mass. It is important not to block air circulation inside the transport vehicle. Although relatively inexpensive, the cooling rate can be fairly slow since the ice only directly contacts the product on the top layer. For this reason, it is recommended that top icing be applied after pre-cooling to crops with lower respiration rates such as leafy vegetables and celery but not for fruit of warm-season crops. Prior to shipping, ice is blown on top of loaded containers to aid cooling and maintenance of higher relative humidity. Ice should be 'tempered' with water to bring the temperature to 0°C to avoid freezing of the product.

*Package icing*: involves distribution of crushed ice within the container. Package icing may be finely crushed ice, flaked ice or a slurry of ice and water called liquid ice (Kitnoja & Kader 1995). Cooling is faster and more uniform than for top icing, but it can be more labour-intensive. Package ice can be used only with water-tolerant, non-chilling-sensitive products (such as carrots, sweetcorn, cantaloupes, escarole, lettuce, spinach, radishes, broccoli and green onions) and with water-tolerant packages (waxed fibreboard, plastic or wood).

A modified version of package icing utilises a slurry of refrigerated water and finely chopped ice drenched over either bulk or containerised produce, or injected into side-hand holds. This 'slush ice' method has been widely adopted for commodities tolerant to direct contact with water and requiring storage at 0°C. The water acts as a carrier for the ice so that the resulting slush, or slurry, can be pumped into a packed container. The rapidly flowing slush causes the product in the container to float momentarily until the water drains out of the bottom. As the product settles in the container, the ice encases the individual vegetables by filling air voids, thus providing good contact for heat removal. Slush icing is somewhat slower than forced-air cooling, but it reduces pulp temperatures to 0°C within a reasonable time and maintains a high relative humidity environment.

*Room cooling*: This simply involves placing the product in a cool store; some products are misted or sprayed with water during room cooling. While this is not always very effective or efficient, best results can be achieved by:

- provision of adequate refrigeration;
- an air velocity of 0.3–0.45 m/s;
- spaced stacking of containers; and
- provision of ventilated containers.

The higher air velocity required for cooling should be reduced to the lower value when the product has reached the required temperature.

*Forced air cooling*: Produce to be cooled by this method is also placed in a cool store and cool air is forced through the container vents, thus achieving closer contact with the warm produce. The technique is versatile and can be very effective. Existing cool stores can be modified to optimise on this technique. Rapid cooling can be accomplished with adequate refrigeration and a large volume of airflow per unit of produce. As cooling is fairly rapid, the maintenance of high relative humidity is less important in a forced air cooling system than in the subsequent storage environment.

Many types of forced-air coolers can be designed to move cold moist air over the commodities. An example of a fixed unit, where a fan is housed inside the wall of a cold room, is given below.

A portable forced-air cooler can be constructed using a canvas or polyethylene sheet. The sheet is rolled over the top and down the back of the boxes of produce to the floor, thus sealing off the unit. Air is then pulled through the vents (vent area should be at least 5% of the surface area of the carton) of the cartons stacked against the cooler. This unit is designed for use inside a refrigerated storage room.

Forced-air cooling is recommended for crops, such as peppers, which are susceptible to water-borne decay organisms.

*Hydro-cooling* requires intimate contact between the product and cold water for rapid heat removal. There are two basic types of hydro-cooler: a conveyor hydro-cooler and a batch hydro-cooler. Both systems place a relatively heavy demand on refrigeration capacity over a short time period and consequently ice is frequently used. Care must be taken to ensure strict sanitation measures are followed to avoid the spread of decay organisms.

Hydro-cooling provides fast, uniform cooling for some commodities such as sweetcorn, snap bean, cucumber and summer squash. The commodity as well as its packaging materials must be tolerant of wetting, chlorine (used to sanitise the hydro-cooling water) and water beating damage.

The simplest version of a hydro-cooler is a tank of cold water in which produce is immersed as it moves along a conveyor. A batch-type hydro-cooler can be constructed to hold entire pallet-loads of produce. Conveyors can be added to help control the time produce stays in contact with the cold water.

*Vacuum cooling* is a very rapid and uniform method for cooling produce, particularly items with a large surface-to-volume ratio such as leafy crops (e.g. iceberg lettuce, cabbages, cress) and in some cases, mushrooms and celery (Longmore 1974). Vacuum cooling operates on the principle that water boils at a lower temperature when under reduced pressure. As the vapour pressure of the chamber is reduced, water evaporates from the product and cooling is achieved as heat is used to evaporate the water. Wetting of some products is practised to reduce the water loss from the product and increase the efficiency of cooling. There is a degree of weight loss from the produce which varies considerably from one commodity to another, but the rule of thumb is: the larger the surface area the less noticeable the moisture loss and weight loss.

### *Market considerations for pre-cooling*

The technical merits of rapid cooling after harvest (pre-cooling) may be clearly apparent, but the benefits will be fully realised only when this is the first step in a cool chain extending from field to market. However the technical merits alone may not be sufficient to justify investment in infrastructure for produce pre-cooling and/or placing demands upon persons along the marketing chain to modify their handling procedure. There are two basic criteria to justify investment in a partial or complete cool chain: (1) the relatively high value of the product, and (2) the perishable nature of the product.

With increases in consumer prices for many fresh products, the demand for optimal quality will increase. This increasing quality market has been seen in many regions of the world as an incentive to invest in pre-cooling and subsequent cool chain handling.

Having satisfied the basic criteria, a detailed examination of the distribution pattern becomes necessary. It is important to understand the financial responsibility of participants in the distribution chain and examine the entire product handling system when planning any pre-cooling and cool chain improvements. Points of the chain that require particular consideration include:

- commodity type and characteristics;
- period of production;
- harvesting method and schedule;
- time and type of packaging material;
- transport and storage facility; and
- consumer demands.

One frequently overlooked critical factor is the need for pre-cooling of products, to remove field heat, to be carried out as soon as possible after harvest. There is little justification for incurring the expense of cool chain handling for produce already inferior in product quality due to poor or lengthy field handling operations.

Careful consideration should be given to the time of cooling in the handling chain. There are both advantages and disadvantages in cooling before and after packing into the market container (Table 5.7).

Having reviewed the organisation and management of the chain, the economics of the operations must be carefully investigated. These include plant ownership, capacity, projected utilisation, capital investment of plant and ancillary services and operating costs, including labour.

**Table 5.7** Advantages and disadvantages in cooling produce before packing.

| Advantages | Disadvantages |
|---|---|
| More effective cooling in field containers compared with market containers | Cost of cooling culls and reject fruit |
| Sooner after harvest, thus reducing rate of deterioration | Rewarming during packing |
| | Higher overall cost of cooling |

## Storage structures

The primary objective of storage is to maintain the quality of the commodity from the time of loading for as long as is required. In practice, many commodities will begin to deteriorate after harvest and so the objective of storage will be to minimise the rate of deterioration. This is achieved by providing protection against external elements (including temperature and humidity as well as pests such as rodents, birds and theft by man) and protection against the products of respiration of the commodity (heat, moisture, and gases such as carbon dioxide and ethylene). There is no single, ideal design of store. This will be determined by the particular storage requirements of the commodity and the potential for loss.

### Store size

The storage space required within a storage structure generally will be determined by the volume and type of produce to be stored. In most instances the usable capacity of the store will be less than the nominal capacity. A number of factors must be taken into consideration when calculating store capacity.

When a commodity is stored in bulk, reference needs to be made to its bulk density, or its inverse, the stowage factor (Tables 5.8 and 5.9). If storing in containers such as bags or cartons, the number of containers (of standard capacity) to be stored may have to be determined before assessing the total quantity of produce that can be stored. Alternatively a 'reduced' bulk density that allows for packaging can be used (compare the figures for bagged and bulk durable commodities in Table 5.8).

Allowance must also be made for the physical and physiological characteristics of the commodity and its packaging. For example, cartons may be crushed and there may be an increased risk of stack collapse if cartons of heavy produce are stored more than 2 m high (Table 5.10). Onions and potatoes may be crushed if they are heaped in excess of 4.6 m and 6 m, respectively (Hardenburg *et al.* 1986). Managerial and operational requirements will affect store capacity. Gangways around, and space above stacks may be needed for inspection, in-store pest control treatments or to allow access to other stacks in the store.

A consideration of store size is particularly important when planning a controlled atmosphere (CA) store for horticultural produce. For effective operation these stores should be filled and emptied as quickly as possible. Ideally, the store should be completely filled with produce of the same or compatible cultivars, but as this may not always be possible the store may be divided into small units to offer greater flexibility. Typically controlled atmosphere stores in the UK have an average capacity of about 100 tonnes (ranging between 50 and 200 tonnes). In continental Europe the capacities of CA vegetable stores range between 100 and 500 tonnes

**Table 5.8** Stowage factors for selected durable commodities. Source: Bridger & Watts 1930.

| Commodity | Stowage factor | |
|---|---|---|
| | Bagged ($m^3$/t) | Bulk ($m^3$/t) |
| Barley | 1.54–1.96 | 1.48–1.79 |
| Oats | 2.12 | 1.82–2.10 |
| Rice: clean | 1.26–1.4 | |
| milled | 1.4–1.5 | |
| paddy | 1.9 | |
| Wheat | 1.48 | 1.20–1.40 |
| Maize | 1.58–1.68 | 1.54–1.60 |
| Sorghum, millet | 1.26–1.43 | |
| Beans | 1.2–1.8 | |
| Peas | 1.4 | |
| Lentils | 1.4 | |
| Soybeans | 1.31–1.60 | |
| Cowpeas | 1.31 | |
| Haricot beans | 1.82–1.90 | |
| Groundnuts kernels | 1.9 | 1.68–1.96 |
| Rape seed | 1.29–1.57 | |
| Linseed | 1.68–1.93 | 1.43 |
| Sesame | 1.68–1.90 | |
| Millet seed | 1.26–1.43 | |
| Cocoa beans | 2.52 | |
| Coffee: washed beans | 1.51 | |
| dry parchment | 2.84 | |
| green beans | 1.54 | |

Note: The bulk density (t/$m^3$) of these commodities will be the reciprocal of the stated values, i.e. stowage factors ($m^3$/t) = 1/bulk density (t/$m^3$).

**Table 5.9** Stowage factors for selected perishable commodities.

| Commodity | Stowage factor | |
|---|---|---|
| | Containerised ($m^3$/t) | Bulk ($m^3$/t) |
| Carrots | | 1.7–1.86 |
| Onions | | 2.25–2.5 |
| Potatoes | 2.0–2.2 (pallet boxes) | 1.5–1.65 |
| Red beetroot | | 1.38 |
| Winter white cabbage | | 2.55 |

**Table 5.10** Suggested stack heights for durable and perishable commodities.

| Commodity | Suggested stack height (m) | Special features |
|---|---|---|
| Whole wheat, milled rice, maize, sorghum, bulgur wheat | 5–6 | Stack height will be limited in smaller stacks to: 3 m for stacks < 100 t; 4 m for stacks < 400 t; 5 m for stacks < 800 t |
| Peas, beans, lentils | 3 | Beans may become hard to cook if storage unduly prolonged. |
| Wheat flour, maize meal | 3 | Avoid strong odours, high temperatures |
| Dried fruits | 3 | Should not be applied to humid areas – keep dry, in separate stores from odour-sensitive foods |
| Potatoes | 4.6 | Maximum for bulk storage |
| Onions | 6.0 | Maximum for bulk storage |

(average about 200 tonnes) whereas larger stores averaging 600 tonnes are found in North America (Bishop 1996). Consideration must also be given to the bin layout and spacing within stores. Bin size will vary with the commodity and area of operation, but the ideal layout and stacking pattern will be one which enhances air flows and store performance.

A typical design process will include:

- determining the type of commodity to be stored and the method of storage (containerised or bulk);
- determination of the physical factors involved (bulk density or the capacity and size of the containers);
- calculation of the space required to contain the commodity;
- adding additional space to allow for store operations (e.g. access space for inspection or treatments), flexibility (higher yields next year?), etc.;
- allowing for any limitations that may be imposed – for example, some commodities cannot be stacked too high, to avoid damage to the lower levels; and
- calculation of store size required or capacity of given store.

## Location and site selection

Many factors will affect the location and specific siting of storage structures including government planning policies, transport infrastructure, cultural requirements and topography. Ultimately, factors may be considered as economic or technical and, although the following discussion is concerned primarily with large storage complexes in mind, many of the factors are equally valid at the small-scale trader or farm level.

### *Economic aspects of site selection*

The location of a storage facility has an important influence on the overall costs of a storage, marketing and distribution system. Locating a store in an area where there is a significantly large marketable surplus or where there is a large consumer demand will minimise the costs of procurement and collection of produce. Produce can be stored promptly after harvest and demands for transport at the peak time after harvest can be minimised. Transportation costs can also be reduced if storage facilities are located on a major highway or railway line, or at a route centre. A consideration of climatic factors which favour a good storage environment, i.e. a cool, dry climate in the case of durable commodities, or a more humid one for certain perishable commodities, will also be important in relation to slowing the rate of deterioration and minimising loss. Finally, consideration may have to be given to local costs such as rent, local taxation/rates of storage structures, labour costs, and service charges (electricity, water, etc.).

### *Technical aspects of site selection*

Local planning rules may determine where storage buildings may or may not be built, but ideally stores must be easily accessible. It would be unwise to build large warehouses, with attendant heavy traffic, in, for example, densely populated residential suburbs, with narrow streets.

The site on which the store is to be built should have an adequate load-bearing capacity. Weak soils substantially increase the costs of a building's foundations. The site should remain fairly dry and firm all the year round, or at least permit easy, low-cost drainage. Flat sites are preferable to sloping ones, although a slight slope (up to 5%) is good for drainage. If stores are built on a hillside, the site will either have to be cut into the hill or built up to create a level area large enough for the warehouse floor. Special drainage will be required and steps may have to be taken to control erosion.

The orientation of the store may be important. If placed with its major axis on an east–west line (± 10°), only the end walls will be warmed by the rising or setting sun, thus reducing the solar gain (heating) inside the building. However, it may be preferable to orientate the building in a way that facilitates ventilation by prevailing winds through ventilators along the long sides of the building. In this respect wind can be important for natural ventilation, in which case sheltered positions should be avoided. However, it is inappropriate to position the store in a natural wind funnel, for example, between two hills, since the wind may carry dust and debris into the store or high winds may lead to structural damage.

The floors of the store must be above ground level. A floor approximately 0.3 m above the surrounding ground level should be sufficient to prevent water entering the building during times of high rainfall. Floors raised above this level add to the cost of construction, make it difficult to enter the store, and can lead to settlement problems.

Access and manoeuvring space for loading and unloading must be adequate. Attention must be paid to the sizes and positions of doorways, and access within the store. Trucks must be able to enter the site, pass over the weighbridge (where appropriate) and approach the store without hindrance. In large stores, trucks may need to enter and manoeuvre inside the store. In busy periods, trucks may need to wait on-site. Generous proportions should, therefore, be allowed for roadways, the radii (sharpness) of curves, cambers (slopes to the side to dispose of rainfall), aprons (additional space to the side of roadways for parking and manoeuvring) and queuing lanes. If vehicles are to enter the stores the floors should be built with these additional loads in mind. Using concrete or steel posts as buffers adjacent to the doorway should protect doorways against damage from vehicles.

### Construction materials for stores

A wide range of materials may be used in the construction of stores and the commonly available and used materials are summarised in Table 5.11. Construction materials are often classed as traditional and imported or modern materials although traditional and non-traditional are perhaps better terms since the important distinction is whether or not materials are in current use. A comparison of the relative advantages and disadvantages is listed in Table 5.12.

The choice of materials for store construction will depend on a number of factors including specific characteristics required of the materials used (e.g. great strength, water resistance, durability; availability of materials and skilled labour; transport costs; and cultural acceptability).

### Design elements

All stores from the smallest of traditional grain stores to the largest warehouse or silo must meet certain basic design objectives to provide protection against rain,

**Table 5.11** Types of materials commonly used in store construction.

| Material | Notes on use |
|---|---|
| Concrete | May be pre-cast or cast on site. Strong in compression but little strength in tension and is therefore reinforced, usually with a steel mesh. Various proportions of sand, cement, and aggregate mix are used for different situations |
| Brick and masonry | Includes bricks, blocks (high and low density), and stone |
| Earth/mud | Includes fired or unfired bricks and stabilised soil blocks (mud with cement added to improve strength and durability). Compacted mud walling is often strengthened with plant material such as grass or straw |
| Metals | Steel section ('I', 'H', 'L' and 'T' shaped sections), steel sheeting (flat or corrugated sheets), aluminium sheeting (corrugated sheets), stainless steels (occasionally used for fixtures and fittings), brass and bronze (fixtures and fittings), copper (pipework) |
| Timber | Beams and cladding (for internal or external covering of buildings). Seasoning of timber is important. Different types of hardwood and softwood available. Structural features vary considerably (e.g. some timbers are more resistant to termite damage) |
| Slates/tiles | Rarely used on stores unless locally available and therefore cheaper than alternatives. Possible high maintenance cost |
| Glass | Toughened/tempered glass, preferably with embedded wire mesh |
| Plastics | Use UV stabilised (chemical added to the plastic during manufacture to prevent the ultraviolet light (in sunlight) from breaking down the plastic). Cladding, fixtures, and pipework |

**Table 5.12** Advantages and disadvantages of traditional materials versus non-traditional materials.

| Traditional | Non-traditional |
|---|---|
| Generally cheaper<br>Local labourers/craftsmen familiar with and therefore equipped to work these materials<br>Craftsmen and materials will be available locally for maintenance and repairs after construction has been completed<br>Material may have been specifically adapted to local conditions<br>Locally acceptable<br>Improved local and national economy | Possibly higher quality than traditional materials due to higher technology and better quality control leading to reduced failure and subsequent maintenance requirements<br>Not restricted to local resources, raw materials or methods of construction. However, maintenance may be difficult if skilled craftsmen are not available<br>Increase options for building type and design<br>Up-to-date, able to adopt new or improved technologies<br>May or may not be locally acceptable |

ground water, adverse temperature, pests and theft. The store should also facilitate the management of both the structure and its contents. Factors to consider include: arrangements for loading and unloading; stock control according to the principle of 'first-in, first-out' (FIFO); commodity inspection and pest control procedures; cleaning and maintenance of the structure, including access to the roof and drains; and good general working conditions for safety, comfort and efficiency, including adequate lighting.

The important design considerations can be identified by considering each design objective in turn. However, in the optimum building design there may be a conflict of desirable characteristics and inevitably a compromise will have to be made. For example, better security is achieved with few doors and windows yet large doors may be necessary for optimum loading and windows and ventilators may be required to ensure adequate ventilation.

The diagrams accompanying the following discussion of important design elements represent a small warehouse as an example, but the principles will apply to any type of store.

### *Protection against moisture*

A well-designed roof and water disposal system is required (see Fig. 5.13). This may be either a wide roof overhang or gutters and downpipes to prevent water running down the walls or being blown into the store. Drains are needed to direct the water away from the building. In large warehouses, valley gutters (where two stores with pitched roofs are built side-to-side and adjoin) should be avoided as they are difficult to clean

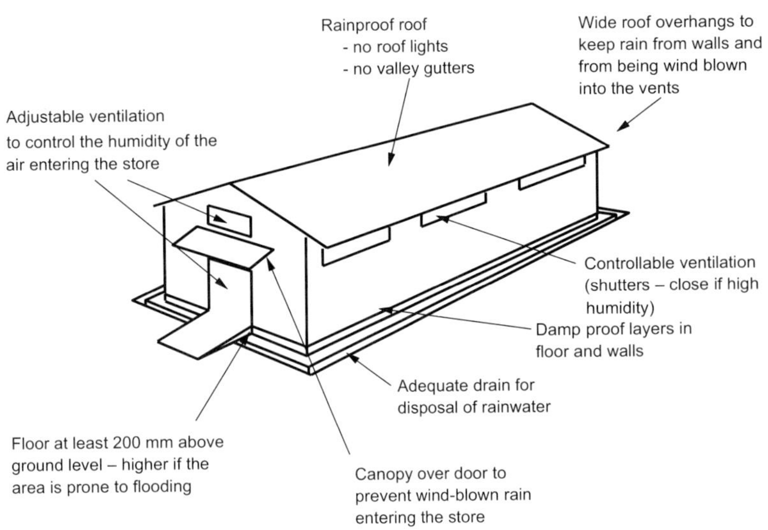


**Fig. 5.13** Store design with respect to moisture.

and maintain, and, in the event of a blockage, water will run into the store.

The floor of the store should be raised above ground level and adequate site drainage must be provided to protect against the risk of flood. Entry of ground water into the store can be prevented by incorporating waterproof barriers (such as plastic sheeting) in the floor and walls.

Ventilation of the store should be controlled so that the exchange of air from the outside only occurs when required to avoid undesirably high or low humidities. Controllable ventilation is also required to remove excess humidity arising from the respiration of the commodity itself or organisms (insects, moulds) within the commodity.

### *Protection against temperature*

Although more difficult to control (compared to moisture) in simple stores, temperatures within a store can be modified by including a number of design elements (see Fig. 5.14). Heat generated by the commodity itself, or by organisms within the commodity, can be limited through the use of controlled ventilation. The cooling effect of ventilation during the night, when ambient temperatures are lowest, can be increased if the store is insulated (to avoid heat gain through the structure during the day).

Heat gain from the surrounding air can be minimised if shiny or light-coloured heat-reflecting materials are used in store construction. Building the store in an east–west direction is claimed to reduce heat gain although little effect is noticed during the middle of the day. However, with the ends of the building pointing in an east–west direction, less wall surface is exposed to the sun, and therefore to heating, during the beginning and end of the day. Wide roof overhangs will shade the walls, and reduce the temperature of the structure. Thick walls (and the use of insulation) further reduce heat gain during the day.

Where temperatures are modified through the use of heaters or, more usually, refrigeration equipment, buildings will need to be insulated. The degree of insulation required will depend on the difference in temperature between the inside and outside of the store. Insulation may also be required to prevent low temperature damage to commodities in cold climates. A well-insulated refrigerated store will require less electricity to keep the produce cool.

### *Controlled atmosphere*

Controlled atmosphere storage (see Fig. 5.15) is rarely used for grains or roots or tubers, but is often essential for the storage of many perishable commodities where it is used in conjunction with refrigerated storage. Controlled atmosphere stores for horticultural produce may be constructed in a similar manner to conventional cold stores but special attention is given to making the store totally gas-tight. Air-tight grain stores are sometimes used to facilitate whole-store fumigation, but such stores are usually specially designed, as it is rarely possible to make an existing store gas-tight.

### *Protection against theft*

Security to guard against theft is enhanced in well-maintained, clean and tidy stores. In some situations

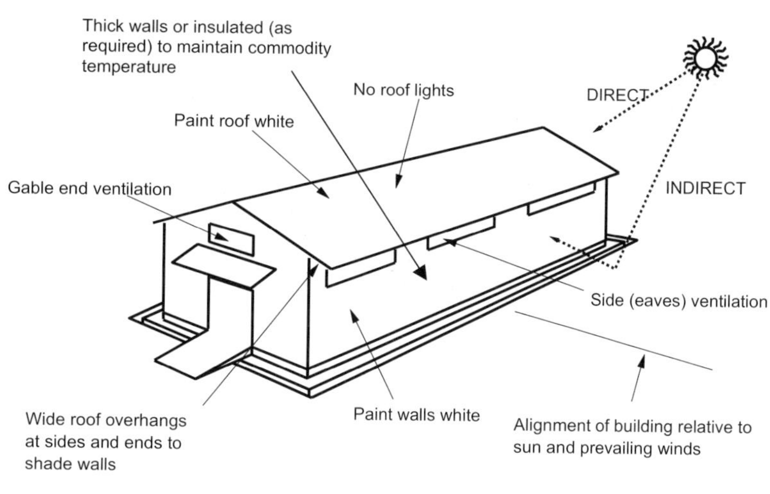


**Fig. 5.14** Store design with respect to temperature.

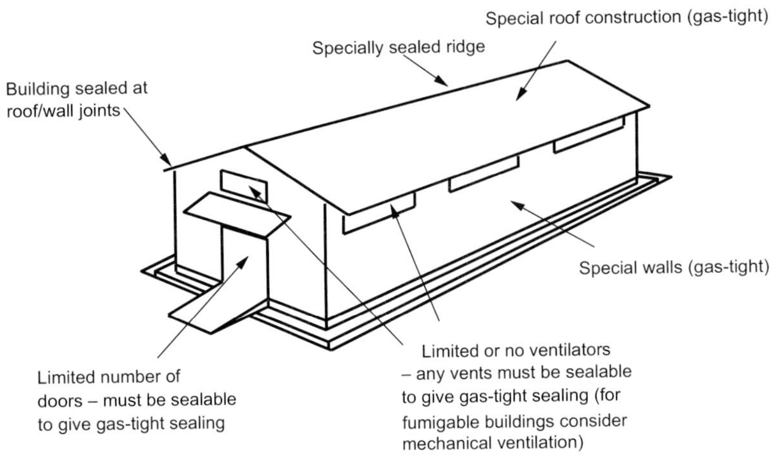


**Fig. 5.15** Store design with respect to the atmosphere.

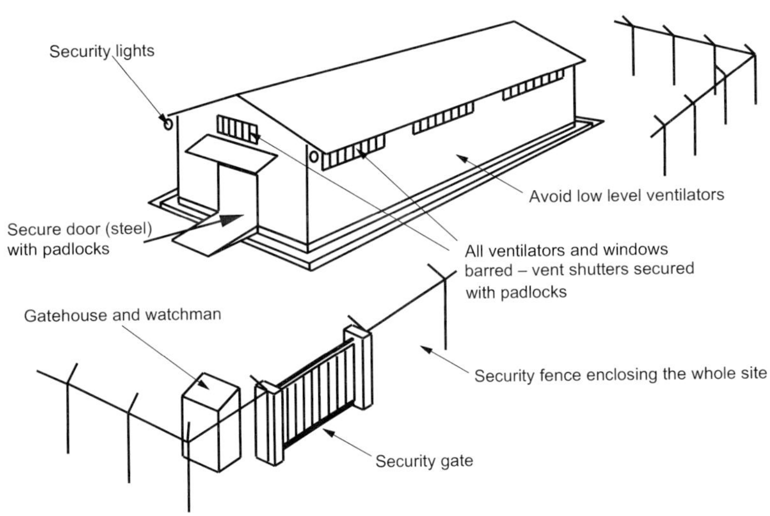


**Fig. 5.16** Store design with respect to security.

a site security fence may be necessary and lighting to illuminate the store and the compound may be desirable (see Fig. 5.16).

### *Protection against birds and rodents*

Stores should be designed to exclude birds and rodents and this requires that any gaps in the structure should be filled or screened to prevent access (see Figs 5.17 and 5.18). Rodents are good climbers, and external drainpipes and overhead cables (from which rodents might gain access to the store) should be avoided, or fitted with rodent guards. Rodent guards are commonly in the form of cones or tight-fitting sleeves made from sheet metal, however in traditional storage structures thorn bush twigs may be tied to store supports as barriers to rodents. As rodents are able to climb rough vertical surfaces it is usual to recommend that walls be rendered smooth; painting the walls with a gloss paint may also help deter climbing rodents. Rodents are capable of tunnelling under walls to enter stores and so attention needs to be paid to foundations to ensure that they are rodent-proof.

### *Protection against insects*

The ability of a storage structure to provide protection against insects (see Fig. 5.19) should be considered from two viewpoints: protection against insects that will damage the commodity (storage insects), and insects that will

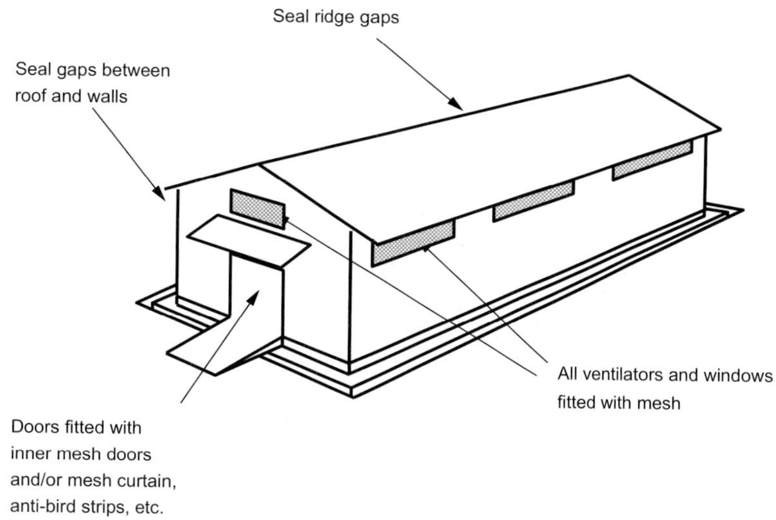


**Fig. 5.17** Store design with respect to bird control.

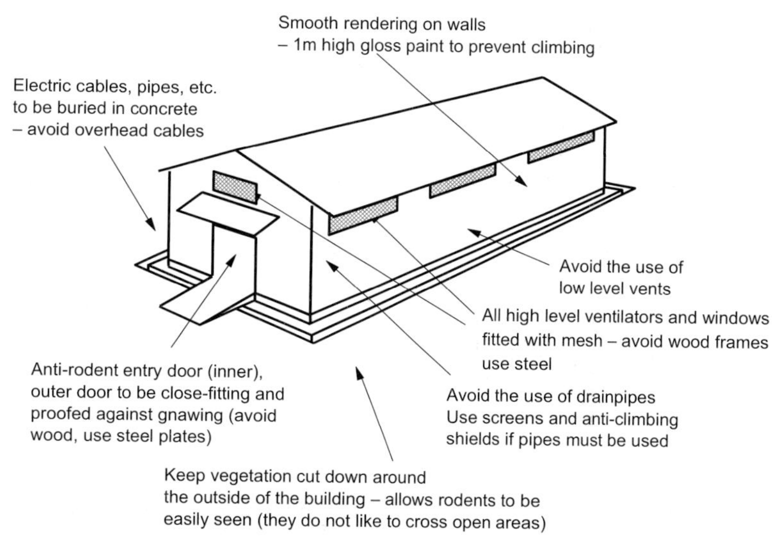


**Fig. 5.18** Store design with respect to rodent control.

damage the structure itself (structural pests such as termites and wood-boring beetles).

*Storage insects*

Because of their small size, insects are able to pass through extremely small gaps and it is very difficult to make stores insect-proof. Screens over doors, windows and ventilators may help to reduce the number of insects entering a store, but it is unlikely that they will exclude them entirely. It is usual to accept that insects cannot be prevented from entering stores, and to include design features that will facilitate cleaning and insect control. The requirement is therefore for smooth wall and floor surfaces, with no cracks or crevices in which insects can hide. Ideally, the wall–floor angle should be curved to eliminate spaces in which insects may hide and to facilitate cleaning. Angled ledges below windows and ventilators are less likely to accumulate dust and are easy to clean. Similarly, if vertical pillars in the walls are built outside the store rather than protruding into the store, sweeping of the store is easier and so insect control is im-

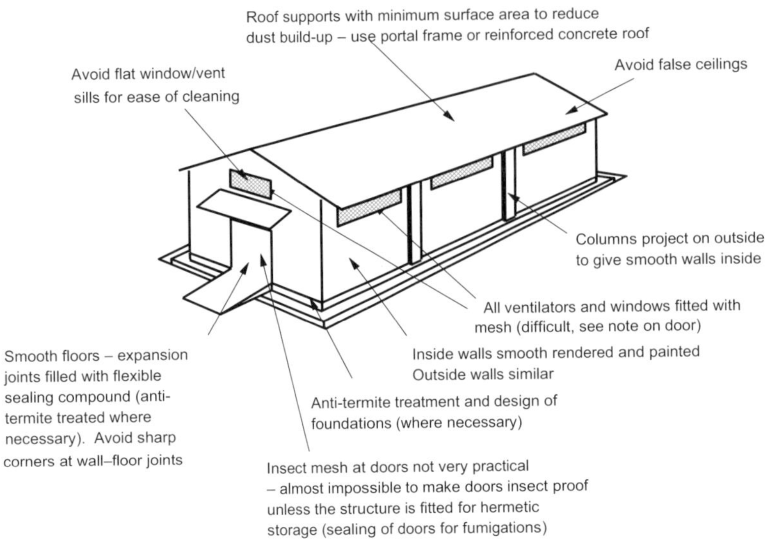


**Fig. 5.19** Store design with respect to insect control.

proved. False ceilings should be avoided as they provide areas where dust (and therefore insects) can collect.

*Structural pests*

A number of insect species may attack the wooden structure of a store. Some beetle species bore into wood but termites are a more common problem. Termites can cause tremendous damage to both wood and mud storage structures in tropical climates. A variety of treatments are available, some being more effective than others. If the termites cannot be removed from the area (by destroying the colonies), thought should be given when planning a new store to the type of materials to be used. If materials susceptible to termite attack have to be used then consideration must be given to ways in which they can be protected. Wherever possible, termite-proof materials, such as stone, concrete or fired bricks, should be used in preference to non-resistant materials such as wood, thatch and mud.

Protection methods include synthetic insecticides (either applied to the ground or foundations during construction, or at regular intervals after construction), plant materials with insecticidal properties, old engine oil or physical barriers (e.g. using brick or concrete supports for wooden structures). If timber has to be used, as in the case of traditional farm stores, it may be possible to find trees that are resistant to termites. Methods such as the charring of the outside of structural timbers are claimed to provide some protection against termites but further research is required.

***Management issues***

As well as protecting the commodity against damage or attack from various sources, the store must be easy to manage (see Fig. 5.20). Running costs, primarily maintenance, can be minimised through the use of good-quality building materials. Although more expensive initially, they will generally last longer and require less attention. Gutters and downpipes (for disposal of rainwater) need to be cleaned, and are susceptible to damage from wind and from operations around the building. If the roof overhangs the walls, gutters and downpipes may not be required. Equipment to control temperature, humidity and atmosphere needs to be well designed and easily maintained.

Large stores with high roofs, wide doors, good lighting, and flat concrete floors are easier to operate. Small stores, low roofs, and narrow doors often restrict movement and use of the building. Thus, operations take longer, often require more people, and are therefore more costly.

## Storage of horticultural produce

Storage facilities for horticultural produce may range from a simple room or container, to systems that incorporate sophisticated cooling facilities. When selecting

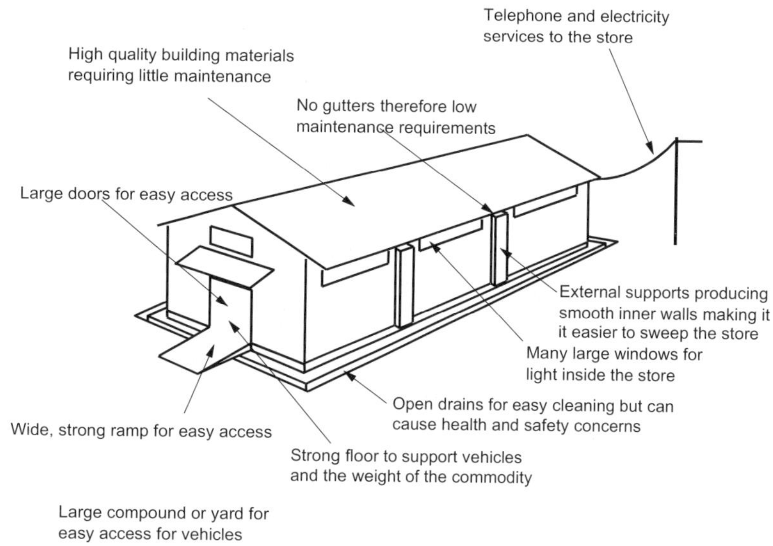


**Fig. 5.20** Store design with respect to management.

a suitable storage system it is important to consider the type of produce that is being handled, as different commodities may have widely differing temperature and humidity requirements.

The successful storage of horticultural produce must begin with a high-quality product that is free from pests, diseases and mechanical damage. However, other factors, namely, the varying sensitivity of commodities to storage temperatures, humidity and ethylene must be taken into account, particularly when considering the storage of different commodities together.

## Product requirements

### *Temperature considerations*

Horticultural crops may be grouped into two broad categories based on sensitivity to storage temperatures. There are those crops – mainly of tropical and subtropical origin – which are chilling sensitive and should be held at a temperature generally above 10–12°C. Storage below this threshold will give rise to a physiological disorder known as *chilling injury*. Symptoms may include surface pitting of the skin, watery breakdown, inability to ripen normally, discoloration of the flesh, off-flavours and increased susceptibility to invasion by micro-organisms. Often the visible effects of chilling injury are delayed until the product is offered for retail sale. The degree of chilling sensitivity, and therefore the lowest safe storage temperature is crop-specific. Skog (1998) gives examples of products susceptible to chilling injury and their symptoms. Care must be taken not to pre-cool or store the products below the recommended temperature. Crops that are not as sensitive to chilling injury may be stored at temperatures as low as 0°C. The optimal storage temperatures for a number of commodities are listed in Table 5.13.

### *Moisture loss*

Fruits and vegetables lose water constantly to the environment. After harvest this cannot be replaced and weight loss will occur. This will be accompanied by a deterioration in visual appearance and consumer acceptability, leading to a direct loss of marketable produce. Moreover, the susceptibility of the product to infection by disease will be increased. Produce loses water because of a water vapour gradient between the internal saturated atmosphere and the less-saturated external atmosphere. The rate of water movement is largely controlled by the vapour pressure difference between the fruit and its environment that is governed by temperature and relative humidity (see Chapter 3 for further details). Produce at 25°C and 30% humidity will lose water 36 times faster than at 0°C and 90% relative humidity. The relative humidity in the storage facility must be within the optimum range for the commodity. For most fruits and vegetables, the optimum relative humidity is 90–95%. However certain commodities, such as onions and garlic, store better in lower relative humidity environments. Table 5.13 lists

**Table 5.13** Recommended storage conditions for selected horticultural produce. (The exact conditions will vary with crop variety.) Source: FAO *Storage of Horticultural Crops*.

| Crop | Temp. (°C) | Temp. (°F) | % Relative humidity | Approx. storage life |
|---|---|---|---|---|
| Apples | −1 to 4 | 30–40 | 90–95 | 1–12 months |
| Asparagus | 0–2 | 32–35 | 95–100 | 2–3 weeks |
| Avocados | 4–13 | 40–55 | 85–95 | 2–8 weeks |
| Bananas, green | 13–14 | 56–58 | 90–95 | 14 weeks |
| Broccoli | 0 | 32 | 95–100 | 10–14 days |
| Carrots, mature | 0 | 32 | 98–100 | 7–9 months |
| Corn, sweet | 0 | 32 | 95–98 | 5–8 days |
| Grapes | −0.5 to 0 | 31–32 | 85 | 2–8 weeks |
| Kiwifruit | 0 | 32 | 90–95 | 3–5 months |
| Lychees | 1.5 | 35 | 90–95 | 3–5 weeks |
| Mangoes | 13 | 55 | 85–90 | 2–3 weeks |
| Onions, dry | 0 | 32 | 65–70 | 1–8 months |
| Papayas | 7–13 | 45–55 | 85–90 | 1–3 weeks |
| Peaches | −0.5 to 0 | 31–32 | 90–95 | 2–4 weeks |
| Potatoes, early | 10–16 | 50–60 | 90–95 | 10–14 days |
| Spinach | 0 | 32 | 95–100 | 10–14 days |
| Tomatoes, firm ripe | 13–15 | 55–60 | 90–95 | 4–7 days |
| Yams | 16 | 61 | 70–80 | 6–7 months |

the recommended relative humidity requirements for a selected number of commodities.

### *Ethylene effects*

Gast and Flores (1992) discuss the effects of ethylene gas on horticultural produce. Ethylene gas, in addition to effecting ripening of climacteric fruits, can also affect non-climacteric fruits and vegetables through a breakdown of chlorophyll and induce the development of physiological disorders. For example, cucumbers and celery turn yellow, while lettuce will turn brown in the presence of ethylene. Therefore, it is important to ensure that ethylene-sensitive and ethylene-producing commodities are not stored together (Table 5.14). The inclusion of potassium permanganate pads can be used to absorb ethylene during transit and storage.

**Table 5.14** Ethylene-sensitive and ethylene-producing crops.

| Ethylene-sensitive crops | Ethylene-producing crops | |
|---|---|---|
| Apples | Carrots | Apples |
| Apricots | Cucumbers | Apricots |
| Bananas | Eggplant | Bananas |
| Kiwifruit | Leafy greens | Melons |
| Melons | Lettuce | Nectarines |
| Pears | Okra | Peaches |
| Plums | Peas | Pears |
| Asparagus | Peppers | Plums |
| Broccoli | Spinach | Mushrooms |
| Cabbage | Sweet potatoes | |

### *Odour compatibility*

Odour compatibility is a concern when storing commodities together. Many commodities produce odours that are easily absorbed by others. For example, if they are stored together, the grower may end up having 'oniony' apples and 'appley' cabbage. Table 5.15 lists a number of odour producing and odour absorbing commodities.

## Storage requirements

### *Temperature control*

An important function of any cooling and storage facility for horticultural produce is to remove field heat and to provide a cold (or cool) storage space. For commercial operations, if a ready supply of electricity is available, mechanical refrigeration (including room cooling, forced-air cooling and evaporative cooling) provides the most reliable system for keeping produce cool. However, a variety of simple methods for cooling produce that do not depend on an electricity supply are available. At the small-scale producer level, cool storage facilities may include covered pits, lined with straw as insulation, root boxes or cellars (essentially pits or trenches lined with straw and wooden planks) or ventilated bins or barns in shaded and well-ventilated locations. Other

**Table 5.15** Odour compatibility between commodities. Source: Gast & Flores 1992.

| Odour producer | Odour absorber |
|---|---|
| Apples, pears | Cabbage, celery, carrots, onion, potatoes |
| Leeks | Grapes |
| Onions, dry and green | Mushrooms, corn, rhubarb, apples, pears, celery |
| Carrots, pepper | Celery, beans, avocados |

techniques include evaporative cooling, the use of ice and storage at high altitude (Thompson 1992). Facilities located at high altitudes can be effective, since air temperature decreases as altitude increases. As a rule of thumb, air temperatures decrease by 10°C for every 1 km increase in altitude.

In cooler regions, including high altitudes, suitable storage temperatures can be maintained in commercial facilities by ventilating at night when outside air is cool. The store should be well insulated and air vents located at the base. Vents can be opened at night and exhaust fans at the top of the storage structure used to pull the cool air though the store. This overhead distribution of air simplifies the store design. Where electricity is not available, wind-powered turbines can help keep storerooms cool by pulling air up through the building.

Ventilated floor systems, similar to those for the storage of bulk grain, can be used for certain perishable commodities. Moreover such systems can be modified for refrigerated storage. However, as refrigerated storage facilities using outside air for ventilation are wasteful of energy, it is better to incorporate a simple recirculatory system with a fan below floor level and a free space at one end of the store to allow cool air to return to the inlet vents.

Small-scale commercial operators may use cold rooms (either purpose-built or prefabricated units), or modified refrigerated transport containers (trucks, railway wagons or marine freight containers). Low-cost cold rooms can be constructed using concrete for floors and polyurethane foam as an insulator. Temperature management during storage can be aided by building a square rather than a rectangular store. Rectangular stores have more wall area per unit area of store space, so more heat is conducted across the walls, making them more expensive to cool. Square structures have lower construction and refrigeration costs. All joints should be caulked and the door should have a rubber seal. An example of a basic plan for a self-built cold room is given by Thompson and Spinoglio (1994) (Fig. 5.21).

#### *Insulation*

Any type of building or facility used for storage of horticultural crops should be insulated for maximum effectiveness. A well-insulated refrigerated building will require less electricity to keep produce cool. If the structure is to be cooled by evaporative or night air ventilation, a well-insulated building will hold the cooled air longer.

Thermal energy always flows from warm objects to cold ones. All materials offer some resistance to the flow of heat. Insulation, however, is any material that offers high resistance to the flow of energy.

Insulation R-values for common building materials modified from data by Boyette *et al.* (1989) are given in Tables 5.16 and 5.17. 'R' refers to resistance and the higher the R-value, the higher the material's resistance to heat conduction and the better the insulating property of the material. In addition to the product's R-value, its cost and the effects of moisture on it should be considered when deciding which material to choose. The cost of insulation varies considerably with the type of material. Of the insulation materials commonly used for refrigerated rooms, loose-fill cellulose is usually the least costly, followed by batts and blankets, then various foam sheet materials, and finally sprayed or foamed-in-place materials, which are the most expensive.

Insulation should be kept dry at all times. When it absorbs moisture, the air is replaced by water and the insulating value is greatly reduced. With the exception of most plastic foam insulation, a suitable vapour barrier (e.g. 4-mm polyethylene sheets) should be installed on the warm side (outside) of the insulation. Doors should have as much insulation as the walls and should be well weather-stripped to reduce the infiltration of warm air.

### Refrigerated storage

Large-scale commercial operations for horticultural produce often require refrigerated storage as part of a cold or cool chain system through which produce is carried from

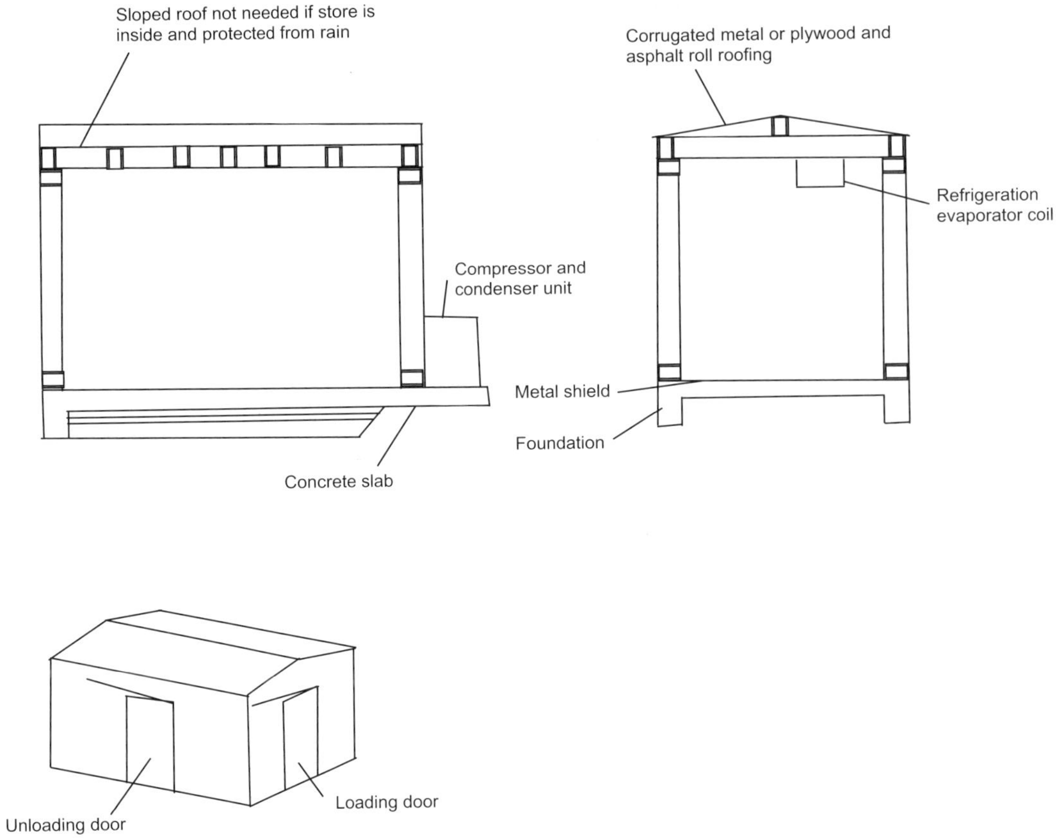


**Fig. 5.21** Design for a small-scale cold room store. Source: Thompson & Spinoglio 1994.

**Table 5.16** Insulation R-values for common insulating materials. Source: Boyette *et al.* (1989).

| Material (2.54 cm thick) | R-value |
|---|---|
| *Batt and blanket insulation* | |
| Glass wool, mineral wool, fibreglass | 3.50 |
| *Fill-type insulation* | |
| Cellulose | 3.50 |
| Glass or mineral wool | 2.50–3.00 |
| Vermiculite | 2.20 |
| Wood shavings or sawdust | 2.22 |
| *Rigid insulation* | |
| Plain expanded extruded polystyrene | 5.00 |
| Expanded rubber | 4.55 |
| Expanded polystyrene moulded beads | 3.57 |
| Aged expanded polyurethane | 6.25 |
| Glass fibre | 4.00 |
| Polyisocyanurate | 8.00 |
| Wood or cane fibreboard | 2.50 |
| *Foamed-in-place insulation* | |
| Sprayed expanded urethane | 6.25 |

**Table 5.17** Insulation R-values for common building materials. Source: Boyette *et al.* (1989).

| Building material (full thickness) | R-value |
|---|---|
| Solid concrete | 0.08 |
| 8-inch concrete block, open core | 1.11 |
| 8-inch lightweight concrete block open core | 2.00 |
| 8-inch concrete block with vermiculite in core | 5.03 |
| Timber (fir or pine) | 1.25 |
| Metal siding | < 0.01 |
| Plywood (1.27 cm) | 1.25–0.62 |
| Masonite particleboard | 1.06 |
| 2 cm insulated sheathing | 2.06 |
| Sheetrock (1.27 cm) | 0.45 |
| Wood lapsiding (1.27 cm) | 0.81 |

producer to market and retailer. This has now become a complex operation requiring considerable technical and managerial skill. The size of refrigeration unit required will be determined by a number of factors:

- the weight of produce to be cooled – since most produce is sold by volume (crates, boxes) one might have to determine its weight per unit of volume;
- the minimum time required from start of finish of cooling – ideally cooling should be just fast enough to prevent significant degradation of the produce (cooling produce faster than necessary incurs extra costs for electrical energy and requires a larger refrigeration system); and
- the nature of the refrigerated space – size, degree of insulation and how it is to be operated.

The choice between cooling and shipping produce immediately or placing it in store often depends not only on the type of produce and market conditions but also on the availability of space in the storage facility. The type of produce to be handled and the volume will determine how much storage space is needed. For example, highly perishable produce requires less storage space than less perishable items simply because it cannot be held for long periods. Adequate storage space offers marketing flexibility while excess storage space is a waste of energy and money. The amount of refrigerated space required can be determined using the following formula:

$$V = 2.5 \times (C + S)$$

where:
V = volume of the refrigerated space in cubic feet;
C = maximum number of bushels to be cooled at any one time; and
S = maximum number of bushels to be stored at any one time.

After V has been determined, divide it by the ceiling height in feet to obtain cooling room floor area in square feet, keeping in mind that the ceiling height should be no more than 18 inches greater than the maximum stacking height of the produce to be cooled. For produce packaged in bulk, volume must be converted to bushels before applying the above equation.

A thermostat is important in the refrigerated facility in order that a predetermined temperature can be maintained in storage. However, the thermostat's thermometer should not be used to monitor storage temperature. Instead, a separate recording thermometer should be installed for this purpose. The most important temperature to control is that of the produce itself. Hence, it is necessary to check the pulp temperature of the produce. A comparison of the produce pulp temperature, the amount of time the produce has been in storage and the thermostat setting will indicate when the thermostat needs resetting or the equipment needs adjusting.

### *Calculation of heat load and sizing of the refrigeration system*

The optimal storage temperature must be continuously maintained during operation of the facilities using an appropriate size of refrigeration system. The correct refrigeration unit will depend on the total amount of heat that the unit must remove from the cooling room. This is known as the heat load. Boyette *et al.* (1989) summarise the sources of leaks and splashes of heat entering a cooling room and which constitute the heat load as follows:

*Heat conduction* – heat entering through the insulated walls, ceiling and floor. The amount of heat flowing through these surfaces is a function of their thermal resistance (R-value), their area, and the temperature difference between one side and the other.

*Field heat* – heat extracted from the produce as it cools to the storage temperature. The amount of field heat is the product of the specific heat of the crop (the amount of heat energy it holds per degree), the difference between the field temperature and the storage temperature, and the weight of the produce. The specific heat of water is 1 calorie per gram per degree Celsius (or 1 Btu per pound per degree Fahrenheit). Since fruit and vegetables are mostly water, their individual specific heat is directly related to their water content and can for practical purposes be estimated as 1.

*Heat of respiration* – heat is generated by the produce as a natural by-product of its respiration. The amount of heat produced depends on the temperature, the crop and the conditions and treatment the crop has received. Clearly, less refrigeration is required to remove the heat of respiration when produce is cool than when it is warm.

*Service load* – heat from lights, equipment such as fans, people, and warm, moist air entering through cracks or through the door when opened. The amount of heat contributed by these sources is usually very difficult to determine accurately. The service load is therefore dealt with collectively and estimated as 10% of the heat from the other three sources: conduction, field heat, and heat of respiration.

### *Condensation and humidity control in refrigerated storage*

Another aspect to consider when handling fruits and vegetables is the relative humidity of the storage environment. The relative humidity in refrigerated storage must be within the optimum range for the commodity (see above). If using mechanical refrigeration for cooling, the larger the area of the refrigerator coils, the higher the relative humidity in the cold room will remain. For fresh market produce, any method of increasing the relative humidity of the storage environment (or decreasing the vapour pressure deficit (VPD) between the commodity and its environment) will slow the rate of water loss. The best method of increasing relative humidity is to reduce temperature. Another method is to add moisture to the air around the commodity as mists, sprays or, as a last resort, by wetting the store room floor.

When relative humidity inside the refrigerated storage is less than 90%, the facility should use a humidifier. The relative humidity of the storage can be monitored with a recording hygrometer. The air-to-coil temperature differential of the refrigeration unit must be no greater than –5°C. The temperature differential equals the difference between the temperature of the refrigerant entering the coil and the temperature of the air in the refrigerated facility. Air-to-coil temperature differentials greater than –5°C condense water vapour in the air, reducing relative humidity in the facility and producing ice accumulations on the evaporator coils. When the humidity is lowered, air tends to absorb water from the stored produce, causing it to wilt. If the air-to-coil differential cannot be adjusted, the installation of a humidifier in the refrigerated store is necessary.

## Controlled atmosphere and modified atmosphere storage

Consumers are increasingly demanding a year round supply of high quality fruit and vegetables and this demand has been met in part by the use of refrigerated storage. However, the development of controlled atmosphere (CA) and modified atmosphere (MA) storage offers further opportunities for extending the storage life of seasonal perishable produce, when refrigeration alone is not sufficient. Controlled atmosphere storage supplements cold storage and is only successful when applied at low temperatures (Kidd & West 1927a, b). Standard refrigeration units are therefore integral components of controlled atmosphere stores. The extended storage life of produce is achieved by reducing the respiration rate, browning reactions, sprouting, softening and decay of fruit and vegetables. Although the precise details of how CA and MA work are not fully understood and are the subject of continuing research, it is accepted that the effects that varying the levels of oxygen and carbon dioxide in the atmosphere have on crops will vary with factors such as:

- the crop species and cultivar;
- the growing conditions before harvest;
- the degree of ripeness of the climacteric fruit;
- the concentration of the gases in the store or package;
- the crop temperature; and
- the presence of ethylene in the store.

There are also interactive effects of the two gases, so that the effect of the carbon dioxide and oxygen in extending the storage life of a crop may be increased when they are combined. Moreover, as well as maintaining the post-harvest quality and extending the shelf-life of produce, CA and MA are also beneficial in reducing or eliminating insect and pathogen damage during transit and storage – this is particularly important as consumers demand fewer chemicals in food crops.

While the technical benefits of CA and MA storage have been demonstrated for a wide range of fruit and vegetables, the economic implications of using this comparatively expensive technology have often limited commercial application. However, with technological developments, more precise control equipment and the reducing cost, the technology is finding wider commercial use for an increasing range of crops.

### *Terminology*

#### *Controlled atmosphere*

While there is no formal definition of CA storage, it can be assumed to be the control of the levels of certain gases around, and therefore within, fresh fruits and vegetables (Thompson 1998). Controlled atmosphere storage requires the constant monitoring and adjustment of the carbon dioxide and oxygen levels within gas-tight stores or containers. The gas mixture will constantly change due to the metabolic activity of the respiring fruits and vegetables in the store and leakage of gases through doors and walls. Hence, they must be measured periodically and adjusted to the predetermined level by introducing fresh air or nitrogen or by passing the store atmosphere through a chemical to remove carbon dioxide (Thompson 1998).

There are different types of CA storage depending mainly on the method or degree of control of the gases. The two most commonly used systems are:

- *static* CA storage – where the product generates the atmosphere; and
- *flushed* CA storage – where the atmosphere is supplied from a flowing gas stream which purges the store continuously. Systems may be designed to allow: (a) initial flushing to reduce the oxygen content, then (b) either injecting carbon dioxide or allowing it to build up through respiration, and then (c) the maintenance of this atmosphere by ventilation and scrubbing.

Scrubbing is the selective removal of carbon dioxide from the atmosphere. In some cases this is referred to as product-generated CA or injected CA.

### *Modified atmosphere*

Modified atmosphere (MA) storage (or modified atmosphere packaging; MAP) is a condition where the fruit or vegetable is enclosed within sealed plastic film, which is slowly permeable to the respiratory gases. The gases will change within the package, thus producing lower concentrations of oxygen and higher concentrations of carbon dioxide than exist in fresh air (Thompson 1998). Modified atmosphere may also be defined as 'an atmosphere of the required composition created by respiration, or mixed and flushed into the product enclosure' (Bishop 1996). This mixture should be maintained over the storage life and no further measurement or control will take place.

## *Current use of CA and MA*

Historically, apples have ranked as the pre-eminent crop stored under CA conditions. This remains true today; however, studies of other fruits and vegetables have shown it to have wide application. An increasing number of crops are being stored and transported under CA and MA conditions although the number that are commercially stored or transported is still relatively small (Table 5.18). The number of crops suitable for CA and MA storage continues to expand as commercial interest in research findings increases.

While CA technology has largely focused on the long-term storage of bulk fruit and vegetables, more recently both CA and MA have been applied to the short-term storage of certain commodities. These include some high-value, speciality crops where limited production in locations distant from consumption areas justify the use of CA and MA to extend storability, and some lightly processed fruits and vegetables (e.g. chopped and shredded lettuce, sliced and shredded carrots and sliced celery, cucumbers, radishes and green peppers).

In most cases, perishable commodities are stored under constant conditions, especially low temperature. However, even under these conditions, the commodity is experiencing physiological and metabolic changes. For this reason, researchers are looking at the storage process as a dynamic system and the use of flexible control rather than constant-value systems in order to obtain a qualitative improvement. Studies have looked at neural networks and genetic algorithms for realising the optimal control of fruit storage (Morimoto *et al.* 1997).

## *Recommended CA storage conditions for selected crops*

Optimum storage conditions for fruit and vegetables depend upon a number of factors, including the cultivar, stage of maturity at harvest and growing conditions. There are considerable variations in local recommendations for the CA storage of Golden Delicious apples grown throughout the world (Bishop 1996). For example, in Australia recommended storage conditions for this variety of apple are 1.5% oxygen and 1.0% carbon dioxide at 0°C, whereas, in Spain, recommendations are for storage at 3% oxygen and 2–4% carbon dioxide at 0.5°C. While the differing recommendations reflect different factors such as soil and climate types, local post-harvest treatment and acceptance criteria, it is also likely

**Table 5.18** Fruits and vegetables for which controlled atmosphere (CA) and modified atmosphere (MA) are being commercially used.

| Fruits | Vegetables |
|---|---|
| Avocado, banana, blueberry, cherry, kiwifruit, mango, nectarine, plum, raspberry, redcurrant, strawberry | Cabbage, onion |

that the choice of research methods and experiences of the researcher have had an impact. Exhaustive studies to define the most appropriate CA regime for a particular cultivar growing under certain conditions have only been undertaken for a select number of crops, of which apple is a prime example. For the majority of other crops, recommended CA storage conditions are based on the results of one or two years' experience at particular gas levels without any work being done to optimise and select the best possible combinations. The recommendations for the optimum storage conditions have varied over time, due mainly to improvements in the control technology over the levels of gases within the stores. Storage recommendations for the cultivar Cox's Orange Pippin, for example, have evolved between 1920 and 1986 (Bishop 1994). Originally, apples stored at 3.5°C under 16% oxygen and 5% carbon dioxide had a storage life of 16 weeks, yet storing apples at the same temperature but under 1% oxygen and less than 1% carbon dioxide more than doubled the storage life to 33 weeks. An exhaustive review of recommended storage conditions for over 100 fruits and vegetables has been produced by Thompson (1998). Table 5.19 provides recommended storage conditions for a small selection and is drawn from a number of sources (Bishop 1996; Thompson 1998). Typically, the recommended storage conditions refer to the maximum time that the fruit or vegetable could be stored. It is usually assumed that the crop is of good quality, free from pests and diseases, and in the case of climacteric fruit, at an appropriate stage of maturity at harvest, before being placed in storage. Moreover, the recommendations may assume that pre-storage treatments such as use of fungicides have been applied correctly and that the produce has been maintained in the cool chain from harvesting.

### *Store construction*

Modern CA stores are made from metal-faced insulated panels (usually polyurethane foam) which are locked together with proprietary locking devices, incorporating seals at the mating face. The joints between panels are usually taped with gas-tight tape or painted with flexible plastic paint to ensure they are gas-tight. This process is also used on floor–wall and wall–ceiling joints.

Leakage is most likely to occur around the doors. Conventional cold store doors even though designated 'leak-tight' are not suitable, and doors specified as suitable for CA rooms must be used. Proper sealing is achieved by rubber gaskets around the edge of the doors, which correspond to similar rubber gaskets around the door jamb or frame. In addition a flexible, soft rubber hose may be hammered into the inside between the door and the frame to give a double rubber seal to ensure gas-tightness. Other door designs incorporate screw jacks around the periphery of the door or inflatable rubber seals.

To protect the store structure from damage due to excessive air pressures – arising from constant changes and adjustments in the store temperature and concentrations of the various gases inside the sealed store – it is essential to install a pressure relief valve, either a simple, water-based vent system or a mechanical valve relying on the

**Table 5.19** Typical controlled-atmosphere storage conditions for selected fruit and vegetables.

| Crop | Variety | Storage temp. (°C) | $CO_2$ (%) | $O_2$ (%) | Storage time |
|---|---|---|---|---|---|
| Apple | Cox's Orange Pippin | 3–4.5 | < 1–3 | 1–3 | up to 30 weeks |
| | Gala | –0.5–2 | 1–5 | 0.8–2.5 | up to 5 months |
| | Golden Delicious | -0.5–2 | 0–8 | 1–4 | up to 10 months |
| | Jonagold | 0–2 | < 1–6 | 1–3 | up to 9 months |
| Banana | various | 11.5–16 | 2–8 | 2–5 | < 8 weeks |
| Broccoli | | 0–15 | 5–15 | 1–2 | < 4 weeks |
| Cauliflower | | 0–5 | 0–10 | 2–10 | 47 days |
| Cucumber | | 12.5–14 | 5–10 | 3–5 | |
| Grape | | 0–5 | 1–10 | 2–5 | < 7 months |
| Kiwifruit | | 0–5 | 5–1 | 1–2 | < 6 months |
| Lychee | | 5–12 | 3–5 | 5 | |
| Mango | | 10–15 | 1–10 | 2–7 | < 6 weeks |
| Papaya | | 10–15 | 5–10 | 1–5 | < 29 days |
| Peach | | 0–5 | 3–5 | 1–2 | < 4 weeks |
| Pineapple | | 8–13 | 5–10 | 2–5 | |
| Tomato | | 12–13 | 2.5–9 | 2.5–5.5 | 3 months |

weight of a sealed disk to obtain the correct pressure. The valve should limit the pressure on the store structure to the designed safe limit, typically this could be ± 190 Pa. However, these valves can make the maintenance of the precise gas level difficult, especially of oxygen in ultra-low oxygen stores. The use of an expansion bag can overcome the problem of pressure differences (Thompson 1998). These gas-tight bags are partially inflated and are located outside the store with their inlets inside the store. As the in-store air pressure increases, the bag automatically further inflates; conversely, when the pressure in the store is reduced, air flows from the bag into the store. The inlet of the expansion bag should be situated before the cooling coils of the refrigeration unit to ensure the air from the expansion bag is cooled before returning to the store. An expansion bag can also buffer pressure changes due to the operation of some types of carbon scrubbers.

### *Gas control*

Central to the design of a CA storage system is the selection of the most appropriate equipment for creating and maintaining the selected gas concentrations in the store. In summary, the following functions need to be undertaken:

- removal of ambient oxygen;
- removal of carbon dioxide by respiration;
- addition of air to replace oxygen consumed by respiration;
- removal of ethylene; and
- addition of carbon dioxide.

The choice of equipment depends on the product being stored and the storage regime. The size of the equipment will also depend upon the respiration rate of the product. Several examples of typical equipment used for the commercial storage of apples, onions, cabbage, kiwi and cherries can be found in Bishop (1996). The atmosphere in a modern CA store is constantly analysed for carbon dioxide and oxygen levels using an infrared gas analyser and a paramagnetic analyser, respectively.

#### *Oxygen control*

In-store oxygen levels can be reduced by the following methods (Bishop 1996):

*Nitrogen flushing*: Nitrogen is discharged slowly into the store (with fans running and the refrigeration turned off) to prevent low-temperature damage to the stored commodity. The temperature in the store must be closely monitored and flushing stopped if it falls too low.

*Nitrogen pressure swing adsorption (PSA)*: These systems consist of two beds of carbon molecular sieve which are used alternately. Pre-treated compressed air enters the bottom of the first bed and passes across the carbon sieve where oxygen and other trace gases are preferentially adsorbed, allowing nitrogen to pass through. After a predetermined interval, when the on-line bed is nearly saturated with adsorbed gases, the system automatically switches the airflow to the second bed, while the first is regenerated by rapidly reducing the pressure inside the column, thus allowing the captured gases to escape to atmosphere.

*Membranes* employ the principle of selective permeation to separate gases. 'Fast' gases such as oxygen, carbon dioxide and water vapour can be separated from 'slow' gases such as nitrogen. Modern membrane separators use hollow fibre technology. High pressure, pre-treated air is required at a typical pressure of 9–12 bar. The oxygen concentration in the output is again dependent upon the output flow.

Once the oxygen in the store has reached the level required for the particular crop being stored, it is maintained by frequently introducing fresh air from outside the store. Usually tolerance limits are set at ± 0.15% for oxygen levels below 2% and ± 0.3% for oxygen levels of 2% and above (Sharples & Stow 1986). Recent developments in the control systems mean that it is possible to control oxygen levels close to the theoretical minimum. This is because modern systems can achieve a much lower fluctuation in gas levels, and ultra-low oxygen storage (levels around 1%) is now common.

#### *Carbon dioxide control*

Similarly precise control can be applied to carbon dioxide. The carbon dioxide level in the store should be maintained at ± 0.5% of the recommended level. When a predetermined level is reached the atmosphere is passed through a chemical that removes carbon dioxide. This is called 'active scrubbing'. Two types of scrubber are commonly used.

*Calcium hydroxide scrubbers*: Calcium hydroxide reacts irreversibly with carbon dioxide to produce calcium carbonate, water and heat. When the carbon dioxide in the store is above the required level, a fan draws the store atmosphere through the room containing bags of freshly hydrated calcium lime (with at least 95% purity of calcium hydroxide) until the required level is reached.

The lime can be loaded in paper sacks into a lime box external to the store and its effect on the store carbon dioxide level controlled by regulating the flow rate of the store atmosphere through the box. The amount of lime required depends on the respiration rate of the fruit, the storage temperature and the length of storage time required. For example, for 1 tonne of Golden Delicious apples, 7.5 kg of high calcium lime is needed every 6–10 weeks (Koelet 1992). The use of carbon hydroxide scrubbers is a common practice, especially for ultra-low oxygen stores.

*Renewable scrubbers*: Renewable scrubbers are compact and are particularly suitable for use in CA transport systems. They consist of two containers of material that can absorb carbon dioxide, such as activated charcoal (active carbon) or a molecular sieve (aluminium calcium silicate). Store air is passed through one of these containers until it becomes saturated with carbon dioxide, when the airflow is automatically switched to the second container. The first container then has fresh air blown through it to detach the carbon dioxide so that it is available for reuse. Where a molecular sieve is used it is necessary to heat it during the purging cycle (Thompson 1998).

### *Ethylene control*

Ethylene removal is important if ethylene-sensitive crops are to be held in CA storage. Ethylene can be adsorbed by potassium permanganate crystals and these have been successfully used during transport of fresh produce but are uneconomic in long-term CA storage (Bishop 1996). Commercial ethylene scrubbers which remove the ethylene by passing the atmosphere over precious metal catalysts running at high temperatures are available, but are expensive in terms of capital costs and the energy needs and therefore they are not commonly used in CA stores except for kiwifruit storage.

## ***Interrupted controlled atmosphere storage***

Controlled atmosphere storage has been shown to have detrimental side-effects on fruits and vegetables, and the possibility of alternating controlled atmosphere storage with air storage has been studied. Results have been very variable. For example, storage of bananas at high temperature causes physiological disorders, and an investigation of the possible benefits of interrupted controlled atmosphere has been carried out. In trials with the cultivar Poyo (from Cameroon), storage at 30–40°C was interrupted by one to three cooling periods (20°C) of 12 hours either in air or in atmospheres with 50% oxygen or 5% oxygen. Cooling periods were found to reduce high temperature damage, especially when fruits stored at 30°C received three cooling periods in 50% oxygen (Dick & Marcellin 1985). Intermittent exposure of the avocado cultivar Haas to 20% carbon dioxide increased storage life at 12°C and reduced chilling injury during storage at 4°C compared to those stored in air at the same temperatures (Marcellin & Chaves 1983).

## ***Total nitrogen or high nitrogen storage***

The storage of some fruits in total nitrogen or nitrogen-rich atmospheres can be beneficial (Thompson 1998). Ripening of plums stored in total nitrogen has been reported to be almost completely inhibited (Anon. 1920). Plums were able to tolerate, for a considerable period, an almost complete absence of oxygen without being killed or developing an alcoholic or unpleasant flavour. For strawberries, 100% and 99% nitrogen atmospheres reduce mould growth during 10 days of storage at 1.1°C with little or no effect on flavour (Parsons *et al.* 1964). Similarly, for peaches, decay was reduced during storage in either 100% or 99% nitrogen at 60°F; off-flavours were detected in fruit after 4 days in 100% nitrogen, but none in those stored in 99% nitrogen.

## ***Controlled atmosphere transport***

Large quantities of fresh fruit and vegetables are transported by sea-freight refrigerated (reefer) containers. It was estimated that the world fleet increased four-fold from 1993 to 1997 when there were around 38 000 reefer containers (Dohring 1997). Relatively small but increasing numbers of containers (approximately 1000 containers) are being used for the CA transportation of fresh fruit and vegetables. These commodities include apples, apricots, asparagus, avocados, bananas, broccoli, cantaloupes, cauliflowers, cherries, eggplants, kiwifruit, limes, mangoes, peaches, nectarines, pears, pineapples, plums, sweetcorn and tomatoes.

A number of technical problems hindering the use of CA transportation in the past (the lack of gas-tight containers, suitable systems for gas control and analysis, and adequate CA-generating systems) have now largely been overcome. A common problem was the maintenance of gas-tight conditions. The leakage from some early systems was as much as 5 $m^3 h^{-1}$, but leakage from current systems can be below 1 $m^3 h^{-1}$ (Garrett 1992). Much of this leakage is through the door (Thompson 1998) and can be reduced by the use of plastic curtains inside the door. However, these curtains are difficult to fit and maintain under working conditions. A system introduced in 1993 had a single door instead of the double doors of reefer containers, and was easier to make gas-tight.

There are a number of commercial systems available and these can be broadly classified as either CA or MA systems. The latter require the injection of an appropriate mixture of gases into the sealed container at the beginning of the journey with no subsequent control, so the atmosphere will constantly change during transport. Controlled atmosphere containers have a mechanism for measuring the changes in gases and adjusting them to a pre-set level (Thompson 1998).

Standard reefer containers can be converted to CA containers either permanently or temporarily. An example is a CA unit measuring 2 m × 2 m × 0.2 m that operates alongside the container's refrigeration system and is capable of controlling, maintaining and recording the levels of oxygen, carbon dioxide and relative humidity to the levels and tolerances pre-set in a programmable controller (Thompson,1998). Ethylene can also be removed from the container by scrubbing. Such level of control is greater than any comparable CA storage system, and can increase shipping range and enhance the quality of fresh produce. The system is easily attached to the container floor and bulkhead, and takes power from the existing reefer equipment. A menu-driven, programmable controller provides the interface to the operator who simply has to pre-set the required percentages of each gas to levels appropriate for the produce in transit.

The systems used to generate the atmosphere in the containers fall into three categories (Garrett 1992):

- The gases can be carried in a compressed liquid form in steel cylinders at the front of the container, with access from the outside. This system involves injecting nitrogen into the container to reduce the level of oxygen. Oxygen levels are maintained by injection of nitrogen if the leakage into the container is greater than the utilisation of oxygen through respiration by the stored crop. If the respiration of the crop is high the oxygen can be replenished by ventilation.
- Membrane technology may be used to generate the gases by separation. The carbon dioxide is generated by the respiration of the crop and nitrogen is injected to reduce the oxygen level. Nitrogen is produced by passing the air through fine porous tubes, made from polysulphones or polyamides, at a pressure of about 5–6 bar. These will divert most of the oxygen through the tube walls, leaving mainly nitrogen which is injected into the container (Sharples *et al.* 1989).
- The gases can be generated in the container and recycled with pressure absorption technology and swing absorption technology. This system uses ventilation to control oxygen levels and a molecular sieve to control carbon dioxide. The molecular sieve will also absorb ethylene and has two distinct circuits that are switched at predetermined intervals so that, while one circuit is absorbing, the other is being regenerated. The regeneration of the molecular sieve beds can be achieved when they are warmed to 100°C to drive off the carbon dioxide and ethylene. This system of regeneration is referred to as 'temperature swing', where the gases are absorbed at low temperatures and released at high temperature. Regeneration can also be achieved by reducing the pressure around the molecular sieve, which is called 'pressure swing'. During the regeneration cycle the trapped gases are usually ventilated to the outside, but they can be directed back into the container if this is required.

Ethylene can be removed from the container air if necessary. A single ethylene absorption bed containing a mixture of activated alumina and Hisea material (a clay mineral-based system patented by the British Oxygen Company) may be used. Air from the container is routed to the bed that remains pressurised for several hours and then depressurised via the oxygen venting lines. The oxygen flow is then routed through the bed for a short period of time in order to scrub it before the process is repeated.

A number of containers use a humidity injection system to increase the relative humidity within the container; an atomised spray of water is injected as required into the main airflow through the reefer. The water supply is taken from a reservoir, typically located on the mainframe, and is fed from the reefer defrost system and the water trap. Air from the instrument air buffer is used to form the spray by drawing up the water as required. The control valve is operated for a short period, and once the additional water spray has been mixed in with the air in the container, the humidity level is measured and the valve operated again if required. In some systems a further stage of moisture removal is carried out using drier beds that are charged with activated alumina. From here the air is routed either directly to nitrogen and carbon dioxide beds, or via an ethylene bed depending on the type of conditioning needed.

### *Modified atmosphere packaging*

Modified atmosphere packaging (MAP) involves the sealing of commodities in plastic films so that the composition of gases of respiration and transpiration in and around the produce is altered. For fresh fruits and vegetables this is commonly achieved by wrapping individual items, boxes of produce or a pallet of boxes in a plastic film. The films can be slowly permeable to carbon dioxide, oxygen and other gases, even water vapour, depending on their thickness and composition.

Modified atmosphere packaging has been known for several decades to have great potential in extending the post-harvest life of a number of fruits and vegetables (Macfie 1956; Scott 1975; Brown *et al.* 1985; Kader 1989). For example, MAP using polyethylene bags extends the storage life of bananas by 6 days compared to non-wrapped fruit at 20°C (Scott *et al.* 1971). In a subsequent trial in Australia, bananas in 13.6 kg commercial packs inside polyethylene film bags remained in good condition during 48 hours of transport at ambient temperatures. Furthermore, the trials showed that the fruit could be held in a commercial ripening room for several days, with minimal weight loss, and good flavour and appearance when subsequently ripened (Scott *et al.* 1971). Modified atmosphere packaging is also advantageous in maintaining some minimally processed tropical fruits such as durian, jackfruit, mangosteen, papaya and pineapple (Powrie *et al.* 1990; Siriphanich 1994). However, overall, reported results for MAP are very variable (Kader 1989) and thought to be due to lack of experimental control (Yahia 1998). Some of the reported beneficial effects of MA may be due to maintaining a humid atmosphere around the commodity, and not to gas modification. Many different types of polymers are used, although the most common are different types of polyethylene. Different thicknesses of the same type of film or different conditions (temperature and r.h.) surrounding the package result in different permeabilities and therefore different in-package atmospheres. It is likely that the differing recommendations for a given crop have in part limited the uptake of this technology by industry. In addition, the limited control of the gases around and within the produce can result in unpredictable effects on post-harvest life and quality of the commodity, which has also had an impact on its uptake. Moreover, there have been some recent concerns about the safety of products in MAP. However despite these apparent constraints, advances in MAP of fresh produce has led in part to the rapid expansion of what is often referred to as the 'fresh-cut revolution'. Modified atmosphere packaging works by slowing down the deterioration of the 'fresh-cut' or minimally processed fruit and vegetables. An example of this technique is the use of low-density polythene film of 27 μm thickness, which creates a steady state atmosphere of 10°C inside the bags of 5.5–5.7% carbon dioxide and 2.0–2.1% oxygen for the preservation of a mixture of carrot, cucumber, garlic and green pepper (Lee & Lee 1996). The potential advantages and disadvantages of MAP are summarised in Table 5.20.

#### *MAP effects*

'The basic premise of MAP technology is that once produce is placed in a package and hermetically sealed, an environment different from ambient conditions will be established. The key to successful use of MAP technology is knowing what type of environment will be most beneficial to the produce inside the package and then determining which packaging materials should be used to create such an environment' (Gorny 1997). Packaging crops into plastic film bags can lead to a build-up of water vapour and changes in the levels of the respiration gases – carbon dioxide, oxygen and ethylene. When fresh produce is sealed inside a polymeric or plastic film package the produce consumes oxygen and releases carbon dioxide, generating a lower in-package oxygen level and/or elevated carbon dioxide level. The effects of these gas changes may be beneficial or detrimental to the crop. If the optimum atmosphere for a product is achieved, its storage life can be increased by many times that which can be expected using conventional

**Table 5.20** Potential advantages and disadvantages of modified atmosphere packing. Source: adapted from Farber 1991.

| Advantages | Disadvantages |
|---|---|
| Potential shelf life increased | Added cost |
| Reduced economic loss | Temperature control required |
| Products can be distributed over longer distances | Different gas formulations for each product type |
| Reduced distribution costs | Special equipment and training required |
| | Product safety to be established |

refrigerated air storage. However, if incorrect packaging materials are chosen, atmospheric conditions that are detrimental may be generated and this can actually result in shorter shelf-life. The concentration of these gases inside the crop will depend upon a number of innate crop characteristics such as type, variety, growing conditions, maturity at harvest, etc., and also such characteristics as:

- the mass of fruit or vegetable within the pack or container;
- the temperature of the fruit or vegetable and the surrounding air;
- the type and thickness of plastic film or membrane used;
- whether moisture condenses on the film or membrane surface;
- external airflow around the film or membrane; and
- activity of micro-organisms.

*Gases used in MAP*

The three main gases used commercially in MAP are oxygen, carbon dioxide and nitrogen (nitrogen is used as a filler gas to prevent pack collapse, which may occur in atmospheres containing high carbon dioxide). Recently, there has been some interest in the potential applications for the use of novel gas mixtures (e.g. argon and nitrous oxide), although these are largely at the experimental stage. The gases and their concentrations should be tailored for each individual product (specific variety). Modified atmosphere may be produced naturally by respiration (passive MA) and by the application of gas flushing techniques (equilibrium MA).

*Film types*

The actual concentration of gases in the fruit will be affected to a limited degree by the amount of space between the fruit and the plastic film, but mainly by the permeability of the film (Thompson 1998). There are a number of different plastic films which have been used for fruit and vegetable packaging including cellulose acetate, ethylene vinyl acetate copolymers, high-density polyethylene (HDPE), linear low-density polyethylene (LLDPE), low-density polyethylene (LDPE), medium-density polyethylene (MDPE), polyolefin, polypropylene and polystyrene. A range of high shrink, multilayer, high speed, machineable shrink films made from polyolefin are available. Polyethylene is also used for shrink film packaging.

*Film permeability*

When fresh produce is sealed inside a polymeric or plastic film package, respiration will reduce the in-package oxygen level and increase the carbon dioxide level. A major challenge in designing MA packages is to match the rate of oxygen uptake and carbon dioxide production of the produce with the oxygen and carbon dioxide permeability of the package. If the package is well designed, gas levels inside the package will equilibrate within a range that benefits the products. For most products, the therapeutic range for carbon dioxide and oxygen is between 2% and 10%. However, optimum $CO_2$ and $O_2$ concentrations are product-specific and vary enormously between products. Examples of optimum $CO_2:O_2$ levels are asparagus 10 : 10, lettuce 2 : 2, peaches 6 : 1.5, pears 2 : 2, carrots 0 : 21 and blueberries 6 :1.5. The transmission of $CO_2$ and $O_2$ through plastic films will vary with film type, but generally films are four to six times more permeable to $CO_2$ than $O_2$. The permeability of films to gases (including water vapour) varies with the type of material for which they are made, temperature and in some cases humidity, the accumulation and concentration of the gas and the thickness of the material. A range of permeabilities can be obtained from films with the same basic specifications. A selection of these is summarised in Table 5.21.

Where the permeability of the film is too low the crop may be damaged because of the accumulation of water or carbon dioxide or depletion of oxygen. In those cases, holes can be punched in the film to improve ventilation.

Film permeability to gases is by active diffusion where the gas molecules dissolve in the film matrix and diffuse through in response to the concentration gradient (Kester & Fennema 1986). A formula to describe film permeability (Crank 1975) is:

$$P = \frac{Jx}{A(p1 - p2)}$$

where J = volumetric rate of gas flow through the film at steady state, x = thickness of film, A = area of permeable surface, p1 = gas partial pressure on side 1 of the film, p2 = gas partial pressure on side 2 of the film (p1 > p2).

*Gas flushing in modified atmosphere packaging*

The levels of carbon dioxide and oxygen can take some time to change inside the modified atmosphere pack. In order to speed up this process the pack can be flushed

**Table 5.21** Permeability of polyethylene film according to density. Adapted from Schlimme & Rooney 1994.

| Film type (polyethylene) | Comment | MW | Density range (g cm–1) | Permeability specifications ($cm^3\ m^{-2}\ day^{-1}$) | |
|---|---|---|---|---|---|
| Low density | Produced by polymerisation of the ethylene monomer which produces a branched chain polymer | 14 000–140 000 | 0.910–0.935 | $O_2$: 3900–13 000<br>$CO_2$: 7700–77 000 | at 1 atm for 0.0254 mm thick at 22–25°C at various or unreported r.h. Water vapour: 6–23.2 $g\ m^{-2}\ day^{-1}$ at 37.8°C and 90% r.h. |
| Medium density | | | 0.926–0.940 | $O_2$: 2600–8293<br>$CO_2$: 7700–38 750 | at 1 atm for 0.0254 mm thick at 22–25°C at various or unreported r.h. Water vapour 8–15 $g\ m^{-2}\ day^{-1}$ at 37.8°C and 90% r.h. |
| High density | 75–90% crystalline structure with an ordered linear arrangement of the molecules with little branching | 90 000–175 000 | 0.995–0.970 | $O_2$: 520–4000<br>$CO_2$: 3900–10 000 | at 1 atm for 0.0254 mm thick at 22–25°C at various or unreported r.h. Water vapour 4–10 $g\ m^{-2}\ day^{-1}$ at 37.8°C and 90% r.h. |
| Linear low density | This combines the properties of low density and high density polyethylene film giving a more crystalline structure than low density polyethylene film but with a controlled number of branches which makes it tougher and suitable for heat sealing. It is made from ethylene with butene, hexene or octene with the latter two co-monomers giving enhanced impact resistance and tear strength | | | $O_2$: 7000–9300 | at 1 atm for 0.0254 mm thick at 22–25°C at various or unreported r.h. Water vapour 16–31 $g\ m^{-2}\ day^{-1}$ at 37.8°C and 90% r.h. |

with nitrogen to reduce the oxygen rapidly, or the atmosphere can be flushed with an appropriate mixture of carbon dioxide, oxygen and nitrogen. In other cases the pack can be connected to a vacuum pump to remove the air so that the respiratory gases can change within the pack more quickly.

*Modelling*

Mathematical models have been developed with the aim of allowing researchers to predict the atmosphere around fresh and minimally processed produce sealed in film bags with a view to selecting the most appropriate packaging (Cameron *et al.* 1989; Lopez-Briones *et al.* 1993; Day 1994; Ben-Yehoshuo *et al.* 1995; Evelo 1995). A number of these models take a systems-oriented approach which allow the user to assess the optimal MAP in realistic distribution chains.

*Effect of product on gas content*

The quantity of produce inside the sealed plastic film bag has been shown to affect the equilibrium gas content, but the levels of carbon dioxide and oxygen do not always follow what would be predicted from permeability data and respiration load of the crop. It has been shown that there was a negative linear relationship between weight of fresh chillies in a sealed Cryovac SSD-310 film package and carbon dioxide and oxygen levels, and carbon dioxide levels were reduced with increasing product in the bag while oxygen levels were proportionately higher (Zagory 1990). Furthermore, a linear relationship, though with a high level of variation (up to four-fold in produce weight) was demonstrated between the weight of produce in a package (20 cm × 30 cm) and its oxygen and carbon dioxide content for one type of plastic film (Zagory 1990).

The number of fruit packed in each plastic bag can alter the effect of MAP. An example of this is that plantains stored at 26–34°C packed with six fruits in a bag ripened in 14.6 days compared to 18.5 days when fruits were packed individually (Thompson *et al.* 1972).

*Perforation*

Punching holes in the plastic can maintain a high humidity around the produce, but it may be less effective in delaying fruit ripening because it does not have the same effect on the carbon dioxide and oxygen content of the atmosphere inside the bag. The holes may be very small and in these cases they are commonly referred to as micro-perforations.

*Absorbents in modified atmosphere packaging*

'Active packaging' of fresh produce has been carried out for many years. This system usually involves the inclusion of a desiccant or oxygen absorber within or as part of the packaging material.

*Polyethylene bags with adjustable diffusion leak*

A simple method of controlled atmosphere storage was developed for strawberries (Anon. 1920) in which the gaseous atmospheres containing reduced amounts of oxygen and moderate amounts of carbon dioxide were obtained by keeping the fruit in a closed vessel fitted with an adjustable diffusion leak.

Polyethylene bags with silicone rubber panels which allow a certain amount of gas exchange have been used for storing vegetables (Marcellin 1973). Good atmosphere control within the bags is obtained for globe artichokes and asparagus at 0°C and green peppers at 12–13°C when the optimum size of the bag and of the silicon gas-exchange panel is determined.

Fruits of *Citrus unshiu* can be stored at 3–8°C in plastic films with or without a silicone window (Hong *et al.* 1983). Storage trials showed that after 110–115 days, 80.8–81.9% of fruits stored with the silicone window were healthy with good coloration of the peel and excellent flavour, while of those without a silicone window, 59.4–76.8% had poor coloration of the peel and poor flavour. Controls stored for 90 days had 67% healthy fruit with poor quality and shrivelling of the peel, calyx browning and a high rate of moisture loss. The size of the silicone window was 20–25 $cm^2\ kg^{-1}$ fruit, giving less than 3% carbon dioxide with greater than 10% oxygen.

*Safety*

The use of MAP has health and safety implications. One important factor that should be taken into account when using MAP is that the gases in the atmosphere can have an impact on the growth of food-borne micro-organisms. The growth of aerobic micro-organisms is optimum at about 21% oxygen and falls off sharply with reduced oxygen levels; optimum growth of anaerobic micro-organisms is at 0% oxygen and this falls as the oxygen level increases (Day 1996). With many MAs containing moderate to high levels of carbon dioxide, the aerobic spoilage organisms, which usually warn consumers of spoilage, are inhibited, while the growth of pathogens may be allowed or even stimulated, which raises safety issues (Farber 1991).

*Humidity*

The gas permeability of some plastics used for film packaging is sensitive to environmental humidity. Gas transmission of polyamides (nylons) can increase by about three times when the relative humidity is increased from 0 to 100% and with ethyl vinyl alcohol copolymers the increase can be as high as 100 times over the same range (Roberts 1990).

Moisture given out by the produce can condense on the inside of the pack. This is especially a problem where there are large fluctuations in external temperatures because the humidity is high within the pack and easily reaches dew point where the film surface is cooler than the pack air. Antifogging chemicals can be added during the manufacture of plastic films. These do not affect the quantity of moisture inside the packs, but cause the moisture that has condensed on the inside of the pack to form sheets rather than discrete drops. They can eventually form puddles at the bottom of the pack. The high humidity provided by individual film wrapping was beneficial, particularly to crops such as citrus fruits and cauliflowers (Zong 1989). The physiological responses to individual film wrapping vary with commodity, and the respiratory pattern of the commodity cannot be used as a basis for predicting commodity responses to individual film wrapping.

## Selection of stores for durable commodities

In comparison to roots, tubers and perishable commodities, durable commodities such as grains are generally considered to be of lower value. The options for storage of grains are therefore limited to relatively simple systems and in many cases the first decision is whether to store in bag or bulk.

### Bag versus bulk

The question of whether grain is best stored in bags or in bulk is often not considered when examining an existing system or developing a new storage system. Failure to consider the various options in the context of the whole handling and storage system can lead to the development of totally inappropriate systems. For example, if freshly harvested dry grain is collected, transported and marketed in bags, it is clear that bag storage would be appropriate. Difficulties can arise, however, when there is a change in the method of handling between field and market, for example, when grain is harvested and transported to collection centres in bags where it may be stored in relatively large quantities before being delivered to a mill where it is held in bulk before processing. The individual circumstances in each case must be examined since the point at which a change from bag to bulk handling needs to occur may not always be obvious. Examples of factors to consider may include:

- The scale of the operation – bulk handling is generally more suited to larger quantities of grain. In the above example, a transfer to bulk storage at the collection centre may be appropriate.
- Transport between field and collection centre (store), and store and mill – in the above example storage in bulk would not be viable if the transport to the mill is only suitable for bagged grain.
- Availability of a reliable electrical supply to power the conveyors, elevators, grain driers and cleaners usually associated with large bulk stores.

Some of the important factors that need to be examined when considering a bag or bulk system are given in Table 5.22.

### Bag storage

Storage systems are generally very simple. Conventional (and preferably clear-spanned) warehouses are usually used although outdoor stacks may be used where the climate allows. The advantages and disadvantages of outdoor and indoor stacks are summarised in Table 5.23.

Loading and unloading may be:

- entirely by hand – trucks unloaded by hand, bags carried into the store and then stacked manually;
- fully mechanised – bags from the production line automatically stacked onto pallets, trucks loaded, unloaded and stacks built by forklift trucks; and
- a mixture of the two – bags from the truck passed into the store and up onto the stack using conveyors/elevators and then stacked by hand.

The choice of system will depend on the availability and cost of labour, as well as the general practices including health and safety aspects.

### Bulk storage

The main choice of storage system for bulk grain is between horizontal (on-floor stores) and vertical stores (silos or bins). On-floor stores may be simple clear-span warehouses or specially constructed warehouses incorporating ventilation in the floor and strengthened retaining walls to allow the grain to be piled against the walls. Silos and bins are specially built stores, either square or round, free-standing or grouped together, incorporating loading, unloading and, usually, aeration systems.

Other options which are less frequently used include below-ground or partly below-ground storage (Cyprus bins) or enamelled, sealed silos, specially designed for storing high moisture-content grain. The advantages and disadvantages of horizontal and vertical storage are summarised in Table 5.24.

## Small-scale farm storage in the developing world

Traditional methods of on-farm storage have evolved over long periods and are generally well suited to agro-climatic regions and social needs. They are sometimes criticised because, it is variously claimed, they are inefficient, the cause of high loss, inflexible, and insecure when compared to 'modern' structures. Nevertheless, they have served well and will continue to meet the needs of many small-scale farmers and will be the obvious choice for farmers where practicable.

However, there are groups of farmers who may need to improve their existing method of storage or change it completely for various reasons including: a need to hold more produce for longer periods on the farm; to provide better protection against pest attack; a shortage of local traditional construction materials as supplies of timber or thatch for roofing become exhausted; availability of

**Table 5.22** Characteristics of bulk and bag storage systems.

| Factor | Bulk storage | Bag storage |
|---|---|---|
| Space utilisation | Excellent, but alternative use of space depends on store type (vertical versus horizontal storage) | Not good – uses only about ¾ of enclosed volume, less in small stores. Stores suitable for other purposes |
| Labour utilisation | Not labour-intensive – requires fewer staff, some of high calibre; and training. High efficiency possible | Can be labour-intensive, and can be inefficient unless labour is well managed |
| Supervision | Wide technical experience/training necessary. Elementary aspects require less constant attention | Elementary aspects, especially hygiene/housekeeping, require constant attention |
| Costs, capital | Higher capital costs to build special stores and install handling machinery. Some existing buildings may be converted | Lower capital costs for a new store. Existing buildings may be suitable, often without conversion |
| Costs, operational | Lower handling costs/tonne for large quantities. No bags needed | Higher operating costs (labour and bags) except for small quantities |
| Versatility of use | Less versatile. Separate units needed to keep different commodities or batches segregated | More versatile. Different commodities or batches more easily kept separate |
| Quality control | Changes in temperature can be monitored using permanently installed instruments | Deterioration may be difficult to detect. Installation of instruments rare |
| Moisture content | Usually require 1% lower m.c. than bag storage. Changes in storage less extensive, if initial m.c. correct. In-store drying or removal for mechanical drying possible | Changes during storage often extensive. In-store drying. Removal for mechanical drying requires a lot of handling – expensive |
| Insect infestation | Reinfestation easier to prevent if store well designed and managed. Inherently less suitable conditions for most pests* | Reinfestation occurs readily and is generally difficult to prevent. All insect pests likely to be troublesome |
| Insect control | When store is gas-tight, easily done using fumigation*, less costly at all times, especially with proper equipment. However, if not gas-tight, extremely difficult to fumigate by conventional means. Specialist equipment/procedures required | Rate of reinfestation faster than some types of fully-enclosed bulk silos. Fumigation under sheets possible |
| Rodent infestation | Depends on type of store – rarely a problem in vertical stores (silos), but can be a problem in on-floor bulk stores | May be a serious problem |
| Rodent control | Problems unlikely, but easy to control and prevent* | More difficult and may be costly |
| Security | Generally less of a problem. Theft rare from silos, less convenient from on-floor stores – time required to bag the grain | Easy – grain ready packaged, but now is more obvious (ref. Stocktaking below) |
| Stocktaking | From volume and bulk density, not straightforward, less accurate | By counting bags and average bag weight– good store management necessary |
| Spillage | Lower losses if properly managed | Higher losses, especially following rodent infestation |
| Wastage | Normally low, but high if a serious problem occurs | Loss levels variable, estimated 2% per annum in best managed stores |
| Commodity damage | Mechanical damage through handling. Heat damage possible | Pest damage. Heat damage possible |
| Integration with associated operations | Very suitable for a smooth flow-pattern | Drying and cleaning before, and milling after storage requires bags to be opened and bulk created |

*Comments apply to vertical bulk (silos); horizontal bulk may suffer higher risk of problems and be more difficult to control.

**Table 5.23** Advantages and disadvantages of outdoor and indoor stacks.

| Outdoor stacks | Indoor stacks |
|---|---|
| *Advantages* | |
| Cheap storage system | High degree of pest control possible |
| Little maintenance of the base required when compared to that of a building (although maintenance of tarpaulins can be expensive) | Good protection from the weather and so storage period is independent of weather conditions |
| Produce can be fumigated for insect control using fumigation sheets over the stack | High level of security possible<br>Mechanisation is not required |
| Flexible in that additional stacks may be built after relatively basic preparations (levelling of the site and the digging of drainage channels around the edges) | Easier to calculate quantity than with bulk in the open |
| *Disadvantages* | |
| Only advisable for temporary or short-term storage | Higher initial and running costs than outdoor storage |
| Only advisable in areas with a reliable dry season over the storage period | More maintenance (i.e. of the store) needed when compared to outdoor storage |
| Little protection against pests | Higher labour requirement than for bulk storage |
| Security is difficult unless in a guarded compound | |
| Problems occur in making/breaking stacks in wet weather if bags are permeable (e.g. paper) | |
| The useful life of tarpaulins is limited | |

**Table 5.24** Advantages and disadvantages of horizontal and vertical storage.

| Horizontal storage | Vertical storage |
|---|---|
| *Advantages* | |
| Specially designed stores are not necessarily required (portable retaining walls can be used to stack the grain against) | Fully mechanised, minimising labour requirements |
| Can be used to store bagged grain if bulk storage is no longer required | Monitoring of temperatures in the grain usually built-in |
| The store can be used for other purposes after the material has been dispatched | If specifically designed to be airtight, grain can be easily fumigated |
| The store can be divided into compartments using internal walls, to separate different types and qualities of material | Easier to segregate different types and batches of grain |
| Can be highly mechanised for fast handling rates | Efficient use of ground space (important when land is expensive) |
| Suitable for materials that tend to cake in vertical storage (for example, fine materials such as flour and processed grains) | Can easily turn a sample of grain (move from one silo cell to another) to sample, treat with pesticide, or prevent compaction |
| Less pressure on the retaining walls than in vertical storage | Most efficient use of space – approaching 100% |
| *Disadvantages* | |
| Handling of large quantities of grain must be mechanised | Most expensive to build |
| Poorer use of storage space if retaining walls (designed to withstand lateral forces from the bulk grain). Especially true when storing different types and qualities of produce (however, if the grain can be piled against the walls, space utilisation is high) | Difficult to detect problems (e.g. hot spots) if instrumentation is not fitted |
| Difficult to ensure 'first-in, first-out' (bulk on-floor require doors at both ends of store) | Handling using conveyors and elevators damages the grain |
| Difficult to estimate quantity (when stored in bulk there are no bags to count – have to rely on volumetric estimate) | Requires dependable electrical supply |
| Insect control difficult (can only access the top of a bulk) | Difficult to control insects if the store is not gas-tight (impossible to cover with fumigation sheets) |
| Materials may contaminate one another in store, either by actual mixing, or by taints and odours | Not suitable for damp commodities or those (such as flour) which can cake or bridge (consolidate), making unloading difficult<br>Alternative uses very limited |

alternative construction materials such as plastic sheets and metal roofing sheets; and the loss of traditional store construction skills, when these are not passed on to the younger generation.

Improved storage systems include modifications to traditional structures and the introduction of new store types that incorporate or are made entirely from industrially produced materials such as prepared timber, cement and galvanised iron sheets. Farmers may be reluctant to accept improved storage systems unless they can see the benefits for themselves and are able to afford them. The individual farmer's level of production and dependence upon storage will influence their willingness or ability to change and the types of storage adopted.

Subsistence farmers rely on their own production to feed their families. They are the most vulnerable group and most dependent upon storage. They are likely to grow the more traditional varieties of grain (rather than improved, high-yielding ones) and the risk of insect damage to stored grain may be relatively low. Subsistence farmers' storage systems will be the simplest and cheapest possible and may include bags, small baskets and other containers made from natural materials. Only simple, low-cost improvements will be possible and the farmer will have to rely on basic good storage management and use of traditional methods of pest control to minimise losses.

Subsistence farmers who are also able to engage in some cash-cropping are likely to grow a mix of traditional and high-yielding varieties of grain and possibly some other cash crop. The traditional grain varieties will be mostly for home consumption and the high-yielding ones for sale. If surpluses are not sold immediately after harvest some additional storage will be necessary. The farmers' storage options will be similar to those of the subsistence farmer but, since they will have some cash to invest in storage, the options will also include construction of additional improved stores and storage of surpluses in bags either in the house or in a separate building. Some will have to invest in insecticides to protect their grain, especially if they are storing high yielding varieties that are more susceptible to insect attack.

Cash-crop farmers will depend on only a minor part of their production to feed their family and will produce a large surplus for sale. This surplus, particularly of grains, may be stored for some time to take advantage of seasonal price rises. Traditional storage systems may be used but these farmers are likely to invest in new, improved, long-term storage systems and use grain storage insecticides.

Farmers who can harvest more than one grain or pulse crop each year or have the opportunity of growing an alternative staple, such as a root crop, will be less dependent upon storage from one season to the next and their storage requirements for surpluses may only be medium term. Multiple cropping may mean that one crop has to be harvested in the wet season and this may pose problems in drying, handling and storage. These farmers will require a flexible storage system that can cope with crops harvested under wet or dry conditions, conserve grains and pulses in the medium term and be adapted for uses other than grain storage, for example, storage of roots and tubers.

### Timber platforms and frames

Platforms and frames can be made at minimal cost from local materials and may be used for storage and drying of produce. Platforms constructed in the open may be four-cornered or circular racks, usually raised about 2 m above the ground. The grain may be placed on the platform soon after harvest and a fire may be lit under the platform to facilitate drying and to deter insects and other pests. Platforms may be covered with a thatch roof that can be lifted off from time to time to aid drying or to remove produce. In humid areas, for example, southern Ghana, Benin and Togo, husked maize on the cob may be stacked in layers to form a cylindrical stack which is then covered with a cone-shaped roof. Cone-shaped, rather than flat platforms are common in humid areas and facilitate drying. Platforms or racks may also be constructed inside the house, often over the cooking fire, or to provide a storage loft below the roof (Fig. 5.22).

Frames consisting of narrow timber or bamboo poles fixed horizontally to heavy upright poles embedded in the ground may be used for storage of maize cobs or sorghum heads, but their main purpose is to facilitate drying of the crop.

Platforms and frames may be improved by fitting rodent barriers around the supporting posts to protect them against rodents. To provide some protection against termites, the posts should be coated with bitumen or used engine oil.

### Yam barns

Yam barns are common among traditional yam farmers of West Africa. The barns consist of timber racks supported on a frame of vertically erected posts about 3 m high and set about 50 cm apart. Horizontal bars are

Fig. 5.22 Platform storage.

attached to the posts to stabilise them. Sometimes, use is made of growing trees as uprights as they strengthen the structure and the foliage can provide natural shade. Where trees cannot be incorporated in the structure, the barn is provided with a roof. Siting of the barn to ensure maximum ventilation is important.

Yams stored in traditional barns may be vulnerable to attack by rodents or even domestic animals. Attempts are therefore made to exclude animals by surrounding the barn with a fence. Alternatively yams may be stored on simple shelving systems with rodent guards fitted to the legs.

### Drying/storage cribs

In humid countries, where grain cannot be dried thoroughly before storage and needs to be ventilated, traditional cribs are used. These are circular or rectangular structures with a framework of wooden poles, with a roof of thatch or corrugated iron sheets. The walls may be made of raffia, bamboo, poles, sawn timber or wire netting and should allow good ventilation and drying. Grain dries best in a narrow crib and, ideally, it should be between 0.5 m and 1.5 m wide and erected in the open, with the long side across the prevailing wind. The floor of the crib should be at least 0.7 m above the ground and the legs should be fitted with rodent guards and treated as necessary to protect against termite attack (Fig. 5.23)

Fig. 5.23 Drying/storage crib.

Cribs provide flexibility in use. They are primarily intended for drying produce, especially maize on the cob. However, they may be used for storage of shelled maize in bags after drying is complete, but in this case the walls should be covered to provide protection against driving rain. They may also be modified for storing other commodities such as root crops and onions.

Traditional cribs usually require a high level of maintenance and may have to be replaced completely after 2–5 years. Improved cribs made from sawn timber, wire netting and metal roof sheets may last for 10–15 years.

The quality of grain stored in cribs will be influenced by the local climate, the presence of pest infestation, design and construction, and the variety of grain being stored. It will usually be necessary to treat stored maize cobs of high-yielding varieties with an insecticide at the time of storing.

### Ventilated structures for perishable crops

Ventilated structures such as cribs can be used for storage of produce with long storage potential such as roots, tubers, and onions. The following essential features should be observed:

- the structure should be located at a site where low night-time temperatures prevail over the storage period;
- if the structure is subject to cold night-time temperatures, movable louvres should be fitted and adjusted to limit the flow of warm air into the structure during the day;
- the structure should be oriented to make maximum use of the prevailing wind for ventilation;
- ventilation spaces below the floor and between walls and roof should be provided to encourage good airflow;
- the roofing material should provide insulation, and the roof should extend to shade the walls of the structure; and
- double-skinned walls will provide better insulation;
- external walls painted white will reflect the heat of the sun.

### Storage baskets

Baskets with an open weave are suitable for drying maize cobs or sorghum heads. Shelled or threshed grain and other commodities such as dried cassava chips can be stored in close-weave baskets or baskets that have been lined with a layer of mud. Mud-plastered walls will provide protection from rain, strengthen the structure and prevent uptake of moisture by dry grain. Baskets may be used for both drying and storage of grain (Fig. 5.24).

Basket stores can be kept inside the house or outside in the open but in both cases they should be raised off the ground to prevent uptake of ground moisture. When used in the open they should be sited under a shelter or be fitted with an extended thatch roof to provide protection against rain and shade from the sun.

Well-constructed baskets may have a life of up to 15 years as long as they are properly maintained. Baskets give no protection against insects, and treatment of the grain with an insecticide is recommended if infestation is expected.

### Solid wall bins

Solid wall bins are often associated with dry climatic conditions under which produce can be dried satisfactorily by simple sun-drying. The bins may be made of clay plaster, clay blocks, stones or burnt bricks and are raised

**Fig. 5.24** Storage basket.

**Fig. 5.25** Solid wall bin.

off the ground on timber frames, on large stones or on a clay pedestal. A base of packed earth is not recommended because it may permit termites and rodents to enter the store. Improved solid wall bins made of stones or bricks may have a concrete pedestal or foundation incorporating a moisture-proof barrier to prevent uptake of moisture. The structures may be round, cylindrical or rectangular in shape and designs are often characteristic of communities or localities. The clay used for construction of solid wall bins commonly comes from termite mounds and is strengthened by the addition of straw. Thin-walled structures are often used for storage of unthreshed grain and pulses whereas the stronger, thick-walled structures are more suitable for storage of threshed grain and dried root crops such as dried cassava chips (Fig. 5.25).

A modified mud-walled bin with thatch roof may be used for storage of yams. These structures, which resemble small mud huts since they are provided with a door or opening, are to be found in the savannah areas of the Yam Belt in West Africa. Good protection is provided against direct sunlight and rain. However, as yams are piled on top of each other, ventilation is restricted and development of rot in the tubers can be a serious problem.

Some solid wall bins have internal dividing walls making several compartments, thus giving some flexibility in the different types or quantities of grain that can be stored. Separate openings for filling and emptying are usually included at the top and bottom of the structures. As in the case of basket stores used in the open, solid wall bins may be sited under a shelter or be fitted with a thatch roof to provide protection against rain and shade from the sun.

The life of solid wall bins will depend upon the construction and local climatic conditions. With good maintenance, bins may last for 20 years or more; bins using mud in their construction will be susceptible to attack by termites.

Insect infestation in stored grain may be dealt with by the application of an appropriate insecticide. Some solid wall structures may be suitable for fumigation, however, most traditional stores will require considerable modification to ensure that they are gas-tight (Brice & Golob 1998).

### Metal storage bins

Metal storage bins may be made from flat or corrugated galvanised metal sheets. They are usually cylindrical in shape with a flat top and bottom. Most bins used for small-farm storage have capacities of around 1 tonne. Grain is commonly loaded through a hatch on the top and removed from an outlet in the side or at the bottom of the bin (Fig. 5.26).

Bins should be raised off the ground on a platform to allow air to circulate around the base to prevent corrosion by ground moisture. They should also be kept in the house or placed under a roof to provide protection against rain and, more particularly, shade to help reduce moisture migration and heating of the grain inside.

There is little experience of the use of metal storage bins in Africa outside of Swaziland. However, they have been used successfully on the Indian subcontinent, in Central America and Turkey. The cost of bins will vary according to the capacity, charges made by skilled metal

**Fig. 5.26** Metal grain storage bin.

workers and the distance they have to be transported. The durability and security offered by the system will appeal to the more affluent farmers. Adoption of metal grain bins is more likely in areas where metal containers are already used as water tanks and they can be made locally by sheet metal workers. Large bins are difficult and expensive to transport to rural areas and are easily damaged in transit over rural roads.

Grain must be very dry for storage in metal bins and so the system is more suited to areas where grain is harvested in a distinct dry season, followed by storage through a rainy season where good protection is desirable. Well-constructed and well-sealed metal bins will provide good protection against insects, rodents and birds, and against mould, if the grain is properly dried. Insect control using insecticides or fumigants is essential. The metal bin is one of the most suitable structures for fumigation.

### Underground storage

Pit storage is an ancient method of grain storage and is still practised in Ethiopia, some Sahelian countries, and parts of southern Africa, India and Turkey. Pit stores used by farmers may have capacities ranging from less than 1 tonne to more than 20 tonnes. The shape of pits varies from region to region; they may be straight sided, square, cylindrical, spherical or amphoric in shape (Gilman & Boxall 1974). The entrance to the pit may be closed with a large stone or strips of wood, covered by a layer of soil. An absorbent lining to the pit made from grass matting and straw, or chaff and grain husks, will help to reduce the damage from moisture that might seep through the walls of the pit (Fig. 5.27).

Improvements to traditional pits have addressed the problem of moisture penetration, mixing of soil and grain and subsequent mould damage. A plaster lining made from a mixture of mud and cow dung, or a single concrete lining, will restrict ingress of water and termites as well as prevent the intermixing of soil with grain. More sophisticated concrete or ferrocement linings are built up in a number of layers and incorporate bitumen or other waterproof coatings. Lining pits with plastic sheets also helps to restrict moisture damage to the stored grain (Boxall 1974).

Underground storage is popular because the entire store can be concealed so grain is unlikely to be stolen and there is no risk of fire. Well-sealed pits will prevent entry by grain insect pests. Some initial mould growth is likely but this may reduce oxygen levels which in turn can asphyxiate insects and inhibit further mould growth. Termites sometimes damage grain stored in pits and burrowing rodents can occasionally gain access to the stored grain.

Underground storage may also be used for relatively short periods for yams and for potatoes. For example, if labour at harvest time is limited, yams may be held temporarily in trenches or pits dug in or close to the field. The pits are usually lined with straw and the tubers stored on a layer of straw either horizontally on top of each other or with the tip vertically downwards beside each other. In areas where there is a pronounced dry season the pits may be constructed and filled in such a way that part

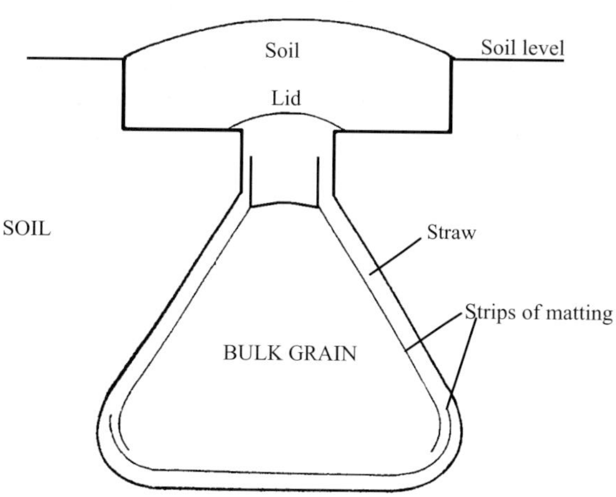


**Fig. 5.27** Underground storage.

of the stored crop is above ground level. The exposed tubers are covered with straw and in some cases a layer of soil may also be added. Although the system provides protection against respiration and transpiration weight losses, it has the disadvantage of reduced ventilation and direct contact between tubers. Consequently the tubers may become warm and conditions may be conducive to formation of rot (Knoth 1993).

Root boxes and root cellars are other forms of underground storage used for potatoes. A root box lined with cloth and straw, buried to the top edge in soil, will keep potatoes cool, while providing protection from freezing. The wooden lid can be lifted for easy access and if necessary additional straw may be placed on top to provide more insulation. Root cellars are constructed by digging out a pit to the depth of about 2 m and framing the sides with wooden planks. A wooden 'chimney' can be fitted as a roof vent.

### Clamps

Clamps are simple structures used for the storage of root crops, especially potatoes, in Europe and Latin America. Potatoes are piled on a bed of straw and over a ventilating duct. The pile of potatoes is then covered with straw before finally being covered with a layer of soil. In tropical and subtropical climates the layer of soil may be replaced by an additional layer of straw to allow better ventilation.

### Bag storage

Bag storage is a convenient way of keeping threshed grain and pulses. Bags provide the flexibility to store different types and different quantities of grains and pulses and the commodity can easily be removed for consumption, inspection or sun-drying. It is also immediately available for sale. Storage capacity is limited only by the number of bags available and the size of the storeroom. Small numbers of bags may be kept in the house or in a separate storeroom. Ideally bags should be raised off the floor on dunnage or on storage platforms to prevent them from getting wet from uptake of moisture from the ground. Successful bag storage depends on adoption of good storage management, especially as they provide little protection against insects, rodents and moisture.

## Transit or trader storage

Throughout the developing world, the role of the grain trader varies tremendously. In many parts of Africa traders have limited storage facilities as they buy and sell produce quickly, and have little need for interseasonal storage. A lack of capital often makes long-term storage of staple grains insufficiently profitable to traders who can earn more money by investing in fast-moving consumer goods. Asian traders on the other hand have a much larger involvement in interseasonal storage, especially of rice and wheat, commodities which must be milled before reaching the consumer. In contrast, in much of Africa where the main staples are coarse grains such as maize, millet and sorghum, consumers buy the grains whole, and either grind them at home or have them ground at small custom-mills.

The requirements of private grain traders have tended to be neglected in favour of the development of storage facilities for government or quasi-government grain marketing organisations (Picard & Procter 1994). Many private grain traders, especially in Africa, rarely store more than a few tonnes of grain at any given time. They tend to dispose of stocks as quickly as possible, thereby minimising losses associated with pest infestation and avoiding the expense of pest control. Their storage facilities, therefore, are small (200 tonnes or less) and of very basic design. Typically a trader's store will consist of little more than a single large room, with bare brick or mud-plastered walls, an earth floor and a corrugated metal roof. Often the store is one of several rooms in a long building sharing a common roof.

Traders suffer storage losses as a result of insect pest damage. Storage space is usually confined and even the most basic storage hygiene is often lacking, which makes cross-infestation (from neighbouring stores), and re-infestation from one season to the next, unavoidable.

Grain traders in parts of Somalia and Sudan store large quantities of grain in big pits. The grain heap may extend above ground level and a layer of matting or straw is placed over it before being covered with a layer of soil. When trading conditions are favourable the grain is bagged up and sent to urban wholesale or retail markets where it will be held temporarily in typical traders' stores.

## Medium and large-scale storage

### Bag storage

Medium or large-scale stores for bagged grain are basically multi-purpose warehouses. They tend to be either trussed or frame designs (in both cases the structure of the building comprises a timber, concrete or steel frame

**Fig. 5.28** Typical indoor bag stack.

with the walls acting as in-fill rather than being load bearing (Fig. 5.28). Designs in which the roof is supported by internal columns should be avoided since bag stacks must be free-standing to facilitate inspection and stock accounting and to allow them to be covered with fumigation sheets.

### *Floor*

The floor of a warehouse should be designed for easy construction without expensive finishing. It must be able to support the weight of the grain that will be stacked upon it and may need to be strengthened if vehicles are to enter the store. The best floor construction is reinforced concrete, 150–200 mm thick, with expansion joints (to prevent cracking) laid on top of consolidated hardcore. A moisture-proof barrier should be incorporated in the floor and made continuous with the damp-proof course in the walls to prevent moisture rising through the floor and being absorbed by the commodities.

The floor level must be sufficiently above ground level to ensure that water will not enter the warehouse and to ensure drainage of water away from the building in all weather conditions. Warehouses are sometimes erected on a plinth raised about 1.2 m above ground level to facilitate loading and unloading of vehicles; but this arrangement is expensive and may add considerably to the cost of construction.

### *Walls*

Most modern warehouses are constructed with a framework, which carries the load from the roof, and the walls, which are not load bearing, are built between the supporting frame pillars.

The best materials for the construction of walls are brick masonry or concrete blocks. The walls should be rendered smooth on both sides. They should be painted white on the inside to facilitate detection of insects. The outside walls may also be painted white or be rendered with a 1 : 1 : 6 mortar of cement, lime and sharp sand to produce a light-coloured, reflective finish, that will keep the store cool. Alternatively the walls may be made of lightweight sheet materials, but these are easily damaged and have higher maintenance requirements.

A vapour-proof barrier (damp-proof course) should be incorporated into the base of the walls to prevent dampness rising and causing damage to the structure and the stored commodities. A concrete strip about 1 m wide should be laid around the outside of the warehouse, to prevent rain from eroding the base of the walls below the damp course.

### *Roof*

The roof should be designed to shed water quickly without leaking, to provide shade and to keep out pests, dust and radiant heat. For large stores, a concrete or steel portal frame is preferred because it gives a clear span and is most likely to be cost effective. Steel frames are most suitable for spans greater than 15 m. Roof frames made of wood are only suitable for warehouses up to 5 m wide. The wood should always be well dried and treated with a preservative.

The pitch of the roof (i.e. the angle of the roof to the horizontal) should be between 17° and 22°. If it is less than 17° rain will not run off effectively, and if greater than 22° additional materials will be required, thus increasing cost with no added benefit. Flat roofs (very slightly sloped to ensure that water still runs off) are more difficult to maintain than pitched roofs and are not to be recommended.

Roof cladding can be aluminium, galvanised steel or asbestos cement. Tiles are not recommended because of the additional cost, weight and maintenance, especially for large warehouses.

The roof should overhang the gables by 0.7–1.0 m and the eaves by at least 1 m. This ensures that rainwater is shed well clear of the building, perhaps into a drain at ground level; it also overcomes the need for gutters and drainpipes, which can become blocked and may provide a means of access to the warehouse for rodents. The overhang also helps to keep the store cool by shading the walls and protects ventilation openings from rain. However, large overhangs can increase the cost of the roof and may be susceptible to wind damage in some areas.

*Doors*

Doors should be large enough and in the best positions to facilitate intake and dispatch of commodities and to permit effective use for supplementary ventilation. All doors must be made rodent-proof if possible and they must be secure against theft. An internal security door, of welded steel mesh, will be useful when the main doors are left open for ventilation. Bollards placed next to the corners of the doors will prevent vehicles from damaging the structure of the store.

### Bulk storage

#### *Horizontal (on-floor stores)*

A conventional clear-span warehouse, of the type used for bag storage, is the simplest bulk on-floor store. Although cheap, space utilisation is low; as the walls will not have been designed to withstand the considerable lateral forces, the building cannot be filled with grain (Fig. 5.29).

Specially designed bulk on-floor stores often incorporate aeration systems. A fan, usually located in a fan house at the rear of the store, is connected to a main air duct or plenum chamber that runs either along the side or along the middle of the store. Smaller lateral air ducts run from the plenum chamber under the stored commodity and have openings or grills along their length to force air up through the crop.

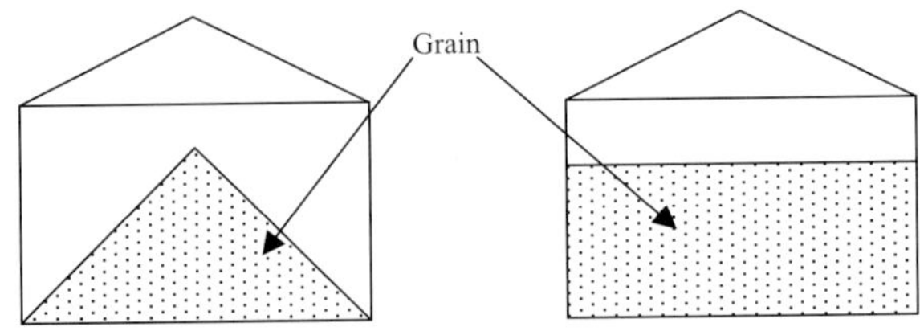


**Fig. 5.29** Space utilisation without and with retaining walls.

Produce may be dumped from a truck either directly in the store or into a reception area outside from where it is then either conveyed into the store or is carried by a loader or tractor-mounted bucket. Grain should be stacked to a uniform height throughout the store to ensure even aeration. Depending on the size of store, the grain may be piled up using an elevator, tractor/loader-mounted bucket, tractor/loader-mounted grain pusher, or a grain thrower (Fig. 5.30). The grain pusher consists of a steel frame, 10–15 m long, attached to the tractor loader with a bulldozer-type blade attached to the front of the beam, and enables far greater depths to be achieved when compared to the conventional tractor-mounted loader.

#### *Vertical stores, silos or bins*

Storage bins may be round or square, within an enclosed store, under an open-sided shelter or totally weatherproofed for siting outdoors (the latter bins are more usually called silos). They may be separate from each other or grouped together to form a block. The simpler designs are generally constructed of galvanised steel, but wood and weld mesh with an inner lining of hessian may be used for indoor bins. Capacities vary from a few tonnes to 3000 tonnes or more.

Round bins are the strongest and use the least amount of sheeting in their construction. Square bins maximise the use of available space and tend to be used where space is limited. Since square bins can easily be secured together, they often form the structure of the store with the roof of the whole store being mounted directly on top of the block of bins.

Bin systems may consist of one or more bins with drying floors with the remaining bins used solely for storage.

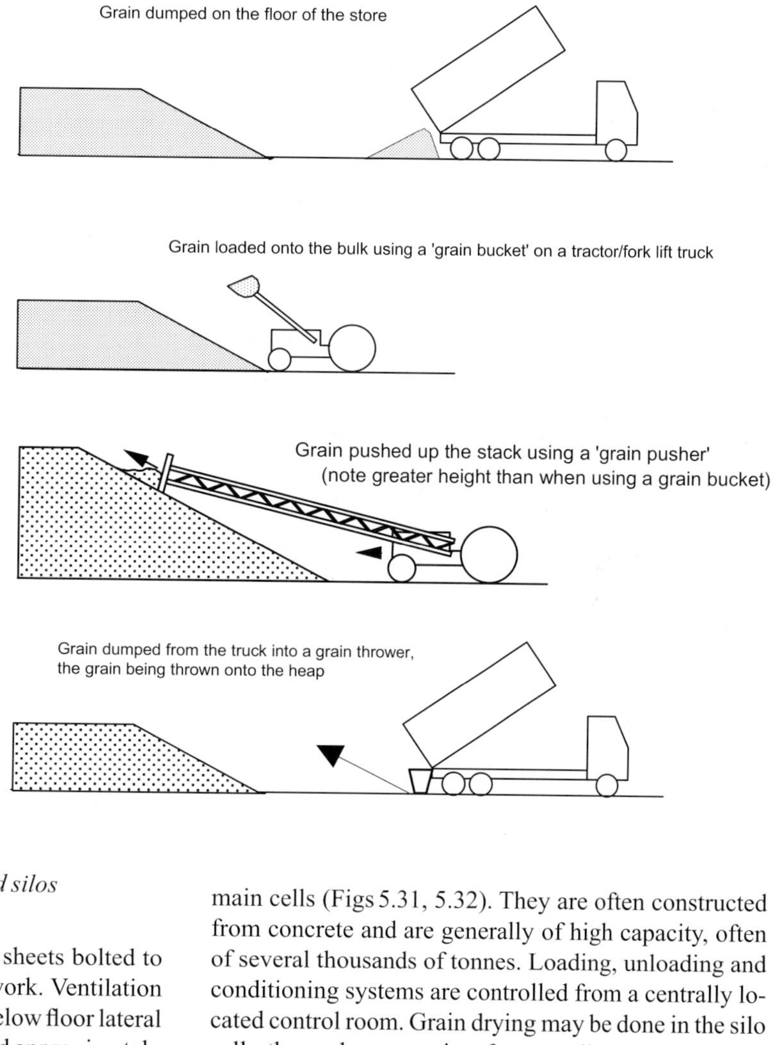


**Fig. 5.30** Options for mechanised loading of grain into bulk on-floor grain stores.

*Free-standing, corrugated metal round silos*

These are constructed from corrugated sheets bolted to either an internal or an external framework. Ventilation may be achieved using either above or below floor lateral ducts or a fully ventilated floor supported approximately 0.5 m above the base.

Bins are loaded from the top, using either a portable inclined conveyor or a fixed overhead conveyor (usually fixed into position), directing grain into the central hatch at the top of the bin. Care must be taken when loading the bins to ensure that the grain is distributed evenly within the silo, avoiding uneven loading against the side walls. Unloading is usually through a central sump in the floor, via a horizontal conveyor in the floor to an external conveying system.

*Large round silo complexes*

These usually comprise at least two rows of round silos, with star-shaped cells in the spaces between the larger main cells (Figs 5.31, 5.32). They are often constructed from concrete and are generally of high capacity, often of several thousands of tonnes. Loading, unloading and conditioning systems are controlled from a centrally located control room. Grain drying may be done in the silo cells themselves or using free-standing high capacity continuous flow driers.

*Medium or large square bin complexes*

These complexes provide a very compact and efficient system. Common in the 1960s and 1970s, square bin storage systems (Fig. 5.33) are less frequently built now in the UK, where they have largely been superseded by round corrugated metal silos which are less expensive to construct.

Grain is tipped directly into a reception pit at the front of the store (inside or outside) where it is elevated to the top of the store. Diverter valves at the top of the elevator direct the grain to a cleaner, drier (if in-bin drying is not possible) or to the storage bins via a top conveyor.

**Fig. 5.31** Large round silo complex.

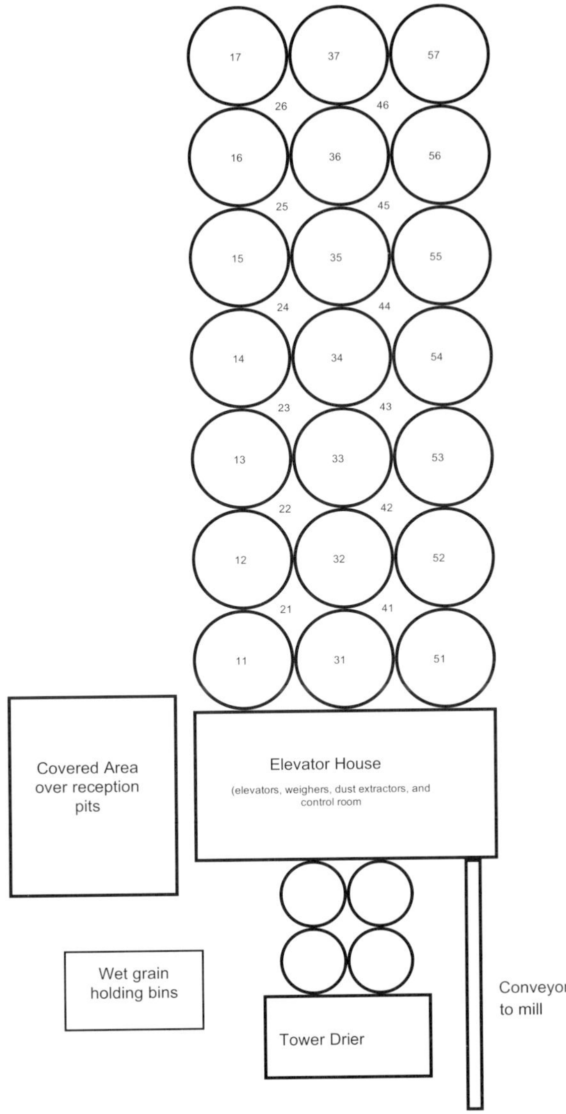


**Fig. 5.32** Layout of a typical large round silo complex.

### Fumigable stores

Gas-tight stores are required if whole-store fumigation is to be implemented. It is extremely difficult to make conventional stores sufficiently gas-tight. Conventional warehouses, corrugated round bins and square bin complexes are rarely suitable. When considering the design of a fumigable store such as a warehouse, all the design elements must be examined, for example, conventional rendered block walls are porous to fumigants, roof sheets must be sealed at the joins, and all doors, ventilators, eaves and ridges must be specially designed. Even if a store has been specifically designed for whole-store fumigation, maintenance is crucial to maintain the desired level of gas-tightness. Round concrete silo complexes may be suitable but checks should be made (visual and pressure decay tests). Some silo complexes may have specially designed built-in fumigation systems (Fig. 5.34).

## Storage management

Storage management involves the preparation or selection of items for storage, arranging suitable storage procedures and ensuring that the commodities stored can be kept as safely as possible to meet particular needs. Thus, storage management requires that the commodities to be stored are in good condition and fit for storage; the site and facilities to be used are fit for the purpose; and the quality of the produce should be maintained sufficiently for the intended use.

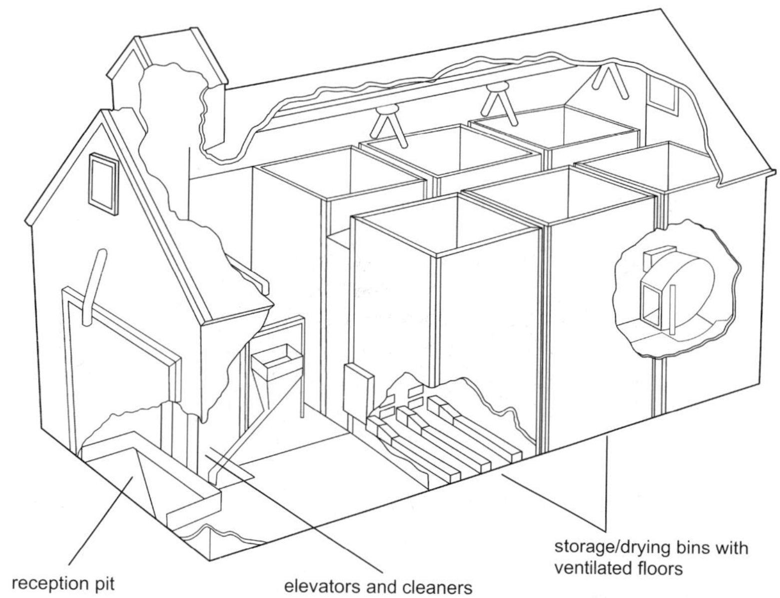


**Fig. 5.33** Block of square metal bins.

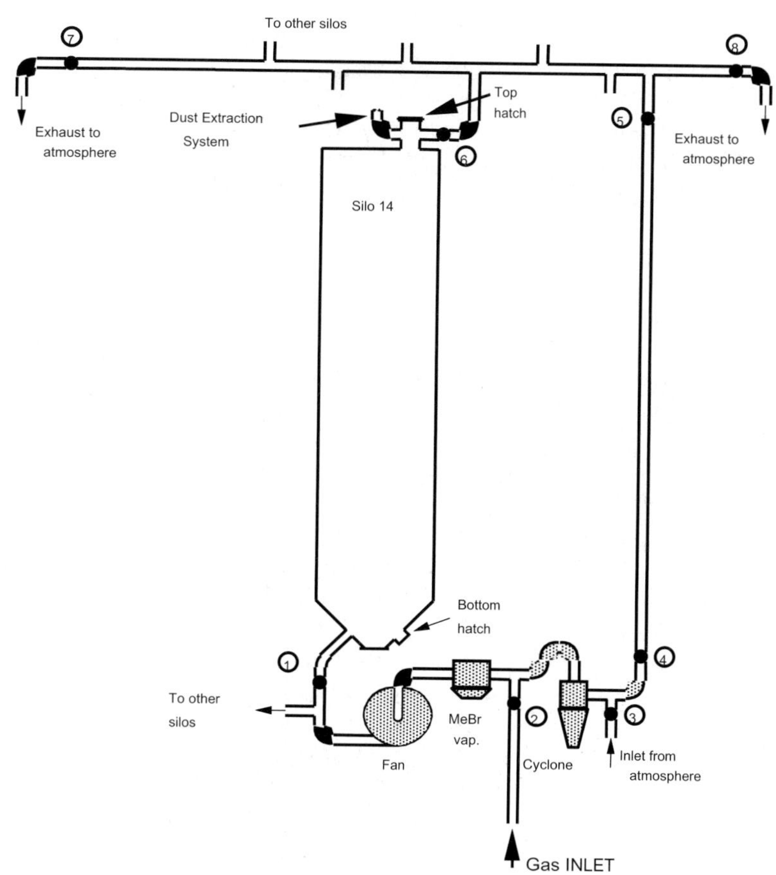


**Fig. 5.34** Fumigation system in a large round silo complex.

Once the commodity has been stored, storage management is largely a matter of common sense. It requires that the storage environment is maintained in such a way that deterioration of the commodity by micro-organisms and damage by insects is minimised and its shelf-life is maintained, and that the storage structure is sufficiently secure against theft and rodent damage.

An important aspect of storage management involves regular and detailed inspection and monitoring of all or part of any consignment of produce, the methods of handling and transport, the storage building and the standards of warehouse-keeping, including the operation of special systems such as refrigeration and controlled atmospheres. The overall objective of such inspections is to provide information as a basis for management action and future planning to ensure that deterioration and loss of produce during storage is kept to a minimum. Inspection calls for the exercise of judgement based mainly on experience and a thorough knowledge of the nature of the commodities and the conditions under which they are to be stored. Those responsible for inspection must have a thorough knowledge of the storage system and the defects or faults that are likely to occur. They must be familiar with the commodities, and must understand how they may be influenced by, for example, changes in temperatures, moisture content or relative humidity, and the potential for damage by micro-organisms, insects or other pests.

A knowledge of the origin and history of the commodity and future requirements is also required, since this will help in deciding what to look for and the action needed should signs of deterioration be detected. Similarly, it is useful to know something of the history of the storage building. In summary, an inspection must consider:

- the source of the commodity;
- past treatments (e.g. any special pre-storage treatment, insecticide applications, fumigation);
- any systems for temperature, humidity and gas control;
- duration of storage – to date and for the future;
- future processing requirements; and
- the previous use of the storage building – especially concerning storage of non-food commodities.

### Quality assessment and grading

Before commodities are accepted for storage they must be inspected to ensure that they are of good quality and fit for the purpose. This usually means that grain must be dry, clean, with few damaged grains and free from pest infestation, and perishable commodities must be free from damaged, diseased or rotten produce. Product temperature is important for perishable commodities and those kept in cold storage should be pre-cooled to a temperature slightly above the storage temperature.

Determining the suitability of a commodity for storage will begin with a gross inspection to determine the apparent condition. If acceptable, more intensive sampling of the consignment may be necessary to assess objectively its quality or grade. Samples will be assessed against grades or standards based on the qualities or properties regarded as important in determining the commodity's suitability for storage or fitness for market, also with regard to local conditions and consumer requirements.

Grades and standards are defined parameters that segregate similar products into categories and describe them with consistent terminology that can be commonly understood by all concerned. In particular, standards are rules of measurement established by regulation or authority and the grades provide a system of classification based on quantifiable attributes. A clear set of descriptive guidelines or standards help to determine grades and these together provide the common language and terminology for defining product value. A system of grades and standards is critical when products are handled in large volumes and are traded over long distances. They convey valuable information about the products that determines prices and helps to define contracts for delivery.

Agricultural products can display a vast array of characteristics such as weight, size, appearance, shape, colour, taste, density, firmness, tolerance to insect damage and moisture content; therefore, a system for clear communication between producer and receiver is vital. A common grading system and terminology will have several benefits such as:

- facilitating the purchase of a product that one has not seen;
- improving incentives for quality and safety;
- making market information meaningful;
- facilitating price and quality comparisons;
- reducing the risk of deception and fraudulent marketing;
- enabling diverse market mechanisms such as futures trading, commodity exchanges, inventory credit or warehouse receipts schemes, and letters of credit; and

- facilitating resolution of disputes regarding quality and/or composition of cargoes.

***Supply chain management for horticultural produce***

The movement of food between the producer and the consumer requires a level of management to ensure that produce, of an acceptable standard, flows to the consumer and that income flows to the producer and all those involved in the supply chain.

The supply chain covers all the steps in the supply of produce from farm to market, i.e. production planning, production, harvesting, assembly, sorting and packing through to transport, storage, processing and distribution. In order to establish a viable commercial supply chain the following are required:

- equitable returns for all players;
- flows of relevant information and technology; and
- efficient management systems.

A successful supply chain may be defined as one in which all the individual goals (economic and non-economic) of the supply chain participants (farm and packhouse labourers, farmers, packhouse managers, truckers, marketing organisations, etc.) have been achieved.

In establishing market linkages, it is very important that the roles of the retailers, farmers and intermediaries (where relevant) are clearly defined and well understood by all partners so that each player has realistic expectations of each other's roles.

One of the keys to achieving a successful supply chain is the need for a pivotal organisation (e.g. export promotion council, exporter, importer) that can bring together the players and stakeholders in the chain. Such an organisation is particularly important in identifying potential export markets, shipping agents, market requirements and importers. For example, in the last decade the supermarket sector, particularly in the UK, has become the driving force in influencing the supply chain of perishable fruits and vegetables.

Prior to assuming this prominence, the supply chain involved a larger number of small producers and exporters supplying wholesalers, often making use of family linkages (Fig. 5.35). The supermarket sector was a relatively minor outlet, sourcing fruits and vegetables from wholesalers.

However, now the horticultural supply chain has become a buyer-driven chain (Fig. 5.36) which:

- bypasses the wholesale market;
- makes use of contracted importers who are category managers of certain produce lines;

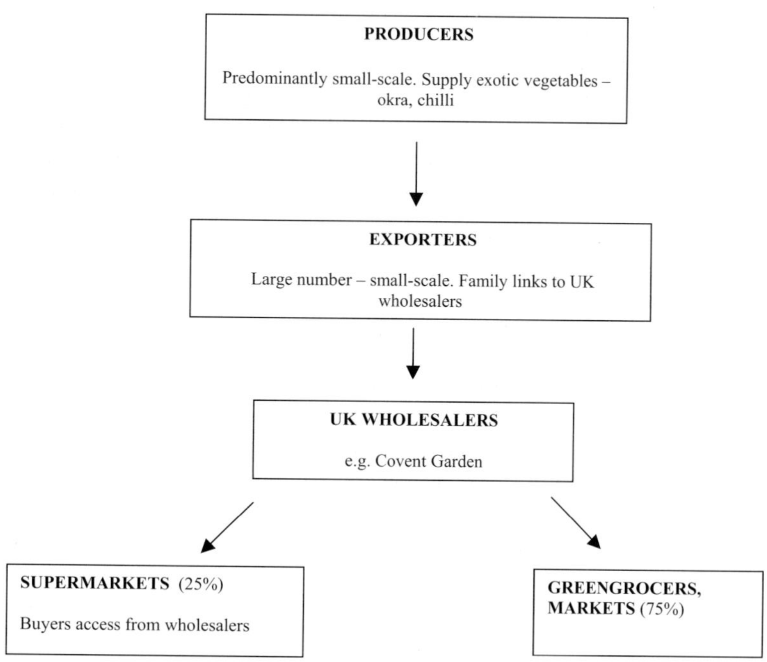


**Fig. 5.35** Supply chain – Kenya to UK in the 1980s.

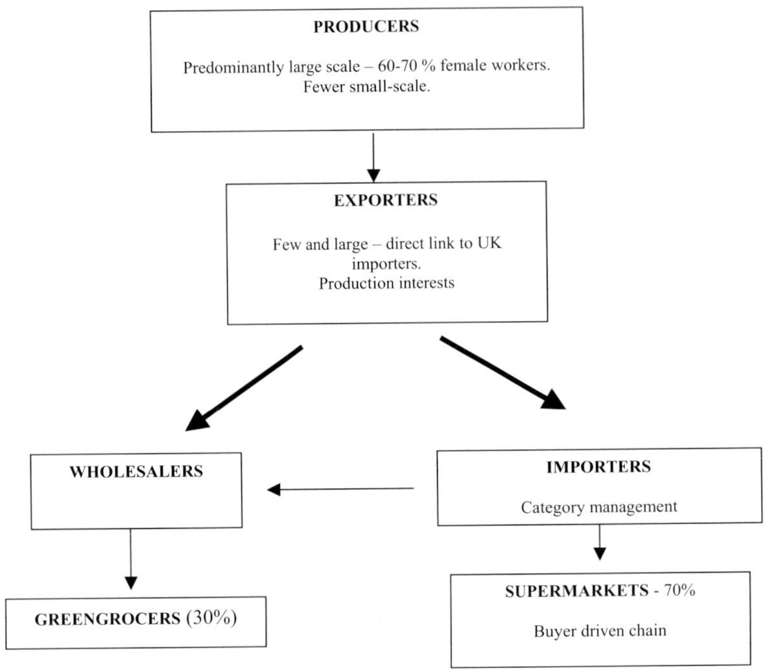


**Fig. 5.36** UK supply chain for horticultural produce in the 2000s.

- demands auditing and compliance throughout the chain;
- uses in-house non-negotiable standards and codes;
- develops national and pan-European alliances to produce codes, e.g. in food safety, such as British Retail Consortium codes for packhouses, and good agricultural practice standards (EUREPGAP); and
- increases opportunities and competitiveness among importers, exporters and growers.

The supermarkets have had to adopt this approach in order to meet customer and legal requirements in relation to:

- The UK Food Safety Act 1990 – requires retailer to show due diligence in supplying food;
- EU harmonised regulations on pesticide residue levels and the UK Department for Environment, Food and Rural Affairs' 'name and shame' policy;
- environment and social welfare – under the influence of the fairtrade NGOs and ethical trade initiatives (making use of core codes of the International Labour Organization);
- quality and consistency – particularly in self-service retailing where customers make visual assessments of the produce;
- frequent, reliable and efficient delivery of high-quality produce all year round;
- price;
- increased produce range – including semi-processed produce; and
- food hygiene and safety needs.

In developing a reliable supply chain, importers have become pivotal organisations, that take responsibility for sourcing supplies, liaising with exporters and growers, ensuring compliance with supermarket and other standards. To achieve these goals, the importing companies have:

- made linkages with large farms and well-organised smallholders' associations with proven reliability and track record (usually developed over more than 20 years of association) and with product experience;
- developed good logistics, with efficient cold chains, particularly at packhouses and airports, with reliable and competent personnel and export officials; and
- implemented standards and codes of practice which meet market and regulatory requirements, with an effective monitoring system that can be verified by independent auditors.

The development and implementation of standards and

codes of practice covering a range of issues have been a part of supply chain management.

***Standards and codes of practice***

Standards and codes of practice are one of the most common instruments used to manage the supply chain. Many private, national and international bodies may be involved with defining quality of produce, such as marketing boards, exporting countries, importing countries, processing industries, national authorities and international agencies.

*Purpose of standardisation*

The following are the principal objectives and purposes for the adoption of standardisation of horticultural produce quality:

- standards serve as a basis for measuring quality – they introduce a common language and understanding at all stages of marketing;
- the introduction of standards brings about a level of discipline, which in turn assists the establishment of sales systems, and particularly assists in the adoption of more advanced marketing methods;
- with the expansion of international trade, standards have accelerated transactions and assisted in the adjustment of supply and demand situations;
- standards are invaluable in assisting in the settlement of disputes between buyers, handling agents and sellers; and
- the establishment of standards, and grades within these standards, forms a basis for market prices, as comparisons of price can readily be given where quality criteria have been previously set.

*Types of standards*

The supply chain must ensure that the marketed produce adheres to legal standards – those which are commonly established by national governments and generally relate to safety. These standards are often mandatory and represent minimum standards of quality. Their major purpose is to ensure that products are not adulterated or do not carry dangerous contamination, such as undesirable micro-organisms, insects, pesticides or potentially toxic additives. The European Union has established maximum residue levels (MRLs) pesticide legislation, and quality standards for horticultural produce (Box 5.2).

**Box 5.2 Outline of European Union common quality standards for horticultural produce.**

| | | |
|---|---|---|
| Extra | Top quality | Approx. 5% of market |
| Class I | Good commercial quality | Approx. 60% of market |
| Class II | Reasonably good commercial quality | Approx. 35% of market |
| Class III | Fit for consumption | When applicable |

**Minimum requirements for all classes**

Produce must be:

- whole
- sound
- clean
- free from foreign exchange moisture
- free from foreign taste or smell

**Additional requirements**

Definition of level of defects permitted in each class, e.g. mechanical damage, shape, insect damage, colour, maturity.

**Size requirements**

Definition of minimum size for each class and size grading requirement within that class.

**Quality and size tolerance**

Definition of level of tolerance of defects outside of a class for each class.

**Packaging, presentation and labelling**

Recommendation and definitions given.

However, these are viewed as minimum standards and there are many types of standards or codes governing practices, for instance:

*Social codes of practice* – such as those produced by: fairtrade organisations (e.g. Max Havelaar, Social Accountability 8000 and the Ethical Trading Initiative);
*National and international sector-specific codes* – such as the Kenya Flower Council code for the cut flower industry and the Horticulture Promotion Council code for the Zimbabwe horticulture industry;
*In-house, company-specific codes* – such as those of the UK supermarkets; and
*Environmental codes* – such as ISO 14000 series, EUREP (Euro-Retailer Produce Working Group) good agricultural practice framework, IFOAM (International Federation of Organic Agriculture Movements) organic standards.

Many of the major European supermarkets as well as other standard-setting bodies are developing codes of practice as a result of growing consumer concern about food production methods and their impact on poor people and the environment. These codes of practice cover:

*Food quality* – there has to be complete traceability from the field to the fork with an agreed set of standards at every step along the production chain, and an audit system for verification;
*Social impact* – there must be due reward for smallholders, and conditions for workers must be reasonable; and
*Environmentally friendly production* – food must be produced in farming systems that are sustainable and do not cause damage to the environment.

The common feature of these codes is that they are auditable. How the auditing is conducted and by whom varies. There is no template for implementing a code of practice, but there is a growing consensus about what constitutes best practice and the key features required.

A recent feature in the process of developing standards has been the approach to develop harmonised codes, which are acceptable to all parts of the supply chain.

Industry standards are sometimes established by an organised industry association in order to establish a reliable identity for a particular product. Normally, such standards become effective because the majority of producers agree to them. They are seldom related to safety, but more to a characteristic quality, which the industry feels is useful to establish credibility for the market. Products such as wheat gluten, corn starch and peanut butter all conform to a set minimum standard established by the industry. These standards are commonly referred to as commodity standards or standards of identity.

In the horticultural sector, a significant initiative in the harmonisation process is being undertaken by the Euro-Retailer Produce Working Group (EUREP), which has produced a code of good agricultural practice (GAP). Suppliers in producing countries aiming to enter or maintaining their place in the EU market are looking to either implement directly EUREP GAP or benchmark national codes with a view to their being recognised as equivalent.

### *Quality characteristics of grains*

Every type of grain possesses many characteristics or qualities that contribute to its overall quality. Different grain types and varieties may be distinguished and identified by variation in intrinsic or natural properties (e.g. grain shape, size, structure and natural colour). These properties also determine the natural quality of the grain. However, during production, or at any of the stages from field to store, the grain invariably acquires a number of properties that are generally detrimental to the overall quality of the grain. Some of the more important of these acquired properties as they affect the safe storage of grain (and to some extent consumer requirements) are considered below.

#### *Foreign matter*

At several stages in the producer/consumer system grain is liable to acquire certain amounts of foreign matter, such as chaff, seed coats, stems, pods, leaves, other crop and weed seeds, insects, mites, dead animals, animal excreta, earth, sand, dust, stones, metal and glass. The presence of this foreign matter can have far-reaching effects on the overall quality and value of the grain and may result in changes in the physical characteristics such as bulk density, changes in thermal conductivity leading to changes in temperature and moisture content, and damage to handling equipment. Organic foreign matter is ideal food for many insect pests and growth of micro-organisms and, even if it is not a source of cross-infestation to the grain, it may shorten its storage life through the translocation of heat or moisture generated by its own infested condition.

#### *Moisture content*

Moisture content is of crucial importance in the storage and processing of grain. A high percentage moisture content can be the cause of several problems. Firstly, if grain is able to dry further after storage it can be a direct cause of weight loss. This is the principal factor responsible for what is commonly referred to in the trade as 'shrinkage', the cost of which is passed on to buyers or consumers even though there may be no significant loss in the food value of the grain. Secondly, high moisture content encourages infestation by insects and micro-organisms. Thirdly, it allows the grain to metabolise more rapidly than is desirable, resulting in unwanted chemical changes in the grain, germination, premature ageing, or increases in temperature, any of which may lead to heat damage and even self-destruction of the grain.

Grain with a low moisture content is unlikely to become badly infested by insect pests, and the risk of microbiological infection during storage is negligible. However, low moisture content may make some grain (particularly paddy rice, pulses and some oilseeds) very brittle, so that it is liable to split, crack, or break easily during handling and storage. This is undesirable if the consumer prefers whole grain.

#### *Broken grain*

Grains are normally marketed as whole grains and consignments showing evidence of broken grains will be considered to be of inferior quality. Breaking, splitting or cracking of the grain renders it more prone to attack by insects and micro-organisms, and exposure of the internal parts of the grain to air induces changes in the composition of the grain, such as the production of free fatty acids, which may lead to rancidity, particularly in oilseeds.

#### *Discoloured grain*

High proportions of discoloured grains are undesirable and indicate inferior quality. Grain may become discoloured as a result of heat damage, caused by over-exposure to the sun, excessively high artificial drying temperatures, or association of the grain with 'hot spots' in bulks.

#### *Infested and infected grain*

Grain damaged by insects and rodents will suffer a loss in weight and will be less attractive to consumers. Moulds and bacteria will spread into and through grains, causing unhealthy discoloration and lesions. The contamination associated with damage by pests and micro-organisms has direct food hygiene implications, and pathogenic micro-organisms spread by rodents and some insect species, and toxins produced by certain moulds may lead to acute or chronic illness.

### *Grain standards*

Grain standards set out the minimum quantities of desirable characteristics required and the maximum quantities of undesirable components that the grain should contain in order for it to be acceptable. When each characteristic is given both minimum and maximum values, a series of intermediate values may be selected to define the limits of various grades.

National grain standards may be drafted by national standards institutions in consultation with interested groups; others may be based on international standards issued by, for example, the International Organization for Standardization (ISO) or the United Nations CODEX Alimentarius Commission (CODEX), established under the FAO/WHO Food Standards Programme.

## Basic principles of sampling and terminology

### *Sampling grain*

Grain delivered to store in bags can be sampled using sampling spears, and grain in small bulks (e.g. in trucks, lorries or barges) can be sampled with long grain probes. Large bulks (e.g. in ships) are more difficult to sample and special mechanical samplers may be used to collect samples from the grain as it is unloaded or conveyed from the unloading site to a storage bin.

Consignments of grain must be sampled in accordance with strict sampling rules to ensure that the resulting sample submitted for analysis is representative of the whole lot. The procedures for selecting and sampling bags at random and for sampling grain transported or stored in bulk are set out in international and national standards or official guidelines based on such standards (ISO 1986a, b, c).

Grain is rarely completely uniform in quality even when it is regarded as acceptable. If a consignment is subjected to poor handling or poor storage, parts of it may become very different from the rest in character. All consignments of grain will contain some damaged grains and foreign matter, and some consignments may contain live or dead insects. However, foreign matter, damaged grains and insect pests are usually spread unevenly throughout the grain.

The only sure way of obtaining complete and accurate information about the commodity would be to carry out an examination of every part of the consignment. This may be possible if the quantity to be examined is small, but it is usually neither practical nor economical for large amounts of grain. The choice then is between not examining the commodity at all and obtaining no information about it, or taking samples and obtaining *some* information while acknowledging that anything less than a total examination will affect the accuracy of results.

The results of sample analyses can be expressed in precise terms. However, precise analytical results may be of little practical value, and may even be misleading, if the samples are obtained without taking into account possible non-uniform, or non-random distribution of foreign matter, defective commodity, insects, and other components of the consignment. The samples must, therefore, be representative, i.e. the various components of the consignment must be proportionally included in the samples.

### *Principles of representative sampling*

To ensure that samples are representative of the consignment from which they are collected, certain basic principles of sampling need to be followed:

- The consignment should be divided into primary units of equal size or status, any or all of which may be sampled. In the case of bagged grain, each bag can be regarded as a primary unit. For loose grain being handled in bulk, the primary unit may be expressed in terms of weight – if the commodity is being moved – or volume, when it is static, as in a truck or bin.
- All primary units should have an equal opportunity of being sampled. Access to any chosen part of a consignment to take a sample is possible only when a stack is being built or dismantled, when a vehicle or vessel is being loaded or unloaded, or when a bulk commodity is being moved.
- The method of sampling should select, without bias, a representative number of primary units from the consignment. In the case of bulk grain that is being moved, at least one primary sample (i.e. a small quantity of grain taken from a single position) is taken from every primary unit. This is usually carried out using automatic sampling equipment (see below). However, grain in bags, or other relatively small containers, must always be sampled by hand. This could be very time-consuming if the number of primary units is very large so a scheme for selecting a proportion of the primary units must be used.

The minimum number of primary units needed to represent a consignment can be determined statistically and depends upon the total number of units present. In the former Soviet Union the recommended number of primary units to be sampled is based on the approximate cube root of the number of units in the consignment. Other countries take samples from 10% of the primary units in a consignment. However, most countries follow the recommendations of the International Organization for Standardization, the International Association for Cereal Chemistry (ICC) or the American Association of Cereal Chemists (AACC), all of which have produced standards incorporating the schedule in Table 5.25.

More intensive sampling is practised if the commodity is particularly valuable and quality is very important, or if it is essential to detect components which may be present in small but significant quantities. For example, cocoa is often sampled in lots of not more than 100 bags, and every bag may be sampled. Similarly, special sampling rules are applied in procedures for detecting aflatoxin in grain and other commodities.

Sometimes during storage, handling, transportation or processing, a part or parts of a consignment may acquire undesirable characteristics. If the affected part is readily distinguishable it should be segregated from the main body of the consignment as far as is practicable and sampled separately. Two or more sampling schemes may have to be adopted to satisfy requirements.

**Table 5.25** Schedule for determining the number of bags to be sampled from consignments of different sizes.

| Bags in consignment | Bags to be sampled |
|---|---|
| Up to 10 | Every bag |
| 11–100 | 10 bags, drawn at random |
| More than 100 | Square root (approximately) of the total number of bags, drawn at random |

*Types of sample*

Samples collected from primary units are referred to as primary samples. These are usually mixed together to form a composite or bulk sample. This composite or bulk sample will be too large to examine and analyse and must be subdivided or reduced to the required number of submitted or laboratory samples, so called because they are the samples that are submitted for examination or analysis. If the submitted sample is still too large for analysis it may be further divided or reduced to one or more working samples. It is the working sample that is examined or analysed.

*Sample size*

The representative nature of a sample depends not only upon the way in which it is obtained but also upon its size. Quality factors that may be present in only small quantities in the consignment, particularly if they are non-randomly distributed, are less likely to be represented in small samples than in large ones. If such a factor happens to be a considerable health hazard, for example, aflatoxin or ergot, or it is likely to adversely affect the acceptability of the commodity (e.g. garlic in wheat), then the working sample needed for its detection with reasonable accuracy should be large. The minimum sample size for mycotoxin analysis for a range of commodities is given in Table 5.26. The sample size for grains is based on data which indicate that the number of grains contaminated in a lot is often very low (below 0.1%).

The European Commission (EC) regulation and directive (1525/98) on maximum permitted levels of aflatoxin in foodstuffs describes the sampling methods and stipulates a 30 kg composite sample for each lot or sub-lot up to 500 tonnes of cereals. The composite sample is composed of 100 primary (or incremental) samples collected from throughout the lot. This is a more rigorous protocol than the ISO 'square-root' protocol for lots of up to 10 000 bags (500 tonnes). The composite sample is then ground before being divided into smaller working samples for analysis.

The detection of very light insect infestations similarly involves the examination of large samples. For the determination of the type, class and factors that are normally randomly distributed in the commodity, relatively small samples are sufficient. It is necessary, however, to quantify the terms 'small' and 'large' as used in the description of samples.

Three categories of sample need to be considered separately, but also in relation to each other: the primary sample, the submitted sample and the working sample. Since the analysis for quality, insect infestation, damage, etc. is carried out on the working sample, it is appropriate to consider first the size of the working sample.

*Working sample*

The bigger the sample the better the results. But, as the size of the sample increases so will the time needed to complete the analysis. Good results will be needed as quickly as possible. It is generally accepted that working samples for determination of common defects such as insect damage, broken grains and discoloured grains should be equivalent to about 500–1000 grains.

The equivalent minimum working sample weights for a selection of commodities are shown in Table 5.27. Samples of this size can be analysed in about 10 to 20 minutes.

Samples for analysis of foreign matter and live insects should be as large as possible. It is common to analyse the whole of the submitted sample by sieving, which can be done quickly.

**Table 5.26** Minimum sample size required for a single 'representative' sample, based on at least 100 sub-samples, for a batch of up to 50 tonnes.

| Commodity | Minimum sample size |
|---|---|
| Very small particles: | |
| milk, milk products, oils | 500 g |
| Intermediate size particles: | |
| meals and flours, peanut butter, lime-treated maize | 3 kg |
| Small grains: | |
| wheat, oats, barley, sorghum, rice, lentils | 5 kg |
| Intermediate-sized grains: | |
| maize, cottonseed, cottonseed cake | 10 kg |
| Large grains: | |
| groundnuts, tree nuts, figs | 20 kg |

**Table 5.27** Equivalent minimum working sample weights for a selection of commodities.

| Commodity | Sample weight (g) |
|---|---|
| Maize (small grain) | 200 |
| Maize (large grain) | 250 |
| Sorghum | 25 |
| Cowpeas | 150 |
| Wheat | 25 |
| Bulrush millet | 10 |
| Paddy rice | 15 |

*Submitted sample*

To determine the size of the submitted sample, it is necessary to consider first the number of separate tests that will have to be carried out and how many times each test must be repeated. If many replicated tests are to be carried out, then the submitted samples may need to be as large as 5 kg, but usually, samples of between 1 kg and 2 kg will be sufficient.

*Bulk sample*

The size of the bulk sample will depend on how many submitted samples are needed. However, the bulk sample must be large enough to meet at least the minimum requirements. For example, if one submitted sample of 1.5 kg is required, then the bulk sample must not be less than this.

*Primary sample*

The minimum weight of primary samples can be determined by dividing the minimum bulk sample weight required by the number of primary samples to be taken. However, if the primary samples are being taken exclusively for the detection and assessment of insect infestations and are not going to be bulked together, each sample should weigh not less than 1 kg.

**Sampling bagged commodities**

Bagged commodities are commonly sampled using a simple bag spear. These are available in a wide variety of sizes and configurations and include simple 'general purpose' spears for various grains, special spears for sampling small grains (rice) or flours and meals, and compartmented spears (Fig. 5.37c) for sampling at more than one point of a bag.

Two common designs of simple bag spears are the cylindrical and tapered types (Fig. 5.37a, b). Sampling spears with a maximum external diameter of about 12 mm are designed for small grains such as wheat, while 25 mm diameter spears are suitable for larger grains. Good penetration into a bag is required and so the sampling spear should be 40–45 cm in length. Long tapering spears are used for sampling flour. Short sampling spears are not good enough for obtaining material from deep inside bags.

For ease of penetration and minimal damage to bag material, a tapered sampling spear is preferred, but because of its shape, this type takes unequal portions of grain from along the line of penetration, which could result in errors in the assessment of grain quality, for example. Larger and more even samples are possible with a cylindrical sampling spear, but this will be harder to push into a bag and it tends to leave large holes in the bag material. This can be a particular problem when sampling

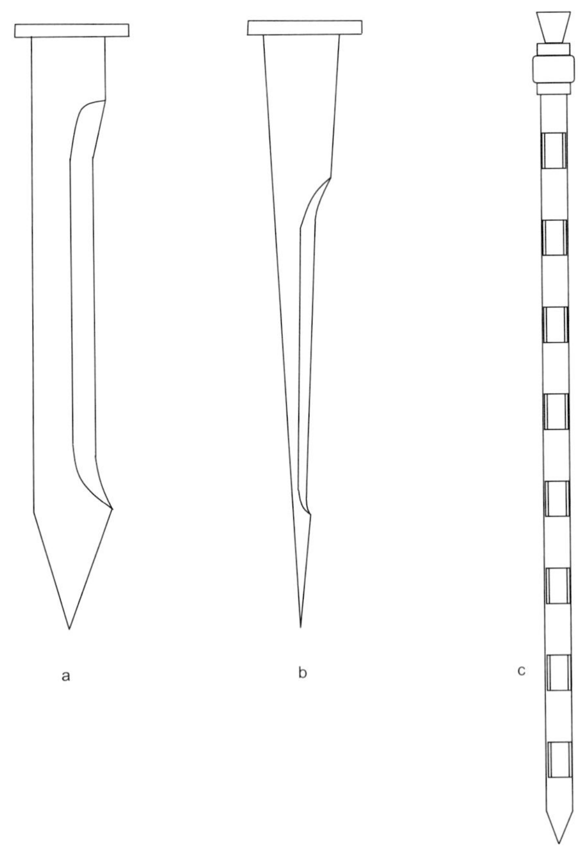


**Fig. 5.37** Sampling spears (not to scale). (a) Cylindrical spear; (b) tapered spear; (c) compartmented spear.

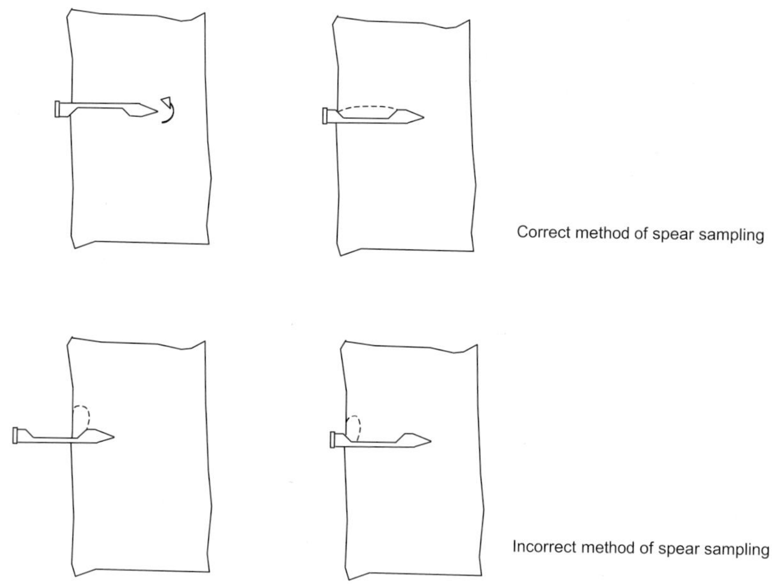


**Fig. 5.38** Correct and incorrect methods of taking a sample with a bag spear.

woven polypropylene bags as the threads will be broken, resulting in excessive spillage during handling.

The correct way to sample with a bag spear is to insert the spear with the open side facing downwards and then to twist the spear so that the open side faces upwards, once it is fully inserted (Fig. 5.38). If a spear is inserted with the open side facing upwards, it will be filled with material from the outer few centimetres of the bag, thus preventing material deeper in the bag from being sampled. It is better to take several small horizontal samples, rather than simply inserting the spear at an angle and allowing grain to flow through the open handle end of the spear. All sampling spears damage the bag material, so they must be used with care. After sampling, the hole made by the spear should be closed by gently pulling the weave of the bag material together.

A disadvantage of sampling with simple bag spears is that the method does not conform to the basic principles of representative sampling. For example, if foreign material is unevenly distributed in a bag, the haphazard nature of spear sampling could lead to distorted quality assessments (Fig. 5.39).

Bagged commodities can also be sampled using small compartmentalised double-tube sampling spears (Fig. 5.37c) although they are primarily used for sampling grain bulks. They are superior to the simple bag-sampling spear since they extract a large sample (or series of samples) from across the width or through the length of a bag. They consist of two metal tubes, one fitting closely inside the other and each with several slots corresponding to similar slots in the other tube. The openings of the spear are opened or closed by turning the inner tube. The spear is inserted into the commodity with the tubes closed, then the inner tube is turned to open the slots and the commodity sampled from several positions along the line of penetration. The slots are then closed and the spear removed. Spears vary in length from 45 cm (for bags) to 3.5 m (for bulks) and in width from 12 to 50 cm. An advantage of these spears is that small batches of commodity corresponding to the original sampling positions can be extracted. However, they are still instruments of haphazard rather than representative sampling.

Sample material can be removed from the spear either by pouring it through the open handle end or by holding

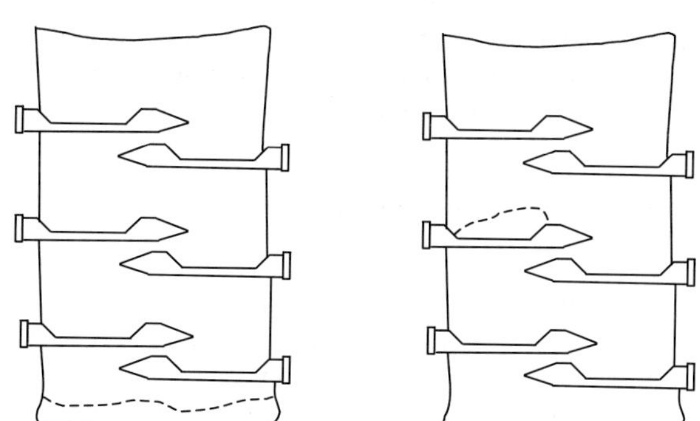

**Fig. 5.39** Disadvantage of sampling with a bag spear.

the spear in the horizontal position with the intake apertures facing downwards and opening them, thus extracting small heaps of material corresponding to the original sampling positions.

## Sampling bulk grain

Bulk grain is sampled either when it is static (in a truck, barge or storage bin) or when it is on the move (when discharged from a spout or carried on a conveyor belt). A wide range of sampling equipment has been developed to meet the special requirements of these various situations, some for small-scale operations and others for situations where grain is handled in very large quantities.

### *Sampling static bulk grain*

#### *Manual sampling*

Large double-tube samplers can be used for sampling commodities stored in bulk and are useful when small batches of commodity, corresponding to the original sampling positions, are required. When a large sample from a single point within the bulk is required it is better to use a probe with a sample collection cup with a capacity of around 250–500 g. These are sometimes referred to as deep-bin samplers, cargo samplers or pod samplers. They usually have a number of extension rods to allow sampling of trucks, deep bins, etc. up to a depth of about 5 m. Before samples are taken from a grain bulk, the surface should be trimmed or levelled to within 30 cm of the mean level. Sampling should then proceed in a standard manner. International Standard ISO 950–1979 recommends that trucks should be sampled at the points shown in Fig. 5.40 (ISO 1979).

This recommendation may be taken as a guideline for determining sampling positions in barges and in other flat bulks.

If standard probing reveals grain which is 'out of condition', additional probes will be necessary in order to determine the boundaries of the inferior portion. The sampling spear should always be inserted into bulk grain at a slight angle from the vertical, with the slots facing upward. The slots are opened only when the spear has reached the sampling position, and it is important to ensure that they are closed before it is removed.

The simplest type of deep bin probe (Fig. 5.41a) consists of a hollow spear head, which serves as a sample cup, with a spring-loaded cap attached to a metal or wooden rod about 1 m long. As the probe is pushed into the grain, extension rods may be attached to it to increase the depth of penetration.

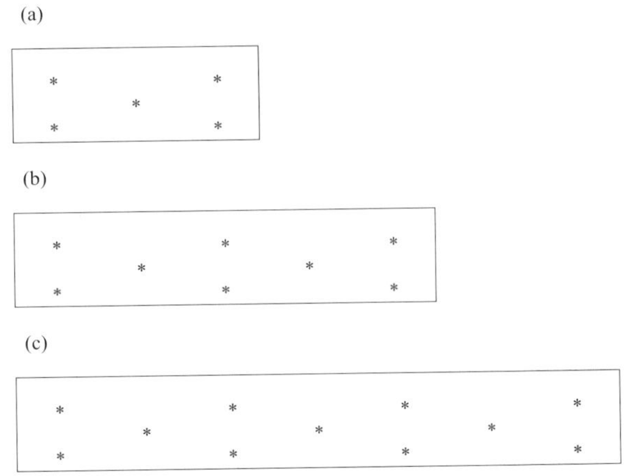


**Fig. 5.40** Recommended sampling sites for grain in trucks. (a) Truck containing up to 15 tonnes: 5 sampling points (middle and approximately 50 cm from sides); (b) truck containing 15–30 tonnes: 8 sampling points; (c) truck containing 30–50 tonnes: 11 sampling points.

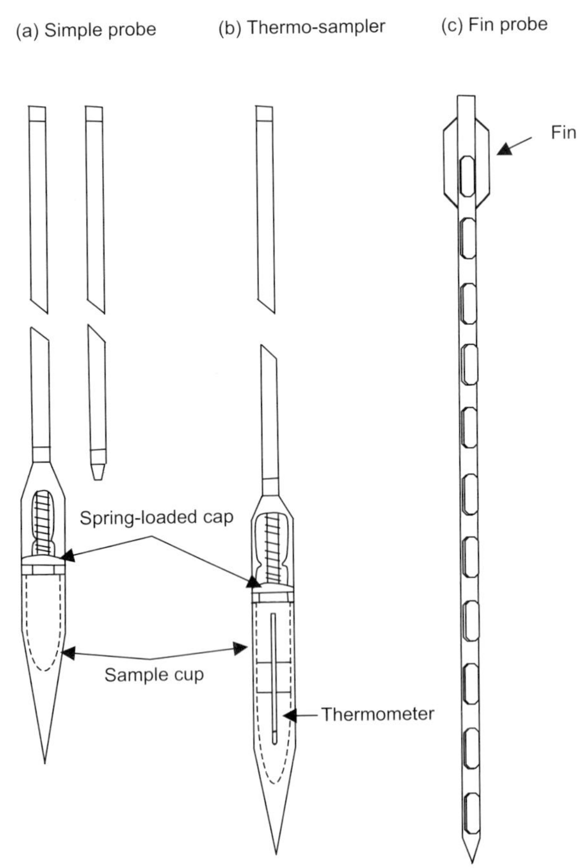


**Fig. 5.41** Deep bin probes (not to scale). (a) Simple probe; (b) thermo-sampler; (c) fin probe.

When the sampling point has been reached, a slight upward pull on the rod lifts the cap of the spearhead, allowing grain to fill the cup. The probe is then withdrawn completely and the sample removed. A single probe yields up to 300 g of sample. The thermo-sampler (Fig. 5.41b) is similar in design and capacity, but has a thermometer in the base of the sample cup, useful for checking grain temperatures. The deep bin fin probe (Fig. 5.41c) consists of a double-tube sampler with a set of extension rods. This probe can deliver up to 600 g of sample material representing a long vertical 'cut'. A considerable amount of physical effort is required to push any of these probes into grain and so none can be expected to penetrate more than 5 m deep.

*Pneumatic grain samplers*

Pneumatic grain samplers overcome the main disadvantages of manual operation by using powered-suction to penetrate to almost any depth of a static bulk of grain, and by taking a continuous sample.

Pneumatic samplers are quicker to operate than manually operated ones, and they can be used to obtain samples from directly against the sides and floors of bulk grain containers.

These samplers (Fig. 5.42) have an electric motor to power a cyclone air pump, which is mounted over a sample collection chamber. The sampling probe is connected to the pump by a flexible hose and a shutter at the base of the collection chamber allows sample material to be discharged into a container as required. The probe consists of a number of aluminium tubes that can be joined together, and a tip section through which sample material is obtained. The tubes are double walled so that the sucking action inside the inner tube combines with a downward flow of air between the two tubes to 'lift' rather than suck the sample material upwards. Pneumatic grain probes have been developed specifically for rapid sampling of bulk grain trucks as they stand at the entrance to a silo (Fig. 5.43). The probe of the sampler is suspended from the end of a hydraulically operated arm that can be raised and lowered, and swung through a 180° arc. Flexible hosing connects the probe to a sample-receiving chamber, which may be located in a grain inspection laboratory overlooking the weighbridge.

## Sampling moving grain

While it may seem convenient to collect samples

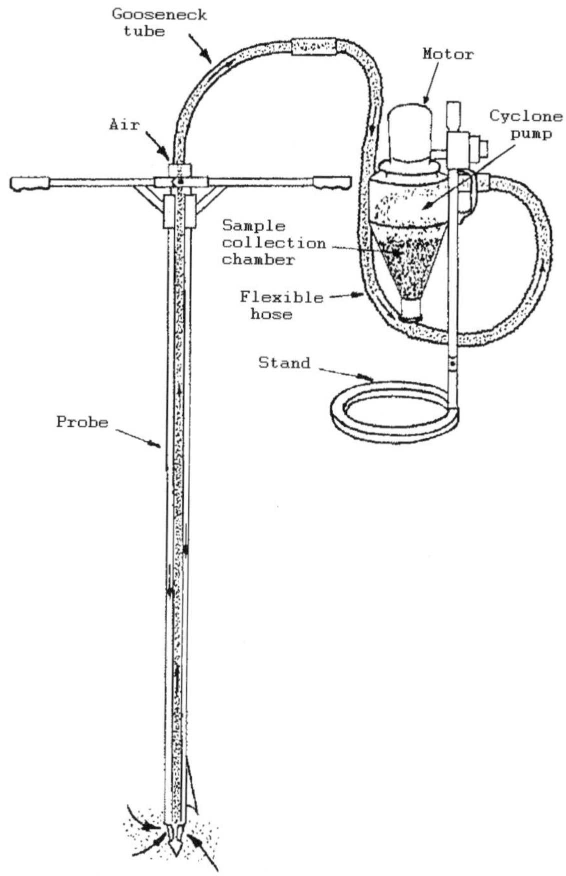


**Fig. 5.42** Probe-A-Vac sampler.

from moving flows of grain, local health and safety regulations may prohibit direct manual sampling from conveyor belts and spouts. This restricts the options for sampling streams of grain. However, manual sampling may be permitted with specially designed scoops such as the 'Ellis cup' and the 'pelican' sampler. The alternative is to install mechanical, diverter, or in-line samplers.

*The Ellis cup sampler*

This is a special kind of hand scoop (Fig. 5.44) designed for obtaining small samples from bulk grain on a moving conveyor belt. It is normally made of high tensile strength aluminium.

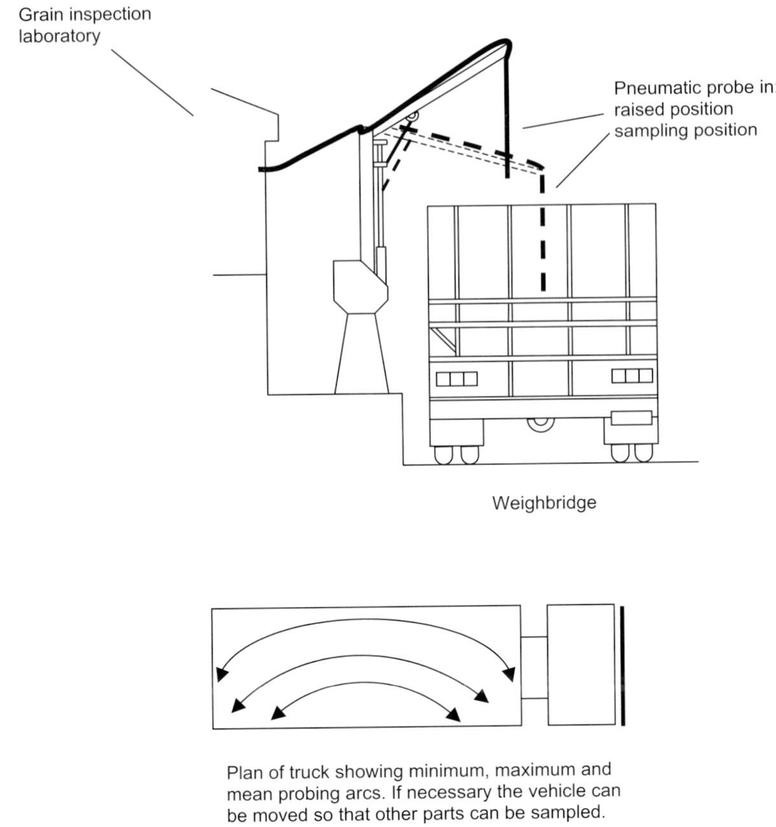


**Fig. 5.43** Pneumatic grain probe.

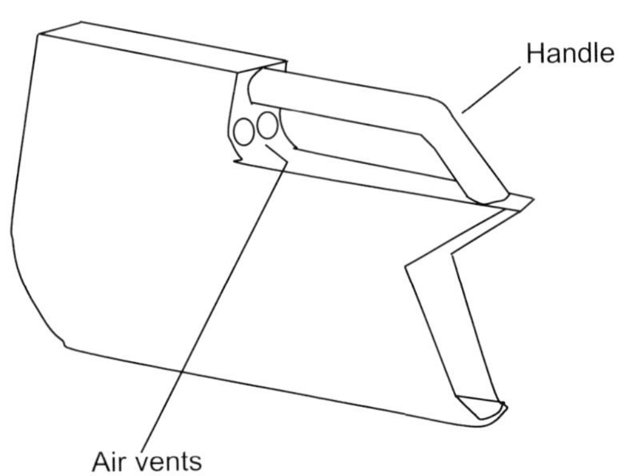


**Fig. 5.44** Ellis cup sampler.

The cup will obtain a vertical section of the flowing grain at the point where it is inserted into the stream. Samples taken in this way are used for making 'spot' checks on the condition of the grain (smell, presence of insects, signs of heating, etc.) and are not intended as substitutes for representative sampling systems.

Sampling with the Ellis cup is hazardous: the operator is dangerously close to moving equipment, and there is a risk that the force of the grain stream could pull the cup out of his grip and carry it into machinery. For these reasons stringent safety regulations are enforced wherever the Ellis cup is used.

*The pelican sampler*

This device consists of a leather pouch attached to a metal frame at the end of a pole (Fig. 5.45) and is used for obtaining samples from free-falling grain, e.g. from a spout discharging into a truck or the hold of a ship.

When grain is flowing vertically from a spout the pelican sampler can be swept across the stream of grain from any side. If the spout is sloping to one side the components of the grain stream (particularly the foreign matter) are likely to be stratified. It is important in this case, therefore, to pass the sampler through the grain stream from side to side and not from under the stream outwards or vice versa. Ideally, the sampler should be filled just

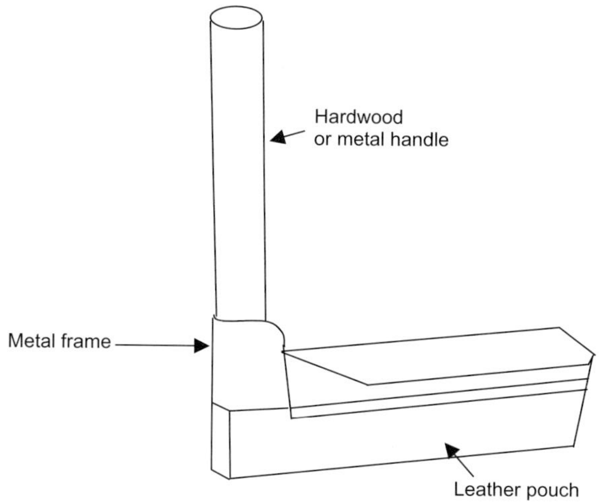


**Fig. 5.45** Pelican sampler.

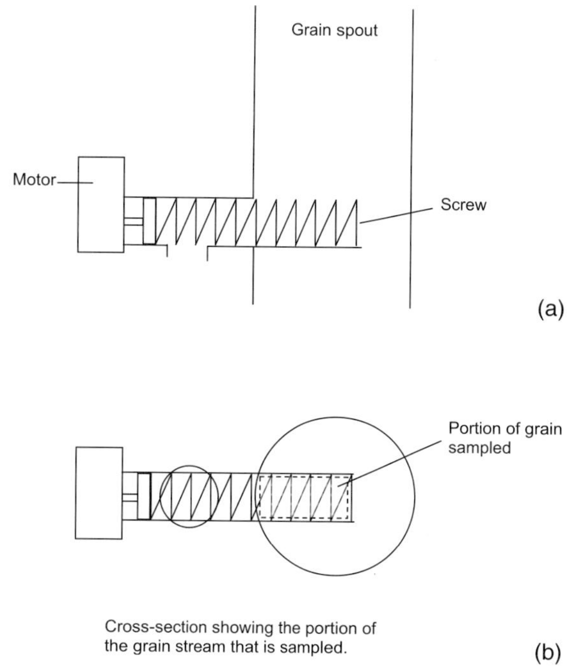


**Fig. 5.46** Limpet sampler with worm-screw extractor. (a) Longitudinal section; (b) cross-section showing the portion of the grain stream that is sampled.

as it leaves the stream of grain. If it fills before the cut is completed the composition of the sample could be biased and therefore unrepresentative of the whole grain stream. A competent user of the pelican sampler can obtain samples as reliably representative as any mechanical sampler. However, it must be remembered that the force behind a stream of grain could throw the user of a pelican sampler off balance or snatch it out of their hands. Mechanical samplers have now largely superseded the pelican sampler where there is a need to regularly sample large quantities of grain flowing at a very high rate.

*Automatic samplers*

In commercial bulk grain handling operations, moving grain may be sampled by a mechanical device fitted to delivery spouts or conveying systems. Limpet samplers, for example, consist of a tube inserted through a hole drilled into the spout wall. The tube is usually open at both ends and has an inlet slot in the upper side projecting into the grain stream. Sample material is collected either by gravity (when fitted to an inclined spout), motorised worm-screw (Fig. 5.46) or a plunger extractor, operated by air under pressure.

Worm-screw extractors can operate continuously or for set intervals. Plunger sample extractors remove samples of fixed size, at rates determined by a timing mechanism. The sample material may be collected in a container mounted under the discharge end of the extractor tube, or directed via piping to a more convenient location for sample collection and examination.

*Diverter samplers*

Diverter samplers are probably the best devices for obtaining representative samples from moving grain. They are commonly standard pieces of equipment in grain silos and mills. There are different models of diverter sampler but the mode of action is basically the same in all cases. The device, operated by a motor-activated timer, takes a complete cross-section of a stream of grain in a standard manner. Grain extracted from the main stream by the diverter sampler is usually fed directly into a secondary sampler, which reduces the sample to a manageable size before it is delivered via spouting to the grain inspection laboratory.

**Dividing and reducing samples**

A composite or bulk sample will usually contain more grain than is needed and will therefore have to be divided or reduced in size. Similarly, the submitted or laboratory samples may contain more grain than is practical to use for analysis and may also have to be divided or reduced several times. When dividing or reducing a sample, care must be taken to ensure that the reduced sample remains truly representative.

### *The coning and quartering method of reducing samples of grain*

This is a simple but very effective method of reducing the size of a sample while ensuring that the final sample is still representative of the original batch from which it was drawn. It can be used for relatively large batches, for example, one or two bags of grain, or for samples submitted to a laboratory. The only requirements are as follows:

- for 50 kg or more of grain: a clean smooth floor, preferably covered with a tarpaulin, a shovel and a suitable length of flat timber or sheet metal (quartering board);
- for smaller quantities of grain: a smooth table top, a scoop and a suitable piece of flat timber or sheet metal (quartering board).

The sample to be reduced should be poured into the centre of the working area. The shovel or scoop is then used to turn the heap of grain from the edge towards the centre. The operator should work round the heap several times to ensure that the mixing is as thorough as possible. The top of the cone should then be flattened, using the board, so that its height is less than the width of the quartering board. The heap is divided into quarters using the board, first by dividing it into two approximately equal portions and then into quarters (Fig. 5.47).

Any two diagonally opposite quarters of the grain are then removed from the working area, and the remaining grain is thoroughly mixed as before. If the reduced sample is still too large, then the procedure is repeated as many times as are necessary to reduce the sample to the required size.

### *Grain sample dividers*

#### *The riffle (multiple slot) sample divider*

The riffle divider is the simplest of the sample dividers recommended by the International Organization for Standardization. It consists of several rectangular-mouthed funnels arranged side by side, with alternate funnels leading to opposite sides of the apparatus (Fig. 5.48). The funnel assembly is fitted inside a box with a shallow hopper at the top and which is open at the bottom. Three identical rectangular sample boxes form part of the apparatus and are designed so that the funnel assembly box can sit on any two of them, while the third is used to pour a sample through the hopper.

#### *Tullgren funnel*

The original Tullgren funnel was designed for extracting living organisms from soil samples, but in modified form it has proved useful for separating free-living insects

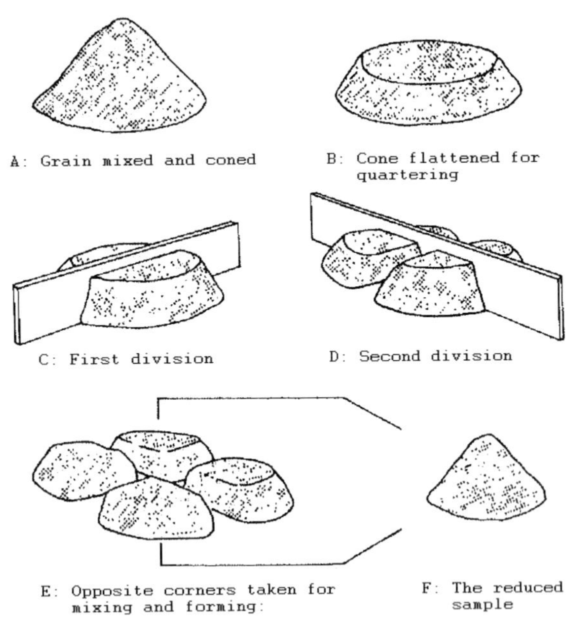


**Fig. 5.47** Coning and quartering technique.

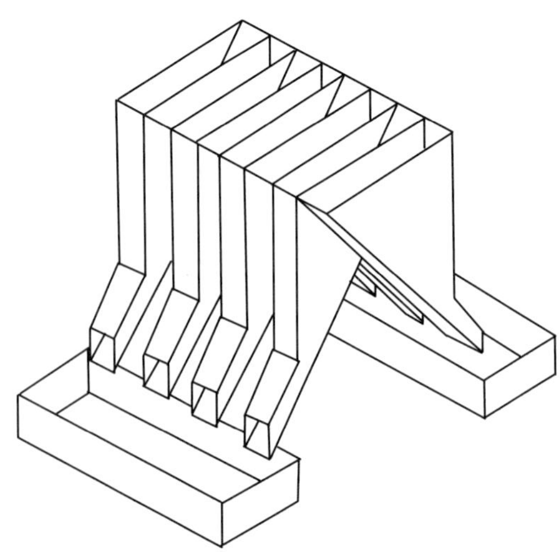

**Fig. 5.48** The riffle divider.

from some durable commodities. The apparatus consists of an electric lamp under which is placed a sample container, with a perforated base, resting on a collecting funnel. When the lamp is switched on the heat generated by the bulb warms the sample and causes free-living insects present to move downwards. The insects eventually pass through the perforated base of the sample container and fall down the funnel into a collecting tube.

The Tullgren funnel is particularly useful for extracting insects from finely divided commodities such as flours and meals, and small grains like wheat, sorghum and millets. However, 100% recovery of the free-living insects in a sample may take 24 h or more; some insect species are more difficult to extract than others (Golob *et al.* 1975).

*The Boerner divider*

Also recommended by the International Organization for Standardization, the Boerner divider (Fig. 5.49) splits a sample into two approximately equal portions. Around the base of the cone are 36 slots – the mouths of funnels that alternately direct the grain into an inner or an outer collecting funnel, from which the grain flows into two receiver pans.

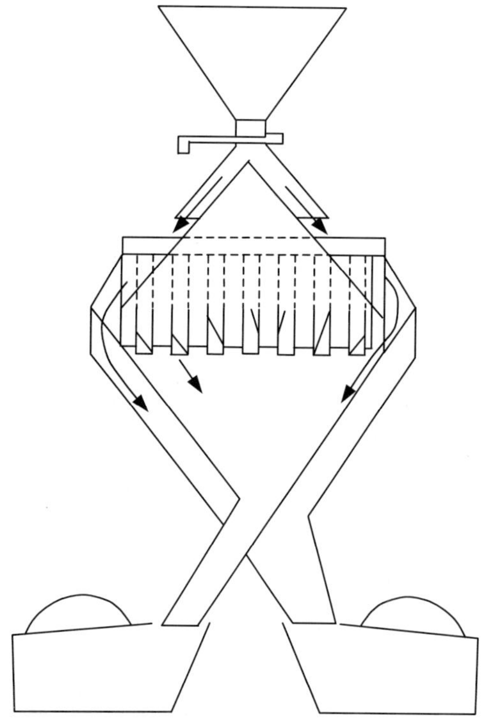

**Fig. 5.49** The Boerner divider.

*Rotary sample dividers*

Some mechanical samplers extract large samples so rapidly that it is necessary to connect them to mechanical sample dividers. Such dividers usually operate on a rotary principle, although other principles (e.g. centrifugal) are sometimes employed.

In the rotary divider illustrated (Fig. 5.50) the sample material is fed in through the top and falls down a bent tube rotated by a geared motor. As the tube rotates, some of the material falls into one or more sample collectors, while the surplus material is guided back into the main stream of grain.

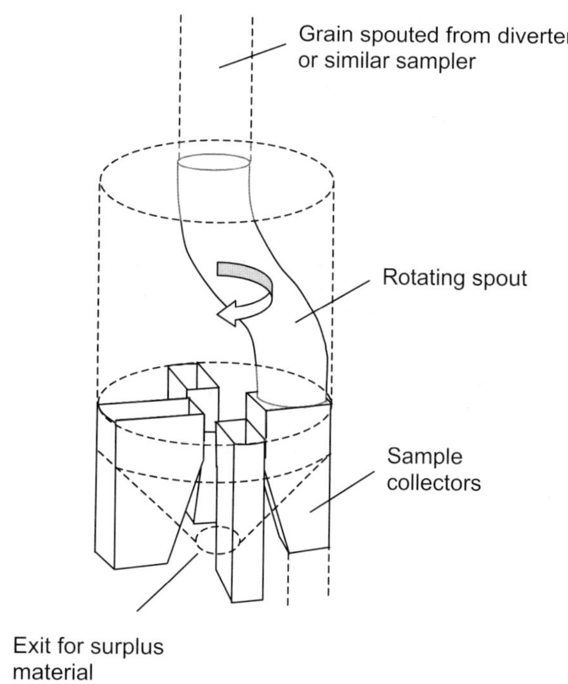


**Fig. 5.50** Rotary sample divider.

## Packaging, transportation and storage of samples

If a sample of grain has to be transported or has to be stored for some time before being examined, it should be packed in such a way that changes in its condition or composition cannot occur or are at least kept to a minimum.

Samples for moisture content determination, or for other tests in which it is important to avoid the loss of volatile matter, should be packed in airtight and moisture-tight containers. The containers should be filled completely to minimise the risk of interchange of moisture between the commodity and the air in the free space and to prevent damage to the sample caused by movement during transportation.

Samples for quality determination are often dispatched in polythene bags, but ideally they should be packed in closely woven bags of unglazed, unbleached cotton, or similar material. If there are several samples to be transported over some distance, then some form of additional packaging will be needed to protect the sample and the primary packaging material.

During transportation, individual samples should be separated from one another with suitable packaging, insulating and shock-absorbing material. Samples of different commodities should be packed in different containers to avoid possible cross-contamination or cross-infestation. All cartons containing samples should be well secured and protected from direct sunshine, sources of extreme heat and away from any form of wetting or contamination.

Samples waiting for examination need to be stored under equable conditions best suited to the commodity concerned and the purpose for which they were obtained. If analysis of the microflora, pesticide residues or other chemical constituents is required, the samples could be stored in a refrigerator or freezer but, for most purposes, a well-ventilated darkened room is adequate.

All samples must be carefully labelled at the time of packaging so that they can be readily identified whenever they are subsequently handled. The following information should always be recorded:

- the commodity;
- the origin of the sample;
- how, when and where it was collected; and
- the purpose for which it was collected.

## Inspection and maintenance of stores

Inspection of stores may be carried out for several reasons, for example: to determine the suitability of the structure for storage of agricultural produce; as part of a pest management programme, to identify possible sources of deterioration; or as part of a planned maintenance programme. Inspection for whatever purpose should be undertaken in a logical manner, starting with an inspection of the site and outside of the building and then continuing inside.

### *Site*

The location and current use of storage building and its relationship to other buildings on the same and adjacent plots is important since there may be potential risks that can affect the safe keeping of produce. Risks may include neighbouring sources of pollution such as sewage or factory discharges, rodent infestation from rubbish dumps and insect infestation from produce stored close by. Attention should also be paid to drainage, the cleanliness of the site, freedom from harbourage for pests and the general ease of maintenance.

### *General structure*

Inspection of the general structure of the store will include an examination of the roof, floor, windows, ventilators and doors to establish that the facility is structurally sound and will provide adequate protection to the stored commodity. Attention must be paid to the general design considerations described in the earlier part of this chapter devoted to storage structures.

## Commodity inspection

One of the major causes of loss in storage, especially in the tropics, is insect infestation in durable commodities. The description of methods of inspection that follows, therefore, concentrates upon detection of insects, but the principles are basically the same for inspections for other forms of qualitative deterioration.

### *General methods of inspection*

Insect pests of all durable commodities are mostly small and dull-coloured and tend to avoid light. Only when they are present in large numbers or have moved out of the commodity, perhaps because of a physical disturbance, are they readily noticeable.

Insects can often be induced to move and show themselves when suddenly exposed to bright light, so that illumination of the sides of a stack or the spaces between bags with a strong torch beam will usually reveal insects if they are present. Some insects may react to the light by 'freezing', or feigning death, and may not be noticed. Therefore, if at first no insects are seen, other methods of detection should be considered.

Physical disturbance, such as sweeping likely resting-places with a hand, or turning bag ears over, may reveal larvae, pupae or adult insects. The stitched overlap of the mouth of the bag should be carefully examined for hidden insects.

Moving some of the bags in the top layer of the stack often reveals free-living insects in previously completely dark or hidden places. If the bags are moved vigorously enough, insects will move out of the bags regardless of light intensity.

Supplementary evidence of infestation should also be sought. Residual infestations are likely to be found in cracks and crevices in the building structure and in piles of sweepings left in the store. Disturbing suspect material is usually sufficient to reveal insects, but samples should be sieved if infestation is not obvious. Other evidence of insect infestation includes webbing produced by migrating moth larvae, or cast larval skins on the commodity or the fabric of the building.

### *Sampling for insects*

Although a general inspection may provide information about the nature and distribution of an insect infestation on and near a stored commodity, it cannot yield details about the infestation within the commodity itself, and samples may have to be drawn for examination. Knowledge of the history of a consignment can help in determining how sampling for insect infestations should be carried out. Well-kept stock record cards will provide the best source of information. If it is known that the consignment was free from infestation when taken into store, or that it was fumigated recently, it can reasonably be assumed that any live insects subsequently infesting the commodity will be found in the outermost layers of stacks. In the case of grain bulks, infestation may be restricted to the top 100 mm of the bulk. Sampling, therefore, can be restricted to these superficial areas.

If the history of the consignment is unknown, it may be necessary to conduct more extensive sampling of inner parts as well as exposed parts of the stack or bulk. This is relatively easy for bulk grain, since a probe sampler can be used to draw samples from parts below the surface. Sampling deep within a stack of bagged grain will involve moving some of the bags and is satisfactorily achieved only when the stack is dismantled. If this is not practicable a compromise of sampling only the top three or four layers of bags has to be accepted.

Free-living insects can be extracted from samples of grain by using either a sack sieve or a hand-held sieve. In order to obtain the maximum extraction of insects, it is important that the sieve has a screen of the right specification and that it is properly used (Table 5.28). As a general guide, 'fine' sieves used for determining foreign matter are suitable for extracting insects from samples of large-grained commodities such as maize. Appropriately fine woven wire screens may be used for separating insects from flour and similar commodities. It is not so easy to extract free-living insects from commodities such as sorghum and bulrush millet using sieves, because the pests associated with these commodities are often similar in size to individual grains. In these cases the grain may have to be hand-sorted to check for free-living infestation.

Accurate estimates of free-living insect populations in granular commodities can be obtained by sieving the contents of whole sacks. Although more time-consuming than hand sieving, it may be particularly important if the commodity is valuable, for example, cocoa or coffee, or when even small populations of insects need to

**Table 5.28** Examples of sieve specifications for different grains.

| Grain type | Shape | Screen* | Diameter (mm) | Breadth × length (mm) |
|---|---|---|---|---|
| Maize | Round | C | 14.0 | |
| Maize | Round | F | 5.0 | |
| Sorghum | Round | C | 6.5 | |
| Sorghum | Round | F | 2.5 | |
| Bulrush millet | Round | C | 4.0 | |
| Bulrush millet | Round | F | 1.0 | |
| Wheat | Slot | C | | 4.5 × 25.0 |
| Wheat | Slot | F | | 2.4 × 20.0 |
| Paddy rice | Slot | C | | 4.5 × 20.0 |
| Paddy rice | Slot | F | | 2.5 × 20.0 |

* C = coarse screen, for removing material larger than the grain; F = fine screen, for removing material smaller than the grain.

be detected. Large sack sieves may be used for cleaning grain, but they are easily adapted for detection of insects. Sack sieves commonly have inclined screens about 1 m wide and 2 m long under which a tray is fitted to collect the insects and foreign material extracted. In the simplest type of sack sieve, the sample is fed directly from the sack onto the top of the screen and then worked across it towards the bottom end where it is guided into another sack. More efficient sack sieves have a hopper for feeding the grain gradually onto the screen and a screen that moves during the sieving operation. The slope of the moving screen ensures that the grain is kept in motion towards the discharge end, while assisting with the emptying of the hopper.

### *Detecting internal insect infestations*

During the course of an inspection few or no free-living insects may be found in the commodity, and there may be no external signs of infestation. However, several insect species spend most of their developing stages inside the commodities they infest and these internal infestations may not become apparent until the adults emerge. By then the damage will have been done and the commodity may have to be downgraded or even rejected as unfit for human consumption. There are several methods of detecting such internal (or 'hidden') infestation, some of which are mentioned below. Methods of determining hidden insect infestation are described in the International Standard ISO 6639 Parts 1–4 (ISO 1986a, b, c, 1987). The standard includes a number of rapid methods and a reference method. Interestingly, the British Standards Institution, a contributor to the development of the International Standard, did not approve ISO 6639 Part 4 – Rapid methods, since it considered that the methods gave insufficiently reliable results. The British Standard therefore combines the parts of the International Standard relating to general principles, sampling and reference method (BSI 1988). The reference method entails keeping a sieved sample of the commodity under conditions suitable for the optimal development of any insects within the commodity under investigation. Ideally, the sample should be kept in a constant climate room in which the temperature is maintained at 25–30°C and the relative humidity is within the range 60–70%. In tropical countries where diurnal variations in temperature are minimal the sample might be kept at room temperature, in a desiccator in which the correct relative humidity can be maintained.

The length of the incubation period depends upon ambient temperature and relative humidity, and also upon the species of insect present. The sample should be examined every 3 or 4 days. Any insects found should be removed before the sample is resealed in its incubation container. The minimum lengths of time that should be allowed for the species concerned to emerge from the commodity are indicated in Table 5.29.

This method of determining internal insect infestation is far too slow for it to be of practical use. However, it is invaluable as a method for providing precise baseline data against which other, less accurate, methods may be calibrated.

If the grain is relatively soft-textured, internal infestation can be detected by dissecting the sample material with a sharp scalpel or knife. This is already part of the standard procedure for determining the quality of cocoa beans. It can be a slow and tedious operation and there is a risk that some insects present, particularly small larvae, may be destroyed.

Radiographic examination of samples is relatively quick but expensive (Sharifi & Mills 1971; Dobie 1973). It is an excellent method for detecting most insect stages in grains, although eggs and small larvae may be missed. The image of a grain on an X-ray negative plate (radiograph) appears white or grey. Cavities within the grain, whether caused by insects or not, show up as relatively

**Table 5.29** Incubation period (days) at 60–70% relative humidity for a selection of insects.

| Insect species | Commodity infested | Temperature | |
|---|---|---|---|
| | | 25°C | 30°C |
| *Acanthoscelides obtectus* | Red kidney beans | 56 | 42 |
| *Araecerus fasciculatus* | Coffee | 84 | 56 |
| *Callosobruchus maculatus* | Cowpeas | 49 | 35 |
| *Rhyzopertha dominica* | Cereals | 70 | 49 |
| *Sitophilus* spp. | Cereals | 56 | 42 |
| *Sitotroga cerealella* | Cereals | 49 | 42 |
| *Zabrotes subfasciatus* | Red kidney beans | 56 | 42 |

dark patches. Insects appear white or grey against the background darkness of cavities, and can generally be recognised by their shape. Larvae, pupae and adults can be distinguished and it is often possible to identify the species or at least the genus.

Another rapid method for detecting hidden insect infestation involves the crushing of grain samples against white paper impregnated with ninhydrin (Dennis & Decker 1962; Ashman *et al.* 1969). If living or freshly killed insects are present inside some of the grains they are also crushed and amino acids in their body fluids react with the ninhydrin to produce purple marks on the paper. The method is good for detecting all insect stages except eggs and very small larvae in small grains such as wheat, rice and sorghum. Large grains such as maize need to be partially broken up or kibbled before they can be tested but this could cause the loss or fragmentation of insects, leading to unreliable results.

A staining technique can be used to determine whether *Sitophilus* spp. have laid eggs on certain cereal grains (Frankenfeld 1948). This involves covering the grain for a short time with acid fuchsin solution. This differentially stains the egg-plugs a cherry-red colour, which is retained when the grain is washed to remove excess stain. Observation of the brightly stained egg-plugs when the grains are examined under a low-powered microscope does not necessarily mean that an equivalent number of *Sitophilus* spp. larvae have infested the grains. However, it does provide a rapid indication that the sample should be further tested to determine whether a living internal infestation is present.

Acoustic methods of detecting hidden insect infestation, involving the conversion of insect noises into amplified electric signals that can be measured electronically, have met with some success but are prone to interference by background noises. However, one piece of equipment consists of a special microphone mounted on a short probe that can be inserted into the surface of a grain bulk or into a bag of grain. The microphone is connected to a high quality amplifier and a set of earphones. It is claimed that the noise made by a feeding larva within a grain can be detected soon after the larva hatches from the egg (Hagstrum *et al.* 1988). Methods of detecting internal infestations and measuring feeding rates via ultrasonic emissions are also being developed.

### Insect trapping methods

If a store has to be inspected frequently for insect infestations it is worth considering the use of insect traps in selected parts of the store. Traps, if properly located, provide a means of continuously monitoring a store between inspections and can provide the first evidence of an infestation. There are many different types of trap; broadly, they can be divided into:

- flight traps which insects will fly into;
- refuge traps which insects will walk into; and
- pitfall traps which insects will fall into.

To increase their efficiency, traps are often provided with bait that will attract insects and enable their detection at times when populations are still relatively small. The bait can be a suitable food or food extract or the specific attractants that the insects release themselves. These attractants are called pheromones, and are used to attract the opposite sex (in the case of sex pheromones) or to attract both males and females (in the case of aggregation pheromones). The pheromones of quite a number of storage pests have been identified and can be synthesised in the laboratory. At the time of writing, aggregation pheromones are available commercially for the beetles *Rhyzopertha dominica*, *Oryzaephilus surinamensis, Stegobium paniceum, Prostephanus truncatus*, *Lasioderma serricorne*, *Trogoderma granarium* and *Tribolium castaneum* and *T. confusum*, and sex pheromones for the moths *Cadra cautella*, *Ephestia elutella, E. figulilella, E. kuhniella, Plodia interpunctella* and *Sitotroga cerealella.* Plastic or rubber capsules are impregnated with pheromone. These capsules are stored in the freezer or refrigerator in sealed sachets (Fig. 5.51). Once the capsules are removed from the sachets and placed in a trap, pheromone will be released for a period of 1–2 weeks. There is no need to open the plastic capsule as the pheromone can diffuse directly through its walls.

#### *Flight traps*

There are a number of designs of flight traps. One of the most common for use in storage systems is the delta trap (Fig. 5.52), so-called because in cross-section it looks like the Greek letter 'Δ'. These traps are coated on the inner surface with a non-drying glue, which will retain insects that fly into them. Funnel traps (Fig. 5.53) are another common design. Flying insects hit a baffle and fall down a funnel into a collecting jar. The jar holds a piece of insecticide-treated filter paper, which will kill any insects that come into contact with it. Delta traps and funnel traps are both used for monitoring the *Prostephanus truncatus*. Catches of *P. truncatus* in the funnel trap are higher than in delta traps, but delta traps are easier to transport and

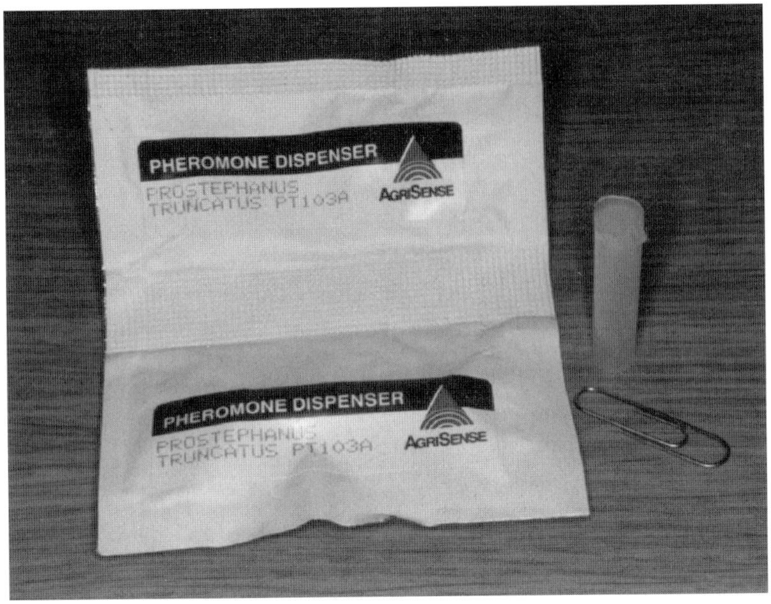


**Fig. 5.51** Pheromone capsule and packets.

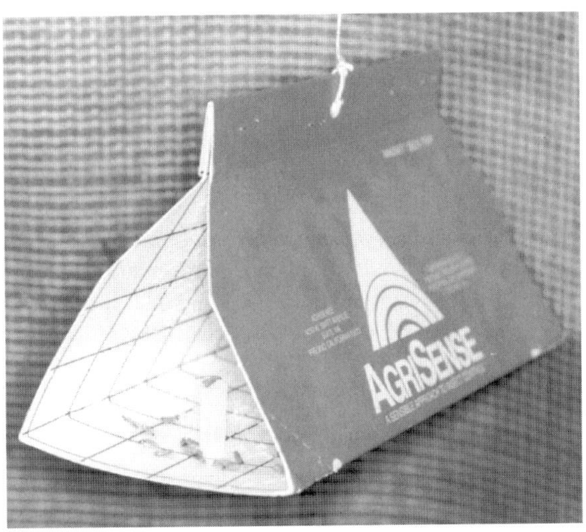


**Fig. 5.52** Delta flight trap hanging in a store to catch *Prostephanus truncatus*.

**Fig. 5.53** A funnel trap hanging in the branches of a tree to catch *Prostephanus truncatus*.

handle, so are favoured for use in large-scale monitoring programmes. Funnel traps are used more often in research studies where fewer traps are required.

### *Refuge traps*

Refuge traps can take a range of forms and exploit behaviour that leads insects to hide in cracks and crevices. One of the simplest of refuge traps consists of three or four small squares of jute sacking sown together at one edge; another simple trap can be made from a short piece of corrugated cardboard rolled up into a tube with a diameter of about 1.5–2.0 cm. Both are ideal refuges for insects. They can be placed in a store and after some days examined for insects. More sophisticated designs of refuge trap can be made by folding corrugated cardboard into multi-layered squares (Fig. 5.54). A recess is cut internally to house a pheromone capsule and/or a food lure, and in some cases, the inner surface is treated with insecticide to kill, and so retain, those insects that are attracted into the trap. Traps like this are available commercially for *Trogoderma granarium* and *Tribolium castaneum*.

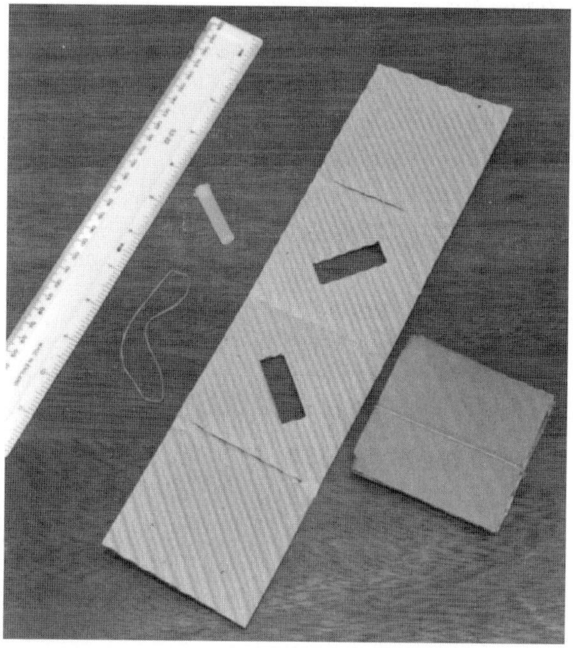

**Fig. 5.54** Multi-layered refuge trap, made from corrugated cardboard, folded (top) and unfolded to show recess for placing pheromone capsule.

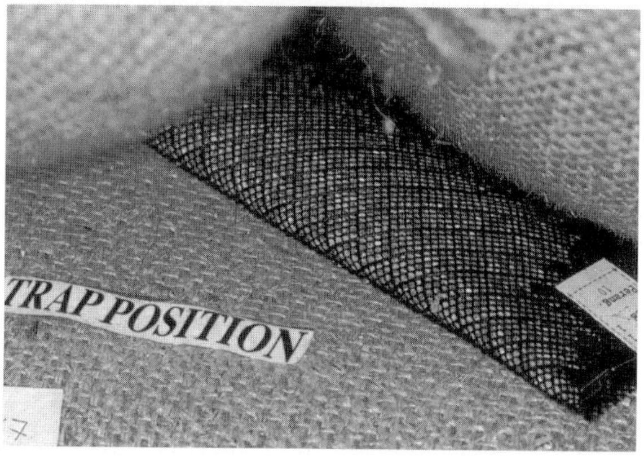


**Fig. 5.55** A bait bag trap located in the crevice between two layers of jute bags.

An alternative refuge trap is the nylon bait bag (Fig. 5.55). This consists of a nylon net of mesh size 2 mm with 20 apertures/cm$^2$, formed into an envelope 10 cm × 20 cm. The envelope is filled with suitable bait. In wheat stores in the UK, bait consisting of groundnuts, wheat and kibbled carobs has been used successfully (Pinniger 1975) while in milled rice stores in Indonesia bait consisting of brown rice (dehusked but not milled rice) worked well (Hodges *et al.* 1985). To monitor infestation in bag stacks the bait bags are placed in the crevices between sacks on the sides of bag stacks and left there for 1 week. After this time they are retrieved and shaken over a tray. The insects that fall out are counted. In the case of the brown rice bait it has been found that the catch in the traps bears a good relationship to the actual population of pests, specifically *Tribolium castaneum*, found in the stores (Hodges *et al.* 1985).

### *Pitfall traps*

Pitfall traps are mostly used in grain bulks. A simple design has been prepared for use on the bulk surface and consists of a hollow plastic cone that can be pushed into the grain (Fig. 5.56). The top of the cone is a perforated lid, which lies flush with the grain surface. Insects fall into the trap through the perforations. A more sophisticated pitfall trap that is pushed below the surface of the grain is the so-called probe trap (Fig. 5.57). This is a hollow cylinder 30 cm long and 2.5 cm wide, usually made of plastic, with all except a short lower portion, perforated. The short lower portion houses a tube for collecting the insects that fall through the perforations above. The collecting tube is painted on the inside with 'fluon', a substance that prevents insects gripping the surface of the tube and climbing out of the trap. Probe traps are pushed into the grain to depths of 10–50 cm and after several days retrieved by pulling on a string

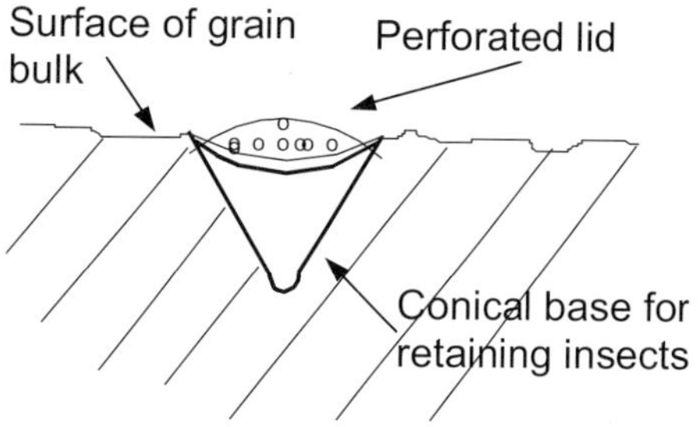


**Fig. 5.56** A cone-shaped pitfall trap for use on the surface of bulk grain.

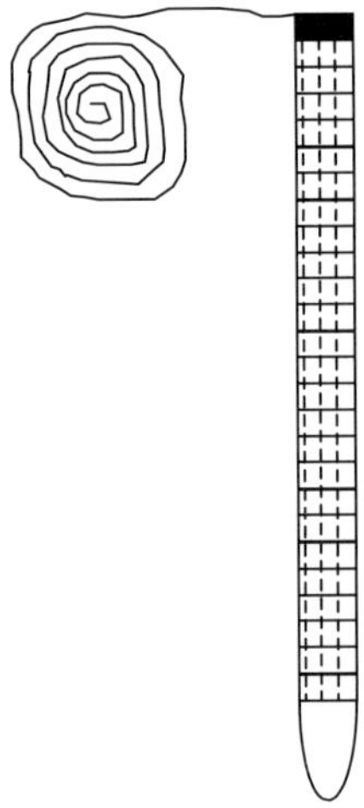

**Fig. 5.57** A probe trap for insertion below the surface of a grain bulk.

fastened to the top of the trap. An automatic version of this trap is currently being developed in the US, the Electronic Grain Probe Insect Counter (Shuman & Epsky 2001), that will be capable of counting insects as they are captured deep inside silos. The counts will be fed directly into a computer, where a decision support system can help the manager decide on the significance of the trap catch and what pest management action, if any, should be taken.

#### *Water traps*

Water traps are useful for catching adult moths, especially gravid females that need to drink water to increase their egg-laying potential. A water trap need only consist of a shallow tray filled to a depth of 1–2 cm with water. If excessive evaporation is expected a small beaker of water can be used. The addition of a few drops of detergent to the water will improve the efficiency of the trap by 'wetting' the insects as they land in the water so they drown quickly and sink to the bottom of the container.

### Investigating fungal attack in stored grain

Increasing temperature and moisture content of stored grain may be an indirect indicator of mould growth as well as insect infestation. Further indicators include a musty odour, staining of bags, caking of grains and, at a later stage of mould attack, obvious fungal webbing on grain.

When examining the stored commodity it will be necessary to look for localised areas of visible mould growth and to examine individual grains for signs of mould growth, discoloration of germ or grain. Visible mould growth on the surface of the grain indicates an advanced state of deterioration. Mould growth inside grains, usually concentrated in the embryo, may have been going on for some time and so grains should be cut in half lengthways from grain tip to base, to expose the embryo and other tissues for close examination.

In order to determine the cause and extent of a particular mould spoilage problem, separate samples should be taken from various points of the consignment and examined to 'map' the distribution of fungi within the lot. Samples should be taken from, for example, the front, middle, back, and top and bottom of a stack. Often a succession of fungi can be isolated, ranging from those preferring the driest conditions, to those able to grow in wetter grain. Samples must be clearly labelled and traceable back to their location for this exercise.

In order to assess the likely pattern of fungal growth, it is also useful to take samples from visually mouldy or caked sections of grain. The identification of species can provide important information on the likely cause of the problem.

The distribution of storage fungi and any mycotoxins they may produce is usually patchy, with the distribution of mycotoxins even more skewed than the mould that produces them. Sampling for moulds and mycotoxins therefore presents more of a problem than sampling for insect infestation, and requires a special sampling protocol (Coker 1998). The sampling protocol adopted will depend on the type of inspection involved. Generally, an inspection to determine an acceptable quality level for grains will only require a small composite or bulk sample of about 1 kg, whereas an exercise which includes analysis for mycotoxins will require a much larger composite sample (10–30 kg for cereals).

## Further reading

Koelet, P.C. (1992) *Industrial Refrigeration*. Macmillan, London, UK.
Wiley, R.C. (Ed.) (1994) *Minimally Processed Refrigerated Fruits and Vegetables*. Chapman & Hall, New York NY, USA.

## References

Adams, J.M. & Harman, G.W. (1977) The evaluation of losses in maize stored in a selection of small farms in Zambia with particular reference to the development of methodology. Tropical Products Institute **G109**, Slough, UK.

Adams, J.M. & Schulten, G.G.M. (1978) Losses caused by insects, mites and micro-organisms. In: K.L. Harris & C.J. Lindblad (Eds) *Post-Harvest Grain Loss Assessment Methods*. American Association of Cereal Chemists, USA.

Anon. (1920) Food Investigation Board. *Department of Scientific and Industrial Research Report for the Year 1920*. pp 16–25.

Anon. (1969) Rapport d'activité de al Commission d'Evaluation des pertes dans les denrées stockées crées a l'issue du Congrès de Marseille sur la protection des cultures tropicales. *L'Agronomie Tropicale*, **24**, 872–876.

Ashman, F., Elias, E.G., Ellison, J.F. & Spratley, R. (1969) Ashman-Simon infestation detector: an instrument for detecting insects within food grains. *Milling*, **151**, 32–36.

Ben-Yehoshuo, S., Fabg, D.Q., Rodov, V., Fishman, S. & Fang, D.Q. (1995) New developments in modified atmosphere packaging. Part II. *Plasticulture*, **107**, 33–40.

Bishop, D.J. (1994) Application of new techniques to CA storage. In: *Commissions C2, D1, D2/33 of the International Institute of Refrigeration International Symposium*, 8–10 June 1994, Istanbul, Turkey.

Bishop, D.J. (1996) *Controlled Atmosphere Storage. A Practical Guide*. David Bishop Design Consultants, Sussex, UK.

Boothumjinda, S., Exell, R.H.B, Rongtawan, S. & Kaewnikon, W. (1983) Field tests of solar rice dryers in Thailand. In: *Proceedings of Solar World Forum*, Perth. Parkville: International Solar Energy Society.

Boxall, R.A. (1974) Underground storage of grain in Harar Province, Ethiopia. *Tropical Stored Products Information*, **28**, 39–48.

Boxall, R.A. (1986) A critical review of the methodology for assessing farm-level grain losses after harvest. Tropical Products Institute **G191**, Slough, UK.

Boxall, R.A. & Gillett, R. (1982) Farmer level storage losses in eastern Nepal. Tropical Products Institute **G157**, Slough, UK.

Boxall, R.A., Greeley, M., Tyagi, D.S. *et al*. (1978) *The prevention of farm level storage losses in India: a social cost benefit analysis*. IDS Research Report, Institute for Development Studies, Brighton, UK.

Boyette, M.D., Wilson, L.G. & Estes, E.A. (1989) Design of room cooling facilities: Structural and energy requirements. AG-414–2. North Carolina Cooperative Extension Service, USA.

Brice, J. & Golob, P. (1998) Fumigation of grain in farmer-level mud stores and metal grain tanks using phosphine. In: Jin Zuxum, Liang Quan, Liang Yongshen, Tan Xianchang & Guan Lainghua (Eds) *Proceedings of the 7th International Working Conference on Stored Products Protection*, Beijing, China.

Bridger, H.H. & Watts, O.M. (1930) *Practical Cargo Stowage for Ships' Officers*. Imray, Laurie, Nome & Wilson Ltd., London, UK.

Brown, B.I., Wong, L.S. & Watson, B.I. (1985) Use of plastic film packaging and low temperature storage for postharvest handling of rambutan, carambola and sapodilla. In: *Proceedings of the Postharvest Horticulture Workshop*, Melbourne, Victoria. Australia.

BSI (1988) *British Standard Methods of Test for Cereals and Pulses*. Part 18. *Determination of Hidden Insect Infestation*. BS 4317:Part18. British Standards Institute, London, UK.

Cameron, A.C., Boyaln-Pett, W. & Lee, J. (1989) Design of modified atmosphere packaging systems: modelling oxygen concentrations within sealed packages of tomato fruits. *Journal of Food Science*, **54**, 1413–1421.

Chaudhry, M.A. (1980) *Food Grain Losses in Pakistan*. 6 Volumes. Faisalabad University of Agriculture, Pakistan.

Coker, R.D. (1998) Design of sampling plans. In: K.K. Singh & D. Bhatnaga (Eds) *Mycotoxins in Agriculture and Food Safety*. Marcel Dekker, New York NY, USA.

Compton, J.A.F. & Sherington, J. (1999) Rapid assessment methods for stored maize cobs: weight losses due to insect pests. *Journal of Stored Products Research*, **35**, 77–87.

Compton, J.A.F., Floyd, S., Ofosu, A. & Agbo, B. (1998) The modified count and weigh method: an improved procedure for assessing weight loss in stored maize cobs. *Journal of Stored Products Research*, **34**, 277–285.

Crank, J. (1975) *The Mathematics of Diffusion*. Second Edition. Oxford University Press, New York NY, USA.

Day, B.P.F. (1994) Modified atmosphere packaging and active packaging of fruits and vegetables. In: VTT-Symposium, *Minimal Processing of Foods*, 14–15 April 1994, Kirkonummi, Finland.

Day, B.P.F. (1996) High oxygen modified atmosphere packaging for fresh prepared produce. *Postharvest News and Information*, **7**, 31N-34N.

De Breve, M., Raboud, G., Sieber, J., Perdomo, J.A. & Velasquez, J.E. (1982) Proyecto post-cosecha: informe sobre los primeros resultados. Tecucigalpa, Honduras: Ministerio de Recursos Naturales, Co-operacion Suiza Al Desarrollo.

De Lima, C.P.F. (1979) The assessment of losses due to insects and rodents in maize stored for subsistence in Kenya. *Tropical Stored Products Information*, **38**, 21–26.

De Lima, C.P.F. (1983) *The assessment of post-harvest losses in Swaziland*. Project SWA/002/PFL, Food and Agriculture Organization of the United Nations, Rome, Italy.

Dennis, N.M. & Decker, R.W. (1962) A method for detecting living internal insect infestation in wheat. *Journal of Economic Entomology*, **55**, 199–203.

Dick, E. & Marcellin, P. (1985) Effect of high temperatures on banana development after harvest. Prophylactic tests. *Fruits*, **40**, 781–784.

Dobie, P. (1973) An investigation into the use of an X-ray technique in the study of pre-emergent stages of *Sitophilus oryzae* (L) developing in Manitoba wheat. *Journal of Stored Products Research*, **9**, 7–12.

Dohring, S. (1997) Over sea and over land: Putting CA research and technology to work for international shipments of fresh produce. In: Seventh International Controlled Atmosphere Research Conference, 13–18 July 1997. University of California, Davis CA, USA.

Elias, D.J. & Fall, M.W. (1988) The Rodent Problem in Latin America. In: I. Prakash (Ed.) *Rodent Pest Management*. CRC Press, Boca Raton FL, USA.

Evelo, R.G. (1995) Modelling modified atmosphere systems. COST94. The post-harvest treatment of fruit and vegetables: systems and operations for post-harvest quality. In: *Proceedings of a Workshop*, Milan, Italy, 14–15 September 1993.

FAO Storage of Horticultural Crops. http://www.fao.org/Wairdocs/X5403E/x5403e09.htm

Farber, J.M. (1991) Microbiological aspects of modified atmosphere packaging technology – a review. *Journal of Food Protection*, **54**, 58–70.

Frankenfeld, J.C. (1948) Staining methods for detecting weevil infestation in grain. Publication No. ET-256 (July). USDA Bureau of Entomology, Washington DC, USA.

Garrett, M. (1992) Applications of controlled atmosphere containers. BEHR'S Seminaire Hamburg, 16–17 November 1992, Munich, Germany.

Gast, K.L. & Flores, R.A. (1992) MF-1033. Storage operations. Fruit and vegetables. Postharvest management of commercial horticultural crops. Cooperative Extension Service, Kansas, USA.

Gilman, G.A. & Boxall, R.A. (1974) The storage of food grains in traditional underground pits. *Tropical Stored Products Information*, **28**, 19–38.

Golob, P. (1981) A practical assessment of food losses sustained during storage by smallholder farms in the Shire Valley Agricultural Development Project Area of Malawi, 1978/79. Tropical Products Institute **G154**, Slough, UK.

Golob, P., Ashman, F. & Evans, N. (1975) The separation of live stored product insect larvae from flour or sievings using a modified Tullgren funnel. *Journal of Stored Products Research*, **11**, 17–23.

Gorny, J.R. (1997) Modified atmospheres packaging and fresh-cut revolution. *Perishables Handling Newsletter*, **90**, 2–4. University of California, USA.

Hagstrum, D.W., Webb, J.C. & Vick, K.W. (1988) Acoustical detection and estimation of *Rhyzopertha dominica* (F.) larval population in stored wheat. Symposium on Agroacoustics. *Florida Entomologist*, **71**, 411–447.

Haines, C. (1991) (Ed.) Insects and arachnids of tropical stored products: their biology and identification. Natural Resources Institute, Chatham, UK.

Haque, A.K.M., Anwarul, Jabber, Quasem, M.A., Choudhury, N.H. & Arboleda, J.R. (1988) Post-harvest rice loss: farmers' valuation in Bangladesh. Bangladesh Journal of Agriculture, **13**, 115–122.

Hardenburg, R.E., Watada, A.E. & Wang, C.Y. (1986) *The Commercial Storage of Fruits, Vegetables and Florist Nursery Stocks. Agriculture Handbook* **66**. United States Department of Agriculture, Washington DC, USA.

Harris, K.L. & Lindblad, C.J. (Eds) (1978) *Post Harvest Grain Loss Assessment Methods*. American Association of Cereal Chemists, New York NY, USA.

Hernandez, A. & Drummond, D.C. (1984) A study of rodent damage to food in some Cuban warehouses and the cost of preventing it. *Journal of Stored Products Research*, **20**, 83–86.

Hodges, R.J., Dunstan, W.R., Magazini, I. & Golob, P. (1983) An outbreak of *Prostephanus truncatus* (Horn) (Coleoptera: Bostrichidae) in East Africa. *Protection Ecology*, **5**, 193–194.

Hodges, R.J., Halid, H., Rees, D., Meik, J. & Sarjono, J. (1985) Insect traps tested as an aid to pest management in milled rice stores. *Journal of Stored Products Research*, **21**, 215–229.

Hong, Q.Z., Sheng, H.Y., Chen, Y.F & Yang, S.J. (1983). Effects of CA storage with a silicone window on Satsuma oranges. *Journal of Fujian Agricultural College*, **12**, 53–60.

Huq, A.K.F. (1980) Some estimates of farm-level storage losses in rice in Bangladesh. *Tropical Stored Products Information*, **39**, 5–12.

ISO (1979) Cereals sampling (as grain). ISO 950. International Organization for Standardization, Geneva, Switzerland.

ISO (1980) Fresh fruits and vegetables – Sampling. ISO 874. International Organization for Standardization, Geneva, Switzerland.

ISO (1986a) Cereals and pulses. Detection of hidden infestation. Part 1. General principles. ISO 6639–1. International Organization for Standardization, Geneva, Switzerland.

ISO (1986b) Cereals and pulses. Detection of hidden infestation. Part 2. Sampling. ISO 6639–2. International Organization for Standardization, Geneva, Switzerland.

ISO (1986c) Cereals and pulses. Detection of hidden infestation. Part 3. Reference method. ISO 6639–3. International Organization for Standardization, Geneva, Switzerland.

ISO (1987) Cereals and pulses. Detection of hidden infestation. Part 4. Rapid methods. ISO 6639–4. International Organization for Standardization, Geneva, Switzerland.

ISO (1999) Cereals, pulses and milled products – Sampling of static batches. ISO 13690. International Organization for Standardization, Geneva, Switzerland.

ISTA (1966) International rules for seed testing. *Proceedings of the International Seed Testing Association*, **31**, 49–91.

Kader, A.A. (1989) A summary of CA requirements and recommendations for fruits other than pome fruits. In: International Controlled Atmosphere Conference, Fifth Proceedings, 14–16 June 1989. Wenatchee WA, USA. Vol. 12. *Other Commodities and Storage Recommendations*.

Kester, J.J. & Fennema, O.R. (1986) Edible films and film coatings: a review. *Food Technology*, **40**, 47–59.

Kidd, F. & West, C. (1927a) A relation between the concentration of oxygen and carbon dioxide in the atmosphere, rate of respiration, and the length of storage of apples. Report of the Food Investigation Board London for 1925, 1926.

Kidd, F. & West, C. (1927b) A relation between the respiratory activity and the keeping quality of apples. Report of the Food Investigation Board London for 1925, 1926.

Kitnoja, L. & Kader, A.A. (1995) *Small-scale Postharvest Handling Practices. A Manual for Horticultural Crops*. Third Edition. Postharvest Horticulture Series **8**. March 1995. University of California, Davis CA, USA.

Klein, J.D. & Lurie, S. (1992) Heat treatments for improved postharvest quality of horticultural crops. *Horticultural Technology*, **2**, 316–320.

Knoth, J. (1993) *Traditional Storage of Yams and Cassava and its Improvement*. Deutshe Gesellschaft für Technische Zusammenarbeit (GTZ), Eschborn, Germany.

Koelet, P.C. (1992) *Industrial Refrigeration*. Macmillan, London, UK.

La Gra, J., Martinez, E. & Martinez, J.F. (1982) Indentificacion, evaluacion y reduccion de perdidas de post-cosecha de arroz en la Republica Dominicana *Publicacion Miscelana IICA 358*. Instituto Interamericano de Co-operacion para la Agricultura, Santo Domingo, Republica Dominicana.

Lee, K.S. & Lee, D.S. (1996) Modified atmosphere packaging of a mixed prepared vegetable salad dish. *International*

*Journal of Food Science and Technology*, **31**, 7–13.
Longmore, A.P. (1974) Vacuum cooling of fresh produce. *Food Industries of South Africa*, **25**, 2, 3 & 7.
Lopez-Briones, G., Varoquaux, P., Bareau, G. & Paascat, B. (1993) Modified atmosphere packaging of common mushroom. *International Journal of Food Science and Technology*, **28**, 57–68.
Macfie, G.R., Jr. (1956) Packaging and handling of fresh lychee fruit. *Florida Lychee Growers' Association Yearbook*.
Marcellin, P. (1973) Conservation des fruits et légumes en atmosphère contrôlée à l'aide des membranes de polymères. *Revue Générale du Froid*, **3**, 217–236.
Marcellin, P. & Chaves, A. (1983) Effects of intermittent high carbon dioxide treatment on storage life of avocado fruits in relation to respiration and ethylene production. *Acta Horticulturae*, **138**, 155–163.
Mian, Y., Ahmed, S. & Brooks, J.E. (1987) Small mammals and stored food losses in farm households in Bangladesh. *Crop Protection*, **6**, 200–203.
Morimoto, T., De Baerdemaeker, J. & Hashimoto, Y. (1997) An intelligent approach for optimal control of fruit-storage process using neural networks and genetic algorithms. *Computers and Electronics in Agriculture*,**18**, 205–224.
Nakasone, H.Y. & Paull, R.E. (1998) *Tropical Fruits. Crop Production Science in Horticulture 7*. CAB International, Wallingford, UK.
Parsons, C.S., Gates, J.E. & Spalding, D.H. (1964) Quality of some fruits and vegetables after holding in nitrogen atmospheres. *American Society for Horticultural Science*, **84**, 549–556.
Picard, J. & Procter, D.L. (1984) Storage at farm/village level and in warehouses. In: *Grain Storage Techniques – Evolution and Trends in Developing Countries. FAO Agricultural Services Bulletin* **109**. Food and Agriculture Organization of the United Nations, Rome, Italy.
Pike, V. (1994) Impact of milling degree on *Liposcellis paetus* population growth rate and assessment of milled rice weight loss due to infestation. *Crop Protection*, **13**, 425–428.
Pinniger, D.B. (1975) A bait trap technique for assessment of stored product beetle populations. *Tropical Stored Products Information*, **19**, 9–11.
Powrie, W.D., Chiu, R., Wu, H. & Skura, B.J. (1990) Preservation of cut and segmented fresh fruit pieces. US Patent 4, 895,729. United States Patent Office, Washington DC, USA.
Ratnadass, A. & Fleurat-Lessard, F. (1991) Improvement in the practical methods of assessment of losses caused by insects in grain stored at the village level in tropical Africa. In: F. Fleurat-Lessard & P. Ducom (Eds) *Proceedings of the 5th International Working Conference on Stored Product Protection, Bordeaux*. INRA, Bordeaux, France.
Ratnadass, A., Berté, S., Diarra, D. & Cisé, B. (1994) Insect losses in sorghum stored in selected Malian villages, with particular emphasis on varietal differences in grain resistance. In: E. Highley, E.J. Wright, H.J. Banks & B.R. Champ (Eds) *Proceedings of the 6th International Working Conference on Stored Product Protection, Canberra*. CAB International, Wallingford, UK.
Roberts, R. (1990) An overview of packaging materials for MAP. In: *International Conference on Modified Atmosphere Packaging. Part 1*. Campden Food and Drinks Research Association, Chipping Campden, UK.
Rowley, J.Q. (1984) An assessment of losses during handling and storage of millet in Mali. *Tropical Stored Products Information*, **47**, 21–33.
Schlimme, D.V. & Rooney, M.L. (1994) Packaging of minimally processed fruits and vegetables. In: R.C. Wiley, (Ed.) *Minimally Processed Refrigerated Fruits and Vegetables*. Chapman & Hall, USA.
Scott, K.J. (1975) The use of polythene bags to extend the life of bananas after harvest. *Food Technology, Australia*, **27**, 481–482.
Scott, K.J., Blacke, J.R., Strachan, G., Tugwell, B.L. & McGlasson, W.B. (1971) Transport of bananas at ambient temperatures using polythene bags. *Tropical Agriculture Trinidad*, **48**, 245–253.
Sharifi, S. & Mills, R.B. (1971) Radiographic studies of *Sitophilus zeamais* Mots. in wheat kernels. *Journal of Stored Products Research*, **7**, 195–206.
Sharples, R.O. & Stow, J.R. (1986) Recommended conditions for the storage of apples and pears. Report of the East Malling Research Station for 1985, 165–170.
Sharples, R.O., Kidd, F. & West, C. (1989) The gas storage of fruit II. Optimum storage temperatures and atmospheres. In: J. Janick (Ed.) *Classical Papers in Horticultural Science*. Prentice Hall, New Jersey, USA.
Shuman, D. & Epsky, N.D. (2001) Commercialisation of the electronic grain probe insect counter. In: *Proceedings of the International Conference on Controlled Atmospheres and Fumigation in Stored Products*. 29 October – 3 November 2000, Fresno CA, USA.
Siriphanich, J. (1994) Minimal processing of tropical fruits. In: B.R. Champ, E.J. Highley & G.I. Johnson (Eds) *Postharvest Handling of Tropical Fruits*. In: *Proceedings of the International Conference, Chiang Mai, Thailand*, 19–23 July 1993. ACIAR Proceedings **50**, 127–137.
Skog, L.J. (1998) Chilling injury of horticultural crops. Fact Sheet, Ministry of Agriculture, Food and Rural Affairs, Ottawa, Ontario, Canada.
Thompson, A.K. (1992) Storage systems. In: A. Kader (Ed.) *Postharvest Technology of Horticultural Crops*. Pub. 3311ANR. University of California, USA.
Thompson, A.K. (1998) *Controlled Atmosphere Storage of Fruits and Vegetables*. CAB International, Wallingford, UK.
Thompson, A.K., Been, B.O. & Perkins, C. (1972) Handling, storage and marketing of plantains. *Proceedings of the Tropical Region of the American Society for Horticultural Science*, **16**, 205–212.
Thompson, J.F & Kasmire, R.F. (1981) An evaporative cooler for vegetable crops. *California Agriculture*, **35**, 20–21.
Thompson, J.F & Spinoglio, M. (1994) Small-scale cold rooms for perishable commodities. Family Farm Series. Small Farm Centre, University of California, Davis CA, USA.
Tyler, P.S. (1981) Report on a survey of grain losses. FAO Action Programme for the Prevention of Food Losses. Report PFL/CYP/001. Food and Agriculture Organization of the United Nations, Rome, Italy.
Tyler, P.S. & Boxall, R.A. (1984) Postharvest loss reduction programmes: a decade of activities: what consequences?

*Tropical Stored Products Information*, **50,** 4–13.

Tyler, P.S., Golob, P., Compton, J. & Bickersteth, S. (1990) Study on phytosanitary requirements to promote maize trade in eastern and southern African countries at risk from the larger grain borer. Report for the European Commission. Natural Resources Institute, Chatham, UK.

Wright, M.A.P. (1995) Loss assessment methods for durable stored products in the tropics: appropriateness and outstanding needs. *Tropical Science*, **35**, 171–185.

Yahia, E.M. (1998) Modified and controlled atmospheres for tropical fruits. *Horticulture Reviews*, **22**, 123–183.

Young, T.W. & Koo, R.C.C.J. (1974) Increasing yield of 'Parvin' and 'Kent' mangoes on Lakewood sand by increasing nitrogen and potassium fertilizer. *Proceedings of the Florida State Horticulture Society*, **87**, 380–384.

Zagory, D. (1990) Application of computers in the design of modified atmosphere packaging to fresh produce. In: *International Conference on Modified Atmosphere Packaging, Part 1*. Campden Food and Drinks Research Association, Chipping Campden, UK.

Zong, R.J. (1989) Physiological aspects of film wrapping of fruits and vegetables. In: *International Controlled Atmosphere Conference, Fifth Proceedings*, 14–16 June 1989. Wenatchee WA, USA.

# Chapter 6
# Pest Management

**P. Golob, L. A. Birkinshaw, T. E. Stathers, A. N. Meyer, S. R. Belmain, G. Farrell and R. J. Hodges**

Post-harvest deterioration is principally caused by biological spoilage organisms including insects, fungi and small vertebrates. The degree of spoilage can be exacerbated by various physical and chemical attributes, such as climate and crop characteristics, but these rarely cause deterioration in their own right. Of course, controlling the underlying factors that influence pest development will assist in maintenance of quality but still, in order to prevent loss of quality and of market value, it is necessary to address the 'front-line' effectors of spoilage.

During the past half-century pest management techniques have become ever more sophisticated. As world population soared and agricultural production increased to meet the new demands, so the technological revolution was able to develop pest management tools that could keep abreast of the need to store and conserve huge quantities of grain and to keep fresh produce in the peak of condition demanded by western urban markets. The development of synthetic insecticides, particularly organophosphates and fumigant gases, enabled countries to insist on 'zero' tolerance for spoilage in imported produce. These chemicals enabled 250 billion tonnes of cereals traded worldwide (http://www.fao.org) to be preserved with very little loss to either weight or value.

Populations continue to grow, particularly in the developing world. The need for sophisticated methods of post-harvest protection would appear to be a greater requirement than ever. However, consumers are becoming ever more aware of the potential hazards that might accrue from using synthetic chemicals, such as evinced by the fumigant methyl bromide's effects as an ozone depletor. The deleterious effects on both consumers and the environment are causing a reassessment of the use of compounds that were previously taken for granted. At a time when more and more food is being produced, and so stored, researchers are searching for methods of protection that significantly reduce or do not require the use of complex synthetic chemical compounds. There has begun a reversion to types of methods that were in common practice in the early part of the twentieth century – for example, the application, from *Lonchocarpus* or *Derris* species, of rotenone as an insecticide which has traditionally been used as a fish poison – or as part of an even earlier cultural tradition.

Natural plant products used in Asia and Africa for hundreds of years are being investigated to see whether they might have the same appeal and popularity in the future. The chemicals produced by the pests themselves, pheromones, juvenile hormones and effectors of chitin deposition, are being exploited to effect control. Methods that rely on physical rather than chemical traits, such as inert dusts, heat application and barriers are being introduced.

Nevertheless, the growth in the production and application of conventional synthetic products continues unabated. Conventional methods will continue to be popular until alternatives are shown to be as effective in curtailing pest problems. This is unlikely to occur through a single non-chemical method used in isolation but may come about if such methods are used in combination. Such a scenario has yet to occur on a large scale. Where it might happen, such as on small-scale farms in developing countries, consumers are able to tolerate a degree of post-harvest damage and do not have stringent demands for pest-free products. In more sophisticated markets, the presence of any poor quality produce is still not tolerated, even in the organic sector. Universal replacement of conventional synthetics might occur when genetic engineering techniques are used to introduce pest resistance into the crop, either directly or by the introduction of secondary substances such as Bt endotoxins. However, although this has already occurred to protect crops during their growth cycle, it has not yet been attempted for stored crops. Furthermore, the transfer of genes into non-traditional plants has potentially more political and ethical implications than has the continued use of conventional synthetic pesticides.

This chapter provides a description of those methods that are currently used for pest management in the post-harvest sector and of methods that are still being developed, and which will become important in the future.

## Safety

Pesticides are toxic to both target species and non-target organisms. Among non-target organisms that may be affected are beneficial insects, such as pest predators and bees, domesticated animals and man. The effects on non-target organisms may be acute, caused by either mishandling of the pesticide during its transportation, storage or application, or chronic as the result of continuous ingestion of chemicals in foodstuffs. There are many examples of acute poisoning caused by synthetic insecticides but very few attributed to chronic exposure.

Most countries routinely record incidences of acute poisoning as a result of insecticide contamination. Poisoning is usually the result of accidental spillage or exposure during spray application but there are many examples of insecticides being used in both suicide and homicide attempts.

It is very difficult to determine the toxicological effects of chronic ingestion of pesticides in man. Long-term dietary experiments have not been performed, for obvious reasons. However, chronic effects of some insecticides have been observed in small mammals, especially rodents and dogs, in order to ascertain the 'no observable effect level' (NOEL). Studies have demonstrated changes in haematology and in the activity of some enzymes, though how pertinent these observations are in relation to human health is open to discussion.

Chronic effects have been observed after feeding animals, frequently rats, with $^{14}$C-labelled active ingredient as part of the daily diet. Residues from these compounds are either absorbed as grain-bound or non-bound compounds. Dietary levels of 1.17 ppm and 7.5 ppm of bound pirimiphos-methyl residues in diets fed to rats for 3.5 months have been shown to cause an increase in alkaline phosphatase, which is a symptom of acute hepatitis, and a reduction in lymphocytes and monocytes, cells essential for the effective operation of the immune system (Akay *et al.* 1992). Nevertheless, the rats in these trials did not exhibit differences in body or organ weights when compared with controls, nor did they show any toxic manifestations.

Insecticidal residues either remain on the surface of the treated commodity or penetrate into the internal structure of the grain. Surface residues can be readily removed by washing and cooking, and they also decay as the storage period continues. Residues inside the grain may initially increase as a result of continuous penetration of active ingredient from the grain surface. However, as time progresses, generally after 2 months or so depending on the relative humidity and temperature, these residues also decline. This rate of degradation depends on whether the compound is bound to the commodity or remains unbound; bound compounds degrade much more slowly.

Studies with $^{14}$C-labelled pirimiphos-methyl applied to a variety of grains and grain products have shown that the insecticide decays gradually with time, but only slowly. For example, in maize 180 days after treatment only 16% of the active ingredient was recoverable (Pedral Sampaio *et al.* 1990) but in wheat and barley more than 80% remained after 12 months, although for non-labelled carriers the active ingredients declined to 78% and 59% respectively (Hadjidemetriou 1990a). Similar losses of activity occurred after eight months' storage of beans and raisins, loss being greater in the former as most of the insecticide was concentrated on the surface (Hadjidemetriou 1990b).

The safety of a pesticide treatment depends upon:

- the toxicity and stability of the active ingredient;
- the formulation in which the chemical is used;
- the method of application;
- the training level of the person applying the chemical;
- the protective measures taken during application (clothing); and
- how excess chemical and chemical containers are disposed of.

## Insecticide toxicity

All pesticides are toxic chemicals and can cause harmful effects, both lethal and sub-lethal. The degree of toxicity varies between different substances. As pesticides are used in a variety of different ways, against many different organisms, it is not immediately obvious how the toxic nature of different chemicals can be compared. In order to be able to make such a comparison, chemical effects are measured in one or more standard scenarios. The most common test is to assess the effect of a single dose given by mouth to rats. This produces the acute, oral $LD_{50}$ value, i.e. that single dose that, when given by mouth, kills 50% of the test population (LD = lethal dose). A similar test using skin (dermal) exposure is also used to compare toxicities. On the basis of these tests pesticides have been divided into a series of categories by the World Health Organization (Table 6.1).

For most pesticides it is normal to quote the oral $LD_{50}$ as this is the speediest, therefore most acutely hazardous, route into the body. Nevertheless, data on dermal toxicity are essential in order to assess the potential effects of exposure to which spray operators are likely to be subjected.

The potential effects of environmental contamination of pesticides can also be assessed by examining $LD_{50}$ against other animals, particularly fish and birds. Utilising $LD_{50}$ values provides a method by which the acute toxicity of compounds can be compared. However, the data should not be used to interpret the 'degree of safety' of a particular compound as neither the laboratory environment under which the tests are conducted nor the use of rodents and other small mammals as test specimens can be directly correlated to the effects on humans and domestic animals under practical conditions. Certainly, the information cannot reflect what happens during chronic exposure, as occurs when treated commodity is ingested over a prolonged period. Furthermore, these data do not provide a clear indication of the biological efficacy of the active ingredient concerned because efficacy is dependent on many factors, including the target species, its age and sex, the geographical strain, the type of formulation and its mode of application, and so on.

Toxicity is also related to the route by which the chemical enters the body. There are three main routes:

(1) Oral (taken in through the mouth, generally by swallowing) – this would apply to consumers eating grain, vegetables or fruit containing insecticide residues.
(2) Dermal (absorbed through the skin) – likely to occur during application through skin exposure to spray particles, or by accidental spillage while a formulation concentrate is being diluted before application. Pesticides can also enter the eyes and cause severe injury or irritation.
(3) Inhalation (breathed in through the mouth or nose and absorbed in the lungs) – particularly likely during application of sprays and dusts.

The toxicity data in Table 6.2 were primarily derived from tests using highly concentrated, technical grade chemicals, which contain more than 95% active ingredient. However, for practical purposes, the insecticide operator will be more interested in the toxicity of the formulation being applied whether this be, for example, a 2% dust or a 25% emulsifiable concentrate. Manufacturers usually publish technical leaflets giving toxicity data of their formulations as well as that of their active ingredients. Where such information is not provided, an approximate indication of the toxicity of the formulation can be obtained from the following formula, which is based on the concentration of active ingredient in the formulation:

$$\text{Toxicity of formulation} = \frac{LD_{50}\text{ of active ingredient} \times 100}{\%\text{ active ingredient in the formulation}}$$

For example: for a 2% dust of methacrifos the toxicity would be:

$$678 \times 100/2 = 33\,900\text{ mg/kg}$$

**Table 6.1** World Health Organization classification of pesticide substances.

| | | $LD_{50}$ for rats (mg/kg body weight) | | | |
|---|---|---|---|---|---|
| | | Oral | | Dermal | |
| Class | Hazard | Solids | Liquids | Solids | Liquids |
| Ia | Extremely hazardous | ≤5 | ≤20 | ≤10 | ≤40 |
| Ib | Highly hazardous | 5–50 | 20–200 | 10–100 | 40–400 |
| II | Moderately hazardous | 50–500 | 200–2000 | 100–1000 | 400–4000 |
| III | Slightly hazardous | >500 | >2000 | >1000 | >4000 |

Note: The WHO classification also lists separately from Class III substances, those pesticides with an oral $LD_{50}$ greater than 2000 mg/kg (solids) or 3000 mg/kg (liquids) under the heading 'products unlikely to present acute hazards in normal use'.

**Table 6.2** $LD_{50}$ of chemicals commonly used as grain protectants and some other substances in common use.

| Common name | WHO class | Acute oral $LD_{50}$ rat (mg/kg body weight)[1] | No observable effect level (NOEL) for rat (mg/kg diet/day)[2] | Comment |
|---|---|---|---|---|
| *Organophosphates* | | | | |
| Pirimiphos-methyl | III | 2050 | 8; 20 (dog) | NOEL tested over 90 days |
| Chlorpyrifos-methyl | II-III | 1630–2140 | 0.1 (rats and dogs) | |
| Fenitrothion | II | 1700 | 10 | |
| Malathion | III | 1375–2800 | 100 | NOEL tested over 21 months |
| Dichlorvos | Ib | 56–108 | 10 | |
| Dimethoate | II | 290–325 | 1 (rat) | |
| Phosmet | II | 113–160 | 40 (rat) | |
| *Pyrethroids* | | | | |
| Permethrin | II-* | 430–4000 | 100 | |
| Deltamethrin | Ib-* | 135–5000 | 2.1; (1 mg/kg b.w. for dogs) | |
| Fenvalerate | II | 451 | 250 | |
| Bioresmethrin | * | 5000–8000 | 50 | |
| Bifenthrin | II | 54.5 | 1.5 (dogs) | NOEL tested over 1 year |
| *Other storage chemicals* | | | | |
| Pyrethrins | II | 584–900 | 100 | |
| Piperonyl butoxide | * | 7500 | Not assessed | |
| Methacrifos | II | 678 | 5 (mg/kg b.w.) | |
| Methoprene | * | >34600 | 5000 | |
| Carbaryl | II | 850 | 200 | |
| *Other compounds* | | | | |
| Aflatoxin B | Ia | 0.37 | | $LD_{50}$ for ducklings, mycotoxin |
| Nicotine | Ia | 1 | | Tobacco |
| Methamphetamine | Ib | 70 | | $LD_{50}$ for mice, stimulant |
| Paracetamol | II | 338 | | $LD_{50}$ for mice, analgesic |
| Warfarin | II | 374 | | $LD_{50}$ for mice, anticoagulant, Rodenticide |
| Aspirin | II | 1500 | | Analgesic |
| Penicillin | III | 2300 | | Antibiotic |
| Acetic acid | III | 3530 | | Vinegar |
| Sodium chloride (common salt) | III | 3750 | | Seasoning |
| Ethanol (alcohol) | III | 10000 | | Beverage |

[1]Where a range of values is given this is the result of conducting tests with different carrier substances, animal species, the sex and age of the animals, the degree of fasting and chemical isomers. Reported values can differ markedly: for example, the $LD_{50}$ for rats for permethrin with a *cis/trans* ratio of 40 : 60 is 430–4000 mg/kg but for a *cis/trans* ratio of 20 : 80 is 6000 mg/kg.
[2]NOEL is the highest dose in an animal toxicology study at which no biologically significant increase in frequency or severity of an effect is observed. Unless otherwise stated, data were derived from 2-year feeding studies; data from shorter studies are only included where no 2-year studies have been undertaken.
*Classified by WHO as products unlikely to cause a hazard.

Then the 2% dust would be classified as a 'product unlikely to present acute hazards in normal use' whereas the active ingredient itself is moderately hazardous (class II). Similarly, other dilute dust formulations are also unlikely to be hazardous, which is why this type of formulation is so appropriate for use by poor farmers in the developing world who generally have no training in pest control techniques or are unable to use the type of protective clothing that would be mandatory in the developed world.

## Maximum residue limits

Few countries in the developing world have rigorous legislation that regulates the use of insecticides. In order to verify manufacturers' advertised efficacy and safety levels, extensive testing may be required, which would be very expensive and require resources most countries do not possess. Instead, countries rely upon recommendations made by United Nations organisations to identify acceptable compounds. International regulation pertaining to the use of insecticides on foodstuffs is governed by the UN Codex Alimentarius Commission (CODEX). CODEX specifies chemicals and residue limits that are permissible in traded grains and other food commodities. Its specifications are based upon recommendations provided by the FAO and WHO Joint Meeting on Pesticide Residues (JMPR). This body accumulates and examines scientific evidence on biological efficacy, mammalian toxicology and environmental contamination of compounds and establishes, in an eight-stage process, maximum permissible residues for different food commodities. These maximum residue limits (MRLs), expressed as quantities of active ingredient (a.i.) of the chemical, take into account the acceptable human daily intake (ADI) of each compound. Only a small number of the insecticides marketed for the protection of agricultural produce have received clearance by CODEX and have MRLs recommended for durable food crops.

Compounds that are not recommended for use by JMPR must not be used for treatment of food grain. It is therefore essential that insecticides and fungicides which are commonly applied as seed treatments, such as thiodan, must not be applied to grain destined for food, as they are too toxic. A few chemicals, for example pirimiphos-methyl (Actellic) are used for both grain and seed treatment.

This system of clearance for additives is not necessarily foolproof in guaranteeing the safety of any chemical used to treat food, but it certainly limits any active ingredient that is likely to be unsafe or a cause for concern. In principle, only those compounds that have passed through the JMPR/CODEX system should be used to treat food grain. Compounds that have not been approved must not be applied to grain under any circumstances.

Many countries among the developed nations have produced their own regulations regarding application of foreign substances to grain and other foods. In general, these are more stringent than those established by JMPR/CODEX. Most developing countries use the data and tolerances established by JMPR/CODEX, but conduct efficacy trials in order to assess the performance of the compound under prevailing agro-climatic conditions.

Few compounds are approved for use on raw cereals and grain legumes. These are primarily organophosphates and pyrethroids. A list of compounds approved for use on raw cereal grain, together with their MRLs, is shown in Table 6.3. Properties of compounds are contin-

**Table 6.3** Active ingredients approved by JMPR/CODEX for use on raw cereal grains, maximum residue limits and acceptable human daily intake.

| Common name | MRL on raw cereals (ppm or mg/kg) | ADI (mg/kg body weight) |
|---|---|---|
| *Organophosphates* | | |
| Pirimiphos-methyl | 10 | 0.03 |
| Chlorpyrifos-methyl | 10 | 0.01 |
| Fenitrothion | 10 | 0.003 |
| Malathion | Withdrawn (0.05 for maize)* | 0.02 |
| Dichlorvos | 5 | 0.004 |
| *Pyrethroids* | | |
| Permethrin | 2 | 0.05 |
| Deltamethrin | 1 | 0.01 |
| Fenvalerate | 2 | 0.02 |
| Bioresmethrin | 1 (wheat) | 0.03 |
| Bifenthrin | 0.5 (wheat) | 0.02 |
| Pyrethrins | 3 | 0.04 |
| *Others* | | |
| Piperonyl butoxide | 10 | 0.03 |
| Methoprene | 5 | 0.1 |
| Carbaryl | 5 (under review) | 0.01 |
| *Fumigants* | | |
| Phosphine | 0.01 | — |

*Until 1999, the MRL for malathion was 8 mg/kg on cereal grains. The MRL was withdrawn and limits were only established for maize and beans (0.5 mg/kg).
Note: MRLs for several compounds including the organophosphates, bromophos, etrimphos and methacrifos have been revoked in recent years.

ually being reassessed and inclusion on the list does not guarantee indefinite approval. For example, some chlorinated hydrocarbons, such as gamma hexachlorohexane, commonly known as lindane, were removed some years ago because of their long persistence, after having been approved for use as a grain additive at 2 mg/kg. Others, such as bromophos, iodophenphos and etrimphos, which are no longer commercially available, have had their tolerances revoked. Absence of a chemical from the list must preclude its use as a grain additive.

There are specific MRLs established for other food commodities. Those designated for other crops, mostly durables, are shown in Table 6.4.

Perishables are also protected against pests and diseases. However, insect pests cause little damage to fresh produce after harvest so that only fungicides are approved as additives. The MRLs are shown in Table 6.5 and the ADI, $LD_{50}$ and NOEL value for each of these fungicides in Table 6.6.

Further restrictions on available chemicals may be imposed through limited formulations that are commercially available. Chlorpyrifos-methyl is very effective in controlling weevils, flour beetles and stored product moths. However, its use is restricted because the chemical is not available in a dilute dust formulation and cannot, therefore, be generally recommended for use by small-scale farmers. It may be appropriate for those farmers who have access to spraying equipment as it could then be applied as a water-based spray.

**Pesticide application: the pesticide label**

It is essential that the manufacturer's instructions specified on the label attached to the pesticide container are always followed in order to ensure that the application procedure and the resultant treated commodity are safe. In many countries, the content of the label descriptions are laid down in law; deviating from the instructions can then be regarded as breaking the law.

The pesticide label should contain the following information:

- the commercial name of the formulation;
- the name and address of the manufacturer;
- the type of active ingredients;
- the concentration of active ingredient in the formulation;
- the date of manufacture of the formulation;
- the quantity;
- what the contents are used for;
- instructions on how to use the contents;
- safety precautions to be taken before, during and after use;
- instructions for disposal of unwanted content;
- any health side effects both on man, domestic and wild animals where these are appropriate; and
- antidotes to the active ingredient.

The packet or bottle should also have a date of manufacture and a use-by date, as well as a production batch

**Table 6.4** Insecticides approved for use on crops other than whole cereal grain.

| | Wheat bran (unprocessed) | Wheat bran (processed) | Wheatgerm | Wheat flour | Wheat (wholemeal) | Rye bran | Maize oil | Beans (dry) |
|---|---|---|---|---|---|---|---|---|
| Fenitrothion | 20 | 2 | | 2 | 5 | | | |
| Chlorpyrifos-methyl | 20 | | | 2 | | | | |
| Malathion | 20 | | | 2 | 2 | 20 | | 8 |
| Pirimiphos-methyl | 20 | | | 2 | 5 | | | |
| Permethrin | 5 | | 2 | 0.5 | 2 | | | |
| Bioresmethrin | 5 | | 3 | 1 | 1 | | | |
| Bifenthrin | 2 | | | 0.2 | 0.5 | | | |
| Deltamethrin | 5 | | | 0.2 | 1 | | | 1 |
| Fenvalerate | 5 | | | 0.2 | 2 | | | |
| Carbaryl* | 20 | | | 0.2 | 2 | | | |
| Methoprene | 10 | | | 2 | 5 | | 0.2** | |

Note: MRLs are also permitted for: pyrethrins on oilseeds (general) and dried fruit (general) at 1 mg/kg; pirimiphos-methyl on whole groundnuts at 25 mg/kg, groundnut kernels at 2 mg/kg and groundnut oil at 15 mg/kg; for dried dates at 0.5 mg/kg and for dried fish at 8 mg/kg; dimethoate on tomato and sweet pepper at 1 mg/kg; phosmet on sweet potato at 10 mg/kg; and for pyrethrins on dried vegetables at 1 mg/kg.
*Clearance allowed to 2003.
**At or about the limit of detection.

**Table 6.5** Fungicides approved by CODEX for post-harvest application to food commodities. Source: http://apps.fao.org/csv_down/ CODEX Pesticide Regulations Committee, September 1999.

| Fungicide | Commodity | MRL (mg/kg) |
|---|---|---|
| Carbendazim | Banana | 1 |
| | Irish potato | 5 |
| | Melon (not watermelon) | 2 |
| Dicloran | Carrot | 10 |
| | Grape | 10 |
| | Onion (bulb) | 10 |
| | Peach | 15 |
| | Plum | 10 |
| Diphenylamine | Apple | 5 |
| Ethoxyquin | Pear | 3 |
| Imazalil | Banana | 2 |
| | Citrus | 5 |
| | Irish potato | 5 |
| | Melon (not watermelon) | 2 |
| | Persimmon | 2 |
| | Pome fruit | 5 |
| Iprodione | Carrot | 10 |
| | Pome fruit | 5 |
| Matalaxyl | Citrus | 5 |
| | Pome fruit | 1 |
| 2-Phenylphenol | Citrus | 10 |
| | Pear | 25 |
| Prochloraz | Avocado | 5 |
| | Orange | 5 |
| | Mango | 2 |
| | Papaya | 1 |
| Tecnazene | Irish potato | 20 |
| Thiabendazole | Banana | 5 |
| | Citrus | 10 |
| Thiophanate-methyl | Carrot | 5 |
| | Celery | 20 |
| | Peach | 10 |
| | Pear | 5 |
| | Citrus | 10 |
| Triadimefon | Pineapple | 2 |
| Triadimenol | Pineapple | 1 |
| Triforine | Peach | 5 |
| Vinclozolin | Cherries | 5 |
| | Peach | 5 |

number. These pieces of information may be either on the label or on the container itself. An example of a typical, well-designed label is shown in Plate 3.

From a safety perspective the instructions on the label must always be followed. However, it is important to remember that label instructions will not always guarantee efficacy. This may be because: the formulation is old and so no longer contains active ingredient at the original concentration, it having decayed with time; the local climatic conditions are not conducive to the action of the chemical; and, especially with dusts, the carrier material may not be fully compatible with the active ingredient and so may cause a loss of activity. This is particularly the case with aluminium phosphide tablets used for fumigation as the manufacturers routinely advocate exposure periods (3 days) that are insufficient to ensure effective disinfestations, which normally require 5–7 days.

### Protective clothing and personal hygiene

The hazard potential of a pesticide is a function not only of the toxicity of the compound but also of the amount and time of contamination. Reducing contamination can be achieved by minimising contact of chemical with the body, usually by wearing protective clothing and washing effectively.

People handling pesticides should understand the reasons for wearing protective clothing. Dirty, contaminated or defective protective equipment is a source of contamination; therefore all safety equipment should be washed or cleaned after each day of use.

Choice of protective clothing is a matter of compromise between efficiency of protection and comfort, acceptability and convenience.

- Materials should be as light-weight as possible and permit maximum air permeability to allow as free body movement as possible and to prevent sweating. A pair of light, durable cotton overalls, which can be washed after use, or alternatively a long-sleeved cotton shirt and trousers, kept separate from other clothing, should be worn. Wearing a hat during spray operations is advisable.
- Gloves should be worn wherever there is a risk of skin contamination. Gloves should be long enough to be worn under the sleeves. Neoprene, PVC or nitrile rubber are preferred. Natural rubber, cotton or leather gloves are not recommended as they offer little protection against the insecticide in certain formulations and to wear such gloves may be more dangerous than wearing no gloves at all. Cuffs of overalls and shirt sleeves should be worn outside the top of the gloves. Spills on gloves should be washed off immediately, as insecticide can penetrate very quickly.
- Protective footwear must always be worn, even when farmers are applying dilute dusts to their own grain. When handling and applying liquids, neoprene or rubber boots are recommended. Trousers should be worn outside the boots.

**Table 6.6** Toxicological characteristics of fungicides approved for use for post-harvest application. Source: http://apps.fao.org/csv_down/ CODEX Pesticide Regulations Committee, September 1999.

| Fungicide | ADI (mg/kg body weight) | Acute oral $LD_{50}$ rat (mg/kg body weight) | No observable effect level (NOEL) | NOEL feeding trial conditions |
|---|---|---|---|---|
| Carbendazim | 0.01 | >15 000 | 300 mg/kg diet | 2-year, dog |
| Dicloran | 0.01 | | 2.5 mg/kg body weight | 2-year, dog |
| Diphenylamine | 0.02 | | | |
| Ethoxyquin | 0.06 | 1 920 | | |
| Imazalil | 0.03 | 227–343 | 2.5 mg/kg body weight | 2-year, rat and dog |
| Iprodione | 0.2 | >2 000 | 150 mg/kg diet | 2-year, rat |
| Metalaxyl | 0.03 | 633 | 2.5 mg/kg body weight | Rat |
| 2-Phenylphenol | 0.02 | 2 700 | 2 g in diet | 2-year, rat |
| Prochloraz | 0.01 | 1 600 | 30 g in diet | 2-year, dog |
| Tecnazene | 0.01 | 1 256–2 047 | 50 g in diet | Long-term, rat |
| Thiabendazole | 0.1 | 3 100 | 40 mg/kg diet | 2-year, rat |
| Thiophanate-methyl | 0.08 | 7 500 | 160 mg/kg diet | 2-year, rat |
| Triadimefon | 0.03 | c. 1 000 | 300 mg/kg diet | 2-year, rat |
| Triadimenol | 0.05 | c. 700 | 125 mg/kg diet | 2-year, rat |
| Triforine | 0.02 | >16 000 | 625 mg/kg diet | 2-year, rat |
| Vinclozolin | 0.07 | >10 000 | 450 mg/kg diet | 90-day |

- A face shield will protect the eyes, face and mouth against pesticide spills, splashes and drift. However, shields give no protection against toxic fumes. Small, lightweight disposable face masks give adequate protection when spraying class III or unclassified chemicals. It is always advisable to wear such face masks when handling dusts and fine powder formulations. When spraying class II chemicals, face masks fitted with filter canisters or respirators should be used, particularly when spraying in confined spaces.

If protective clothing is not available, wear sensible footwear, place a clean cotton cloth over the mouth and nose, and wear a hat or scarf on the head. Wash all items carefully after each day's use and destroy if items become heavily contaminated. Suitable washing and cleaning procedures can be the most fundamental and effective line of defence against pesticide hazards. The following points should be strictly observed:

- always have a plentiful supply of clean water and soap available whenever pesticides are in use;
- always wash immediately when accidental contamination of the body has taken place;
- remove dry contaminated protective clothing immediately;
- wash protective clothing after use; gloves should be washed inside and out (all oil-soluble pesticides are readily absorbed and retained in rubber and plastics); and
- protective clothing should be washed separately from other clothing.

Other general safety instructions when handling pesticides include the following:

- never leave pesticides unattended in a place where security is poor;
- never transfer pesticides to other containers, especially beer, soft drink or other beverage bottles;
- never re-use empty pesticide containers;
- never work alone when handling toxic pesticides;
- never eat, drink, smoke, rub your eyes or touch your mouth while working with pesticides and keep food, drink and tobacco separate from pesticides;
- destroy heavily contaminated clothing and faulty protective clothing, especially gloves and respirators;
- if spills and leaks occur, decontaminate immediately; and
- keep unauthorised persons, especially children, away from pesticides.

### Pesticide disposal

Disposal of pesticides will require to be undertaken when:

- excess diluted spray solution remains after a store or commodity has been treated;
- a formulation becomes out-of-date; and
- spills occur.

Problems of excess spray volumes can be avoided if the correct volume of the formulation needed for the job is calculated in advance and just that amount is made up. Any small excess quantity can then be used up by slightly increasing the quantity applied. The best way to dispose of diluted pesticide solution is to use it for the job being undertaken.

Liquid or solid formulations of grain protectants bought from retail outlets should be disposed of by mixing them with soil or sand and the mixture then buried. This mixing is best done in a small trench and the contaminated material should be covered with about 50 cm of soil. It is important to ensure that the active ingredient has no opportunity to leach into the surrounding area where it could cause environmental hazards, especially if it leaches into watercourses. If available, the following chemicals can be added to the soil or sand and pesticide mix to facilitate the inactivation of the active ingredients:

- alkalis such as sodium hypochlorite (bleach), sodium hydroxide (caustic soda) or sodium carbonate (washing soda) for deactivating organophosphates and N-methylcarbamates;
- acids such as hydrochloric or sulphuric for treating thiocarbamate fungicides; and
- possibly an improvement on sand or soil is to mix the concentrate with farmyard manure; this has shown to be particularly effective in tropical conditions for dealing with pirimiphos-methyl and fenvalerate (Cox & Kilminster 1995).

Acids and alkalis should be diluted to about 10% before mixing with pesticides as they are less hazardous to use in a diluted form. Dilutions must not be carried out in buckets or a drum, because a rapid, uncontrolled rise in temperature or evolution of gas may eject the decomposing mixture with consequent danger to the operator. Protective clothing must be worn during these procedures.

When spills occur, for example, as pesticide concentrates are being diluted, they must be dealt with immediately. If action is not taken quickly the active ingredient may be absorbed into the floor, especially if the surface is porous. Liquids must be soaked up with absorbent material such as sawdust, sand or soil. After sweeping up, the contaminated area can be flooded with a 1% solution of household bleach and left until the next day if convenient. Any excess can be removed by scrubbing with soap or detergent and by washing with clean water.

In countries where the resources and services exist, chemicals must be disposed of by an organisation registered for this purpose. Chemicals are rendered harmless most effectively by incineration. In the US conditions for incineration require a temperature of 1200°C with a 3% oxygen surplus and a residence time of 2 seconds (in the UK the conditions required are 1800°C, 6% excess oxygen and 1.5 seconds residence). Incomplete incineration can be not only ineffective but positively dangerous: burning 2,4,5-trichlorphenol (related to the herbicide 2,4,5-T) at 600°C can produce the extremely toxic compound dioxin at a yield of about 0.5%.

### Symptoms of poisoning

Organophosphates cause acute toxic effects by inhibiting the enzyme, acetylcholinesterase (AChE), which leads to the accumulation of acetylcholine at all cholinergic transmission sites in the central and peripheral nervous systems. Symptoms of mild acute poisoning include dizziness, headache, nausea, vomiting, miosis (small pupils), excessive sweating and tracheobroncheal and salivary excretions (He & Chen 1999). As the poisoning increases, patients suffer shortness of breath and muscular fasciculation, and with severe poisoning pulmonary oedema, respiratory depression and coma result. Some organophosphates induce a polyneuropathy after a delay of 2–4 weeks, when symptoms include weakness of the feet and hands with accompanying burning sensations. A further outcome of severe acute organophosphate poisoning is intermediate myasthenia syndrome (IMS) (He & Chen 1999), which is characterised by muscular weakness affecting the neck, proximal limb muscles, muscles innervated by motor cranial nerves and respiratory muscles. IMS leads to death in about 20% of cases and it occurs some days after the patient has recovered from the acute cholinergic crisis.

Repeated absorption of organophosphate at significant dosage, but in amounts not sufficient to cause acute poisoning, may cause persistent weakness, anorexia and malaise (Morgan 1989).

Carbamates are also cholinesterase inhibitors and have a high acute toxicity. Symptoms of poisoning are similar to those produced by organophosphates. However, because of spontaneous hydrolysis of the carbamylated AChE, the symptoms are less severe and of shorter duration. Carbamates penetrate the blood–brain barrier poorly and, therefore, produce minimal effects on brain cholinesterase activity and few central nervous system symptoms.

Synthetic pyrethroids have high insecticidal activity but relatively low mammalian toxicity. They are not cholinesterase inhibitors. Dermal contact may result in skin irritation such as burning, itching and tingling, progressing to numbness. Exposure to large doses results in stomach pain, nausea, vomiting, dizziness, headache, anorexia, fatigue and muscular tremors. In very severe cases convulsive attacks and other symptoms indicating pulmonary oedema may occur. However, reports of symptoms other than skin irritation occur extremely rarely and even oral ingestion of pyrethroids presents little risk.

### Long-term effects of pesticide exposure

Long-term health effects from prolonged exposure to pesticides have been studied in many countries. Suppression of spermatogenesis and increased follicle-stimulating hormone and luteinising hormone have been found as a result of exposure to dibromochloropropane. However, firm conclusions on other adverse effects of chronic exposure, whether to chemical vapours during the manufacturing process or as a result of continuous ingestion of treated food, are difficult to draw at present and require much more research (He & Chen 1999).

### Treatment of poisoning

First aid can be administered when acute poisoning first occurs. The patient should be moved away from the pesticide source and contaminated clothing removed immediately. The body should be washed thoroughly with soap and water, but not scrubbed as this might abrade the skin and provide entry points for the contaminant. If the eye has been splashed with pesticide it should be bathed with clean water for at least 10 minutes. If poison was taken orally the patient should be made to vomit by giving saline solution or by tickling the back of the throat. This should only be done if the patient is conscious. The patient should be kept warm and quiet. If convulsions occur, prevent the patient from hurting him- or herself; biting the tongue can be prevented by placing a tightly folded handkerchief or piece of wood between the teeth. If unconsciousness occurs, ensure the patient can breathe by clearing the airway, turning the person on their side and removing any food or vomit from the mouth, and removing loose dentures.

Symptoms of acute poisoning usually occur within 12 hours of exposure; symptoms of chronic poisoning will be less severe and may be mistaken for other stress reactions. The patient must be taken to a doctor as soon as possible. The doctor must be told the name of the active ingredient and given as much information about the product, preferably by showing the pesticide container and any other relevant literature.

Breakdown of organophosphates occurs mainly by hydrolysis in the liver. However, some compounds, which break down relatively slowly, may be stored in body fat. The primary treatment is to administer atropine sulphate intravenously in order to counteract the effects of excessive concentrations of acetylcholine at neuroreceptors. Atropine does not antagonise the cholinesterase inhibitor enzyme itself. Where poisoning is relatively severe, especially when symptoms are prolonged such as when organophosphate is continuously released from body fat into the blood, pralidoxime, a cholinesterase reactivator, may need to be administered along with atropine. However, pralidoxime is ineffective and may be hazardous in carbamate poisoning.

In severe cases of poisoning when the chemical has been ingested it is necessary to flush out the contents of the stomach and intestine. A slurry of activated charcoal in isotonic saline or solutions of sodium or magnesium sulphate can be used to cause regurgitation (Morgan 1989).

Poisoning from anticoagulant rodenticides should be treated with vitamin A injections. Vitamin C may be a useful adjunct in therapy (Oudejans 1982).

## Insect control

### Contact insecticides

Contact insecticides are solid or liquid formulations which are toxic to insects and which exert their effect when insects come into direct contact with them. Generally, people think of synthetic chemicals when they hear a reference to an insecticide, but of course there are many other substances that have insecticidal properties such as some botanicals and some inert dusts. A number of the natural products used as grain protectants are in fact insect repellents or antifeedants exerting their effect by preventing colonisation of the commodity rather than killing insects that have arrived.

Most of the man-made contact insecticides available for post-harvest use were originally developed to protect field crops and then found to be useful for stored products. The development and registration of pesticides is an expensive process; it costs \$50–100m to develop a new insecticide (Casida & Quistad 1998). These costs can only be recovered through the very widespread use of a pesticide in agriculture or public health; only 4% of total insecticide usage is for the protection of stored produce. The number of insecticides available for use on stored food commodities is also limited by the obvious requirement for very low toxicity to humans when consumed, as some insecticide active ingredients may remain after the commodity is processed.

### Historical perspective

Most insecticide residues will be removed during processing of the raw stored product before reaching the consumer. However, consumer pressure for 'zero' tolerance of chemical residues is one force that is promoting the use of alternative methods. In addition, it is often possible for insects to develop resistance to contact insecticides and the spread of such resistance, together with concerns regarding environmental contamination by pesticides, have inspired a search for more natural and environmentally friendly protectants and more efficient ways of managing their application (Pimentel *et al.* 1993). In consequence, during the last two decades a great deal of research has been undertaken to identify materials or methods which might be used as alternatives.

The popularity of some insecticides has, in a sense, come full circle. Some of the substances first used, such as dusts and botanicals, are currently being either re-evaluated or developed (see later sections). In terms of synthetic chemicals, the trend has been from simple, highly toxic, relatively non-specific chemicals to more complex organic chemicals, of ever increasing specificity to the target pest and of lower mammalian toxicity (see Fig. 6.1). Increasingly, features of the biology of the target pest are being studied and used to identify exploitable weaknesses. For example, to develop, insects need to moult. Some of the hormones that control this

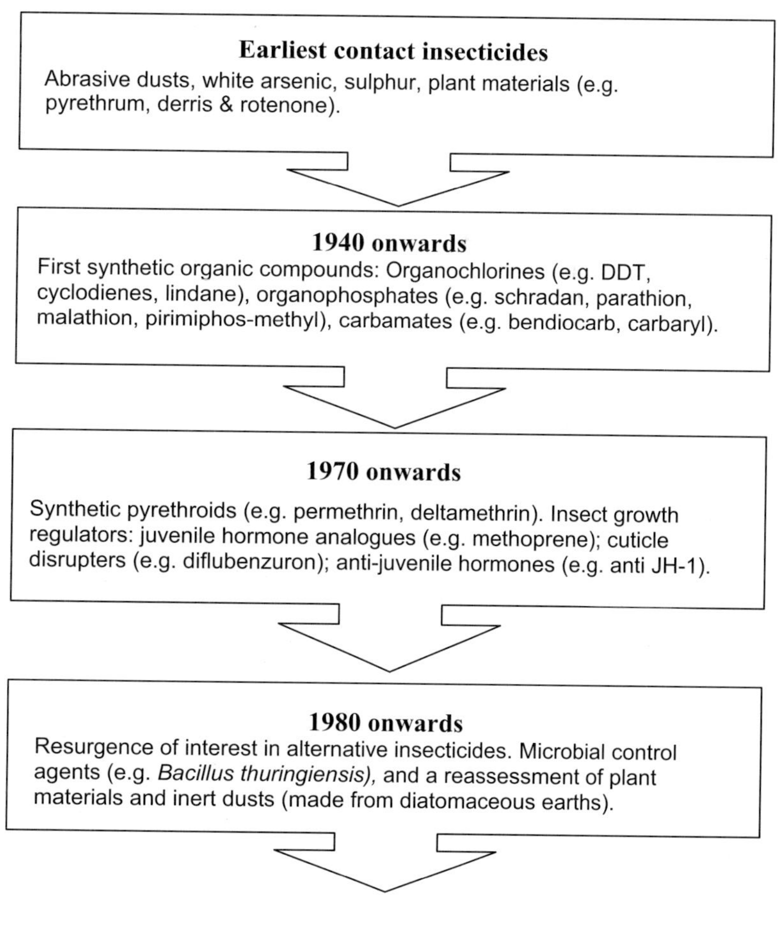


**Fig. 6.1** Historical sequence of different contact insecticide use.

process have been studied, artificially produced and used to disrupt this process, resulting in death of developing juvenile stages. The use of biological control has taken this approach a step further and considers the exploitation of the target pest's relationship with its immediate biotic environment.

In medium to large-scale grain-storage systems, contact insecticides have mostly been applied as residual treatments on the commodity as the grain is loaded into store (Longstaff 1994; Arthur 1996). Such treatments are mostly preventive measures, applied even if there is no evidence of any infestation prior to storage. Contact insecticides are not particularly effective at eliminating well-established insect pest populations (Snelson 1987) and fumigation should be used for this purpose. However, if treated soon after harvest, when pest infestations are relatively light, it may well be advantageous to use insecticides that have fast action, especially when it is inappropriate to use fumigants. Conventional chemical insecticides such as organophosphorus, pyrethroid and carbamate insecticides, and indeed some newer grain protectants such as inert dusts (Golob 1997), can be used in this way. Such fast-acting 'quick fix' insecticides may fit well into emerging pest-management regimes that employ decision-making tools where treatments are dependent on detection of insects through monitoring. Many of the newer treatments such as insect growth regulators and various biological control options may not be fast acting enough to be used as 'quick fixes', and may be used in the kind of initial preventive role previously filled by conventional chemical treatments (Arthur 1996). For example, methoprene (an insect growth regulator) could be used as part of a mixture of active ingredients that may replace malathion as a storage protectant (Daglish *et al.* 1995).

Innovative use of insecticides does not fit particularly well with storage management in developing countries and progress to change has been slow. In East Africa, for example, research in the 1960s had already pointed to problems with malathion stability as a result of the inappropriate use of local fillers for the widely used dust formulation (McFarlane 1969). Nevertheless, despite the short half-life of the dust, malathion is still formulated locally using the same diatomite filler. It is perhaps ironic that this filler is now commercially exploited as an insecticide in its own right. Recommendations entrenched in local legislation, whether produced pre- or early post-independence, are still followed despite being inappropriate; an example is the application of highly volatile dichlorvos as a residual spray treatment to store surfaces of stores in Ghana and Tanzania. Notwithstanding, change has taken place. Gamma BHC, widely used as a protectant of maize cobs until the 1970s, has been replaced by pirimiphos-methyl because of insect resistance problems and fears that its long persistence would lead to health problems. It will, no doubt, be some while before developing countries move away from well entrenched methods of pest management that are dependent on conventional synthetics, though the displacement of government control of crop marketing in favour of the private sector will facilitate change. An additional stimulus will be the increasing globalisation of markets and accompanying procedures essential for the maintenance of quality and safety necessary to enable countries to export, a stimulus which will enforce conformity on the developing world.

Some desirable properties for insecticides are:

- fast action;
- highly toxic to target species (i.e. insect pest);
- non-toxic (or at least of low toxicity at expected levels of exposure) to beneficial and non-target species, e.g. humans, other mammals, birds, fish, non-target insects;
- cost-effective;
- absence of smell, taint or toxic breakdown products;
- non-irritant, non-corrosive;
- easy to formulate/apply;
- long shelf-life;
- effective against resistant strains of insect;
- effective under a wide range of environmental conditions;
- non-flammable; and
- relatively persistent after application (should remain effective for entire storage season).

The perfect insecticide that exactly meets all the criteria above remains to be discovered. All insecticides have some drawbacks. Some of the earliest pesticides used in storage, such as arsenic-containing compounds, were very toxic to non-target species. The organochlorines seemed to provide a cheap solution, but they are now known to be dangerously persistent in the environment, and bioaccumulate in the tissues of living organisms, ultimately reaching very high concentrations in the bodies of top predators. The exact nature of the toxic consequences of such doses are not known. However, human fatalities from acute doses through accidental consumption of undiluted product have occurred and some chronic effects have been detected such as the thinning of eggshells in predatory birds and effects on child development (Carson 1962; Van Straten 2001).

Very often there is a trade-off between desirable characteristics of the insecticides where, for instance, cheaper ones may be more toxic to non-target species. Different features will be of particular importance to different markets. For example, dichlorvos is highly volatile and so has a significant vapour action that can be very effectively used for space application in stores to control moths and flying stages of beetle pests, but its general level of toxicity makes its use inappropriate for direct application to grain. Different insecticides might fulfil different roles in an integrated approach where sometimes a quick-fix immediate 100% kill might be required, and at other times a more preventive treatment might be needed.

#### *Impact on the environment of the use of chemical insecticides in post-harvest*

> 'It seems to be generally assumed that since the compounds are used in small quantities in confined areas, and since most of them decompose into non-active substances there are no adverse environmental effects. Although this is a reasonable assumption, there is little concrete evidence either for or against this widely held opinion.'
>
> *http://www.fao.org*

### Currently used conventional chemical insecticides

#### *Chemistry of the active ingredients*

All conventional chemical insecticides that are currently approved for use in the post-harvest system contain synthetic organic compounds as their active ingredients (a.i.) (Fig. 6.2). We have already said something about the organochlorines such as DDT, that are now generally banned for use in food storage, although lindane, another organochlorine, is still available and used in some tropical countries even though it may not be officially approved for this use.

Synthetic organic chemicals that are currently approved for use fall into one of three groups: organophosphates, carbamates and synthetic pyrethroids. A good overview of the chemistry of these and other groups can be found in Ware (1989). Up-to-date information on the toxicity, to both target and non-target species, and uses of pesticides is regularly collated in *The Pesticide Manual* (Tomlin 2000) and in reports from the Joint Meeting on Pesticide Residues as submitted to the committee on pesticide residues of the Codex Alimentarius Commission.

#### *Organophosphates (OPs)*

These compounds are chemically relatively unstable compared to the very persistent organochlorines that preceded them. They all contain phosphorus and are derived from phosphoric acid.

The organophosphates can be divided up into three groups according to which type of molecules branch off from the phosphorus atom: the aliphatic, phenyl and heterocyclic derivatives:

*Aliphatic derivatives*: These all have straight carbon chains off the phosphorus atom, rather than any kind of ring structures. Malathion is one of the oldest aliphatic organophosphates and was registered for use on stored grain because it is one of the least toxic to mammals. Dichlorvos is both an insecticide and acaricide that was introduced in the mid-1950s. Dichlorvos is available in many formulations from emulsifiable concentrates (ECs) to fogs, but has short residual stability.

> 'The most spectacular development came in the early sixties, with the introduction of malathion, a residual insecticide that soon became the basis of all grain protection in Australia. The euphoria associated with the initial success did not last long, however, and its demise as a grain protectant in Australia began as early as 1968, when a resistant strain of *Tribolium castaneum* was found on treated peanuts in Queensland.'
>
> *Longstaff (1994)*

*Phenyl derivatives*: All phenyl derivatives contain a benzene ring that has one of its hydrogens displaced to allow linkage to the phosphorus component. Some of the other hydrogens of the benzene ring may also be displaced by other groups such as $NO_2$ and $CH_3$ in fenitrothion. Fenitrothion is a broad-spectrum insecticide that is used on stored products, but, in common with many organophosphates, it is not particularly effective against Bostrichidae beetles such as *Rhyzopertha dominica* and *Prostephanus truncatus*. For this reason, some formulations of fenitrothion include a synthetic pyrethroid that has good action against bostrichids.

*Heterocyclic derivatives*: These all have ring structures whose ring is made up of two or more types of

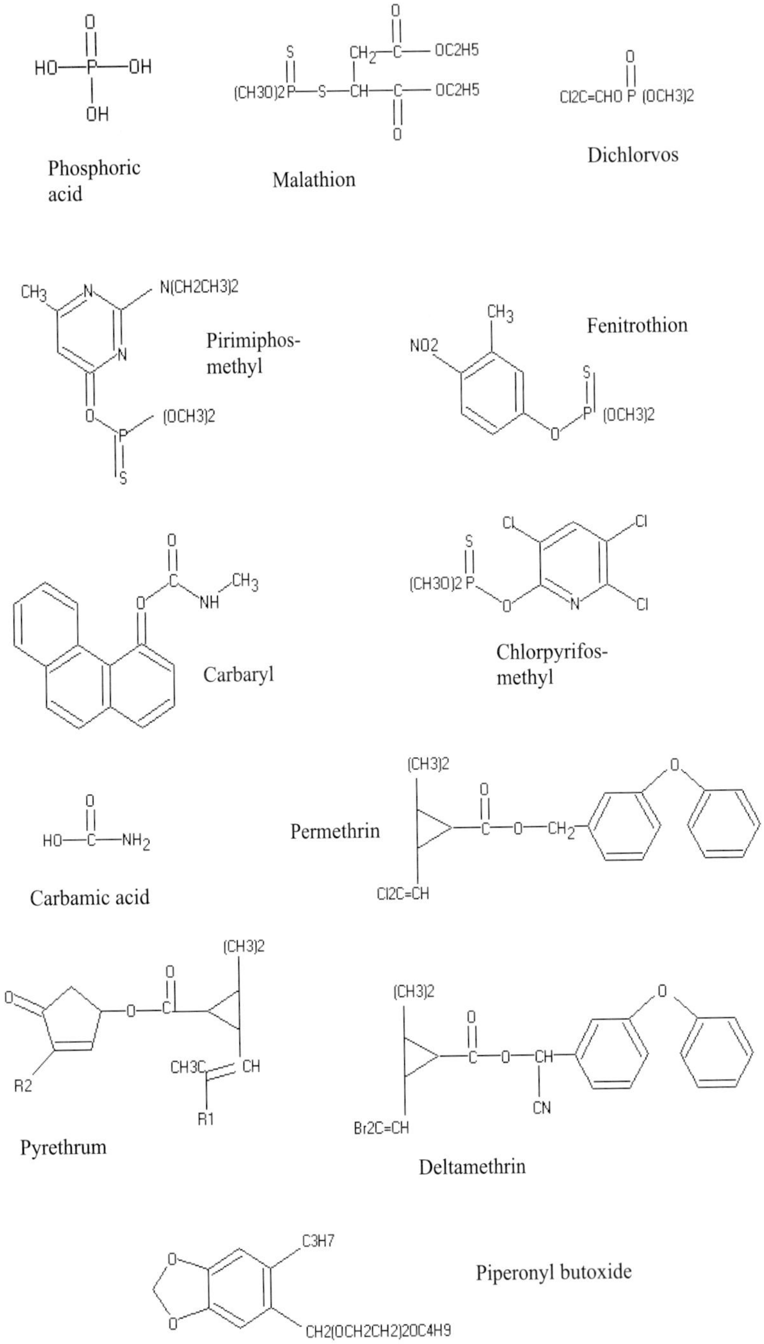


**Fig. 6.2** Structures of active ingredients of current, conventional insecticides.

atoms. For example the chlorpyrifos-methyl molecule has a six-atom ring where five of the atoms are carbon and the sixth is nitrogen (N). Pirimiphos-methyl has two nitrogen atoms in its ring structure. Both chlorpyrifos-methyl and pirimiphos-methyl are broad-spectrum insecticides of relatively low mammalian toxicity.

*Carbamates*

In the same way that organophosphates are derivatives of phosphoric acid, carbamates are derivatives of carbamic acid. Carbaryl was the first very effective carbamate to be used as an insecticide and its structure is described above. It was one of the first insecticides to show good action against *Rhyzopertha dominica* before synthetic pyrethroids took over (Snelson 1987). Carbaryl has only limited approval by CODEX and its use is due to be phased out in 2005.

*Natural and synthetic pyrethroids*

Pyrethrum is the oily extract (containing approximately 2% active ingredient) from the pyrethrum flower *Tanacetum cinarariifolium* (formerly *Chrysanthemum cinerariaefolium*) that has good insecticidal effect on a broad range of insects and has low mammalian oral toxicity. Pyrethrum is actually a mixture of six compounds that all have the same basic structure of a pentose ring, but have different groups attached as $R_1$ and $R_2$. Pyrethrin I and II, making up more than 70% of the mixture, are highly insecticidal whereas cinerins I and II (19%) and jasmolins I and II (8%) are much less active but contribute to the efficacy of the mixture. One constraint of this natural product is that insects, which are rapidly immobilised by pyrethrum and 'knocked down', often recover after several hours as they have the ability to detoxify the small amount of pyrethrum they receive. Pyrethrum is also fairly unstable and breaks down quickly after application, particularly if exposed to sunlight. However, when combined with synergists such as piperonyl butoxide, some of these constraints are reduced.

Cheaper and more stable pyrethrin-like compounds have been developed, the synthetic pyrethroids. Various 'generations' of synthetic pyrethroids have been produced of generally increasing insecticidal activity and increased photo-stability, making them more useful as residual treatments on grain (Ware 1989). Bioresmethrin is one of the second-generation pyrethroids that has been extensively used in Australia for post-harvest applications in wheat stores (Snelson 1987). Permethrin and deltamethrin are structurally similar synthetic pyrethroids that are classed as third and fourth generation respectively. Deltamethrin is active at lower doses than permethrin (Golob *et al.* 1985); the cyano group in the deltamethrin molecule is thought to contribute to this difference (Davies 1985). Synthetic pyrethroids such as permethrin and deltamethrin are particularly effective against Bostrichidae beetles such as the grain borers, *Rhyzopertha dominica* and *Prostephanus truncatus* (Golob *et al.* 1985). The synthetic pyrethroids are often formulated with an organophosphate component such as pirimiphos-methyl, to provide good activity against both grain borers and other storage pests such as *Sitophilus* spp.

*Synergists or activators*

These are compounds that are not particularly toxic in themselves but whose presence can enhance the toxicity or stability of some insecticides. They represent an additional cost and so are only used where they have most impact. As we have already mentioned, some pyrethroids are more effective with the synergist piperonyl butoxide. Piperonyl butoxide is thought to increase the toxicity of pyrethroids by disrupting the breakdown or metabolism of insecticides by mixed-function oxidase enzymes in microsomes in the insects' body cells (Keserü *et al.* 1999). Piperonyl butoxide is therefore particularly useful in killing those individuals who are resistant to pyrethroids because they possess higher than normal capacity to metabolically detoxify these insecticides.

### *Modes of action*

*Entry into the insect*

Conventional synthetic chemical insecticides used as grain protectants all function as a result of direct contact with the insect body. Insecticides can enter insects by mouth, by penetration of the cuticle, particularly through the feet, and through the spiracles (air passages) (Fig. 6.3). Different types of formulation may be taken up in different ways by the insect. For example, a surface treatment like an emulsifiable concentrate (EC) applied to a wall may be taken up by through the insect's feet, but a dust admixed into the commodity may be ingested (they are not, however, stomach poisons). Oil-based formulations are particularly good at penetrating the waxy cuticle. Once the insecticide has entered the insect it is transported to the site within the insect body where poisoning occurs. Inert dusts, such as diatomaceous earths, are an interesting exception. Diatomaceous earths exert their influence by disrupting the cuticle itself and therefore do not need to penetrate into the insect body to be effective.

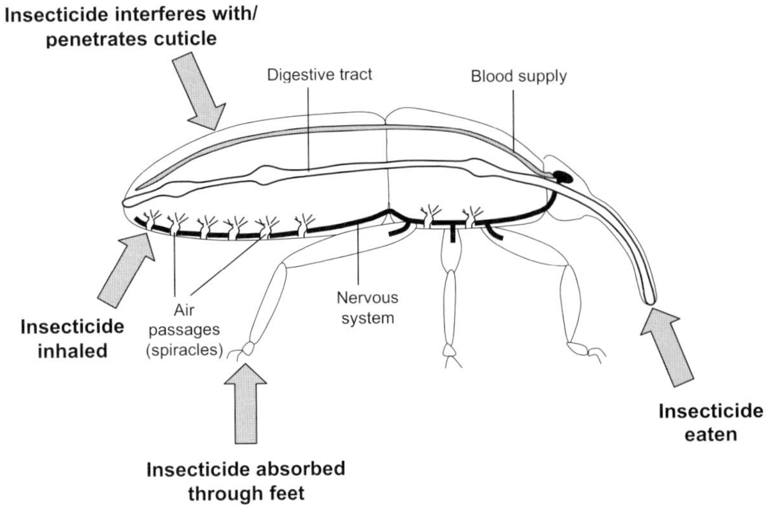


**Fig. 6.3** Cross-section through a simplified insect showing common points of insecticide adsorption/ entry into insects and the transport systems (e.g. blood supply) that can transport insecticides to their site of action.

### *Nerve poisons*

Almost all chemical contact insecticides in use at the moment are nerve poisons. These cause disruption of the passage of nerve impulses from the central nervous system to the muscles. Nerve impulses are transmitted by nerve cells in two ways: as an electrical signal down the axon, and as a chemical movement across a synapse between two cells (see Fig. 6.4). Different types of contact insecticides have different effects on the nervous system. Indeed, it is thought that some insecticides may act in more than one way and for many insecticides the mode of action, from uptake to chemical mechanism, is not fully understood.

#### *Axonic poisons*

Pyrethroids and DDT-type chlorinated insecticides act by interfering with the electrical signal that passes down the axon of nerve cells (see Fig. 6.4). Insects poisoned by pyrethroids are initially excited, often described as 'buzzing' and then this leads to loss of co-ordination, convulsions and ultimately death. The electrical signal (electrical nerve impulse) that normally passes down the axon is reliant on sodium ions moving across the lipid membrane of the axon into the cell contents. Pyrethroids appear to influence the opening and closing of these channels (see Ware 1989 for more detailed explanation).

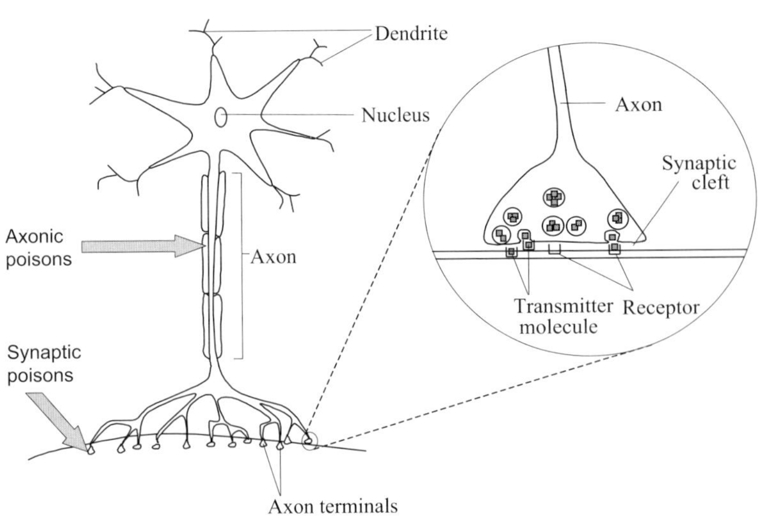


**Fig. 6.4** Site of action of axonic and synaptic poisons, including detail of release of transmitter substances across a synapse. Source: Ware 1989.

Pyrethroids also appear to have some effect on synapses and therefore also act as synaptic poisons.

*Synaptic poisons*

Synapses are the junctions between a nerve cell and other cells (usually other nerve cells or muscle cells) that allow the nerve impulse to be transmitted between these cells (Fig. 6.4). The electrical signal that passes down the axon initiates the release of a chemical (transmitter substance) at the end of the nerve, that moves across the synapse and is taken up by the cell at the other side of the synapse. This process can initiate another electrical signal if the second cell is another nerve cell, or some other appropriate response in muscle or gland cells. Acetylcholine and noradrenaline are two common chemical transmitter substances.

After acetylcholine transmits the nerve impulse it is broken down by an enzyme, acetylcholinesterase (Fig. 6.5). Organophosphorus insecticides and carbamate insecticides interfere with this breakdown of acetylcholine and therefore the transmitter substance

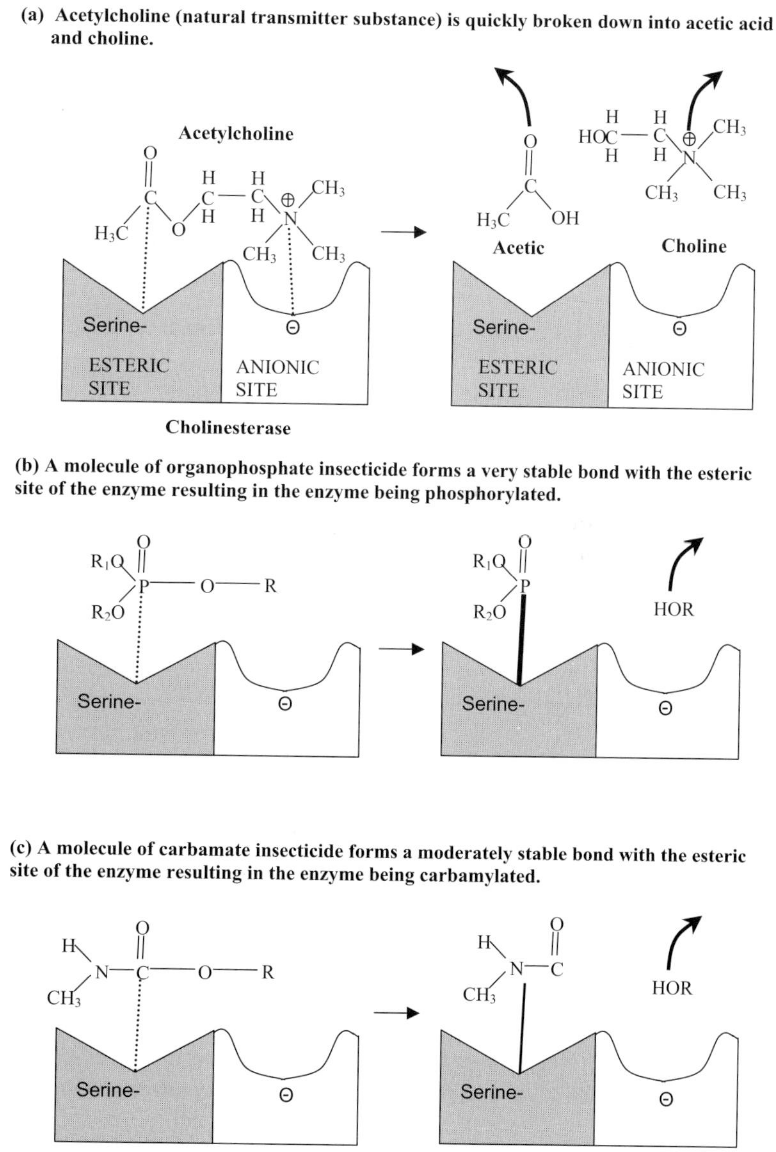


**Fig. 6.5** Interactions on the surface of the enzyme, acetylcholinesterase, with (a) acetylcholine, (b) an organophosphate insecticide molecule, and (c) a carbamate insecticide molecule. Source: redrawn from Ware 1989.

accumulates in the synapse and effectively masks new impulses, causing muscle rigor.

Part of the structures of both organophosphates and carbamates fits extremely well into one half of the acetylcholinesterase enzyme. Whereas acetylcholine, the natural transmitter substance, locks onto the enzyme, is quickly broken into two halves, and then immediately leaves the enzyme, the insecticide molecules break into two halves but one half remains much more firmly attached to the enzyme.

> 'It has been estimated that 300 000 molecules of acetylcholine are destroyed by one molecule of enzyme per minute at 37°C.'
>
> *Hassell 1990*

In the case of organophosphates, the phosphate part of the molecule chemically bonds to the enzyme and can remain there for up to approximately a million times longer than the natural transmitter substance. Carbamates also fit in a stable way into the enzyme, but do not form a chemical bond. This link is therefore more easily broken and the enzyme can resume its natural function much faster (hours rather than days) (Ware 1989).

**Fig. 6.6** Some packets of insecticide dust.

### *Formulations and application*

Contact insecticides are formulated into preparations that enable them to be applied efficiently with appropriate equipment and that enhance their toxic effects on insects. A good detailed account of the differences between formulations and their uses in a wider context beyond post-harvest is given by Matthews (1992).

#### *Dilute dusting powders*

These usually contain between 0.5% and 3% concentration of active ingredient (a.i.), the remaining 97% or more of the dust being an inert filler, for example, talc, which permits the insecticide to be dispersed throughout the commodity. They are particularly used for the protection of grains stored on-farm in developing countries because they are often available packaged in small quantities and the formulation sold is already diluted (Fig. 6.6). However, because of their very nature dilute dusts are prone to being adulterated or even completely replaced with powders that are ineffective. This problem can be overcome by improving the packaging and by good publicity, but in countries where supply of pesticides is uncontrolled the sale of inferior quality products remains a problem.

Dusts with acceptable shelf-lives have been difficult to formulate, so that few of these products are commercially available. The active ingredients most commonly available are pirimiphos-methyl and fenitrothion. In areas where *Prostephanus truncatus* has become established these organophosphates are mixed with either permethrin or deltamethrin dusts, to which this pest is particularly susceptible (Taylor *et al.* 1992).

#### *Emulsifiable (miscible) concentrates*

Emulsifiable concentrates (EC) make use of an emulsifier, an ingredient that is partly hydrophilic and partly lipophilic. This enables the usually lipophilic active ingredient to disperse in water by the formation of very small (usually less than 10 μm in diameter) oily droplets. Commercially available EC is usually sold with a particularly high percentage of active ingredient, commonly varying between 20% and 80% of the formulation, according to supplier and insecticide type. One exception to this is deltamethrin EC, which can be bought at 2% concentration of active ingredient. Emulsifiable concentrates consist of the insecticide in a suitable solvent which is usually then diluted with water for use at 1% or 2% concentration. This type of formulation is most suitable for application to non-absorbent surfaces such as metal or painted wood. It is also used for applying low dosages of insecticides directly onto grain to be stored in flat bulks or silos. The high concentrations of active ingredient in these formulations prior to dilution can make them a hazard if not used correctly.

*Wettable powders (water dispersible powders)*

Wettable powders (WP) consist of very small solid particles of insecticide, together with substances on the surface that enable the powder to mix well with water. These contain between 25% and 80% of active ingredient and, for use, are dispersed in water (initially in a small quantity to form a cream) to attain the required concentration for application, usually 1% or 2%. This type of formulation is more effective than emulsifiable concentrates for spraying structural surfaces which are absorbent in nature, for example, cement, brick, and the outside of bags. However, they are more difficult to use than EC formulations because they are harder to dilute, and the solid particles often block sprayer nozzles. The solid particles also tend to settle out of suspension, and so are often applied with sprayers that incorporate agitators in the spray tank. Wettable powders can leave a noticeable white powder on dark surfaces, which can be unsightly, yet an advantage if the pest controller needs to easily see which areas have been treated.

*Suspension concentrates (flowables)*

Whereas wettable powders are packaged in powder form ready to be mixed with water, suspension concentrates (SC), sometimes called flowables, consist of solid particles of insecticide already suspended in a liquid. Initially, settling out of the solid particles during storage limited the shelf-life of this kind of formulation, however progress is being made on solving this problem (see references in Matthews 1992). One advantage of suspension concentrates over wettable powders is that pest controllers generally find liquids easier to measure out and work with than powders.

*Space treatments*

Space treatments must not be confused with fumigations. Space treatments are not undertaken with gases and do not penetrate well into grain. For this reason their main application is to kill flying insects and some of those on the surface of grain. The application of space treatments can be particularly hazardous and therefore requires specifically trained staff.

*Smoke treatments*

Smoke treatments consist of solid particles of chemical insecticide released when the formulation is burnt (seen as large quantities of white smoke), therefore the insecticide must not be one that is readily destroyed by heat. The formulation contains an oxidant and combustible material that generates a lot of hot gas very quickly. High speed of burning is important to reduce the time that the active ingredient is exposed to potentially damaging high temperatures. The insecticide smoke is carried to all parts of an enclosure by the movement of the hot gas produced and eventually settles on the upper parts of horizontal surfaces but not on the vertical surfaces or the underside of horizontal surfaces.

*Fog formulations*

In fog formulations the insecticide is dissolved (or sometimes suspended) in a solvent and, for use, is diluted with odourless kerosene or a heavier technical white oil. Oil sprays penetrate the bodies of insects more rapidly than water-based sprays and may have a quicker visible effect. They are suitable for application as aerosols, mists or fogs. They must not be employed in situations in which naked flames are present because of the fire risk.

*Formulations containing more than one active ingredient*

Insecticide formulations sometimes contain more than one active ingredient, perhaps one to control bostrichids and another to control other families of insect. In Australia, mixtures of pyrethroids, such as synergised bioresmethrin or (1R)-phenothrin, with OPs including pirimiphos-methyl, chlorpyrifos-methyl or fenitrothion, were found to be effective in protecting wheat against *Rhyzopertha dominica* and non-bostrichids including *Sitophilus* (Bengston *et al.* 1980a, b). Mixtures of pirimiphos-methyl and permethrin have been recommended for use in both East and West Africa for treatment of grain infested with *Prostephanus truncatus* and other storage insects (Dales & Golob 1997; Richter *et al.* 1998).

Mixtures of active ingredients may reduce the risk of the development of insecticide resistance. This is because an insect would need to combat more than one active ingredient to survive. This is less likely than an insect being resistant to just one insecticide, especially if the mixture is made up of chemicals with different modes of action.

The active ingredients in mixtures could interact as protectants and produce lower rates of mortality than that produced by the active ingredients alone (antagonistic effect) or higher mortalities than the active ingredients alone (additive or synergistic effects). No such interactions between active ingredients in insecticide mixtures were demonstrated between four different organophosphorus insecticides and methoprene or deltamethrin (Daglish 1998).

As a general rule, insecticides should not be mixed unless recommended on the label. Local regulations also need to be checked before mixing.

## The main uses of insecticides in storage

### *Surface treatment of store structures*

In this usage insecticide is applied to the structure of buildings or the surface of bags in a bag stack so as to kill any insects which may be present and to prevent re-invasion (Plate 4). This is a very common use of insecticides for the protection of stored products, and requires insecticides to be both effective and sufficiently stable to have a residual action for an adequate period of time after application. Often the insecticide is not stable enough on surfaces to give lasting protection, particularly in hot climates (Gudrups 1996). There is also concern that some insects may not spend long enough on the treated surface to pick up a lethal dose (Hodges *et al.* 1992). Further, once insects have penetrated the insecticide barrier, they may build up in numbers under the surface and will be out of reach of any subsequent surface treatment. On a more positive note, surface treatments of a store can be very useful as an effective method of killing the last pests in an empty store which are missed by sweeping, before it is loaded with fresh commodity. Surface treatment of stores is also recommended for use in conjunction with fumigation. Insect numbers around an infested commodity are likely to be relatively high. Surface treatment around the commodity either before or immediately after its fumigation (before the sheet is removed) will reduce the risk of immediate reinfestation of the fumigated commodity.

### *Direct admixture with commodities*

The commonest use of insecticides in this manner is admixture with raw grains (Plate 5). Insecticides generally decay by oxidation at a slower rate when admixed with the grain compared to when they are applied to store surfaces. A major constraint in direct application of insecticides to food commodities is that of ensuring that there is no hazard to the consumer due to toxic residues.

### *Application to air spaces*

This type of treatment is aimed at the control of flying insects (Plate 6). It has been found to be of particular use in certain situations in warehouses where flying insects are a particular problem. However, since space treatments do not penetrate well into a stack or bulk of commodity, their effect is mostly superficial. Use is usually limited to warehouses containing high-value commodities such as tobacco or cocoa, and where visibility of even a light moth infestation could cause downgrading in value of the commodity. Space sprays are applied at regular, short intervals, frequently daily or weekly, and are timed to catch the insects' maximum flight activity, usually at dusk. Treatments are often used in combination with fumigation to ensure that immigrating adults do not have an opportunity to become established.

### *Application equipment and procedures*

Many dust formulations can be applied directly to commodities, however, formulations like emulsifiable concentrates and wettable powders need to be diluted before application. The recommended proportion of formulation to diluent should be given on the insecticide label. Labels often give a range of diluent quantities; this reflects the variation needed depending on the absorbency of the area to be sprayed.

#### *Hand dusters*

Very simple yet effective hand dusters for small-scale use can be made using a tin or a piece of sacking (Plate 7). Often a couple of small stones are placed in the tin to dislodge any dust that is stuck in the corners. The tin dispenser can also be used to sprinkle liquid formulations. Simple bellows or rotary-operated dusters are also available for applications of dust by hand.

*Motorised dusters*

Motorised knapsack spraying machines can be fitted with an attachment for dispensing dusts and are suitable for use in situations where the hand-operated dusters are too small or tedious to use. Electrically operated vibrating dust dispensers can be used to trickle dust onto grain on a moving conveyor loading silos or flat storage. The rate of dust feed is set to match the grain flow in order to treat at the required dosage level.

### *Dusting procedures*

*Shelled grain (Plate 8)*: The simplest method of applying insecticide dusts to relatively small quantities of grains is to mix it into a heap with a shovel or to tumble the grain with the dust in an adapted oil drum. If the floor is concrete the mixing can be done on the floor itself. If the floor is earth the grain must be emptied onto some kind of covering like a metal sheet, tarpaulin or polythene sheet. For amounts above 100 kg it is useful to mix the insecticide in by shifting the heap of grain from one spot to another, repeating three or four times until the insecticide is properly mixed and no patches of powder can be seen. For larger quantities, dust would usually be dispensed over the commodity as it is mechanically loaded into store.

*Grain on the cob or unthreshed commodities (Fig. 6.7)*: Dusts are best applied to cob grain or to unthreshed commodities such as maize, sorghum and cowpeas by applying between each layer of commodity as a store is filled.

### *Spraying equipment*

*Hand operated hydraulic systems (Plate 9)*: A hand pump giving a spray at high pressure is useful in treating small areas with insecticide. Such equipment is suitable for farmers treating cribs and village stores. Extra spray liquid can be contained in a bucket linked to the sprayer with tubing. The operator carries no load on their back.

*Hydraulic knapsack sprayers (Fig. 6.8)*: A variety of simple hand-operated knapsack sprayers are manufactured. The hydraulic type requires constant pumping action of a forward projecting handle. The force generated by pumping directly expels liquid from the sprayer using the hydraulic pressure exerted on the liquid since liquids are relatively incompressible. A long lance with a trigger control allows the sprayer to treat reasonably high targets. Capacity of these sprayers is usually 15–20 litres. Hand-operated knapsack sprayers are suitable for dispensing water-based emulsions and suspensions or light oil sprays, in order to surface treat walls, floors, sacks, and so on. Pressure-regulating valves should be fitted to these and to pneumatic sprayers in order to allow a uniform spraying rate throughout the operation.

*Pneumatic pre-pressurised sprayers (Plate 10)*: Pneumatic knapsack sprayers have the advantage in that it is possible to pressurise the tank by a hand pump before the sprayer is mounted on the back of the operator. The energy from pumping is stored by pressuring an airspace above the liquid insecticide. Usually such sprayers cannot be filled more than two-thirds full to enable a large enough airspace to provide enough pressure to empty most of the tank. Spraying can then be undertaken using the trigger-fitted lance without further pumping action. This allows the operator to retain one hand free during the spraying procedure. However, due to the high pressures involved, sprayers usually have metal tanks that can be very heavy. Capacity is usually 10–15 litres.

*Motorised knapsack sprayers (Plate 11)*: For the application of residual films of insecticide to large areas, high walls and ceilings, a motorised knapsack sprayer is useful. These are powered by a two-stroke engine, the insecticide entering the air-stream from the engine at the nozzle, being broken into fine droplets and forced out in a continuous stream. In some models, the rate of insecticide dispensed is reduced as the nozzle is raised to treat elevated areas, and account must be taken of this factor to ensure the proper treatment rate is achieved. In practice it is nearly impossible to spray a wall accurately with this type of sprayer, and the operator is liable to be contaminated with drifting and bouncing spray particles. The capacity of motorised knapsack sprayers is usually about 10 litres and, when full, the complete sprayer weighs about 12 kg.

*Portable large-scale spray units (Plate 12)*: For very large-scale surface treatments, insecticide can be dispensed from a powered hydraulic pump unit that may be driven by an engine or electric motor. For convenience and mobility large sprayers may be mounted on wheels. These are much more controllable and therefore accurate than the motorised knapsack sprayers.

*Space sprayers*: These are usually electrically operated and produce a fine mist or aerosol for control of fly-

**Fig. 6.7** Instructions on how to apply dust formulations of insecticide onto maize on the cob, intended for small-scale farmers in Africa.

ing insect pests. Battery-operated and petrol-driven misters allowing full portability are available. Some machines can be adjusted to produce a much coarser spray and can be used to apply surface residual treatments.

*Thermal fog generator*: Control of flying insect pests in storage buildings can be achieved by the use of fogs containing chemicals such as pyrethrins or pirimiphos-methyl. The heat of the exhaust from a resonant jet engine vaporises the chemical contained in the oil. Fogs are created when the heavy oil condenses giving very fine droplets that completely fill any enclosure very rapidly. There is little deposition or wetting of surfaces by oil. These are best used in closed warehouses where the fog can be retained well for several hours. Machines may be portable, or trolley or vehicle mounted. Large machines are used to treat ships' holds and tobacco warehouses.

### *Liquid application to grain during loading*

The application of residual insecticide to grain on loading into a silo or other bulk store can be conveniently

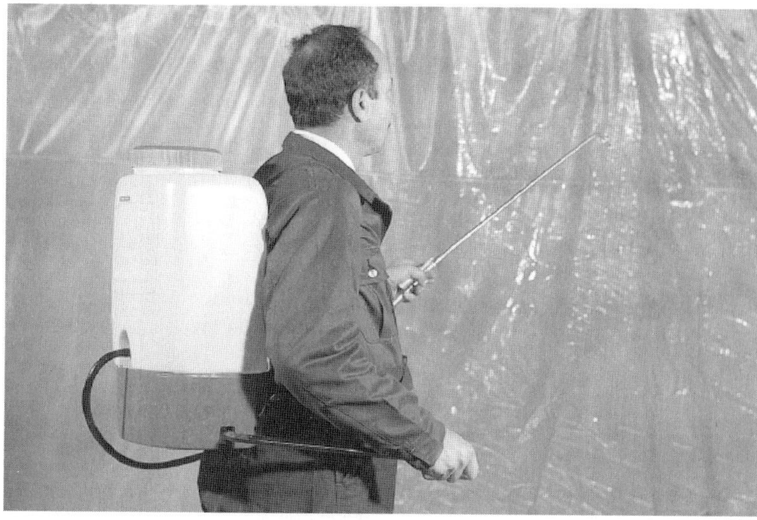

**Fig. 6.8** A knapsack sprayer in use.

undertaken using an emulsion spray in water. An electrically operated pump unit can be employed, the spray being applied at the appropriate rate to a grain stream on a moving conveyor. Sprays can also be applied to large flat bulks using a boom sprayer mounted onto a tractor. Spraying is often preferred to dusting since there is less hazard to operators than when dust is applied to a similar intake system. At the recommended insecticide application rates, the moisture content of grain treated with a water-based emulsion spray need not be increased by more than 0.1–0.2%.

### *Losses during application*

Some methods of pesticide application in food storage are more efficient than others. Any pesticide lost to the environment is not only an economic waste but a potential environmental hazard. Windy conditions during spraying or sprinkling should obviously be avoided. However, even when extra care is taken, loss of active ingredient will occur. This is particularly important when small quantities of grain are being treated because a small loss of dust, for example, can result in a proportionally large loss of the active compound. Even under well-controlled experimental conditions the application of dilute dust insecticide invariably results in at least 40% loss of active ingredient (Golob & Hanks 1990), which may still be effective immediately after treatment but will not remain so for very long. More than 30% of the active ingredient of permethrin and pirimiphos-methyl applied to bulk stored wheat or sorghum stored on commercial silos in Australia was lost within a week of application, which could have been due to immediate metabolism of the compounds on contact with the grain or more probably the result of loss during application (Bengston *et al.*, quoted in Snelson 1987). Similarly, in the UK applications of synergised permethrin to wheat on a conveyor achieved only 35–55% of the nominal dosage (Halls & Periam, quoted in Snelson 1987).

Sprays applied to stores are not prone to losses as treatments of walls are carried out to the point of 'run-off', that is the spray is applied until the liquid particles coalesce and begin to run down the wall. Care should be taken to avoid overdosing in this situation.

### *Environmental influences on the efficacy of insecticide treatments*

Of course, once the insecticide has been applied, this is not the end of the story. Environmental conditions can have far-reaching impact on the success of any treatment. Parameters such as air humidity, grain moisture content and temperature can all influence the biological activity of insecticides in a variety of ways (Fig. 6.9). Mostly conventional chemical insecticides are more toxic at higher temperatures. The exceptions are pyrethrum and synthetic pyrethroids that are often more toxic at lower temperatures (Harris & Kinoshita 1977).

Diatomaceous earths (DEs) may become either more, or less, toxic to insects with increasing temperature, depending on species. The increased toxicity may be because of increased insect movement, therefore more

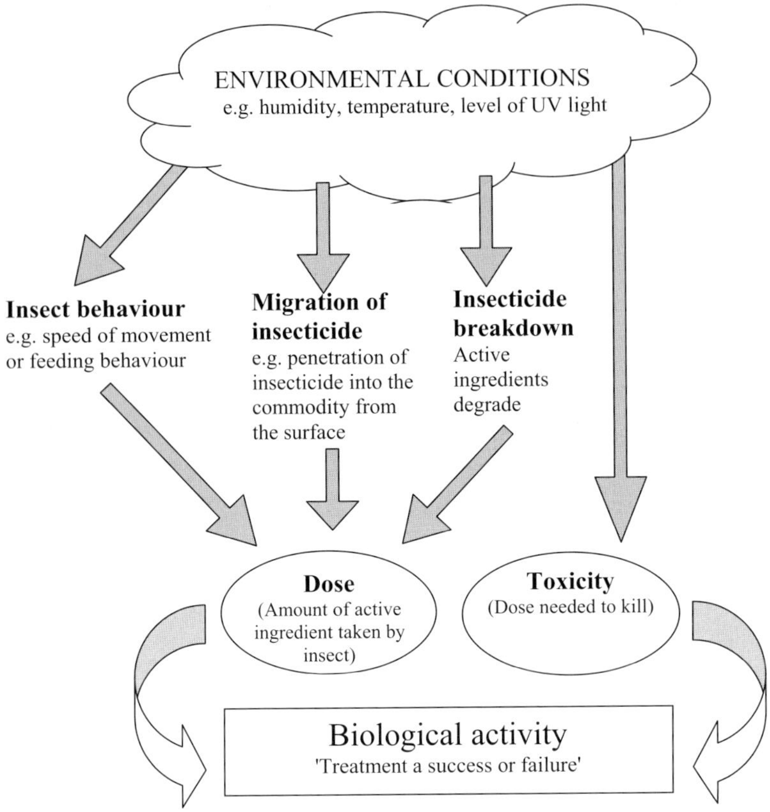


**Fig. 6.9** Influence of environmental conditions on the biological efficacy of an insecticide treatment.

contact with DE, and greater movement also means more water loss, especially through the spiracles. But increased temperature also means more feeding, which could replace water loss. With DE there is no confounding effect of degradation of the insecticide (Fields & Korunic 2000).

As well as altering toxicity, different environmental conditions can alter the rate at which insecticides degrade. This is obviously a particularly important consideration for treatments that are intended to remain effective over extended periods of time. Even before pirimiphos-methyl was approved for use in stores, it was shown that compared to malathion, bromophos and iodofenphos, it was chemically more stable over time, particularly at higher grain moisture contents (Mensah & Watters 1979) (Fig. 6.10).

Fleurat-Lessard *et al.* (1998) conducted an experiment to study how two factors, insecticide breakdown and biological activity, change with time under different environmental conditions in relatively small (2 kg) lots of wheat. They found that the breakdown of chlorpyrifos-methyl increased with increasing temperature and water activity. The toxicity of the grain against a primary and secondary beetle pest of wheat was not, however, closely correlated with insecticide residue analysis. The authors suggest that as the moisture content of grains increases, insecticide on the surface of grains tends to migrate to the inside of each kernel away from the surface, thereby changing the rate of exposure of insects to the insecticide. This suggests that such 'availability' of insecticide may be more important in determining the protection afforded to the grain than the total quantity of insecticide as measured by residue analysis on the grains as a whole.

## Resistance

Resistance is the acquired ability of some insect species to survive insecticide treatments, which has arisen through artificial selection of the most resistant individuals of a population through pesticide use (Fig. 6.11).

Insecticide tolerance is the inherent ability of some insect species or stages of insects (for example, larvae)

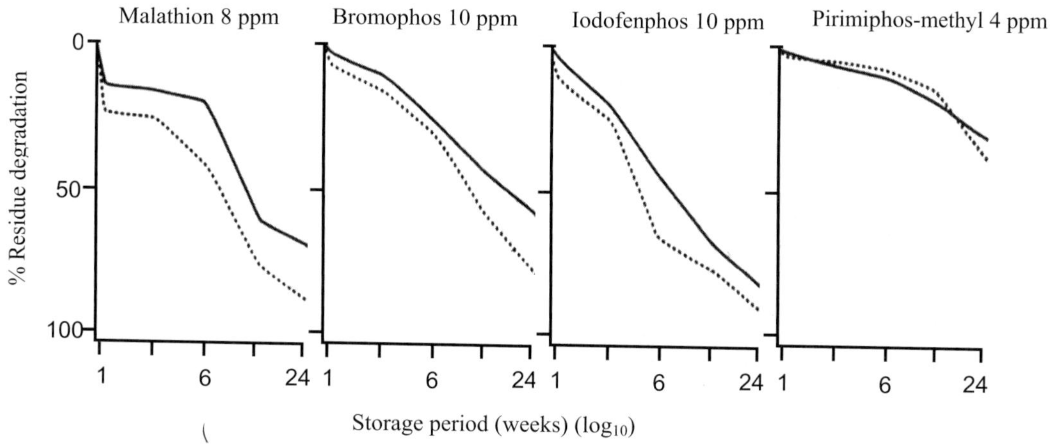


**Fig. 6.10** Degradation of insecticides. Source: redrawn from Mensah & Watters 1979.

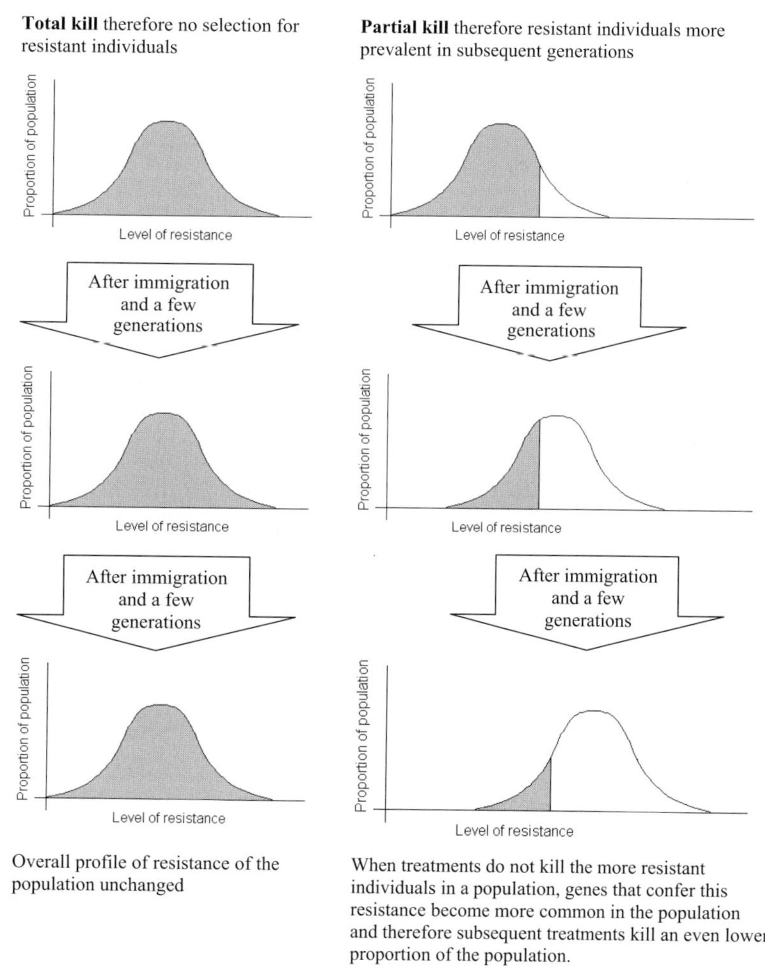


**Fig. 6.11** Selection for insecticide resistance in a population exposed to faulty insecticide applications that leave the resistant portion of the population alive. This simplified scheme describes the possible case where resistance is conferred by the additive action of many genes. NB. Shaded area under the graph indicates those individuals killed by the treatment.

to survive insecticide treatments, that is, without natural selection following exposure to insecticides.

It is important to remember that insect resistance should not be suspected immediately as the most likely cause of an insecticide control failure. It should only be considered as a likely potential cause after eliminating other possibilities such as sub-standard insecticide, mistakes in insecticide dilution, or faulty application.

In the early 1970s a worldwide survey of contact insecticide and fumigant resistance was carried out by the FAO (Champ & Dyte 1976). It was apparent at that time that resistance to the then commonly used insecticides, lindane and malathion, was widespread.

Two types of malathion resistance are known throughout the world. One form is specific to malathion, whereas the other involves cross-resistance to various other organophosphorus insecticides. Both types are widely distributed (Giga & Mazarura 1990).

Fast development of insect resistance to insecticides is often associated with a history of poor quality pest management, where insecticide has been applied at the wrong dose and/or in the wrong situation (Fig. 6.11).

### Types of resistance

#### Morphological

Changes in the structure of the insect can potentially confer resistance to insecticides, for example, the insects may develop a thicker cuticle that enables them to absorb less of the insecticide (Fig. 6.12a).

#### Vigour tolerance

The overall size of the insects may be increased, thus allowing them to reduce the amount of insecticide absorbed per unit body weight (Fig. 6.12b).

#### Biochemical

The insects may develop internal chemical defences against the insecticide. This could include a novel enzyme that detoxifies the toxin, or perhaps a new waste product that happens to combine with the toxin making it easy to expel it from the insect's body. An important family of enzymes involved in insecticide detoxification by metabolism are the P450 group. Resistance to insecticides is often conferred by increased activity of P450 enzymes (Feyereisen 1999) (Fig. 6.12c).

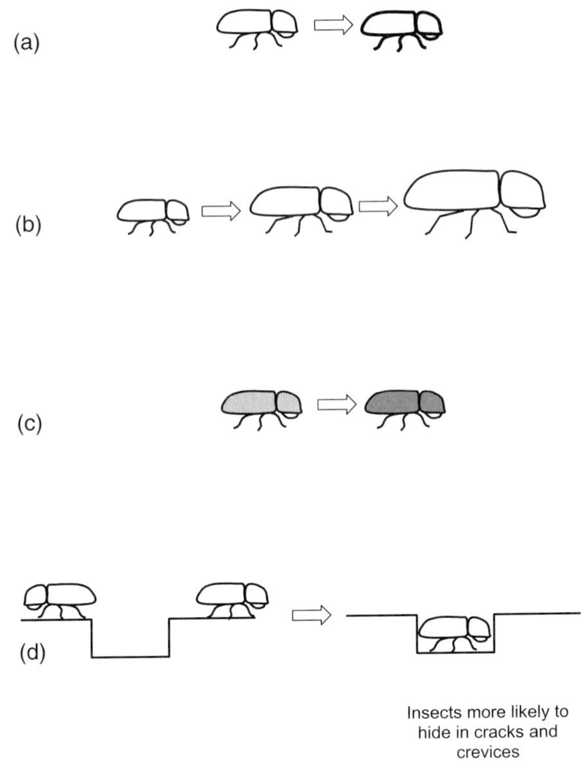


**Fig. 6.12** Types of insect resistance.

#### Behavioural

In common with other types of resistance, behavioural traits can be heritable (passed on to the next generation via genes). Some behaviours may result in insects reducing the chance that they will be exposed to the insecticide, for example, spending more time within grains rather than moving between grains. Avoidance behaviour (repellency) has been demonstrated in some strains of *Oryzaephilus surinamensis* against permethrin (Watson & Barson 1996) (Fig. 6.12d).

Mechanisms of pyrethroid insecticides (and DDT) have since been studied in detail using the cotton bollworm (a pre-harvest pest of cotton). Pyrethroid insecticides are toxic because they are axonic poisons that result in continuous firing of the nerve impulse. At least three mechanisms have been identified for counteracting pyrethroids: delayed penetration of insecticide through the cuticle, nerve insensitivity and enhanced metabolism of the active ingredient with mono-oxygenases and esterases (Gunning *et al.* 1991; McCaffery *et al.* 1997).

Synergists such as piperonyl butoxide have been shown to enhance the toxicity of pyrethroids by slowing the insects' metabolic breakdown of the active ingredient (Keserü *et al.* 1999).

*Cross-resistance*

This occurs when resistance gained to one insecticide also confers resistance to another insecticide. Obviously some forms of resistance, such as behaviours that make insects less likely to come into contact with surface applications of insecticide, will confer some protection against many such applications. Even biochemical resistance gained by a population from poor treatment with one chemical may result in a population with increased capacity to detoxify similar chemicals.

Laboratory insecticide-resistance selection studies over 25 generations of insects have shown that populations of *Sitophilus oryzae* that are selected for permethrin resistance then also exhibited increased resistance for deltamethrin, even without any exposure to this chemical. The reverse situation was also true: insects selected for deltamethrin resistance were then also resistant to permethrin and both strains were also found to possess increased resistance to a range of pyrethroids (Heather 1986). Interestingly, the resistance genes appeared to be sex-linked with males more likely to be resistant. Furthermore, this extensive cross-resistance may mean that the use of one pyrethroid could confer resistance to others (Heather 1986).

### *Managing pesticide resistance*

Any pest management practice that exposes insects to sub-lethal doses of insecticides will select for resistant individuals. Uneven coverage during application will result in some under-dosing. Poor store maintenance and hygiene can lead to harbourages for insects which may then only be exposed to the insecticide some time after the treatment, after it has degraded to a sub-lethal concentration.

In theory, at least, the use of more than one type of insecticide against a population of insects could slow the appearance of resistance by exposing insects to conflicting selection pressures. The hope is that those individuals that are able to survive one type of insecticide treatment would be killed by a different one. Therefore, resistance would not spread through the population. Alternating treatment types, applying insecticides as mixtures or treating different volumes of commodity with different chemicals were suggested strategies (Tabashnik 1989). Research so far suggests that alternating treatments does not delay the onset of resistance, compared with continuous use of one insecticide and then another. The use of mixtures of insecticides is thought to be more likely to delay the onset of resistance. However, strict conditions need to be met that may prove to be unrealistic. For example, the two (or more) active ingredients need to be formulated such that they degrade at similar rates, ensuring that one does not remain active in the absence of the other (Denholm & Rowland 1992).

The existence of a large immigrant population of insects that are not exposed to treatment can slow the spread of resistance by reducing the proportion of the next generations of insects that are made up of survivors of an insecticide treatment. Management of resistance may therefore be most effectively conducted on a relatively large geographic scale where the pest control regime operating in one region may influence another. In the broader context of all insecticide usage there has been an increase in the implementation of IRM (insecticide resistance management) through national networks of insecticide providers and users (Denholm & Rowland 1992). The range of influence will depend on the distances over which the particular pest species commonly disperse. In addition to this, some pests exploit a range of hosts and are not restricted to one commodity and may even have considerable populations in the natural environment. For example, *Prostephanus truncatus* has considerable populations away from storage, living in woody plant hosts (Nang'ayo *et al.* 1993; Borgemeister *et al.* 1998) and therefore we would predict that these relatively insecticide-free refuges would increase the chances that resistant individuals will be diluted with susceptible individuals, if individuals commonly migrate between these host types.

Of course the development of novel methods of killing insects will potentially allow future pest managers to keep one step ahead of the evolving pest populations. An ambitious idea is described by Zlotkin (1999) who suggests that insecticides could be developed to particularly target those individuals who are resistant to existing chemicals by exploiting the characteristic that is conferring resistance. Zlotkin suggests that this strategy may be possible for those insecticides that target the voltage-gate, sodium-ion channels in insect nerves.

One of the challenges for large-scale insecticide resistance management is to consider the effects of in-

## Box 6.1 The changing and unchanging face of malathion resistance

Sometimes insecticide-resistant individuals also have lower overall fitness (capacity to increase in number) than susceptible individuals. Therefore, in the absence of insecticide treatments, they tend to become rarer and rarer in the population. Muggleton (1983) demonstrated this in populations of *Oryzaephilus surinamensis* containing some resistant individuals. After two years with no exposure to insecticide the population contained proportionately fewer resistant individuals and Muggleton calculated that some resistant strains were less than 70% as efficient at providing offspring for each subsequent generation as susceptible individuals.

Malathion resistance was, however, found to persist in pest populations in New South Wales in Australia even after pest controllers ceased using it. One type of malathion resistance is known to convey cross-resistance against other active ingredients, including some of the compounds recommended to replace malathion. Therefore, the use of replacement insecticides is thought to have ensured that a high proportion of insect individuals remain resistant (Herron 1990).

secticide applications in many different contexts. Some insecticides have a relatively broad spectrum of activity against many different species and may therefore be used in many contexts. Add to this the phenomenon of cross-resistance and it is evident that pre-harvest applications could influence post-harvest pest populations. Pesticide applications made to combat vectors of disease such as mosquitoes or tsetse flies could also be an important selection pressure.

Some insect species will be inherently faster than others at developing resistance because of their biology. Genes conferring insect resistance will spread more quickly in populations of insects with shorter life cycles. Also, those species with larger populations exposed to sub-lethal doses are more likely to contain individuals that happen to have characteristics which make them less susceptible to the insecticide.

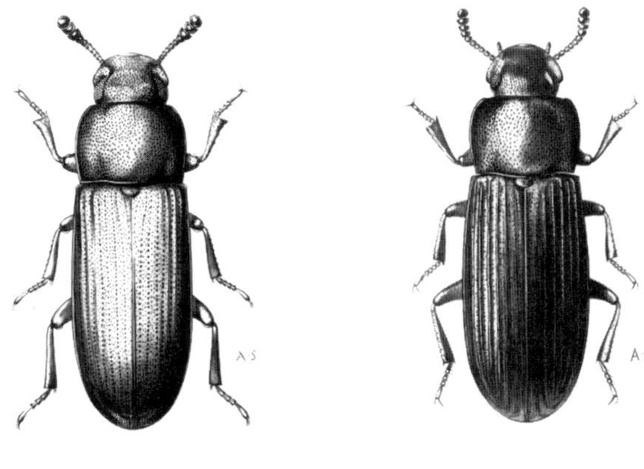


*Tribolium castaneum* *Tribolium confusum*

**Fig. 6.13** Not all *Tribolium* beetles are equal (courtesy of Central Science Laboratory, UK).

Even very similar species can develop insecticide resistance at different rates (Fig. 6.13). The levels of malathion resistance exhibited by several strains of these two *Tribolium* species collected from flour mills in the United States were found to be very different (Zettler and Arthur, 1997). Populations of *Tribolium castaneum* were found to be far more resistant to malathion than populations of *T. confusum*, even though malathion has been used extensively as a control strategy against both these species. The authors propose that the selection pressure against *T. castaneum* may have been higher since this pest predominates in warmer climates than *T. confusum,* where pest control activities are necessarily more intense since these conditions are more favourable for insect development.

### *Detecting resistance to insecticides in an insect population*

Choice of the best method to measure insecticide resistance in a population of insects depends on what information is required. Different levels of precision are required for different possible outcomes (Roush & Miller 1986):

*Low precision*: Determine whether an insecticide treatment has failed because the insects are resistant.

*High precision*: Quantify the degree of resistance in a field population of insects already known to have some insecticide resistance.

*Very high precision*: Detect areas where insect populations contain the 1–4% of resistant individuals that may then quickly spread throughout the population unless action is taken.

Generating dose/response curves and comparing populations using the gradient of these slopes is an inefficient way of detecting differences (Roush & Miller 1986). Far greater accuracy can be obtained from the same sample size if a discriminating dose test is used, that is, test insects against a single dose of insecticide that will kill susceptible individuals.

Strict protocols should be followed when screening for resistance to insecticides since repeatable and comparable results are needed. Generally, test-insects of unknown resistance level are compared to a baseline population of susceptible insects. Insects can be scored for their mortality, or the percentage of beetles that show 'knockdown', after a defined exposure time to a specific dose of insecticide. Knockdown is defined as the inability of insects to stand and walk, but some may recover after time.

There is a range of standardised protocols used for testing for insecticide resistance in collections of insects. FAO have recommended both direct application of specific doses of insecticides onto adult insects (called topical application), and also exposure of insects to discriminating dosages of insecticides applied to filter papers. Details of the features of a 'good' resistance-testing protocol are given by Champ and Campbell-Brown (1970).

Such simplified tests will not detect all forms of resistance. For example, theoretically a population of insects that is relatively resistant to an insecticide because of its increased refuge-finding behaviour will have no opportunity to exhibit this behaviour in such an assay. In practice, in one case, resistance to chlorpyrifos-methyl in different strains of *Rhyzopertha dominica*, assessed by holding them on a treated surface, was not a reliable predictor of the chance of the strain surviving in storage conditions (Arthur 1992).

Rather than indicate resistance in an insect population, standard discriminating dose tests are perhaps more useful in allowing simple comparisons of efficacy to be made between different active ingredients. However, even for this purpose, complications arise. Pyrethroids cannot be used in filter paper tests as the active ingredient becomes bound to the paper fabric and is then unavailable to the insect; aluminium-backed cardboard acts as an effective substitute (R.W.D. Taylor, pers.comm.).

### Insect growth regulators

In the search for more insect-specific pesticides, compounds have been developed which directly interfere with the insect life cycle. Synthetic forms of insect hormones or their agonists and antagonists have been shown to prevent the normal development of juvenile stages (Hoffmann & Lorenz 1998). Such insect growth regulators (IGRs) have been used to protect stored products from insect damage (Oberlander *et al.* 1997). During their pre-adult life, insects have an exoskeleton that cannot expand sufficiently to allow for growth, therefore at frequent intervals during the larval phase the cuticle is renewed by the process of moulting or ecdysis. During larval and pupal stages ecdysteroids and juvenile hormones are used in the control of moulting and metamorphosis. Synthetic analogues of juvenile hormone known as JHAs (e.g. methoprene and fenoxycarb) have been developed which produce severe abnormalities and premature death of treated insects. Ecdysteroid agonists (e.g. tebufenozide and methoxyfenozide) have not been fully evaluated but they are known to be very effective against *Plodia interpunctella* (Oberlander & Silhacek 2000). These two types of insect growth regulator therefore act against the larval stages which, when affected, usually fail to survive the next moult.

Chitin is a major component of the exoskeleton of the insect. It is a linear polymer of N-acetyl glucosamine. Several compounds, benzoyl phenyl ureas, have been synthesised that interfere with chitin synthesis. These include chlorfluazuron (currently the most active), teflubenzuron, hexaflumuron, flufenoxuron, triflumuron and diflubenzuron, the first and least active. As well as affecting the enzymatic control of chitin production by interfering with chitin synthetase, they also have ovicidal activity so preventing hatching of fully developed larvae from eggs. Their effective rates of application (0.5 mg/kg or less) are generally lower than for conventional insecticides (1–10 mg/kg) (Elek & Longstaff 1994). Furthermore, the chitinase inhibitors are much more stable, providing protection to stored grain for as long as 2 years.

Both groups of IGR do not kill adult insects but prevent juvenile stages from completing their development. Not surprisingly therefore, there is some evidence that insect growth regulators are more effective against species whose larvae develop on the commodity surface, such as *Tribolium* spp, compared to internal feeding larvae such as those of *Sitophilus* spp. (Smet *et al.* 1989; Kostyukovsky *et al.* 2000). Those larvae completing development inside the commodity may avoid a treatment applied on the grain surface. However, pre-exposure of adult insects to a chitin synthesis inhibitor subsequently increased progeny mortality in some species of Coleoptera (Elek 1998). Therefore, insect growth regulators may influence the next generation before eggs are laid.

When the idea of using insect hormones as insecticides was first proposed it was thought that insects would not be able to develop resistance against them since they play such a crucial role in the normal development of juvenile stages. However, insects have indeed developed resistance to some IGRs, notably methoprene and diflubenzuron. Currently it is thought that most IGR resistance is conferred by increased barriers to uptake of the chemicals and possibly increased metabolism of the compounds that do enter the body (Hoffmann & Lorenz 1998).

IGRs were first tested on a large scale in the mid-1980s (Vick *et al* 1985). Methoprene caused a greater than 90% reduction in adult emergence during a 2-year test with stored in-shell peanuts, which were initially infested with *Ephestia cautella*. Subsequently, methoprene came into wide use as a protectant for stored tobacco against *Lasioderma serricorne* and *Ephestia elutella*, achieving protection for as long as 2 years (Oberlander *et al.* 1997). Mixtures of IGRs and conventional synthetic insecticides may be used in an attempt to slow resistance. However, no additional gain has been demonstrated through synergism between methoprene and conventional chemical protectants (Daglish *et al.* 1995, 1996). The specificity of some insect growth regulators means that they may be particularly compatible with the use of biological control if the beneficial organism remains unaffected by the treatment.

## Biological control

### *Principles*

The term 'biological control' was first used by Smith (1919) to describe the introduction of exotic natural enemies of insects for the permanent suppression of insect pests. Both the pest to be controlled and the natural enemy to be used can come from a wide range of taxa. One of the first examples was the use of mammals (cats) to control other mammals (rats and mice). Currently, many interactions are exploited. For example, fungi can be used to control insects, bacteria used to control fungi, insects to control other insects or plant weeds, and mites can be used to kill other mites.

The way that biological control agents are used varies greatly. At one extreme they are sometimes viewed as a direct substitute for chemicals, as 'biopesticides', being applied in a similar manner (Rodgers 1993). Such is the case for many bacterial treatments against insects. On the other hand, a broader ecological approach can be taken and the treatment viewed as a long-term manipulation of the species composition of a store. Some of the common approaches are listed and defined below (Waage & Greathead 1988):

*Classical biological control*: the introduction of exotic natural enemies;

*Inoculation*: the periodic establishment of agents in conditions where they cannot persist all the year round, hence each inoculation provides control over a number of pest generations;

*Augmentation*: the supplemental release of indigenous natural enemies to increase control of a pest, often strategically timed for a vulnerable stage of the pest population growth; and

*Inundation*: the release of large numbers of agents to control a single pest generation, with no anticipation of effects on subsequent generations (biological pesticides).

### *Background*

In common with some of the newest contact insecticides like diatomaceous earths and insect growth regulators, biological control has recently been under particular investigation as a potential alternative to conventional chemical pesticides. The term 'biological control' is used very broadly by some to include control techniques such as the use of resistant crop varieties and the release of sterilised male insects that rely on the biology of the pest or crop system. Here, however, we will use the term to describe the use of one or more species (the biological control agents) to suppress the damage inflicted by pest species. In food storage, biological control often occurs naturally, particularly in stocks that have not been treated with pesticides (that are often better at killing biological control agents than the pests themselves). Reviews of the potential of biological control in post-harvest systems can be found in Wilson *et al.* (1991), Brower *et al.* (1995), Arthur (1996) and Cox and Wilkin (1996).

Each biological control agent has its own advantages and disadvantages that are described in more detail later. In very broad terms, however, the benefits of biological control that specifically address the constraints posed to conventional chemical use are listed below:

- they usually pose a particularly low risk of toxic effect on mammals and therefore the end consumer;
- some techniques are currently exempt from many of the registration costs required for the use of new chemical pesticides (see Rodgers 1993);
- they provide another tool that might be called upon as part of an integrated pest management system; and
- development of resistance is mostly thought to be slower than to chemical agents (Cox & Wilkin 1996).

However, there are potential barriers to the use of biological control. Again in broad terms (particular details of exceptions and other caveats are given later) the main disadvantages of biological control are:

- biological control techniques rarely eliminate pests or result in zero damage or pest contamination;
- the biological control agent itself is often an unwanted extra residue on a commodity;
- generally the use of biological control techniques is management intensive, requiring monitoring of populations for the best results; and
- often biological control is relatively slow-acting (Cox & Wilkin 1996).

In a survey of stakeholders in 1996 in the UK grain storage system, storekeepers were open-minded about the use of biological control, but did not feel there was a need at the time, since conventional methods were cheaper and more effective (Cox & Wilkin 1996). Flour millers were against the use of biological control, mostly because of concerns over the residues they might leave in the grain.

> 'In 1988 a grain storage company was prosecuted in the USA under Federal Regulations for releasing beneficial insects into grain, even though they were added for the purpose of reducing pests and pesticide usage.'
>
> *Cox (1990) cited in Cox and Wilkin (1996)*

The level of consumer acceptance of produce protected using biological control is hard to predict. On the one hand, since biological agents rarely achieve 100% control, consumers would generally be presented with produce with higher levels of cosmetic damage, particularly noticeable for unprocessed products such as fruit and vegetables. Conversely, small amounts of damage are seen as a positive feature by an increasing market of consumers (both in developed and developing countries) who value indications that the produce has not been treated with conventional pesticides. Indeed, some consumers will pay extra for 'organic' produce that has been produced without the aid of chemical additives. The reaction of consumers to the presence of insect body parts or micro-organism remains in food is also difficult to gauge. In the UK, where public reactions to food scares can be particularly extreme, there was little resistance to the introduction of bio-yoghurt containing live cultures of bacteria, and little public outcry when significant amounts of arthropod remains were reported in UK breakfast cereals. Consumer tolerance of the residues from biological control may therefore be tolerated in exchange for reduced chemical pesticide residues.

Cox and Wilkin (1996) suggested that the bulk of research into biological control in grain storage has been aimed at specific pest issues in developing countries. The US (funded largely by the Environment Protection Agency to reduce the dependence on chemical control) and Australia were credited with having the most extensive research programmes in this field applicable to the UK situation. Other topics include research in the Czech Republic (then Czechoslovakia) on the predatory mite *Cheyletus eruditus* (Žd'árková & Horák 1990) and work done in Vietnam on the use of the toxin-producing bacterium, *Bacillus thuringiensis* (Cox & Wilkin 1996).

A code of practice has been produced that outlines procedures and responsibilities for the importation and release of exotic biocontrol agents (FAO 1996). The code outlines the responsibilities of three groups of users (government organisations or their representatives, exporters, and importers) and describes three responsibility phases in the process (before export, before and upon importation, and after importation). In the post-harvest sector there has been one example requiring the implementation of this code, these being all concerned with the use of *Teretrius nigrescens*, the predator of *Prostephanus truncatus*.

### *Characteristics of the 'ideal' biocontrol agent*

The ideal biocontrol agent:

- is relatively host-specific and will always choose the intended target over other organisms if the target is present – in practice, often it is deemed acceptable if the biocontrol agent feeds at a low level on other species during times when the pest species is absent;
- is not harmful to economically important organisms;

- does not have a significant impact on populations of native species, particularly rare species;
- has a relatively fast rate of population increase (to provide a fast numerical response to the presence of the target species);
- has excellent host-finding and handling adaptations (to provide an effective functional response to the presence of the target species);
- is easy to rear, preferably requiring only very basic equipment; and
- has biology that is very well understood, so the chance of unforeseen outcomes is low.

### *Predicting the outcome of biological control*

Environmental parameters such as humidity and temperature generally appear to be even more influential on biocontrol efficacy than on chemical insecticides (Rodgers 1993). However, the relative stability of the post-harvest environment and protection from high levels of ultraviolet light may make some biocontrol agents more effective post-harvest rather than pre-harvest.

Many early attempts at biological control made use of generalised predators (those that could feed on a range of hosts) introduced as classical biological control agents. Host switching was common and many of these schemes resulted in devastating environmental damage to the indigenous flora and fauna.

The stored-product environment has features that have particular implications for the design of biocontrol strategies. A store of food commodity usually represents a fairly discrete area where some types of biocontrol are relatively easy to manage such that they should have little or no impact on the wider environment, particularly if the biological control agent can be found naturally in stores. For example, strains of pathogenic bacteria and fungal pathogens that could potentially be used as biological control agents have been originally collected from stores (Bernhard *et al.* 1997; Oduor *et al.* 2000). From the perspective of classical biological control programmes (where the biological control agent is not a native species), the patchiness of the stored commodity potentially facilitates the persistence of refuges of prey that may ensure that the predator is unlikely to become extinct. The outcome of classical biological control is harder to predict than that for inoculation or inundation since the implications of classical biological control are much longer term and geographically more widespread.

There are concerns that pest species might evolve to be less and less influenced by a biocontrol agent in a similar manner to the development of resistance to chemical pesticides. This may be true, particularly for strains of *Bacillus thuringiensis* (toxin-producing bacteria) (McGaughey & Johnson 1992). It was hoped that using a blend of more than one strain of Bt in a formulation would slow the development of resistance by exposing insects to a range of selection pressures. Laboratory studies on *Plodia interpunctella* have demonstrated that the benefit of using mixtures may not be great (McGaughey & Johnson 1992). However, unlike chemical pesticides, biocontrol agents can evolve. There is some evidence that biocontrol agents after introduction to a new location can become more efficient as time progresses; possibly they also have to adapt to features of the new environment.

### *Types and scope of biological control agents*

In terms of pesticide sales at least, biological control is still a relatively small market. Of course, only some types of biological control are 'marketable' in the sense that the user needs to repeatedly reapply them. However, it has been estimated that world biopesticide (inundation, inoculation and augmentation products) sales were 'approximately \$120m in 1990, which represented less than 0.5% of the world agrochemical market' (Rodgers 1993). Over 90% of these sales were of products containing *Bacillus thuringiensis.* To truly assess the impact of biological control in agricultural production, either pre- or post-harvest, an analysis would have to quantify the loss reductions against money spent rather than simply the market value of the industry. Classical biological control agents are in theory at least only 'sold' once and thereafter that inoculation should be self-sustaining. Therefore, changes in the use of this kind of biocontrol will not really show up in sales data.

Research in the area is constantly throwing up new options, and an increasing range of species are being identified as potential biological agents against pests in the food storage environment. Their relationship with the pest varies from that of predator–prey, to pathogen–host. A range of examples is given in Table 6.7 that are classified in terms of the relationship between the biological control agent and its intended pest target. It can be seen that many types of biological control are still under development, some have been mostly rejected, and some have been promising enough to be developed into com-

**Table 6.7** Different biological control systems that are in various stages of development.

| Biological control species | Target pest species | Commodity | Proposed method of control | Stage of development | Reference |
|---|---|---|---|---|---|
| **Predators** | | | | | |
| *Teretrius nigrescens* | *Prostephanus truncatus* | Maize, dried cassava | Classical | Predator released in several African countries, established well in all but Tanzania, impact debatable | Giles *et al.* 1995 |
| *Xylocoris flavipes* | Broad range predator of most stages of many beetles and moths | Old grain (including nuts), residues before new harvest | Augmentation | Tested and used in USA | Cox & Wilkin 1996 |
| *Cheyletus eruditus* | Mites and young stages of some beetles, moths and psocids | Grain | Augmentation | Commercialised in the Czech Republic | Žd'árková & Horák 1990, 1990; Cox & Wilkin 1996 |
| **Parasites** | | | | | |
| *Steinernema feltiae* (also known as *Neoaplectana carpocapsae*) (Nematode) | Various adult or larvae stages of some beetle pests | Trialled in a range of situations including almond orchards | Augmentation/ inoculation | Commercially available for pre-harvest, but post-harvest environment likely to be too dry | Cox & Wilkin 1996 |
| **Parasitoids** | | | | | |
| *Bracon hebetor* | Some moth pests | Range of commodities including sultanas and peanuts | Augmentation | Commercialised in the USA | Cox & Wilkin 1996 |
| *Uscana lariophaga* | *Callosobruchus maculatus* | Cowpea | Augmentation | Laboratory studies | Van Alebeek 1994 |
| *Anisopteromalus calandrae* | Larvae and pupae of a range of beetle pests | Grain | Augmentation | Commercialised in the USA | Cox & Wilkin 1996 |
| **Pathogens** | | | | | |
| *Pseudomonas fluorescens* | *Gibberella pulicaris* | Potato tubers | Inoculation/ augmentation | Control demonstrated in the laboratory | Schisler *et al.* 2000 |
| *Bacillus thuringiensis* | *Plodia interpunctella, Ephestia cautella* | Wheat and maize | Inoculation/ augmentation | 50–60% control (wheat) 80% control (maize) demonstrated in grain silos and bins in Midwestern USA. Developed commercially as 'Dipel' in the USA | McGaughey 1980, 1985 |
| *Beauveria bassiana* | *Oryzaephilus surinamensis, Sitophilus zeamais* | Grain | Augmentation/ inoculation | Experimental work demonstrated the importance of relative humidity to efficacy | Lourenção *et al.* 1993; Adane *et al.* 1996 |
| *Beauveria brongniartii* | *Sitophilus zeamais* | Maize | Augmentation/ inoculation | Control demonstrated experimentally | Rodrigues & Pratissoli 1990 |
| *Metarhizium anisopliae* | Termites | Wooden buildings | Inoculation | Commercialised in US and Australia<br>Patent numbers:<br>US 5595746<br>AU 9229365<br>WO 93/0962 | |

mercial products. Notice also that many biological control systems have been developed for fairly high-value commodities such as nuts, fruit and vegetables. It may be these specialised markets that are best suited to the high monetary and time investments that are required for some biological control strategies.

*Insects and mites*

Figure 6.14 shows examples of an insect and a mite. Insect biological control agents include predators (e.g. *Teretrius nigrescens* against *Prostephanus truncatus*) and parasitoids, for example, wasp parasitoids *Dinarmus basalis* and *Eupelmus vuilleti* have been shown to have some influence over the population growth of a beetle pest, *Callosobruchus maculatus*, of cowpeas *Vigna unguiculata* (Sanon *et al.* 1999). (See Haines (1999) for a review of the use and research effort into arthropod biological control against stored-product pests.)

The predatory mite *Cheyletus eruditus* attacks other species of mite that can be pests in storage. *Cheyletus eruditus* eats other mites by feeding on their body fluids and has been used in commercial grain storage in the former Czechoslovakia. A bag containing 2000–3000 *C. eruditus* has been used per treatment of empty warehouse stores to reduce the mite pest population that remains even after cleaning, before fresh commodity is loaded (Žd'árková & Horák 1990).

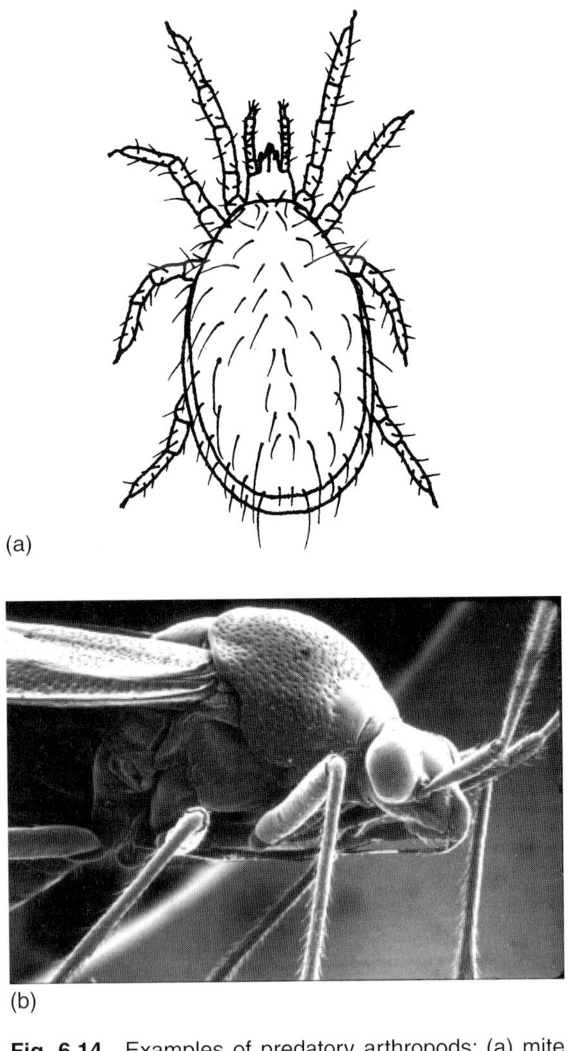

(a)

(b)

**Fig. 6.14** Examples of predatory arthropods: (a) mite (female adult of *Blattisocius tarsalis*; (b) hemipteran bug. Source: University of Greenwich.

*Protozoa*

The biological control potential of some protozoan pathogens such as *Mattesia trogodermae* has been investigated (Cox & Wilkin 1996) although there are no reports of any instances of protozoa actually being used on-farm. Limitations identified included slow action, low virulence and poor persistence. If they are to be used it is likely that it will be alongside other complementary measures.

*Fungi*

Most of the entomopathogenic fungi, for example, *Beauvaria bassiana* and *Metarhizium anisopliae* (Plate 13), are facultative parasites. The infection unit is a spore that on germination penetrates the integument with mechanical pressure or enzymatic action. Hyphae can also initiate infection. Metabolites of fungal pathogens are involved in the infectious process. Pigments like biochromes, such as bassianin and tenellin, or dibenzo-quinones, such as oosprein, are responsible for the colour change in the insect body. Many of the entomopathogenic fungi produce toxins that act as poisons for the insects (Ananthakrishnan 1992). So far, fungi have not yet been used against insects on a commercial scale in the stored-product environment, and work is still at the exploratory stage (Searle & Doberski 1984; Sheeba *et al.* 2001).

*Bacteria*

*Bacillus thuringiensis* (Bt) (Fig. 6.15) is a bacterial insect pathogen that is arguably the biological control agent that has had the largest impact on crop pest popu-

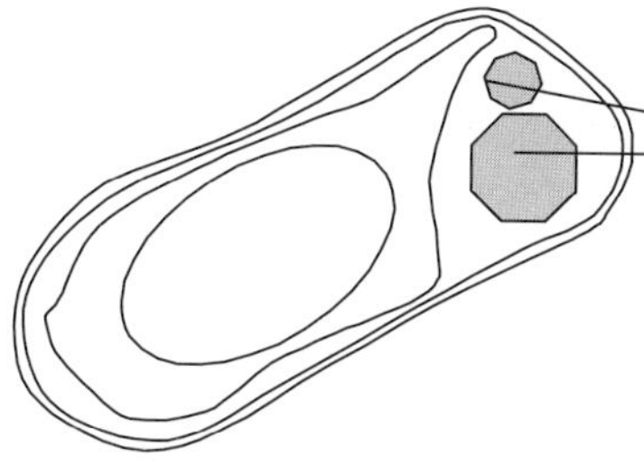


**Fig. 6.15** Diagram of *Bacillus thuringiensis* showing regular shaped crystals of endotoxin.

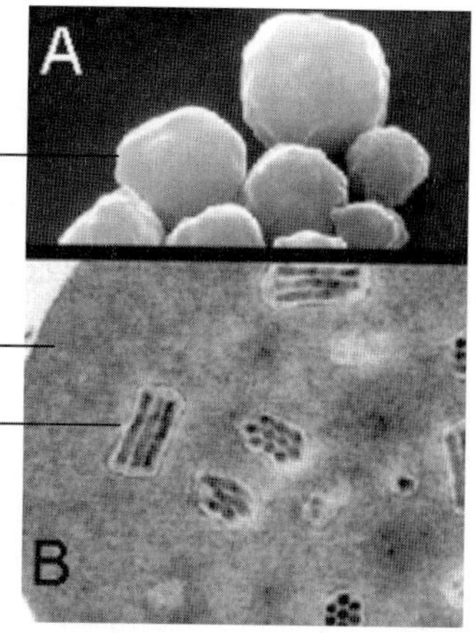


**Fig. 6.16** Electron micrographs of Nuclear Polyhedrosis Virus. Source: University of Greenwich.

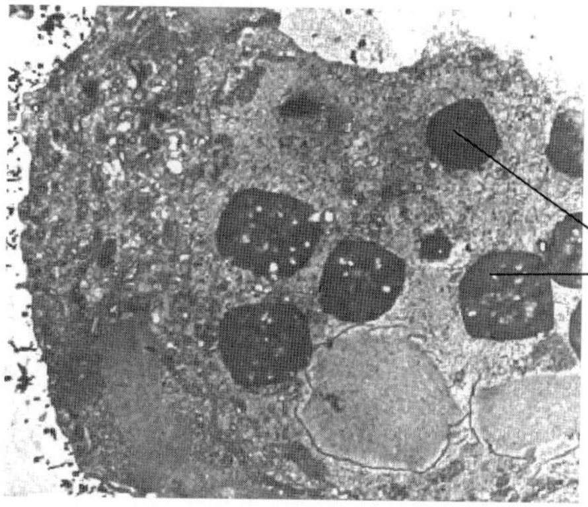


**Fig. 6.17** Electron micrograph of Nuclear Polyhedrosis Virus, × 10 000. Source: University of Greenwich.

lations. Bt exerts its toxicity when endotoxins that it produces are ingested by the target insect. The toxins dissolve in the midgut and are modified into activated toxins that bind to the cell membrane of gut lining and effectively punch holes in it that then leads to salt imbalances. Ultimately the cells split open. The insect usually stops eating and dies from starvation (Feitelson *et al.* 1992). There are many different strains of Bt that produce many different toxins. Different insect species are susceptible to different toxins. Bt is widely distributed in the environment (Bernhard *et al.* 1997) and libraries of strains have been collected for screening against different pests (Meadows *et al.* 1992).

Most of the published work is on the use of Bt as a protectant of crops in the field. However, see Table 6.7 for an example of the use of a commercial preparation of Bt against moth pests of stored products. Bt is relatively easy to produce by simple fermentation techniques and therefore could, in theory, be produced locally on a fairly small scale where there was a demand (Salama *et al.* 1993). The genes that are responsible for the production of the Bt toxins can be moved from the Bt bacteria into the crop plant itself to produce genetically modified grains that make their own toxin(s).

### *Viruses*

The development of biological control through the use of viral pathogens (Figs 6.16, 6.17) against pests of stored-products is largely at the research stage, with most of the work still at a fairly upstream level (see review by Moore *et al.* 2000). Often viral pathogens are fairly focused in the species that they affect and also in the stage of the insect, for example, many viruses only attack larvae (Moore *et al.* 2000). Baculoviruses are naturally influential insect pathogens, especially against Lepidoptera. However, they do not seem to be widespread antagonists against Coleoptera. In common with other pathogens, there is the theoretical potential at least to use the insect pests themselves to incubate the pathogen and spread it among their population. Initial inoculation of some individuals may be facilitated by the use of pheromone-baited traps in a 'trap, inoculate and release' system.

### *Commercial exploitation*

One commercial operation in the US, Biofac in Texas, breeds and distributes ten arthropod biocontrol agents. These include *Anisopteromalus calandrae, Bracon herbetor, Xylocoris flavipes* and *Trichogramma petrosium*, which are recommended for control of beetles and moths in silo bins, trucks and similar closed containers. Apart from the obvious difficulties with grain contamination caused by the presence of insect bodies, another drawback is that applications have to be made every 2 weeks or even at more frequent intervals. None of these parasites is particularly effective at low temperatures, and below 18°C application rates would have to be markedly increased.

### Varietal resistance

The innate ability of stored grain crops to resist insect attack could eventually result in there being no requirement for the use of insecticides. However, even though many plant varieties have shown to be relatively resistant to insects that normally attack the species, no cultivated line has been shown to be completely resistant; insect pests have been able eventually to overcome the resistance factors. Nevertheless, varietal resistance does offer a significant element for incorporation into any integrated pest management programme. A comprehensive review of varietal resistance is given by Throne *et al.* (2000).

Resistant varieties have been developed in classical plant breeding programmes for all major crops and by transgenic technology. In maize it has long been known that 'local' varieties produced by African farmers are much more resistant than much higher yielding composites and hybrids (Golob *et al.* 1985). This resistance has partly been attributed to grain hardness (Dobie 1974), which is correlated to the presence of phenols, particularly ferulic acid (Arnason *et al.* 1992). Another feature that potentiates resistance in maize is the presence of a tight, elongated husk, which acts as a mechanical barrier against pre-harvest *Sitophilus zeamais* infestation and delays subsequent population development in store; cobs with loose or open husks are much more readily infested (Giles & Ashman 1971).

Husk (or hull) tightness in rice (paddy) is also responsible for conferring protection against stored grain insects. Primary pests are unable to infest rice if the husk is intact but even with small cracks *Rhyzopertha dominica* larvae are able to attack the grain. Grain hardness is positively correlated with resistance but grain size is negatively correlated (Morallo-Rejesus *et al.* 1982) and more eggs tend to be laid on larger, smoother grains (Zhang & Deng 1993).

Grain hardness in wheat is inversely related to insect progeny production (e.g. McGaughey *et al.* 1990) but this expression of resistance is only evident when there are large differences in hardness. However, grain hardness is the best character for expressing susceptibility in wheat varieties (Ram & Singh 1996).

Resistance in barley and oats is dependent on the presence of the seed hull and on its thickness. Barley tends to be more susceptible than wheat, maize or rice but oats are more resistant than wheat.

Sorghum resistance is related to grain hardness, the presence of the glume (husk), and the tannin content of the testa. On the head, glumes are intact and the grain can resist attack by *Sitophilus* spp but it is susceptible to *Sitotroga cerealella* and to a lesser extent to *Rhyzopertha dominica.* When glumes are removed by threshing the grain becomes very susceptible to *Sitophilus oryzae* (Wongo & Pedersen 1990; Mvumi 2001, in press). High tannin content, which gives the testa a red or brown colour, tends to prevent bird attack in those areas of Africa where severe damage can be caused by birds, particularly *Quelea quelea.* However, the evidence linking tannin content to insect resistance is contradictory (Throne *et al.* 2000). Nevertheless, sorghum contains other compounds such as proteinase inhibitors that might confer some degree of resistance.

Almost all pulses developed from plant breeding programmes are very susceptible to at least one bruchid pest. Physical characteristics of the seed, such as roughness of the testa, can restrict insect oviposition but a single characteristic such as this is generally not sufficient to prevent significant damage to the grain. As bruchids infest grain legumes in the field once the crop is mature, the thickness and dryness of the pod can play a part in restricting infestation (Nahdy *et al.* 1999); dry pods which have cracked allow adults to enter and oviposit directly on the seed, but intact dry pods are more difficult to attack as the emerged first instar larvae must be able to cross the air space between the pod wall, on which the egg was deposited, and the seed testa. Thickness and hardness of the testa may cause exhaustion or starvation of larvae before they are able to reach the cotyledons (Kitch *et al.* 1991).

Pulses contain a wide range of secondary metabolites that can contribute to resistance, including saponins, non-protein amino acids and alkaloids. However, the most probable biochemical sources of resistance are lectins, arcelins, α-amylase inhibitors and protease inhibitors, all of which are structurally related. Although the evidence is rather contradictory it appears that arcelins are primarily responsible for resistance in common beans, *Phaseolus vulgaris*, to *Zabrotes subfasciatus* and possibly *Acanthoscelides obtectus*, and α-amylase inhibitors confer resistance in cowpeas to *Callosobruchus* spp. α-Amylase inhibitors, which interfere with carbohydrate digestion, do not account for all the resistance in *Callosobruchus*. Other storage proteins, vicilins, confer resistance by binding chitin in the midgut and by their low digestibility (Yunes *et al.* 1998). Resistance to *Callosobruchus* in small grain legumes such as green gram may be caused by alkaloid vignatic acids (Kaga & Ishimoto 1998). The presence of other antimetabolites confuses the picture, and the recent discovery of a variant α-amylase inhibitor that inhibits midgut enzymes in *Zabrotes subfasciatus* (Grossi de Sa *et al.* 1997) suggests that a combination of compounds might be an effective weapon to protect pulses as a group.

### *The role of transgenic crops*

The technology now exists to replace some chemical solutions to the problem of insect damage with genetic ones by the use of genetically modified (GM) crops (Estruch *et al.* 1997; Gatehouse & Gatehouse 1998). Transgenic plants that possess the genes that code for some *Bacillus thuringiensis* toxins have been produced and shown to have high insect resistance (Estruch *et al.* 1997). Like most of the chemicals that are approved for use in post-harvest systems, the funding for the development of transgenic plants has come from their use to protect the growing crop in the field. (A comprehensive review of the prospects for using transgenic resistance to insects in crop improvement is given by Sharma *et al.* 2000.)

Depending on the particular toxins produced by the plant and their range of toxicity, genetically modified crops may also be protected against insect attack in store. Primary insect pests that feed mostly within grains often only have limited exposure to contact insecticides that are applied to the surface of the grain. The use of a GM crop could ensure that even those pests that develop entirely within one grain would ingest the toxin.

There have been seven reports of the use of transgenic technology for storage insect control (Throne *et al.* 2000). These are:

(1) Pea (*Pisum sativum*) seeds expressing common bean (*Phaseolus vulgaris*) α-amylase inhibitor exhibited a high level of resistance to *Callosobruchus* and *Bruchus pisorum* (Shade *et al.* 1994; Schroeder *et al.* 1995). Field trials in Australia showed that these peas were 99.5% resistant to bruchid attack and required no insecticide application during production or storage (Throne *et al.* 2000). Enhanced levels of resistance to bruchids have also been found in transgenic adzuki beans (Ishimoto *et al.* 1996).

(2) Transgenic hybrid maize containing Bt endotoxin CryIABb caused high larval mortality of the lepidopteran *Plodia interpunctella* but had virtually no effect on two beetles, *Tribolium castaneum* and *Sitophilus oryzae*. Bt products with action against stored product beetles have yet to be identified.

(3) Transgenic potato plants containing Cry1A(b) gene Bt 884 and a truncated gene Cry1A(b)6 resulted in less damage to leaves by *Phthorimaea operculella*, the potato tuber moth. In potato tubers containing the Cry1A(b)6 gene stored for up to 6 months moth mortality was 100%, there being no damage during this period (Jansens *et al.* 1995).

(4) One of the glycoproteins found in chicken egg, avidin, has been shown to prevent the action of the vitamin, biotin, that is essential for insect development. Maize has been engineered to contain avidin and has shown good toxicity against storage insects including both internal and external feeders (Kramer *et al.* 2000).

(5) *Ceutorhynchus assimilis*, the cabbage seed weevil, exhibited differential susceptibility to oilseed rape that expressed oryzacystatin, the cysteine protease inhibitor of rice (Girard *et al.* 1998). Of two beetle strains tested one was unaffected while the other showed increased growth, even though oryzacystatin inhibited the beetle digestive proteases *in vitro*. Clearly, other, as yet unknown, metabolic processes occurred in the insects to overcome the effects of the protease inhibitor.

(6) Rice transformed with a bifunctional enzyme inhibitor from barley, subtilisin/α-amylase inhibitor, caused increased mortality in *C. pusillus* (Ohtsubo & Richardson 1992).

(7) Survival of early instar larvae of *Sitotroga cerealella* on wheat transformed with barley trypsin inhibitor was significantly reduced compared with controls (Altpeter *et al.* 1999).

Transgenic crops that include inhibitors of major insect digestive enzymes, such as lectins, chitinases, peroxidases and immunoglobulins, are also being developed. However, to date no single transgenic crop has been commercially produced, specifically to deter stored product insect pests.

## Inert dusts

Inert dusts are dry powders of different origins that are chemically unreactive in nature. They can be divided into five categories differentiated by their chemical composition or level of activity (Golob 1997):

- non-silica dusts (include katelsous (rock phosphate and ground sulphur), lime (calcium hydroxide), limestone (calcium carbonate) and common salt (sodium chloride);
- sand, kaolin, paddy husk ash, wood ash and clays;
- diatomaceous earths (or diatomite);
- synthetic silicates and precipitated silicates; and
- silica aerogels.

The use of inert dusts as grain protectants is not new. Observations of birds and mammals taking dust baths to rid themselves of mites and parasites is believed to have led the Chinese to start using diatomaceous earths for pest control more than 4000 years ago (Allen 1972). The Aztecs of ancient Mexico are said to have mixed maize with lime to preserve their grain (Golob 1997). In the Philippines farmers layer lime with maize cobs to protect against insect damage, while in Honduras lime is mixed with the grain (Golob & Webley 1980). Many small-scale farmers in the developing world still use traditional methods of mixing sand, kaolin, paddy husk ash, wood ash and clays with grain as a protectant. However, despite these materials being locally available, the large quantities (>20% by weight) which are characteristically required to exert an effect put many farmers off (Golob & Webley 1980). Farmers are not keen on this level of adulteration of their grain and the cleaning of these huge quantities of ash and sand from the grain is tedious and time-consuming.

Unlike insecticides, inert dusts function through their physical properties and are, therefore, generally slower acting (Maceljski & Korunic 1972). Synthetic silicates and diatomaceous earths are active at much lower rates of application than sand, ash and lime and similar materials traditionally used by small-scale farmers. However, synthetic silicates, which are manufactured for industrial uses, have a very high silicon dioxide content, and are very expensive and therefore inappropriate for use as grain protectants. In this section we will concentrate on the use of sand, ashes and clays and diatomaceous earths in stored grain protection.

### *Ashes, sands and clays*

The use of ashes and minerals in which physical barrier effects are responsible for the control of insects is clearly quite distinct from the use of whole plants or parts of plants such as powdered seeds or roots where there may be some chemical or repellent effect (Golob & Webley 1980). Their use in large quantities to fill up the interstitial space in grain bulks or to provide a barrier to insect movement is a widespread grain protection technique among small-scale farmers in the developing world (Davies 1970; Golob 1997). These materials impede entry and movement of insects within intergranular spaces (Katanga Apuuli & Villet 1996; Chinwada & Giga 1997), which in turn affects reproduction, oviposition and population growth. Wood ash and sand may have a desiccating effect on the adult insects and crawling larvae with which they come into surface contact (Chiu 1939; Wolfson *et al.* 1991; Baier & Webster 1992). There are also suggestions that the ash particles may clog insect spiracles and trachea, causing suffocation (Wolfson *et al.* 1991). Ashes are available in areas where synthetic chemicals are often difficult to obtain, and the ash residue which covers the grains can be easily seen and washed off before consumption (Katanga Apuuli & Villet 1996).

Generally, the need to apply these inert protectants at such high rates makes smallholder farmers reluctant to use them, especially on high-value grains (Subramanyam & Roesli 2000). Although farmers want to minimise insect damage, they do not want to add large quantities of extraneous matter to their grain. Protectants such as cereal husks, though not as effective as ash, would be more acceptable as they can be easily removed by winnowing and do not make the grain look unsightly (Chinwada & Giga 1997). There is potential to use ashes and sands in conjunction with other non-chemical methods such as host plant insect resistance, storage in sealed containers or in the smoke of a fire, which may enable reduced quantities of these materials to be used, thereby increasing their attractiveness to farmers.

Although much research has been conducted to systematically validate traditionally used storage methods, this research has frequently been carried out in laboratories. In order to fully understand a storage method's potential applicability from all perspectives it needs to be evaluated on-farm within the socio-economic context in which it is to be used.

There is wide discrepancy about the necessary application rates for ash, and results from numerous trials are contradictory. It is possible that some of this variation is a result of the different origins and particle sizes of the ashes, in combination with varying levels of susceptibility of the varieties of commodity tested, insect behaviour and climatic conditions. Ash from different sources has been demonstrated to vary in its effectiveness (Wegmann 1983; Javaid & Ramatlakapela 1995). Although ash from cooking fires is a readily available resource to small-scale farmers in developing countries, ash from animal dung requires preparation. However, ashes are non-polluting and are known to be economically and environmentally acceptable to a majority of resource-poor farmers in developing countries (Talekar 1987; Stoll 1988).

#### *Use of ash and sand against bruchid pests*

In northern Cameroon, farmers traditionally use ash to protect cowpeas from bruchid damage during storage. After threshing, cowpeas are combined with sieved ash and mixed, the mixture is then put into a mud granary or a clay jar, tapped down to compress the mixture and often covered with an additional layer of ash. Alternatively, the cowpeas and ash are layered in the storage container, finishing with a layer of ash. The farmers described the process as time-consuming but effective and inexpensive, and confirmed that after the ash is rinsed off, neither taste nor seed germination is affected by the storage method (Wolfson *et al.* 1991). In the Northern Province of South Africa, farmers mix ash from *Aloe marlothi* with maize grain prior to storing it in drums; as only a low dosage of ash is applied it is not washed off but milled with the grain (Achiano *et al.* 1999). Aloe ash is also particularly useful for protecting seed (Ofuya 1986; Katanga Apuuli & Villet 1996), it does not affect germination and there is some suggestion that the cations in the ash may even enhance growth (Philogene 1972).

When field infested cowpeas were mixed with an equal quantity of ash, and then covered with a layer of several centimetres of ash, during on-farm storage trials in Cameroon, both invasion by and development of the bruchid pest *Callosobruchus maculatus* was prevented (Wolfson *et al.* 1991). If the proportion of ash was reduced, *C. maculatus* larvae were able to develop into adults but were unable to reproduce. In these experiments, cowpeas were protected by being covered with a 3 cm layer of ash, which prevented *C. maculatus* access.

Laboratory trials in Zimbabwe found mixing ash (from mixed tree species) with stored beans at concentrations of 1 : 1, 1 : 2 and 1 : 4 v/v reduced both the damage and the number of F1 bruchids emerging (Chinwada & Giga 1997). However, sand was only effective at reducing F1 bruchid emergence at the higher application rates of 1 : 1 v/v. In laboratory studies in Botswana, cow-dung ash and sand, applied to cowpeas at 30% and 100% w/w respectively, reduced infestation by *C. maculatus* (Javaid & Ramatlakapela 1995). Most of the studies on the use of ashes to protect cowpeas recommend the use of high doses of ash (50–100% w/w) (Dovlo *et al.* 1976; Zehrer 1985; Wolfson *et al.* 1991).

In Columbia, the mixing of kitchen ash 20% w/w with beans led to effective control of the bruchid *Acanthoscelides obtectus* during a 39-week period of on-farm storage (Baier & Webster 1992).

#### *Use of ash and sand against* **Prostephanus truncatus** *and* **Sitophilus** *spp.*

Ashes, particularly paddy husk ash and kitchen (wood) ash effectively reduced insect damage to maize grains in three seasons of field trials in Tabora, Tanzania (Golob & Hanks 1990). Again, the higher application rates, 5% w/w paddy husk ash and 30% w/w wood ash, were found to be the most effective against the main pests *Sitophilus zeamais*, *S. oryzae* and *Prostephanus truncatus* during a 40-week storage period (Fig. 6.18). However, high variation in treatments was observed between years. Sand was also effective, but only when applied at a rate of 20% v/v. Similar results were obtained in field trials in Malawi, where wood ash mixed with maize grain at 30% w/w was found to be almost as effective at controlling insect damage as pirimiphos-methyl (Golob *et al.* 1982). The Malawian trials also demonstrated the importance of relating storage trials to the actual storage season and prevailing climate, to realistically understand the effect of periods of the storage season on insect behaviour and the efficacy of the control method. In Mexico, volcanic ash from Mount Chichonal produced good control of *P. truncatus* when applied to maize at 1% w/w (Sanchez-Arroyo *et al.* 1989).

The use of ash, sand and clays to protect stored commodities will undoubtedly continue to be important to rural communities, especially in areas where farmers lack access to other grain protection methods. However, to date the research findings suggest that application rates and use of ashes of different origins may be highly location-specific.

### *Diatomaceous earths*

Diatomaceous earths (DEs) consist of the fossils of phytoplanktons (diatoms) (Fig. 6.19), which are com-

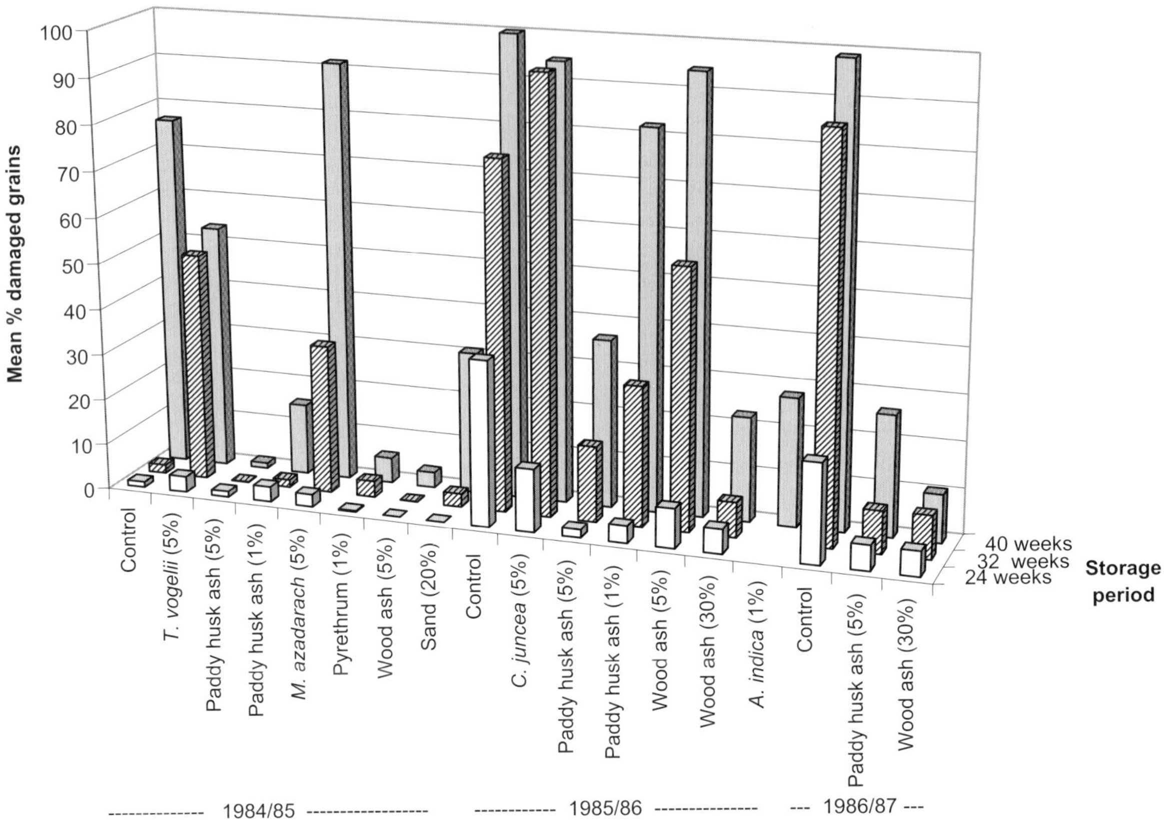


**Fig. 6.18** Mean percentage damaged maize grain admixed with locally available materials during storage.

**Fig. 6.19** Scanning electron micrograph of diatoms. Source: Cereal Research Centre, Agriculture and Agri-Food Canada.

posed mainly of amorphous hydrated silica (~90% $SiO_2$) and other minerals including aluminium, iron oxide, magnesium, sodium and lime. Diatoms are unicellular organisms found in both fresh and marine water. They extract silicic acid from the water and incorporate it into their shells. When they die they sink down into a sedimentary layer. Over many centuries a thick layer builds up, which becomes compressed and fossilised into a soft, chalky rock called diatomite. This layer of diatomite can be quarried, dried and ground in order to reduce both the particle size and moisture content, resulting in a fine talc-like dust (diatomaceous earth) considered to be non-toxic to mammals (Quarles 1992). The high porosity of diatomite has resulted in its use in filters to help clarify fruit juices, beers, wine, pharmaceuticals, swimming pool waste, and dry cleaning solvents, amongst others (Subramanyam & Roesli 2000), as a filler in paints, plastics, asphalt, coating agent in fertilisers, carrier for pesticides (Jefferson & Eads 1951), as a mild abrasive and as a particle aggregate in industrial absorbents. Diatoms are the dominant phytoplankton in areas where dissolved silicon concentrates, which are typically located at equatorial and subpolar latitudes as well as along the western continental margins (Libes 1992). There are more than 25 000 species of diatoms, and as many as 7–8 billion diatoms can exist per square metre of ocean (Round *et al.* 1992). Much of the DE being used today originated more than 20 million years ago in the lakes and seas of the Miocene era.

### *Mode of action*

Diatomaceous earth also has insecticidal properties, exerting its effect through physical means and, although not affecting metabolic pathways by chemical action, may well be chemically active under some circumstances. When particles of DE come into contact with insects they absorb wax from the cuticle, resulting in water loss, desiccation and death (Ebeling 1971). Death occurs when 28–35% of the body weight (about 60% of the water content) is lost (Ebeling 1971). Many dusts including DEs have a repellent effect against insects (White *et al.* 1966) and, it has been suggested, against rodents as very low numbers of rodent hairs and faecal matter were found in grains, cereals and dried fruits treated with DE (cited by Allen 1972).

Stored product species show variation in their susceptibility to diatomaceous earth (Carlson & Ball 1962; Desmarchelier & Dines 1987; Korunic 1998; Subramanyam *et al.* 1998; Fields & Korunic 2000). The most susceptible tend to be those:

- with large surface to volume ratios (small insects);
- with body hair (DE particles collect on the hair) (Carlson & Ball 1962);
- with thin cuticles (Bartlett 1951);
- protected by low-melting grease as opposed to a hardened waxy cuticle (Ebeling 1971); and
- that feed on dry grain, as opposed to sucking insects (Flanders 1941).

Although results are conflicting, there is a general consensus that the most sensitive stored product species are in the genus *Cryptolestes*; *Sitophilus* spp are less susceptible, followed by *Oryzaephilus*, *Rhyzopertha* and *Tribolium* spp which appear most resistant (Maceljski & Korunic 1972; Desmarchelier & Dines 1987; Korunic & Fields 1995; Fields & Muir 1996). However, much of the DE research work has focused on a very limited number of insect species important in large-scale storage. Insects such as *Prostephanus truncatus*, the larger grain borer, and moth species devastating to small-scale farmers in developing countries, have been largely ignored.

The different insect life-stages also vary in their susceptibility to DE. First instars of *Plodia interpunctella* are more susceptible to the DE 'Insecto' than third and fifth instars (Subramanyam *et al.* 1998). *Tribolium confusum* larvae survive seven times as long as *T. confusum* adults on wheat admixed with DEs (Mewis & Reichmuth 1999). This larval tolerance might be linked to the ability of the larvae to regenerate their cuticle frequently, preventing the DE particles from breaking the water barrier of the continuously growing new wax layers (Mewis & Reichmuth 1999). Those insects which develop and feed internally within grains are less likely to come into contact with DE particles applied to the surface of grains than insects which develop externally or are highly mobile within commodities. This fact necessitates the need for DE treatment of grain either prior to infestation or immediately following the destruction of insect populations by fumigation, particularly in commodities commonly attacked by boring beetles such as the bostrichids *P. truncatus* and *Rhyzopertha dominica*.

Most of the work undertaken to investigate insect response to inert dusts has been conducted to observe acute effects on adult mortality under rigidly controlled experimental conditions in the laboratory. A list of these studies is provided by Korunic (1998). There is little information on the effect on insect fecundity, the persistence of the treatments or the effects on mixed populations of insects.

### *What affects the efficacy of DEs in storage?*

The efficacy of a DE is reduced by factors that affect its ability to absorb the wax from the insect cuticle directly.

Relative humidity affects the rate of water loss and therefore determines the effectiveness of inert dusts. For example, a 37-fold increase in silica dust dosage was required on wheat grain at 16% m.c., compared to 11% m.c., in order to obtain the same 7-day $LD_{50}$ of *Sitophilus granarius* (Le Patourel 1986). Because the mode of action of DE is due to desiccation, increased grain moisture content reduces DE efficacy (Carlson & Ball 1962; La Hue 1965; Maceljski & Korunic 1971; Desmarchelier & Dines 1987; Aldryhim 1990, 1993). DE efficacy is considerably reduced if the moisture content of the commodity is more than 14% or relative humidity exceeds 70% (Korunic 1994), and in these circumstances higher DE application rates will be required for effective stored pest control (Korunic 1998). Synthetic insecticides also degrade faster on grain with higher moisture content, rendering them less effective (Snelson 1987).

The effect of temperature on DE efficacy is less well studied, and results of different studies contradict each other. A progressive increase in mortality of *Tribolium castaneum* and *T. confusum* occurs as temperatures increase from 22°C to 27°C and 32°C, which could be linked to increased insect activity at the higher temperatures (Arthur 2000). This activity, resulting in contact with and adherence of a greater number of DE particles, would also result in increased respiration and water loss (Korunic & Fields 1999). Researchers have also found that *T. confusum* and *T. castaneum* were less sensitive to DEs at higher temperature (Aldryhim 1990, 1993; Korunic & Fields 1999) though not *Cryptolestes ferrugineus* (Korunic & Fields 1999). An increase in toxicity with temperature was also reported for studies with organophosphate insecticides (Turnbull & Harris 1986), while the effect on pyrethroids was inversely related (Snelson 1987). If DE is being used to treat fabric of mills, and the facilities are not climate controlled, the environmental differences between seasons could affect the success of the treatment.

Diatomaceous earths do not adhere as well to maize as to wheat or sorghum (La Hue 1972). Furthermore, DE efficacy rapidly reduces in wheat containing broken kernels, which is probably due to the absorption of fatty acids from the broken kernels by the DE, reducing its efficacy (Nielsen 1998). Similarly, DE is less effective on milled rice than rough rice and on ground feed in comparison with whole seed wheat (Cotton & Frankfeld 1949; McGaughey 1972). An explanation may be that the grain dust present in broken and milled commodities may coat the DE particles, reducing their effectiveness. These findings have serious implications for the use of DEs as a structural treatment in mills.

The efficacy of DE is also related to the physical and morphological characteristics of diatoms (Korunic 1998) which vary depending on their source (Fig. 6.20) (Snetsinger 1988; Katz 1991; McLaughlin 1994). The most effective DE samples have lower tapped densities (< 300 g/L), good adherence to the grain (> 70%), a $SiO_2$ content greater than 80%, a pH below 8.5 and reduce grain bulk density by more than 2.5 kg/hl when applied at a concentration of 50 mg/kg (Korunic & Ormesher 1999). Mean particle size (below 15 μm), and diatom shapes are not correlated with insecticidal activity.

DEs also vary in their oil absorption capacity (Subramanyam & Roesli 2000), probably as a result of the different diatom morphologies. After DEs have absorbed lipids from the surfaces of grains or insects, their potential to absorb further lipids is reduced. Aged deposits of the DE Perma Guard are less effective than fresh deposits on rice (McGaughey 1972). This will affect the persistence of DEs on both treated commodities and surfaces.

### *Commercial uses of diatomaceous earth products*

During the 1960s and 1970s experiments in the US found DEs had insecticidal activity (Strong & Sbur 1963; La Hue 1965, 1967, 1977; Quinlan & Berndt 1966; Redlinger & Womack 1966; White *et al.* 1966). However, the DEs these researchers worked with required dosages as high as 0.35% w/w to ensure 12 months' protection. These high DE application rates were not acceptable commercially as they were believed to abrade grain handling machinery and increase wear (Golob 1997). As a result, research into the use of DEs in grain storage ceased. The development of organophosphate resistance in stored product insects led to the reappraisal of DEs, and in 1984 the US Environmental Protection Agency registered Insecto, a new DE which could be effectively applied to grain at dosages as low as 0.05–0.1% w/w (Subramanyam *et al.* 1994).

Many DE dusts are now commercially available, and are registered for use as grain protectants in Australia, Brazil, Canada, China, Croatia, Germany, Indonesia,

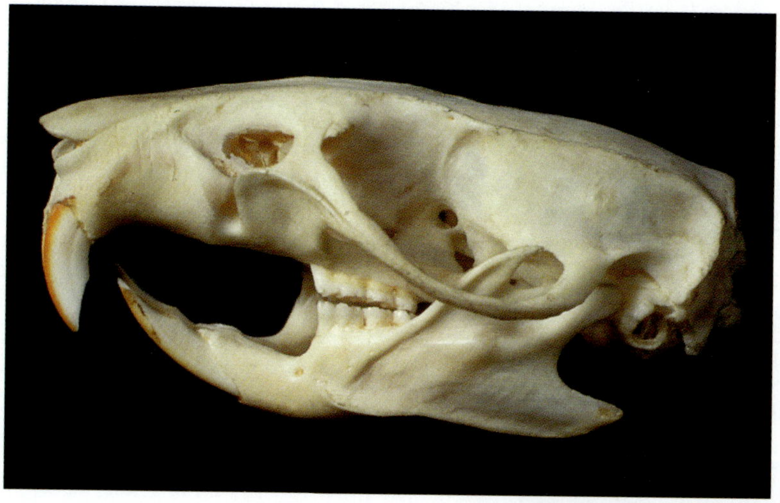

**Plate 1** Side view of a rodent skull showing the diastema and incisors (courtesy of Natural Resources Institute).

(a)

(b)

**Plate 2** Typical 'best' and 'worst' cobs for classes 2–5 of visual scale: (a) Class 2; (b) Class 3; (c) Class 4; (d) Class 5.

(c)

(d)

**Plate 2** *(Continued.)*

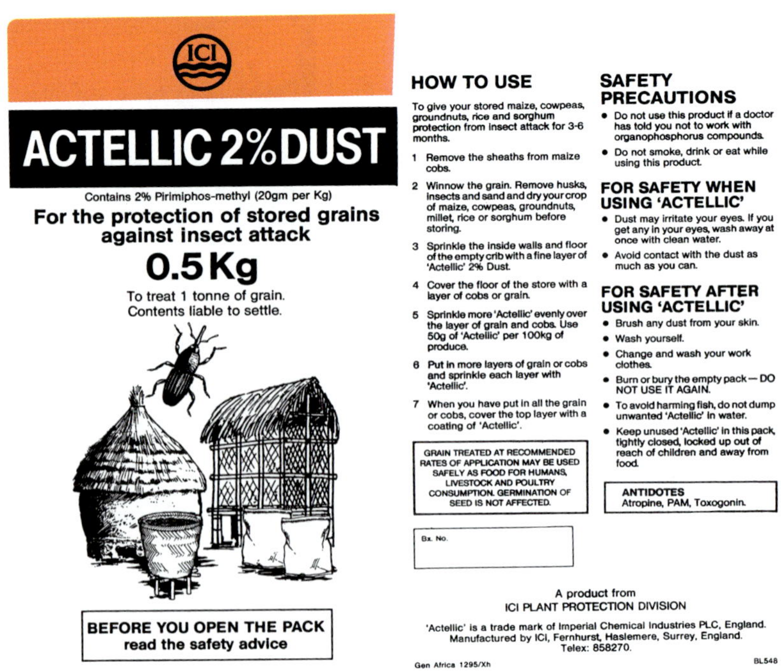


**Plate 3** Example of a typical pesticide container label.

**Plate 4** Liquid spray application onto a store structure using a hydraulic knapsack sprayer.

**Plate 5** Admixing a dust formulation with grain. Source: Golob 1977.

**Plate 6** Thermal fogger being used to apply a space treatment.

(a)

(b)

(c)

**Plate 7** Three types of hand duster. (a) Simple tin and bag hand dusters; (b) bellows-type hand duster; (c) rotary-operated hand duster.

(a)

(b)

**Plate 8** Applying dilute insecticide dust to maize grain: (a) insecticide being added to a pile of shelled maize ready for mixing; (b) insecticide dust being mixed in evenly into shelled maize using a shovel.

**Plate 9** Small hand-sprayer.

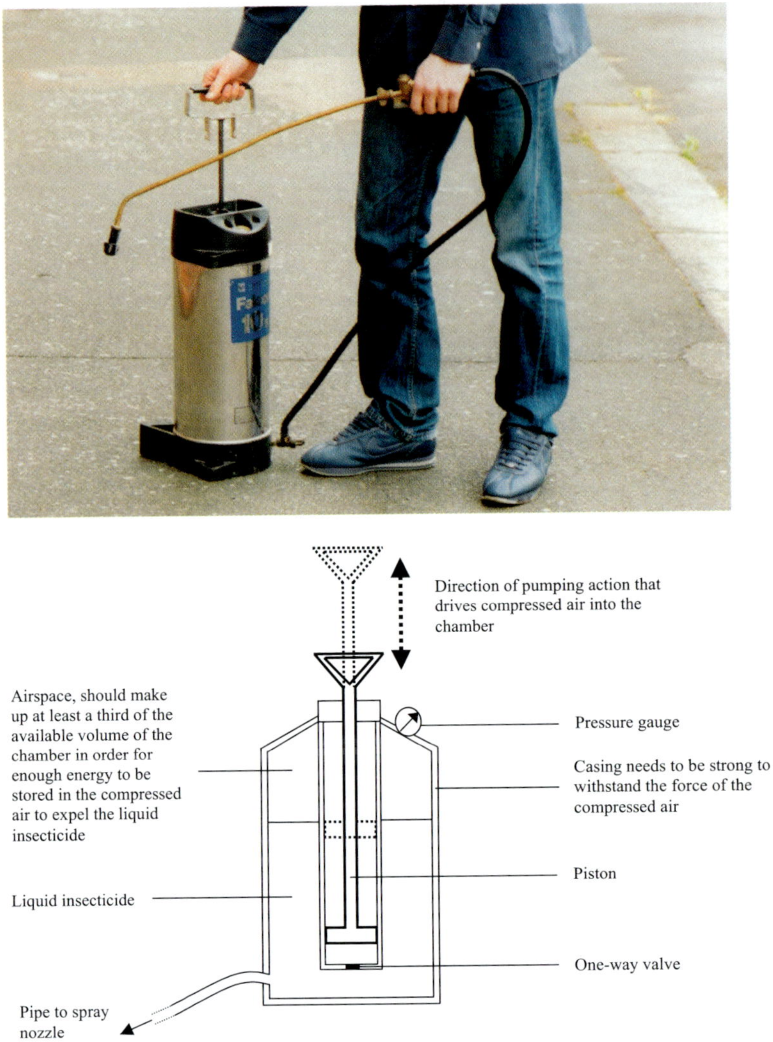


**Plate 10** Pre-pressurised sprayer. Photograph and diagram of cross-section showing pump mechanism.

**Plate 11** A motorised knapsack sprayer.

**Plate 12** Portable large-scale sprayer unit.

**Plate 13** *Oryzaephilus surinamensis* infected with entomopathogenic fungus (courtesy of Central Science Laboratory, UK).

**Plate 14** Dry diatomaceous earth application. Source: S. Allen, CSIRO.

**Plate 15** *Tanacetum cinerariifolium* (*Chrysanthemum cinerariaefolium*) from which pyrethrum is extracted.

**Plate 16** *Cymbopogon schoenanthus.*

**Plate 17** *Azadirachta indica.*

**Plate 18** Setting up farmer participatory research trials to test different botanicals as stored grain protectants. Information on insecticidal efficacy under local conditions as well as on farmer acceptability can be evaluated.

**Plate 19** Extracting plant material into hot water – the indigenous way.

**Plate 20** Treating commodity with a water extract of an insecticidal botanical by dipping the commodity into the extract.

**Plate 21** Pounding plant materials to make a dry powder or to extract oil to treat commodity stored by small-scale farmers is usually done by hand. The extra labour involved will be assessed by the farmer when considering appropriate pest control strategies.

**Plate 22** Pest: *Prostephanus truncatus* (larger grain borer) (courtesy of R.J. Hodges).

**Plate 23** Maize cobs being stacked ready for storage in an Ewe-style store in Ghana, West Africa.

**Plate 24** 'Katchalla' maize-cob store in Ghana, West Africa.

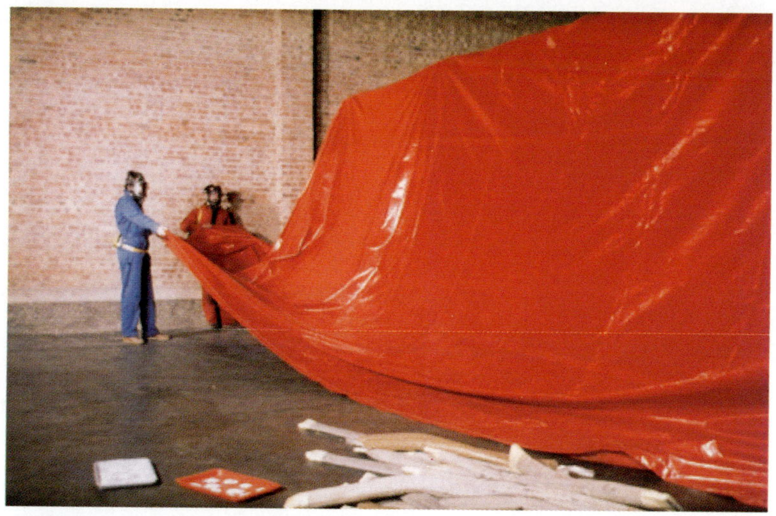

**Plate 25** Bag stack fumigated under gas-proof sheets.

**Plate 26** Application of aluminium phosphide tablets.

**Plate 27** Pounding millet to remove the bran layers.

**Plate 28** Traditional method for milling grain.

**Plate 29** Mixture of millet, maize and groundnuts (prior to milling) to be used as weaning food – Lake Zone, Tanzania.

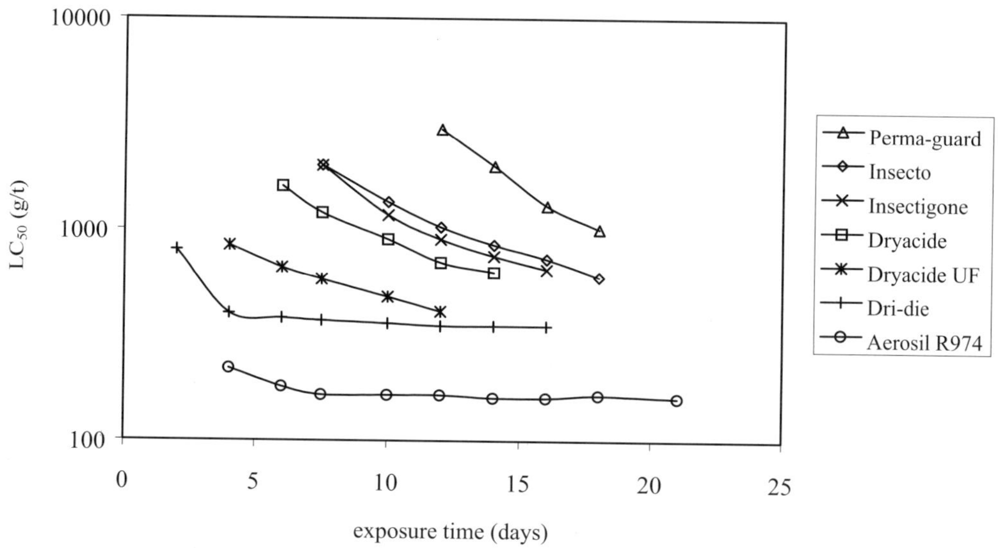


**Fig. 6.20** Relationship between $LC_{50}$ of *Sitophilus granarius* on wheat treated with various desiccant dusts and duration of exposure. Source: McLaughlin 1994.

Japan, Philippines, Saudi Arabia, United Arab Emirates and the US. DEs from different sources vary in their efficacy against insects (Snetsinger 1988; Katz 1991; McLaughlin 1994). This variation is due mainly to the different physical and morphological characteristics of the diatoms rather than their origin (Korunic 1998) and helps to explain why some registered DEs are more effective than others. The extensive amount of research into the use of DEs against various stored product insects has been summarised by Korunic (1998). However, despite research attention, the application of DEs to control stored-product insects remains limited. No figures are currently available on quantities of grain treated with DEs. DEs can be used for the treatment of both grain and structures.

### *Grain treatment*

Diatomaceous earths can be applied directly to dry grain. Historically, very high dosages were required. However, improved formulations which are effective at dosages between 0.5 and 1 kg/t (Insecto, USA; Dryacide®, Australia) or from 0.1 to 1 kg/t (Protect-It®, USA) and innovative combinations with other grain management practices enable reduced dosages to be used. The simplest application method is to admix the DE with the small quantities of grain or seed using a shovel, prior to storage. Uneven mixing and distribution of the DE within the commodity can enable pockets of insect populations to develop. Larger quantities of grain can be treated while on auger hoppers, belt conveyors or bucket elevators using a dust applicator for dry DE or a spray system for aqueous DE slurries.

However, even though dosages have been reduced, the problems affecting the physical properties of the grain mass remain similar. Present regulations defining the quality parameters of the grain in many countries prohibit the addition of any dust to grain intended either for export or for large-scale handling. Grain is also graded for quality, based on bulk density. The presence of DE creates greater friction between grains, which reduces the bulk density and flowability of the grain and the evidence of visible residues on the grains also affects the quality assessment (Johnson & Kozak 1966; Quinlan & Berndt 1966; La Hue 1970; Desmarchelier & Dines 1987; Quarles 1992; Jackson & Webley 1994; Korunic *et al.* 1996; Korunic 1997). Until such regulations are changed, the future of large-scale direct application of DE to grain remains limited. Investigations into the alteration of the regulations to enable DE treated grain to be assessed fairly and further research into reduced DE dosages are ongoing. Nonetheless, the grain industry is likely to accept only DE formulations that can be used effectively at concentrations below 0.3 kg/t of grain, and that can be applied as a spray as well as a dust (Korunic *et al.* 1996).

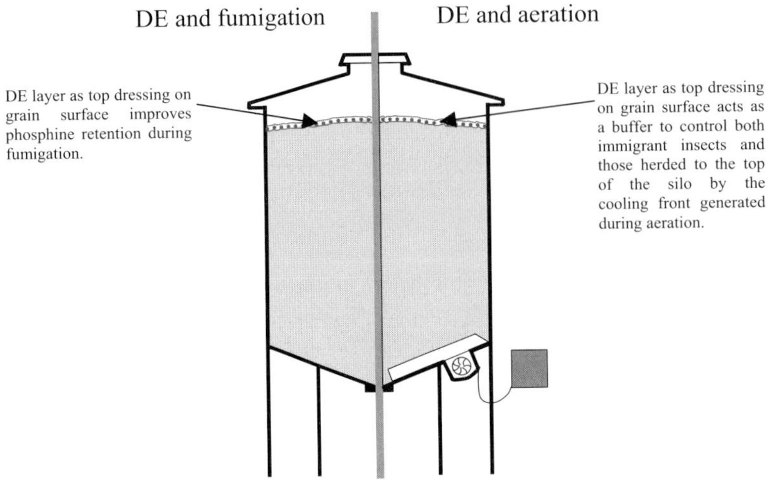


**Fig. 6.21** A silo with the grain surface top-dressed with diatomaceous earth to enhance both fumigation and aeration practices.

An alternative protection system favoured by bulk grain handlers in Australia is to apply the enhanced DE Dryacide® as a top-dressing and fumigate with phosphine every 2–3 months (Fig. 6.21). The use of DE as a dust to cap the grain surface in low-flow phosphine fumigations is an industry practice in bulk handling companies in Eastern Australia (Bridgeman 1999). The DE layer improves phosphine retention, enabling insect-free and residue-free storage in poorly or unsealed structures (Bridgeman & Collins 1994). In this practice Dryacide® is dusted onto the level surface of the grain at a rate of 100 g/m$^2$. Trials have indicated that the use of DE in this manner gives superior results when compared to covering the grain surface with a PVC cover (Winks & Russell 1994). However, if the concentration of the DE on the top of the grain is high it can alter the characteristics of the grain flow, causing problems during silo unloading (Bridgeman 1999).

Controlling insect populations with aeration alone is not completely effective, the addition of DE to the surface layer of the grain reduces insect populations still further (Plate 14) (Nickson *et al.* 1994; Bridgeman 1999). Insects which move to the top of the silo as a result of the cooling front and immigrating insects are controlled by the DE layer (Bridgeman 1999).

### *Structural treatment*

Diatomaceous earths can also be applied as dry powders or wet aqueous slurries to empty storage facilities (Fig. 6.22) and grain handling equipment such as trucks, headers, augers and combines for disinfestation and long-term protection purposes (Anon. 1994a; Bridgeman 1994). Before empty storage facilities or grain handling equipment are treated with DE they must be cleaned of grain dusts and residues. The Dryacide prod-

**Fig. 6.22** Diatomaceous earth slurry spray application.

uct manual recommends that dry DE dust is applied to combine harvesters, augers, aeration fans, ducting and other grain handling equipment to avoid corrosion problem from water exposure. Dry DE dusts can be applied using a venturi blower, power duster, sand blasting gun equipped with compressed air or through aeration fans by workers following appropriate safety precautions. The surfaces of silos and other storage facilities can be sprayed with an aqueous DE slurry; in Australia, this is now a popular practice and has been shown to provide protection for up to 12 months (Desmarchelier & Dines 1987; Bridgeman 1991, 1994). Dryacide recommend a slurry application rate of 6 g/m$^2$, which is higher than the recommended Dryacide® dust application rate of 2 g/m$^2$. As application rates differ between DE formulations, individual label recommendations must be followed. Grainco, the bulk grain handling authority of Queensland, Australia, use a Hydra-Cell slurry pump for application, while the Rice Growers Co-operative Ltd, Australia, developed a special slurry spray unit (Bridgeman 1991; Anon. 1994b). Slurry application of DEs gives more even coverage than dusts and occupational safety is improved as DE dust is only generated during tank mixing. In response to consumer pressure to reduce both chemical treatment of food commodities and pesticide residues in food, the use of DEs that are approved for organic processing, for structural treatment of cracks and crevices in buildings or as a top-dressing in grain stores is likely to increase in the future.

The retail cost per kg of Dryacide®, Protect-It® Insecto and PermaGuard® ranges from about \$3 to \$8. Bulk purchasing will lead to reduced costs. The application procedure will affect the cost, as the admixture of DE with commodity will require larger quantities of DE than a layer of DE applied as a top-dressing for combined use with fumigation or aeration. Similarly, due to different dosage recommendations, wet and dry structural treatments will differ in cost.

Diatomaceous earth can be used not only against storage pests but also against domestic and field pests. Diatomaceous earth is effective against pests that live in close association with humans, such as cockroaches, silverfish, mites, ants, houseflies, spiders, bedbugs, fleas and crickets (St Aubin 1991). It can be used to treat cracks, wall crevices, wall voids and attics to repel insects and deny harbourage in these areas (Quarles 1992). Of the 44 DEs registered in the United States, eight are registered for household use (Subramanyam & Roesli 2000). There are claims that DE is deadly to a wide range of field pests including gypsy moth, codling moth, pink boll weevil, lygus bug, twig borer, thrips, mites, slugs, snails, nematodes and mildew (Allen 1972). Diatomaceous earth can be applied directly to the soil, or to moist foliage using an electrostatic applicator (Quarles 1992). However, there are few data to support the effective use of DEs against domestic or field pests.

### *Resistance*

The high reproductive potential of stored product insects, with many generations occurring in a short period of time, can select for resistance if populations are exposed to continuous insecticide application. The persistence of DE on grain and structures could result in exposure of numerous generations of insects to the same material (Desmarchelier & Allen 1999). Originally, it was thought that because the mode of action of these inert dusts did not depend on affecting metabolic pathways, it was unlikely that insects would be selected genetically by their action, so that physiological resistance would not occur (Ebeling 1971). Although insects may develop a behavioural response to avoid contact with inert dusts, avoidance of DEs is known to occur in cockroaches (Desmarchelier & Allen 1999). Recently, researchers have shown that *Tribolium castaneum*, *Rhyzopertha dominica*, *Cryptolestes ferrugineus* and *Sitophilus zeamais* developed tolerance after five to seven generations of exposure to DEs (Korunic & Ormesher 1999; Stathers *et al.* 2000). However, because of the very different mode of action of inert dusts compared to synthetic chemical grain protectants and fumigants, cross-resistance is unlikely to develop, making it potentially a very useful non-selective tool for the management of insecticide resistance (Longstaff 1994).

### *Safe use of diatomaceous earths in storage*

When considering health and safety aspects of DE use in storage, there are two main areas: consumer safety and worker safety.

*Consumer safety*: DEs have extremely low toxicity to mammals. For example, Insecto® has a rat oral $LD_{50}$ > 5000 mg/kg (Subramanyam *et al.* 1994); silicon dioxide (the major constituent of DE) has a rat oral $LC_{50}$ = 3160 mg/kg (NIOSH 1977). DEs are considered 'generally regarded as safe' by the US Environmental Protection Authority (Anon. 1991). The Food and Drug Authority has exempted DE from requirements of fixed residue levels when added to stored grain (Anon. 1961).

Cattle, poultry and dog owners commonly use DEs as a feed mix to combat internal parasites (Allen 1972). Silica occurs naturally in vegetables and grains such as rice, and the average human intake from natural sources is about 200 mg per day (Quarles & Winn 1996). Silica does not accumulate in mammals as it is excreted as silicate in the urine (Desmarchelier & Allen 1999). Silica is used as a thickener in ointments and suppositories, as a filler in tablets, as an anti-caking agent in processed foods, in toothpaste, and to prevent clogging in hygroscopic powders (Blacow 1975; Budavari 1989; FDA 1995). Since protective amounts of DE on grain are often less than 0.1% w/w, and as 98% of DE is removed during processing, DE is not likely to become a health problem for consumers (Desmarchelier *et al.* 1996; Quarles & Winn 1996). The traditional method of cleaning grain by washing with water is also effective in removing DE (Desmarchelier & Paine 1988).

*Worker safety*: The only possible negative health effect comes from long-term chronic exposure to quantities of inhaled dust, and workers involved in DE application and/or handling of DE-treated grain should take appropriate safety precautions. The important parameters include the amount of dust, its particle size and the crystalline silica content of the DE (Desmarchelier & Allen 1999). During the process of sedimentation, geological forces can convert amorphous silica into forms of crystalline silica including the highly dangerous cristabolite. Exposure to crystalline silica dust is a known cause of lung disease (Hughes *et al.* 1998) and in 1997 the International Agency for Research on Cancer (IARC) classified it as a group 1, human carcinogen. This recent decision has caused much debate, details of which can be found in Goldsmith (1999) and Hessel *et al.* (2000). Fortunately, most DEs are mainly composed of amorphous (non-crystalline) silica which is classified by the IARC as group 3, not carcinogenic (Korunic 1998), and average < 3% crystalline silica (Quarles & Winn 1996). It should be noted that DE used in swimming pool filters can contain up to 60% crystalline silica and only DEs specifically registered for use as grain protectants should be used on stored grain or in storage structures.

The particles of a DE, or any type of dust, can be divided into two fractions: the inspirable fraction, which can be inhaled, and the non-inspirable fraction which cannot. The inspirable fraction contains large particles which are usually deposited in the nose, smaller particles which may reach the tracheobronchial tree or even the alveolar region of the lung, and very fine respirable particles (particles with an aerodynamic diameter < 7 μm) that can enter the lower region of the lung. Crystalline forms of silica are dangerous if the particles are small enough to enter the alveolar region of the lung where they cause scar (fibrotic) tissue (Desmarchelier & Allen 1999). Most DEs are amorphous (non-crystalline) silicas. The US Occupational Safety and Health Administration (OSHA) established limit for DE containing less than 1% of crystalline silica is 6 mg/m$^3$. Above these limits workers are required to wear dust masks (OSHA 1991). Why exposure standards vary between countries is not clear. A comparison of the Australian time weighted average (TWA) maximum exposure levels of workers for different dusts based on continuous exposure during an 8-hour day for 5 days per week are shown in Table 6.8. These figures suggest that DEs are potentially less hazardous to workers than wood or cotton dusts. However, in order to minimise risk, anyone involved in handling or applying any quantity of DEs should wear protective dust masks.

Safety precautions include reducing the amount of dust in the workplace, wearing masks to prevent inhalation, and ensuring the DE meets the regulatory specifications in terms of particle size and absence of crystalline silica. In broad terms exposure safety limits for amorphous DEs are similar to those for such common materials as cement and lime (Desmarchelier & Allen 1999). Interestingly, the use of DEs could reduce worker exposure to grain dust because small respirable particles of grain dust can attach themselves to a non-respirable particle of DE, actually reducing the amount of respirable dust in the workspace (Desmarchelier & Allen 1999).

Protective clothing (hats, overalls and gloves) should also be worn to prevent DEs from drying out the skin

**Table 6.8** A comparison of Australian time-weighted average maximum exposure levels (8 hours per day and 5 days per week) for a range of dusts. Adapted from National Occupational Health and Safety Commission (NOHSC 1995), cited by Desmarchelier & Allen (1999).

| Material | TWA maximum exposure levels (mg m$^{-3}$) |
|---|---|
| Uncalcined DE | 10 |
| Silica gel | 10 |
| Kaolin | 10 |
| Starch | 10 |
| Lime | 5 |
| Wood dust | 1–5 (depending on type) |
| Cotton dust | 0.2 |
| White asbestos | 1 fibre per ml of air |
| Blue asbestos | 0.1 fibre per ml of air |

(Desmarchelier & Allen 1999). A moisturiser with sun block should be worn if working outside. Safety glasses should be worn for eye protection. Protective clothing can be washed in water to remove DE particles. If a person is exposed to excessive concentrations of dust, they should be removed from the dusty atmosphere into fresh air, and should then wash their nose, face and exposed skin with clean water (McDonald 1989; Miles 1990).

### *Integration of diatomaceous earths with other storage pest management technologies*

The integration of DEs with aeration and fumigation is already finding application in Australia. Further innovative techniques for integrating DEs with other storage pest management technologies are likely to be used in the future.

The use of DEs in combination with heat sterilisation is more effective in controlling both *T. confusum* and *T. castaneum* (Table 6.9) than the use of heat sterilisation alone (Fields *et al.* 1997; Dowdy 1999). Heating may increase the rate of desiccation due to the control of water regulation being compromised by the DE (Dowdy 1999). These findings suggest that the temperature necessary for insect control in a processing facility may be reduced when diatomaceous earth is used in combination with heat sterilisation, as might the time period (Fields *et al.* 1997; Dowdy 1999).

Diatomaceous earths have historically been used as carriers for pesticides (Jefferson & Eads 1951). Some of the DE products on the market contain a percentage of insecticide, such as pyrethrum (0.1 to 0.2%) and piperonyl butoxide (1.0%) (Korunic 1998). As previously mentioned, insects differ in their susceptibility to DEs, and it may be more effective to use a mixture of DEs with synthetic or botanical pesticides against those insects that are difficult to control with DEs alone. Slightly higher *T. castaneum* and *S. oryzae* mortality was achieved when both the DE Fossil Shield® (1 g/kg) and a commercial neem product Neem Azal-T/S® (1 g/kg) were admixed with rice, in comparison to either treatment alone (Ulrich & Mewis 2000). Combining DEs with high concentrations of insecticides which knock down insects can, however, result in the reduced pick-up of DE particles by insects (Ebeling 1971; Le Patourel & Singh 1984). Recent work has found that the addition of DE to concrete surfaces, prior to applying insecticides, increases insect mortality and reduces the degradation of deltamethrin and piperonyl butoxide (S. Allen, pers. comm.).

It is advantageous if parasitoids and predators can be integrated with other pest management protocols. However, a recent study of the DE Protect-It® on the efficacy of the hymenopteran parasitoid, *Anisopteromalus calandrae*, found the parasitoid was very sensitive to direct contact with the DE (Perez-Mendoza *et al.* 1999). This confirms field observations in Zimbabwe (Stathers *et al.* 2000). Direct contact with the DE particles caused mortality, the particles adhered tightly to the cuticle surfaces and wasps were observed grooming extensively, attempting to remove the particles. Parasitisation of *Sitophilus oryzae* weevils was also reduced by the DE. However, during two-choice bioassays, the adult parasitoids were seen to avoid the diatomaceous earth-treated wheat, suggesting that some biological control may occur in stored grain even in the presence of DE (Perez-Mendoza *et*

**Table 6.9** Percentage mortality* of *Tribolium castaneum* adults following exposure to insecticidal dusts at 34°C and 50°C for 15 or 30 minutes. Source: Dowdy 1999.

| Exposure temperature (°C) | Exposure time (min.) | Control | Pennac-P® | Natural Guard® | Concern® | Insecto | Protect-It® |
|---|---|---|---|---|---|---|---|
| 1-day mortality | | | | | | | |
| 34 | 15 | 0.0 | 0.0 | 1.3 | 2.5 | 1.3 | 2.5 |
| | 30 | 0.0 | 2.5 | 0.0 | 1.3 | 1.3 | 2.5 |
| 50 | 15 | 0.0c | 7.5b | 12.5b | 8.8b | 8.8b | 36.3a |
| | 30 | 36.3bc | 38.8c | 35.0c | 51.3b | 53.8b | 100.0a |
| 7-day mortality | | | | | | | |
| 34 | 15 | 1.3 | 0.0 | 2.5 | 2.5 | 1.3 | 3.8 |
| | 30 | 0.0 | 5.0 | 3.8 | 6.3 | 3.8 | 6.3 |
| 50 | 15 | 3.8c | 8.8bc | 16.3b | 13.8b | 8.8bc | 41.3a |
| | 30 | 56.3bc | 50.0c | 51.3c | 63.8b | 62.5b | 100.0a |

*Within a row, means followed by the same letter are not significantly different ($P > 0.05$). Rows with no letters following means indicate no significant difference among treatments. $N = 4$.

*al.* 1999). Scale insect outbreaks along roads in citrus orchards were found to be linked to the effect of the mineral-based inert dusts being blown off the roads onto the parasitic Hymenoptera (Bartlett 1951). Parasitoid species differ in their host locating activities, a number of species penetrate grain masses while others remain mainly on the grain surface. Searching behaviour among these species could affect the degree of their exposure to DEs. It would appear that, like many insecticides, effective formulations of DE are non-discriminatory in their action against both pests and beneficial insects.

Preliminary laboratory work on the use of DEs in combination with entomopathogenic fungi *Beauveria bassiana* against *R. dominica* found that synergy existed (Lord 1999). When *B. bassiana* spores are admixed with wheat treated with 200 ppm DE, the *B. bassiana* $LC_{50}$ for *Rhyzopertha dominica* is reduced from 0.102 g/kg of wheat to 0.025 g/kg. DEs may induce changes in cuticular lipids that may affect attachment, germination and penetration of entomopathogenic fungi (Lord 1999).

## Botanicals

Mankind has achieved much in agricultural production, largely thanks to pesticides. Indeed, it is impossible to imagine modern agriculture could meet the demands placed upon it by a rapidly growing human population without effective chemicals to control pest damage. It would, however, be slightly pretentious to claim that humans invented pesticides when plants have evolved a variety of compounds that help protect them against the ravages of herbivorous pests and diseases. The most successful example of the commercial exploitation of plant compounds for agrochemical use is the pyrethroid group of insecticides. Pyrethroids are synthetically modified versions of pyrethrum originally extracted from *Tanacetum cinerariifolium* (*Chrysanthemum cinerariaefolium*) (Plate 15).

People have known about the insecticidal properties of certain plant species for millennia. Analysis of grain stored in Oriental (3000–30 BC), Greek (2000–200 BC) and Roman antiquity (500 BC–AD 476) has shown that various plants such as *Cymbopogon schoenanthus* (Plate 16) were used to protect stored food against insect damage (Levinson & Levinson 1999). Plants such as *Cymbopogon* spp are still widely used today in traditional farming systems (Srivastava *et al.* 1988; Swidan 1994). Other insecticidal plant compounds such as rotenone from *Derris elliptica* and nicotine from *Nicotiana tabacum* were widely used in agriculture until the arrival of synthetic organochlorines such as DDT during the 1940s. The neem tree, *Azadirachta indica* (Plate 17), is one of the more famous botanical pesticides and is still widely used today in South Asia and parts of Africa (Puri 1999).

The purpose of this section on insecticidal plant materials is to summarise the global information currently available about botanical usage in post-harvest pest management. More importantly, the shortcomings of the existing research base will be highlighted, giving a clear indication of what scientists in the future should focus upon. The use of botanicals is obviously not restricted to the post-harvest sector, with most research actually focusing upon the medicinal properties of plants or their pre-harvest pest management uses, where some good texts are available (Arnason *et al.* 1989; Prakash & Rao 1997; Wrigley *et al.* 1997; Chauhan 1999). Conversely, there are relatively few works that have focused upon the post-harvest use of botanicals (Dales 1996; Golob *et al.* 1999; Weaver & Subramanyam 2000), which is perhaps related to the general differences in food storage practices between developed and developing countries. Expectations in developing and developed countries about botanicals is a major subject divide which will be addressed in this section.

### Indigenous use and rediscovering the past

Farmers living in poor rural farming villages in Africa or Asia often use plants to protect their crops. Although the range and scale of plant species used for insecticides will vary from community to community, poor village communities in the developing world are familiar with the properties of local plants used not only for pest problems but as traditional medicines. Current usage of plant materials for the protection of stored products is largely at the small-scale level by poor rural farmers. Many general ethnobotanical surveys have been conducted to assess the use of plants by indigenous communities (Smith 1991; Chabra *et al.* 1993; Torto & Hassanali 1997). However, there are very few surveys that have been specifically conducted to assess the usage of plants for stored product protection (Cobbinah *et al.* 1999). Dissemination of ethnobotanical information may become increasingly problematic due to changes to national legislation enacted after the UN Convention on Biological Diversity (Wrigley *et al.* 1997).

Because of the lack of historical information, it is difficult to formally assess whether indigenous botanical usage is in overall decline. However, the use of insecticidal plants has more than likely declined since the advent of synthetic chemicals. Associated ethnobotanical information on the uses of plants may also be under threat as farmers increasingly rely upon commercial products, leading to a breakdown in the passing on of local knowledge between generations. On the other hand, scientific research on plant-derived compounds with pesticidal properties has been continually increasing throughout the twentieth century. Figure 6.23 indicates the rapid increase in research published on post-harvest botanical usage since the 1960s. Similar research trends on botanicals can be found in grey literature produced by government and international institutions (http://www.fao.org/inpho/).

Although difficult to assess, botanical research in the commercially sensitive private sector continues. Many companies already have large chemical libraries containing plant compounds still waiting to be screened for biological activity. Notwithstanding, in the days of combinatorial chemistry, many agrochemical companies claim that they are no longer investing heavily in bio-prospecting research. It remains to be seen, however, whether scientific tools such as combinatorial chemistry will be superior to evolutionary pressure in creating useful compounds for the twenty-first century.

Botanicals are typically pesticides, working in the same fashion as commercial synthetics. However, because of their natural source, many practitioners believe botanicals to be safer than synthetic alternatives. This is not entirely true, and some botanicals such as *Nicotiana tabacum* can be more toxic than many commercial alternatives. The difference between botanicals and synthetic pesticides is that botanicals often have several modes of action. Toxicity against insects may be expressed by (1) directly killing particular life stages of the insect, (2) interfering with mating or suppressing reproduction, (3) acting as a repellent or affecting host finding and selection in a way that prevents infestation or (4) reducing or preventing feeding, i.e. an anti-feedant. Many botanicals have more than one mode of action, and these can all contribute to the efficacy of the botanical in reducing insect damage. Other advantages could be that the potential for insects to develop resistance to a botanical with many modes of action is reduced. Environmental or vertebrate toxicity may also be lower when botanicals rely more upon anti-feedancy and repellency than, say, direct insect toxicity.

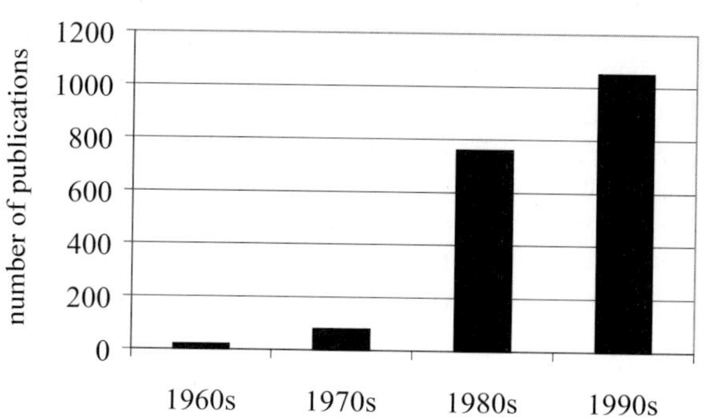


**Fig. 6.23** Number of published references from the latter half of the twentieth century citing research work on the use of plant materials tested against a range of stored products.

### On-farm or in the supermarket?

Most post-harvest usage of plant materials is done on a small scale to protect against insects attacking grain and legumes during storage. Generally, wild growing plant materials are collected from the local surrounding area, processed in some way and then mixed with the commodity. As sufficient plant material must be collected, there are practical limitations on the quantity of grain that can be treated in this way. Most farmers who use botanicals in their store use them to treat small amounts of high-value crops such as cowpea. Medium- to large-scale farm operations that have several tons of stored produce to protect would find it enormously difficult to collect enough plant material to treat the grain at a suitably high enough concentration. Treatment of stored produce with botanicals under the large-scale commercial operations found in developed countries would, therefore, be impossible without the equally large-scale cultivation of insecticidal botanicals. There are, however, some promising developments using the volatile properties from essential herb extracts to overcome the practical constraints of treating large amounts of commodity (Shaaya *et al.* 1991; Gbolade & Adebayo 1993; Don-Pedro 1996; Keita *et al.* 2000). As herbs such as sweet basil, *Ocimum basilicum,* and lemon grass, *Cymbopogon citratus*, are already widely used as culinary spices, their safety and registration is less problematic. Using essential oils as a fumigant for stored grain and legumes could be particularly relevant as methyl bromide is removed from use. However, there is only very limited evidence (Shaaya *et al.* 1997) demonstrating the penetrative power of oils, i.e. their ability to treat empty spaces is much greater than their ability to penetrate through grain bulks, which must occur if plant extracts are to emulate fumigant gases. Another potential method for the large-scale application of botanicals could involve the use of vegetable oils. Particularly with respect to leguminous pests such as bruchid beetles, many different edible oils have been shown to offer good protection (Ivbijaro *et al.* 1985; Don-Pedro 1989; Singal & Singh 1990; Cockfield 1992).

Even if the practical constraints to using botanicals on a large scale can be overcome, such as by cultivation of botanicals, there are many problems when it comes to the safety assessments and official registration of insecticidal additives required in many countries. For these reasons, post-harvest use of plant materials will continue to remain predominantly an indigenous practice at the subsistence level farm store for many years yet. In this context, botanicals have many advantages because they are normally gathered locally by farmers and can provide an inexpensive method of pest control during storage. For the majority of farmers in the developing world, commercial insecticides are often too costly or unavailable. Similarly, many uneducated farmers use synthetic pesticides inappropriately, leading to environmental and human safety hazards as well as promoting insecticide resistance. Poor farmers will continue to use plant materials for stored product protection, and so research on post-harvest botanicals should focus upon optimising small-scale farmer usage of these materials.

Small-scale use of plant protectants for storage has been well researched in the context of testing plants against a range of stored product pest species in laboratory trials. Dales (1996) describes in detail the work which has been done in this field and provides a good list of published references dealing with botanical efficacy in stored product entomology. Similar lists of plants can be found in Prakash and Rao (1997). Work continues to identify plants with activity against stored product insects (Obeng-Ofori *et al.* 2000; Owusu 2000). However, much of this laboratory work is not of immediate practical value without greater research on the acceptability and availability of the botanical treatments. There are relatively few published works referring to field trials or farmer usage of botanicals in stored product protection (Baier & Webster 1992). Thus, there is a real need to increase field-based research so that the impact of differences in the susceptibility of insect species and varying climatic effects can be measured on botanical efficacy under local conditions. Stored product pests are well-known to have regional biotypes that can vary quite considerably in their behaviour and physiology (Ofuya & Credland 1995). Similarly, the insecticidal effect is only one of many criteria that require assessment before understanding which plant material may be the best one to use. Research on the economic effectiveness of botanicals and their acceptability by farmers will help clarify whether the plants adversely affect the flavour, colour or germination characteristics of treated grain (Plate 18). More importantly, adverse effects of insecticidal plants upon the consumer need to be assessed so that any potential vertebrate toxicity is understood.

#### It's all in the technique

The processing of plant materials to be used for treating commodity to be stored varies enormously. The part of plant used is quite important and can range from the leaves, fruit, root, flower, seed, bark and sometimes the entire plant in the case of perennial weeds. One of the better-known examples of variable application technology can be found with regard to the neem tree, *Azadirachta indica*. It is well known that most of the active ingredient found in neem, the compound azadirachtin, is found in the oil extracted from the seed (van der Nat *et al.* 1991). Many farmers will use neem oil as a stored grain protectant, but others choose to use neem leaves. Indigenous practice results in a myriad of subtle differences in application technology. For example, some farmers use fresh neem leaves, others use dried neem leaves. The leaves may be whole or ground into a powder, and the material may be placed in layers or admixed with the commodity. Some farmers also make a water extract of the neem leaves (Plate 19). The fresh leaves may be pounded before adding them to hot or cold water. The commodity is then dipped into the water extract or the extract is poured or sprayed over the commodity (Plate 20).

As highlighted in Fig. 6.24, the range of choices faced by a farmer using an insecticidal plant is quite considerable. Many of the methods chosen may affect the availability of the active ingredient(s), the type of active ingredient present and the concentration ultimately applied to the commodity.

The reasons why a farmer chooses a particular application method hinge upon several practical constraints. As with most pest management activities, there is a trade-off between the amount of invested time and resources for an expected reward or outcome. Although many farmers familiar with neem recognise that the oil from the seed is most effective, they choose to use leaves because they require less effort to use. Neem seeds often ripen some months before most agricultural crops are ready to harvest, meaning farmers have to collect and store the seeds

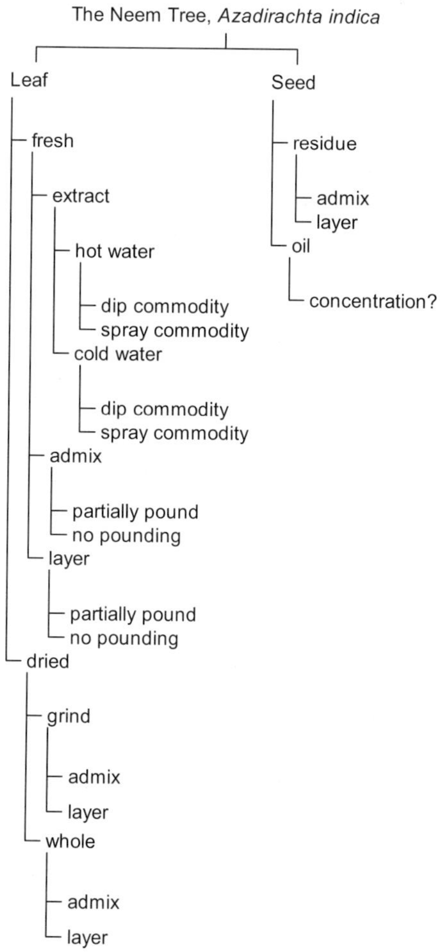


**Fig. 6.24** Application choices of a typical botanical used in post-harvest protection.

until required. More importantly, oil extraction is done by pounding the seeds by hand (Plate 21). Oil extraction from neem is, therefore, laborious. The oil is also smelly and bitter, making the job unpleasant, and can lead to tainting of the wooden mortar and pestle used as well as the treated commodity. The analysis of the farmer's decisions for choosing particular plant materials and application methods is, therefore, an important issue to address for researchers involved in screening plant materials as suitable grain protectants. The most efficacious plant material is not necessarily the best one to use.

The most common way of using plants in post-harvest protection is the admixture of powdered plant material. In addition to the admixture of oils and making water extracts of plants, more sophisticated methods include the use of essential oils and organic solvent extracts. Essential oils of plants are obtained via steam distillation, and essential oil extracts of some plants have been found to act as effective fumigants (Shaaya *et al.* 1991; Rahman & Schmidt 1999; Tunç *et al.* 2000). Such sophisticated methods are inappropriate for small-scale subsistence farmers but they may eventually have commercial applications for large-scale storage.

### Resource-poor or resource-rich?

The United Nations Convention on Biological Diversity (CBD) was agreed in Rio de Janeiro in 1992 and has now been ratified by over 120 countries. The treaty reaffirms the sovereign rights of States over their own biological resources, and it is designed to protect indigenous knowledge systems, such as how plants are used, as a form of intellectual property. The CBD states that benefits from biological resources, including ethnobotanical information, should be shared equitably among all those involved. It is, therefore, in the interest of signatories to the CBD to invest in the conservation of their biological diversity and to add value to their biological resources by establishing current knowledge systems of indigenous populations. The CBD has given many countries the incentives and the means to implement appropriate legislation to protect indigenous knowledge. The process of scientifically validating ethnobotanical information will add value to a State's biological resources and, in conjunction with the CBD, promote sustainable ecosystem management (Kate & Laird 1999).

There are, however, many potential caveats with the CBD, and its impact upon further research to understand the usage and safety of insecticidal botanicals is not clear (Rosendal 2000). Subsistence farmers in the developing world are often termed 'resource poor'. However, with respect to botanicals and protecting the harvest, subsistence farmers could be considered to be 'resource rich' in their understanding and use of indigenous plants. Farmers are usually quite happy to share their knowledge with others, and there is a danger that national governments enact legislation which severely restricts communication and collaboration in order to protect knowledge systems (Lacy 1995). Such legislation could severely slow down the process of scientific research, ultimately reducing any benefits that may be achieved through pesticidal botanical research (Steinberg 1998).

Bio-prospecting and research on bio-active plants is likely to continue into the twenty-first century for a number of reasons:

- Increasing government regulation on food standards and food safety, particularly in the United States and the European Union, is restricting the number of commercial products available for pest control.
- Consumers are increasingly demanding reduced pesticide usage because of perceptions that pesticides are dangerous to human health and the environment. This may only be marginally based upon factual evidence, and more likened to widespread chemophobia. Well-publicised cases of pesticide poisoning and the withdrawal of commercial products cultivates often false perceptions about what is natural or safe, leading to scepticism and fear of science.
- Organic, more natural and less intensive agriculture is considered by many societies to be more environmentally and ethically sustainable as part of an integrated pest management approach. Pest control with natural products from plants fits in with this ethos despite the fact that plant extracts may contain highly toxic compounds.
- Plant materials can be more cost-effective than synthetics. This is particularly relevant to resource-poor farmers in developing countries where indigenous use of ethnobotanicals is widespread.
- The need for new and even safer alternatives to current commercial pesticides will continue to encourage scientific investigation on natural products.
- Industry will carry on the search for new modes of action resultant from problems associated with the development of insecticide resistance.

The main constraint to increased promotion of plant materials for stored product protection is the assessment of their potential toxicity to vertebrates, and safety concerns (Copping 1996). In many countries existing registration procedures for pesticides require a barrage of tests which will be difficult to apply to a botanical that is a mixture of compounds with various levels of toxicity, differing modes of action and synergistic effects. Scientifically understanding these complexities will be expensive and laborious if existing procedures for synthetic compounds are used for the licensing of botanicals. There is, as of yet, no new registration pathway developed which could be used for botanical products (Isman 1997). However, there is a widespread perception that registration data will need to be modified if botanical usage is to increase (Rejesus 1995). Endorsement of existing, albeit optimised, indigenous practices should be a relatively simpler task than commercial registration. Countries where farmers currently use botanicals in post-harvest protection still require some level of safety assessment before existing practices are institutionally promoted. It is hoped that the scientific community begins to use the existing knowledge base of potential insecticidal botanicals to focus efforts upon developing suitable vertebrate toxicity data.

## Rodent control

### Identification of rodent activity

Rodent control operations should always set objectives to obtain 100% control. Lower levels of mortality will allow the residual population to recover rapidly to pre-control levels as a result of the high breeding potential of rodents.

The successful application of control measures to a rodent infestation requires a number of components. Of prime importance is the need for pest control operatives to have a comprehensive understanding of the nature of the infestation. It is, therefore, essential to identify the species present. The efficacy of control techniques employed is dependent on the behaviour, biology and susceptibility of the specific rodent species present, and without this knowledge control techniques are unlikely to be successful.

The most reliable way to identify the species is to obtain either live or dead specimens of the rodents involved, or at the very least to have had a good sighting of them, sufficient to be able to make a reliable identification. The nocturnal and secretive nature of rodent activity means that such sightings do not always occur. Alternatively, identification can be derived by utilising the presence of the many signs and traces of infestation that rodents inevitably leave behind, and which are species-specific. In addition, the traces will provide information on the scope of the infestation.

It is also essential to know how extensive the infestation is and to have some idea as to the size (numbers) of the population concerned. Without this additional knowledge it is unlikely that the control techniques being applied will either be placed sufficiently extensively or intensively to be likely to achieve successful control.

#### *Survey, signs and traces*

While rodents infest a particular area they will inevitably leave behind traces of their activity. These should be used in a survey to determine whether or not an area is infested and the extent and severity of infestation.

*Smears*: Greasy smears are left behind by rodent fur as it rubs regularly against surfaces. These can not only provide information as to where the rodents are active, but can also be species-specific. Examples include the continuous and discontinuous loop smears left behind by *Rattus norvegicus* and *Rattus rattus*, respectively, when negotiating overhanging obtrusions such as beams in roof spaces.

*Runs*: Compacted earth or dust and dirt-free runways are formed by continual rodent activity along a particular route. Many commensal rodents prefer to utilise familiar and safer routes when moving around their territory. This inevitably can lead not only to the development of smears left by the grease on their fur, but also to other signs of regular usage.

*Droppings*: Rodents can produce up to 60–70 droppings of faeces per day depending on the species. These inevitably remain in the environment they infest, unless removed. Their size and shape can be species-specific and their distribution can indicate areas of activity.

*Damaged goods and structures*: Materials may be found damaged or partially eaten by rodents. When surveying an infestation a search should be made for such damaged materials. The nature of the damage, for example, the way in which grains have been damaged, or the width of the incisor grooves on damaged materials, can be indicative of specific rodent species and the extent of the infestation.

*Urination pillars*: These are caused by house mice urinating regularly on a particular spot. The pillars are usually found at what seem to be the edges of territories and are probably indicative of intensive territory marking, particularly on spots where immigration or invasion from adjoining territories is likely.

*Smell*: Different rodent species, particularly the global commensals, seem to generate their own particular smell when infesting closed areas. These smells are difficult to describe other than as 'musty', but are recognisable to those who have been undertaking control for some time.

*Burrows or holes*: Burrows or holes are signs of rodent access to either subsurface harbourage or to other areas of the infested environment. These can be species indicative and are a clear sign of infestation.

*Footprints*: Rodents passing over wet, dusty or even granular surfaces can leave behind footprints. These can be species indicative, particularly when more clearly defined.

The objective of the survey must be to build up a picture of the infestation which is as complete as possible and should ideally be recorded, at least on a simple map. Subsequent application of management and control techniques can then be specific to the areas of activity and the susceptibilities of the particular environment. Storage environments are by their very nature three-dimensional and therefore all survey and subsequent control must take account of the degree to which the three-dimensional habitat is being used by the species of rodent involved.

It is perfectly possible for two or more species to be infesting the same storage facility. If this occurs it is usually not because the species are competing for any resource that they require, usually either food, water or harbourage. If competition does occur, the larger and more aggressive species will usually exclude or even predate upon the weaker.

### Integrated rodent management

The effective control or management of a rodent problem can be achieved by the combined use of the following three elements:

(1) Management of the environment: to reduce the availability of food, water and harbourage – as these are reduced, so will be the carrying capacity of that environment, making control easier;
(2) Exclusion and proofing: taking measures to prevent rodents accessing the food store; and
(3) Control: the use of rodenticides and traps, to kill and remove the rats directly.

#### *The food store*

If they are to survive and increase in numbers rodents require food, harbourage (somewhere to live and breed) and water. House mice (*Mus* spp) are able to survive without free water, but will certainly use it if it is available and will certainly breed more successfully if water is present.

Control in food stores is clearly complicated by the existence of the one factor above all others, food, in plentiful and, usually, unlimited quantity. In addition, particularly in food stores that have not been specifically constructed to exclude rodents, harbourage is usually readily available either within or nearby the store. However, in these stores, unlike most other situations in which rodents cause problems, those responsible for the control of the rodents are also responsible for most other aspects of store management. This makes the food store an ideal place to apply a fully co-ordinated integrated rodent management (IRM) strategy. Such a strategy must incorporate each of the following elements: management of the environment around and outside the store; management and rodent proofing of the storage structure; management of the food and the storage practice within the store; application of rodent control measures; monitoring of pest populations and the efficacy of control measures.

Any food store must lie within an environment that is likely to be infested by rodents, whether this is in a rural agricultural area or within a more built up or urban area. It will inevitably be, therefore, at continual risk of being invaded by rodents from these adjoining habitats. Figure 6.25 indicates the likely continuous movement of rodents between and within the habitats involved, including the store (the store matrix).

The objective of the first two elements of the IRM strategy must therefore be to prevent the rodents entering the store at all. This can be achieved, firstly, by making the environment around the store as large and as inhospitable to rodents as possible, so reducing the number of rodents that actually reach the store; secondly, by proofing the store, so that those rodents that do reach the store cannot enter.

### *Management of the environment outside the store*

The external environment must be as clean, tidy and free from vegetation and other encumbrances as is possible. In practice, there should be an absolutely clear zone around the outside of the store of at least 10 m and ideally 30 m. This zone will not act as a barrier to rodent movement, but will act as a deterrent. Rodents do not like crossing open spaces where they may be seen and exposed to predators. The space must be kept clear of any other material that might act as a harbourage, such as old packing crates and sacks, and vegetation must be kept as short as possible. It is also essential not to allow any trees or other structures to overhang the store or for vegetation to grow around the store.

In one case where a colony of *R. norvegicus* was living under uncut vegetation around a food store in Sri Lanka, grazing by a herd of goats for a few days each month not only reduced all the vegetation, but also destroyed the rat burrows. This resulted in a significant reduction of the rat infestation, at no cost!

### *Rodent-proofing the store*

Stores are proofed to prevent rodents gaining access. The following actions or precautions should be employed to ensure stores are adequately protected.

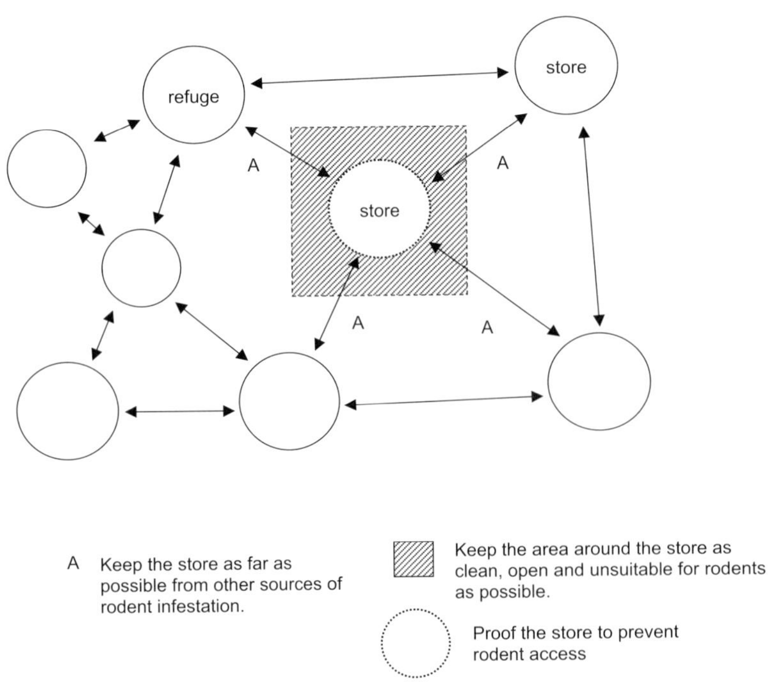


**Fig. 6.25** The store matrix. Source: Buckle 1994.

- All holes need to be blocked. If ventilation is required then galvanised metal sheet with perforation holes of less than 5 mm should be fixed across the access point.
- Doors must be rodent-proof. This means that there should be no gap larger than 5 mm under or behind any doors. If this is not possible to achieve, for instance behind sliding doors, then a movable metal barrier should be fixed in place behind the doors when they are closed or not in use. The barrier should be at least 1 m high and should include a back projection (lapping) of at least 15 cm. The barrier should lie within a groove in the floor surface and should be fixed flush with all adjoining surfaces.
- The outlets of all rainwater downpipes should be covered with a wire mesh or at their top with a wire balloon. Both will need to be removable for cleaning and must be maintained. Where the downpipes access directly into drainage systems and cannot be proofed at their base, they must be proofed at their apex.
- All pipes running down the side of the building should be proofed with a metal cone guard.
- A smooth strip not less than 10 cm wide should be painted around the entire outside wall of the building. This will help prevent rodents climbing the wall. Several coats of gloss paint may be required.
- A trench, at least 30 cm depth and 20 cm wide, should be dug around all exterior walls and filled with fine pea gravel. This should be maintained free of vegetation.
- Sound foundations should be constructed.
- A concrete curtain wall some 100 mm thick, extending not less than 600 mm below ground with the base turned out some 300 mm away from the building, in the shape of an L, should prevent rodents digging under a building.
- Rat guards on downpipes should be fixed high enough above ground level to clear passers-by or vehicles, but should not be sited above features such as mouldings, sills or branch pipes near enough to the downpipes to give rats an alternative route into the building.
- Horizontal pipes and cables running between buildings at high level should be proofed with circular metal guards projecting 23 cm all round, similar to the guards used on ships' mooring lines.
- All electrical cables and telephone cable access points must be rendered rodent-proof. Where these are aerial they should be proofed with a cone guard.
- All potential sources of water must be proofed, turned off, isolated or otherwise removed. If this is not possible then they need to be regularly monitored for rodent activity.
- All food spillage should be removed immediately and disposed of so that it does not provide a food source or attraction for rodents.

Where the store is not a permanent structure and the above elements are not relevant then they should be adapted in whatever may be appropriate. For example, small structures should be raised off the ground by a minimum of 1 m and metal cone guards or anti-climbing bands fitted to the supporting legs. Wherever possible the legs supporting food stores should be of smooth materials and kept as smooth as possible throughout their functional lives. Where raised storage is used, only removable step access should be provided. Steps should be removed as soon as access is completed. If at all possible food storage facilities should be kept separate from domestic and living accommodation and should be built at least 5 m from these buildings. Grass, palm or thatch should not be used in the construction of roofs for storage facilities if they can possibly be avoided, since they provide ideal harbourage for rodents.

All buildings are different and it is not possible to identify all potential proofing measures that will be required for each store. A comprehensive proofing strategy should be developed and maintained for each store.

#### *Management within the store*

It is essential that high levels of store management are maintained within the store. Food should be stored in relatively small stacks, which can be individually inspected and monitored. These stacks should be at least 1 m apart and 1 m from any walls to facilitate inspection, monitoring and the application of rodent and other pest control measures. Where there is evidence of rodent activity in the roof areas, the stacks should be kept at least 2 m from adjacent walls and at least 2 m from supporting beams and other structures. All stacks should be stored off the ground on pallets or similar supports. Only food should be stored in a food store! The storage of any other materials should not be permitted but should be kept in a separate building. All stacks should contain food material that entered the store at the same time. Different aged stocks should not be stored together. Returned goods should be stored separately. Good records of all stacks, their contents and their history should be maintained. Good stock rotation records should be maintained.

If rodent activity is found in any stack then rodent control measures should be applied immediately. If such measures are not considered likely to provide 100% control within the stack then the stack should be dismantled and removed. Measures must be taken to prevent rodents escaping into adjacent stacks as dismantling takes place. All potential rodent harbourages must be removed from the store. All potential access to free water sources must be prevented. For example, access to water in toilets must be prevented by proofing the toilet or enclosing the water. It is essential to ensure that at all times any spillage and other problems that may result in food lying around are removed and disposed of immediately.

### Rodent control methods

While the maintenance of high standards of proofing and store management are essential to maintain a rodent-free store, infestation will still occur and it will be necessary to have effective rodent control measures available. In addition, the application of good housekeeping will help reduce the availability of spilt and damaged food and will hence increase the efficacy of rodent baiting. Nevertheless, direct rodent control will still be a necessary component of most IRM strategies, though control will be helped by the successful application of the other elements of the IRM strategy.

#### *Chemical methods*

Rodenticides to be effective must have a high level of toxicity to rodents. However, rodent are mammals as are human beings, domesticated livestock and many other animals that share our world. Inevitably, rodenticides can be toxic to humans and other non-target animals and great care must be taken during their application. It is essential, therefore, that rodenticides are only ever used by those who are trained and familiar with their use, that they are used with maximum attention to safety at all times and that the recommendations for use on rodenticide labels are always followed. Furthermore, in food storage environments it is not only the non-target animals that need to be protected, but it is also essential that the food in the store does not become contaminated with these chemicals.

Rodenticides can be divided into three groups: the acute, subacute and chronic rodenticides. Historically, the acute rodenticides are the oldest: many have been used for the control of rodents for hundreds and, in some cases, thousands of years.

##### *Acute rodenticides*

These are frequently cheap to produce and to buy and they are often readily available. In the developing world acute rodenticides can often be purchased from petrol filling stations, retail chemical outlets and even from market traders. Unfortunately, they have two characteristics that restrict their efficiency. Firstly, they usually work very rapidly, sometimes in minutes and certainly in a few hours, and secondly they tend to cause death in a relatively painful way. Toxicities of these compounds are shown in Table 6.10.

The first of these characteristics might be regarded as an advantage. However, the problem lies not in the acute rodenticide itself, but more with the feeding behaviour of the rodents that are being controlled. Many rodents are neophobic, suspicious of and avoiding new objects. This neophobia extends to a reluctance to feed immediately on new foods, particularly when placed in unusual places.

While this neophobia may be gradually overcome, it means that the first time a rodent feeds on the bait it is more likely to be a small, tentative feed, consuming only a small amount of bait that is unlikely to be sufficient to deliver a lethal dose of the rodenticide. The subsequent rapid action of the rodenticide and the development of unpleasant, but not lethal symptoms, will prevent further

**Table 6.10** Toxicity (acute oral $LD_{50}$ in mg $kg^{-1}$) of some acute and subacute rodenticides to commensal rodents. Source: Buckle 1994.

| Compound | *Rattus norvegicus* | *Rattus rattus* | *Mus musculus* |
|---|---|---|---|
| Zinc phosphide | 27.0–40.5 | 21.0 | 32.3–53.3 |
| Scilliroside | 0.4–2.6 | | <0.2–0.47 |
| Sodium monofluoroacetate | 0.2–5.0 | 0.1–1.0 | 6.3–16.5 |
| Alphachloralose | 200–400 | | 190–300 |
| Thallium sulphate | 16.0–25.0 | | 16.0–27.0 |
| Bromethalin | 2.0–2.5 | 6.6 | 5.3–8.1 |
| Calciferol | 43.6–56.0 | | 23.7–42.5 |

feeding on the rodenticide bait. The rapid onset of the symptoms enables the rodent to associate cause and effect and it will become 'bait shy' or 'poison shy'; it will avoid both the bait base and the taste and smell of the rodenticide in the future.

Prebaiting (Fig. 6.26), allowing the rodent to feed first on unpoisoned bait, may overcome problems due to neophobia, at least in part. This may take a few days in house mice and up to 2 weeks or more for rats. When consumption has been maximised, the unpoisoned bait is removed and is replaced with the same bait base, but containing the acute rodenticide.

Unfortunately there remains some neophobic response in the rodents, particularly when they detect the taste and or smell of the acute rodenticide in the bait. It is thus very difficult to obtain the target 100% mortality with acute rodenticides. Average mortalities of about 80% are obtained if thorough prebaiting has been applied. This is not ideal, but may be acceptable if there are no available alternatives.

For those rodents that do not exhibit neophobia the problem persists. Such rodents, house mice (*Mus* spp) for instance, tend to be inquisitive and ready to feed on new baits, but feed only very sparingly from any particular bait point. Thus, a sublethal dose is consumed, symptoms develop soon afterwards, feeding stops and shyness follows recovery. Here it is necessary to use large numbers of bait points, to increase the chance that a lethal dose of acute rodenticide is consumed.

The efficacy of the acute rodenticides is often overestimated, because the rapid onset of symptoms and death (for those rodents that have consumed a lethal dose) results in the rodents dying in the open, rather than in their burrows and harbourages as with the slower-acting chronic rodenticides. It is important to remember that it is not the number of rodents killed that matters, but the number left behind. In general, an average mortality of some 70% might be expected from the acute rodenticides, even when prebaiting is used.

The acute rodenticides can, however, have their uses, particularly when a rapid reduction in population is required or where there is resistance to the alternative available chronic poisons. Some of the more commonly used acute rodenticides include the following:

*Alphachloralose,* applied at 2–4%, is a narcotic and slows metabolic processes when consumed. The rodent produces insufficient heat to be able to maintain its body temperature and dies of hypothermia. It is, therefore, most effective against small rodents, which have a high surface area to volume ratio, and at low temperatures (below about 14°C).

*Sodium (mono) fluoroacetate (1080)* is highly toxic to rodents, and works by blocking the tricarboxylic acid cycle, causing accumulation of citric acid and leading to convulsion and respiratory or circulatory failure. Due to its high toxicity, it is used at 0.08–0.5%, and in the absence of any antidote it is strictly regulated in most countries.

*Fluoroacetamide (1081)* is an analogue of 1080 and is very similar in most respects. However, it is slightly less toxic and is therefore applied in baits at slightly higher concentrations, usually at 1–2%.

The very high toxicity of both 1080 and 1081, as well as their relatively odourless and tasteless char-

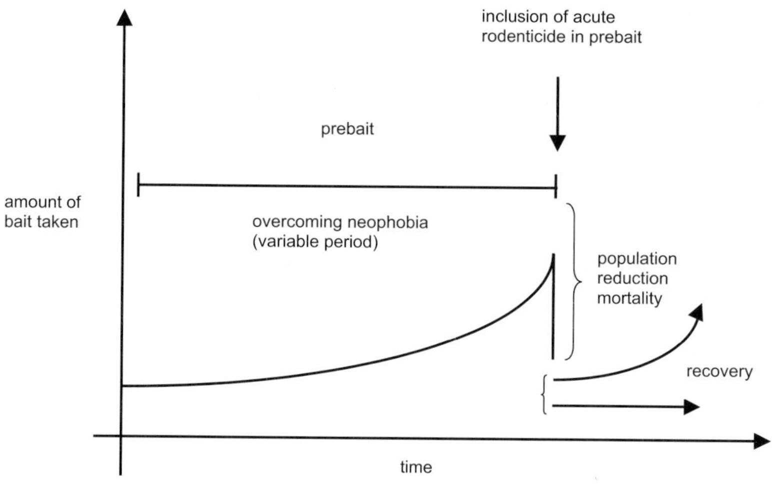


**Fig. 6.26** Use of an acute rodenticide requiring prebaiting.

acteristics, means that they can be used with some success in the absence of prebaiting. These same characteristics, however, restrict their use on grounds of non-target risk.

*Red squill* is an extract from bulbs of the Mediterranean plant, *Drimia* (*Urginea*) *maritima,* which causes death by convulsion. It is available in a purified form as 'Silmurin' which is used from 0.015% to 0.05% in baits. However, it is banned is some countries as being a cruel poison. Both unpalatability and bait shyness have been reported. Its efficacy against *R. norvegicus* is similar to other acute rodenticides but efficacy against other commensal rodents is questionable (Meehan 1984).

*Thallium sulphate* is recommended for use in baits at 0.5–1.5% and seems not to induce bait shyness, possibly because it is relatively tasteless. However, largely as a result of its toxicity to non-target species and the lack of an antidote, it is no longer widely available.

*Zinc phosphide* is probably the most commonly used acute rodenticide worldwide and is widely available to non-professional users. It is usually available as a dark greyish powder, smelling of garlic, and is applied at 1–5%, although most commonly at about 2%. Zinc phosphide acts by liberating phosphine into the stomach, which subsequently enters the bloodstream, causing heart failure and organ damage. In terms of its use and efficacy and availability it is probably the most effective acute rodenticide currently available.

The acute rodenticides are not as available as they have been in the past and many are no longer available, either because of their low efficacy and hence low usage or because their use may have been restricted. Examples of acute rodenticides that are now rarely used include ANTU, crimidine, gophacide, norbormide, silatrane and pyriminyl.

### *Subacute rodenticides*

These compounds are similar to the acute rodenticides, but differ in one main respect in that the symptoms are slower to appear and they may not be as dramatic as those exhibited by the acutes. Hence, a lethal dose need not necessarily be consumed in a single feed and feeding may continue for 24 hours or more. Death is normally delayed for several days (Table 6.10).

A period of anorexia may be induced in animals that have taken both lethal and sub-lethal doses. This is the stop-feed action claimed for these compounds. This is only a benefit if a lethal dose has been consumed prior to cessation of feeding. If a sub-lethal dose has been consumed, death will not occur.

*Bromethalin* is a pale yellow crystalline solid, which works by uncoupling oxidative phosphorylation in cells of the central nervous system, causing tremors, paralysis of the hind limbs and convulsion. It is used in baits at either 0.005% or 0.01% and is relatively rapid acting, death usually occurring in 24 hours. This compound is effective against many rodent species and prebaiting is not considered necessary.

*Calciferol* is a naturally occurring, fat-soluble vitamin in two forms: ergocalciferol (Vitamin D2) and cholecalciferol (Vitamin D3). In high doses, calciferol promotes the absorption of calcium, which leads to calcium deposition in the kidneys and cardiovascular system and death through hypercalcaemia. The process can be slow, taking some 2–3 days. The stop-feed action of calciferol is apparent in rats but less so in house mice where effective control can be achieved without prebaiting. Against *R. norvegicus*, however, the stop-feed action precludes its use without prebaiting.

### *Chronic rodenticides*

The development of anticoagulant rodenticides in the early 1950s revolutionised rodent control. Up to this time, the limitations of acute rodenticides had made it virtually impossible to achieve a 100% clearance of a rodent population and repeated applications of acute rodenticides to rapidly recovering and 'shy' rodent populations were required.

All the chronic rodenticides are anticoagulants. They inhibit effective blood coagulation, which results in the animal bleeding to death. Anticoagulants are much more effective than alternative rodenticides for several reasons. Firstly, in contrast to acute rodenticides, where symptoms are apparent within minutes or hours of bait consumption, anticoagulants take days to take effect. Minimum time to death is 2–3 days, with average times being 7–8 days; in some cases death may not occur for 14 days.

Secondly, anticoagulants have a relatively painless mode of action. While it would not be correct to say that the process is always painless, the level of discomfort is not comparable to the actions of the acute and subacute rodenticides.

The combination of these two factors enables rodents to keep feeding on the anticoagulant rodenticide baits until a lethal dose has been consumed. If other aspects of the rodenticide treatment have been applied correctly, then 100% mortality can be achieved.

In any chronic rodenticide treatment it is essential to ensure that baits are placed correctly, as would be a requirement for any rodenticide treatment, and that the rodents are able to feed on the baits on a daily basis, possibly for several weeks, even though the daily consumption of bait may be very low. This is termed 'surplus' or 'saturation' baiting. It is essential that the baits are visited and replaced frequently, daily if necessary (Fig. 6.27).

The first of the anticoagulants to appear on the market in the early 1950s was warfarin, named after the Wisconsin Alumini Research Foundation that introduced the rodenticide. Within a few years additional anticoagulants were available, including diphacinone, pindone, coumatetralyl, coumachlor and chloropacinone, now known collectively as the first-generation anticoagulants. They all have broadly similar levels of toxicity to the commensal rodents although there is some variation between them in their toxicity to other species. Unfortunately, the initial widespread success of these early anticoagulants was not maintained. Resistance not only to warfarin but to all the first-generation compounds was detected in some *R. norvegicus* and *Mus musculus* populations by the late 1950s and early 1960s.

> 'Anticoagulant resistance is a major loss of efficacy in practical conditions where the anticoagulant has been applied correctly, the loss of efficacy being due to the presence of a strain of rodent with a heritable and commensurably reduced sensitivity to the anticoagulant.'
>
> *Greaves 1994*

*Rattus norvegicus* resistance to the first-generation anticoagulants was found first in the United Kingdom and subsequently in many other industrial countries. While resistance may be present in some areas this does not mean that it is present in all areas, and first-generation resistance in *R. norvegicus* continues to be very localised. *Rattus rattus* have lower susceptibility to the first-generation anticoagulants and resistance has also been found in some populations. Resistance to these compounds in house mice is more widespread; in the United Kingdom, for example, resistance is now so widespread that there is little point in using a first-generation anticoagulant.

More potent compounds became available in the mid-1970s. Difenacoum and bromadiolone were marketed as an effective method of control of resistant populations of both *R. norvegicus* and house mice. Most recently, difethialone has been released. These three compounds, the second-generation anticoagulants, are

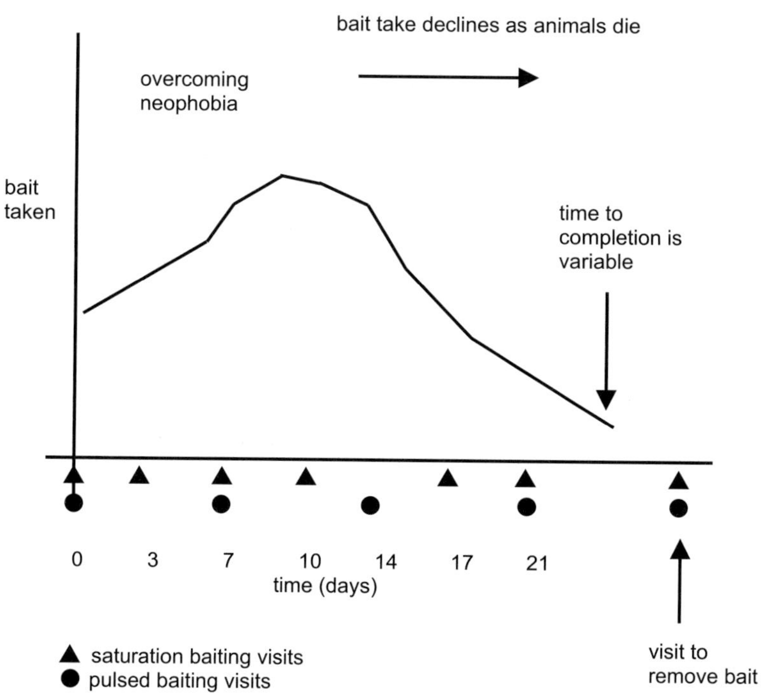


**Fig. 6.27** Progress of anticoagulant treatment against *Rattus norvegicus.*

| Compound | *Rattus norvegicus* | *Ratus rattus* | *Mus musculus* |
|---|---|---|---|
| Defenacoum | 1.8 | | 0.8 |
| Bromadiolone | 1.1–1.8 | | 1.75 |
| Brodifacoum | 0.22–0.27 | 0.65–0.73 | 0.4 |
| Flocoumafen | 0.25–0.56 | 1.0–1.18 | 0.79–2.4 |
| Difethialone | 0.56 | | 1.29 |

**Table 6.11** Acute oral $LD_{50}$ (mg $kg^{-1}$) data for the second-generation anticoagulants against warfarin-susceptible *Rattus norvegicus, R. rattus* and *Mus musculus.* Source: Buckle 1994.

all significantly more toxic than the first-generation ones (Table 6.11). All three, however, require application using the saturation or surplus baiting techniques discussed earlier. For some while these compounds seemed to be effective against the resistant populations of both *R. norvegicus* and house mice. In recent years, increasing resistance has been detected in some countries to both difenacoum and bromadiolone.

Resistance to anticoagulants is not to be assumed as being present in an area unless there is evidence of treatment failure. If there is little history of anticoagulant use in an area or in a storage facility, then it is likely that all anticoagulants, including first-generation products, should continue to be effective.

The development of brodifacoum and flocoumafen enabled the development of an alternative baiting strategy. The new strategy, 'pulsed' baiting, allows some flexibility with the bait application regime. It is not so essential that presence of these baits is maintained continuously throughout the treatment period (Fig. 6.27). Visits to replace baits may be made less frequently, perhaps at weekly intervals, and the quantity of bait placed can be reduced.

Interestingly, the increased toxicity of the anticoagulants does not result in faster control. Both the minimum time and average times to death remain the same as for the first-generation anticoagulants. Similarly, the time taken to control typical *R. norvegicus*, *R. rattus* or *M. musculus* infestation remains much the same at about 21 days.

The disadvantage of the increased toxicity of both brodifacoum and flocoumafen is that it has brought with it the increased risk to non-target species. For this reason, the use of these two compounds is restricted in certain countries.

### *Rodenticide formulations*

To be effective, rodenticides must be formulated in such a way that they become acceptable to and ingested by the target rodent population. There are three main types of formulation: edible baits, contact rodenticides and liquid rodenticides.

#### *Edible baits*

There is an increasing range of ready-made, edible bait formulations available. These include loose cereals and other foods, pellets, wax blocks, gels and pastes. It is difficult to choose between these formulations and there is little comparative data on palatability and efficacy. The most effective way to decide which to use in any particular situation is to try out those that are available and select the one that gives the best results.

It is generally accepted that loose baits are the most palatable of the formulations and are likely to provide the best opportunity for control. However, concerns relating to the safety of loose cereal bait formulations, particularly in an area where food is stored, will require careful consideration.

Wax block baits may be a safer alternative because they can be fixed in one position and so reduce the chance that the food being stored might become contaminated. However, they may not compete well with the possibly more palatable and familiar food that is being stored. Wax block formulations may also be more costly.

An alternative to ready-made formulations is to purchase a rodenticide concentrate and mix an edible formulation using a locally available, possibly cheaper, loose food bait. Advice and instructions on the concentrate label must be followed if maximum efficacy and safety are to be preserved. The temptation to increase the concentration at which the rodenticide is used may be counterproductive because palatability, and therefore efficacy, may be reduced.

There may be an advantage in using a bait base on which the rodents are already feeding; the rodents will be familiar with this food and it is likely to compete well with the alternatives. It is essential to ensure that it does not become mixed with the stored material. Factors that affect feeding behaviour in rats were investigated by Berdoy and Macdonald (1991).

*Placement of rodenticide baits*: It is essential to place baits correctly. Incorrectly placed baits can nullify the effect of even very attractive bait (Quy *et al.* 1996).

Rats are particularly neophobic and are reluctant to take risks. They do not therefore like to encounter new objects and are reluctant to feed on unfamiliar foods. Placing baits often involves placing unfamiliar foods at unfamiliar points in unfamiliar containers. It is important to try and make it as easy as possible for the rodents to find and then feed on the bait.

Hoarding behaviour, characteristic of most rodents, involves their taking food from where they find it, to somewhere where they feel they can hide it or feed on it more safely. *Rattus norvegicus* feel most secure feeding in their burrows or in their harbourages and if they can do this they will be more likely to feed on baits that are not particularly palatable. It appears that in stable environments (such as a food store) they may be particularly neophobic and reluctant to take risks, such as feeding on unfamiliar baits or entering unfamiliar bait boxes. The nearer a bait can be placed to those areas where they do feel safe, their harbourage, the more likely they are to feed on it. The objective in placing baits must therefore be to identify where the rats are living and to ensure that baits are placed in these areas, if it can be done safely.

Most effective control of *R. norvegicus* was obtained on stable English farm environments by placing the bait inside the rat burrows. The use of similar baits in bait boxes away from the burrows failed to achieve even reasonable control (Quy *et al.* 1996).

*Bait boxes and baiting containers*: Bait may be laid in many different ways: directly on the ground, on trays or boards or in containers of varying descriptions. In recent years there has been a significant move to the use of specially constructed, tamper-resistant, bait containers, designed not only to protect the bait from spillage, but also from interference. However, if baits can be laid safely without such containers, then there is no reason to use them.

Unfortunately there are few comparative data available on the efficacy of these containers, nor are there many data available on the response of the rodents to them. The few data available suggest that rats in stable situations may be very reluctant to enter them.

The establishment of permanent points at which bait or traps might be laid if an infestation were to occur can be useful. These points become part of the environment and are less likely to elicit neophobic responses as a result.

#### *Contact rodenticides*

Dust formulations are the main forms of contact rodenticide. Rodents coming in contact with the dust pick it up on their fur. Rodents spend some 20% of their time grooming (Meehan 1984) and will ingest the rodenticide as they do so. Rodenticidal contact dusts are, however, not suitable for use in areas where food or the surfaces against which food is stored may become contaminated by the dust. Such contamination may occur directly or through rodents running over the dust and then over the food, contaminating the food with the dust as they do so.

A contact gel formulation is also available, although restricted, which was specially formulated for the control of *R. rattus* in high and inaccessible areas. Additional forms of contact rodenticide include a 'contact wick' technique for the control of house mice. The wick is impregnated with brodifacoum and the mice rub against the wick as they run through the tube. Subsequent grooming results in ingestion of the brodifacoum.

#### *Liquid rodenticides*

These, as their name suggests, are simply soluble forms of the rodenticide presented to the rodent in a drinkable form. They can be very effective in food stores, particularly the larger ones, where water and access to water may be in short supply. Regular replenishment of drinking points is necessary to compensate for evaporation as well as liquid drunk by the rodents. The concentration of liquid should not be allowed to increase too much as it is likely to reduce palatability.

An alternative to toxic liquid formulations is to use plain, unpoisoned water in a drinking point adjacent to rodenticide bait points. Work undertaken in Sri Lanka against *R. rattus* in food stores showed that the presence of unpoisoned water adjacent to an edible bait point significantly increased bait consumption at that point (Meyer 1989).

### *Non-chemical and non-lethal methods of control*

There is a range of non-chemical methods of controlling and managing rodents.

#### *Trapping*

There are two forms of trap, kill traps and live capture traps. Killing traps are 'active' traps usually working on the 'break back' principle. A spring, released by the activity of the rodent on the trigger mechanism, drives down a metal bar, breaking the back of the rodent. The mechanism can work very well if the spring is strong enough, the design of the trap suitable and the rodent is in the correct position. In these circumstances the rodent is killed. If, however, all these criteria are not met and the trap is triggered before the rodent is in the correct position, or if the spring is not strong enough to kill the rodent, the rodent will survive. In surviving it will have gone through an unpleasant experience (pain, shock, fright) and in exactly the same way as with shyness induced by the acute rodenticides, it will learn to avoid the trap in the future and will become trap-shy and difficult to kill in similar traps. It is also likely that some individual rodents are naturally trap-shy and reluctant to enter traps.

For these reasons it is very difficult to eliminate a rodent population of any size by trapping alone. It may, however, be possible to eliminate a small population. Intensive break-back trapping with a well-designed trap suitable for the species being caught can have a major impact on the size of a rodent group in a food store or house. Work done in Mozambique has shown that intensive trapping, on a daily basis, in home food stores can reduce the population by some 70%, at low cost and with significant reduction in food storage losses (S. Belmain, pers. comm.).

Live capture traps work on a variety of designs, the principle being to entice the rodent into a container from which it is not permitted to leave. These traps are usually passive in action, in that the mechanism plays little part in the retention, it merely prevents the animal from escaping. Thus the rodent is either trapped or not, and if it is not trapped it will probably not have undergone an unpleasant experience, will not have learned to avoid the trap and is no less likely to be trapped on the next occasion. However, some animals are naturally trap-shy.

Some live capture traps incorporate spring mechanisms that help to retain the rodent, usually mice, by some form of trip mechanism triggered by the animal. In these cases, if the mouse is not properly retained but escapes, it may develop induced shyness. Traps should contain some form of food and should be visited at least once a day to prevent undue suffering and starvation. The food may act as an attractant.

There is a third form of trap: these are 'glue' or 'sticky' traps. The idea is that the rodent runs into the glue and becomes caught. The technique is cheap and easily applied and can be effective. It is not, however, a very efficient form of control. Many rodents will avoid the glue and others learn to avoid it. Nevertheless, there is no doubt that as long as glue traps are used with these constraints in mind they can help reduce the rodent population, particularly where other techniques are not proving successful or where they cannot be used. The use of this technique is prohibited or controlled in some countries on the grounds of being inhumane.

Traps have a clear part to play in the control of rodents in the food storage environment, if for no other reason than that they are non-toxic and will not contaminate the food in store. They will be most effective if they are used intensively, placed carefully in areas where the rodents are active and if thought is given as to how to use the traps to best advantage.

*Predators*

In some situations predators may have a part to play in the control of rodent infestations. This is unlikely to be the case in food stores, which are extremely artificial environments and where predator levels would have to be maintained artificially. Furthermore, the very high breeding potential of the rodents in stores means that it is very unlikely that predators would provide any form of effective control; predators may also contaminate the food themselves with their own droppings and urine.

*Ultrasound*

Some rodent communication is undertaken with ultrasonic frequencies. This ability has been exploited to produce devices that claim to be able to drive rodents from areas by producing very loud ultrasound – which humans are not able to hear. There is absolutely no evidence that these devices work, on the contrary, there are good data (Meehan 1984) to demonstrate that they are ineffective. Ultrasound should not be used as a means of protecting food stores from rodent infestation.

*Electromagnetic devices*

A range of devices are sold throughout the world that claim to produce electromagnetic fields that deter rodents. These devices have not proven effective.

*Chemical repellents*

While many thousands of potential repellents have been tested for their repellency effects, none have so far proved sufficiently effective to be used as a means of managing rodent populations. There are currently no repellents available that have been shown to be cost-effective.

*Reproduction inhibitors*

Extensive work has been undertaken on the use of chemosterilants as a means of eliminating rodent populations, so far without practicable success. While it is possible to reduce infestations by continued use of chemosterilants, populations are not eliminated. There are currently no effective chemosterilant formulations available.

### *Monitoring and record-keeping in rodent control*

One of the most important and perhaps the least well-applied aspects of rodent control is the monitoring of rodent management operations and the recording of the data which go with this monitoring. The extent and degree to which monitoring is necessary will depend upon the nature of the problem. A large food storage facility will require a more comprehensive monitoring programme than a small, single-household rural store.

Monitoring the activity of rodents in an infestation will provide evidence relating to the progress of the control operation. It starts with the initial survey of activity. The initial survey is perhaps the most important part of the control programme, which often fails because either insufficient time was given to the survey, or it was not undertaken thoroughly. The first survey should involve a search of the infested area for all the signs and traces of rodent infestation. From a record of the findings, the surveyor will be able to determine the species present and size and extent of the rodent infestation. Decisions can then be made as to the control techniques that will be most appropriate to achieve control.

The next stage of the programme will be to apply the control techniques. Details of this application, including numbers and position of control points (baits, traps, etc.), type and quantity of rodenticide, amount of bait laid and dates, must be recorded if the programme is to be monitored correctly. Control points must be visited at regular intervals, dependent on the technique being used. At each visit a record should be kept of the rodent activity and the actions taken as a result. With time, of course, the activity at each control point should reduce, as the control begins to have an effect.

The problem with a single method of monitoring is that it may not provide the information required. Rodents may not be eating bait from a bait point, the absence of a bait take does not mean that there are no rodents, the baits may not be in the correct place or the rodents may simply not want to feed on the baits. Similarly, failure to catch rodents in traps does not mean that there are no rodents.

It is, therefore, essential that not only do visual surveys continue throughout the treatment, but that a secondary method of survey is employed if there are any suggestions that the programme is not progressing well. A good method of monitoring for rodents is to use unpoisoned dust, such as sand or chalk dust. Rodents may not have to eat baits or go into traps, but they usually have to move around. If these dusts are placed carefully in rodent runways, they will show footprints or other marks if the rodents move over them. This technique can be used easily when monitoring stacks of food. Bands of dust can be laid around the base of the stack and any movement into or out of the stack will be evident.

Similarly, rodents may enter bait boxes used for presenting rodenticide baits safely but they do not necessarily have to eat the bait. In such cases there is no evidence that the rodent has entered the box. A new technique for monitoring *M. musculus* activity is now available, incorporating a small plastic strip in the entrance to the bait box. The position of the plastic strip will indicate whether or not a mouse has entered the box.

Effective monitoring permits those managing the control programme to evaluate progress. Changes can then be made to improve the situation if the control regime is not working. Detailed monitoring can also provide additional information, on proofing and hygiene requirements, for example.

### *Managing rodent control*

To be effective any large-scale rodent control programme requires good management. It is essential that if rodent control programmes are set up they are managed through an unambiguous chain of command. There must be managers who know what their objectives are and have unequivocal responsibility for the control operations. Similarly, technicians who apply the rodenticides must be clear as to their objectives and certain as to where their responsibilities lie. Above all, all staff must be well trained to carry out the functions and tasks for which they are responsible.

## Bird control

The management and control of bird problems in and

around a food store should not be undertaken without first understanding the nature of the problem. It is, therefore, necessary to identify the species of bird(s) involved, the nature and extent of the damage, loss and contamination that is being caused and to obtain some idea as to the cost implications of these problems. With this information, solutions to the problem can be more readily identified. Then, all the available options must be reviewed so that the most appropriate can be selected for application.

The options for the management of a bird problem in a store are culling, exclusion (proofing), anti-roosting techniques and scaring, and environmental and store management.

## Culling

Culling is essentially the killing of the birds causing the problem and it is used in many bird management strategies.

### *Poisons (avicides)*

These are chemicals that can be used to kill birds. There are, in most countries, strict legal constraints against the use of most, if not all, avicides. However, among those that are available are alphachloralose, seconal, strychnine, Avitrol (contains 4-aminopyridine and is used more as a bird frightening agent), endrin (9.4%) and fenthion (11%). While most avicides are used in edible bait formulations, the latter two are used on perches used by the birds and the toxicants are absorbed through the bird's feet.

There are a number of problems with the use of toxicants for bird control. The first is that all are also very toxic to other non-target species, including birds and mammals. While it may be possible to put poisoned bait down for the target species, it may be difficult to stop other birds, or even other animals in the area, from obtaining it. Non-target species may ingest the pesticide either directly, by eating a poisoned bait (primary poisoning), or by eating birds or other animals that may have been poisoned (secondary poisoning). It is essential that these chemicals are used safely and the methods employed reduce the risk of non-target species ingesting the toxicant. Poisoned baits should only be used after very careful consideration and then only if all precautions are taken to safeguard the bait.

It is even more difficult to control access to the birds that have eaten the poisoned bait. Their mobility, and the time taken for the poison to act, can mean that the poisoned birds have travelled some distance from the source of the bait before they succumb. They then die or become ill and may quickly be taken by other animal predators and scavengers. People may take the birds in the hope that they can be eaten.

Thus, avicides are not generally considered safe and should not be used except under very unusual circumstances.

Perhaps the best reason for avoiding the use of toxicants is that they rarely solve the problem. There are two reasons for this. Firstly, putting down bait actually increases the amount of food in the area. Some of the birds that feed on the poisoned bait may be birds that have been attracted in to what is effectively a new food resource. This is particularly true if prolonged prebaiting is used to condition the birds to a bait base and poisoning site prior to the treatment. Thus, many of the birds that are killed may not have been the cause of the problem. The technique is not very specific.

Secondly, the use of poison baits frequently does not address the reason for the birds being present. As a result, the same food and the same resources remain available and the remaining birds breed rapidly to replace those that have been killed. In addition, those killed are rapidly replaced by birds from outside the area that migrate to occupy the vacant niches.

### *Chemosterilants*

Some attempts have been made to use chemosterilants to reduce the breeding potential of birds. Many different types have been tried, but perhaps the most widely used has been 0.1% aza-steroid (20,25-diazacholestenol dihydrochloride). However, the need to reapply the chemosterilant on a regular basis, the potential for non-targets to take the bait and the continuing problem with migrants coming in to replace any vacant niches, reduces the effectiveness of this compound.

### *Trapping*

There is a range of live trapping devices for taking birds. The use of trapping techniques is again strictly controlled in many countries. Spring traps, which kill the birds, should not be used because it is rarely possible to prevent non-target species going into the trap.

Live trapping devices are able to trap many of the bird species that frequently cause problems around food stores, but not all birds are easily trapped. Feral pigeons, for example, are relatively easily trapped, whereas starlings are much more difficult to trap. The main problem with traps is that in almost all cases the birds will only enter if there is food inside. Thus, in placing the food the carrying capacity of the environment is increased. Those birds that already have a food source (those that are already at the site) are less likely to be tempted into the trap than those that do not (those that are from outside the site). Thus, many of the birds trapped will not have been part of the original problem. For this reason, trapping is not as efficient as it may appear at first sight.

Nevertheless, prolonged and intensive trapping can reduce the number of birds at a site, but trapping will have to be maintained if the population is to be maintained at a lower level. Immediately the trapping stops the population will return to previous levels if the levels of food availability and other environmental factors remain the same.

Trapping can be very expensive in labour costs and it is essential that the traps are visited regularly to avoid stress to trapped birds. They should then be humanely destroyed.

#### *Shooting*

Shooting is perhaps the most effective culling technique. The main benefits of shooting are that the technique is specific and only birds that are causing a problem need be shot. There is no non-target risk as long as the operation is undertaken carefully. It is essential, however, that the facility is not damaged by the shooting and that those undertaking the work are thoroughly trained.

Once again, any short-term gains from shooting will not last for long. The shot birds will be replaced by breeding or immigration, if birds continue to have access to the store.

Culling or killing does not, therefore, have as important role in the control of birds at food storage sites as it does with insect and rodent pests and rarely provides an effective solution to a problem.

### Physical exclusion (proofing), anti-roosting techniques and scaring

These techniques are designed to protect the interior of the store where feeding, nesting and roosting may occur. Any reduction in bird activity within the store will reduce the amount of food eaten and contaminated, and will reduce damage and fouling of the store interiors and structures.

There is a range of measures for preventing access to stores. A guide to rodent-proofing is provided by Jenson (1979).

#### *Filling or blocking*

This is perhaps the most obvious technique and simply involves identifying the points at which birds gain access to the store and physically filling them with whatever materials are available, such as concrete or metal sheeting. In some situations this is not possible nor practicable and more specific techniques must be used.

#### *Netting*

This involves placing bird-proof nets across openings and access points into stores, such as gaps beneath roof eaves, to prevent bird access. The netting needs to be applied correctly and gaps must not be left between the edges of the net and the frame. There are many techniques for doing this, but the use of straining wires around the edge of the opening provides easy attachment points.

It is essential that the mesh size is correct for proofing against the bird concerned. Appropriate mesh sizes are 50 mm for feral pigeons, 28 mm for starlings and 19 mm for house sparrows.

Netting can be a convenient and cost-effective method of preventing access but must be well maintained to provide maximum cost benefit. An advantage of netting is that it does not interfere with ventilation.

#### *Curtains*

A range of curtain techniques are available. These include plastic strips, metal chains, air curtains and others. Curtains provide a barrier to bird entry, without interfering too much with human or vehicle access. They are usually hung on hinges that allow the curtain to be moved as required. Since they do not provide a fixed barrier, birds can sometimes learn to obtain access. For instance, both feral pigeons and house sparrows have been observed hopping through fixed plastic curtain sheets behind a fork-lift truck entering the building. Curtains can be coloured or transparent to enhance traffic management into and out of the store.

#### *Anti-roosting techniques*

Birds may cause a problem because they use a particular

ledge or beam for roosting and consequently foul the areas below. When this occurs, special anti-roosting techniques can be applied to prevent further roosting. These techniques are most effective against feral pigeons and birds of similar size, and are less effective and may not work against smaller birds such as house sparrows and larger species such as some of the gull species, although specific techniques have been developed in some cases to overcome these problems.

- Bird repellent gels – these are soft gels that are applied along the leading edges of ledges and surfaces used for roosting. Their use has reduced in recent years as more effective wire and spike systems have been developed. The gels have a limited life, become very dirty, and are expensive to remove.
- Bird wires – special sprung wires supported along the edges of ledges and beams.
- Spike systems – fixed wire spikes on plastic bases that physically prevent the birds from having the space to roost on the ledges on which they are applied. Plastic spike systems can be found, but are not generally as effective as metal wire spikes.

In addition, there is a wide range of other anti-roosting techniques that may be useful in unusual situations. All these techniques require a level of skill and an understanding for their use to be effective.

#### *Scaring techniques*

There is a range of scaring methods available. In general, these will not be very effective around food stores because the attraction of the food outweighs the effect of the scarer. They have to be used intelligently if they are to have greatest chance of success.

- Visual scarers – these include a range of devices, some static and some moving, which are designed to scare through their visual impact. In almost all cases the birds rapidly accommodate to the devices and their effect wears off within a few days. For maximum benefit, these scarers should be used in a manner that is unpredictable to the birds. However, this will be a costly operation.
- Acoustic scarers – most of these scarers simply rely on the production of loud noises, sometimes randomly emitted, to scare the birds. Birds rapidly become accustomed to their presence and their position needs to be changed or the noise needs to be reinforced with other actions for maximum benefit.

Perhaps the most effective acoustic devices are those that are called ‘distress calls’. These mimic the distress calls emitted by specific bird species and can certainly be more reliable, particularly if used in short periods and reinforced with other sounds and visual scaring. It is important to use the call specific to the species being scared, although there can be some scaring effect on other species. Some birds do not seem to have distress calls, or it is possible that they have not yet been identified.

There is also a range of ultrasonic devices on the market for bird repulsion, particularly sold for use inside buildings. These are not effective.

#### *Predators*

In some situations, birds may be scared away by flying predatory birds around the store. This usually has only a temporary effect and the birds rapidly return. The flying of predators needs to be undertaken regularly, perhaps several times a day, if the scaring effect is to be maintained.

### The role of store management in controlling bird populations

Food availability is the overriding factor that determines the size of a bird population at any one point. The available food is the reason why the birds are present, and also, together with the ecology of the birds concerned, why so many of the alternative control and management techniques are not effective.

There are, therefore, two principles associated with the effective management of birds at a food store. Firstly, ensure the birds do not have access to any spillage of food at any time. The absence of spillage will reduce the availability of food and will reduce the number of birds far more effectively than any control technique. It is very unlikely that spillage of food can be avoided in an active food storage facility and so there is a need to maintain high levels of hygiene and to have the spillage removed as soon as it occurs. The collected food can then be returned to the system or disposed of in some other appropriate way.

Secondly, even if there is no spillage there will still be food on site, even though it may be in containers or packaged. If the birds can gain access to the storage facility and can then gain access to the packaging, then a population may build up on site. For example, Hessian sacking allows easy access for feral pigeons to the contents of the sack; pigeons simply peck the food out through holes in the sacking that they make with their beaks. The option in these circumstances is to identify the access routes and block them by using the range of proofing and exclusion techniques available. If necessary and exclusion is not possible, bag stacks should be covered with netting or some bird-proof material.

Bird problems at a food storage facility are very much related to the standards of store management, i.e. the standards of hygiene and maintenance of the store structures. If management is maintained to a high level, bird problems of any significance are unlikely to occur.

## Moulds and bacteria

Fungi and bacteria gain easy access to produce through wounds or natural openings and so avoidance of damage is important in minimising their attack. Produce can be damaged both before and after harvest and, to ensure high quality, damage during handling should be minimised. Where infections begin in the field, chemical control is best directed before harvest. However, with latent infections that occur before harvest, such as anthracnose in bananas and mangoes, field treatments may not be economic and so post-harvest management needs to be considered (Waller 2001).

Physical damage to the produce occurs during harvesting and subsequent handling. Every time the commodity is moved from silo to silo, in and out of cold stores or from ship to shore, the risk of damage increases. Grain that is too dry is at greater risk from cracking and shattering than grain at the optimum moisture content; there is, therefore, a risk from overdrying grain. Damaged grain cannot be stored for as long as good-quality grain, under the same conditions, before fungal growth occurs. Perishable produce may deteriorate very quickly once pathogens gain ingress, particularly in the humid tropics.

Insects, rodents and birds also damage stored produce, which reduces storage life. In addition, insects often carry fungal spores, and so they can act as vectors of pathogens to and within the commodity. Insect frass and dust produced by feeding activities provide a food source for fungi and bacteria, and their faeces are often contaminated with viable fungal spores. Insects are metabolically highly active and increase the temperature and moisture content of grain within the storage environment, again improving conditions for fungal growth.

Some insects are attracted to fungi by the volatiles that the fungi produce. For example, short-chain alcohols and ketones known to be produced by fungi were strongly attractive to *Cryptolestes ferrugineus, Oryzaephilus surinamensis*, *O. mercator* and *Ahasverus advena* (Pierce *et al.* 1991). *Carpophilus hemipterus*, a generalist insect herbivore, was attracted to volatiles produced by the yeast *Saccharomyces cerevisiae* growing on the surface of bananas. Host location by *C. hemipterus* was thought to occur as a response to a variety of blends of common fruit volatiles, whose concentrations were enhanced by fungi (Phelan & Lin 1991). Other species of insects and mites also feed on mycelium, spores or yeasts on the surface of fruits, spreading the pathogen as they move within the store.

### Drying

The most common means of controlling fungi and bacteria in stored products is to dry them to 'safe' moisture contents. The lower limit of $a_w$ that will permit growth of storage moulds is 0.7, over a period of about 2 years at a temperature of 21–27°C. This is equivalent to an inter-seed equilibrium relative humidity of 70%. As mould growth may occur slowly above the $a_w$ level, this represents the maximum level to which the crop should be dried for prolonged storage. Some grains, in particular corn that is harvested at high moisture contents, must be artificially dried to prevent spoilage or may be ensiled and used for animal feed.

### Low temperature

Reduced temperatures in combination with low moisture are more effective for preventing fungal biodeterioration than drying alone. These conditions can be achieved by aeration, in which ambient air is blown or drawn through the grain. A low, uniform moisture content coupled with a low uniform temperature reduces the possibility of moisture transfer within the bulk and adds to the storage life of the commodity.

Adequate aeration and lower temperatures in colder climates can be used for crops up to about 17% moisture content. Maize can be safely stored for up to a year at 15% moisture content and 15°C, whereas if stored at 30°C, then the same maize would be spoiled by moulds within 3 months. However, in tropical countries with higher temperatures the moisture content of the grain should not be more than about 12.5% for long-term storage.

### Chemical control

Relatively few resources have been allocated to the development of post-harvest chemical control methods, in comparison to those devoted to field problems, because the size of the market for post-harvest preparations is much smaller than that for field fungicides. Consequently, post-harvest fungicides were originally developed for field use (Edney 1983). However, the two areas are quite distinct, and require different chemical properties (Waller 2001). Chemical preservatives have limited application in controlling the spoilage of moist stored produce. Many preservatives for this purpose have not proved acceptable to industry and several are also unacceptable to health authorities. Some, which have general potential, are propionic acid or calcium propionate, sorbate or sorbate with carbon dioxide, acetic acid and propylene oxide.

As the effectiveness of fungicides depends on their dissociation in water, the compounds may not be fungicidal against storage fungi that grow under relatively dry conditions, for example, at $a_w$ of 0.7–0.8. Levels of fungicides high enough to inhibit fungal growth during storage usually kill the seeds. Fungicides for use against storage moulds often have serious limitations, such as toxicity to animals, excessive cost, difficulty of application, undesirable effects on processing quality of the commodity, lack of toxicity to storage fungi and lack of penetrative ability to reach latent infections beneath the epidermis. For example, borax was discontinued as a dip treatment because of its low solubility and although sodium salicylanide controlled *Gloeosporium* on bananas it was not commercially economic in use. However, biphenyl remains the fungistat of choice for control of post-harvest disease of citrus, despite the risk of imparting an odour to the fruit and rendering the 'button' (calyx and receptacle) susceptible to *Alternaria* rot. Many other chemicals have been tried on post-harvest produce. They can be grouped under three broad headings: fumigants, treated packaging and dips, sprays and dusts.

Fumigation helps to control mould growth on grains by killing insects that damage kernels, but the effect is not long-lasting. Some gases such as ammonia, sulphur dioxide and methyl bromide destroy moulds and insects. Phosphine at low concentrations (0.1 g/m$^3$) can retard development of storage fungi that are actively growing and can also limit mycotoxin production. However, phosphine has minimal effect on spores and mycelia.

Citrus and apples have been protected by chemically treated wraps, which help to prevent fungal development, sporulation and spread so that a single diseased fruit does not infect the whole batch. This type of treatment is most effective when the fungicide has some vapour action. Wrappers impregnated with biphenyl have been used extensively but may adversely affect the flavour of the produce and give rise to unpleasant odours. Other wrap chemicals include copper sulphate, some active halogen compounds, pine oil, sodium-*o*-phenyl phenate and *o*-phenylphenol esters.

Dips and sprays include antibiotics, borax, dithiocarbamates, organic acids and halogen and phenolic and sulphur compounds. Systemic fungicides such as benomyl, dicarboximides (iprodione and vinclozolin), thiabendazole and other benzimidazole derivatives have also been used. However, resistance to benzimidazole compounds has developed in some post-harvest fungi. The broad-spectrum dicarboximides are used on field crops and in post-harvest protection of vegetables and fruits, though resistant strains of *Botrytis cinerea* have appeared. Metalaxyl, alone and in combination, controlled potato blight rots when applied in the field. Some other phycomycete storage diseases have been controlled by aluminium tris-*o*-ethyl phosphonate. In addition, tecnazene is used for control of *Fusarium caeruleum* and suppression of sprouting on stored potatoes, iprodione for control of *Alternaria* and *Botrytis* on stored cabbages, and carbendazim and metalaxyl for control of *Phytophthora*, also on cabbages (Whitehead 2000).

Propionic acid was first employed in the 1960s in the UK to preserve damp grain. Dose rates depend on the moisture content of the grain (for example, 10.5 litres per tonne at 25% mc) but treating grain above 30% mc is not economic. Treated grain will not germinate and cannot be used for human consumption, but can be fed to livestock. Uniform treatment may protect stored grain for up to 1 year. However, inconsistent treatment will allow fungal growth to occur. In addition, some fungi can metabolise propionic acid at low doses, and spread into otherwise adequately treated bulks (Stables 1989).

Aqueous suspensions or solutions are used for protecting fruit and vegetables, whereas dusts are more applicable in grain bulks. Water-based protectants are simple to prepare and apply, and often penetrate the commodity as efficiently as fumigants, so that latent pathogens can be targeted. Treatments may be incorporated into mechanical handling procedures. For example, dusts can be automatically added to grain on conveyors as it enters silos. Perishable produce may need to be wetted for cleaning or cooling purposes, and this provides an opportunity for the fungicide or bactericide to be added to the wash water. It may even be necessary to incorporate protectants at this stage to prevent cross-contamination of the produce from water-borne spores or mycelium.

However, care should be exercised because wetting of produce that does not normally require it may induce a more rapid decay. Perishable produce can be dipped or sprayed, but dipping is advantageous because the whole of the commodity is exposed to the pesticide, though it is more expensive since a greater quantity of pesticide is usually required.

## Approaches to pest management in stored grain

### General principles

For a short period in the twentieth century, from 1940 to the 1960s, it appeared that the application of synthetic pesticides, particularly in North America and Europe, was the answer to pest control problems. However, it has since become clear that excessive pesticide application may pose a threat to health and the environment and that in any case many insect species are able to become resistant to these chemicals. It is against this background that efforts have been made to move away from the 'blanket' application of pesticides and instead to adopt an approach termed integrated pest management (IPM). This is a pest management system which takes into account the environment, the population dynamics of the pest and uses all suitable techniques and methods in the most compatible manner possible to maintain the pest population below levels that would cause economic injury (Smith & Reynolds 1966). IPM is now a well-developed approach for the protection of crops before harvest (Dent 1991). However, it is much less well-developed for the protection of stored grain and we will be considering why this is so.

Pest management in grain storage is not a synonym for pest control in grain stores. Pest management is a decision-making process that leads to pests being controlled in a cost-effective manner (McFarlane 1989; Mumford & Norton 1991; Subramanyam & Hagstrum 1995). Besides this, good pest management should reduce to a minimum the dangers inherent in the use of insecticides and fumigants to consumers, pest control staff and the environment and limit the chances of the development of pest resistance.

To understand the principles of pest management, especially those relating to cost-effectiveness, it is necessary to define what is meant by a pest. The definitions often vary according to the objectives of the storage or marketing system. However, it is generally agreed that a pest is anything that can cause a financial loss. Financial losses result when there is a reduction in either the quality or quantity of stored food. The loss in quality may include a reduction in nutritional value of the food or the creation of a potential health hazard.

A useful concept in pest management is pest status. This concept takes into account the fact that the importance of a pest is dependent on the circumstances in which it occurs. The following two examples of common storage pests illustrate pest status.

*Lasioderma serricorne* is well-known as the tobacco beetle even though it reproduces much better on maize. Its pest status on maize, however, is relatively low as it causes little damage, and small numbers of insects on maize may make little difference to its value. In contrast, on tobacco it has a high pest status as the burrowing of just a few beetles can ruin a large consignment of tobacco because, for marketing, this commodity needs to be completely pest-free. This was illustrated in Malawi in 1981 where the presence of the beetle in tobacco bales at the auction floors resulted in the closing down of the export market for that year (P. Golob, pers. comm.).

*Latheticus oryzae* is an important pest of cereal products and able to develop at temperatures ranging from 26°C to 40°C. If a grain stock is kept at 35°C then the amount of damage and hence pest status of this species is potentially high. For stocks maintained at less than 26°C, pest status is potentially very low.

### How to achieve cost-effective pest management

All those responsible for grain storage should aim to ensure that the benefits of pest control outweigh the costs (Hebblethwaite 1985), that is, the potential losses, if pest control was not applied, should amount to a value equal to or more than the cost of pest control. To achieve this requires a knowledge of the economic injury level (EIL), which is the pest density causing a reduction in market value equal to the cost of pest control (Fig. 6.28).

It is clearly not cost-effective to apply pest control too early, before it is actually needed. Much benefit is lost if it is applied too late, as is the case when the loss in value approaches or exceeds the cost of pest control. Pest control should be applied when the insect population reaches an intermediate density, called the economic control threshold (ECT). This is where the insect population is large enough for a reliable estimate to be made of it and to show benefit from the costs of control procedures against it. Pest management action at this pest density will prevent damage or contamination reaching the EIL. However, if stocks are going to be consumed before the

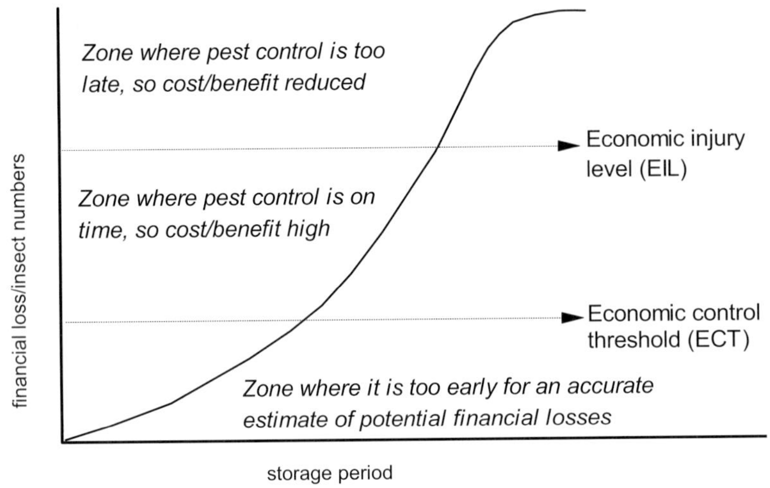


**Fig. 6.28** How to optimise the timing of pest control to obtain a favourable cost/benefit ratio.

pest population has risen to the EIL, then pest control action may not be justified even though the ECT has been reached. In setting the ECT, account must be taken of typical delays that may occur before control measures can be applied (often called the lead time to pest control). Hence the pest density representing the ECT may have to be set at a rather low value in view of the fact that the lead time may be several weeks. In such a period, pest populations may increase rapidly as insects have high reproductive rates.

To establish ECTs it is important to be able to assess the losses associated with particular pest densities. Once the relationship between losses and pest density is known then a measure of pest density may be all that is required for decision-making. The use of traps and sampling probes for the estimation of pest density is discussed elsewhere.

## Practical approaches to pest management

There are broadly three approaches that can be taken to the management of pests. These are listed below and illustrated with examples taken from the conventional protection of stores holding bagged grain using fumigation and insecticide spraying. These approaches can apply to a number of grain storage systems, except controlled atmosphere storage where grain is protected by the continuous use of a single technique. However, this method must still demonstrate that it is cost-effective compared to other options for stock protection.

### *Mixed pest control*

Two or more pest control techniques are applied together in a rigid system without consideration of their cost-effectiveness. A good example of this would be the application of fumigation and spraying of residual insecticide in stores on a fixed routine as in Fig. 6.29(a).

Here the two activities are applied whether they are needed or not and no consideration is given as to whether the system is actually cost-effective.

### *Integrated pest control*

In this case, one or more techniques are integrated into a specific storage management system in a cost-effective manner. An example of this might be whether or not to use insecticide spraying on the basis of cost comparisons with fumigation. In Fig. 6.29(b), where the cost of four spray treatments is less than the cost of a single fumigation, it is worth continuing to use insecticide treatment (although recent research suggests that routine insecticide spraying is unlikely to be effective). Alternatively, where the cost of four spray treatments is more than the cost of one fumigation, then fumigation alone, albeit at a slightly greater frequency, can provide equally good grain quality preservation at a lower cost.

### *Integrated pest management (IPM)*

This differs from integrated pest control in that control

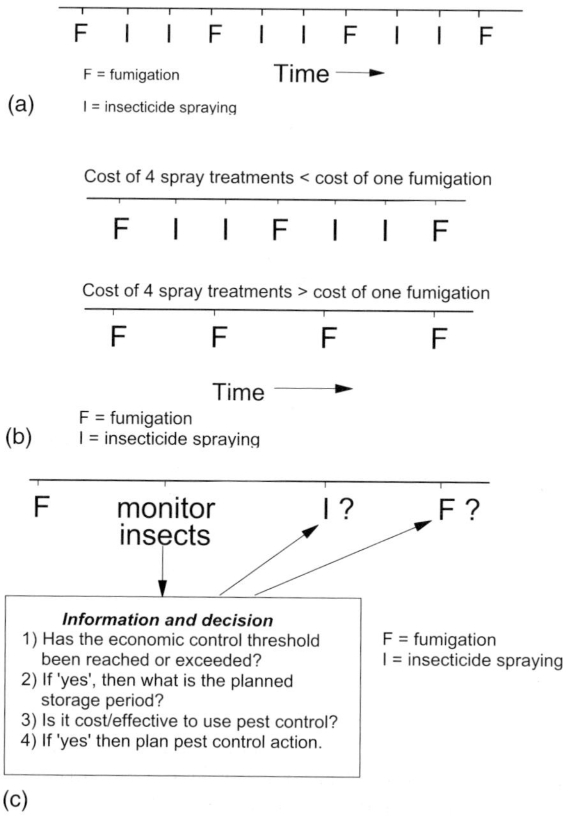


**Fig. 6.29** Combining fumigation and insecticide spraying.

procedures are applied in response to a decision-making process that refers to an economic control threshold. An integral part of this process is an insect monitoring procedure to estimate the pest population (Fig. 6.29(c)).

It is true that to date such IPM approaches are rarely taken in grain stores as they present both practical constraints and conceptual challenges. Some of these are listed in Table 6.12.

### Decision support systems

The consideration of IPM above has shown that effective storage pest management may require complex decision-making. Over the last 10 years several computer-based decision support systems have been developed to help managers improve the decision-making process. Data are entered into a computer, the best course of action is offered and then the manager chooses whether or not to act on the advice. Good examples of decision support systems are *Stored Grain Advisor* in the US (Flinn & Hagstrum 1990), *Grain Pest Advisor* in the UK (Wilkin *et al.* 1990) and *PestMan* in Australia (Longstaff 1997). These contain advice on a range of pest management issues. The *Stored Grain Advisor* is designed to help farmers in the US who store bulk grain. It can do the following:

- help in the recognition of a limited range of insects;
- help with insect sampling, designing a sampling programme using spear sampling or probe traps, estimating the accuracy of trap catch and estimating the probability of insect detection;
- give advice on the significance of a given population of storage pests; and
- simulate pesticide breakdown and population growth of a pest population depending on length of storage, initial grain temperature and moisture, aeration/fumigation and identity of pest species present.

Besides entering real data into the decision support system and receiving a suggested course of action, it is possible to enter several sets of hypothetical figures and learn what might happen to grain stocks under a

**Table 6.12** Challenges to the use of integrated pest management in grain stores. Source: Haines 1999.

| Constraint to IPM | Effect |
|---|---|
| Duration of storage is often not specified in advance | Difficult to apply an economic control threshold |
| In many situations it is very difficult to make an accurate assessment of insect populations | Difficult to operate an economic control threshold |
| The future marketing route through the post-harvest system is often uncertain | Creates spatial and logistic uncertainties |
| The link between where in the marketing chain the critical pest damage occurs and where it has economic impact may be weak | Lack of economic incentive to apply pest control at the critical time |
| The end use of food (e.g whether for human or animal consumption) or consumer perceptions about the significance of pest damage are often not known, i.e. the pest status is unknown | Inability to choose appropriate pest management regime |

range of different circumstances. This makes the system very useful for educational purposes, since by comparing the outcomes under different conditions the student can learn a great deal about grain store management. To date these systems have been used almost entirely for educational purposes; a version of *PestMan* has been developed for use in Indonesia where it has been renamed the *Training Workbench.*

Two simple decision support systems, each dealing with a single aspect of grain store management, have been developed to deal with specific problems. In Francophone Africa, *Cereanly* has been developed to improve the management of food security stocks. Grain quality data can be entered into the system and advice is provided on storage options. In Indonesia, a *Fumigation Decisions Support System* (FDSS) has been developed to assist pest management staff in deciding the optimum timing for the fumigation of milled rice stocks (Hodges 1999). To do this, insect populations, particularly those of *Tribolium castaneum*, are estimated using traps such as bait bags. The estimated pest numbers are entered into the computer program and future pest population growth predicted. From this prediction, the FDSS can indicate at what time in the future the pest population will reach the economic control threshold (Fig. 6.30), and it will be at this time that pest control should be implemented. In this particular case, pest control action is required at the time the numbers of *Tribolium castaneum* reach about 10 per kg in the rice. From experience, fumigation at this time will prevent an unacceptable degree of contamination since the insect population will be destroyed before it reaches the fastest period of growth, the exponential phase (in this phase of growth, damage and contamination increase most rapidly). The ECT is based only on unacceptable degrees of contamination and not rice weight losses. This is because weight losses are very small relative to the significance of contamination. However, the FDSS allows the user to determine what weight losses would be expected from different insect densities.

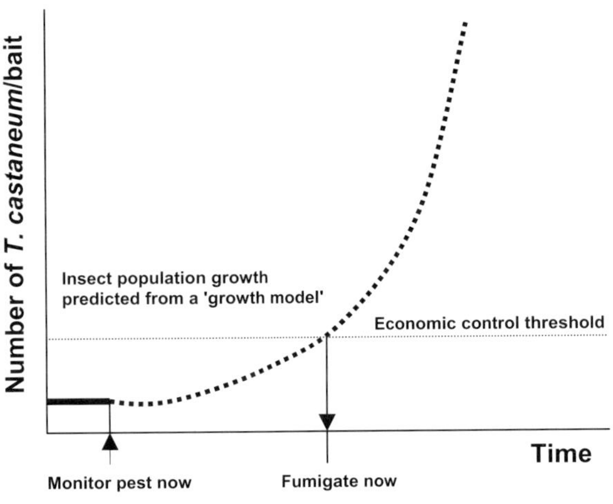


**Fig. 6.30** How the Fumigation Decision Support System helps advise on the most cost-effective timing for fumigation.

Health issues sometimes need to be taken into account when considering what pest density could be tolerated before pest control is implemented. In the case of *Tribolium castaneum*, adults release mutagenic chemicals, called quinones, from thoracic and abdominal defence glands. The quinones impart a strong odour to the food and could present a serious health hazard. Fortunately, research has shown that they do not need to be considered in the storage of milled rice since it does not accumulate significant amounts of these chemicals (Hodges *et al.* 1996). However, they do accumulate in wheat flour (and presumably other flours) where pest management operations should take this into account.

Decision support systems do not have to be computer-based. A simple printed decision tree can be effective. The decision tree shown in Fig. 6.31 was developed to help farmers in Kenya manage their stores to reduce losses caused by *Prostephanus truncatus* (Farrell *et al.* 1996). Field work in both Tanzania and Kenya indicated that if maize was to be stored for less than 5 months in either country then use of pesticide was probably not justified (Henckes 1992). This conclusion was based on the break-even point between the value of losses and the costs of treating grain with a dilute dust insecticide. As the value of maize and cost of pesticide treatment changes from year to year then the break-even for pesticide treatment needs to be recalculated from time to time. This would be a job for the extension services that advise subsistence farmers.

Looking at the decision tree (Fig. 6.31), it can be seen that if the storage period was to be greater than 5 months then knowledge of the previous history of *P. truncatus* infestation in a store is important. If there was *P. truncatus* in the store during the previous year then the risk of infestation again in the current year is higher and the admixture of a dilute dust insecticide is recommended. However, if no *P. truncatus* were observed last year then regular inspection is required. If the pest is found subsequently, then grain shelling and insecticide treatment are required.

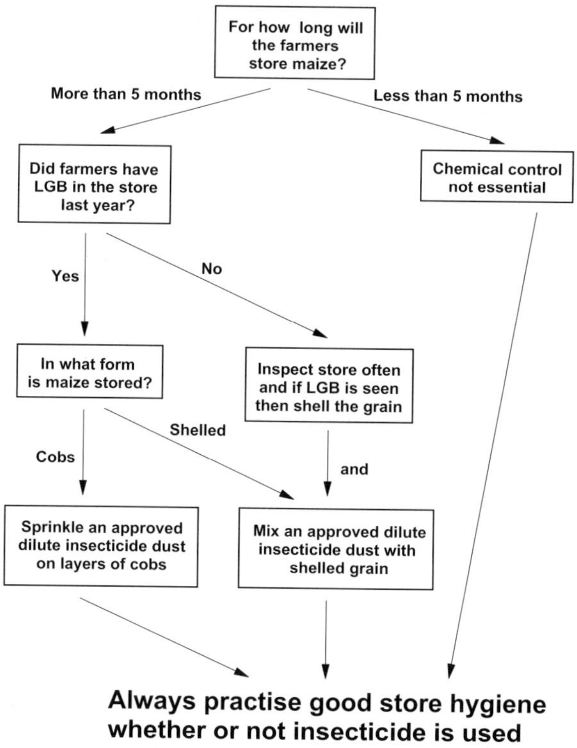


**Fig. 6.31** A decision tree to help extension services advise farmers on the protection of their maize against *Prostephanus truncatus*. Source: Farrell *et al.* 1996.

## IPM for limiting insecticide resistance

The sources and mechanisms for insecticide resistance have already been discussed. A major incentive for the use of IPM is that it offers an approach that can help to avoid the development of insecticide resistance. As far as possible IPM uses non-chemical means of control and a decision to use pesticides is only made when it is clear that this option is needed and would be cost-effective. This all leads to a reduction in insecticide usage and so limits the opportunity for the development of resistance. Additional measures can be taken, for instance occasional rotation between the pesticides used so that the opportunities for the development of resistance are reduced (Roush 1989).

## Future improvements to conventional pest management systems

Conventional pest management systems can be made more effective by integrating compatible methods of control, and the best results can be achieved through integrated pest management. The following examples consider possible improvements in pest management for small-scale farm storage and larger-scale trader/central storage.

### *Farm storage*

A flow diagram of typical farm production and storage of maize is shown in Fig. 6.32. For most small-scale farmers, maize production and storage has to be a very cheap, low-input activity. Nevertheless, improved protection need not necessarily involve a very large increase in costs. An example of how traditional farm storage of maize might be improved through the integration of several means of storage protection is shown in Fig. 6.33. Two suggestions for initial improvements to the traditional system are the choice of a maize variety more resistant to insect pests and the use of a drying crib. The crib enables earlier harvesting, thus reducing deterioration on the plant. After drying, the grain is shelled and treated with pesticide if storage is to be long-term. Alternatively, the grain could be stored without pesticide treatment in sealed containers, such as metal bins or good-quality mud silos.

Farmers in Ghana have moved some way towards these improvements under the guidance of Sasakawa Global 2000, and treatment of the dried grain with pesticide. However, farmers still do not have access to maize with increased resistance to post-harvest damage. Although maize cultivars have been screened for such resistance for some years (Dobie 1974; Kossou *et al.* 1993), varieties are not yet available to farmers that would reduce their problems with post-harvest pests. Almost invariably, local unimproved varieties with good husk cover offer the greatest resistance to storage pest attack. The International Centre for Maize and Wheat Improvement (CIMMYT) and the International Institute of Tropical Agriculture (IITA) are both screening maize cultivars for resistance to post-harvest pests and sources of resistance of varying degrees have been identified. Similarly, cowpea cultivars are known with resistance either in the pods or in the pulse but have yet to result in varieties available to farmers with both good agronomic characteristics and resistance to bruchid beetles (Lienard & Seck 1994). In the future, it may prove possible to develop resistant varieties more rapidly. Gene transfer techniques can offer new opportunities and reduce the efforts involved in

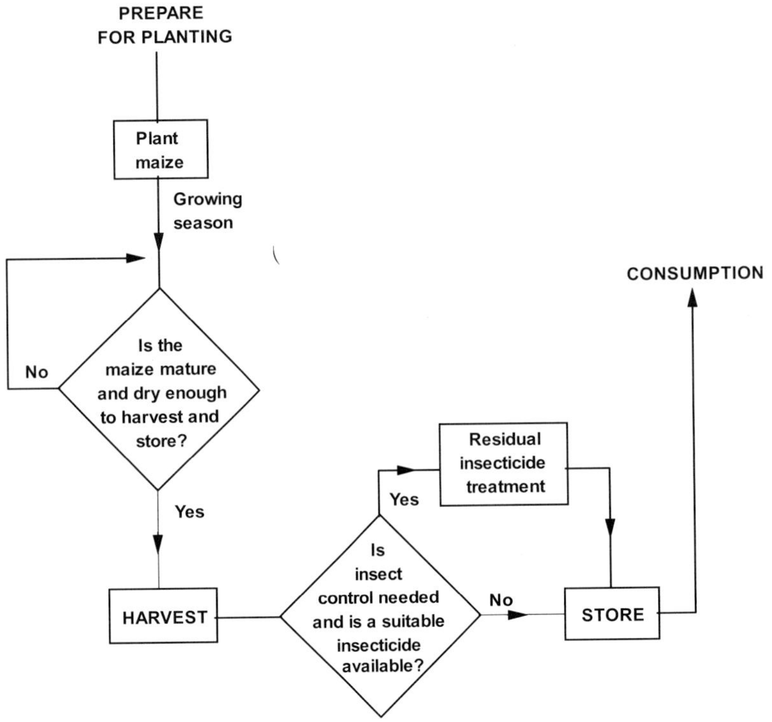


**Fig. 6.32** Conventional production and on-farm storage of maize in the tropics. Modified from Dobie 1984.

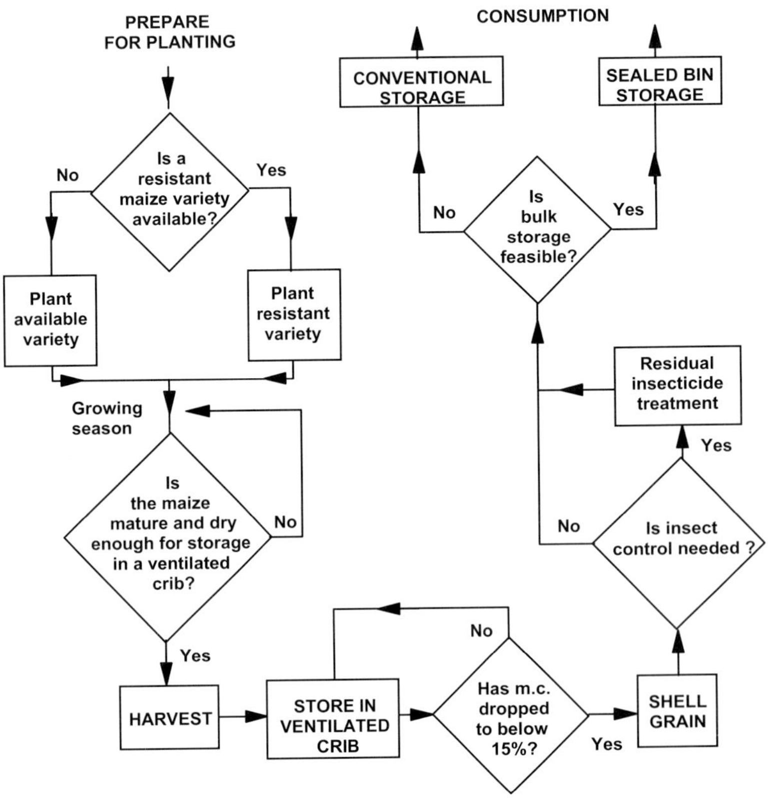


**Fig. 6.33** A modification of the production and storage of maize on-farm in Fig. 6.32, including alternative pest control options. Modified from Dobie 1984.

plant breeding, while more rapid screening of cultivars for potential insect resistance should be possible through the use of a biomonitor first developed in the US (Shade *et al.* 1990) and now undergoing development at the Natural Resources Institute (Devereau *et al.* 1999). This equipment detects the ultrasonic emissions generated by insect feeding in individual grains. Resistant grains can be identified rapidly as they are associated with reduced feeding rates.

Predators and parasites may also play a role in farm storage. This is discussed in more detail elsewhere.

### *Trader and central storage*

In many developed countries grain is stored large-scale in silos or flat stores that can be sealed. Pest management in such situations can be achieved conveniently using controlled atmospheres. However, in many other circumstances trader and other types of large-scale warehousing are dependent on keeping grain in bags and it is in this situation where most improvements in pest management can be achieved. In bag storage, the commodity is conventionally protected using an initial fumigation on arrival in store followed by residual insecticide treatment (Fig. 6.34). Thereafter, inspection by spear sampling warns of the need for further pest management action. Possible improvements might be to use a pheromone or food-baited trapping system to provide a more timely warning of pest attack (Fig. 6.35) and possibly an estimate of the size of the pest population. Advances in pest management could also be provided by an appropriate decision support system. Where pest

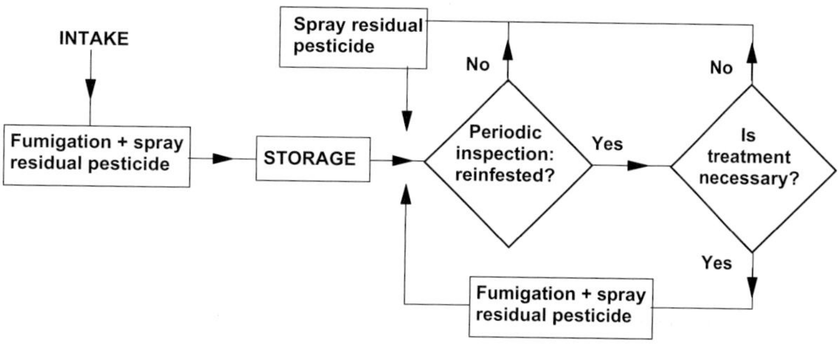


**Fig. 6.34** A common management pattern, for the storage of a bagged commodity in a medium or large warehouse, with dependence upon chemical pest control and conventional inspection methods. Modified from Dobie 1984.

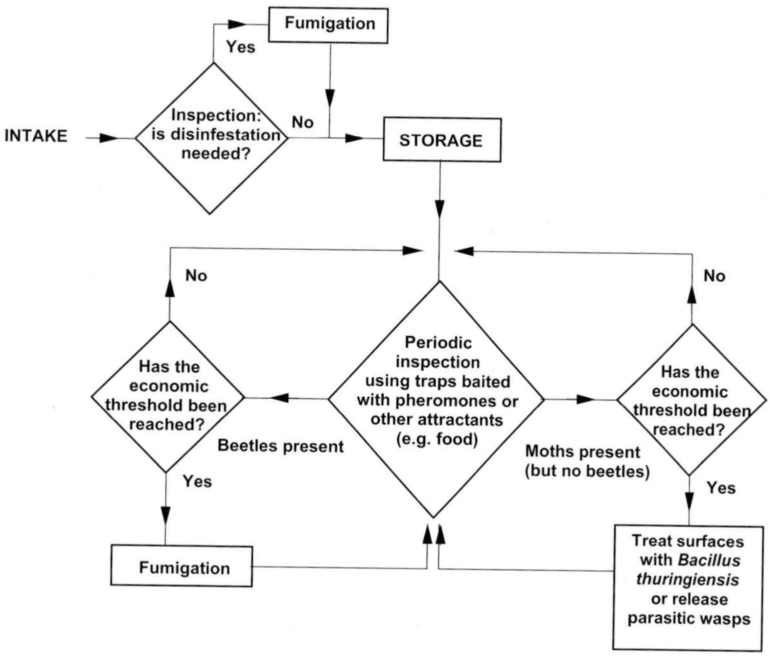


**Fig. 6.35** Possible modifications to the management pattern for the storage of bagged grain in a medium or large warehouse. Modified from Dobie 1984.

attack is by beetles, then fumigation is likely to be the most appropriate action if the commodity is to remain in store for a considerable period. If attack is by moths, then control might prove possible by use of pheromones for mating disruption (Hagstrum & Davis 1982; Hodges *et al.* 1984; Mafra-Neto & Baker 1996), application of formulations of *Bacillus thuringiensis* or the release of parasitic wasps.

### Box 6.2 Case study – use of a predatory beetle, *Teretrius nigrescens*, against larger grain borer, a beetle pest of stored maize and cassava in Africa

When the larger grain borer (LGB; *Prostephanus truncatus* (Coleoptera: Bostrichidae); Plate 22) was first introduced in the African continent in Tanzania in the early 1980s it was utterly devastating to affected subsistence farmers. Intact maize cobs are quickly reduced to dust if infested with LGB. Reducing this new threat to the food security of rural populations in Africa became the focus of much effort, both from the affected countries and from international donors.

The pest-status of LGB in large-scale storage is relatively low since it can be controlled fairly easily by fumigation. Small-scale farmers however are particularly vulnerable to the threat of LGB attack. It is not easy to reduce the damage inflicted by LGB once it has entered a small store of maize or cassava (Plates 23, 24). One recommended strategy that has had a significant impact is to shell the maize as soon as possible after harvest and then to treat the grain with a cocktail of insecticide containing a synthetic pyrethroid to kill the LGB and an organophosphate to kill other pests such as *Sitophilus* spp. Many farmers in affected areas are unable to afford these chemicals or cannot buy them when they are needed, because of poor distribution networks, and furthermore, the quality of the chemicals may be low because of adulteration (W. Riwa, pers. comm).

For any one farmer the risk of a store becoming infested is also very unpredictable. There is large variation in the overall levels of damage caused between years, and also, in any one year some stores will become very heavily damaged whereas others will remain untouched. This spatial and temporal patchiness of damage is in contrast to the largely predictable levels of damage a farmer can expect from another main insect pest, *Sitophilus* spp.

A farmer is then confronted with a dilemma. Either treat grain against LGB every year, even though for many years treatment will not be necessary, or wait until an infestation is observed before treating, and risk substantial loss of the harvest if an infestation is not detected and dealt with quickly enough. Classical biological control (the introduction of an exotic natural enemy) was pursued in the hope that populations of the pest would be reduced sufficiently to prevent 'bad years' of LGB damage with no need for any inputs from the farmers themselves.

Classical biological control was also favoured as a strategy since LGB is an unusual storage pest because it has considerable populations living in the bush and woodland away from stores, that are thought to provide reservoirs of sources of infestation. It would be both impractical and environmentally devastating to try and reduce these wider populations with chemicals.

In 1992, a native predator of LGB from Central America, *Teretrius nigrescens* (Coleoptera: Histeridae) (formerly known as *Teretriosoma nigrescens*) was introduced into Togo and subsequently into most other LGB-infested countries in Africa. *Teretrius nigrescens* (TN) adults and larvae prey on the eggs and larvae of LGB (Rees 1994). *Teretrius nigrescens* (Fig. 6.36) is thought to be a relatively specific predator of LGB since dispersing TN will follow the particular chemical signals emitted by male LGB that are thought to be used by males to attract mates (Cork *et al.* 1991; Scholz *et al.* 1998; Birkinshaw & Smith 2000).

Before releasing TN into the African continent, checks were made that it would not attack economically important insects such as bees, and an evaluation of the host specificity of the predator was made (Boeye *et al.* 1992). It was found that given a choice TN would always tend to prey on LGB, but would eat other insects and some grain if LGB were not available. The ability to subsist during times when the pest is unavailable was

**Fig. 6.36** Predator: *Teretrius nigrescens* (TN). Bar = 1 mm.

seen as a strength, since this would result in a higher chance that, once introduced, TN would be self-sustaining in its new environment and there would be no need for additional introductions. Indeed, in most countries where TN has now been introduced, the population has grown and shows no sign of local extinction (Giles *et al.* 1995; Borgemeister *et al.* 1997; Hodges & Birkinshaw 1999). One puzzling exception to this is in Tanzania, where although TN has thrived in neighbouring Kenya, few TN have so far been caught in LGB-pheromone baited flight-traps (W. Riwa, pers. comm.).

Predicting the likely impact of TN on LGB populations and the damage they cause in stores was severely hampered by several gaps in our knowledge of the population dynamics of LGB. The relative importance of biotic and non-biotic factors (such as climate) in LGB population fluctuations was not known. Nobody knew the relative importance of LGB populations in the bush and in store, and even what the common woody hosts of LGB are, and whether or not TN can be effective in these environments. So really TN was released on the basis of, 'well let's hope it does work' rather than any sure evidence that impact would be achieved.

Assessing the actual impact of classical biological control programmes is extremely difficult. The damage from the pest may go down after the introduction of the control agent, but it is extremely difficult to be sure that this is not due to other reasons, such as changes in farmer practices. Encouraging falls in LGB trap catches were recorded as TN numbers increased in Togo and Kenya (Giles *et al.* 1995; Borgemeister *et al.* 1997), and some in-store work indicated that losses had dropped over 80% following release of TN (Richter *et al.* 1997). However, TN numbers are high in some parts of Ghana but this did not prevent a dramatic increase in LGB catches five years after TN was introduced (Hodges & Birkinshaw 1999), and researchers in Benin predict that TN will be unable to reliably prevent the build-up of an LGB infestation in store (Holst *et al.* 2000).

In summary, classical biological control has been attempted to limit the damage of a particular storage pest whose distribution is more widespread than the storage environment itself. Almost all other current examples of the use of biological control against storage pests involves augmentation or inundation techniques that are relatively management intensive and explicitly target large-scale stores. The hope was that TN would have sufficient impact on the population dynamics of LGB in and away from stores to prevent 'bad years' of LGB damage. So far the evidence of success is patchy and may vary between different situations of climate and timing of TN introduction. The challenge for the future is to determine whether the impact of TN can be enhanced by changing farm storage practices, to produce a more integrated management system.

## Suggested reading

Buckle, A.P. & Smith, R.H. (Eds) (1994) *Rodent Pests and their Control*. Commonwealth Agricultural Bureau, Wallingford, UK.

Haines, C.P. (1983) Pest management in stored products. *Protection Ecology*, **4**, 321–330.

Haines C.P. (2000) IPM for food storage in developing countries, 20th century aspirations for the 21st century. *Crop Protection*, **19**, 825–830.

Jayas, D.S., White, N.D.G. & Muir, W.E. (Eds) (1995) *Stored Grain Ecosystems*. Marcel Dekker, New York, USA.

Meehan, A.P. (1984) *Rats and Mice: Their Biology and Control*. Rentokil Ltd., East Grinstead, UK.

Prakash, I. (Ed.) (1988) *Rodent Pest Management.* CRC Press, USA.

Subramanyam, Bh. & Hagstrum, D.W. (Eds) (1995) *Integrated Management of Insects in Stored Products.* Marcel Dekker, New York, USA.

## References


Achiano, K.A., Giliomee, J.H. & Pringle, K.L. (1999) The use of ash from *Aloe marlothii* Berger for the control of maize weevil, *Sitophilus zeamais* Motschulsky (Coleoptera, Curculionidae), in stored maize. *African Entomology*, **7**, 169–172.

Adane, K., Moore, D. & Archer, S.A. (1996) Preliminary studies on the use *of Beauvaria bassiana* to control *Sitophilus zeamais* (Coleoptera, Bostrichidae) in the laboratory. *Journal of Stored Products Research*, **32**, 105–113.

Akay, M.T., Yilmazoğlu, G., Yaşacan, S., Tūrk, H. & Kolanyak, D. (1992) Bioavailability and toxicological potential of wheat-bound pirimiphos-methyl residues in rats. *Xenobiotica*, **22**, 293–302.

Aldryhim, Y.N. (1990) Efficacy of amorphous silica dust, Dryacide®, against *Tribolium confusum* Duv. and *Sitophilus granarius* (L.) (Coleoptera, Tenebrionidae and Curculionidae). *Journal of Stored Products Research*, **26**, 207–210.

Aldryhim, Y.N. (1993) Combination of classes of wheat and environmental factors affecting the efficacy of amorphous silica dust, Dryacide®, against *Rhyzopertha dominica* (F.). *Journal of Stored Products Research*, **29**, 271–275.

Allen, F. (1972) A natural earth that controls insects. *Organic Gardening and Farming*, November, 50–56.

Altpeter, F., Diaz, I., McAuslane, H., Gaddom, K., Carbonero, P. & Vasil, I.K. (1999) Increased insect resistance in transgenic wheat stably expressing trypsin inhibitor Cme. *Molecular Breeding*, **5**, 53–63.

Ananthakrishnan, T.N. (1992) *Dimensions of Insect–Plant Interactions.* Intercept, Andover, UK.

Anon. (1961) *Federal Register* **26**, 10228. USA.

Anon. (1991) EPA R.E.D. FACTS. Silicon dioxide and Silica Gel. 21T-1021, **1–4**. Environmental Protection Agency, Washington DC, USA.

Anon. (1994a) *Dryacide® and Integrated Pest Management in the Grain Industry.* Dryacide Product Manual. Dryacide Australia Pty. Ltd, Maddington, Western Australia.

Anon. (1994b) Dust application workshop, Temora, New South Wales, Australia. Workshop held in conjunction with the 6th International Working Conference on Stored-product Protection, 17–23 April 1994, Canberra, Australia. Organised by Graincorp Operations Limited, New South Wales, Australia.

Anon, (1999) Biotechnology in crop protection, facts and fallacies. *International Pest Control*, January/February, 7–11.

Arnason, J.T., Philogene, B.J.R. & Morand, P. (Eds) (1989) *Insecticides of plant origin.* ACS Symposium Series **387**. American Chemical Society, Washington, DC, USA.

Arnason, J.T., Gale, J., de Beyssac, B.C., *et al.* (1992) Role of phenolics in resistance of maize grain to the stored grain insects, *Prostephanus truncatus* (Horn) and *Sitophilus zeamais* (Motsch). *Journal of Stored Products Research*, **28**, 119–126.

Arthur, F. H. (1992) Control of lesser grain borer (Coleoptera, Bostrichidae) with chlorpyrifos-methyl, bioresmethrin and resmethrin, effect of chlorpyrifos-methyl resistance and environmental degradation. *Journal of Economic Entomology*, **85**, 1471–1475.

Arthur, F.H. (1996) Grain protectants, current status and prospects for the future. *Journal of Stored Products Research*, **32**, 293–302.

Arthur, F.H. (2000) Toxicity of diatomaceous earth to red flour beetles and confused flour beetles (Coleoptera, Tenebrionidae), effects of temperature and relative humidity. *Journal of Economic Entomology*, **93**, 526–532.

Baier, A.H. & Webster, B.D. (1992) Control of *Acanthoscelides obtectus* Say (Coleoptera, Bruchidae) in *Phaseolus vulgaris* L. seed stored on small farmers – I. Evaluation of damage. *Journal of Stored Products Research*, **28**, 289–293.

Bartlett, B. (1951) The action of certain 'inert' dusts materials on parasitic Hymenoptera. *Journal of Economic Entomology*, **44**, 891–896.

Bengston, M., Connell, M., Davies, R.A.H., *et al.* (1980a) Chlorpyrifos-methyl plus bioresmethrin; methacrifos; pirimiphos-methyl plus bioresmethrin; and synergised bioresmethrin as grain protectants for wheat. *Pesticide Science*, **11**, 61–76.

Bengston, M., Connell, M., Davies, R.A.H., *et al.* (1980b) Fenitrothion plus (1R)-phenothrin, and pirimiphos-methyl plus carbaryl, as grain protection combinations for wheat. *Pesticide Science*, **11**, 471–482.

Berdoy, M.L. & Macdonald, D.W. (1991) Factors affecting feeding in wild rats. *Acta Oecologia*, **12**, 261–279.

Bernhard, K., Jarrett, P., Meadows, M., *et al.* (1997) Natural isolates of *Bacillus thuringiensis*, worldwide distribution, characterization, and activity against insect pests. *Journal of Invertebrate Pathology*, **70**, 59–68.

Birkinshaw, L.A. & Smith, R.H. (2000) Function of aggregation pheromone in the larger grain borer *Prostephanus truncatus*, variation in response to individuals as evidence for a role in sexual selection. *Journal of Chemical Ecology*, **26**, 1325–1339.

Blacow, N. W. (Ed.) (1975) *The Extra Pharmacopoeia.* Pharmaceutical Press, London, UK.

Boeye, J., Wright, M. & Laborius, G.A. (Eds) (1992) Implementation of and further research on biological control of the larger grain borer. Proceedings of an FAO/GTZ coordination meeting Lomé, Republic of Togo, 5–6 November 1990.

Borgemeister, C., Djossou, F., Adda, C., *et al.* (1997) Establishment, spread, and impact of *Teretriosoma nigrescens* (Coleoptera, Histeridae), an exotic predator of the larger grain borer (Coleoptera, Bostrichidae) in southwestern Benin. *Environmental Entomology*, **26**, 1405–1415.

Borgemeister, C., Tchabi, A. & Scholz, D. (1998) Trees or stores? The origin of migrating *Prostephanus truncatus* collected in different ecological habitats in southern Benin. *Entomologia Experimentalis et Applicata*, **87**, 285–294.

Bridgeman, B.W. (1991) *Grainco Structural Treatment Dryacide® SYMBOL 210/f 'Symbol' Manual.* Internal Grainco Publication, Grainco Queensland, Australia.

Bridgeman, B.W. (1994) Structural treatment with amorphous silica slurry, an integral component of GRAINCO's IPM strategy. In: E. Highley, E.J. Wright, H.J. Banks & B.R. Champ (Eds) *Proceedings of the 6th International Working Conference on Stored Product Protection*, Canberra, Australia. **2**, 628–630. CABI, Wallingford, UK.

Bridgeman, B.W. (1999) Application technology and usage patterns of diatomaceous earth in stored product protection. In: J. Zuxun, L. Quan, L. Yongsheng, T. Xianchang & G. Lianghua (Eds) *Proceedings of the 7th International Working Conference on Stored Product Protection*, Beijing, China. **1**, 785–789. Sichuan Publishing House of Science and Technology, Chengdu, Sichuan Province, P.R. China.

Bridgeman, B.W. & Collins, P.J. (1994) Integrated pest management in the Grainco, Queensland, Australia, Storage System. In: E. Highley, E.J. Wright, H.J. Banks & B.R. Champ (Eds) *Proceedings of the 6th International Working Conference on Stored Product Protection*, Canberra, Australia. **2**, 910–914. CABI, Wallingford, UK.

Brower, J.H., Smith, L., Vail, P.V. & Flinn, P.W. (1995) Biological control. In: Bh. Subramanyam & D.W. Hagstrum (Eds) (1995) *Integrated Management of Insects in Stored Products.* Marcel Dekker, New York, USA.

Buckle, A.P. (1994) Rodent control methods. In: A.P. Buckle & R.H. Smith (Eds) *Rodent Pests and their Control.* Commonwealth Agricultural Bureau, Wallingford, UK.

Budavari, S. (1989) *The Merck Index.* Merck & Co., Rahway, New Jersey, USA.

Carlson, S.D. & Ball, H.J. (1962) Mode of action and insecticidal value of a diatomaceous earth as a grain protectant. *Journal of Economic Entomology*, **55**, 964–970.

Carson, R. (1962) *Silent Spring.* London, Hamish Hamilton, London, UK.

Casida, J.E. & Quistad, G.B. (1998) Golden age of insecticide research, past, present, or future? *Annual Review of Entomology*, **43**, 1–16.

Chabra, S.C., Mahunnah, R.L.A. & Mshiu, E.N. (1993) Plants used in traditional medicine in Eastern Tanzania. VI. Angiosperms (Sapotaceae to Zingiberaceae). *Journal of Ethnopharmacology*, **39**, 83–103.

Champ, B.R. & Campbell-Brown, M.J. (1970) Insecticide resistance in Australian *Tribolium castaneum* (Herbst) – I. A test method for detecting insecticide resistance. *Journal of Stored Products Research*, **6**, 53–70.

Champ, B.R. & Dyte, C.E. (1976) Report of the FAO Global survey of pesticide susceptibility of stored grain pests. *Plant Production and Protection Series* **5**. Food and Agriculture Organization of the United Nations, Rome, Italy.

Chauhan, N.S. (1999) *Medicinal and Aromatic Plants of Himachal Pradesh.* Indus Publishing Company, New Delhi, India.

Chinwada, P. & Giga, D.P. (1997) Traditional seed protectants for the control of bean bruchids. *Tropical Science*, **37**, 80–84.

Chiu, S.F. (1939) Toxicity studies of so-called 'inert' materials with the bean weevil, *Acanthoscelides obtectus* (Say). *Journal of Economic Entomology*, **57**, 1013–1014.

Cobbinah, J.R., Moss, C., Golob, P. & Belmain, S.R. (1999) Conducting ethnobotanical surveys, an example from Ghana on plants used for the protection of stored cereals and pulses. *NRI Bulletin*, **77**. Natural Resources Institute, University of Greenwich, Chatham, UK.

Cockfield, S.D. (1992) Groundnut oil application and varietal resistance for control of *Callosobruchus maculatus* (F.) in cowpea grain in The Gambia. *Tropical Pest Management*, **38**, 268–270.

Copping, L.G. (1996) *Crop Protection Agents from Nature, Natural Products and Analogues*. Royal Society of Chemistry, Cambridge, UK.

Cork, A., Hall, D.R., Hodges, R.J. & Pickett, J.A. (1991) Identification of the major component of male-produced aggregation pheromone of larger grain borer, *Prostephanus truncatus* (Horn) (Coleoptera, Bostrichidae). *Journal of Chemical Ecology*, **17**, 789–803.

Cotton, R.T. & Frankenfeld, J.C. (1949) Silica aerogel for protecting stored seed or milled cereal products from insects. *Journal of Economic Entomology*, **42**, 553.

Cox, J.R. & Kilminster, K.M. (1995) Pesticide and container disposal. Final Technical Report, Project A0192. Unpublished report, Natural Resources Institute, Chatham, UK.

Cox, P.D. & Wilkin, D.R. (1996) The potential use of biological control of pests in stored grain. *Home Grown Cereals Authority Research Review*, **36**. HGCA, London, UK.

Daglish, G.J. (1998) Efficacy of six grain protectants applied alone or in combination against three species of Coleoptera. *Journal of Stored Products Research*, **34**, 263–268.

Daglish, G.J., Eelkema, M. & Harrison, L.M. (1995) Chlorpyrifos-methyl plus either methoprene or synergized phenothrin for control of coleoptera in maize in Queensland, Australia. *Journal of Stored Products Research*, **31**, 235–241.

Daglish, G.J., Eelkema, M. & Harrison, L.M. (1996) Control of *Sitophilus oryzae* (L.) (Coleoptera, Curculionidae) in paddy rice using chlorpyrifos-methyl or fenitrothion in combination with several other protectants. *Journal of Stored Products Research*, **32**, 247–253.

Dales, M.J. (1996) A review of plant material used for controlling insect pests of stored products. *NRI Bulletin*, **65**. Natural Resources Institute, University of Greenwich, Chatham, UK.

Dales, M.J. & Golob, P. (1997) The protection of maize against *Prostephanus truncatus* (Horn), using insecticide sprays in Tanzania. *International Journal of Pest Management*, **43**, 39–43.

Davies, J.C. (1970) Insect infestation and crop storage. In: J.D Jameson (Ed.) *Agriculture in Uganda.* Oxford University Press, Oxford, UK.

Davies, J.H. (1985) The pyrethroids, an historical introduction. In: J.P. Leahey (Ed.) *The Pyrethroid Insecticides.* Taylor & Francis, London, UK.

Denholm, I. & Rowland, M.W. (1992) Tactics for managing

pesticide resistance in arthropods, theory and practice. *Annual Review of Entomology*, **37**, 91–112.

Dent, D. (1991) *Insect Pest Management*. CABI, Wallingford, UK.

Desmarchelier, J.M. & Allen, S.E. (1999) Diatomaceous earths, health, safety, environment, residues and regulatory issues. In: J. Zuxun, L. Quan, L. Yongsheng, T. Xianchang & G. Lianghua (Eds) *Proceedings of the 7th International Working Conference on Stored Product Protection*, Beijing, China. **1**, 758–764. Sichuan Publishing House of Science and Technology, Chengdu, Sichuan Province, P.R. China.

Desmarchelier, J.M. & Dines, J.C. (1987) Dryacide® treatment of stored wheat, its efficacy against insects, and after processing. *Australian Journal of Experimental Agriculture*, **27**, 309–312.

Desmarchelier, J.M. & Paine, C. (1988) Insecticide-retentive carriers. I. Methods of analysis. *Australian Journal of Experimental Agriculture*, **28**, 267–269.

Desmarchelier, J.M., Allanson, R., Allen, S.E. & Moss, R. (1996) *Dryacide® removal during wheat cleaning*. Technical Report **68**. Division of Entomology, CSIRO, Canberra, Australia.

Devereau, A., Credland, P., Appleby, J. & Jackai, L. (1999) Rapid screening of grain for insect resistance. In: J. Zuxun, L. Quan, L. Yongsheng, T. Xianchang & G. Lianghua (Eds) *Proceedings of the 7th International Working Conference on Stored Product Protection*, Beijing, China. **1**, 18–26. Sichuan Publishing House of Science and Technology, Chengdu, Sichuan Province, P.R. China.

Dobie, P. (1974) The laboratory assessment of the inherent susceptibility of maize varieties to post harvest infestation by *Sitophilus zeamais* Motsch. (Coleoptera, Curculionidae). *Journal of Stored Products Research*, **10**, 183–197.

Dobie, P. (1984) Biological methods for the integrated control of insects and mites in tropical stored products. *Tropical Stored Products Information*, **48**, 21–25.

Don-Pedro, K.N. (1989) Effects of fixed vegetable oils on oviposition and adult mortality of *Callosobruchus* (F.) on cowpea. *International Pest Control*, **31**, 34–37.

Don-Pedro, K.N. (1996) Fumigant toxicity of citrus peel oils against adult and immature stages of storage insect pests. *Pesticide Science*, **47**, 213–223.

Dovlo, F.E., Williams, C.E. & Zoaka, L. (1976) *Cowpeas, home preparation and use in West Africa*. Integrated Development and Research Centre, Ottawa. Canada.

Dowdy, A.K. (1999) Mortality of red flour beetle, *Tribolium castaneum* (Coleoptera, Tenebrionidae) exposed to high temperature and diatomaceous earth combinations. *Journal of Stored Products Research*, **35**, 175–182.

Ebeling, W. (1971) Sorptive dust for pest control. *Annual Review of Entomology*, **16**, 123–158.

Edney, K.L. (1983) Top fruit. In: C. Dennis (Ed.) *Post-harvest Pathology of Fruits and Vegetables*. Academic Press, London, UK.

Elek, J. A. (1998) Interaction of treatment of both adult and immature Coleoptera with a chitin synthesis inhibitor, affects mortality and development time of their progeny. *Entomologia Experimentalis et Applicata*, **89**, 125–136.

Elek, J.A. & Longstaff, B.C. (1994) The effect of chitin synthesis inhibitors on stored product beetles. *Pesticide Science*, **40**, 225–230.

Estruch, J.J., Carozzi, N.B., Desai, N., Duck, N.B., Warren, G.W. & Koziel, M.G. (1997) Transgenic plants, an emerging approach to pest control. *Reviews in Biotechnology*, **15**, 137–141.

FAO (1996) *Code of Conduct for the Import and Release of Exotic Biological Control Agents*. Food and Agriculture Organization of the United Nations, Rome, Italy.

Farrell, G., Hodges, R.J. & Golob, P. (1996) Integration of control methods for stored products pests in East Africa. In: G. Farrell, A.H. Greathead, M.G. Hill & G.N. Kibata (Eds) *Management of Farm Storage Pests in East and Central Africa*. Proceedings of the East and Central Africa Storage Pest Management Workshop, Naivasha, Kenya, 14–19 April 1996. International Institute of Biological Control, UK.

FDA (1995) Silicon dioxide silica gel up to 2% additive to human foods as an anti-caking agent. 21 CFR Section 172. 480. Food and Drug Administration, Washington, USA.

Feitelson, J.S., Payne, J. & Kim, L. (1992) *Bacillus thuringiensis*, insects and beyond. *Biotechnology*, **10**, 271–275.

Feyereisen, R. (1999) Insect P450 enzymes. *Annual Review of Entomology*, **44**, 507–533.

Fields, P. & Korunic, Z. (2000) The effect of grain moisture content and temperature on the efficacy of diatomaceous earths from different geographical locations against stored-product beetles. *Journal of Stored Products Research*, **36**, 1–13.

Fields, P., Dowdy, A. & Marcotte, M. (1997) Structural pest control, the use of an enhanced diatomaceous earth product combined with heat treatment for the control of insect pests in food processing facilities. Canada-United States Working Group on Methyl Bromide Alternatives. http://res.agr.ca/winn/Heat-DE.htm.

Fields, P.G & Muir, W.E. (1996) Physical control. In: Bh. Subramanyam & D.W. Hagstrum (Eds) (1995) *Integrated Management of Insects in Stored Products*. Marcel Dekker, New York, USA.

Flanders, S.F. (1941) Dust as an inhibiting factor in the reproduction of insects. *Journal of Economic Entomology*, **34**, 470–472.

Fleurat-Lessard, F., Vidal, M.-L. & Budzinski, H. (1998) Modelling biological efficacy decrease and rate of degradation of chlorpyrifos-methyl on wheat stored under controlled conditions. *Journal of Stored Products Research*, **34**, 341–354.

Flinn, P.W. & Hagstrum, D. W. (1990) An expert system for managing insect pests of stored grain. In: F. Fleurat-Lessard & P. Ducom (Eds) *Proceedings of the 5th International Working Conference on Stored-Products Protection*, Bordeaux, France, **3**, 2011–2018.

Gatehouse, A.M.R. & Gatehouse, J.A. (1998) Identifying proteins with insecticidal activity: use of encoding genes to produce insect resistant transgenic crops. *Pesticide Science*, **52**, 165–175.

Gbolade, A.A. & Adebayo, T.A. (1993) Fumigant effects of some volatile oils on fecundity and adult emergence of *Callosobruchus maculatus* F. *Insect Science and its Application*,

**14**, 631–636.

Giga, D.P. & Mazarura, U.M. (1990) Malathion resistance in *Sitophilus zeamais* (Motsch.) in Zimbabwe. *Tropical Pest Management*, **36**, 320.

Giles, P.H. & Ashman, F. (1971) A study of pre-harvest infestation of maize by *Sitophilus zeamais* Motsch. (Coleoptera, Curculionidae) in the Kenya Highlands. *Journal of Stored Products Research*, **7**, 69–83.

Giles, P., Hill, G., Nang'ayo, F., Farrell, G., Stabrawa, A. & Wekesa, P. (1995) Entomological and socio-economic investigations for the development of integrated pest management strategies against Prostephanus truncatus. *Final Project Report.* Natural Resources Institute, Chatham, UK; National Agriculture Research Laboratories, Nairobi, Kenya.

Girard, C., Metayer, M.L., Bonade-Bottino, M., Pham-Delegue, M.H., & Jouanin, L.(1998) High level of resistance to proteinase inhibitors may be conferred by proteolytic cleavage in beetle larvae. *Insect Biochemistry and Molecular Biology*, **28**, 229–237.

Goldsmith, D.F. (1999) Research and policy implications of IARC's classification of silica as a Group 1 carcinogen. *Indoor Built Environment*, **8**, 136–142.

Golob, P. (1997) Current status and future perspectives for inert dusts for control of stored product insects. *Journal of Stored Products Research*, **33**, 69–79.

Golob, P. & Hanks, C. (1990) Protection of farmer stored maize against infestation by *Prostephanus truncatus* (Horn) and *Sitophilus* species in Tanzania. *Journal of Stored Products Research*, **26**, 187–198.

Golob, P. & Webley, D.J. (1980) *The use of plants and minerals as traditional protectants of stored products*. Tropical Products Institute **G138**, Slough, UK.

Golob, P., Mwambula, J., Mhango, V. & Ngulube, F. (1982) The use of locally available materials as protectants of maize grain against insect infestation during storage in Malawi. *Journal of Stored Products Research*, **18**, 67–74.

Golob, P., Changjaroen, P., Ahmed, A. & Cox, J. (1985) Susceptibility of *Prostephanus truncatus* (Horn) (Coleoptera, Bostrichidae) to insecticides. *Journal of Stored Products Research*, **21**, 141–150.

Golob, P., Moss, C., Dales, M., Fidgen, A., Evans, J. & Gudrups, I. (1999) The use of spices and medicinals as bioactive protectants for grains. *FAO Agricultural Services Bulletin*, **137**. Food and Agriculture Organization of the United Nations, Rome, Italy.

Greaves, J.H. (1994) Resistance to anticoagulent rodenticides. In: A.P. Buckle & R.H. Smith (Eds) *Rodent Pests and their Control.* Commonwealth Agricultural Bureau, Wallingford, UK.

Grossi deSa, M.F., Mirkov, T..E., Ishimoto, M., Colucci, G., Bateman, K.S., & Chrispeels, M.J. (1997) Molecular characterization of a bean alpha-amylase inhibitor that inhibits the alpha-amylase of the Mexican bean weevil *Zabrotes subfasciatus. Planta*, **203**, 295–303.

Gudrups, I. (1996) Surface spraying in tropical warehouses, is it worthwhile? *International Pest Control*, **38**, 58–60, 64.

Gunning, R.V., Easton, C.S., Balfe, M.E. & Ferris, I.G. (1991) Pyrethroid resistance mechanisms in Australian *Helicoverpa armigera. Pesticide Science*, **33**, 473–490.

Hadjidemetriou, D.G. (1990a) Pirimiphos-methyl residues in stored wheat and barley, bread, burghul and parboiled wheat. In: *Studies of the magnitude and nature of pesticide residues in stored products using radiotracer techniques. Proceedings of a research co-ordination meeting*, Ankara, Turkey, 30 May–3 June 1988. International Atomic Energy Authority, Vienna, Austria.

Hadjidemetriou, D.G. (1990b) Persistence of pirimiphos-methyl in stored sultana raisins, common beans and their processed products. In: *Studies of the magnitude and nature of pesticide residues in stored products using radiotracer techniques. Proceedings of a research co-ordination meeting*, Ankara, Turkey, 30 May–3 June 1988. International Atomic Energy Authority, Vienna, Austria.

Hagstrum, D.W. & Davis, L.R. (1982) Mate seeking behaviour and reduced mating by *Ephestia cautella* (Walker) in sex pheromone permeated atmosphere. *Journal of Chemical Ecology*, **8**, 507–515.

Haines, C. P. (1999) Arthropod natural enemies in stored products – overlooked and under-exploited. In: J. Zuxun, L. Quan, L. Yongsheng, T. Xianchang & G. Lianghua (Eds) *Proceedings of the 7th International Working Conference on Stored Product Protection*, Beijing, China, **2**, 1205–1226. Sichuan Publishing House of Science and Technology, Chengdu, Sichuan Province, P.R. China.

Harris, C. R. & Kinoshita, G. B. (1977) Influence of post-treatment temperature on the toxicity of pyrethroid insecticides. *Journal of Economic Entomology*, **70**, 215–218.

Hassell, K.A. (1990) *The Biochemistry and Uses of Pesticides: Structure, Metabolism, Mode of Action and Uses in Crop Protection.* Second Edition. MacMillan Press Ltd, UK.

He, F. & Chen, S. (1999) Health impacts of pesticide exposure and approaches to prevention. *Asian-Pacific Newsletter on Occupational Health and Safety*, **6**. http: //www.occuphealth.fi/e/info/asi/ap399/health02.htman.

Heather, N.W. (1986) Sex-linked resistance to pyrethroids in *Sitophilus oryzae* (L.) (Coleoptera, Curculionidae). *Journal of Stored Products Research*, **22**, 15–20.

Hebblethwaite, M.J. (1985) The application of economics to pre- and post-harvest pest control programmes in developing countries. *Tropical Science*, **25**, 215–230.

Henckes, C. (1992) *Investigations into the Population Dynamics, Damage and Losses of Stored Maize – an approach to IPM on small farms in Tanzania with special reference to* Prostephanus truncatus *(Horn).* GTZ, Hamburg, Germany.

Herron, G.A. (1990) Resistance to grain protectants and phosphine in coleopterous pests of grain stored on farms in New South Wales. *Journal of the Australian Entomology Society*, **29**, 183–189.

Hessel, P.A., Gamble, J.F., Gee, J.B.L., *et al.* (2000) Silica, silicosis and lung cancer. *Journal of Occupational and Environmental Medicine*, **42**, 704–720.

Hodges, R.J. (1999) Decision making tools for more cost-effective pest management in milled rice bag-stores. In: I.F. Grant & C. Sear (Eds) *Decision Tools for Sustainable Development.* Natural Resources Institute, University of Greenwich,

Chatham, UK.

Hodges, R.J. & Birkinshaw, L.A. (1999) *Risk warning to farmers of Larger Grain Borer infestation and reduced pesticide treatment in farm maize stores.* Final Report to the DFID Crop Post-harvest Programme (Project R6684). Natural Resources Institute, University of Greenwich, Chatham, UK.

Hodges, R.J., Benton, F., Hall, D.R. & dos Santos Serodio (1984) Control of *Ephestia cautella* (Walker) by synthetic sex pheromones in the laboratory and store. *Journal of Stored Products Research*, **20**, 191–197.

Hodges, R.J., Sidik, M., Halid, H. & Conway, J.A. (1992) Cost efficiency of respraying store surfaces with insecticide to protect bagged milled rice from insect attack. *Tropical Pest Management*, **38**, 391–397.

Hodges, R.J., Robinson, R. & Hall, D.R. (1996) Quinone contamination of dehusked rice by *Tribolium castaneum* (Herbst) (Coleoptera, Bostrichidae). *Journal of Stored Products Research*, **32**, 31–37.

Hoffmann, K.H. & Lorenz, M.W. (1998) Recent advances in hormones in insect pest control. *Phytoparasitica*, **26**, 323–330.

Holst, N., Meikle, W.G., Nansen, C. & Markham, R.H. (2000) Analysing the impact of *Teretrius nigrescens* on *Prostephanus truncatus* in maize stores in west Africa. Poster presentation at the XXI International Congress of Entomology and XVIII Brazilian Congress of Entomology. http://ace.orst.edu/info/nptn.

Hughes, J.M., Weill, H., Checkoway, H., *et al.* (1998) Radiographic evidence of silicosis risk in the diatomaceous earth industry. *American Journal of Respiratory and Critical Care Medicine*, **158**, 807–814.

Ishimoto, M., Sato, Chrispeels, M.J. & Kitamura, K. (1996) Bruchid resistance of transgenic azuki bean expressing seed alpha amylase inhibitor of common bean. *Entomologia Experimentalis et Applicata*, **79**, 309–315.

Isman, M.B. (1997) Neem and other botanical insecticides, barriers to commercialisation. *Phytoparasitica*, **259**, 339–344.

Ivbijaro, M.F., Ligan, C. & Youdeowei, A. (1985) Control of rice weevils, *Sitophilus oryzae* (L.), in stored maize with vegetable oils. *Agriculture, Ecosystems and Environment*, **14**, 237–242.

Jackson, K. & Webley, D. (1994) Effects of Dryacide® on the physical properties of grains, pulses and oilseeds. In: E. Highley, E.J. Wright, H.J. Banks & B.R. Champ (Eds) *Proceedings of the 6th International Working Conference on Stored Product Protection*, Canberra, Australia, **2**, 635–637. CABI, Wallingford, UK.

Jansens, S., Cornelissen, M., Clercq, R. de, Reynaerts, A. & Perferoen, M. (1995) *Phthorimaea operculella* (Lepidoptera: Gelechidae) resistance in potato by expression of *Bacillus thuringiensis* Cry IA(b) insecticidal crystal protein. *Journal of Economic Entomology*, **88**, 1469–1476.

Javaid, I. & Ramatlakapela, K. (1995) The management of cowpea weevils [*Callosobruchus maculatus* (Fabricius)] in cowpea seeds by using ash and sand. *Journal of Sustainable Agriculture*, **7**, 147–154.

Jefferson, R.N. & Eads, C.O. (1951) Control of aphids transmitting stock mosaic. *Journal of Economic Entomology*, **44**, 878–882.

Jenson, A.G. (1979) *Proofing of Buildings Against Rats and Mice.* Technical Bulletin **12**, Ministry of Agriculture, Fisheries and Food, London, UK.

Johnson, R.M. & Kozak, A.S. (1966) Correction for the effect of diatomaceous earth on moisture content of wheat as determined by capacitance measurements. *Agronomy Journal*, **58**, 135–137.

Kaga, A. & Ishimoto, M. (1998) Genetic localization of a bruchid resistance gene and its relationship to insecticidal cyclopeptide alkaloids, the vignatic acids, in mungbean (*Vigna radiata* L. Wilczek). *Molecular and General Genetics*, **258**, 378–384.

Katanga Apuuli, J.K. & Villet, M.H. (1996) The use of wood ash for the protection of stored cowpea seed (*Vigna unguiculata* (L.) Walp.) against Bruchidae (Coleoptera). *African Entomology*, **4**, 97–99.

Kate, K.T. & Laird, S.A. (1999) *The Commercial Use of Biodiversity, Access to Genetic Resources and Benefit-sharing.* Earthscan, London, UK.

Katz, H. (1991) Desiccants: dry as dust means insect's death. *Pest Control Technology*, **82**, 84.

Keita, S.M., Vincent, C., Schmit, J.P., Ramaswamy, S. & Belanger, A. (2000) Effect of various essential oils on *Callosobruchus maculatus* (F.) (Coleoptera, Bruchidae). *Journal of Stored Products Research*, **36**, 355–364.

Keserü, G.M., Kolossváry, I. & Bertók, B. (1999) Piperonyl butoxide-mediated inhibition of cytochrome P450-catalysed insecticide metabolism, a rational approach. *Pesticide Science* **55**, 1004–1006.

Kitch, L.W., Shade, R.E. & Murdock, L.L. (1991) Resistance to the cowpea weevil (*Callosobruchus maculatus*) larvae in pods of cowpea (*Vigna unguiculata*). *Entomologia Experimentalis et Applicata*, **60**, 183–192.

Korunic, Z. (1994) Dijatomejska zemlja prirodni ineskticid. [Diatomaceous earths as natural insecticides.] In: Z. Korunic (Ed.) Proceedings of ZUPP '94 (Protection of Stored Agricultural Products), 9–11 March 1994, Novi Vinodolski, Croatia. pp 136–148.

Korunic, Z. (1997) Rapid assessment of the insecticidal value of diatomaceous earths without conducting bioassays. *Journal of Stored Products Research*, **33**, 219–229.

Korunic, Z. (1998) Diatomaceous earths, a group of natural insecticides. *Journal of Stored Products Research*, **34**, 87–97.

Korunic, Z. & Fields, P. (1995) Diatomaceous earth insecticidal composition. Canadian and USA Patent Pending, 1995.

Korunic, Z. & Fields, P. (1999) The effect of grain moisture content and temperature on efficacy of six diatomaceous earths against three stored product beetles. In: J. Zuxun, L. Quan, L. Yongsheng, T. Xianchang & G. Lianghua (Eds) *Proceedings of the 7th International Working Conference on Stored Product Protection*, Beijing, China. **1**, 790–795. Sichuan Publishing House of Science and Technology, Chengdu, Sichuan Province, P.R. China.

Korunic, Z. & Ormesher, P. (1999) Evaluation and standardised testing of diatomaceous earth. In: J. Zuxun, L. Quan, L. Yongsheng, T. Xianchang & G. Lianghua (Eds) *Proceedings of the 7th International Working Conference on Stored*

*Product Protection*, Beijing, China. **1**, 738–744. Sichuan Publishing House of Science and Technology, Chengdu, Sichuan Province, P.R. China.

Korunic, Z., Fields, P.G., White, N.D.G., MacKay, A. & Timlick, B. (1996) The effectiveness of Diatomaceous Earth against stored-grain insect pests in farm storage. In: *Proceedings XX International Congress of Entomology*, Florence, Italy.

Kossou, D.K., Mareck, J.H. & Bosque-Perez, N.A. (1993) Comparison of improved and local maize varieties in the Republic of Benin with emphasis on susceptibility to *Sitophilus zeamais* Motschulsky. *Journal of Stored Products Research*, **29**, 333–343.

Kostyukovsky, M., Chen, B., Atsmi, S & Shaaya, E. (2000) Biological activity of two juvenoids and two ecdysteroids against three stored product insects. *Insect Biochemistry and Molecular Biology*, **30**, 891–897.

Kramer, K.J., Morgan, T.D., Throne, J.E., Dowell, F.E., Bailey, M. & Howard, J.A. (2000) Transgenic avidin maize is resistant to storage insect pests. *Nature Biotechnology*, **18**, 670–674.

La Hue, D.W. (1965) Evaluation of malathion, synergised pyrethrum and diatomaceous earth as wheat protectants in small bins. Marketing Research Report **726**. Agricultural Research Service, United States Department of Agriculture.

La Hue, D.W. (1967) Evaluation of four inert dusts on wheat as protectants against insects in small bins. Marketing Research Report **780**. Agricultural Research Service, United States Department of Agriculture.

La Hue, D.W. (1970) Evaluation of malathion, diazinon, a silica aerogel and a diatomaceous earth as protectants on wheat against lesser grain borer attack. Marketing Research Report **860**. Agricultural Research Service, United States Department of Agriculture.

La Hue, D.W. (1972) The retention of diatomaceous earths and silica aerogels on shelled corn, hard winter wheat and sorghum grain. Marketing Research Report **51044**. Agricultural Research Service, United States Department of Agriculture.

La Hue, D.W. (1977) Grain protectants for seed corn, field test. *Journal of Economic Entomology*, **70**, 720–722.

Lacy, W.B. (1995) The global plant genetic resources system, a competition–cooperation paradox. *Crop Science, Madison, Wisconsin*, **35**, 335–345.

Le Patourel, G.N.J. (1986) The effect of grain moisture content on the toxicity of a sorptive silica dust to four species of grain beetle. *Journal of Stored Products Research*, **25**, 65–72.

Le Patourel, G.N.J. & Singh, J. (1984) Toxicity of amorphous silicas and silica pyrethroids mixtures to *Tribolium castaneum* (Herbst) (Coleoptera, Tenebrionidae). *Journal of Stored Products Research*, **20**, 183–190.

Levinson, H. & Levinson, A.T.I. (1999) Pest control of stored grain in antiquity. *Informatore Fitopatologico*, **49**, 13–18.

Libes, S.M. (1992) Biogenic silica. In: *An Introduction to Marine Biogeochemistry*. John Wiley & Sons Inc., New York, USA.

Lienard, V. & Seck, D. (1994) Revue des methodes de lutte contre *Callosobruchus maculatus* (F.) (Coleoptera, Bruchidae) ravageur des graines niebe (*Vigna unguiculata* (L.) en Afrique tropicale. *Insect Science and its Application*, **15**, 301–311.

Longstaff, B.C. (1994) The management of stored product pests by non-chemical means, an Australian perspective. *Journal of Stored Products Research*, **30**, 179–185.

Longstaff, B.C. (1997) Decision tools for grain storage pest management. *Journal of Stored Products Research*, **33**, 99–114.

Lord, J. (1999) Silicaceous dust synergises *Beauveria bassiana* applied to stored grain beetles. 32nd Annual Meeting of Society of Invertebrate Pathologists, 23–27 August 1999. University of California, Irvine, CA, USA.

Lourencao, A.L., Komatsu, S.S. & Alves, S.B. (1993) Controle de *Sithophilus zeamays* em milho com *Beauveria bassiana, Metarhizium anisoplae* e pirimifos metil. *Ecossistema, Espirito Santos Do Pinhal*, **18**, 69–74.

McCaffery, A.R., Head, D.J., Jianguo, T., Dubbeldam, A.A., Subramaniam, V.R. & Callaghan, A. (1997) Nerve insensitivity resistance to pyrethroids in Heliothine Lepidoptera. *Pesticide Science*, **51**, 315–320.

McDonald, J.W. (1989) Silica, silicosis and lung cancer. *British Journal of Industrial Medicine*, **46**, 289–291.

McFarlane, J. (1969) The effects of synergized pyrethrins and lindane on pre-emergence mortality of *Acanthoscelides obtectus* (Say.) (Coleoptera: Bruchidae). *Journal of Stored Products Research*, **5**,177–180.

McFarlane, J.A. (1989) *Guidelines for pest management research to reduce stored food losses caused by insects and mites*. Overseas Development Natural Resources Institute, Bulletin **22**, Slough, UK.

McGaughey, W.H. (1972) Diatomaceous earth for confused flour beetle and rice weevil control in rough, brown and milled rice. *Journal of Economic Entomology*, **65**, 1427–1428.

McGaughey, W.H. (1980) *Bacillus thuringiensis* for moth control in stored wheat. *Canadian Entomologist*, **112**, 327–331.

McGaughey, W.H. (1985) Evaluation of *Bacillus thuringiensis* for controlling Indian meal moths in farm grain bins and elevator silos. *Journal of Economic Entomology*, **78**, 1089–1094.

McGaughey, W.H. & Johnson, D.E. (1992) Indian meal moth (Lepidoptera, Pyralidae) resistance to different strains and mixtures of *Bacillus thuringiensis. Journal of Economic Entomology*, **85**, 1594–1600.

McGaughey, W.H., Speirs, R.D. & Martin, C.R. (1990) Susceptibility of classes of wheat grown in the United States to stored grain insects. *Journal of Economic Entomology*, **83**, 1122–1127.

McLaughlin, A. (1994) Laboratory trials on desiccant dust insecticides. In: E. Highley, E.J. Wright, H.J. Banks & B.R. Champ (Eds) *Proceedings of the 6th International Working Conference on Stored Product Protection*, Canberra, Australia, **1**, 638–645. CABI, Wallingford, UK.

Maceljski, M. & Korunic, Z. (1971) Trials of inert dusts in water suspension for controlling stored-product pest. *Zastita Bilja*, **22**, 377–387.

Maceljski, M. & Korunic, Z. (1972) *The effectiveness against stored-product insects of inert dusts, insect pathogens, temperature and humidity*. Project No. E30-MQ-1. Grant USDA/YU No. FG-YU-130. Final Report of Institute for Plant Protection, Zagreb, Yugoslavia.

Mafra-Neto, A. & Baker, T.C. (1996) Timed, metered sprays of pheromone disrupt mating of *Cadra cautella* (Lepidoptera, Pyralidae). *Journal of Agricultural Entomology*, **13**, 149–168.

Matthews, G.A. (1992) *Pesticide Application Methods*. Second Edition. Longman Group, UK.

Meadows, M.P., Ellis, D.J., Butt, J., Jarrett, P. & Burges, H.D. (1992) Distribution, frequency, and diversity of *Bacillus thuringiensis* in an animal feed mill. *Applied and Environmental Microbiology*, **58**, 1344–1350.

Meehan, A.P. (1984) *Rats and Mice. Their Biology and Control*. Rentokil Ltd, East Grinstead, UK.

Mensah, G.W.K. & Watters, F.L. (1979) Comparison of four organophosphorus insecticides on stored wheat for control of susceptible and malathion-resistant strains of the red flour beetle. *Journal of Economic Entomology*, **72**, 456–461.

Mewis, I. & Reichmuth, C. (1999) Diatomaceous earths against coleoptera granary weevil *Sitophilus granarius* (Curculionidae), the confused flour beetle *Tribolium confusum* (Tenebrionidae) and the mealworm *Tenebrio molitor* (Tenebrionidae). In: J. Zuxun, L. Quan, L. Yongsheng, T. Xianchang & G. Lianghua (Eds) *Proceedings of the 7th International Working Conference on Stored Product Protection*, Beijing, China, **1**, 765–780. Sichuan Publishing House of Science and Technology, Chengdu, Sichuan Province, P.R. China.

Meyer, A.N. (1989) Vertebrate pest control report on a visit to the Food Department, Sri Lanka. Overseas Development Natural Resources Institute, Chatham, UK.

Miles, W.J. (1990) Mining industry responds to crystalline silica regulations. *Mining Engineering*, April, 345–348.

Moore, D., Lord, J. & Smith, S.M. (2000) Pathogens. In: Bh. Subramanyam & D.W. Hagstrum (Eds) *Alternatives to Pesticides in Stored-product IPM.* Kluwer Academic Publishers, Boston MA, USA.

Morallo-Rejesus, B., Javier, J.A. & Juliano, B.O. (1982) Properties of brown rice and varietal resistance to storage insects. *Philippine Entomology*, **5**, 227–238.

Morgan, D.P. (1989) Organophosphate insecticides. In: *Recognition and Management of Pesticide Poisonings*. Fourth Edition. Environmental Protection Agency, Washington, DC, USA.

Muggleton, J. (1983) Relative fitness of malathion-resistant phenotypes of *Oryzaephilus surinamensis* L. (Coleoptera, Silvanidae). *Journal of Applied Ecology*, **20**, 245–254.

Mumford, J. & Norton, G. (1991) Economics and decision making in stored-product pest management. In: F. Fleurat-Lessard & P. Ducom (Eds) *Proceedings of the 5th International Working Conference on Stored Products Protection*, Bordeaux, France, **3**, 1913–1923.

Nahdy, M.S., Silim, S.N. & Ellis, R.H. (1999) Effect of field infestations of immature pigeonpea (*Cajanus cajan* (L.) Millsp.) pods on production of active (flight) and sedentary (flightless) morphs of *Callosobruchus chinensis* (L.) *Journal of Stored Products Research*, **35**, 339–354.

Nang'ayo, F.L.O., Hill, M.G., Chandi, E.A., Chiro, C.T., Nzeve, N.V. & Obiero, J. (1993) The natural environment as a reservoir for the larger grain borer *Prostephanus truncatus* (Horn) (Coleoptera, Bostrichidae) in Kenya. *African Crop Science Journal*, **1**, 39–47.

NIOSH (1977) Registry of toxic effect of chemical substances. US Department of Health, Education and Welfare, DHEW Publication Number (NIOSH) 78–104-A. National Institute of Occupational Safety and Health, Washington, DC, USA.

NOHSC (1995) *Exposure standards for atmospheric contaminants in the occupational environment, guidance note on the interpretation of exposure standards for atmospheric contaminants in the occupational environment* (NOHSC, 3008), *adopted national exposure standards for atmospheric contaminants in the occupational environment* (NOHSC, 1003). Third Edition. National Occupational Health and Safety Commission, Canberra, Australia.

Nickson, P.J., Desmarchelier, J.M. & Gibbs, P. (1994) Combination of cooling with a surface application of Dryacide® to control insects. In: E. Highley, E.J. Wright, H.J. Banks & B.R. Champ (Eds) *Proceedings of the 6th International Working Conference on Stored Product Protection*, Canberra, Australia, **2**, 638–645. CABI, Wallingford, UK.

Nielsen, P.S. (1998) The effect of a diatomaceous earth formulation on the larvae of *Ephestia kuehniella* Zeller. *Journal of Stored Products Research*, **34**, 113–121.

Obeng-Ofori, D., Jembere, B., Hassanali, A. & Reichmuth, C. (2000) Effectiveness of plant oils and essential oil of *Ocimum* plant species for protection of stored grains against damage by stored product beetles. In: J. Zuxun, L. Quan, L. Yongsheng, T. Xianchang & G. Lianghua (Eds) *Proceedings of the 7th International Working Conference on Stored Product Protection*, Beijing, China. Sichuan Publishing House of Science and Technology, Chengdu, Sichuan Province, P.R. China.

Oberlander, H. & Silhacek, D.L. (2000) Insect growth regulators. In: Bh. Subramanyan & D.W. Hagstrum (Eds) *Alternatives to Pesticides in Stored-product IPM.* Kluwer Academic Publishers, Boston MA, USA.

Oberlander, H., Silhacek, D.L., Shaaya, E. & Ishaaya, I. (1997) Current status and future perspectives of the use of insect growth regulators for the control of stored product insects. *Journal of Stored Products Research*, **33**, 1–6.

Oduor, G.I., Smith, S.M., Chandi, E.A., Karanja, L.W., Agano, J.O. & Moore, D. (2000) Occurrence of *Beauveria bassiana* on insect pests of stored maize in Kenya. *Journal of Stored Products Research*, **36**, 177–185.

Ofuya, T.I. (1986) Use of wood ash, dried chilli pepper fruits and onion scale leaves for reducing *Callosobruchus maculatus* (Fabricius) damage in cow-pea seeds during storage. *Journal of Agricultural Science, Cambridge*, **107**, 467–468.

Ofuya, T.I. & Credland, P.F. (1995) Responses of three populations of the seed beetle, *Callosobruchus maculatus* (F.) (Coleoptera, Bruchidae), to seed resistance in selected varieties of cowpea, *Vigna unguiculata* (L.) Walp. *Journal of Stored Products Research*, **31**, 17–27.

Ohtsubo, K. & Richardson, M (1992) The amino acid sequence of a 20 kDa bifunctional subtilisin/α-amylase inhibitor from bran of rice (*Oryza sativa*) seeds. *FEBS Letters*, **309**, 68–72.

OSHA (1991) *Occupational health and safety standards*. Patr. 1910.1000. 7-1-91 Edition. Occupational Safety and Health Association, Washington, DC, USA.

Oudejans, J.H. (1982) *Agro-pesticides, their management and application*. United Nations Economic Commission for Asia and the Pacific, Bangkok, Thailand.

Owusu, E.O. (2000) Effect of some Ghanaian plant components on control of two stored product insect pests of cereals. *Journal of Stored Products Research*, **37**, 85–91.

Pedral Sampaio, M.R.F., Ruegg, E.F., Viliotti, W.M. & Homen de Mello, M.H. (1990) $^{14}$C-pirimiphos-methyl residues in stored maize grains. In: *Studies of the magnitude and nature of pesticide residues in stored products using radio tracer techniques. Proceedings of a research co-ordination meeting*, Ankara, Turkey, 30 May–3 June 1988. International Atomic Energy Authority, Vienna, Austria.

Perez-Mendoza, J., Baker, J.E., Arthur, F.H. & Flinn, P.W. (1999) Effects of Protect-it® on efficacy of *Anisopteromalus calandrae* (Hymenoptera, Pteromalidae) parasitising rice weevils (Coleoptera, Curculionidae) in wheat. *Environmental Entomology*, **28**, 529–534.

Phelan, P.L. & Lin, H. (1991) Chemical characterisation of fruit and fungal volatiles attractive to dried-fruit beetle, *Carpophilus hemipterus* (L.) (Coleoptera, Nitidulidae). *Journal of Chemical Ecology*, **17**, 1253–1272.

Philogene, B.J. (1972) Volcanic ash for insect control. *Canadian Entomologist*, **104**, 1487.

Pierce, A.M., Pierce, H.D. Jr, Borden, J.H. & Oehlschlager, A.C. (1991) Fungal volatiles, semiochemicals for stored-product beetles (Coleoptera, Cucujidae). *Journal of Chemical Ecology*, **17**, 581–597.

Pimentel, D., McLaughlin, L. & Zepp, A. (1993) Environmental and economic effects of reducing pesticide use in agriculture. Agriculture, Ecosystems and Environment, 46, 273–288.

Prakash, A. & Rao, J. (1997) *Botanical Pesticides in Agriculture*. CRC Press, Boca Raton, FL, USA.

Puri, H.S. (1999) *Neem. The Divine Tree* Azadirachta indica. Harwood Academic Publishers, Amsterdam, Netherlands.

Quarles, W. (1992) Diatomaceous earth for pest control. *IPM Practitioner*, **18**, 1–10.

Quarles, W. & Winn, P. (1996) Diatomaceous earth and stored product pests. *IPM Practitioner*, **14**, 1–11.

Quinlan, J.K. & Berndt, W.L. (1966) *Evaluation in Illinois of a few inert dusts on stored shelled corn for protection against insects*. Agricultural Research Service Report ARS 51–56, United States Department of Agriculture.

Quy, R.J., Cowan, D.P., Morgan, C. & Swinney, T. (1996) Palatability of rodenticide baits in relation to their effectiveness against farm populations of the Norway rat. In: R.M.Timm & A.C. Crabb (Eds) *Proceedings of the 17th Vertebrate Pest Conference*. University of California, Davis CA, USA.

Rahman, M.M. & Schmidt, G.H. (1999) Effect of *Acorus calamus* (L.) (Araceae) essential oil vapours from various origins on *Callosobruchus phaseoli* (Gyllenhal) (Coleoptera, Bruchidae). *Journal of Stored Products Research*, **35**, 285–295.

Ram, C. & Singh, V.S. (1996) Resistance to *Sitophilus oryzae* in wheat and associated grain characteristics. *Indian Journal of Entomology*, **58**, 79–90.

Redlinger, L.M. & Womack, H. (1966) *Evaluation of four inert dusts for the protection of shelled corn in Georgia from insect attack*. Agricultural Research Service Report ARS 51–7. United States Department of Agriculture.

Rees, D.P. (1994) Biological Control – Session Summary. In: E. Highley, E.J. Wright, H.J. Banks & B.R. Champ (Eds) *Proceedings of the 6th International Working Conference on Stored Product Protection*, Canberra, Australia, **1**, 1165. CABI, Wallingford, UK.

Rejesus, B.M. (1995) Research and policies needed to promote the use of botanical biocides in Asia. *Philippine Entomologist*, **9**, 449–466.

Richter, J., Biliwa, A., Helbig, J. & Henning-Helbig, S. (1997) Impact of *Teretrius nigrescens* Lewis (Coleoptera, Histeridae) on *Prostephanus truncatus* (Horn) (Coleoptera, Bostrichidae) and losses in traditional maize stores in southern Togo. *Journal of Stored Products Research*, **33**, 137–142.

Richter, J., Biliwa, A. & Henning-Helbig, S. (1998) Efficacy of dust formulated insecticides in traditional maize stores in West Africa. *Journal of Stored Products Research*, **34**, 181–187.

Rodgers, P.B. (1993) Potential of biopesticides in agriculture. *Pesticide Science*, **39**, 117–129.

Rodrigues, C. & Pratissoli, D. (1990) Patogenicidade de *Beauvaria brongniartii* (Sacc.) Petch. e *Metarhizium anisopliae* (Mots.) Sorok. e seu efeito sobre o gorgulho do milho e caruncho do feijão. *Anais da Sociedade Entomologica do Brasil*, **19**, 301–306.

Rosendal, G.K. (2000) *The Convention on Biological Diversity and Developing Countries*. Kluwer Academic, Dordrecht, Netherlands.

Round, F.E., Crawford, R.M. & Mann, D.G. (1992) *The Diatoms: Biology and Morphology of the Genera*. Cambridge University Press, New York, USA.

Roush, R.T. (1989) Designing resistance management programs, how can you choose? *Pesticide Science*, **26**, 423–441.

Roush, R.T. & Miller, G.L. (1986) Considerations for design of insecticide resistance monitoring programmes. *Journal of Economic Entomology*, **79**, 293–298.

Salama, H.S., Morris, O.N. & Rached, E. (Eds) (1993) The biopesticide *Bacillus thuringiensis* and its applications in developing countries. In: *Proceedings of an international workshop*, 4–6 November 1991. NRC-Cairo, Agriculture Canada and IDRC, Cairo, Egypt.

Sanchez-Arroyo, H., Lagunes-Tejeda, A. & Llanderal-Cazares, C. (1989) Satividad de polvos minerales para el combate de *Prostephanus truncatus* (Horn) y *Sitophilus zeamais* Motschulsky, en maiz almacendo. *Agrociencia*, **76**, 47–58.

Sanon, A., Tricault, Y., Ouedraogo, P.A.& Huignard, J. (1999) Biological control of *Callosobruchus maculatus* (F.) (Bruchidae), during storage of cowpea. *Annals de la Societé*

*Entomologique de France*, **35**, 496–501.

Schisler, D.A., Slininger, P.J., Kleinkopf, G., Bothast, R.J. & Ostrowski, R.C. (2000) Biological control of fusarium dry rot of potato tubers under commercial storage conditions. *American Journal of Potato Research*, **77**, 29–40.

Scholz, D., Borgemeister, C. & Poehling, H.-M. (1998) EAG and behavioural responses of the larger grain borer, *Prostephanus truncatus,* and its predator, *Teretriosoma nigrescens,* to the borer-produced aggregation pheromone. *Physiological Entomology*, **23**, 265–273.

Schroeder, H.E., Gollasch, A., Moore, A., *et al.* (1995) Bean α-amylase inhibitor confers resistance to the pea weevil (*Bruchus pisorum*) in transgenic peas (*Pisum sativum* L.). *Plant Physiology*, **107**, 1233–1239.

Searle, T. & Doberski, J. (1984) An investigation of the entomogenous fungus *Beauveria bassiana* (Bals.) Vuill as a potential biological control agent for *Oryzaephilus surinamensis*. *Journal of Stored Products Research*, **20**, 7–23.

Shaaya, E., Kostjukovski, M., Eilberg, J. & Sukprakarn, C. (1997) Plant oils as fumigants and contact insecticides for the control of stored-product insects. *Journal of Stored Products Research*, **33**, 7–15.

Shaaya, E., Ravid, U., Paster, N., Juven, B., Zisman, U. & Pisarev, V. (1991) Fumigant toxicity of essential oils against four major stored-product insects. *Journal of Chemical Ecology*, **17**, 499–504.

Shade, R.E., Furgason, E.S. & Murdock, L.L. (1990) Detection of hidden insect infestations by feeding-generated ultrasonic signals. *American Entomologist*, **36**, 231–234.

Shade, R.E., Schroeder, H.E., Pueyo, J.J. *et al.* (1994) Transgenic pea seeds expressing the α-amylase inhibitor of the common bean are resistant to bruchid beetles. *Bio-Technology*, **12**, 793–796.

Sharma, H.C., Sharma, K.K., Seetharama, N. & Ortiz, R. (2000) Prospects for using transgenic resistance to insects in crop improvement. *EJB Electronic Journal of Biochemistry* (http://www.ejb.org/content/vol13/issue2/full/3/).

Sheeba, G., Seshadri, S., Raja, N., Janarthanan, S. & Ignacimuthus, S. (2001) Efficacy of *Beauveria bassiana* for control of the rice weevil *Sitophilus oryzae* (L.) (Coleoptera, Curculionidae). *Applied Entomology and Zoology*, **36**, 117–120.

Singal, S.K. & Singh, Z. (1990) Studies of plant oils as surface protectants against pulse beetle, *Callosobruchus chinensis* (L.) in chickpea *Cicer arietinum* L. in India. *Tropical Pest Management*, **36**, 314–316.

Smet, H.J., Rans, M. & De Loof, A. (1989) Activity of new juvenile hormone analogues on a stored food insect, *Tribolium confusum*. *Journal of Stored Products Research*, **25**, 165–169.

Smith, H.S. (1919) On some phases of insect control by the biological method. *Journal of Economic Entomology*, **12**, 288–292.

Smith, N.M. (1991) Ethnobotanical field notes from the Northern Territory, Australia. *Journal of the Adelaide Botanic Gardens*, **14**, 1–65.

Smith, R.F. & Reynolds, H.T. (1966) Principles, definitions and scope of integrated pest control. *Proceedings of the FAO Symposium on Integrated Pest Control* **1**, 11–17.

Smith, R.H. (1994) Rodent control methods: non-chemical and non-lethal control In: A.P. Buckle and R.H. Smith (Eds) *Rodent Pests and their Control.* Commonwealth Agricultural Bureau, Wallingford, UK.

Snelson, J.T. (1987) *Grain Protectants*. Ruskin Press, Melbourne, Australia.

Snetsinger, R. (1988) *Report on Shellshock insecticide*. Department of Entomology, Pennsylvania State University, PA, USA.

Srivastava, S., Gupta, K.C. & Agrawal, A. (1988) Effect of plant product on *Callosobruchus chinensis* L. infection on red gram. *Seed Research*, **16**, 98–101.

St Aubin, F. (1991) Everything old is new again. *Pest Control Technology*, **50**, 52–102.

Stables, L.M. (1989) Insect, mite and fungal pests of stored cereals and oilseed rape. In: N. Scopes & L.M. Stables (Eds) *Pest and Disease Control Handbook.* Third Edition. BCPC Publications, Croydon, UK.

Stathers, T., Mvumi, B., Chigariro, C., Mudiwa, M. & Golob, P. (2000) *Grain Storage Pest Management using Inert Dusts. DFID Crop Post-Harvest Programme Final Technical Report*. Natural Resources Institute, University of Greenwich, Chatham, UK.

Steinberg, P.F. (1998) Consensus by design, policy by default, implementing the convention on biological diversity. *Social and Natural Resources*, **11**, 375–385.

Stoll, G. (1988) *Natural Crop Protection in the Tropics*. Margnas Publishers Germany.

Strong, R.G. & Sbur, D.E. (1963) Protection of wheat seed with diatomaceous earth. *Journal of Economic Entomology*, **56**, 372–374.

Subramanyam, Bh. & Hagstrum, D.W. (Eds) (1995) *Integrated Management of Insects in Stored Products.* Marcel Dekker, New York, USA.

Subramanyam, B. & Roesli, R. (Eds) (2000) Inert dusts. In: Bh. Subramanyam & D.W. Hagstrum (Eds) *Alternatives to Pesticides in Stored-product IPM.* Kluwer Academic Publishers, Boston MA, USA.

Subramanyam, Bh., Swanson, C.L., Madamanchi, N. & Norwood, S. (1994) Effectiveness of Insecto, a new diatomaceous earth formulation in suppressing several stored grain insect species. In: E. Highley, E.J. Wright, H.J. Banks & B.R. Champ (Eds) *Proceedings of the 6th International Working Conference on Stored Product Protection*, Canberra, Australia, **2**, 650–659, CABI, Wallingford, UK.

Subramanyam, Bh., Madamanchi, N. & Norwood, S. (1998) Effectiveness of Insecto applied to shelled maize against stored-product insect larvae. *Journal of Economic Entomology*, **91**,280–286.

Swidan, M.H. (1994) Antifeedant activity of 24 plant extracts against larvae of *Spodoptera littoralis* (Boisduval) (Lepidoptera, Noctuidae). *Alexandria Journal of Agricultural Research*, **39**, 363–374.

Tabashnick, B.E. (1989) Managing resistance with multiple pesticide tactics, theory, evidence, and recommendations. *Journal of Economic Entomology*, **82**, 1263–1269.

Talekar, N.S. (1987) Host plant resistance to insect attacking soybean and mung beans in the tropics. *Insect Science and its Application*, **8**, 777–782.

Taylor, R.W., Golob, P & Hodges, R.J. (1992) Current trends in the protection of stored cereals in the tropics by insecticides and fumigants. In: *Proceedings of the Brighton Crop Protection Conference – Pest and Diseases*. pp 281–290.

Throne, J.E., Baker, J.E., Messina, F.J., Kramer, K.J. & Howard, J.A. (2000) Varietal resistance. In: Bh. Subramanyam & D.W. Hagstrum (Eds) *Alternatives to Pesticides in Stored-product IPM*. pp 165–192. Kluwer Academic Publishers, Norwell, MA.

Tomlin, C. (2000) *The Pesticide Manual*. British Crop Protection Council and the Royal Society of Chemistry, UK.

Torto, B. & Hassanali, A. (1997) Progress in the search for anti-arthropod botanicals. *Recent Research Developments in Phytochemistry*, **1**, 475–488.

Tunç, I., Berger, B.M., Erler, F. & Dagli, F. (2000) Ovicidal activity of essential oils from five plants against two stored product insects. *Journal of Stored Products Research*, **36**, 161–168.

Turnbull, S.A. & Harris, C.R. (1986) Influence of post-treatment temperature on the contact toxicity of ten organophosphorus and pyrethroid insecticides to onion maggot adults (Diptera, Anthomyiidae). *Proceedings of the Entomological Society of Ontario*, **117**, 41–44.

Ulrich, Ch. & Mewis, I. (2000) Treatment of rice with Neem and diatomaceous earth for controlling the stored-product Coleoptera, *Sitophilus oryzae* and *Tribolium castaneum*. *Journal of Pest Science*, **73**, 37–40.

Van Alebeek, F.A.N. (1994) The functional response of *Uscana lariophaga* Steffan (Hym., Trichogrammatidae) under different egg distributions of its host *Callosobruchus maculatus* L. (Col., Bruchidae). In: E. Highley, E.J. Wright, H.J. Banks & B.R. Champ (Eds) *Proceedings of the 6th International Working Conference on Stored Product Protection*, Canberra, Australia. **2**, 1152–1157. CABI, Wallingford, UK.

van der Nat, J.M, van der Sluis, W.G., de Silva, K.T.D. & Labadie, R.P. (1991) Ethnopharmacognostical survey of *Azadirachta indica* A. Juss (Meliaceae). *Journal of Ethnopharmacology*, **35**, 1–24.

Van Straten, M. (2001) *Organic Living*. Francis Lincoln Ltd, Italy.

Vick, K.W., Coffelt, J.A., Silhacek, D.L. & Oberlander, H. (1985) Methoprene and sex pheromones as control agents for the almond moth (Lepidoptera: Phycitidae) on peanuts stored in the shell. *Journal of Economic Entomology*, **78**, 258–262.

Waage, J.K. & Greathead, D.J. (1988) Biological control: challenges and opportunities. *Philosophical Transactions of the Royal Society*, **B 318**, 111–126.

Waller, J.M. (2001) Postharvest diseases. In: J.M. Waller, J.M. Lenné & S.J. Waller (Eds) *Plant Pathologist's Pocketbook*. Third Edition. CABI Publishing, Wallingford, UK.

Ware, G. W. (1989) *The Pesticide Book*. Third Edition. Thomson Publications, Fresno CA, USA.

Watson, E. & Barson, G. (1996) A laboratory assessment of the behavioural responses of three strains of *Oryzaephilus surinamensis* (L.) (Coleoptera, Silvanidae) to three insecticides and the insect repellent N,N-diethyl-*m*-toluamide. *Journal of Stored Products Research*, **32**, 59–67.

Weaver, D.K. & Subramanyam, Bh. (2000) Botanicals. In: Bh. Subramanyam & D.W. Hagstrum (Eds) *Alternatives to Pesticides in Stored-Product IPM*. Kluwer Academic Publishers, Dordrecht, Netherlands.

Wegmann, E. (1983) Holzaschen als wirksames Mittel zur Bekaempfung von *Callosobruchus maculatus* in traditionellen Bohnenlagern Westafrikas. *Gesunde Pflanzen*, **9**, 229–234.

White, G.D., Berndt, W.L., Schesser, J.H. & Wilson, J.L. (1966) Evaluation of four inert dusts for the protection of stored wheat in Kansas from insect attack. *ARS Report* **22**, 51–58. Agricultural Research Service, United States Department of Agriculture.

Whitehead, R. (Ed.) 2000. *The UK Pesticide Guide*. BCPC, Farnham & CABI Publishing, Wallingford, UK.

Wilkin, D.R., Mumford, J. & Norton, G. (1990) The role of expert systems in current and future grain protection. In: F. Fleurat-Lessard & P. Ducom (Eds) *Proceedings of the 5th International Working Conference on Stored-Products Protection*, Bordeaux, France. **3**, 2039–2046.

Wilson, C.L., Wisniewski, M.E., Biles, C.L., McLaughlin, R., Chalutz, E. & Droby, S. (1991) Biological control of post-harvest diseases of fruits and vegetables – alternatives to synthetic fungicides. *Crop Protection*, **1**, 172–177.

Winks, R.G. & Russell, G.F. (1994) Effectiveness of Siroflo in vertical silos. In: E. Highley, E.J. Wright, H.J. Banks & B.R. Champ (Eds) *Proceedings of the 6th International Working Conference on Stored Product Protection*, Canberra, Australia, **2**, 1245–1249, CABI, Wallingford, UK.

Wolfson, J.L, Shade, R.E., Mentzer, P.E. & Murdock, L.L. (1991) Efficacy of ash for controlling infestations of *Callosobruchus maculatus* (F.) (Coleoptera, Bruchidae) in stored cowpeas. *Journal of Stored Products Research*, **27**, 239–243.

Wongo, L.E. & Pedersen, J.R. (1990) Effect of threshing different sorghum cultivars on *Sitotroga cerealella* (Oliver) and *Sitophilus oryzae* (L) (Lepidoptera, Gelechidae and Coleoptera, Curculionidae). *Journal of Stored Products Research*, **26**, 89–96.

Wrigley, S., Hayes, M., Thomas, R. & Chrystal, E. (Eds) (1997) *Phytochemical Diversity, a Source of New Industrial Products*. Royal Society of Chemistry, Cambridge, UK.

Yunes, A.N.A., de Andrade, M.T., Sales, M.P., *et al.* (1998) Legume seed vicilins (7S storage proteins) interfere with the development of the cowpea weevil (*Callosobruchus maculatus* (F)). *Journal of the Science of Food and Agriculture*, **76**, 111–116.

Žd'árková, E. & Horák, E. (1990) Preventative biological control of stored food mites in empty stores using *Cheyletus eruditus* (Schrank). *Crop Protection*, **9**, 378–382.

Zehrer, M.W. (1985) Methode traditionaille de stockage des nieby dans le nord Togo. *Comptes rendes colloque sur les Legumineuses Alimenrairs Africaines*. Niamey (Niger) 19–22 Aupelf, Paris, France.

Zettler, J.L. & Arthur, F.H. (1997) Dose–response tests on red flour beetle and confused flour beetle (Coleoptera, Tenebrionidae) collected from flour mills in the United States. *Journal of Economic Entomology*, **90**, 1157–1162.

Zhang, H. & Deng, W. (1993) The resistant mechanism of stored rice grain to rice weevil. *Acta Phytophylacia Sinica*, **20**, 143–146.

Zlotkin, E. (1999) The insect voltage-gated sodium channel as target of insecticides. *Annual Review of Entomology*, **44**, 429–455.

## Websites

National pesticide telecommunications network – a site created by the US government with updated information and discussions, particularly on the toxicity of organophosphate insecticides. http://ace.orst.edu/info/nptn.

CODEX Pesticide Regulations Committee, September 1999. http://apps.fao.org/csv_down/.

COLEACP site for information on MRLs for horticultural produce. http://www.coleacp.org.

# Chapter 7
# Remedial Treatments in Pest Management

**R. W. D. Taylor, A. D. Devereau and P. Golob**

This chapter describes methods that can be used in pest management strategies specifically to disinfest commodities. The main means of rectifying pests established in food commodities is by applying a gaseous fumigant. The main focus of the chapter is on fumigation as a technique and the gases in common use. Alternative techniques using other gases including carbon dioxide are described, as are methods using controlled or modified atmospheres. Radiation and the use of microwaves are also considered.

## Fumigation

The effectiveness of certain gases for pest control has been known since ancient times when the burning of sulphur was used for public health purposes. The development of modern techniques of fumigation was partly stimulated by World War II when it became necessary to conserve large stocks of cereal grains. In the 1950s, the use of under-sheet fumigation resulted in dramatic reductions in grain damage by insect pests, and the subsequent development of fumigation technology, particularly the introduction of phosphine, enabled this method of controlling insects to be adopted in almost every country. The number of available fumigants has never been great and no new gases have been introduced in recent years. On the contrary, many have fallen into disuse, often banned because of risks to human health. From the 1980s onwards problems with insect resistance to phosphine have given increasing cause for concern. During the early 1990s confirmation that methyl bromide, a major fumigant used worldwide, was an ozone-depleting substance resulted in significant research and development into alternative methods of insect control. These two events have probably stimulated a greater volume of research into fumigation technology than had been undertaken previously.

### Definition of fumigants

Fumigants are chemicals that can exist as gases in sufficient concentration under the required conditions of temperature and pressure to be lethal to certain pest organisms. Chemicals regarded as fumigants can remain in the vapour state throughout the pest control treatment period, which may be of many days' duration. The definition, therefore, excludes aerosol, smoke, and mist formulations of insecticides that have particle sizes much greater than those of fumigants, and which prevents them from having significant penetrative properties or from remaining airborne other than for short periods. The number and variety of chemicals suitable for classification as fumigants is restricted by the special properties required of them. They should:

- be gases at normal temperatures;
- have good diffusive properties;
- be toxic to all developmental stages of target pests;
- be not greatly heavier than air;
- not leave harmful chemical residues in products treated.

Bond (1984) listed chemicals that have been considered as potential fumigants in the past; the more important of these appear in Table 7.1. Many of these chemicals have now been excluded from use, either for health and safety reasons, or because in most countries registration has lapsed, other fumigants being found to be more effective.

### Principal uses for fumigants

Fumigation is widely applied in the post-harvest protection of agricultural commodities and timber, and is also used to some extent for disinfesting processed foods. Although fumigation is used mostly to treat durable (low

**Table 7.1** Gases with potential fumigant action. Source: after Bond 1984.

| Chemical name | Use and present status |
|---|---|
| Acrylonitrile | No longer used because of high human toxicity |
| Carbon disulphide | Now used only in Australia for grain treatment at farm level, and on certain products in China |
| Chloropicrin | Used as warning gas with methyl bromide, also a potential soil fumigant |
| Ethylene dibromide | Formerly used for disinfesting fruit, or in mixtures for treating farm-stored grain. Use discontinued due to human health hazards |
| Ethylene dichloride | Formerly used in mixtures for treating farm-stored grain but discontinued because of high human toxicity |
| Ethylene oxide | Use as a grain fumigant discontinued but used as a sterilant in some countries for treating spices |
| Ethyl formate | Now only used on dried fruit; potential for treating grain being re-examined in Australia |
| Hydrogen cyanide | Use on stored products and for flour mills superseded by methyl bromide and phosphine; used for rodent control in some countries |
| Methylallyl chloride | Was used in the former USSR, elsewhere did not proceed far beyond field testing due to high mammalian toxicity |
| Methyl bromide | Used on stored products, for structures, and for quarantine treatment of fruit and vegetables, is a major soil fumigant used worldwide. Will be phased out worldwide by 2015 because of environmental concerns |
| Phosphine | The most widely used fumigant for stored products |
| Sulphuryl fluoride | Used for termite control in buildings in the USA for many years. More recently used in China to treat timber. Potential for application to stored products being evaluated |

moisture content) commodities, some fruits, vegetables and cut flowers are also treated to control insect pests. The fumigation of fresh commodities is, however, almost always conducted for quarantine purposes to prevent the introduction and establishment of insects that may be harmful to the agriculture of importing countries.

In addition to the treatment of commodities stored in bulk, in bags, or in boxes, fumigation is also employed to disinfest buildings, food-processing facilities (including flour mills), and contaminated transport vehicles. Although the majority of commodity fumigations are conducted in warehouses (Plate 25), or even outdoors in some countries, treatments are sometimes carried out during commodity transportation in ships, in freight containers (increasingly) and in rail wagons.

## Theory of fumigation

To control insects and other target pests effectively by fumigation they must be exposed to a lethal dose (concentration) of the chemical for a sufficient time-period. A variety of factors will determine this dose including the type of chemical employed, the species of pest to be controlled and certain physical factors, especially temperature. Fumigants are more effective at high temperatures, and consequently dosage rates can be reduced accordingly. For certain fumigants, the concentration of gas and the period of exposure required to obtain effective insect control are directly related, and reduced dosages (rates of application) can be compensated for by longer periods of exposure (Haber's rule). Under this rule, the concentration of gas ($c$) multiplied by the exposure period ($t$) is a constant. The product of $c \times t$ is known as the *ct*-product and applies, for example, to methyl bromide, but not in the case of phosphine where a minimum time period for exposure to the fumigant is essential, and cannot be compensated for by increasing the amount of phosphine applied. Effective control of pests will be dependent on both the concentration of fumigant gas and the period of exposure.

## Pests controlled by fumigation

In the post-harvest disinfestation of agricultural commodities, storage buildings, and transport vehicles, fumigation is used almost exclusively to control insect pests. These include primarily beetles and moths, and fumigation can effectively control all insect species, and all stages of development. The adult stage is most easily killed by fumigation, the egg and pupal stage usually being the most tolerant, requiring higher concentrations or, in the case of phosphine, longer periods of exposure to achieve complete mortality. Fumigation is also used against mites, although mite eggs are more difficult to kill than insect eggs. The effectiveness of fumigants against mites was reported by Bowley and Bell (1981).

Rodents are easily controlled by fumigation, usually at application rates much lower than those required to kill insects. Control of rats and mice attained during commodity fumigation is almost always incidental to the control of insect pests. Because of the high cost of treatment, fumigation against rodents is usually conducted for public health purposes only, mostly in ships and aeroplanes.

The effectiveness of fumigants for controlling microflora in grain has been investigated. In reviewing reports relating to methyl bromide, Bond (1984) indicates that to control micro-organisms such as *Aspergillus* spp and *Penicillium* spp, much larger *ct*-products are necessary than those needed to control insects. There has been recent interest in the potential of phosphine to control fungi on stored grain (Dharmaputra 1997; Dharmaputra *et al.* 1997; Dharmaputra & Putri in press; De Castro *et al.* in press). Application rates of phosphine evaluated have been as high as 4 g $m^{-3}$.

Methyl bromide has a particularly broad spectrum of activity against pests, which is the principal reason for its use as a soil fumigant where it controls insects, together with soil-borne fungi, nematodes, bacteria and weeds. Application rates are very much greater than those used in post-harvest disinfestation.

### Use of contact insecticides to supplement fumigation

Contact insecticides are often used in conjunction with fumigation for the purpose of reducing the incidence of re-infestation of bag stacks of commodities. In most situations contact insecticides cannot replace fumigation, chiefly because they cannot be used to control insects inside grains, or in the centre of grain bulks or bag stacks, unless applied to these during construction. The period of protection against re-infestation of bag stacks provided by surface application of contact insecticides to store and sack surfaces, particularly in hot climates, has been demonstrated to be much less than once thought (Gudrups *et al.* 1994). The differences in action between fumigants and contact insecticides are given in Table 7.2.

### Individual fumigants

#### *Phosphine*

Phosphine is the only fumigant other than methyl bromide that is registered worldwide for commodity disinfestation. It was first developed as a fumigant in Germany in the 1930s, but only came to prominence in the 1960s, and then mainly in tropical and subtropical regions. This restriction was principally because phosphine is more effective at higher temperatures, often being recommended as best employed at temperatures greater than 15°C. The toxic action of phosphine against insects is relatively slow, and consists of two parts, an absorption phase, followed by a biochemical phase (Price 1984). For this reason, a sufficiently long period of exposure to the gas is essential for fumigation with phosphine to be fully effective.

##### *Physical properties*

Phosphine, also known as hydrogen phosphide, has the chemical formula $PH_3$. Pure phosphine is generally regarded to be odourless, and the 'garlic-like' smell associated with the gas is usually attributed to impurities such as diphosphine. Phosphine is explosive at concentrations by volume in air greater than 1.79%, and spontaneously combusts above 100°C, and at reduced pressures. It has a density only slightly greater than air, diffuses well, and

**Table 7.2** Basic differences between contact insecticides and fumigants.

| Fumigants | Contact insecticides |
|---|---|
| No lasting protection | Lasting protection possible |
| Grain can be treated *in situ* | Normally grain has to be moved in order to apply insecticide |
| Can be used for treating most commodities | In most countries, only permitted on commodities before processing |
| Disinfestation can be completed within 1–15 days, according to temperature | Disinfestation achieved over a longer period, since stages of those species which develop within the grain are not affected until they develop into adults |
| Only skilled certified personnel can apply fumigants | Semi-skilled operators can apply contact insecticides |
| Generally effective against all insect species | Various compounds are selectively effective against different insect species |
| No incidence of substantial methyl bromide resistance known, but development of resistance to phosphine a current concern | Most insect pests develop resistance to particular insecticides or groups of insecticides, with continued use |
| Good penetration of grain bulks | Poor penetration of grain bulks |

readily penetrates most commodities, and materials used for holding commodities including paper, cardboard and sacking materials. The ease with which phosphine penetrates building materials such as sand–cement rendering, and concrete blocks, however, makes it essential that if warehouses are to be fumigated, fully gas-tight materials are used in the construction. Such materials include some plastics, metal, and densely formed concrete. Surface applications such as oil-based gloss paint can provide a gas-proof layer to sand–cement wall rendering.

Phosphine reacts with certain metals, the most important of these, because of its extensive use in electrical systems, being copper. Damage may occur if phosphine is used in circumstances where electrical equipment or wiring may be exposed to the gas. Where phosphine has to be used the best course of action is the removal of equipment likely to be 'at risk', but where this is not possible some form of protection is necessary. It has been demonstrated in Canada that, contrary to formerly held opinion, corrosion by phosphine can take place even under conditions of low relative humidity (Brigham 1999a, b).

### *Formulations of phosphine*

Traditionally, phosphine has been generated on site from solid preparations, the active ingredient being aluminium or magnesium phosphide, which react with air-borne moisture to produce phosphine (Plate 26). Other ingredients of commercial formulations containing aluminium phosphide include paraffin wax that helps to retard the rate of gas evolution, and ammonium carbamate, which produces ammonia and carbon dioxide according to the following reaction:

$$NH_2COONH_4 \rightarrow 2NH_3 + CO_2$$

The purpose of the ammonia is to provide a warning to applicators that the formulation is commencing to react with atmospheric moisture, and phosphine is about to be generated. The function of carbon dioxide is to reduce the risk of phosphine combustion. Formulations containing magnesium phosphide are claimed to release phosphine more rapidly than those containing aluminium phosphide. They do not release ammonia, and were developed originally for their potential for fumigating perishable commodities. Table 7.3 indicates the various commercial formulations generating phosphine. The chemical reactions by which phosphine is generated are:

$$AlP + 3H_2O \rightarrow Al(OH)_3 + PH_3$$

$$Mg_3P_2 + 6H_2O \rightarrow 3Mg(OH)_2 + 2PH_3$$

### *Other sources of phosphine*

During the 1980s, research and development took place, particularly in Australia, into the potential for applying phosphine gas from cylinders. The gas, usually 2–3% by concentration mixed with carbon dioxide, was stored in cylinders ready for direct application to grain or other commodities. The carbon dioxide acts as a fire suppressant (necessary for the higher phosphine concentration used), and also as a carrier for the phosphine. A major factor in the development of a system employing cylinder-based phosphine was the need to fumigate Australian grain silos that were not gas-tight, while avoiding the

**Table 7.3** Commercial formulations of phosphine.

| Formulation | Unit weight (g) | Phosphine released (g) | Pack content |
|---|---|---|---|
| Pellets | 0.6 | 0.2 | 1600 in resealable flask |
| Tablets | 3.0 | 1.0 | 30 per tube, 100 or 250 in resealable flask |
| Mini bags | 9.0 | 3.0 | Individually sealed, contained in tins of various numbers |
| Bag (sachet) | 34.0 | 11.3 | Individually sealed, contained in tins of various numbers |
| Bag (sachet) with strings or chains | 340 | 113 | Ten bags (sachets) in chains, one or two chains per tin |
| Blanket | 3400 | 1130 | 100 bags per strip (several bags connected together) |
| Plates | 117 | 33 | 32 per tin |
| Plate strips | 2340 | 660 | 1 strip of 20 plates per tin |
| Belt | 136 | 45.2 | 1 per tin (for probing into bulk grain) |
| Prepacks | 99 | 33 | 4 per tin |
| Fumicel discs | 10 | 5.0 | 25 per tin |
| Fumi strips | | $16 \times 33$ | $2 \times 16$ per tin |

selection of insects resistant to the gas. In addition, use of the system satisfied the market requirement for grain free of insecticide residues (Winks 1992). The system developed was known as Siroflo® and enabled a constant concentration of phosphine to be maintained over long periods, which might be up to 4 weeks depending on the temperature. These long exposures greatly reduced the risk of selecting phosphine-resistant insects.

The use of cylinder-based phosphine increased during the 1990s, and research and development was encouraged further by the need to find alternatives to methyl bromide. In most countries, aluminium or magnesium phosphide formulations are registered as the source from which phosphine is obtained, and a constraining factor in widening the use of cylinder formulations has been the need to register phosphine gas separately as a fumigant. The number of countries where the gas is registered is increasing slowly, the US being added to the list in 2000. As an alternative to methyl bromide, phosphine in combination with carbon dioxide and heat has been investigated and recommended as a potential method of controlling insects in structural treatments (Mueller 1996).

Recently, a new cylinder-based formulation containing phosphine has been developed in Germany. This formulation contains 1.5% phosphine in nitrogen, and is commercially available under the trade name 'Frissin' but is not registered yet outside Germany.

Other phosphine-generating devices are currently under development, these being based upon the hydrolysis of aluminium or magnesium phosphide. A stream of carbon dioxide may be used to entrain the phosphine produced in order to carry it to the fumigation site, and also to act as a fire suppressant (Horn 1997). It is claimed that phosphine dosages from these generators are more consistent and controllable than when the fumigant is obtained from solid metal phosphide formulations. The most recently announced development of phosphine generators involves the use of a new formulation of aluminium phosphide (Waterford *et. al.* 1994) which is impregnated into wax blocks and currently being evaluated in field trials. This new type of generator is unique in being activated by the addition of water, a practice never used with conventional metal phosphide formulations because of the risk of explosion (Waterford & Asher 2000).

### *Application, exposure periods, and dosage rates using phosphine*

Phosphine distributes well by diffusion and by air movement and there is little, if any, need for forced circulation in many applications. In the fumigation of bag stacks under gas-proof sheets, for example, solid formulations of metal phosphides can be sited at floor level or on top of the stack as convenient. The distribution of gas within a stack should be satisfactory whatever application location is selected. However, in some types of treatment such as those involving bulk grain, injection of the fumigant combined with forced air recirculation may be necessary to ensure complete gas distribution. If this practice is not possible, such as with filled silos requiring treatment, the only alternative may be to transfer the grain to an empty bin applying fumigant to the grain stream during loading.

Phosphine is not absorbed appreciably, except by a very few commodities such as rice in husk. For most applications the dosage rate used does not vary with the commodity but will depend upon the method of storage, and to some extent also upon the temperature. Recommended dosage rates for phosphine are given in Table 7.4, and minimum exposure periods in

**Table 7.4** Recommended dosage rates for phosphine.

| Type of fumigation | Recommended dosage* | |
|---|---|---|
| | g per tonne | g per $m^3$ |
| Bulk fumigation in gas-tight silos | 2–4 | 1.5–3.0 |
| Bagged commodities under gas-proof sheets | 3–5 | 2–3.5 |
| In-bag fumigations | 0.2 g per bag† | |
| Space fumigations, e.g. empty store | | 1.0‡ |

*For control of *Sitophilus* spp, *T. granarium, Ephestia* spp, and mites, the highest dosage in the range recommended will be required, and exposures longer than the minimum given are likely to be necessary to control all developmental stages.
†50-kg bag. Equivalent to one pellet per bag.
‡For complete control of insects, structures being fumigated must be sufficiently gas-tight to retain a lethal concentration of fumigant throughout the exposure period.

Table 7.5. Table 7.6 provides information on the exposure periods required for control of various insect pests.

**Table 7.5** Minimum exposure periods recommended using phosphine.

| Temperature (°C) | Exposure period (days after application of the fumigant) |
|---|---|
| <15 | Do not use phosphine |
| 15–25 | 10 |
| >25 | 7 |

**Table 7.6** Minimum exposure periods (days) required for control of all stages of the stored product pests listed, based on a phosphine concentration of 1.0 gm$^{-3}$ (after ASEAN 1989).

| Species | Temperature (°C) | |
|---|---|---|
| | 15–20 | 20–30* |
| *Oryzaephilus surinamensis* | 3 | 3 |
| *Cryptolestes pusillus* | 5 | 4 |
| *Oryzaephilus mercator* | 5 | 4 |
| *Tribolium castaneum* | 5 | 4 |
| *Lasioderma serricorne* | 5 | 5 |
| *Acanthoscelides obtectus* | 8 | 5 |
| *Corcyra cephalonica* | 8 | 5 |
| *Cryptolestes ferrugineus* | 8 | 5 |
| *Plodia interpunctella* | 8 | 5 |
| *Ptinus tectus* | 8 | 5 |
| *Rhyzopertha dominica* | 8 | 5 |
| *Sitotroga cerealella* | 8 | 5 |
| *Tribolium confusum* | 8 | 5 |
| *Cadra cautella* | 10 | 5 |
| *Ephestia elutella* | 10 | 5 |
| *Ephestia kuehniella* | 10 | 5 |
| *Caryedon serratus* | 10 | 8 |
| *Sitophilus granarius* | 16 | 8 |
| *Sitophilus oryzae* | 16 | 8 |
| *Sitophilus zeamais* | 16 | 8 |
| *Trogoderma granarium* | 16 | 8 |

*All species listed succumb to a 4-day exposure at this dosage level at 30°C or above. This dosage is recommended for good conditions and the dosage applied will usually need to be increased considerably in leaky situations. For certain commodities in long-term storage, where it is necessary to control a mite infestation, two fumigations may be carried out, separated by an interval dependent on ambient temperature, allowing eggs surviving the first fumigation to hatch. This interval varies from 2 weeks at 20°C to 6 weeks at 10°C (Bowley & Bell 1981).

*Insect resistance to phosphine*

Resistance to fumigants was not perceived as a real problem affecting insect control in the field until its detection in relation to phosphine. The first indications of resistance were revealed in a survey conducted by FAO in the early 1970s, which included phosphine among a number of pesticides screened against several insect species damaging stored products (Champ & Dyte 1976). Failures to control insects by phosphine fumigation during field treatments were not reported from the survey but were predicted. Within 10 years, published data were available indicating that insect resistance to the fumigant was widespread (Taylor & Halliday 1986). The mechanism of phosphine resistance in insects involves an active exclusion principle causing insects to be unaffected by gas concentrations that are lethal to susceptible individuals. However, this mechanism is unable to exclude phosphine indefinitely and, provided exposure periods are extended sufficiently, effective insect control with phosphine can be achieved (Price 1984).

Resistance in the field was detected in food warehouses in Bangladesh in 1982 (Mills 1983; Tyler *et al.* 1983). In that country, and in others in South Asia, fumigation of complete warehouses that were not gas-tight had been conducted routinely for many years, and had undoubtedly been a common factor in the selection of resistant insects in the region. Suspected phosphine resistance in *Trogoderma granarium* (Khapra beetle) had been reported from the Punjab as early as 1979 (Borah & Chalal 1979), and some of the most phosphine-resistant insects reported have been from Pakistan (Sardar Alam & Ahmed 1989). Winks (1987) suggested that improvements to the techniques employed in using phosphine were essential to avoid serious problems with insect resistance. Better sealing of fumigated enclosures and longer exposure periods were regarded as major factors in reducing the risks of insect resistance (Taylor & Gudrups 1996). A recent research report from India suggests that resistance to phosphine is widespread in that country, with high levels in some populations of *R. dominica* and *S. oryzae* (Rajendran 1999). In view of the acute problem in India, revised phosphine dosage schedules are reported to be likely to be accompanied by better sealing practices and gas monitoring. The most recent reports from Australia have indicated that the magnitude of resistance in *R. dominica* has risen to a level where the most resistant populations are no longer controlled using the registered dose and time protocols. Strategies to control these higher levels of resistance are being developed including adoption of increased dosage rates (Collins & Daglish, in press).

A comprehensive review of the insecticidal action of phosphine, including the mechanism of resistance, has been given by Chaudhry (1997). This author suggests that the chemical reducing properties of phosphine are likely to play a major role in its insecticidal action. The oxidation of phosphine in producing reactive phosphorylating compounds, and the interactions of phosphine with biological redox systems, appear to be the basis of phosphine toxicity in insects. Phosphine-resistant strains of insects are reported to absorb very small amounts of the chemical compared to susceptible strains, this resulting from respiratory exclusion of phosphine. The reduced uptake of the gas might be due to the presence of a phosphine-insensitive target site or to a membrane system that excludes the gas in resistant insects. Chaudhry concludes that the nature of the phosphine exclusion system is not yet fully understood.

There is little doubt that insect resistance is a continuing and increasing problem requiring careful management. This management often requires the use of alternative insect control technologies, but an additional problem is that most incidences of resistance occur in developing countries where there are often limitations to such management (Zettler 1997).

### *Detection and measurement of resistance*

Following the FAO survey mentioned earlier (Champ & Dyte 1976), more intense screening programmes were conducted to detect and measure insect resistance, particularly of insects from Africa and Asia. The technique commonly employed a laboratory-based method in which insects are exposed to a discriminating dose of phosphine that had been determined for particular species, enabling differentiation between known susceptible strains and potentially resistant insects collected from field situations. This method (FAO 1976) relied upon exposing insects for 20 h to an accurately known concentration of phosphine, but the availability of such concentrations in the past limited the extent to which reliable results were reported. This test method provides information only on the frequency of resistance in a population of insects, but not information on the magnitude of resistance. To determine this, further tests are necessary, employing concentrations above the level of the discriminating dose in order to calculate a regression line for comparison with that for susceptible insects. The FAO test method requires insects to be exposed overnight, and to hasten the test a new same-day method was devised (Savvidou *et al.* 1994). This method is based upon assessment of the knockdown of insects exposed to phosphine for fixed periods of several hours. The method claims to overcome the problems of insect narcosis associated with exposure to high concentrations, and to be capable also of differentiating between heterozygous and susceptible insects. Data enabling use of the same-day test have so far been published only for *C. ferrugineus*, *T. castaneum* and *S. oryzae*.

### *Other potential constraints on phosphine*

In 1998, phosphine came under scrutiny in the US during the process of re-registering the chemical. A wide range of new risk mitigation measures in relation to bystanders has been proposed by the Environmental Protection Agency, some of which, if put into effect, could greatly constrain use of the fumigant. Two of the most controversial of these measures are proposals to impose a buffer zone of 500 feet (150 m) around any treatment with phosphine, and to reduce the threshold limit value by a factor of ten to 0.03 ppm. The new measures proposed were put out for public consultation. The latest information available suggests that many of the less radical mitigation measures proposed, such as better monitoring and incident reporting, will be accepted by the US Department of Agriculture and others using phosphine. A compromise may need to be agreed regarding the two measures that potentially would have the most constraining effects on use of the fumigant.

## Methyl bromide

Methyl bromide is the only chemical other than phosphine that is currently registered worldwide as a fumigant for the post-harvest disinfestation of a range of commodities. It was first used in the 1930s and, because of its very broad spectrum of activity against pests, became increasingly important as a post-harvest treatment, particularly for cereal grains and timber. Methyl bromide is a fast-acting pesticide, and for this reason is also used to fumigate perishable commodities, including fresh fruit, vegetables, and cut flowers, pot plants and bulbs, in treatments generally lasting a few hours only. Its fast action makes methyl bromide very suitable for all fumigations that must be completed quickly, such as those conducted at ports where undue delay may increase costs to unacceptable levels. One of the most important uses of the chemical is for treatments conducted for quarantine purposes, particularly for perishable commodities, and for which there are presently few widely accepted alternative treatment techniques. The fumigation of structures commonly employs methyl bromide, and its fast action is again especially useful in flour mills, enabling the treatment to be completed rapidly and reducing the period of mill closure to a minimum. More detailed accounts of methyl bromide are given by Bond (1984), and by Thompson (1966).

Although methyl bromide is important as a post-harvest fumigant, on a volume basis the major use worldwide for the chemical is as a soil fumigant, effectively controlling nematodes, insects, fungi, bacteria and weeds. Large-scale production of crops such as strawberries and tomatoes has relied heavily upon use of the chemical for many years. More recently, there have been significant increases in the use of methyl bromide as a soil fumigant in developing countries, the majority of this use being in the production of such cash crops as flowers for export.

### *Environmental concerns and the future of methyl bromide*

In 1992, methyl bromide was listed as an ozone-depleting substance under the provisions of the Montreal Protocol, and a programme to control and restrict its use agreed internationally (Taylor 1997). Control of the fumigant is effected through manufacture and distribution and its use will cease in developed countries in 2005, except for certain exempted or emergency uses. Developing countries have an extra 10 years in which to adopt alternative insect control strategies and will be aided in this task with financial and technical assistance from developed countries. It is unlikely that suitable alternatives for all of the uses of methyl bromide, particularly for quarantine treatments, will be available by 2005. Use of the fumigant for quarantine and pre-shipment purposes, for example, and for which currently there are few alternative treatments, is exempt from controls under the Montreal Protocol. This exemption, made because restrictions on the fumigant are seen as posing a threat to international trade in infestable commodities, may continue for some years after the phase-out date until suitable alternatives are found. Exempting use of the fumigant for quarantine and pre-shipment purposes has been seen as a loophole resulting in increased use of methyl bromide in recent years (Taylor 2000). The continued exemption from controls is also seen to be a disincentive for research into alternative technologies, and the reason for the European Union considering imposing a cap on the quantity of the fumigant used for quarantine and pre-shipment purposes. Detailed descriptions of methyl bromide and its effect on the environment are given by Taylor (1994) and by Bell *et al.* (1996). The reports of the Methyl Bromide Technical Options Committee that operates under the United Nations Environment Programme provides detailed information on the current uses of the fumigant and potential alternative pest control technologies (Anon. 1995, 1998).

### *Properties and application*

Methyl bromide is a low-boiling point liquid (boiling point 3.6°C) that is odourless, colourless and non-flammable at normal temperatures. Depending upon the use intended, 2% chloropicrin may be added to methyl bromide as a lachrymatory warning agent to protect those involved in application. However, because of its high boiling point (112°C), chloropicrin often remains in fumigated commodities for very much longer than methyl bromide, and this can be unacceptable in some circumstances. Methyl bromide is more than three times heavier than air and this factor must be taken into account in the design of gas application systems. The gas penetrates relatively easily into most commodities at normal pressure, but vacuum fumigation can be employed where penetration is slow, such as with compressed tobacco bales. For the fumigation of bag stacks of commodities, a technique very common in developing countries, methyl bromide is distributed through a branched piping system placed on top of the stack. Such a system maximises the distribution of gas, ensuring that fumigant reaches all parts of the stack (Figs 7.1, 7.2). In cool climates, it is

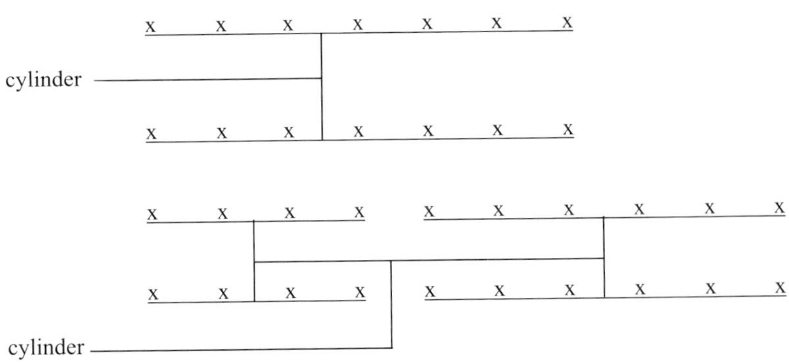


**Fig. 7.1** Piping arrangements for distributing methyl bromide to bag stacks.

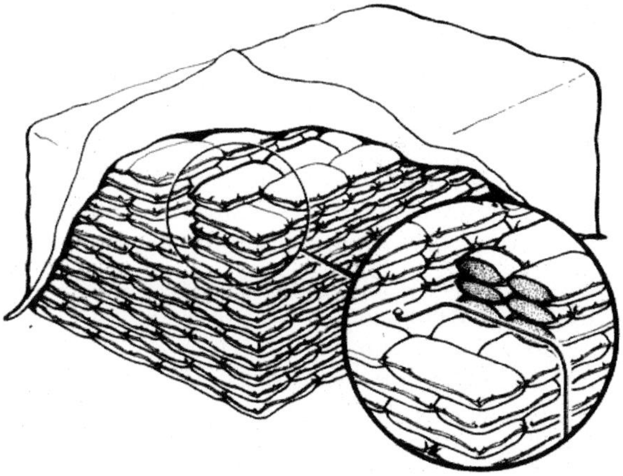

Fig. 7.2 Preparing a bag stack for fumigation with methyl bromide.

often necessary to pass methyl bromide through a heat exchanger to encourage volatilisation, and to prevent freezing of the gas at the nozzles located at the end of the piping distribution system. Unaided, methyl bromide does not distribute readily through bulk grain, and silos to be fumigated with methyl bromide need to be equipped with a recirculatory system. Portable recirculatory systems, originally designed for use in ship fumigation, have been introduced successfully in recent years (Taylor & Locke 1996) (Fig. 7.4).

Methyl bromide is chemically reactive, particularly with sulphur-containing molecules, and in the liquid state reacts with aluminium. It is a powerful organic solvent for natural rubber, though the concentration of gas used in fumigation does not normally affect plastic materials. Liquid methyl bromide can cause the partition of the laminates used in sheeting materials.

*Formulations available*

Methyl bromide is supplied as a liquid under pressure in steel cylinders, usually of about 90 kg capacity; these may be with or without 2% chloropicrin added as warning agent (tear gas). The fumigant is also supplied in small metal cans, typically containing 670 g of liquid fumigant under pressure.

*Residues*

Methyl bromide is strongly sorbed by many commodities, particularly those that are finely divided such as flour, or those that have a high oil content. Much of the absorption is physical, the gas being desorbed at a later stage, but some chemical reaction with treated products occurs. This can result in elevated levels of inorganic bromide residues in commodities, and methylation of protein components of wheat has been demonstrated (Bridges 1955; Winteringham *et al.* 1955). The varying degree to which methyl bromide is sorbed by different commodities has a significant influence on the dosage rates that need to be applied. The greater the sorption, the higher the dosage of fumigant that is required to maintain a concentration of gas in the intergranular spaces that is lethal to insects. Experience has shown that where ab-

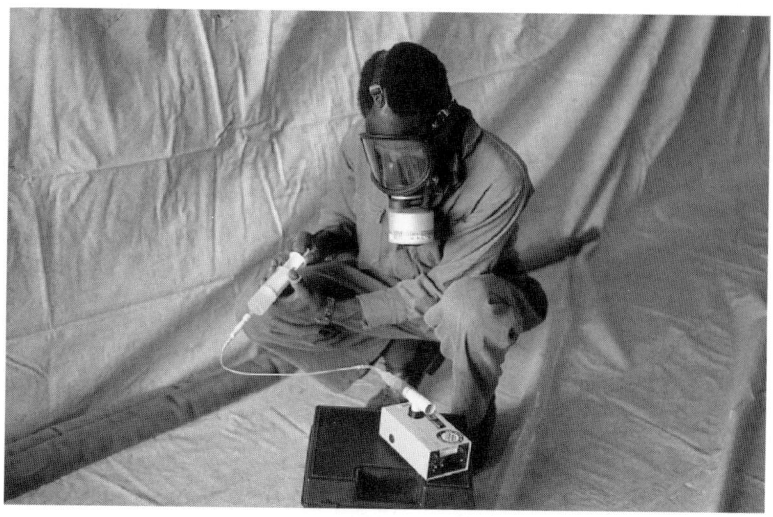

Fig. 7.3 Measurement of phosphine concentration using an electronic meter.

sorption and/or gas penetration render fumigation with methyl bromide difficult it is preferable to use another fumigant, which in recent years has almost always meant employing phosphine. The dosage rates recommended for applying methyl bromide to durable commodities are based upon the degree of absorption of the gas, and it is convenient, therefore, to group together commodities with similar absorption levels. Dosage rates for commodities commonly fumigated with methyl bromide using these groupings are given in Table 7.7.

### *Fumigation of perishable commodities*

Almost all fumigations of perishable commodities are conducted for quarantine purposes to prevent the introduction and establishment of exotic pests that may harm the agricultural production of importing countries (Anon. 1995). For this reason a very high standard of pest control is essential for all such fumigations. Commodities may be treated in either the importing or exporting country, and may include fruits, vegetables, cut flowers, ornamental plants, fresh root crops and bulbs. Methyl bromide is currently the only fumigant available for treating these commodities, the use of ethylene dibromide having been discontinued in the mid-1980s because of concerns over human health hazards. Although its fast toxic action against pests makes methyl bromide an ideal fumigant for perishable commodities, it cannot be used for all commodities, or for all the varieties of some commodities. This is because of phytotoxic effects, or the potential for causing injury to some varieties of fruits and vegetables. Pests that are typically controlled by fumigation include a number of fruit fly species, codling moth, and mites. Most treatments of perishable commodities involving methyl bromide are for the control of surface feeding insects. Control of internal feeders would require much higher fumigant application rates (unless reduced pressure can be applied) and could cause injury to the product treated.

Information on the dosage rates of methyl bromide that are applied to a variety of perishable commodities, and the exposure period recommended are to be found in the publications of FAO (1994) and the US Department of Agriculture (USDA 1998). Typical exposure periods used are between 2 and 4 hours, and dosage rates may vary from 16 g to 64 g m$^{-3}$, depending on the commodity treated (Table 7.8). The majority of perishable commodities are fumigated at atmospheric pressure, but some vegetables are more easily treated under a partial vacuum. A further variation in treatment protocols is that some fruits are found to be less liable to fumigation injury after a period of cold storage. These include apples, pears and grapes treated to control pests such as light brown apple moth (*Epiphyas* spp), Queensland fruit fly (*Bactrocera tryoni*) and Mediterranean fruit fly (*Ceratitis capitata*).

**Table 7.7** Methyl bromide dosage table for durable commodities adopted by the European Plant Protection Organization. Source: EPPO 1993.

| Group | Commodities | Dosage (g m$^{-3}$) | | | Exposure period (h) |
|---|---|---|---|---|---|
| | | <10°C | 10–20°C | >20°C | |
| 1 | Rice, peas, beans, cocoa beans, dried vine fruits | 25 | 15 | 10 | 24 |
| 2 | Wheat, barley, oats, maize, lentils | 50 | 35 | 25 | 24 |
| 3 | Pollards, rice bran | 70 | 45 | 3 | 48 |
| 4a | Sorghum, nuts, figs | 75 | 50 | 35 | 24 |
| 4b | Groundnuts, oilseeds, dates, empty sacks | 75 | 50 | 35 | 48 |
| 5 | Oilseed cakes and meals | 120 | 85 | 60 | 48 |
| 6 | Fishmeal, dried blood, etc. | 140 | 100 | 65 | 48 |
| 7 | Flour | 50 | 50 | 40 | 48 |

These dosage rates apply to fumigations under gas-proof sheets and in freight containers that are usually fully loaded. If this method is to be used for mite control, dosage rates should be doubled accordingly.
Penetration of methyl bromide into commodities in groups 5 and 6 is poor and fumigation may be uneconomic using the recommended dosage rates. In such cases the use of phosphine should be considered and this is the preferred fumigant for group 7 (flour).
To reduce the possibility of taint the dose for flour should never exceed 50 g m$^{-3}$.
Diapausing larvae of *Trogoderma granarium* (Khapra beetle) and *Ephestia elutella* (warehouse moth) are highly tolerant of methyl bromide. Where these insects are present, dosages should be increased by one-half and, where applicable, exposure periods increased to 48 h in order to achieve the requisite *ct*-products (300 mg/L/h).

**Table 7.8** Examples of perishable commodities commonly fumigated with methyl bromide, and dosage rates used at normal atmospheric pressure.

| Commodity | Typical dosage rate and exposure period at 25°C* | Remarks |
|---|---|---|
| Apple | $32\,g\,m^{-3}$ for 2 h§ | Control of external feeding insects, some varieties sustain injury if fumigated |
| Asparagus | 32–48 $g\,m^{-3}$ for 2 h§ | Control of external feeders, rate depends on origin of the crop |
| Cabbage | $32\,g\,m^{-3}$ for 2 h§ | Control of external feeders |
| Cherry | $32\,g\,m^{-3}$ for 2 h§ | If fruit flies not present |
| | $22\,g\,m^{-3}$ for 2 h | Control of fruit flies or codling moth |
| Ginger rhizome | $48\,g\,m^{-3}$ for 3.5 h§ | Control of external feeders |
| | $48\,g\,m^{-3}$ for 3 h in vacuum chamber | Control of internal feeders |
| Grape | $32\,g\,m^{-3}$ for 3.5 h§ | Treatment depends on pest and origin of crop |
| Grapefruit | 24–40 $g\,m^{-3}$ for 2 h§ | Treatment depends on pest and origin of crop |
| Kiwi fruit | $32\,g\,m^{-3}$ for 3.5 h§ | Treatment depends on pests |
| Leafy vegetables | $32\,g\,m^{-3}$ for 2 h | Control of external feeders |
| Melon | $32\,g\,m^{-3}$ for 2 h§ | Control of external feeders |
| Orange | 24–40 $g\,m^{-3}$ for 2 h§ | Some varieties sustain injury |
| Potato | $44\,g\,m^{-3}$ for 2 h§ | Control of tuber moth |
| Sweet potato and yam | $56\,g\,m^{-3}$ for 4 h§ | Control of external and internal feeders |
| Zucchini/squash | $32\,g\,m^{-3}$ for 2 h§ | Control of external feeders |

*Rates quoted are those given in the *USDA Treatment Manual* (USDA 1998).
§Dosage rate requires minimum scheduled gas concentrations to be attained after 0.5 h and at the end of the exposure period.

A variety of other products is sometimes fumigated with methyl bromide, usually, but not always, to control insect pests and mostly for quarantine purposes. Examples of these products, together with fumigant application rates and exposure periods recommended, are given in Table 7.9.

## Controlled atmospheres

Controlled atmospheres, sometimes known as modified atmospheres, are usually based on either carbon dioxide or nitrogen, with low residual oxygen levels, and, under certain conditions, offer an alternative to the more conventional fumigants. A good recent review of the subject is given by Adler *et al.* (2000).

In recent years, the pressure to find alternatives to methyl bromide has stimulated much research to develop control procedures which utilise increased carbon dioxide or reduced oxygen concentrations in air as an integral part of pest management. For example, in California controlled atmospheres containing 0.4% oxygen, together with a microbial insecticide (*Plodia interpunctella* granulosus virus) and cold storage have been used to protect almonds and raisins (Johnson *et*

**Table 7.9** Examples of some other commodities treated with methyl bromide. Source: *USDA Treatment Manual* (USDA 1998).

| Type | Dosage rate at 25°C | Remarks |
|---|---|---|
| Bulbs corms and tubers | $48\,g\,m^{-3}$ for 2–4 h | Treatment depends on plant and pest to be controlled |
| Chrysanthemum cuttings | $12\,g\,m^{-3}$ for 2 h | Control of aphids |
| Cotton and cotton products | 48–64 $g\,m^{-3}$ for 12–24 h | Control of *Pectinophora* spp |
| Oak logs | $240\,g\,m^{-3}$ for 72 h | Control of oak wilt disease, minimum scheduled gas concentrations must be met during treatment period |
| Orchids: plants and cuttings | 32–48 $g\,m^{-3}$ for 1–2 h | Treatment depends on pests to be controlled |
| Wood products | $72\,g\,m^{-3}$ for 16 h | Control of boring insects, minimum scheduled gas concentrations must be met during treatment period |

*al.* 2001). Alternative treatments that have been tried continuously kept oxygen at 5%. These treatments were capable of maintaining the quality of the dry products to acceptable standards.

As alternatives to other pesticide treatments, controlled atmospheres have both advantages and disadvantages, which are illustrated in Table 7.10.

Low oxygen tension has also been combined with heat for the disinfestation of empty structures; in stores containing bulk grain the cost of providing sufficient heat to raise the grain temperature makes the technology prohibitively expensive (Adler *et al.* 2000).

### *Carbon dioxide*

The use of carbon dioxide to control insects is frequently regarded as a controlled atmosphere but, although only weakly insecticidal, the gas possesses a positive toxic action. In this respect carbon dioxide may alternatively be considered as a fumigant. As with all fumigants, the insecticidal action of carbon dioxide increases significantly with increasing temperature, but in the temperature range of 15–30°C, over which grain is likely to be treated, a relatively long exposure period is necessary. Research on carbon dioxide has been reported from workers in a number of countries, the more important of these being Australia, France, Israel, United Kingdom and US. Annis (1987) has provided detailed recommendations for application regimes to control a wide range of stored-product insects using carbon dioxide. Summarising the recommendations, Annis suggested the following to be necessary for grain temperatures between 20 and 29°C:

(1) Under constant concentrations of $CO_2$:
80%, 16 days if *Trogoderma granarium* is present; 8.5 days for all other species
60%, 11 days for all species except *T. granarium*
40%, 17 days for all species except *T. granarium*

(2) Under declining $CO_2$ concentrations:
An initial concentration above 70% declining to 35% in 15 days or longer.

These recommendations have been adopted for many of the field applications of carbon dioxide, an example being the treatment of milled rice in long-term storage in Indonesia (Nataredja & Hodges 1990). Large quantities of carbon dioxide are necessary for disinfestation purposes and the cost-effectiveness of employing the gas depends upon its ready local availability. For example, carbon dioxide is mined locally in the Rift Valley in Kenya and may therefore be a potentially low-cost source for the gas. Trials in Kenya demonstrated that individual concrete silo cells containing 2000 tonnes of grain could be effectively treated with carbon dioxide provided the cells were well sealed (Anon. 1998). Each silo required 10 tonnes of carbon dioxide and, for routine use, large tanks would need to be installed at the silo site.

Carbon dioxide can be used to disinfest grain effectively, provided time is not a constraining factor, and well-sealed enclosures are available. It may be advantageous to routinely monitor the carbon dioxide concentration during the treatment so that topping up with gas can be carried out when necessary. Without this facility some insects may inevitably survive.

Investigation of the potential of applying carbon dioxide under high pressure has been conducted, and this has led to its limited introduction in Germany for treating beverages, nuts and spices. Although the treatment can be completed in several hours the high cost of producing and operating pressure chambers is likely to limit its use (Anon. 1998).

### *Nitrogen*

Nitrogen itself is not toxic and its application to disinfest

**Table 7.10** Advantages and disadvantages of controlled/modified atmospheres. Source: Adler *et al.* 2000.

| Advantages | Disadvantages |
|---|---|
| No residues in treated produce | High gas tightness required |
| Workers' safety for $CO_2$: threshold limit value (TLV) for $CO_2$ 5000 ppm; no TLV for $N_2$ | Longer treatment times required (14+ days for $CO_2$ if treatment is not pressurised) |
| Environmentally safe | Increased production costs |
| Hypoxia has fungistatic effects and conserves product quality through reduced product respiration | Storage atmosphere has to be continuously replenished |
| Low risk for development of resistance | No warning smell |
| | Logistical problems related to material availability and supply |

grain is designed to reduce the level of oxygen below which insects can survive, usually 1% or lower. As with the application of carbon dioxide, extremely gas-tight enclosures are necessary in order to control insects effectively, and long exposure periods. In small-scale treatments, nitrogen can be supplied from cylinders, but on-site generation of the gas is more practical for large-scale use. Provision of a facility for topping up is also considered important to enable compensation for any leakage that may permit oxygen levels to rise above the 1% level. Nitrogen-based systems are in commercial operation in Australia at an export grain terminal originally equipped for application of methyl bromide (Cassells *et al.* 1994).

In addition to the use of nitrogen for reducing oxygen levels in grain, propane combustion units have been designed that can deliver a cool low oxygen-content atmosphere (Bell *et al.* 1991).

### Other fumigants in use or under investigation

Until relatively recently, the availability of methyl bromide and phosphine meant there was little need for additional fumigants for post-harvest use. Although some interest in the search for new fumigants arose as a result of the emergence of insect resistance to phosphine, it was the decision to phase out methyl bromide that highlighted the urgent need for additional fumigants. Table 7.11 summarises the principal gases that are under investigation as possible alternative fumigants.

Of those chemicals listed, carbonyl sulphide and sulphuryl fluoride are two of the most promising that are likely to become commercially available in the future. Carbonyl sulphide is reported to be effective against a wide range of the common insect pests of stored products in exposures of 1–5 days. Considerable laboratory and field evaluation of the chemical has been undertaken, particularly in Australia (Wright, in press), and it is likely that registration and commercialisation will be sought in the future. Sulphuryl fluoride, under the trademark Vikane®, has been used for almost 40 years as a termiticide in buildings and is now being evaluated for its wider potential as a replacement for some uses of methyl bromide. Under the brand name Profume®, the manufacturer of sulphuryl fluoride, Dow AgroSciences, is investigating the possible use of the chemical for control of a range of stored product pests in food storage, processing, milling, and warehousing (Welker *et al.* in press). Data are currently being sought to enable registration of the chemical for use on food commodities in the US and other countries. A major constraint to sulphuryl fluoride is its poor ovicidal action but this may be overcome to some extent by extending exposure periods or raising the temperature. A particular use for sulphuryl fluoride appears likely to be the disinfestation of walnuts and almonds immediately after harvest to control *Cydia pomonella* (codling moth) and *Amyelois transitella* (navel orange worm) which may be present in the egg stage at this time (Zettler & Gill 1999).

### Fumigation on farms

Fumigation was widely practised in the past on farms in industrialised countries using low volatility (liquid)

**Table 7.11** Other gases that may have potential as fumigants for post-harvest application.

| Chemical | Use and present status |
|---|---|
| Carbon dioxide | Fumigant for grain where long treatment periods are possible |
| Carbon disulphide | Fire hazard, used now only in China and Australia |
| Carbonyl sulphide | Experimental use only for grain and some other commodities, patent applied for in Australia |
| Cyanogen | Under investigation as a grain fumigant, patent applied for in Australia |
| Ethyl formate | Used to treat dried fruit, formerly used as a grain fumigant. Renewed interest in Australia as a grain fumigant |
| Methyl iodide | Experimental use only, similar properties to methyl bromide |
| Methyl isothiocyanate | Experimental use only for grain and for perishable commodities |
| Methyl phosphine | Experimental use only, has specific action against phosphine-resistant insects, UK patent application |
| Ozone | Laboratory use only against grain pests and some fungi |
| Propylene oxide | Used a food sterilant, under investigated as a fumigant for stored products |
| Sulphuryl fluoride | Used to control termites in the USA for many years, also in China for treating timber recently to control *Anoplophora glabripennis* (the Asian longhorn beetle). Potential use as fumigant for buildings and some food commodities now under investigation in the USA and elsewhere |

fumigants that were easy to apply to bulk grain stored in silos. The fumigants, including mixtures of ethylene dibromide, ethylene dichloride and carbon tetrachloride, and also carbon disulphide, were applied by simply pouring the liquid on to the top of the grain. The fumigation process was very slow and the silos usually remained sealed for several months, or until the grain was needed. Use of these low-boiling fumigants largely ceased in the 1980s due to perceived health dangers to those applying the chemicals (Taylor 1981), phosphine being used instead. In Australia, however, carbon disulphide continues to be used to treat farm-stored grain and this chemical is reported also to be used in China (Anon. 1998).

In developing countries there was limited use of low volatility liquid fumigants for disinfesting grain stored at farm and village level, chiefly in West Africa, also in Southern Africa (Swaziland), and on the Indian subcontinent. The structures fumigated in West Africa largely consisted of traditional granaries sealed with mud, although in Swaziland, metal water tanks were adapted for grain storage. With the decline in use of low volatility fumigants in developing countries there has been some substitution of phosphine for treating grain at farm and village level.

Factors to be taken into account when considering the advisability of fumigation at farm and village level include the poor gas-tightness of structures being treated, and that phosphine is not marketed in single-dose formulations suitable for application to small structures. It is common practice for farmers to purchase individual metal phosphide tablets in local markets when required. Poor gas-tightness of the structures fumigated can be expected to lead to risks of selecting insects for resistance to phosphine, and health hazards associated with leakage of gas affecting humans, and farm and domestic animals. There have been recommendations that fumigation should not be encouraged at farm level in developing countries (Gwinner *et al.* 1996).

In Ghana, where use of fumigation at farm/trader level has been a common practice for many years, a programme of investigation demonstrated that the mud structures used to store grains and pulses could not be fumigated effectively without costly modifications (Brice & Golob 1999). However, the metal tanks used in Swaziland can be fumigated effectively provided the filling and emptying spouts are well sealed (Brice & Golob 1999). In order to overcome the need for fumigation to be carried out by individual traders, a small fumigation centre has been developed in a town in northern Ghana, permitting communal fumigation under supervised conditions (Golob *et al.* in press). Such centres should encourage the safe and effective use of phosphine, and might be usefully adopted by groups of traders in other countries.

### Fumigation during transportation

The need to ensure that durable commodities moving in trade (particularly international trade) are not infested has resulted in fumigation being adopted as a major method for in-transit pest control in selected situations. Fumigations are conducted in ships, in freight containers, and in rail wagons, mostly for quality maintenance purposes, but also for quarantine purposes if insects such as *Trogoderma granarium* are detected. The importance of shipboard fumigation is shown by reports that, in 1998, approximately 50 million tonnes of commodities were treated (Watson, unpublished data). Both methyl bromide and phosphine have been used for fumigating commodities in ships but the efficacy of many treatments, and the safety aspect to ships' crew, is often in doubt. Problems with efficacy arise because of the difficulties of good gas distribution in ships unless a properly designed system is installed. Hazards to ships' crew can also arise where fumigant escapes into living quarters, or crew enter areas under gas without proper protective equipment. The need for a worldwide standard for shipboard fumigation with respect to both safety and efficacy has been called for (Watson *et al.* in press).

The increasing use of freight containers for moving commodities internationally has led to a large increase in the number that are fumigated, some of these being shipped 'under gas', others being fumigated and aired to release the fumigant prior to despatch. The dangers to ships' crew from fumigant leakage from containers led to recommendations by the International Maritime Organization that containers under gas should always be carried on deck, or in holds that are regarded as holds under fumigation (IMO 1993). As with the fumigation of ships, the treatment of containers is not always effective because they are not always gas-tight. For this reason containers that are not sufficiently gas-tight may be particularly unsuitable where relatively long exposures are necessary such as with phosphine (Taylor 1995). Recommendations have been made that containers should be pressure tested before fumigation, and where a container fails the test it should be rejected or alternatively fumigated under a gas-proof sheet (De Lima 1994).

### Toxic action and factors affecting fumigation

Some fumigants such as methyl bromide act as respiratory poisons, mainly entering the insect body through the spiracles. In the case of insect eggs, the gas diffuses through the outer shell (chorion). The rate of respiration of insects is affected by a variety of external factors and these will also influence the effectiveness of fumigants. Several of these factors are of practical significance in fumigation and must be taken into account when determining whether effective insect control will be achieved under particular conditions. Any factors that increase the rate of respiration will result in an insect becoming more susceptible to fumigation. The most important of these is temperature, and fumigation is normally conducted in the range 10–35°C. Within this range, the greater the temperature, the lower the amount of fumigant required to cause insect mortality. However, when using methyl bromide at very low temperatures, the effect of temperature is less clear and, in addition, the temperature to which certain products have been exposed prior to treatment may also influence fumigant effectiveness (Bond 1984). Relative humidity does not appear to have a strong influence on fumigant effectiveness although in certain situations it may affect the rate at which fumigant gases are evolved (e.g. phosphine from metal phosphide formulations). Carbon dioxide can stimulate the rate at which insects respire and can increase the effectiveness of fumigants, resulting in lower application rates and possibly shorter exposure periods (Navarro 1986; Navarro *et al.* 1999).

The toxic action of phosphine has been investigated recently, particularly in relation to the development of insect resistance. Unlike other fumigants, uptake of phosphine does not appear to require the spiracles to be open. The presence of oxygen has been shown to be essential for the toxic action of phosphine to proceed, and there is no absorption of the gas by insects in the absence of oxygen. The increase in effectiveness of phosphine at elevated temperatures is probably due to an overall increased metabolic rate, in which an increased consumption of oxygen stimulates the uptake of phosphine (Chaudhry 1997).

Methyl bromide is a fast-acting fumigant permitting post-harvest treatments to be completed reasonably quickly, 24 h being the exposure period commonly employed for durable commodities. This rapidity of treatment can be very advantageous in situations where commodity export is involved, and where undue delays could result in increased costs. Durable products that are treated with methyl bromide include cereal grains, pulses, coffee, dried fruit, and timber. Horticultural products that may need to be fumigated include fresh fruit, vegetables, cut flowers, root crops and bulbs, and treatment periods for these do not usually exceed a few hours, making methyl bromide a suitable choice. Almost all fumigation of horticultural products is conducted for quarantine purposes, frequently on arrival in the importing country. Although the majority of durable commodities can be fumigated with methyl bromide without adverse effects (seed grain is best avoided), horticultural commodities vary greatly in their response to fumigation. This is true not only in the type of fruit, for example, but also on the variety and on the temperature conditions. It is generally recommended that, unless it is known that a particular variety can be fumigated without injury, a test sample should be investigated first under local conditions (Bond 1984).

### Safety aspects of fumigation

All fumigants are toxic to humans, and to domestic and farm animals, and should only be used by those who have been adequately trained in their safe and effective application. The principal risk is to those applying fumigants by inhalation of the gas, but there may also be risks to the general public, and also to animals during the course of a fumigation. Hazards may arise because of leakage of fumigant from a treated enclosure that is not gas-tight, or because of unauthorised entry into a fumigated area. Fumigators need to be provided with appropriate personal protective equipment, which commonly takes the form of a full-face respirator (gas mask) fitted with an appropriate canister. Although the type of canister appropriate for phosphine can be used more than once, those designed for use with low-boiling point organic gases, such as methyl bromide, must be discarded after a single use. Gases such as methyl bromide are only loosely absorbed by the activated carbon component of a filter and can pass across this filter, providing a significant hazard to the wearer if used on a second occasion. When using methyl bromide, there is also a risk to fumigators of skin contact with the liquid fumigant that may cause blistering, and is the reason why gloves are not worn during application of this fumigant.

In large-scale treatments such as flour mills, those involved in the application of fumigant may need to be provided with breathing apparatus rather than gas masks. Such equipment may be self-contained with cylinders of compressed air, or attached to lines providing air from an external source.

### *Exposure to gaseous fumigants*

The limits of fumigant concentration in air to which workers are considered safe to be exposed are commonly known in the US and many countries as the threshold limit value or TLV. Other terms may be applied in some countries when referring to these limits, including the occupational exposure standard (OES) and the maximum acceptable concentration (MAC). The TLV is subdivided according to the length of the exposure time of workers. The TLV-time weighted average (TWA) exposure limit applies to exposures calculated over a normal 8-hour working day, for a 40-hour working week. The use of an average value allows for periods when the fumigant concentration may, on occasions, be above or below the limit. The TLV-short term exposure limit (STEL) applies for exposures not exceeding a specified period. Safe use recommendations for phosphine can be found in WHO (1989) and for methyl bromide in WHO (1994).

For phosphine:

TLV-TWA = 0.3 ppm
TLV-STEL = 1.0 ppm for exposures not exceeding 15 min, with a maximum of four exposures per day, and with a period of at least 60 min between exposures.

For methyl bromide:

TLV-TWA = 5 ppm
TLV-STEL =15 ppm for exposures not exceeding 10 min. Exposure for a few hours to concentrations of 100–200 ppm may cause severe illness or death (Bond 1984).

### *Monitoring of fumigants*

Monitoring of fumigant concentrations may be carried out for two purposes, safety or efficacy. For safety purposes, the techniques employed must be capable of accurately detecting and measuring relatively low-level concentrations of fumigant. For efficacy testing, equipment used must be capable of providing accurate and reproducible measurement over the range of gas concentrations expected during the fumigation.

Although the detection and measurement of fumigant gases for safety purposes is quite common in some countries, the monitoring of concentrations for assessing fumigation efficacy is much less common. Monitoring for efficacy is carried out in relatively few countries, and then may only be in relation to the fumigation of commodities to be exported. One important factor constraining the more widespread monitoring of fumigant concentrations is the lack of low-cost and easy-to-use equipment. An example of a method for measuring gas concentrations, and that is widely available, is the detector tube. These tubes are relatively simple to use and have been available for many years, but can be used only once. For use, gas is drawn from the site to be monitored into the tube using a hand-operated bellows supplied by the tube manufacturer. Tubes contain a chemical, or chemical complex, specific to the gas under test, and the reaction taking place with the gas drawn in produces a colour change. The length of the colour change developed along the graduation on the tube enables the concentration of gas to be read directly in parts per million. Several manufacturers produce tubes for most gases, and for measuring concentrations at both the safety and efficacy levels. For more accurate measurement, electronic meters are available that operate on the principle of thermal conductivity for methyl bromide, or by electron transfer for phosphine (Fig. 7.3).

In Australia, in response to increasing insect resistance to phosphine, an indicating device has been developed known as the Phoscard® (Emery & Kostas, in press). The card incorporates a thin metal copper strip that discolours progressively in the presence of phosphine. The card includes a colour indicator that can be used to determine the effectiveness of a phosphine fumigation in which the card has been exposed. The developers of the card stress that it is not a substitute for accurate monitoring of phosphine concentrations.

### *Fumigant residues in food commodities*

The formation of residues in commodities following fumigation will depend upon a number of factors including the commodity, the fumigant, and the physical conditions, particularly the temperature. Residues resulting from fumigation may be permanent due to a chemical reaction between the fumigant and the commodity treated (chemical absorption), or temporary due to physical adsorption. The latter is particularly important in the methyl bromide treatment of commodities having a high oil content or that are finely divided. Adsorbed fumigants will be lost by desorption, although a considerable time period may be necessary for complete loss of all the fumigant sorbed. This is particularly important in situations where treated commodities are kept in closed containers such as airtight grain bins or even in shipping containers. Methyl bromide is a reactive chemical and

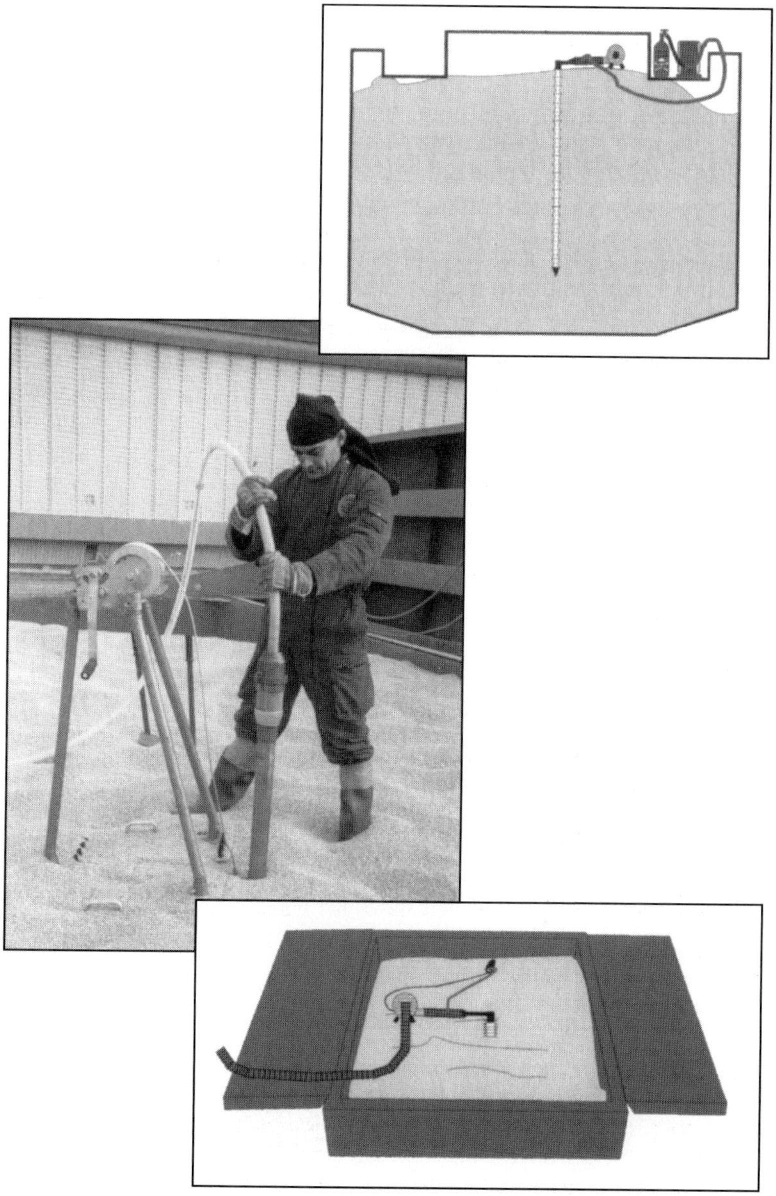

Fig. 7.4 Portable fan-assisted fumigation injection system (courtesy IGROX, UK).

fumigation of certain commodities may result in the formation of permanent residues of inorganic bromide, and methylation of proteins. The magnitude of residue will depend upon the type of commodity treated. Comprehensive summaries of the formation of residues following fumigation with methyl bromide and their effect are given by Bond (1984), and by Thompson (1966). Inorganic bromide is not considered particularly hazardous to human health and is used in medical applications. A maximum residue limit of 50 mg/kg has been established for unprocessed cereals, principally as an indicator that the commodity has not been subjected to excessive fumigation with methyl bromide. A limit of 0.01 mg/kg was established for processed commodities.

Phosphine is much less reactive than methyl bromide and permanent residues in treated commodities are not common. Small amounts of unreacted phosphine can be detected in treated commodities and maximum residue limits have been established. These are 0.1 mg/kg for unprocessed cereals and 0.01 mg/kg for processed foods.

## Radiation disinfestation

Radiation processing of foods, or food irradiation, involves exposure of food to ionising radiation. It is a physical process based on the transfer of energy to the food and its contaminants, just as exposure to solar radiation causes heat energy to be transferred to a material. It leaves no residues and does not induce radioactivity in foods (Lagunas-Solar 1995). It is defined by the World Health Organization (WHO) as a process in which foods are exposed to predetermined levels of radiation energy with either a radioactive source (cobalt-60 or caesium-137), with electron beams (EB) or with X-rays generated in electrically driven machines or accelerators (Lagunas-Solar 1995). Foods are subjected to precise levels of exposure to achieve particular degrees of preservation, but are rarely exposed to levels which achieve complete sterility as this tends to lead to off-flavours developing (Robins 1991).

Irradiation is primarily a preservation technique (Robins 1991), reducing or eliminating micro-organisms and arresting spoilage (Lagunas-Solar 1995), but it can also prolong the shelf-life of foods by inhibiting sprouting in root vegetables, delaying ripening in some fruits and vegetables, and reducing or eliminating pests and parasites (Robins 1991). Applications of irradiation that have been investigated include the following:

- disinfestation of stored grain, quarantine treatment of agricultural commodities, and genetic control and the sterile insect technique (Sethi & Bhatia 1979; Tilton & Burditt 1982; Hallman 1998);
- irradiation of packaging materials and food irradiation (disinfestation, radurisation or pasteurisation, radicidation or sanitisation) (Neijssen 1995; Ignatowicz 1997);
- sprout inhibition of root crops, delay of ripening, elimination of pathogenic organisms and reduction of microbial load (Mitchell 1986);
- inhibition of sprouting in tubers, bulbs and root vegetables, inhibition of post-harvest growth of mushrooms and asparagus, insect disinfestation, alteration of ripening and senescence of various fruits and control of post-harvest disease (Kader 1986).

Other investigations of the application of irradiation include McGivney (1988) and the International Atomic Energy Authority (IAEA 1991).

As a disinfestation technique, irradiation is quicker than fumigation, leaves no undesirable residues and can be as effective as other current procedures (Tilton & Burditt 1982). It has been suggested as an alternative to fumigation with methyl bromide when the latter is withdrawn (Forsythe & Evangelou 1994). It can achieve deep and uniform treatment of fruits and may be applied to packaged commodities (Adamo *et al.* 1996). The International Atomic Energy Agency (1991) and Lagunas-Solar (1995) concluded that irradiation is a technically viable alternative to chemical pesticides and a known and technically desirable solution to the problems of microbial contamination, excess chemical residues in food and the need for insect quarantine control in regional and international markets. Irradiation uses much less energy than canning, freezing, refrigeration or frozen storage for perishable foods, so could lead to considerable energy savings if used widely (Andreski 1984). Extended shelf-lives could result in less spoilage, waste and deliveries.

Despite this, full commercialisation of the technique within the food industry faces several barriers, not technical or scientific in nature, but centring on concerns about economic viability and levels of consumer acceptance (Neijssen 1995; Hallman 1998).

### Principles of radiation disinfestation

During irradiation, the foodstuff absorbs energy from an ionising radiation source. Ionising radiation refers in this context to electromagnetic radiation with a short wavelength. X-rays and gamma rays fall within this category – they have similar wavelengths and in food irradiation the terms may be considered to be interchangeable (Robins 1991). Ionising radiation causes electrons to split away from some atoms or molecules of the food and any contaminants it contains, leaving behind them ions, that is, electrically charged atomic or molecular fragments. The interaction of the energetic electrons and ions with the organic matter leads to chemical changes in the food and contaminants and affects the ability of cells to reproduce (Robins 1991). The growth of insect or arachnid pests and their ability to reproduce are therefore controlled (Lagunas-Solar 1995). The transfer of energy to the food during irradiation causes heating within it, but as the amounts of energy involved are relatively small compared to other processes there are fewer physical and sensory changes than those caused by cooking, freezing or canning (Lagunas-Solar 1995). Irradiation can also affect the ability of insects to move and therefore disperse (Ignatowicz 1996).

The most commonly used sources of radiation are the radioisotopes, caesium-137 and cobalt-60, which decay to produce gamma rays. Caesium-137 is a by-product of nuclear fission and was available in vast quantities from the US energy and weapons programmes (Lagunas-Solar 1995). However, the association with nuclear weapons programmes helped to focus consumer opposition to irradiation, and the cancellation of the US government's by-products utilisation programme in 1988 affected its availability. Cobalt-60 is produced in nuclear fusion reactors and has no association with nuclear weapons. It is used in many applications, including cancer therapy and large-scale industrial processes such as sterilisation and polymerisation, and has therefore gained a reputation for efficacy, reliability and safety (Lagunas-Solar 1995), with well-established procedures and codes for operatives (Robins 1991). There are, however, economic drawbacks to the use of cobalt-60 (Lagunas-Solar 1995): the supply is limited and is practically a monopoly of one company – prices doubled between 1985 and 1993 – and a sufficient future supply is questionable if even modest levels of commercialisation of food irradiation take place worldwide. Radioactive isotopes need to be replaced eventually, require extensive shielding and radiate permanently. These factors together have cost implications: extensive capitalisation is required in building an irradiation plant (Robins 1991).

Electron-beam (EB) accelerators are electrically powered, producing a stream of high-velocity electrons using electric and magnetic fields (Robins 1991). They are more flexible than radioisotopes as they can operate at different beam energies and can be used in two different ways for food irradiation: (1) by irradiating the food directly using the electron (particle) beam, or (2) converting the electron beam into more penetrating X-rays by striking a metal target prior to irradiation (Robins 1991; Lagunas-Solar 1995). Electron beam accelerators have numerous research applications but limited large-scale industrial application, are limited in the thickness of foods for which they may be used, and are unable to match the throughput capacity of megacurie-level, cobalt-60 facilities (Lagunas-Solar 1995). Their size and geometry is also large in comparison to gamma sources (Robins 1991), though advances are expected in their design and performance. Electron beam accelerators compared favourably with other sources economically, offered better safety advantages over permanently radiating sources and, therefore, potentially are of less public concern.

Gamma irradiation has been found to be more effective in some cases than EB irradiation in inducing mortality and sterility of some insect pests (Hayashi 1991). X-ray treatment has also been found to be more effective than gamma radiation from a cobalt-60 source at similar dosages, with the effectiveness tending to increase with increasing wavelength (Wohlgemuth 1973).

Insects showing resistance to chemical pesticides have shown no differences in sensitivity to irradiation, making the technique potentially useful where there are resistant populations (Tilton & Brower 1973; Brower 1974; Tilton & Burditt 1982). The cost of irradiation means that it may be limited to lower doses, which sterilise rather than kill insects, raising the possibility that sub-sterilising doses may be applied, leading to the development of radiation-resistant insect strains. Tests exposing successive generations of the stored product insects to sub-sterilising doses of gamma radiation have however shown no evidence that this occurs (Hossain *et al.* 1972; Brower *et al.* 1973; Tilton & Brower 1973).

### Irradiation facilities

The facilities used for irradiation consist of the radiation source, a water pool for storing it if it is a radioisotope, and handling machinery (Fig. 7.5). Thick concrete shielding is needed around the source to protect workers and the local environment. Facilities to store and handle the food are needed which will depend on the type of food being treated and its packaging.

With EB sources an air stream is often used to move grains past the radiation source at high speed. This has the added advantage of achieving a degree of disinfestation through the impacts of the grains, particularly with insects such as adult moths that are external to the grain. The disadvantage is that some grains will be broken (Tilton & Burditt 1982).

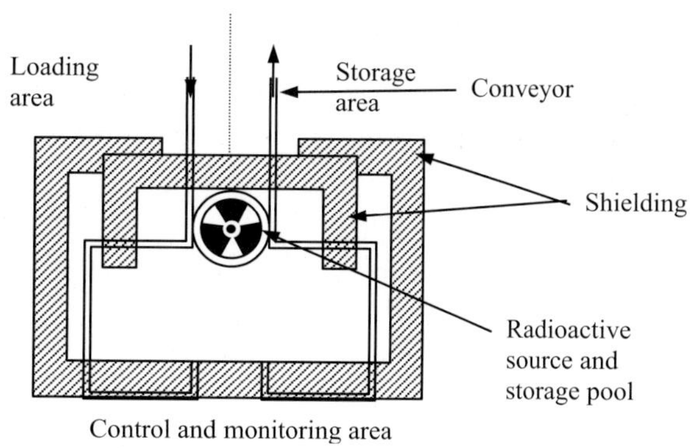


**Fig. 7.5** Representation of an irradiation facility.

### Safety of irradiated foods

Chemical changes occurring in the foodstuff during irradiation are perhaps the most contentious issue in irradiation. Definitions of the wholesomeness of food are difficult (Robins 1991) but centre around taste, texture, nutritional value and absence of harmful additives. Irradiation destroys some vitamin and major nutrient content, but there is little evidence of harmful radiolytic products, that is, products of chemical breakdown induced by irradiation, being formed through irradiation. Foods high in fats are prone to development of rancidity, but some foods like prawns, and low-fat foods such as strawberries, have their flavours enhanced. Irradiation, like heating and cooking, causes softening of food and this can have significant effects on the texture of some fruits and vegetables, making them unsuitable for irradiation (Robins 1991).

The World Health Organization (summarised in Lagunas-Solar 1995) evaluated the wholesomeness of irradiated foods. They noted that radioactivity can occur in foods through either naturally occurring radiation sources or man-made sources reaching foods through accidents or disposal. It has been argued that there are no radiolytic products formed during irradiation that are not already present in the human diet (Lagunas-Solar 1995). Though it is claimed that the exact nature of the chemical changes induced by irradiation are not understood, it is also argued that there is a substantial body of scientific work on the radiation chemistry of a wide range of foods. Chemical changes occur in nutritional properties but it is contended that these are comparable with or less severe than those occurring during other food processing techniques: microbial loads are greatly reduced, and studies have provided justification supporting the toxicological safety of irradiated foods (Robins 1991). Irradiation also has the potential to reduce the use of chemical pesticides for insect control.

Toxicological studies are reported which have shown no toxic effects from irradiated foods or special nutritional problems, and it is suggested that losses of vitamins on irradiation of permitted foods in western countries will be of no nutritional importance at the doses limited by organoleptic changes and legislation (Kilcast 1994). There are difficulties however in conducting trials to determine whether irradiated foods pose any health hazard to consumers, centring on the potential imbalance between animal and human nutritional requirements, and the existence of subtle, long-term effects from prolonged exposure. No adverse effects have been reported from hospital patients requiring bacteria-free diets that have been treated by irradiation. These diets are however limited to the duration of hospital stays and do not include foods containing polyunsaturates. Other trials have reported more equivocal results over which there is some controversy. One major effect of irradiation is the depletion of vitamins – not an addition of toxic compounds but a potential reduction in nutritional value (Robins 1991).

Another aspect of safety is that of workers involved in exposing food to radiation. The deleterious effects of exposure to radiation are well documented (Robins 1991). The dangers of exposure to gamma rays are common to many industries and activities that involve nuclear power, reprocessing or medical uses, so maximum exposure levels are common to these industries and attract considerable weight of legislation, official monitoring and control.

### Dosimetry

Dosimetry is the measurement of exposure to irradiation. Units used to measure the absorbed dose of radiation include the rad and gray (Gy). The roentgen (R) is applicable to X-ray or gamma radiation and was used as a unit of exposure rather than absorption. The rad and gray have replaced the roentgen for clinical work involving X-rays or radioactive sources. The exact equivalence depends on the material being exposed (Jerrand & McNeill 1972).

An absorbed dose is one rad when a mass of 1 kg absorbs $10^{-2}$ J of irradiated energy. 1 gray = 1 J/kg, therefore 100 rad = 1 Gy.

Applied radiation doses can be calculated from the properties of the source used, but it is often necessary to make an independent measurement using a dosimeter. Chemical dosimeters that can be calibrated very accurately against known radiation exposures are used. The simplest devices contain radiation-sensitive dyes which can be inserted into food packages in the form of films (Robins 1991).

Dosimeters have the important job of ensuring that the food product has received a dose within allowable limits in commercial systems, where the need for high throughput can force a compromise to be made with uniformity of treatment. In bulk grain, the dosimeter must be able to withstand being mixed within the grain. One such dosimeter was based on lithium fluoride thermoluminescence (Tilton & Burditt 1982). This involved placing powder into small capsules which were placed into the grain, then recovered after exposure, using a sieve. It was speculated that impregnated polythene rods could be safer and easier to use. Dosimeters can be placed into fruit cartons attached to a string to aid recovery, and reactive labels can be attached to the box or fruit itself.

Detection of the irradiation history of a foodstuff is important, given the need for irradiated foods to be labelled. It is however difficult. Irradiation at levels below 10 kGy has little discernible effect on the food. The products of radiolytic changes tend to be present in tiny, transient quantities, and are not unique to irradiation. While there are viable methods for detecting irradiation in the laboratory, these methods are not viable in industrial quality control (Robins 1991). More practical methods are being investigated.

### Application of radiation disinfestation

Early work on radiation disinfestation included unsuccessful attempts to use X-rays to control *Lasioderma serricorne* in cigars and *Sitophilus oryzae* in rice in 1912 and 1913 respectively (Tilton & Burditt 1982). The principles for disinfestation were established during the 1960s. It was found that the radiosensitivity of insects could not be predicted and, therefore, must be determined empirically, and that irradiation doses must be chosen based on the least radiosensitive insect species present. Although many studies are reported, data on the effects of irradiation on insects infesting cereals, fruits, etc., are mostly scattered and incomplete. Comparison of data is difficult because of the many factors which can affect sensitivity, including insect age, sex and strain, type of food, temperature, differences in the types of radiation used, dosimetry and exposure rates, criteria for assessing death or sterility, and treatments and conditions before, during and after treatment. Assessment of mortality and the effects of radiation needs uniform criteria, and the radiosensitivity of insects changes significantly between and even during different life stages (Tilton & Burditt 1982).

High doses of irradiation can be used to achieve an immediate and complete kill of insects. Such doses are not required though – lower, more economical doses that do not cause undesirable effects like loss of taste can be used to sterilise pupae and adults and prevent emergence of eggs and larvae (Tilton & Burditt 1982). It is assumed that a dose that will sterilise adults will suffice for disinfestation (Hoedaya *et al.* 1973). This means that feeding may continue for some time after treatment, though possibly at a reduced rate. For quarantine of fruit, irradiation doses which prevent insect pupation or egg hatch are too high to be practical, but doses which allow some eggs to hatch and pupae to form are still sufficient to induce sterility and provide quarantine security (Tilton & Burditt 1982). For example, irradiation of packed dried dates airfreighted from Iraq to Poland at doses of 0.715 kGy and 0.455 kGy provided quarantine security if induced sterility was taken into account, though fumigation with methyl bromide was found to be more successful if mortality was used as a criterion for control (Ignatowicz *et al.* 1994). Continuation of development of early insect stages to the pupal stage could result in doubts about the effectiveness of treatments during quarantine inspections despite the adult mortality rate (Heather *et al.* 1991).

Dose rates in the range of 0.2–1.0 kGy are recommended for radiation disinfestation of foods (Hackwood 1991; Robins 1991). Some maximum average radiation doses proposed for authorisation within the EU and for imported goods from tropical or developing countries are shown in Table 7.12 (Ehlermann 1991).

#### *Combination treatments*

Some researchers have looked at the compatibility of irradiation with existing chemical pesticide treatments and the possibility of combining irradiation with other disinfestation methods in order to reduce radiation disinfestation costs.

Combinations of irradiation with malathion and methyl bromide were both found to be compatible and to increase the efficacy of the individual treatments. Combinations of infra-red or microwave disinfestion treatments and irradiation applied to *Sitotroga cerealella*, *Rhyzopertha dominica* and *Sitophilus oryzae* in wheat were likewise more effective than the sum of the individual treatments, prompting speculation that dose

**Table 7.12** Some maximum average radiation doses proposed for authorisation within the European Union. Source: Ehlermann 1991.

| Foodstuff | Maximum average radiation dose (kGy) |
|---|---|
| Bulbs and tubers | 0.2 |
| Dried fruits | 1 |
| Pulses (legumes) | 1 |
| Cereal flakes | 1 |
| Dehydrated vegetables | 10 |

rates and costs of treatment could be reduced without affecting efficacy (Tilton & Burditt 1982).

The possibility that more or less toxic products could be formed by irradiation of pesticides when the two techniques are used in combination has been investigated. Irradiation of malathion at up to 43 kGy did not cause degradation or reduction in toxicity (Tilton & Burditt 1982).

Other investigations have examined hot water treatment and irradiation for disinfestation of Tephritidae in *Averrhoa carambola* (Paull 1992) and also a combination of cold treatment and gamma irradiation for control of the Mediterranean fruit fly *Ceratitis capitata* in oranges and lemons, where a synergistic effect was observed (Ohta *et al.* 1989).

### *Control of specific insects*

Substantial differences have been found in the radiosensitivity of species of Coleoptera that infest stored commodities. In general, sensitivity increases from the highly sensitive bruchids and curculionids, through the cucujids and most of the tenebrionids to the increasingly resistant anobiids, dermestids and ptinids. The most resistant species has been found to be the tenebrionid *Palorus subdepressus* which reproduced after exposure to 0.3 kGy, leading to the conclusion that a dose of 0.5 kGy will control even the most resistant beetles when commodities are infested with a range of species (Tilton & Burditt 1982).

Specific irradiation doses for control of various pest species of Coleoptera, Lepidoptera, Diptera and Acarina are summarised in Tables 7.13–7.16 respectively.

### Effects of irradiation on commodities

It is possible that some or all of the effects observed on insects after irradiation are due to changes caused to their food. Early work showed no effect on insects' development, behaviour or progeny production when they were fed with irradiated diets (Tilton & Burditt 1982). However, other work has shown conflicting indicators: both increases and decreases in fertility and longevity of insects fed on irradiated diets. No significant effects were observed for *Ephestia cautella* and their progeny fed exclusively on dates irradiated at 0.5 kGy or 1 kGy (Al-Hakkak *et al.* 1983).

Many researchers have investigated the effects of irradiation on the nutritional values of the commodities being treated and have found no significant effects. Commodities investigated have included white beans, peas and lentils (Delincee & Bognar 1993), mangoes (Manoto *et al.* 1987; Mitchell *et al.* 1990), red capsicums (Mitchell *et al.* 1990), lychees (McLauchlan *et al.* 1992), blueberries (Miller & McDonald 1996), walnuts (Abdus-Sattar *et al.* 1989) and hazelnuts (Bogunovic *et al.* 1993).

The sensory qualities of peaches, nectarines and plums have been found to be affected by irradiation (Moy *et al.* 1983). At a dose of 0.3 kGy, plums and nectarines were unaffected in comparison to controls but differences in the colour and flavour of peaches were detected. At 0.5 kGy differences in the colour and texture of plums and the aroma and colour of peaches were detected. At 1 kGy differences were found in the texture of all three fruits and in the colour and flavour of nectarines. The age of mangoes when irradiated has been found to be significant; less mature fruit were strongly affected by gamma irradiation while those which were partially ripe and in their climacteric rise when treated were largely unaffected (Boag *et al.* 1990). Kader (1986) has tabulated the effects of gamma rays, X-rays and accelerated electrons on fresh fruit and vegetables.

For seed treatment it is important that their viability and vigour are not adversely affected. This has been investigated for *Phaseolus vulgaris* beans infested with adult *Zabrotes subfasciatus* and irradiated using cobalt-60 gamma radiation. No effects on seed physiology and seedling vigour were found, though there were some abnormal seedlings produced (Moraes-Rego *et al.* 1987).

## Other methods of disinfestation and protection

### Airtight storage

Airtight storage can be distinguished from controlled atmosphere storage in that there is no specific attempt to manipulate the gas content of the environment. In airtight storage the intergranular air (oxygen) is depleted by the respiratory activity of insects and other organisms present in the grain. This leads to hypoxic conditions that result ultimately in the death of all stages of the pest life cycle. Completely hermetic conditions will result with relatively little grain damage occurring. However, the time required for this autodisinfestation and the amount of grain damage caused both depend on the air available in the container, which in turn depends on its degree of filling. With larger head spaces disinfestation requires longer periods and grain damage could become signifi-

**Table 7.13** Irradiation doses for control of Coleoptera.

| Species | Foodstuff | Radiation type/source | Dose (kGy) | Stage of insect development | Level of control | Reference |
|---|---|---|---|---|---|---|
| *Oryzaephilus surinamensis* | Dried fruits (raisins, Zante currants, prunes, dried apricots) | | 0.4 or less | All | Sufficient to control infestation | Tilton & Burditt 1982 |
| | | | 0.2 | Eggs, young larvae | Sufficient to control infestation | Tilton & Burditt 1982 |
| | | | 1–2 | All | Mortality in a few days | Tilton & Burditt 1982 |
| | Dried dates in cardboard boxes | $^{60}$Co | 0.76 ± 0.08 | All | 100% control after 25 days. Surviving immature stages did not develop further and adults were sterile | Ahmed *et al.* 1982 |
| | Dried dates | Gamma | 0.2 | Most damaging stages | Complete kill often achieved | Ahmed 1981 |
| | Rice (Indonesia) | $^{60}$Co | 0.1 | Adults | Sterility | Hoedaya *et al.* 1973 |
| | | | 0.2 | Adults | Lethal in 15 days | |
| | | Gamma | 0.1 | Eggs | Prevented survival to adult stage | Rejesus & Lapis 1973 |
| | | | 0.15 | Adults 0–15 days after emergence | Lethal within 28 days | |
| | Whole and prepacked ground spices; *Capsicum annum, Coriandrum sativum, Piper nigrum,* ginger and turmeric | $^{60}$Co | 1.0 | | Adults did not emerge; stored at 28–30°C | Padwal-Desai *et al.* 1987 |
| Dermestids | | $^{60}$Co | 3, but 0.4 thought sufficient | | Control | Tilton & Burditt 1982 |
| *Sitophilus granarius* | Wheat | $^{60}$Co and EB | 0.06–0.2 | Adults | Mortality significantly higher with gamma than EB after 28 days | Hayashi 1991 |
| | | | 0.1–0.2 | Pupae | Emerging adults sterile regardless of type | |
| | | | 0.06–0.1 | Pupae | Gamma rays more effective at inducing sterility than EB in emerging adults | |
| *Sitophilus oryzae* | Rice (Indonesia) | $^{60}$Co | 0.075 | Adults | Sterility | Hoedaya *et al.* 1973 |
| | | | 0.2 | Adults | Lethal in 20 days | |
| | Lotus (*Nelumbo nucifera*) | $^{60}$Co | 1.5 | | Complete mortality in 5 days | Wu 1991 |
| *Tribolium confusum* | Wpp bags | EB | >2 | Larvae and adults | Completely mortality within 3 weeks | Igantowicz & Migdal 1997 |
| | For use on packaging materials | Gamma | >2 | Larvae and adults | Lethal within a few days | Igantowicz & Zaedee 1995 |
| | | | 1–2 | Larvae and adults | Lethal within 2 weeks | |
| | | | <1 | Larvae and adults | Lethal within several weeks, surviving insects inhibited | |

**Table 7.13** *(Continued.)*

| Species | Foodstuff | Radiation type/source | Dose (kGy) | Stage of insect development | Level of control | Reference |
|---|---|---|---|---|---|---|
| | Cut flowers; Geraldton Wax (*Chamelaucium uncinatum*) cv Purple Pride, *Banksia hookeriana*, red and green kangeroo paw (*Anigozanthos manglesii*) | Gamma | >2 | | Immediate mortality; radiation was found to significantly reduce flower and foliage vase lives | Seaton & Joyce 1992 |
| | Rice (Indonesia) | $^{60}Co$ | 0.1 | Adults | Sterility | Hoedaya *et al.* 1973 |
| | | | 0.25 | Adults | Lethal in 29 days | |
| *Tribolium castaneum* | Walnuts and pine nuts | | 1 | | Completely checked infestation; nuts irradiated at 0.5 kGy or higher retained better sensory scores | Abdus-Sattar *et al.* 1989 |
| | Whole and prepacked ground spices; *Capsicum annum, Coriandrum sativum, Piper nigrum*, ginger and turmeric | $^{60}Co$, gamma | 1 | | Adults did not emerge; stored at 28–30°C | Padwal-Desai *et al.* 1987 |
| | Dried dates and raisins | | 0.25 | | Not effective for 1 year storage in polythene packaging at 10–36°C, but checked infestation at 10–20°C storage | Mohammed *et al.* 1989 |
| | | Gamma | 0.1 | Eggs | Prevented survival to adult stage | Rejesus & Lapis 1973 |
| | | | 0.15 | Adults 0–15 days after emergence | Lethal within 28 days | |
| | | Gamma | 0.5 | Eggs and young larvae | Complete mortality within 1 month | Tilton *et al.* 1978 |
| | | | 0.2 | Eggs and young larvae | Delayed mortality | |
| *Pantomorus cervinus*, fuller rose beetle | Lemons | Gamma | 0.17 | Eggs 10–13 days old | Prevented from hatching: this was the most resistant stage. 0.5–6.1% of fruits were affected by damage due to treatment | Johnson *et al.* 1990 |
| *Lasioderma serricorne* | | Gamma | 0.25 | Eggs | Development to adult prevented at 28°C and 60–70% r.h. | Harwalkar *et al.* 1995 |
| | | | 0.3 | Adults | Sterility induced | |
| | | | 0.5 | Older larvae | Prevention of development | |
| | Whole and prepacked ground spices; *Capsicum annum, Coriandrum sativum, Piper nigrum*, ginger and turmeric | $^{60}Co$ | 1 | | Adults did not emerge; stored at 28–30°C | Padwal-Desai *et al.* 1987 |
| *Rhyzopertha dominica* | | Gamma | 0.5 | Eggs and young larvae | Complete mortality within 1 month | Tilton *et al.* 1978 |
| | | | 0.2 | Eggs and young larvae | Delayed mortality | |

| Pest | Commodity | Source | Dose | Stage | Effect | Reference |
|---|---|---|---|---|---|---|
| | Whole and prepacked ground spices; *Capsicum annum, Coriandrum sativum, Piper nigrum*, ginger and turmeric | $^{60}$Co | 1 | | Adults did not emerge; stored at 28–30°C | Padwal-Desai *et al.* 1987 |
| *Callosobruchus chinensis* | | Gamma | 0.02 | Eggs | Lethal | Huda & Rezaur 1982 |
| | | | 0.02 | Larvae | Sublethal, did not prevent some adult emergence | |
| | | | 0.4 | Pupae | All adults died in 6 days | |
| | | | 0.4 | Adults | Died in 7 days | |
| | Pulses | | 0.25 | Eggs | Complete mortality | Gill & Pajni 1991 |
| | | | 0.3 | Larvae | Complete mortality | |
| | | | 0.35 | Early pupae | Complete mortality | |
| | | | 0.8 | Older pupae and adults | Sterility | |
| | Gram (*Vigna radiata*) | | 0.1 | All | Mortality | Bui-Cong-Hein *et al.* 1997 |
| *Callosobruchus maculatus* | Pulses | | 0.25 | Eggs | Complete mortality | Gill & Pajni 1991 |
| | | | 0.25 | Larvae | Complete mortality | |
| | | | 0.35 | Early pupae | Complete mortality | |
| | | | 0.8 | Older pupae and adults | Sterility | |
| | | | 0.01 | Eggs 0–24 hours old | Prevented hatching | Dongre *et al.* 1997 |
| | | | 0.01 | Late eggs (3 days) | Hatched but further development adversely affected | |
| | | | 0.02 | Early to late larvae | Development arrested | |
| | | | 0.15 | Pupae | 50% adult emergence | |
| | | | 0.5 | Adults | Longevity adversely affected | |
| | | | 1.5 | Adults | Mortality within 3 days | |
| | Cowpeas | | <0.1 | All | Sterilisation of adults and destruction of other stages | Diop *et al.* 1997 |
| *Callosobruchus analis* | Pulses | | 0.20 | Eggs | Complete mortality | Gill & Pajni 1991 |
| | | | 0.25 | Larvae | Complete mortality | |
| | | | 0.30 | Early pupae | Complete mortality | |
| | | | 0.8 | Older pupae and adults | Sterility | |
| *Bruchus dentipes* | Fava beans | Gamma | 0.09 | Larvae | 100% mortality | Mansour & Al-Bacheer 1995 |
| | | | 0.3 | Pupae | 93.8% mortality; emerging adults unable to fly | |
| *Stegobium paniceum* | | Gamma | 0.025 | Eggs | Development to adults prevented | Harwalker *at al.* 1995 |
| | | | 0.03 | Adults | Sterility induced | |
| | | | 0.05 | Older larvae | Prevention of development | |

**Table 7.14** Irradiation doses for control of Lepidoptera.

| Species | Foodstuff | Radiation type/source | Dose (kGy) | Stage of insect development | Level of control | Reference |
|---|---|---|---|---|---|---|
| *Plodia* spp | | | > 1 | All | Sterility | Tilton & Burditt 1982 |
| | | | < 1 | All | Sterile progeny | |
| *Plodia interpunctella* | Dried fruit; raisins, Zante currants, prunes, apricots | | 0.4 | All | Sufficient for control | Tilton & Burditt 1982 |
| | | | 0.2 | Eggs, young larvae | Sufficient for control | |
| | Almond and groundnut | | 1 | All? | Completely checked infestation | Abdus-Sattar *et al.* 1989 |
| | | Gamma | 0.5 | Eggs, young larvae | Complete mortality within 1 month | Tilton *et al.* 1978 |
| | | | 0.2 | Eggs, young larvae | Delayed mortality | |
| *Sitotroga* spp | | | > 1 | All | Sterility | Tilton & Burditt 1982 |
| | | | < 1 | All | Sterile progeny | |
| *S. cerealella* | Walnut and pine nuts | | 1 | | Completely checked infestation | Abdus-Sattar *et al.* 1989 |
| | Whole and prepacked ground spices; *Capsicum annum, Coriandrum sativum, Piper nigrum,* ginger and turmeric | $^{60}$Co | 1 | | Adults did not emerge; stored at 28–30°C | Padwal-Desai *et al.* 1987 |
| *Ephestia* spp | Rice | $^{60}$Co | 0.18 (0.45 males adults only, 0.24 females only) | | Sterility | Hoedaya *et al.* 1973 |
| *E. cautella* | Almond and groundnut | | 1 | | Completely checked infestation | Abdus-Sattar *et al.* 1989 |

| | | | | | | |
|---|---|---|---|---|---|---|
| | Dried dates in cardboard boxes | $^{60}Co$ | 0.76 | All | 100% control after 25 days; surviving immature stages did not develop further, surviving adults were sterile | Ahmed *et al.* 1982 |
| | Dried apricots and figs | | 0.25 | | Not effective in controlling infestation for 1 year in polythene packaging at 10–36°C, but checked infestation at storage temperature of 10–20°C | Mohammed *et al.* 1989 |
| | Dried dates in Iraq | Gamma | 0.2 | Most damaging stages | Complete kill often achieved | Ahmed 1981 |
| *Cydia pomonella* | | Gamma | 0.05 | Eggs 5–6 days old | Reduced numbers developed into adults, these did not mate | Toba & Burditt 1992 |
| | | | 0.1 | Eggs 5–6 days old | Only males developed into pupae | |
| | | | 0.1 | Eggs 0–1 day old | No hatching | |
| | | | 0.2 | Eggs 5–6 days old | Prevented pupation | |
| | | | 0.4 | Eggs 5–6 days old | Prevented development apparently beyond first instar | |
| | | | 1.4 | Eggs 5–6 days old | No hatching | |
| *Corcyra cephalonica* | Dried apricots and figs | | 0.25 | | Not effective in controlling infestation for one year in polythene packaging at 10–36°C, but checked infestation at storage temperature of 10–20°C | Mohammed *et al.* 1989 |
| | | $^{60}Co$ | 0.25 | Immature stages | Most likely dose for disinfestation | Allotey 1985 |

**Table 7.15** Irradiation doses for control of Diptera.

| Species | Foodstuff | Radiation type/source | Dose (kGy) | Stage of insect development | Level of control | Reference |
|---|---|---|---|---|---|---|
| *Anastrepha obliqua* | *Averrhoa carambola* fruit | Gamma | 0.05 | Larvae | Completely inhibited adult emergence at 25ºC and 70% r.h. | Arthur & Wiendl 1994 |
| | | | 0.6 | Larvae | Killed 100% | |
| *Dacus cucurbitae* | | | < 0.05 | Eggs and larvae | Prevented 95% of adult emergence | Tilton & Burditt 1982 |
| | | | 0.05–0.25 | Eggs and larvae | Prevented adult emergence | |
| | | | 0.05–0.25 | Eggs < 50% developed | Did not hatch | |
| | | | 0.15–0.25 | Larvae | Pupated but did not emerge | |
| | | | 0.25–1.2 | Eggs >50% developed | Hatched but did not reach maturity or pupal stage | |
| | | | > 1 | Mature larvae | Prevented pupation | |
| | | Gamma | < 0.05 | Eggs and larvae | Prevented adult emergence | Tilton & Burditt 1982 |
| | Mangoes | $^{60}Co$ | 0.15–1 | All stages | All doses prevented adult emergence, higher doses also prevented further development | Thomas & Rahalkar 1975 |
| *Ceratitis capitata* | | | <0.05 | Eggs and larvae | Prevented 95% of adult emergence | Tilton & Burditt 1982 |
| | | | 0.05–0.25 | Eggs and larvae | Prevented adult emergence | |
| | | | 0.05–0.25 | Eggs < 50% developed | Did not hatch | |
| | | | 0.15–0.25 | Larvae | Pupated but did not emerge | |
| | | | 0.25–1.2 | Eggs > 50% developed | Hatched but did not reach maturity or pupal stage | |
| | | | > 1 | Mature larvae | Prevented pupation | |
| | | Gamma | < 0.05 | Eggs and larvae | Prevented adult emergence | Tilton & Burditt 1982 |
| | Oranges | | 3 | | Disinfestation for quarantine | Adamo *et al.* 1996 |
| | Peaches | $^{60}Co$ | 0.05 | | Completely inhibited adult emergence | Arthur *et al.* 1993 |
| | Peaches, nectarines, plums | $^{60}Co$ | 0.4–0.5 | Eggs | No hatch in peaches and nectarines, inconclusive in plums | Moy *et al.* 1983 |

| | | | | | | |
|---|---|---|---|---|---|---|
| | Oranges and lemons | Gamma | 0.3, then storage for 21 days at 5.5°C (oranges) and 11.1°C (lemons) | Eggs | Nearly total egg mortality | Ohta *et al.* 1989 |
| | | | 0.5, then storage for 14 days at 5.5°C (oranges) and 11.1°C (lemons) | Eggs | Nearly total egg mortality | |
| | Oranges and mandarins | $^{60}$Co, gamma | 0.05 at 18–22°C and 65–75% r.h. | Eggs of all ages | Disinfestation | Costa *et al.* 1996 |
| | Cut flowers; Geraldton Wax (*Chamelaucium uncinatum*) cv Purple Pride, *Banksia hookeriana* and red and green kangeroo paw (*Anigozanthos manglesii*) | Gamma | > 2 | | Immediate mortality; radiation was found to significantly reduce flower and foliage vase lives | Seaton & Joyce 1992 |
| *Anastrepha ludens* | | Gamma | < 0.05 | Eggs and larvae | Prevented adult emergence | Tilton & Burditt 1982 |
| *Bactrocera tryoni* | | Gamma | < 0.05 | Eggs and larvae | Prevented adult emergence | Tilton & Burditt 1982 |
| *Dacus oleae* | | Gamma | < 0.05 | Eggs and larvae | Prevented adult emergence | Tilton & Burditt 1982 |
| *D. zonatus* | | Gamma | < 0.05 | Eggs and larvae | Prevented adult emergence | Tilton & Burditt 1982 |
| *D. ferrugienus* | | Gamma | < 0.05 | Eggs and larvae | Prevented adult emergence | Tilton & Burditt 1982 |
| Oriental fruit fly | Papaya in fibreboard boxes | Gamma | 0.18 | Immature stages | 99.9968% quarantine security | Tilton & Burditt 1982 |
| | | | < 0.05 | Eggs and larvae | Prevented 95% of adult emergence | Tilton & Burditt 1982 |
| | | | 0.05–0.25 | Eggs and larvae | Prevented adult emergence | |
| | | | 0.05–0.25 | Eggs < 50% developed | Did not hatch | |
| | | | 0.15–0.25 | Larvae | Pupated but did not emerge | |
| | | | 0.25–1.2 | Eggs > 50% developed | Hatched but did not reach maturity or pupal stage | |
| | | | > 1 | Mature larvae | Prevented pupation | |
| Mango weevil | Mangoes | Gamma | 0.1–0.2 | Adult females | No eggs laid | Tilton & Burditt 1982 |
| | | | 0.5 | Larvae and pupae | 100% mortality | |
| | | | 0.75 | Adults | Prevented from escaping from the seedpod | |

**Table 7.15** *(Continued.)*

| Species | Foodstuff | Radiation type/source | Dose (kGy) | Stage of insect development | Level of control | Reference |
|---|---|---|---|---|---|---|
| *Planococcus citri* | Chayote (*Sechium edule*) | Gamma | 0.04 | 0–2 week old adults | At 24–27°C and 65–75% r.h. no offspring in 50 days | Arthur & Wiendl 1996a |
| *Anastrepha fracterculus* | Apples | $^{60}$Co | 0.025 | All stages | No adult emergence | Arthur & Wiendl 1996b |
| *Rhagoletis pomonella* | Apples | | 0.015 | 3$^{rd}$ instar | Prevention of adult emergence | Hallman & Thomas 1999 |
| | | | 0.058 | 3$^{rd}$ instar | Prevention of pupae at 99% level | |
| *Rhagoletis mendax* | Blueberries | | 0.024 | 3$^{rd}$ instar | Prevention of pupae at 99% level | Hallman & Thomas 1999 |
| *Ecdytolopha aurantiana* | Oranges | Gamma | 0.4 | Immature stages | Prevention of adult emergence | Faria *et al.* 1998 |
| *D. dorsalis* | Mangoes | $^{60}$Co | 0.15–1 | All stages | All doses prevented adult emergence, higher doses also prevented further development | Thomas & Rahalkar 1975 |
| | Mangoes | Gamma | 0.05 | | Sterilisation of male and female, suppression of orchard populations in Philippines | Manoto *et al.* 1987 |
| | | Gamma | 0.5 | Most resistant stages | Mortality *in situ* | |
| *Bactrocera tryoni* | Mangoes | Gamma | 0.074–0.10 | Eggs (24 h old) larvae (5 day old) | Prevention of adult emergence | Heather *et al.* 1991 |
| *B. jarvisi* | Mangoes | Gamma | 0.074–0.10 | Eggs (24 h old), larvae (5 day old) | Prevention of adult emergence | Heather *et al.* 1991 |
| *Ctenopseustis obliquana* | Fruit | Gamma | 0.056 | Eggs (1 day old) | 99% egg mortality | Lester & Barrington 1997 |
| | | | 0.07, 0.15, 0.215 | Eggs (1 day old)<br>5$^{th}$ instar larvae | Inhibition of development<br>Effective disinfestations | |
| | | | 0.269 | Eggs (5 days old) | 99% egg mortality | |

**Table 7.16** Irradiation doses for control of mites.

| Species | Foodstuff | Radiation type/source | Dose (kGy) | Stage of insect development | Level of control | Reference |
|---|---|---|---|---|---|---|
| *Acarus siro* | | | 0.25 | | Highest dose permitting reproduction in females or mixed populations | Tilton & Burditt 1982 |
| | | | 0.45 | | Sterilisation of mixed populations | |
| | | | 0.5 | | Highest dose permitting reproduction in males | |
| *Tyrophagus putrescentiae* | Packaging materials | | 0.26 | Deutonymphs, adults | Sterilisation | Ignatowicz 1997 |
| | | | 1.5–2.1 | Deutonymphs | Mortality within 10 days | |
| *Tetranychus urticae* | Cut roses | Gamma | 0.3 | All | Eggs (0–24 h) were killed, otherwise females were sterilised | Goodwin & Wellham 1990 |
| | Placed on *Phaseolus vulgaris* leaves after irradiation | | 0.35 | Oviposition | Probit 9 quarantine security for production of sterile eggs from mites in diapause for 1–12 weeks to meet the Probit 9 standard, treatment must kill or sterilise 99.9968% of pests in a test of at least 100 000 pests | Lester & Petry 1995 |
| | | | 2.11 | Oviposition | 99% inhibition of oviposition | |
| | | | 6.58 | Adult | $LD_{99}$ within 10 days | |
| Tea mites | Puer tea | $^{60}Co$ | 0.48, 0.95 | | Mortality | Luo *et al.* 1996 |

cant. Nevertheless, with filled containers the quality of the grain is well maintained, as is its germination potential. If carbon dioxide or nitrogen is introduced into the container we then have a controlled atmosphere.

The cheapest and simplest forms of airtight storage are traditional underground pits, which are used in many parts of the world. These pits, often constructed by digging large holes in the earth, are only partially airtight, but the rate of diffusion of air is so low that oxygen cannot be replenished fast enough to compensate for its rate of use by respiring insects. In Somalia, Ethiopia and the Sudan underground pits are used for storage of sorghum by farmers and traders; in Swaziland they are used for maize storage. Considerable bulks can be stored quite safely for long periods though it is inevitable that moisture redistribution or infiltration will occur to cause local mould damage. This will be exacerbated if the grain is moist and mycotoxin problems may result. Moisture problems are increased by ambient temperature fluctuations which will affect the upper parts of the pit, especially when they are incompletely filled (McFarlane 1989).

Large, semi-underground concrete grain bins have been used in the past for longer storage of strategic grain reserves in Kenya and Cyprus, for example. Theoretically, these provide a good means of long-term storage but they have had operational difficulties. Furthermore, despite the bulk of the commodity remaining in good condition, some of the grain closest to the intake/exit hatches and close to the exposed concrete surfaces has lost quality, and therefore value, due to mould development.

Clay pots and gourds can also be effective storage containers, particularly for small quantities of grain selected and stored for seed. They provide a barrier against insect ingress if the entry point is well sealed with clay. Hermetic conditions can be achieved if the external surfaces are treated with an impermeable coating such as an oil-based paint or varnish. Plastic buckets fitted with tightly sealed lids make excellent hermetic stores without requiring modification, and very large plastic tanks, having capacities of 3–4 t of grain, have been introduced in northern Ghana for the storage of cowpea. Sealed metal drums also make good stores provided the grain is dry and kept cool. Large metal tanks are used extensively for maize storage in Swaziland, Honduras and other Central American countries. These tanks can be made hermetic with simple modifications to the joints where inlet and outlet pipes join the tank; generally, though, farmers use aluminium phosphide to ensure complete disinfestation.

### Cooling and aeration

Pest development is particularly affected by temperature. Manipulation of temperature has been used in many places to exert control and protect grain. In temperate climates and where cold conditions exist for parts of the year, cooling by aeration with ambient air at lower temperatures in bulks is quite common. Because insects are ectotherms they can only survive and develop across a relatively narrow temperature range; outside this range their development slows down and eventually death occurs (Table 7.17).

Aeration occurs by forcing air through grain to reduce its temperature. In mechanically controlled systems, it usually operates to cool the bulk in steps of 5–10°C at a time. It is a very useful tool where the grain moisture content is above the safe level for storage. Aeration can be used as effectively in sealed as in unsealed stores; sealed stores require an air exhaust ventilator (Proctor 1994).

In tropical and subtropical climates cooling is most easily obtained by refrigeration. However, this is a relatively expensive process and is usually not cost-effective in the developing world. Aeration cooling requires periods when the wet bulb temperature is low but in the more humid tropics and in the rainy season generally this is not obtainable very often. In this situation, drying

**Table 7.17** Response of stored product insects to temperature. Source: Fields & Muir 1995.

| Zone | Temperature (°C) | Effect |
|---|---|---|
| Lethal | >62 | Death in <1 minute |
| | 50–62 | Death in <1 hour |
| | 45–50 | Death in <1 day |
| | 35–42 | Population dies out |
| Sub-optimum | 35 | Development stops |
| | 33–35 | Slow development |
| Optimum | 25–33 | Maximum rate of development |
| Sub-optimum | 20–25 | Slow development |
| | 13–20 | Development slows or stops |
| Lethal | 3–13 | Death in days (unacclimatised); movement stops |
| | −10 to −5 | Death in weeks to months |
| | −25 to −15 | Death in minutes, insects freeze |

Note: Some insects can develop outside the optima. *Sitophilus granarius* can develop down to 13°C if the moisture content is sufficiently high and *Trogoderma granarium* can develop well at 40°C. Mites only stop developing at 2°C. If insects are gradually acclimatised to low temperatures they are able to survive for longer periods at lower temperatures.

by removal of moisture is easier to achieve than cooling. However, it may be possible to aerate at certain times of the day. A strategy of aeration using manual or mechanical ventilators, such as fans, which operate at certain times during the day, can be effective in bringing down grain temperatures even under general ambient conditions which do not predispose to effective cooling.

### Heating

Heating is also an effective method of grain disinfestation. In the developing world heating grain can be a simple but effective method for farmers to disinfest it before storage. Spreading grain in thin layers out in the sun can raise its temperature by up to 40°C, to in excess of 60°C. The rate of heat increase can be enhanced by using a solar collector, such as a sheet of black polythene or even a hessian sack, placed beneath the grain. Heat can also be retained in and around the grain, maintaining high temperatures for long periods, by covering it with a sheet of clear polythene, which creates a greenhouse effect.

For large-scale drying, grain can be heated with hot air in a batch drier or in a fluidised bed. In a conventional batch drier, heating and cooling is slow, the temperature variation among kernels is high and unwanted drying may occur. In a fluidised bed, air is passed vertically upward through a moving bed at a velocity sufficient to lift and mix individual kernels. The grain quickly comes into equilibrium with the air temperature and can be cooled just as quickly as the kernels move through the system.

Space heating has been used to disinfest flour mills and food processing factories. The area to be disinfested is closed off and heaters raise the temperature to 50–60°C for at least 24 h; fans distribute the hot air (Fields & Muir 1995). After heating, machinery must be lubricated and checked to ensure that belts have not expanded and plastic and rubber equipment has not been damaged. Recently, heating has been used in combination with carbon dioxide and phosphine for disinfestation of mills and similar buildings (Mueller 1996). Heating for drying purposes is described in Chapter 3.

### Non-ionising radiation

Non-ionising forms of radiation can also be used to control insects and other storage pests. Microwaves are generated by a magnetron tube that converts electrical energy into an electromagnetic field with positive and negative charges that change direction millions of times a second. Ranging from 1 mm to 0.3 m, microwaves have longer wavelengths than infra-red radiation and shorter wavelengths than radio and television waves. They consist of polar and non-polar molecules which continually switch back and forth, disrupting hydrogen bonds between water molecules, causing friction and generating heat. Water, a polar substance, must be present in order to effect microwave heating. The insect body contains sufficient water. Although insect pests can be killed using microwaves there have been no practical applications of this technology for grain disinfestation to date; experiments are being conducted to control codling moth and light brown apple moth using pulsed microwaves.

Radio waves are claimed to be able to control several storage insects. *Sitophilus oryzae, Plodia interpunctella* and *Tribolium castaneum* adults were claimed to be totally killed when subjected, in a partial vacuum, to a radio frequency field for 90 seconds (King 2001). However, no detail is given regarding control effects, nor of the conditions used for the tests. Nevertheless, disinfestation using radio waves deserves further investigation. Devices producing ultrasonic sound, which is claimed to repel flying insects, are commercially available.

### Other methods

Impaction is used extensively in the milling industry to kill insects. For more than 60 years the 'Entoleter' has been used for this purpose. The system operates using centrifugal forces to create impact action and then aspiration is used to remove frass and detritus, including dead insects, to produce clean grain.

Grain stores, but more particularly areas in which processed or part processed foods are stored and served, often rely on electronic grids combined with light traps to kill invading flying insects, whether these be pests or casual visitors. Grids are frequently used with ultraviolet light. Insects attracted to the trap are subjected to low-voltage electric shocks which incapacitate and desiccate them.

Openings into and between stores, whether windows or doors, can be protected against entry by flying insects and other detritus by the presence of curtains. Physical curtains made from vertically hung strips of plastic can be used to prevent rodent and bird entry although they are by no means foolproof. Air curtains, which provide a powerful stream of air, can be very effective in excluding insects. Of course, to maintain the barrier the air curtain must operate permanently.

Simple physical barriers may be effective deterrents to insect attack. The barrier may be a sack, bag, carton, drum or other unit container, a protective stack cover, a bin or even the building itself (McFarlane 1989). However, the complete and effective insect-proofing of a building is very difficult and expensive to achieve and has generally not been attempted when grain stores have been designed.

A woven sack, whether of jute, sisal or polypropylene, is never proof against storage insects and mites because these pests are small enough to penetrate the weave. Multi-wall paper sacks provide very good protection, but this is considerably reduced if the end closures are not specially treated to prevent or repel entry by insects through stitching holes. Sacks with polythene film liners give good protection if carefully sealed. Both paper and polythene film can be penetrated by storage insects but this is only likely to occur by high density populations in the grain or flour trying to disperse to find a new food source; small numbers of individuals on the outside of the sacks are much less likely to do so. Stitching holes can be protected against penetration by early instar larvae by using waxed paper seals or plastic tapes impregnated with pyrethrins or similar repellent insecticide. Unlined paper sacks allow a degree of airflow through the commodity, which can also be fumigated if necessary. Grain stored in polythene-lined sacks must be sufficiently dry to prevent mould and bacterial infection and free from insects because the contents cannot be fumigated. However, the commodity stored in a polythene-lined sack can be fumigated as long as the liner is thick enough to prevent rapid gas diffusion and if the mouth of the sack can be tightly closed.

Closely woven cotton sheets have been used successfully as barriers to insect infestation to protect food security grain stocks in Mali, with an application of fenitrothion to seal the barrier to the floor (McFarlane 1989). Similar protection has been obtained in traders' stores using insecticide-impregnated cotton sheeting in northern Ghana (Golob *et al.* in press). The ambient relative humidity in Mali is generally very low and stored grain has correspondingly low water activity. Where ambient conditions predispose to high diurnal fluctuations in temperature and relative humidity, the use of protective stack barriers can cause problems if these prevent airflow through the stack. The use of barriers of polythene sheeting caused condensation and fungal infection in Indonesia whereas grain was stored safely under cotton sheets. However, with dry grain stored under conditions of high carbon dioxide and low oxygen, storage under impermeable sheets has been successful for periods of a year or more.

## Further reading

There are several useful handbooks available that provide detailed information on the technology of fumigation. These include guidelines for the effective use of fumigants (MAFF 1999), and recommendations for fumigation in the ASEAN region, with comprehensive instructions on methods for application of fumigants (ASEAN 1989; van Graver & Annis 1994).

## References


Abdus-Sattar, Jan, M., Ahmad, A., Wahid, M. & Khan, I. (1989) Irradiation disinfestation and biochemical quality of dry nuts. *Acta Alimentarius*, **18**, 45–52.

Adamo, M., Ilio, V.D., Gionfriddo, F., *et al.* (1996) The technique of ionisation of orange fruits infested by *Ceratitis capitata*. *Informatore Agrario*, **52,** 73–75.

Adler, C., Corinth, H.-G. & Reichmuth, C. (2000) Modified atmospheres. In: Bh. Subramanyam & D.W. Hagstrum (Eds) *Alternatives to Pesticides in Stored-product IPM.* Kluwer, Amsterdam.

Ahmed, M.S.H. (1981) Investigations on insect disinfestation of dried dates by using radiation – a review. *Date Palm Journal*, **1**, 107–116.

Ahmed, M.S.H., Al-Hakkak, Z.S., Ali, S.R., *et al.* (1982) Disinfestation of commercially packed dates, Zahdi variety, by ionising radiation. *Date Palm Journal*, **1**, 249–273.

Al-Hakkak, Z.S., Ali, S.R., Ahmed, M.S.H. & Al-Miliky, S.K. (1983) Wholesomeness studies with a full diet of irradiated dates, using the insect *Ephestia cautella* (Walker): I. Disinfestation doses of gamma radiation. *Date Palm Journal*, **2**, 87–102.

Allotey, J. (1985) Study of radiosensitivity of the immature stages of *Corcyra cephalonica* (Stainton)(Lepidoptera: Galleriidae). *Insect Science and its Application*, **6**, 621–625.

Andreski, R. (1984) Irradiation: a post-harvest treatment process. *Agribusiness Worldwide*, September–October, 6–10.

Angerilli, N.P.D. & Fitzgibbon, F. (1990) Effects of cobalt gamma radiation on San Jose scale (Homoptera: Diaspididae) survival on apples in cold and controlled atmosphere storage. *Journal of Economic Entomology*, **83**, 893–895.

Annis, P.C. (1987) Towards rational controlled atmosphere dosage schedules: a review of current knowledge. In: E. Donahaye & S. Navarro (Eds) *Proceedings of the 4th International Working Conference on Stored Product Protection*, 21–26 September 1986, Tel Aviv, Israel. The Caspit Press, Jerusalem, Israel.

Anon. (1995) *Assessment of alternatives to methyl bromide.* Methyl Bromide Technical Options Committee, Ozone Secretariat, United Nations Environment Programme, Nairobi, Kenya.

Anon. (1998) *Assessment of alternatives to methyl bromide.* Methyl Bromide Technical Options Committee, Ozone Secretariat, United Nations Environment Programme, Nairobi, Kenya. http://www.teap.org/html/methyl_bromide_reports.html.

Arthur, V. & Wiendl, F.M. (1994) Disinfestation of *Averrhoa carambola* infested by *Anastrepha obliqua* (Macquart, 1835)(Diptera – Tephritidae) using gamma radiation. *Scientia Agricola*, **51**, 216–221.

Arthur, V. & Wiendl, F.M. (1996a) Irradiation of *Planococcus citri* (Risso)(Homoptera: Pseudococcidae) with gamma radiation from cobalt 60 to determine the disinifestation dose. *Anais da Sociedade Entomologica do Brasil*, **25**, 345–346.

Arthur, V. & Wiendl, F.M. (1996b) Disinfestation of apples attacked by the fruit fly *Anastrepha fraterculus* (Wied.)(Diptera: Tephritidae) using gamma radiations of cobalt 60. *Anais da Sociedade Entomologica do Brasil*, **25**, 157–159.

Arthur, V., Wiendl, F.M. & Wiendl, J.A. (1993) Disinfestation of *Prunus persica* artificially infested with *Ceratitis capitata* (Wied., 1824)(Diptera, Tephritidae) using gamma radiation from cobalt 60. *Revista de Agricultura Piracicaba*, 68, 323–330.

ASEAN (1989) *Suggested recommendations for the fumigation of grain in the ASEAN Region. Part 1. Principles and general practice.* ASEAN Food Handling Bureau, Kuala Lumpur, Malaysia. Australia Centre for International Agriculture, Canberra, Australia.

Bell, C.H., Llewellin, B.E., Sami, B., Chakrabarti, B & Mills, K.A. (1991) In: F. Fleurat-Lessard, F. & P. Ducom (Eds) *Proceedings of the 5th International Working Conference on Stored Product Protection*, 9–14 September, Bordeaux, France.

Bell, C.H., Price, N. & Chakrabarti, B. (Eds) (1996) *The Methyl Bromide Issue.* Agrochemicals and Plant Protection, Volume 1. John Wiley & Sons, UK.

Boag, T.S., Johnson, G.I., Izard, M., Murray, C. & Fitzsimmons, K.C. (1990) Physiological responses of mangoes cv. Kensington Pride to gamma irradiation treatment as affected by fruit maturity and ripeness. *Annals of Applied Biology*, **116**, 177–187.

Bogunovic, M., Katusin-Razem, B., Duric-Bezmalinovic, S. & Razem, D. (1993) Chemical and organoleptic investigations of hazelnut irradiated with multiple disinfestation doses. *Prehrambeno Tehnoloska I Biotehnoloska Revija*, **31**, 83–88.

Bond, E.J. (1984) Manual of fumigation for insect control. *FAO Plant Protection Paper* **54**, FAO, Rome, Italy.

Borah, B. & Chalal, B.S. (1979) Development of resistance in *Trogoderma granarium* (Everts) to phosphine in the Punjab. *FAO Plant Protection Bulletin*, **27**, 77–80.

Bowley, C.R. & Bell, C.H. (1981) The toxicity of twelve fumigants to three species of mites infesting grain. *Journal of Stored Products Research*, **17**, 83–87.

Brice, J. & Golob, P. (1999) Fumigation of grain in farmer-level mud stores and metal grain tanks using phosphine. In: Jin Zuxun, Liang Quan, Liang Yongsheng, Tan Xiachang & Guan Lianghua (Eds) *Proceedings of the 7th International Working Conference on Stored Product Protection,* 14–19 October 1998, Beijing, P. R China. Sichuan Publishing House of Science and Technology, Chengdu, Sichuan Province, P.R. China.

Bridges, R.G. (1955) N-methylation as a result of fumigating wheat. *Journal of the Science of Food and Agriculture*, **6**, 261–268.

Brigham, R.J. (1999a) *Corrosive effects of phosphine, carbon dioxide, heat and humidity on electronic equipment: Phase 11.* Agriculture and Agri-Food Canada Publication.

Brigham, R.J. (1999b) Corrosion of metals in phosphine. In: *Proceedings of the Annual International Research Conference on Methyl Bromide Alternatives and Emissions Reductions*, 1–4 November 1999, San Diego CA, USA.

Brower, J.H. (1974) Radio-sensitivity of an insecticide-resistant strain of *Tribolium castaneum* (Herbst.). *Journal of Stored Products Research*, **10**, 129–131.

Brower, J.H., Hossain, M.M. & Tilton, E.W. (1973) Radiation sensitivity of successively irradiated generations of *Tribolium castaneum* (Herbst)(Coleoptera, Tenebrionidae), and *Sitophilus oryzae* (L.)(Coleoptera, Curculionidae). *Journal of Stored Products Research*, **9**, 43–49.

Bui-Cong-Hein, Sidik, M., Rejesus, B.M., *et al.* (Eds) (1997) Effect of gamma radiation and bag packaging materials on the development of *Callosobruchus chinensis* (L.) (Coleoptera: Bruchidae) in grams (*Vigna radiata* (L.) Wilczek). *Proceedings of the Symposium on Pest Management for Stored Food and Feed*, 5–7 September 1995, Bogor, Indonesia. BIOTROP Special Publication **59**, 225–230.

Cassells, J., Banks, H.J. & Allanson, R. (1994) Application of pressure-swing absorption (PSA) and liquid nitrogen as methods for providing controlled atmospheres in grain terminals. In: E. Highley, E.J. Wright, H.J. Banks & B.R. Champ (Eds) *Proceedings of the 6th International Working Conference on Stored Product Protection*, 17–24 April 1994, Canberra, Australia. CAB International, Wallingford, UK.

Champ, B.R. & Dyte, C.E. (1976) *Report of the FAO global survey of pesticide susceptibility of stored grain pests.* FAO Plant Production and Protection Series **5**. FAO, Rome, Italy.

Chaudhry, M.Q. (1997) A review of the mechanisms involved in the action of phosphine as an insecticide and phosphine resistance in stored-product insects. *Pesticide Science*, **49**, 213–228.

Collins, P.J. & Daglish G.J. (in press) Combating the evolution of strong resistance to phosphine in Australia. In: *Proceeding of an International Conference on Controlled Atmosphere and Fumigation*, 29 October to 3 November 2000, Fresno, CA, USA.

Costa, N., Arthur, V. & Wiendl, F.M. (1996) Disinfestation with gamma radiation from Cobalt 60 of oranges and tangerines (*Citrus* spp.) artificially infested with eggs of *Ceratitis capitata* (Wied., 1824). *Revista de Agricultura Piracicaba*, **71**, 379–386.

De Castro, M.F.P.M., Leitao, M.F.F., do Vale, J.O., Bragnolo, N., Anichiareo, E.S. & Mills, K.A. (in press) Effects of phosphine in the development of *Aspergillis flavus* aflatoxin production in maize grains stored at different moisture contents. In: *Proceeding of an International Conference on Controlled Atmosphere and Fumigation*, 29 October to 3 November 2000, Fresno, CA, USA.

De Lima, C.P.F. (1994) Improved procedures for fumigation of oaten hay in shipping containers. In: E. Highley, E.J. Wright, H.J. Banks & B.R. Champ (Eds) *Proceedings of the 6th International Working Conference on Stored Product Protection*, 17–24 April 1994, Canberra, Australia. CAB International, Wallingford, UK.

Delincee, H. & Bognar, A. (1993) Effect of ionizing radiation on the nutritional value of legumes. *Bioavailability '93 – Nutritional, Chemical and Food Processing Implications of Nutrient Availability, Part II*. pp 367–371.

Dharmaputra, O.S. (1997) Effect of controlled atmospheres and fumigants on storage fungi – a review of the research activities at Semeo Biotrop. In: E.J. Donahaye, S. Navarro & A. Varnava (Eds) *Proceedings of the International Conference on Controlled Atmosphere and Fumigation in Stored Products,* 21–26 April 1996, Nicosia, Cyprus.

Dharmaputra, O.S., & Putri, A.S.R. (in press). Effectivity of phosphine on the growth of storage fungi *in vitro*. In: *Proceeding of an International Conference on Controlled Atmosphere and Fumigation*, 29 October to 3 November 2000, Fresno, CA, USA.

Dharmaputra, O.S., Putri, A.S.R. & Sidik, M. (1997) Effect of phosphine and bag type on storage fungi of milled rice. In: E.J. Donahaye, S. Navarro & A. Varnava (Eds) *Proceedings of the International Conference on Controlled Atmosphere and Fumigation in Stored Products,* 21–26 April 1996, Nicosia, Cyprus.

Diop, Y.M., Marchioni, E., Ba, D. & Hasselmann, C. (1997) Radiation disinfestation of cowpea seeds contaminated by *Callosobruchus maculatus*. *Journal of Food Processing and Preservation*, **21**, 69–81.

Dongre, T.K., Harwalker, M.R., Nene, S.P. & Padwal-Desai, S.R. (1997) Radio sensitivity of different development stages of pulse beetle (*Callosobruchus maculatus*). *Journal of Food Science and Technology, Mysore*, **34**, 413–415.

Ehlermann, D.A.E. (1991) Current status of food irradiation in Europe. In: S. Thorne (Ed.) *Food Irradiation*. Elsevier Applied Science, London, UK.

Emery, R.N. & Kostas, E. (in press) Phoscards – phosphine fumigation cards for farmers. In: *Proceeding of an International Conference on Controlled Atmosphere and Fumigation*, 29 October to 3 November 2000, Fresno, CA, USA.

EPPO (1993) Methyl bromide fumigation of stored products to control stored-product insect pests in general. Standard 12. *OEPP/EPPO Bulletin* **23**, 207–208.

FAO (1976) Recommended method for the detection and measurement of resistance of agricultural pests to pesticides. Tentative method for adults of some major beetle pests of stored cereals with methyl bromide and phosphine. FAO Method No. 16. *FAO Plant Protection Bulletin* **23**, 12–25.

FAO (1994) *International Plant Quarantine Treatment Manual.* FAO Plant Production and Protection Paper No. 50. FAO Rome, Italy.

Faria, J.T., Arthur, V., Wiendl, T.A. & Wiendl, F.M. (1998) Gamma radiation effects on immature stages of the orange fruit borer, *Ecdytolopha aurantiana* (Lima). *Journal of Nuclear Agriculture and Biology*, **27**, 52–56.

Fields, P.G. & Muir, W.E. (1995) Physical control. In: Bh. Subramanyam & D.W. Hagstrum (Eds) *Integrated Management of Insects in Stored Products.* Marcel Dekker, New York, USA.

Forsythe, K.W. & Evangelou, P. (1994) *Costs and benefits of irradiation versus methyl bromide fumigation for disinfestation of US fruit and vegetable imports.* Staff Report No. 9412, Economic Research Service, US Department of Agriculture.

Gill, M. & Pajni, H.R. (1991) Disinfestation of stored pulses with gamma radiations. *Entomology for Defense Services*, 239–243.

Golob, P., Tran, B. & Andan, H.F. (Prince) (in press) Integrated pest management including fumigation provision, of traders' stocks in Northern Ghana. In: *Proceedings of an International Conference on Controlled Atmosphere and Fumigation*, 29 October–3 November 2000, Fresno, CA, USA.

Goodwin, S. & Wellham, T.M. (1990) Gamma irradiation for disinfestation of cut flowers infested by two-spotted spider mite (Acarina: Tetranychidae). *Journal of Economic Entomology*, **83**, 1455–1458.

Gudrups, I., Golob, P. & Dales, M. (1994) Are residual insecticide applications to store surface worth using? In: E. Highley, E.J. Wright, H.J. Banks & B.R. Champ (Eds) *Proceedings of the 6th International Working Conference on Stored Product Protection*, 17–24 April 1994, Canberra, Australia. CAB International, Wallingford, UK.

Gwinner, J., Harnisch, R., & Mück, O. (1996) *Manual on the prevention of post-harvest grain losses.* Post-Harvest Project. Deutsche Gesellschaft für Technische Zusammenarbeit (GTZ), Eschborn, Germany.

Hackwood, S. (1991) An introduction to the irradiation processing of foods. In: S. Thorne (Ed.) *Food Irradiation.* Elsevier Applied Science, London, UK.

Halfhill, J.E. (1988) Irradiation disinfestation of asparagus spears contaminated with *Brachycorynella asparagi* (Mordvilko)(Homoptera: Aphididae). *Journal of Economic Entomology*, **81**, 873–876.

Hallman, G.J. (1998) Ionizing radiation quarantine treatments. *Anais da Sociedade Entomologica do Brasil*, **27**, 313–323.

Hallman, G.J. & Thomas, D.B. (1999) Gamma irradiation treatment against blueberry maggot and apple maggot (Diptera: Tephritidae). *Journal of Economic Entomology*, **92**, 1373–1376.

Harwalkar, M.R., Dongre, T.K. & Padwal-Desai, S.R. (1995) Radiation disinfestation of spices and spice products. I. Radiation sensitivity of development stages of *Lasioderma serricorne* and *Stegobium paniceum*. *Journal of Food Science and Technology, Mysore*, **32**, 249–251.

Hayashi, T. (1991) Comparative effectiveness of gamma rays and electron beams in food irradiation. In: S. Thorne (Ed.) *Food Irradiation*. Elsevier Applied Science, London, UK.

Heather, N.W., Corcoran, R.J. & Banos, C. (1991) Disinfestation of mangoes with gamma radiation against two Australian fruit flies (Diptera: Tephritidae). *Journal of Economic Entomology*, **84**, 1304–1307.

Hoedaya, M.S., Hutabarat, D., Sastradihardja, S.I. & Soetrisno, S. (1973) Radiation effects on four species of insects in stored rice and the use of radiation disinfestation in their control. In: *Radiation Preservation of Food.* Proceedings of a Symposium Jointly Organised by the International Atomic Energy Agency and the Food and Agriculture Organization of the United Nations, 13–17 November 1972, Bombay, India. International Atomic Energy Agency, Vienna, Austria.

Horn, F. (1997) The Horn phosphine generator and fumigation methods with this new tool for space-stored products and fruits. In: *Proceedings of the Annual International Research Conference on Methyl Bromide Alternatives and Emissions Reductions*, 3–5 November 1997, San Diego CA, USA.

Hossain, M.M., Brower, J.H. & Tilton, E.W. (1972) Sensitivity to an acute gamma radiation exposure of successively irradiated generations of the cowpea weevil. *Journal of Economic Entomology*, **65**, 1566–1568.

Huda, M.S. & Rezaur, R. (1982) Control of pulse pests by gamma radiation: a study of *Callosobruchus chinensis* (Coleoptera, Bruchidae). *Acta Entomologica Bohemoslovaca*, **79**, 344–350.

IAEA (1991) *Insect disinfestation of food and agricultural products by irradiation*: proceedings of the final research co-ordination meeting, 25–29 May 1987, Beijing, China. International Atomic Energy Agency, Vienna, Austria.

Ignatowicz, S. (1996) Locomotor activity of irradiated adults of the grain weevil, *Sitophilus granarius* (L.), and the rice weevil, *S. oryzae* (L.)(Coleoptera: Curculionidae). Roczniki Nauk Rolniczych. Seria E. *Ochrona Roslin*, **25,** 113–118.

Ignatowicz, S. (1997) Guideline for radiation disinfestation of used packaging. *Annals of Warsaw Agricultural University, Horticulture*, **18**, 21–29.

Igantowicz, S. & Migdal, W. (1997) Disinfestation of polypropylene bags with electron beam radiation generated by a linear electron accelerator. Roczniki Nauk Rolniczych. Seria E. *Ochrona Roslin*, **26,** 19–24.

Igantowicz, S. & Zaedee, I.H.M. (1995) Mortality of larvae and adults of the confused flour beetle, *Tribolium confusum* Du Val., after gamma radiation. *Annals of Warsaw Agricultural University, Horticulture*, **17**, 49–54.

Ignatowicz, S., Zaedee, I.H.M., Ahmed, M.S.H., Taweel, A.A.A. & Hameed, A.A. (1994) Evaluation of trial shipments of fumigated and radiation disinfested dates from Iraq to Poland. Roczniki Nauk Rolniczych. Seria E. *Ochrona Roslin*, **23,** 97–101.

IMO (1993) *Recommendations on the safe use of pesticides in ships* MSC/Circ.612/Corr.1, 8 September 1993. International Maritime Organization, London, UK.

Jerrand, H.G. & McNeill, D.B. (1972) *A Dictionary of Scientific Units*. Third Edition. Chapman & Hall, London, UK.

Johnson, J.A., Soderstrom, E.L., Brandl, D.G., Houck, L.G.& Wofford, P.L. (1990) Gamma radiation as a quarantine treatment for Fuller rose beetle eggs (Coleoptera: Curculionidae) on citrus fruit. *Journal of Economic Entomology*, **83**, 905–909.

Johnson, J.A., Vail, P.V., Brandl, D.G., Tebbets, J.S. & Valero, K.A. (2001) Integration of non-chemical treatments for control of post-harvest pyralid moths (Lepidoptera: Pyralidae) in almonds and raisins. http://www.nalusda.gov/ttic/tektran/data/000012/08/0000120803.html.

Kader, A.A. (1986) Potential applications of ionising radiation in postharvest handling of fresh fruits and vegetables. *Food Technology*, **40**, 117–121.

Kaneshiro, K.Y., Ohta, A.T., Kurihara, J.S., Kanegawa, K.M. & Nagamine, L.R. (1983) Gamma radiation treatment for disinfestation of the Mediterranean fruit fly in California grown fruits. I. Stone fruits. *Proceedings of the Hawaiian Entomological Society*, **24**, 245–259.

Kilcast, D. (1994) Effect of irradiation on vitamins. *Food Chemistry*, **49**, 157–164.

King, M. (2001) Insect pest control the non-chemical way for grains, books, papers, files, documents, textiles and wood objects. http://www.midwestfreezedryltd.com/pest_control.htm.

Lagunas-Solar, M.C. (1995) Radiation processing of foods: an overview of scientific principles and current status. *Journal of Food Protection*, **58,** 186–192.

Lester, P.J. & Barrington, A.M. (1997) Gamma irradiation for postharvest disinfestation of *Ctenopseustis obliquana* (Walker)(Lep., Tortricidae). *Journal of Applied Entomology*, **121**, 107–110.

Lester, P.J. & Petry, R.J. (1995) Gamma irradiation for after harvest disinfestation of diapausing two-spotted spider mite (Acari: Tetranychidae). *Journal of Economic Entomology*, **88**, 1361–1364.

Luo, X.M., Yu, Q., Deng, Z.Q., Hou, R.H. & Li, Z.H. (1996) Studies on the effect of irradiation disinfestation on mites in Puer tea. *Journal of South China Agricultural University*, **17,** 93–97.

McFarlane, J.A. (1989) Guidelines for pest management research to reduce stored food losses caused by insects and mites. Overseas Development and Natural Resources Institute Bulletin **22**, Slough, UK.

McGivney, W.T. (1988) Preservation of food products by irradiation. *Seminars in Nuclear Medicine*, **18**, 36–45.

McLauchlan, R.L., Mitchell, G.E., Johnson, G.I., Nottingham, S.M. & Hammerton, K.M. (1992) Effects of disinfestation-dose irradiation on the physiology of Tai So lychee. *Postharvest Biology and Technology*, **1**, 273–281.

MAFF (1999) *Fumigation guidelines*. PB 3988. Ministry of Agriculture Fisheries and Food, London, UK.

Manoto, E.C., Rosario, M.S. & Del-Rosario, M.S. (1987) Mango disinfestation strategies in relation to quarantine requirements in the Philippines. *Plant Quarantine and Phytosanitary Barriers to Trade in the ASEAN*, 9–10 December 1986. ASEAN Plant Quarantine Centre and Training Institute, Selangor, Malaysia.

Mansour, M.Y. & Al-Bacheer, M. (1995) Gamma irradiation as a disinfestation and quarantine treatment for faba bean seeds infested with *Bruchus dentipes* Baudi (Col., Bruchidae). *Journal of Applied Entomology*, **119**, 631–636.

Miller, W.R. & McDonald, R.E. (1996) Quality of 'Brightwell' and 'Tifblue' blueberries after gamma irradiation for quarantine treatment. *HortScience*, **31,** 1234.

Mills, K.A. (1983) Resistance to the fumigant hydrogen phosphide in some stored product species associated with repeated inadequate treatments. In: *Proceedings of the Second European Congress of Entomology*, Kiel, Germany.

Mitchell, G.E. (1986) Food irradiation. *Queensland Agricultural Journal*, **112,** 17–21.

Mitchell, G.E., McLauchlan, R.L., Beattie, T.R., Banos, C. & Gillen, A.A. (1990) Effect of gamma irradiation on the carotene content of mangoes and red capsicums. *Journal of Food Science*, **55,** 1185–1186.

Mohammed, W., Abdus, S., Johammad, J., *et al.* (1989) Effects of combination methods on insect disinfestation and quality of dry fruits. *Journal of Food Processing and Preservation*, **13**, 79–85.

Moraes-Rego, A.F., Rodrigues, Z.A., Oliveira, M.L., Leao-Veiga, A.F.S. & Santana, M.D.L. (1987) Influence of $^{60}$Co gamma radiation on *Zabrotes subfasciatus* (Boheman, 1833) (Coleoptera, Bruchidae) and the bean beetle *Phaseolus vulgaris* (L.). *Anais da Society Entomologica do Brasil*,

**16,** 19–36.

Moy, J.H., Kaneshiro, K.Y., Ohta, A.T. & Nagai, N. (1983) Radiation disinfestation of California stone fruits infested by medfly – effectiveness and fruit quality. *Journal of Food Science*, **48**, 928–931.

Mueller, D.K. (1996) The combination fumigation method – an update. In: *Proceedings of the Annual International Research Conference on Methyl Bromide Alternatives and Emissions Reductions*, 4–6 November 1996, Orlando, FL, USA.

Nataredja, Y.C. & Hodges, R.J. (1990) Commercial experience of sealed storage of bag stacks in Indonesia. In: B.R. Champ, E.J. Highley & H.J. Banks (Eds) *Fumigation and Controlled Atmosphere Storage of Grain*, ACIAR Proceedings **25**, 197–202.

Navarro, S. (1986) Gaseous fumigant mixtures, with special reference to mixtures if carbon dioxide with phosphine or methyl bromide. GASGA Seminar on Fumigation Technology in Developing Countries. Tropical Development Research Institute, London, UK.

Navarro, S., Donahaye, E., Sabio, G. C., Rinder, M., Dias, R., & Azrieli, A. (1999) Reducing methyl bromide dosage or exposure time using carbon dioxide with methyl bromide or carbon dioxide with heat. In: *Proceedings of the Annual International Research Conference on Methyl Bromide Alternatives and Emissions Reductions*, 1–4 November 1999, San Diego CA, USA.

Neijssen, P.J.G. (1995) Practical applications of gamma irradiation in the food industry. *Food Tech Europe*, **2,** 84 -88.

Ohta, A.T., Kaneshiro, K.Y., Kurihara, J.S., Kanegawa, K.M. & Nagamine, L.R. (1989) Gamma radiation and cold treatments for the disinfestation of the Mediterranean fruit fly in California-grown oranges and lemons. *Pacific Science*, **43,** 17–26.

Padwal-Desai, S.R., Sharma, A., Amonkar, S.V. & Desai, S.R. (1987) Disinfestation of whole and ground spices by gamma radiation. *Journal of Food Science and Technology, India*, **24,** 321–322.

Paull, R.E. (1992) Evaluation of possible carambola fruit fly disinfestation methods. *Acta Horticulturae*, **296,** 181–184.

Price, N. (1984) Active exclusion of phosphine as a mechanism of resistance in *Rhyzopertha dominica* (F.) (Coleoptera: Bostrychidae). *Journal of Stored Products Research*, **20**,163–168.

Proctor, D.L. (1994) Grain storage techniques: evolution and trends in developing countries. *Agricultural Services Bulletin*, **69**. FAO, Rome, Italy.

Rajendran, S. (1999) Phosphine resistance in stored grain insect pests in India. In: Jin Zuxun, Liang Quan, Liang Yongsheng, Tan Xiachang & Guan Lianghua (Eds) *Proceedings of the 7th International Working Conference on Stored Product Protection,* 14–19 October 1998, Beijing, P. R China. Sichuan Publishing House of Science and Technology, Chengdu, Sichuan Province, P.R. China.

Rejesus, R.S. & Lapis, E.B. (1973) Disinfestation of stored products by gamma irradiation: I. Defining lethal doses for the Philippine strain of red flour beetle, *Tribolium castaneum* (Herbst) and saw-toothed grain beetle, *Oryzaephilus surinamensis* (Lin.). *Philippine Entomologist*, **2,** 350–358.

Robins, D. (1991) *The Preservation of Food by Irradiation.* IBC Technical Services Ltd, London, UK.

Sardar Alam, M. & Ahmed, M. (1989) *Development of resistance in beetle pests of stored grain against phosphine and contact insecticides in Pakistan. Report No. 7.* Grain Quality Preservation Group, Grain Storage Research Laboratory, Pest Management Research Institute, Pakistan Agricultural Research Council, University Campus, Karachi, Pakistan.

Savvidou, N., Mills, K.A. & Bell, C.H. (1994) A same-day test for detecting resistance to phosphine. In: E. Highley, E.J. Wright, H.J. Banks & B.R. Champ (Eds) *Proceedings of the 6th International Working Conference on Stored Product Protection*, 17–24 April 1994, Canberra, Australia. CAB International, Wallingford, UK.

Seaton, K.A. & Joyce, D.C. (1992) Gamma irradiation for disinfestation damages native Australian cut flowers. *Scientia Horticulturae*, **52**, 343–355.

Sethi, G.R. & Bhatia, P. (1979) Nuclear methods in entomological investigations. *Journal of Nuclear Agriculture and Biology*, **8**, 117–120.

Taylor, R.W.D. (1981) Low volatility (liquid) fumigants. *Tropical Stored Products Information*, **41**, 19–20.

Taylor, R.W.D. (1994) Methyl bromide: Is there any future for this noteworthy fumigant? *Journal of Stored Products Research*, **30**, 253–260.

Taylor, R.W.D. (1995) Fumigation of freight containers – a study in Jamaica. In: *Proceedings, Annual International Conference on Methyl Bromide Alternatives and Emissions Reductions*, San Diego CA, USA.

Taylor, R.W.D. (1997) Methyl bromide – phase-out brought forward. *Pesticide Outlook*, **8**, 6–8.

Taylor, R.W.D. (2000) Methyl bromide – is phase out on schedule? *Pesticide Outlook*, **11**, 54–57.

Taylor, R.W.D. & Gudrups, I. (1996) *Using phosphine as an effective commodity fumigant*. Natural Resources Institute, University of Greenwich, Chatham, UK.

Taylor, R.W.D. & Halliday, D. (1986) The geographical spread of resistance to phosphine by coleopterous pests of stored products. In: *Proceedings of the British Crop Protection Conference, Pests and Diseases*, Brighton, UK.

Taylor, R.W.D. & Locke, M.T. (1996) In: *Phytosanitary Procedures to be Adopted for Maize and Other Food-aid Shipments to Reduce the Risks of Insect Infestation and Damage.* Food and Agriculture Organization of the United Nations, Rome, Italy.

Thomas, P. & Rahalkar, G.W. (1975) Disinfestation of fruit flies in mango by gamma irradiation. *Current Science*, **44**, 775–776.

Thompson, R.H. (1966) A review of the properties and usage of methyl bromide as a fumigant. *Journal of Stored Products Research*, **1**, 353–376.

Tilton, E.W. & Brower, J.H. (1973) Status of US Department of Agriculture research on irradiation disinfestation of grain and grain products. In: *Radiation Preservation of Food.* Proceedings of a Symposium Jointly Organised by the International Atomic Energy Agency and the Food and Agriculture Organization of the United Nations, Bombay, 13–17 November 1972. Vienna, International Atomic Energy Agency, Austria.

Tilton, E.W. & Burditt, A.K. (1982) Insect disinfestation of grain and fruit. In: F.S. Josephson & M.S. Peterson (Eds)

*Preservation of Food by Ionising Irradiation.* Volume III. CRC Press Inc., Boca Raton, FL, USA.

Tilton, E.W., Brower, J.H. & Cogburn, R.R. (1978) Irradiation disinfestation of cornmeal. *Journal of Economic Entomology*, **71**, 701–703.

Toba, H.H. & Burditt, A.K. Jr (1992) Gamma irradiation of codling moth (Lepidoptera: Tortricidae) eggs as a quarantine treatment. *Journal of Economic Entomology*, **85**, 464–467.

Tyler, P.S., Taylor, R.W.D. & Rees, D.P. (1983) Insect resistance to phosphine fumigation in food warehouses in Bangladesh. *International Pest Control*, **25**, 10–13.

USDA (1998) *New Treatment Manual.* United States Department of Agriculture, Animal Health and Inspection Service, Plant Protection and Quarantine, Washington, USA.

Van S. Graver, J. & Annis, P. C. (1994) *Suggested recommendations for the fumigation of grain in the ASEAN Region. Part 3.* Phosphine fumigation of bag-stacks sealed in plastic enclosures: an operations manual. Australia Centre for International Agriculture, Canberra, Australia.

Waterford, C.J. & Asher, P.P. (2000) A new formulation of aluminium phosphide providing controlled release of phosphine. In: *Proceedings of the Annual International Research Conference on Methyl Bromide Alternatives and Emissions Reductions*, 6–9 November, Orlando, FL, USA.

Waterford, C.J., Whittle, C.P. & Winks, R.G. (1994) New aluminium phosphide formulations for the controlled release of phosphine. In: E. Highley, E.J. Wright, H.J. Banks & B.R. Champ (Eds) *Proceedings of the 6th International Working Conference on Stored Product Protection*, 17–24 April 1994, Canberra, Australia. CAB International, Wallingford, UK.

Watson, C.R., Boye, J., Pruithi, N., Bureau, D., Macdonald, C. & Roca. J. (in press) New developments in the fumigation of bulk and bagged goods in transit. In: *Proceedings of the International Conference on Controlled Atmosphere and Fumigation*, 29 October to 3 November 2000, Fresno CA, USA.

Welker, J., Rowley, J. & Hindes, D. (in press) Update on the development of sulfuryl fluoride as an alternative to methyl bromide. In: *Proceedings of the International Conference on Controlled Atmosphere and Fumigation*, 29 October to 3 November 2000, Fresno CA, USA.

WHO (1989) *Phosphine health and safety guide no. 28*, World Health Organization, Geneva, Switzerland.

WHO (1994) *Methyl bromide (bromomethane) health and safety guide no. 86*, World Health Organization, Geneva, Switzerland.

Winks, R.G. (1987) Strategies for effective use of phosphine as a grain fumigant and the implications of resistance. In: E. Donahaye & S. Navarro (Eds) *Proceedings of the 4th International Working Conference on Stored Product Protection*, 21–26 September 1986, Tel Aviv, Israel. The Caspit Press, Jerusalem, Israel.

Winks, R.J. (1992) The development of Siroflo® in Australia. In: *Proceedings of an International Conference on Controlled Atmosphere and Fumigation in Grain Storage*, Israel. Caspit Press, Jerusalem, Israel.

Winteringham, F.P.W., Harrison, A., Bridges, R.G. & Bridges, P.R. (1955) Nature of methyl bromide residues in fumigated wheat. *Journal of the Science of Food and Agriculture*, **6**, 251–261.

Wohlgemuth, R. (1973) Comparative studies on the action of Roentgen and gamma radiation ($^{60}$Co) on the grain weevil (*Sitophilus granarius*). *Zeitschrift fur Angewandte Entomologie*, **74**, 7–24.

Wright, E.J. (in press) *Proceedings of an International Conference on Controlled Atmosphere and Fumigation,* 29 October to 3 November 2000, Fresno CA, USA.

Wu, C.X. (1991) Study on the radiodisinfestation of lotus seed. *Journal of Nuclear Agricultural Sciences*, **12,** 23–25.

Zettler, J.L. (1997) Influence of resistance on future fumigation technology. In: E.J. Donahaye, S. Navarro & A. Varnava (Eds) *Proceedings of an International Conference on Controlled Atmosphere and Fumigation in Stored Products*, 21–26 April 1996, Nicosia, Cyprus.

Zettler, J.L. & Gill, R.F. (1999) Sulfuryl fluoride: A disinfestation treatment for walnuts and almonds. In: *Proceedings of the Annual International Research Conference on Methyl Bromide Alternatives and Emissions Reductions*, 1–4 November 1999, San Diego CA, USA.

# Chapter 8
# Food Processing and Preservation

A. A. Swetman, L. Nicolaides, P. W. Wareing, J. H. New, J. F. Wood and L. Hammond

Food is an extremely perishable commodity and will become unfit for human consumption unless it is preserved and/or stored under suitable conditions. For example, fish will spoil rapidly once captured and left at ambient temperatures. However, if it is chilled immediately and stored in ice, it will have a shelf-life of between 12 and 30 days depending on the species and its origin. All food commodities are sources of a diverse range of micro-organisms, some of which will cause spoilage while others have the ability to cause food-borne illness. Micro-organisms such as bacteria, fungi and some viruses have the ability to reproduce and increase in numbers in food, if it is not preserved or stored correctly. Parasites, however, are unable to reproduce outside a host, and so will not increase in number on food. Nonetheless, once consumed they may cause food-borne infections. Food processes and preservation methods, in combination with packaging methods, have been developed to minimise losses and keep food resources safe and wholesome to eat.

## Food preservation

### Factors affecting microbial growth

#### *Growth rate*

All micro-organisms require certain chemicals and chemical reactions in order to survive and reproduce. The microbial cell must be able to obtain energy and synthesise cellular protoplasm from its environment. Organisms require carbon and nitrogen, and growth factors such as vitamins, minerals and water.

There are three general patterns of microbial growth displayed by bacteria, yeasts and filamentous fungi (moulds). Most bacteria multiply by binary fission, yeasts usually by budding, and moulds usually by hyphal extension and branching. Provided the concentration of a nutrient is above a certain level the growth rate of a micro-organism is constant. Below that level the growth rate declines. As an organism grows in a medium it changes the composition of the medium so that conditions are not always optimal for growth.

Growth of a micro-organism in a batch (closed system) culture is characterised by a sequence of phases:

(1) *Lag phase*: The population remains static, adjusting to its new environment. Metabolic activity is initiated and new cell components are synthesised.
(2) *Exponential or logarithmic phase*: Cells divide and grow at regular intervals at the maximum rate possible, given the conditions under which they are grown and their genetic potential. Cultures are very uniform at this stage and have generation times or doubling times in minutes or hours. Typical doubling times are 20–60 min for bacteria, 90–120 min for yeasts, and 4–8 h or more for moulds. The rate of growth can be described according to the mean growth rate constant, which is the number of generations per unit time, usually per hour, as in the equation below:

$$k = \frac{n}{t} = \frac{\log N_t - \log N_0}{0.301t} \tag{8.1}$$

where:
$k$ = mean growth rate constant
$N_0$ = initial population number
$N_t$ = population at time $t$
$n$ = number of generations within time $t$

(3) *Stationary phase*: Cell numbers reach the maximum that the environment can support, growth

ceases due to nutrient depletion or build-up of wastes or environmental change (low pH, for example). Maximum cell density is usually around $10^9$ or $10^{10}$ cells per ml of culture fluid for bacteria.

(4) *Death phase:* Phase of decline. Build-up of wastes or environmental change is severe enough to cause death. Death rate is also a logarithmic function, sometimes declining when the population has been greatly reduced, due to the survival of resistant cells.

Several interconnected factors can affect growth, for example, water activity, pH, temperature, redox potential, gaseous atmosphere, nutrient content, antimicrobial constituents, antimicrobial structures and microbial interactions.

### *Water activity*

Water is an essential requirement for microbial growth. It has four functions – as a chemical reactant, solvent, maintainer of cell turgidity and in supporting cell structure.

Solutes or particles in an aqueous (water) medium are surrounded by a shell or layers of water molecules which are oriented by a variety of physical forces. This partial binding and organisation of the water molecules reduces the availability of water to take part in those functions that support microbial growth. A similar situation occurs at low temperatures when water is removed from the solution as ice crystals. Therefore, water has to be present in a medium and also be available to support microbial growth.

The development of a micro-organism has a maximum, optimum and minimum $a_w$ for growth. The values vary between organisms. No microbial growth can take place below $a_w$ =0.60. As the $a_w$ decreases from the growth optimum, the growth rate and final cell population decrease and the lag phase increases. Below the minimum the lag phase is infinite.

Generally, bacteria require a higher $a_w$ than do yeasts. Growth of most bacteria, moulds and yeasts occurs at $a_w$ above 0.90. Most spoilage and pathogenic organisms do not grow well below $a_w$ 0.93. *Staphylococcus aureus* is a notable exception, growing at $a_w$ of 0.86 (Table 8.1).

The water activity of a foodstuff is related to the water content and the composition of the food. For example, a water activity of 0.70 would be exhibited by pears and apples at approximately 30–32% moisture content, wheat, maize and sorghum at 13.5–14.5%, soybeans at 13%, sunflower at 9.5% and groundnuts at 7.0% water content, on a wet weight basis. The lipid (fat) content of foods greatly influences the water activity; the higher the lipid content, the higher the water activity at a given moisture content.

**Table 8.1** Approximate minimum water activities for growth of micro-organisms.

| Organism | Minimum $a_w$ |
|---|---|
| Most spoilage bacteria | 0.90–0.91 |
| *Acinetobacter* spp | 0.95–0.98 |
| *Aeromonas* spp | 0.95–0.98 |
| *Alcaligenes* spp | 0.95–0.98 |
| *Arthrobacter* | 0.95–0.98 |
| *Bacillus* spp | 0.90–0.99 |
| *Bacillus cereus* | 0.92–0.95 |
| *Citrobacter* spp | 0.95–0.98 |
| *Clostridium botulinum* | 0.90–0.98 |
| *Clostridium perfringens* | 0.93–0.97 |
| *Corynebacterium* spp | 0.95–0.98 |
| *Enterobacter* spp | 0.95–0.98 |
| *Escherichia coli* | 0.94–0.97 |
| *Klebsiella* spp | 0.95–0.98 |
| *Lactobacillus* spp | 0.90–0.96 |
| *Leuconostoc* spp | 0.96–0.98 |
| *Micrococcus* spp | 0.90–0.95 |
| *Pseudomonas aeruginosa* | 0.96–0.98 |
| *Salmonella* spp | 0.93–0.96 |
| Halophilic bacteria | 0.75 |
| *Staphylococcus aureus* | 0.83–0.92 |
| *Streptococcus* spp | 0.92–0.98 |
| *Vibrio parahaemolyticus* | 0.94–0.98 |
| Most yeasts | 0.87–0.94 |
| Osmophilic yeasts | 0.60–0.78 |
| Most moulds | 0.85–0.98 |
| Xerophilic moulds | <0.85 |
| Extreme xerophilic moulds | 0.60–0.70 |
| *Aspergillus* spp | 0.68–0.90 |
| *Aspergillus flavus* | 0.78–0.90 |
| *Aspergillus halophilicus* | 0.68 |
| *Aspergillus niger* | 0.80–0.84 |
| *Botrytis cinerea* | 0.93 |
| *Fusarium* spp | 0.82–0.92 |
| *Hansenula* spp | 0.89–0.90 |
| *Mucor* spp | 0.80–0.93 |
| *Penicillium* spp | 0.78–0.93 |
| *Saccharomyces cerevisiae* | 0.90–0.94 |
| *Xeromyces* spp | 0.60–0.61 |

The level of added solutes is another factor that affects the water activity. Salt and sugar (glucose, sucrose and fructose) are frequently added to foods to reduce the water activity and improve its microbial stability. Typically, salt is added to vegetables, meat and fish, whereas sugars are added to fruits and processed fruit products.

Below $a_w$ 0.93 the growth of *Bacillus cereus*, *Clostridium botulinum*, *Clostridium perfringens*, *Salmonella* spp and *Vibrio* spp is prevented or severely restricted. There are, however, groups of organisms that can grow at low $a_w$ values and these are important in the spoilage of low $a_w$ foodstuffs.

### *Halophiles*

Halophiles are organisms that require substantial quantities (around 20% w/w) of salt to be present before growth can occur. They are often found in sun-dried salt. Halophiles are usually responsible for the spoilage of products such as salted fish. Examples include *Halobacterium* spp and *Halococcus* spp. *Vibrio parahaemolyticus* is a halophilic bacterium that can also cause human illness.

### *Xerophiles*

Xerophiles are organisms that can grow at water activities of less than 0.85. These are always yeasts and moulds, which are important in the spoilage of so-called intermediate moisture foods, such as dried fruits, cereals, molasses and nuts.

Some sugar-tolerant yeasts which are xerophiles are referred to as *osmophiles*, that is, micro-organisms that can grow in the presence of high levels of sugar in the foodstuff. These are responsible for the spoilage of products where sugar is the solute reducing the $a_w$, for example, jams, jellies and dried fruits.

Table 8.2 shows the water activity of selected foods.

**Table 8.2** Approximate water activities of selected foods.

| Food | $a_w$ |
|---|---|
| Fresh fruit | 0.97–1.00 |
| Fresh vegetables | 0.97–1.00 |
| Fresh poultry | 0.98–1.00 |
| Fresh fish | 0.98–1.00 |
| Fresh meat | 0.95–1.00 |
| Bread | 0.95–0.96 |
| Cheese | 0.91–1.00 |
| Cake | 0.90–0.94 |
| Dried rice | <0.70 |
| Nuts | 0.66–0.84 |
| Molasses | 0.76 |
| Dried fruits | 0.51–0.89 |
| Dried noodles | 0.50 |
| Biscuits | 0.30 |
| Sugar | 0.10 |

Besides the minimum water activity necessary for growth, $a_w$ also influences the germination of spores, toxin production, resistance to heat and survival of micro-organisms.

The storage environment also has an important influence on $a_w$. If the relative humidity of the environment is higher than the $a_w$ this may allow surface growth to take place. If the relative humidity is lower than the $a_w$ then the stored product will tend to dry out, lowering the $a_w$ but possibly producing undesirable quality changes in the food.

## *pH*

The pH of a medium is an important factor affecting microbial growth. Most organisms function over a range of about four pH units with an optimum of one to two units. Most bacteria grow best at pH values around 7, although there are some notable exceptions, for example, *Lactobacillus* spp (5.5–6.0) and *Acetobacter* spp (5.4–6.3). Table 8.3 shows the minimum, optimum and maximum pH levels for growth of selected food-borne pathogens.

Moulds and yeasts grow best at acid pH values less than 7, yeasts usually between pH 3 and pH 6 and moulds usually at pH 3–7.The pH level of a specific food will influence the type of spoilage and pathogenic microflora associated with it. For example, both fruit and vegetables have pH values less than 7. Fruits are generally more acidic than vegetables (Table 8.4). Fruits are usually subject to yeast and mould spoilage while bacterial spoilage is important in vegetables, e.g. soft rots due to *Erwinia carotovora*.

Decreasing the pH level of a food by an acid-producing fermentation or the addition of acid, for example, vinegar, is one of the traditional food preservation tech-

**Table 8.3** Minimum pH for growth of some food-borne pathogens.

| Organism | pH Minimum | pH Optimum | pH Maximum |
|---|---|---|---|
| *Bacillus cereus* | 5.0 | 6.0–7.0 | 8.8 |
| *Salmonella* spp | 3.8 | 7.0–7.5 | 9.5 |
| *Escherichia coli* | 4.4 | 6.0–7.0 | 9.0 |
| *Clostridium botulinum* | 4.6 | — | — |
| *Clostridium perfringens* | 5.0 | — | 9.0 |
| *Staphylococcus aureus* | 4.0 | 6.0–7.0 | 10.0 |
| *Vibrio parahaemolyticus* | 4.8 | 7.8–8.6 | 11.0 |

**Table 8.4** Approximate pH ranges of selected foods.

| Fruit | pH range | Vegetable | pH range |
|---|---|---|---|
| Bananas | 4.5–5.2 | Carrots | 4.9–6.3 |
| Pineapple | 3.2–4.1 | Onions | 5.3–5.8 |
| Oranges | 2.8–4.0 | Sweet potatoes | 5.3–5.6 |
| Lemons | 2.2–2.4 | Tomatoes | 3.7–4.9 |
| Apples | 2.9–3.3 | Lettuce | 6.0 |
| Limes | 1.8–2.0 | Broccoli | 6.5 |
| Grapes | 3.4–4.5 | Beans | 4.6–6.5 |

nologies. The effect of low pH on microbial growth will vary depending on the acid used. Lipophilic acids such as acetic acid and lactic acid in their undissociated form can pass across the plasma membrane to the cell interior where they dissociate, causing a drop in interior pH, a breakdown of transport and energy-yielding processes, and a build-up of toxic levels of anion. Strong acids such as hydrochloric, sulphuric and phosphoric acids produce a very low external pH and high proton concentration that disrupts membrane processes. These effects are usually unacceptable in foodstuffs other than carbonated beverages.

A low pH in combination with other factors inhibitory to microbial growth can have a complementary effect. For example, a low pH inhibits the germination of spores surviving a heat treatment, therefore reducing the heat treatment necessary to produce commercial sterility in acidic foodstuffs.

### *Temperature*

Micro-organisms can grow over a wide range of temperature from –10°C to more than 90°C. Probably the most important determinant is that water should be in the liquid state. In the case of temperatures below freezing, suspended solutes can allow water to be present in the liquid state in a food. The range of temperatures over which a single organism will grow seldom exceeds 35°C. For fungi it is usually lower, around 30°C.

Organisms that will grow below 20°C are known as psychrophiles or psychrotrophs. Those growing in the range 10–45°C (optimum 30–40°C) are known as mesophiles. Those growing above 45°C are known as thermophiles. Table 8.5 shows the temperature ranges, and maxima and minima of several groups of organisms. At growth temperatures below the optimum there is a slow decline in growth rate ascribed to three factors:

**Table 8.5** Approximate temperature ranges and growth for arbitrary classes of micro-organisms.

| | Temperature °C | | |
|---|---|---|---|
| | Minimum | Optimum | Maximum |
| Psychrophilic | –15 to –5 | 10–30 | 20–40 |
| Obligate | –15 to 0 | 10–20 | 20–22 |
| Facultative | –5 to 0 | 20–30 | 30–40 |
| Psychrotrophic | –5 to 5 | 25–30 | 30–40 |
| Mesophilic | 5–25 | 25–40 | 40–50 |
| Thermophilic | 35–45 | 45–65 | 60–90 |
| Obligate | 40–45 | 55–65 | 70–90 |
| Facultative | 36–40 | 45–55 | 60–80 |

- the decreasing rate of chemical reactions with decreasing temperatures;
- the breakdown in control mechanisms at low temperatures; and
- changes in membrane fluidity.

All of these factors are generally reversible, and so storage of a food below the minimum growth temperature of its usual spoilage microflora will prevent them spoiling the product although it will not necessarily kill them. If the temperature rises then growth can resume. At growth temperatures higher than the optimum the growth rate declines sharply. This is due to the degradation and denaturation of essential cell components. This is not a reversible process and is, therefore, an effective way of killing micro-organisms. Vegetative cells of bacteria and fungi are destroyed by temperatures 10–15°C above the maximum for growth. Death of a population is not, however, instantaneous.

### *Redox potential ($E_h$)*

The redox potential or $E_h$ is a measure of the tendency of a system to reduce or oxidise and is expressed in mV. In its broadest sense, oxidation and reduction can be defined as the losing or gaining of electrons by an element or compound. A high positive $E_h$, for example, +300 mV, indicates an oxidising aerobic environment suitable for the growth of aerobic organisms such as *Bacillus* spp and *Pseudomonas* spp. A low negative $E_h$ (–420 mV) indicates a reducing environment suitable for anaerobic organisms such as *Clostridium* spp.

Many organisms have the capacity to grow under both aerobic and anaerobic conditions and are termed facultative, for example, Enterobacteriaceae and yeasts. Some organisms grow best under slightly reduced conditions and are known as microaerophilic, such as the lactic acid bacteria.

The $E_h$ of a foodstuff reflects the balance between oxidising and reducing agencies and is determined by:

- the characteristic $E_h$ of the food, for example, the presence of reducing compounds such as –SH groups in proteinaceous foods, reducing sugars, and so on;
- the poising capacity (resistance to change in potential) or buffering ability in pH measurement ($E_h$ is a measure of intensity, not capacity);
- the oxygen tension in the atmosphere around the food, for example, vacuum packaging or the use of carbon dioxide or nitrogen atmospheres will tend to decrease the $E_h$; and
- the accessibility of the foodstuff to the atmosphere – degree of division, mincing, etc.

Generally, there is an inverse relation between redox potential and shelf-life. The higher the $E_h$, the more rapidly a food will spoil, because aerobic growth of organisms is usually faster than anaerobic growth. The $E_h$ is difficult to measure accurately and reproducibly but the fact that microbial growth lowers the $E_h$ is used as the basis of some rapid methods for the assessment of total microbial load, for example, methylene blue and resazurin tests in milk microbiology. The $E_h$ of plant products varies between +383 mV and +436 mV in fruit juices, +74 mV in spinach, +225 mV in barley, +179 mV in cherries and +175 mV in peaches.

### *Gaseous atmosphere*

The type of gas in the atmosphere surrounding the food may determine the types of organisms that become dominant. Oxygen in the atmosphere will favour the growth of aerobic types. The lack of oxygen, or a vacuum, will allow facultative anaerobes to become dominant.

Micro-organisms vary widely in their tolerance to carbon dioxide. In a carbon dioxide atmosphere, the growth of some micro-organisms is completely suppressed, while others are less affected. On the other hand, it is well documented that the presence of low concentrations (1% or less) of carbon dioxide stimulates oxygen uptake by bacteria. The amount of stimulation varies with the species.

### *Nutrient content*

The nutrient content of a medium is one of several factors that affect the growth of micro-organisms. It is sometimes known as an intrinsic factor, in that it is a property of the growth medium. Intrinsic factors include macronutrients, for example, carbon, hydrogen, oxygen, nitrogen, sulphur and phosphorus (the building blocks of carbohydrates, proteins, nucleic acids and lipids). Other macronutrients include potassium, sodium, calcium, iron and magnesium. The latter have a variety of roles in enzyme structure, enzyme cofactors, membrane structure and cytochrome synthesis.

Micronutrients such as zinc, copper, manganese, molybdenum, cobalt and nickel are required in much smaller amounts. These are mainly needed for enzyme structure and by some cofactors.

### *Antimicrobial constituents*

Many foods contain natural microbial inhibitors which confer some protection (Table 8.6).

### *Antimicrobial structures*

Many foods possess physical barriers to micro-organisms when intact, for example cuticle, shell and membrane of an egg, skin of an animal, peel or skin of fruit and the shell of a nut.

**Table 8.6** Antimicrobial factors in foods.

| Foodstuff | Antimicrobial compound | Effect |
|---|---|---|
| Eggs | Lysozyme | Lysis of cell walls of Gram-positive bacteria |
| | Avidin | Binds to biotin |
| Dairy products | Lactoferrin | Iron-binding protein from milk |
| Cranberries | Oleuropein – phenolic | Inhibitory to *Lactobacillus*, *Leuconostoc,* yeasts and moulds |
| Hops | Lupulone, humulone and isohumulone (hop oil) | Inhibitory to Gram-positive bacteria |
| Onions and garlic | Thiosulphinates | Inhibitory to bacteria |

### *Microbial interactions*

Micro-organisms can inhibit or stimulate the growth of one another. Although pure cultures are usually used in laboratory studies, mixed cultures are found in nature and on food products. Since the main goals of all living forms are self-preservation and perpetuation of the species, micro-organisms have to compete for food and other resources. As a result, there are various actions and reactions (or coactions) that may be harmful or beneficial when organisms are in a mixed population:

*Predation*: in this relationship, the strong predators feed on the weak prey;
*Parasitism*: the weak parasite benefits at the expense of the strong host;
*Commensalism*: the coaction results in the weak benefiting, and the strong is unaffected;
*Symbiosis*: mutual aid, with both organisms benefiting from the relationship; and
*Neutrality*: neither organism benefits or loses from the relationship – there is no coaction between them.

## Food preservation by physical methods

The mechanisms of inhibition of microbial growth are reviewed to understand the relative merits of physical methods of preservation and to indicate the nature of the microflora that can survive these methods.

Physical methods for food preservation include heat treatments (sterilisation, pasteurisation, blanching, canning), refrigeration and freezing, removal of water (dehydration or additives), irradiation (ultraviolet or ionising), controlled atmospheres and modification of acidity (fermentation, additives).

### *Heat treatments*

Micro-organisms have been characterised in terms of their thermal death times. This value, $D$, represents the time (in minutes) at a specified temperature required to decrease the population by 90%. The thermal death of micro-organisms is a first order process. Consequently, if the number of survivors (log scale) is plotted against time a linear relationship is produced.

Another important index is the $Z$ value: the number of °F that the temperature must be changed in order to effect a ten-fold change in $D$. The $Z$ values are obtained by plotting log $D$ against temperature.

Micro-organisms are classified as thermophiles, mesophiles or psychrophiles. The micro-organisms that are most difficult to remove by heat treatments are the thermophilic spore-formers. For example, the $D_{121°C}$ of *B. stearothermophilus* exceeds 4 minutes and the value for *Clostridium botulinum* is about 0.1–0.2 min. Non-spore formers are much more sensitive, with $D_{121°C}$ values usually 1 min or less.

#### *Non-genetic factors that affect heat resistance*

Resistance is greatest at neutral pH and increases as the water activity of the food decreases. Vegetative cells growing near their upper temperature limit of growth are more resistant than those growing near their lower temperature limit. Cells growing in the exponential phase of growth are more sensitive to heat than cells growing in the other growth phases.

#### *Pasteurisation*

Pasteurisation is the term given to a relatively mild heat treatment that has only a slight effect on the physical and chemical properties of food. Food that is usually preserved in this way includes fruit juices and purees, milk and preserved meats such as hams. All potential pathogens encountered in raw food, with the exception of the heat-resistant spore formers *Bacillus cereus*, *Clostridium perfringens* and *C. botulinum*, are killed by the mild heat treatment of pasteurisation. The index micro-organism for pasteurisation is *Mycobacterium tuberculosis*. This means that any treatment that will have destroyed all strains of this bacterium will also kill other pathogens. The present pasteurisation standard provides a safety margin of 28.5 min at 61.7°C and 15 s at 71.7°C.

The low temperature holding (LTH) method (61–62°C for 30 min) was most widely used in the 1920s and 1930s. Technological improvements have led to the high temperature short time (HTST) process (71.7°C for 15 s). Processes that employ temperatures above those of pasteurisation are known as ultra-high temperature (UHT) treatments (135–150°C for 2 s). Ultra-high temperature treatment will destroy all but the most heat-resistant spores and produce a product that is commercially sterile. If packed aseptically into sterile containers it can be marketed without refrigeration.

#### *Blanching*

Blanching involves heating the produce to a temperature of about 90°C or above for very short periods of time, typically ranging from 20 s to 2–3 min. It is a pretreatment that is usually applied to selected fruit and vegetables before the drying stage. As well as facilitating drying, blanching will also help in the retention of colour of the commodity or retain certain nutrients. Blanching will almost certainly reduce the levels of viable micro-organisms. The objective of blanching is to deactivate enzymes (Arthey & Dennis 1991) that may cause the deterioration of colour, flavour and texture during drying and subsequent storage (Rocha *et al.* 1993). It may be carried out by hot water scalding, steam or microwave heating.

With some commodities it is advantageous to add sodium metabisulphite or sodium carbonate to give a slight alkalinity to green vegetables in order to prevent undesirable colour changes caused by the breakdown of chlorophyll to phaeophytin. Salt and the pH level may also be adjusted for water blanching. Products are cooled rapidly after blanching to avoid excessive heating that may damage the product.

### *Canning*

Canning is a relatively modern technology which enables food to be preserved in an edible condition under a wide range of storage conditions for long periods – from a few months to several years. There are three different stages to the process: hermetically sealing the food in a container, heat 'sterilising' the sealed unit, and cooling it to ambient temperature for subsequent storage.

Canned foods have the potential of keeping almost indefinitely; the bacteria which cause deterioration are killed during the canning process. The canning process should aim at placing food in the freshest possible condition into the can, removing the air by a heat treatment and hermetically sealing the lid and sterilising the can by further heat processing. The second heat treatment must be carried out at temperatures around 120°C, using steam under pressure or in autoclaves. The preparation of the food and the canning operation itself vary according to the type of food and size of can.

The headspace is the space left in the top of the can to allow for expansion of the contents during the heating process. However, leaving air in this headspace causes considerable internal pressure during processing and leads to oxidation of the contents (surface discoloration and rancidity) and of the container (corrosion) during subsequent storage. The lid (or filler's end) is attached to the body of the can in a double-seaming operation. It is vitally important that the side seam and the double seams are completely hermetic. The former uses solder (98 parts lead to 2 parts tin) to complete the seal, while the latter is sealed by the melting and re-setting of a plastic sealing compound on the inside curl of the end-piece.

It is important that equipment used to form double seams is monitored regularly both for visible faults and for slackness, by measurement of can overlap.

Canning is always an expensive operation and it is therefore essential that careful consideration be given to the economics of production. In many less developed countries, the high cost of cans has hindered the development of canning industries.

### *Sterilisation*

For the purpose of determining the degree of heat treatment which is needed to sterilise food within a can, three pH groupings are recognised:

*Acid foods of pH less than 4.5* cannot support the growth of heat-resistant spore-forming pathogens such as *C. botulinum*. To effectively preserve such foods (e.g. most fruits and pickles), it is necessary only to destroy the relatively heat-sensitive, acid-tolerant micro-organisms that could otherwise grow and cause spoilage. A mild heat process is sufficient (e.g. the coldest point in the can should receive a minimum treatment of 5 min at 100°C).

*Medium to low acid foods of pH between 4.5 and 5.3* will support the growth of pathogenic heat-resistant spore-formers like *C. botulinum*, and are therefore processed to reduce the chance of such spore survival to virtual insignificance (e.g. the coldest point in the can should receive a minimum treatment of 10 min at 121°C).

*Low acid foods of pH greater than 5.3* will support the growth of organisms such as *C. botulinum*, as well as the germination and growth of highly heat-resistant spores like those of *Bacillus stearothermophilus*, which cause flat-sour spoilage. Fortunately, these spores will only germinate and grow at temperatures greater than 37°C and therefore often do not germinate in food stored at cooler temperatures. The severity of the heat process required would render the food unpalatable.

Most sterilisation heat processes are designed to eliminate *C. botulinum* spores. However, it can be seen from Table 8.7 that a process achieving 12 decimal reductions of *C. botulinum* spores (i.e. $12 \times 0.2 = 2.4$ min at 121°C) would only achieve approximately half a decimal reduction in the number of *B. stearothermophilus* spores.

**Table 8.7** Bacterial groups and their heat resistance.

| Organism | *D* value (°C) and time (min) |
|---|---|
| Thermophiles: | $D_{121°C}$ |
| *Bacillus stearothermophilus* spores | 4–5 |
| *Clostridium thermosaccharolyticum* spores | 3–4 |
| *C. nigrificans* spores | 2–3 |
| *C. botulinum* spores (types A and B) | 0.1–0.2 |
| Mesophiles: | |
| *B. coagulans* spores | 0.01–0.07 |
| | $D_{66°C}$ |
| Non-spore-forming mesophilic bacteria, yeast and moulds | 0.5–1.0 |

Once the temperature history of the process of a canned food has been plotted and the main spoilage organism identified and its *Z* value found, a graph of L versus time of processing may be plotted throughout the process. (The L-value or equivalent killing power is a function of *Z*-value and temperature, related to a reference temperature, Î,, which is 121.1°C or 250°F.) The area beneath this graph must exceed *m D* for commercial sterility, where *m* is the process factor. This value has been called the 'equivalent time' and is the *F* value.

Packs with pH 4.5 are generally processed to commercial sterility with reference to *C. botulinum*, the minimum order of process factor *m* being taken as 12. Thus, *m D* should be 12 × 0.3 = 3.6 min at 121°C at the cold spot. However, in other foods there are often spoilage organisms with more heat labile characteristics.

Pressure is applied by closing off the drain and steam exit valves in the retort while still allowing steam. Various petcocks are left open to allow any air, which may be admitted with the steam, to escape. Common processing temperatures are 115.5°C and 121°C. The pressure, and hence the retort temperature, is controlled by an automatic steam pressure control valve; this opens when the set pressure is exceeded and closes again when the pressure falls below that set.

### *Cooling*

The pressure in the retort is maintained after closing the steam inlet valve by admitting compressed air. If this was not done the high pressure inside the can would cause the cans to distort outwards (peaking), possibly damaging the integrity of the seams. While the retort pressure is being maintained with compressed air, chlorinated cooling water is admitted.

The cooling water is chlorinated because, at this stage, the sealing compound in the double seams is still molten and the vacuum forming in the headspace, due to condensing steam, could pull drops of cooling water through the double seams. If this cooling water contains viable micro-organisms this leakage may lead to 'leaker spoilage'. This type of spoilage is by far the most common in food poisoning attributed to canned food. Cooling water is generally recirculated and dosed automatically with chlorine. A residence time of at least 20 min between dosing and utilisation is necessary to allow the chlorine to take effect. The free residual chlorine content of the cooling water should be measured in water draining from the retort rather than in that entering the retort. Common chlorination levels lead to 5–20 ppm free residual chlorine in the drain water. Too high a level can lead to corrosion of the can.

As cooling proceeds, it becomes necessary to reduce the air pressure in the retort since the pressure inside the can falls with the temperature of its contents, eventually becoming a partial vacuum. If the pressure outside the can far exceeds the pressure inside, the can may buckle inwards (panelling) which could also damage the can seam. Improper or incomplete cooling of the cans could lead to 'stack burn', where hot cans are stored and the contents overcook.

When the cooling process has been completed (the can contents have reached a sufficiently low temperature and the retort pressure has been reduced to atmospheric pressure), the retort is opened and the wet cans lifted out. It is essential that the wet cans are not handled at this stage: the danger of contaminating the can contents via a leaking seam still exists. Cans should be conveyed mechanically to a can drier along chlorinated runways before they are labelled and packed into cases or shrink-wrapped.

In warm climates it is important to ensure the heat process is sufficient to reduce the levels of thermophiles to $<1$ in 100, or that the storage temperature is unfavourable for their growth. Canned food products should be stored under conditions which avoid sweating caused by extreme temperature fluctuations. This may result in rusting of the exterior of the can. This is particularly a problem in humid conditions.

### *Problems related to canned foods*

#### *Can lacquers*

Fish and meat proteins are rich in sulphur amino acids which, on heat processing, release hydrogen sulphide. This can react with iron in the tinplate producing black ferrous sulphide ('sulphur staining'). To avoid these unsightly black deposits, a special lacquer incorporating zinc oxide or zinc carbonate is used to coat the internal can walls. The hydrogen sulphide released now reacts preferentially with the zinc oxide or carbonate, producing white zinc sulphide which remains embedded in the lacquer so that an attractive internal appearance is maintained.

#### *Struvite*

In some canned fish products, glass-like crystals of calcium struvite may form on storage and become the reason for many 'foreign body' complaints in canned fish. This phenomenon may be avoided by the addition of small amounts of citric acid to the product prior to filling and processing. Citric acid complexes available calcium ions, thus preventing them from forming calcium struvite.

### *Refrigeration and freezing*

The lag phase of growth is extended as the temperature drops and growth becomes very slow below 4°C. The common food pathogens are mesophilic and are therefore sensitive to refrigeration, but growth of *Listeria monocytogenes* is an exception. Freezing causes cell death because of the rupture of cell membranes by ice crystals; reduced water activity is also a factor. Gram-negative bacteria are more sensitive to freezing than Gram-positive species. Yeasts and moulds are relatively resistant to low temperatures. The quality of frozen foods is good and the packaging cost low. However, the processing cost and the cost of maintaining a cold-chain between the factory and the consumer are high.

### *Drying*

The water concentration in foods is most usefully represented by the water activity, as detailed above. On this basis one can calculate the maximum safe water content (at 20°C) in terms of the moisture contents of common foods corresponding to an $a_w$ of 0.70.

One means of achieving these low moisture contents is by drying (forced-air, solar drying, drum-drying, spray-drying, etc.). The heat treatments involved cause some of the quality problems already referred to, but since the product is stable, it can be stored in a simpler package.

### *Salting and sugaring*

Another method of reducing water activity is to add low-molecular weight compounds (humectants) such as salt or sugar to the food. For example, a 3.4% salt solution has an $a_w$ of 0.98, and the $a_w$ values of 11.9% and 16.3% solutions are 0.92 and 0.88, respectively.

Gram-positive bacteria are more tolerant of reduced $a_w$ than are Gram-negative bacteria. Thus, as $a_w$ decreases from 0.98 to 0.93, Gram-negative bacteria are succeeded by Gram-positive species. The growth of most food-borne pathogens is prevented at the lower end of this range. The key exception is *Staphylococcus aureus*.

### *Modification of pH*

Most bacteria are sensitive to acidity; low pH is used to achieve direct inhibition, decrease in *D* and reduction in the viability of bacteria at low $a_w$.

The growth of many food-borne pathogens is inhibited at pH below 5.0 (e.g. *Bacillus cereus*). Neither *C. botulinum* nor *Vibrio parahaemolyticus* can grow below pH 4.5, while *Salmonella* spp and *Staphylococcus aureus* are controlled by ensuring the pH is below 4.0.

Most bacterial spoilage of foods of pH 4.5 is due to Gram-positive species (especially Lactobacillaceae). Acidification is often used in conjunction with food preservatives to prevent spore germination. Spore germination in canned foods can be predicted from the pH of the product. The pH of most foods cannot be changed without grossly affecting product acceptability. The principal example of the use of pH as a physical method of control is with fermented foods, the majority of which are lactic fermentations.

### *Smoking*

Smoking probably originated as a natural development of drying foodstuffs practised by early man. Smoke from fires flavoured and dried food hung up for storage. When the improvements in flavour and preservation time were noted, then smoking began to develop as a method of food preservation.

In Europe during the Middle Ages, food products were heavily salted and smoked for some weeks. These products owed their long storage life, at normal temperatures, to a high salt concentration and to very long smoking and drying times and, therefore, low water content. Modern transport and distribution facilities in industrially developed countries have greatly reduced the need for long-term storage of smoked and dried products and most are now available for the consumer within a few days of processing. Smoke curing, as a method of food preservation, has also lost importance in many countries due to rapid advances in freezing and cold storage techniques; most smoked products are only lightly cured in order to give them a mild savoury flavour. They will not remain in a wholesome condition much longer than fresh products when stored at normal temperatures and so should be stored under refrigeration.

In most tropical developing countries, however, smoking is used not only to impart desirable flavours but more importantly to accelerate the drying process. Smoked products in tropical countries have storage properties that enable them to be marketed without the use of sophisticated refrigeration systems. Smoking is often combined with a period of sun drying and/or preliminary brining. The temperature of smoking varies from place to place depending on consumer requirements and the type of smoking kiln or oven used. Most products, however, are hot smoked, that is, the temperature of smoking is sufficiently high to cook the product.

#### *How smoking preserves*

Smoking is a method of preserving food which combines three effects:

*Cooking*: If food is smoked at a high temperature, as with hot smoking, the flesh will be cooked and this will destroy the enzymes and kill bacteria;

*Drying*: The fire which produces the smoke also generates heat and this will not only cook but dry the food;

*Preservative value of the smoke*: The smoke produced from burning wood contains a large number of compounds, such as phenols, which will kill bacteria.

The long storage life of some smoked food products is due more to drying and cooking than to the preservative value of the chemical compounds deposited from the smoke. The burning of wood or sawdust to produce smoke is extremely complex since the smoke is the result of incomplete combustion and this will vary with the source of the fuel and the ventilation of the fire. A slow burning fire will produce much more smoke than an intense fire. Wood smoke is a mixture of gases, vapours and droplets. Droplets form the visible part of the smoke although the invisible vapours contribute to the characteristic smell. It has been shown that mainly the vapours are taken up by fish or meat during smoking. The substances in the vapours dissolve in the liquid on the surface of the product and the rate of uptake depends on the moisture on the surface and the rate of flow of the smoke. In the processing of traditional hot-smoked fish products in developing countries it is the heat of the fire rather than the smoke which is of most importance.

Smoked fish can be divided into two general categories:

*Cold smoked*: During the smoking process, the temperature at no time rises to a level where the flesh is cooked sufficiently to denature protein. In practice, this means a maximum temperature of approximately 30–40°C. It is mainly carried out in temperate climates. Tropical fish species can be cold smoked at higher temperatures than temperate species due to the fact their proteins are denatured at higher temperatures.

*Hot smoked*: During the smoking process, the flesh is cooked. Traditional smoking in tropical countries falls within this category.

### ***Irradiation***

#### *Ultraviolet light*

This is only a partial method of control used, for example, in conjunction with refrigeration to extend the shelf-life of fresh meat (but not fatty meat such as pork, because of rancidity problems). Gram-negative bacteria are more susceptible than Gram-positive species. Yeasts have a similar resistance to bacteria, whereas moulds are much more resistant.

#### *Ionising radiation*

This has four advantages:

- highly lethal even at low levels such as 5 kGy at which organoleptic changes are undetectable;
- no residues of non-food material;
- negligible heat production; and

- penetration of the irradiation is uniform and instantaneous and so the process is easier to control.

However, irradiation equipment is expensive and requires costly shielding, possible mutagenic effects of irradiated foods have been reported, and further processing may be necessary to inactivate endogenous enzymes. Gram-negative bacteria are more sensitive than Gram-positive ones. Yeasts are more resistant than moulds, which have a similar resistance to vegetative bacteria. Bacterial spores are more resistant than their corresponding vegetative cells by a factor of about 5–15.

### *Controlled atmosphere*

Elimination of oxygen favours anaerobes and facultative anaerobes such as Lactobacillaceae, Enterobacteriaceae, Corynebacteriaceae and *Clostridia* spp, and prevents mould growth.

The inclusion of carbon dioxide inhibits the growth of some micro-organisms by a mechanism which is not clearly understood. Moulds are sensitive while yeasts are relatively resistant. Bacterial susceptibility varies considerably, but Gram-negative species are generally more sensitive than Gram-positive types. Packaging of fresh meats in gas-impermeable plastic is an important application of this method.

Modification of atmosphere can occur at three levels.

(1) Controlled atmosphere: mainly bulk storage or transport. Gas composition, humidity and temperature are controlled to give optimal storage conditions for long-term storage of fruit, meat and other foods.
(2) Gas packing: bulk and retail packs. Packs are flushed and filled with nitrogen or other gas mixtures. Gas composition may be changed as a result of pack permeability, chemical and biological activities.
(3) Vacuum packing: mainly retail packing. The original air atmosphere is evacuated and the atmosphere that develops is mainly the result of biological activity occurring within the food.

### *Food preservation by chemical methods*

While the human economy and way of life remained agricultural and rural, traditional methods of food preservation met our needs, such as drying, salting, syruping, pickling, fermentation and smoking. These methods were established by early civilisations, having evolved empirically from chance discoveries.

The situation was changed dramatically in Europe and North America by the Industrial Revolution that stimulated the development of large-scale food manufacturing and distribution. Initially, this encouraged widespread adulteration of foods with chemicals to add bulk and colour and to disguise poor-quality materials and unhygienic processing. The legal position on food additives was not clearly defined in the UK until 1925. Prior to this date chemicals in common use included boric acid, carbolic acid, creosote, formaldehyde, salicylic acid, thymol and alum.

Great technical advances have occurred over the last century with physical methods of food preservation, but these methods are not satisfactory for all types of food for three reasons:

(1) Organoleptic and nutritional changes caused by these methods limit their applicability.
(2) The cost of these methods limits their application in many tropical countries in which microbial growth rates are increased by the favourable climatic conditions.
(3) The advent of convenience foods which need minimal preparation and cooking. Here the use of preservatives may help to extend the shelf-life of refrigerated or mildly heat-processed or intermediate moisture foods.

Chemical preservatives are nearly always used after or in combination with physical methods of preservation. Consequently we have to bear in mind the nature of the micro-organism likely to survive these physical methods before selecting the appropriate chemical preservative.

Although chemical preservatives have been used in the past to cover up the use of spoiled foods and unhygienic processing, the types and levels presently permitted would not prevent rapid decomposition if the growth of spoilage organisms is already advanced. In other words, chemical preservatives cannot replace good manufacturing practice.

### *Food additives*

Preservatives are food additives that check or prevent the growth of micro-organisms in food, and the term usually only refers to particular chemicals such as benzoate, sorbate, sulphur dioxide, nitrite, propionate and parabens (para-hydroxy benzoic acid series).

#### *Legislation on the use of preservatives*

Legislation aims to protect the health of the public and

to prevent fraud. Legislation has followed investigation of the toxicity of preservatives in terms of their acceptable daily intake. This is usually one hundred times less than the 'no effect level' in long-term feeding trials. Most countries have a list of permitted preservatives and no other preservative may be used.

Most countries also restrict the use of permitted preservatives by specifying that only certain named foods may contain them, and by laying down maximum use levels. In the UK for a new preservative to be allowed it must be demonstrated that there is a need for it and that it is safe at the concentrations used.

### *Desirable properties of food preservatives*

There are seven requirements for food preservatives:

- no toxicity problems;
- microbiocidal rather than microbiostatic properties;
- must be stable in foods (especially if only microbiostatic);
- the spectrum of activity should correspond to the spectrum of micro-organisms likely to appear in the food;
- must not stimulate the development of resistant strains of micro-organisms;
- chemicals used therapeutically are not recommended as food additives; and
- an assay procedure should be available.

### *Selection of preservatives*

Table 8.8 shows how permitted preservatives match these criteria.

### *Factors affecting preservative action*

No single preservative matches up to the ideal requirements. Therefore, for a particular application one must attempt to optimise the desirable characteristics of the available preservatives. Effectiveness depends not only on its inherent properties (antimicrobial activity, solubility dissociation constant and reactivity) but also on its microenvironment ($a_w$, pH, fat content, type of packaging, process and storage). The number, type and condition of the food microflora also affect the efficacy of preservatives.

The pH of the food is an important factor, not only because it determines the types of micro-organisms present, but also because it has a direct effect on the efficiency of preservatives. Efficiency is dependent upon the proportion of undissociated acid present, which is determined by the dissociation constant of the preservative as well as the pH of the food.

### *Adverse effects*

Concern has been expressed about the adverse effect of various preservatives on health. In the UK this has led to some demand for preservative-free foods, but this is not possible for every product.

*Benzoate*: People who suffer asthma or who have recurrent urticaria are likely to be sensitive. It may cause gastric irritation if consumed in large quantities.

*Parabens*: Allergic reaction in people who suffer asthma or who have recurrent urticaria or are sensitive to aspirin. May be skin sensitivity and/or numbing effect in mouth.

**Table 8.8** Suitability of some preservatives.

| | Preservative | | | | | |
|---|---|---|---|---|---|---|
| Criterion | Benzoate | Parabens | Sorbate | Propionate | Nitrite | Sulphite |
| Toxicology | + | + | + | + | ? | + |
| Microbiocidal activity | — | — | — | — | (+) | (+) |
| Stability | + | + | + | + | — | — |
| Activity spectrum | | YBM | YBMb | YBMb | MB | B |
| YBMb | | | | | | |
| Appearance of resistant strains | — | + | — | + | + | — |
| Not used in medicine | + | + | + | + | + | + |
| Assay available | + | + | + | + | + | + |

Y = yeasts; M = moulds; B = Gram-positive bacteria; b = Gram-negative bacteria.
+ = suitable; – = not suitable; ? = not known; (+) = marginally suitable.

*Sulphur dioxide*: Causes irritation of alimentary canal. Can be dangerous to asthmatics.
*Propionate*: Recently associated with carcinoma in rats.
*Nitrite*: Reacts with amines to form nitrosamines, which are potentially carcinogenic. Unsuitable for babies (blue baby syndrome).
*Sorbic acid*: Generally acceptable.

*Typical applications*

*Benzoate and parabens*: jams, beer, dessert sauces, fruit pulp, fruit juice, marinated fish, pickles, fruit yoghurt, coffee essence.
*Sorbate*: yoghurt, fermented milks, sweets, soft drinks, various cheeses, certain cakes, candied peel, wine and cider, soup concentrates.
*Propionate*: bakery and dairy products.
*Nitrite*: various meats and cheeses.
*Sulphur dioxide*: many foods including fruit juices and purees, soft drinks, dried banana and apricots, beer, wine, cider, desiccated coconut, tinned crabmeat, powdered garlic, frozen mushrooms, sausage meat.

Dependence on a single preservation method is not advisable for many foods, because of adverse effects on organoleptic quality, cost or toxicity. This has led to extensive research into the interactions between preservatives and physical methods of food preservation, and between the different preservatives. In many cases the effects of each component preservative/process are additive, each comprising a hurdle to growth (the 'hurdle effect').

*Minor food preservatives*

The principal food preservatives are the ones listed above, but other chemicals that have a preservative role under some circumstances include antibiotics (e.g. nisin), sequestrants and chelators (EDTA), phenols and related compounds (antioxidants), alcohol, carbon dioxide (fruit and vegetables, drinks, packed meats), lactic acid bacteria (lactic acid, acetic acid, diacetyl, carbon dioxide, bacteriocins, hydrogen peroxide) and sterilising agents (ozone for water, ethylene oxide for spices).

## Packaging

The discussion that follows is concerned with commodities traded on a commercial scale. It is not meaningful to analyse packaging in relation to the small-scale farmer, whose product may be largely stored and consumed on the farm, or exchanged or sold in a local market by the farmer him- or herself. Although containers of some kind will be used in such cases, these will usually be multipurpose and recycled; for the short times and distances involved, the properties of the containers are of little importance.

Bulk handling and transport is technically feasible for many raw commodities, and for some products of primary processing. For efficient operations, bulk handling usually requires specialised facilities and appropriate vehicles, perhaps dedicated to specific products. Consequently an adequate scale of throughput is required to make bulk handling cost-effective, and these systems are used particularly for international trade and for supply to large-scale (industrial) customers. Examples of the use of bulk systems include the delivery of cereal grains to millers, flour to bakeries, edible oils to food manufacturers, and fresh fruits to processors. Apart from the constraints of scale, the application of bulk systems may be restricted by the characteristics of products.

Packaging provides a more generally applicable method of facilitating the exchange of goods. Three basic functions of a package are generally recognised:

(1) To contain the product, enabling the required quantity to be handled as one unit without loss, throughout transport and storage; effective performance of this function aids good management and is a necessary precondition for other functions.
(2) To protect the product during distribution; effective protection from the hazards of transport and storage reduces the mechanical stresses to which the product is exposed, and reduces the rate of loss of quality through biological, physical and chemical effects of the environment.
(3) To inform, by enabling the indication of nature, quality, grade, origin, destination, and other features of the product. This is particularly significant in relation to advertised branded products, especially when sold in supermarkets, but can also aid management and marketing in other situations.

The relative importance of these functions depends on the product and the market concerned.

Packages cover a wide range of sizes; it is convenient to consider three classes, although their boundaries overlap:

*Intermediate bulk* – units of 200–1000 kg, necessarily handled with mechanical aids.

*Transport, market or wholesale pack* – for distribution to retailer; the weight is usually in a range suitable for manual handling, typically 10–50 kg, but may extend from 2 kg to 100 kg. Packs for the institutional market usually also fall into this range.

*Consumer or retail pack* – units of retail sale, about 10 g–15 kg according to product.

Furthermore, these categories are not mutually exclusive; in most instances, retail packs are collated into a transport pack for delivery to the shop and, in industrialised countries, transport packs are commonly assembled into a unit load on a pallet, requiring mechanical handling in the same manner as an intermediate bulk container.

In selecting a package, or package system, several requirements should, in principle, be taken into consideration. The characteristics of the product to be packed are of primary importance. Its physical state determines the need for containment; its fragility determines the need for protection against the mechanical hazards of impact, vibration and compression. The chemical and physiological properties of the product, in particular the relationship with water vapour and oxygen, are critical in determining appropriate packaging to maintain its quality. The process or packing operation may be a major influence on the choice of package, especially when a new product is to be handled by an existing facility. Whether a simple manual activity or a highly mechanised system is involved, some constraints are likely to apply. Special requirements are imposed on the package where a preservation process is closely linked to the packing operation, such as in food canning and aseptic packaging of heat-sterilised foods.

The specific market in which the product is to be sold can impose many requirements. Usually the market will have some expectation of the general type of package appropriate for a product, and a new supplier will often have to conform to this. There will frequently be both legal and customary requirements for the information to be given on the package. The graphic design on the package is usually important, especially for processed products, and particularly so where marketing of the product is linked to an advertising campaign. In recent years the issue of recycling or disposal of packages after use has become important especially in European markets. Conditions during transport and storage may also impose constraints on the package; unusually rough handling, high humidity or long storage time might require special attention.

The cost of packaging is also a constraint. To the farmer, the cost of a package may be seen as simply an expense, to be minimised as much as possible; to the manufacturer of branded processed foods it is perceived as one component of marketing costs. The objective should be to seek not the lowest cost package but the package which will minimise the total costs of package filling, storage, transport and distribution in a particular operation, taking account of the value of lost or damaged product, while also optimising sales of the product. This optimum package will (in principle) be optimum only for a specific system – there is no universal 'best' package. The cost of a package can be very variable, depending on source, the detailed specification, and the quantity of packaging material purchased in a single order. With manufactured foods, package costs falling in the range of 5–10% of the value of the product are common; however, with certain products, in particular fresh fruits and vegetables, the package cost is usually greater, perhaps exceeding the value of the produce to the grower.

### Packaging materials

Many different raw materials are used to make packages, and a number of these are used in several different ways. Industrial processes on a substantial scale are involved at some stage of manufacture, for most packages. Consequently, to obtain the benefits of scale, there is commonly some degree of standardisation, whether through official standards or trade practices. Nevertheless, many packages are supplied according to the specification of individual packers, if only in surface design. In general, the manufacture of packages is carried out separately from the packing operations of filling and closing. The principal materials and the ways in which they are used for packages are discussed below.

#### *Wood*

Wood is rigid, strong, and (in the short term) relatively unaffected by moisture, and consequently is used to make boxes (nailed, wire-stitched and wire-bound) and pallets. Species used include softwoods such as pine and low-grade hardwoods such as poplar. Plywood, hardboard and wood chipboard also find some application in box manufacture. Coopered barrels were formerly important general-purpose packages but now their use is limited to certain alcoholic beverages.

#### *Natural fibres*

Mainly jute, kenaf and sisal are spun to yield coarse

strong yarns. These are woven into cloth which is then fabricated into textile sacks; the spun yarn is also used as sewing thread. Cotton fibre results in finer yarns and fabric which can also be formed into sacks.

### *Paper and board*

Paper is used in several different forms. Strong papers (especially 'kraft') are used principally for making bags of various styles and multiwall sacks; paper weights are typically in the range 30–90 g/m$^2$. Fine papers are used for labels, and as a component in flexible laminates to provide stiffness, opacity and printability. Plain cartonboard is used for general-purpose folding cartons. For special applications, such as cartons for liquids, board may be coated with polyethylene or incorporated into multilayer laminates. Corrugated fibreboard, in the form of boxes and trays, is the most important material used for transport packaging for all kinds of goods. The component materials and construction of corrugated board may be varied to provide a wide choice of specifications, with individual layers ranging from 112 to 400 g/m$^2$. The majority of the commonly used styles of box are described in the International Fibreboard Case Code. Solid fibreboard has similar uses to corrugated fibreboard, and is also used to make drums with fibreboard or steel ends.

### *Steel*

Steel is used in the form of mild steel sheet with coatings of paint or lacquer as required. It is used for drums and pails and for closures for jars and bottles; steel is also used for staples and nails. Cans for processed foods have traditionally been made from tin-coated mild steel but alternative inorganic coatings are now also used; both are often further coated with lacquers. Stainless steel is used occasionally for specialised drums. Aluminium is used for barrels, drinks cans and for flat food cans, especially for fish, also for closures for jars and bottles. Rolled into a thin sheet (foil), aluminium is used for semi-rigid trays and, in gauges of 0.006–0.012 mm, as a component of flexible laminates to provide opacity and a high barrier to water vapour and gas transmission. The process of vacuum deposition is used to apply an extremely thin coating of aluminium to certain plastic films to provide a mirror-like appearance and enhance their barrier properties when incorporated in laminates.

### *Glass*

Glass is used for jars and bottles; these require a closure of metal or plastic to form a complete package. Green or brown glass is usually used when the package contents require protection from light, although other colours can be produced.

### *Plastics*

Plastics are widely used for package manufacture. Most are thermoplastics, although there is limited used of thermosetting materials for bottle closures. Processes for forming plastics materials into packages include:

*Injection moulding*, in which the plastic in a fluid state is forced under high pressure to fill a mould cavity of the required shape; used to produce pails, crates, closures and minor components;

*Blow-moulding*, in which the plastic in a fluid state is first formed into a tube closed at one end and is then inflated to conform to the inner surface of a mould cavity of the required shape; used to produce drums and bottles;

*Extrusion*, in which the plastic in a fluid state is forced under high pressure through a straight or ring slot die to yield a semi-rigid sheet or a flexible film. Multilayer sheets can be produced by feeding two or more plastics into a suitably designed die. The sheet or film may be subjected to a variety of further processes, including:

- biaxial orientation, in which a film is stretched in both directions under carefully controlled temperature conditions, thereby improving many of its physical properties;
- coating with other polymeric compounds in solution or aqueous dispersion (in particular, vinylidene dichloride copolymer, PVDC) to confer heat sealability and increase the resistance to gas and water vapour transmission;
- laminating, by joining the film to one or more other components, which may be other films, aluminium foil or paper, using some kind of adhesive (which might itself be a plastic);
- heat-sealing to form a film into a package, by softening a small area of the film by heating, bringing two surfaces together, and allowing the plastic to cool. For coated films and multilayer flexible packaging the seal will usually be made using

only one surface layer, whereas for single films the seal will usually be a weld involving the whole thickness of the film; different sealing equipment is used in these two cases; and

- thermoforming, in which a sheet is heated to soften it and then forced by air pressure (in some instances with mechanical assistance) to conform to the inner surface of a mould of the required shape.

The principal plastics, their leading properties and major uses in packaging are listed below.

*Polyethylene, low density (LDPE)*: cheap, fairly tough, low water vapour but high gas transmission rates, excellent heat sealability – used for sacks, bags, shrinkwrap, stretchwrap, and as a component in most multilayer flexible packaging, especially for sealing, also for some bottles and caps.

*Polyethylene, high density (HDPE)*: rigid, tough, low water vapour and gas transmission rates, good heat sealability – used for drums, pails, crates, bottles, tubs, caps, bags, especially boil-in-the-bag packs, sacks woven from film tapes.

*Polypropylene (PP)*: cheap, rigid, tough, good appearance, low water vapour and gas transmission rates – used for pails, crates, bottles, tubs, punnets, caps, bags and wraps (several different types of film are available), component of laminates, sacks woven from film tapes.

*Polyester (PET)*: tough, good appearance, low water vapour and gas transmission rates – used for bottles, punnets, blister packs, bags (usually as component of laminates).

*Polyvinylchloride (PVC)*: rigid, low water vapour and gas transmission rates but can be formulated as soft film with high transmission rates – used for bottles, punnets, nestpacks, blister packs, and the soft plasticised film is used as shrinkwrap and stretchwrap.

*Polystyrene and styrene co-polymers*: rigid, good appearance (most types) – used for tubs, punnets and nestpacks. As expanded polystyrene the plastic has low density, low thermal conductivity, moderate rigidity – used for boxes, trays, nestpacks.

*Polyamide (nylons)*: tough, low gas transmission rates – used in flexible laminates.

### *Mixed materials: plastics and laminates*

The retort pouch is perhaps the best-known application of laminate packaging. The retort pouch is a heat-resistant flexible replacement for the metal can. The pouch is a laminate of the following three layers:

(1) An outer layer – gives strength and protection to the pouch and allows for labelling – polyester film is used as the outer layer.
(2) The core – acts as the barrier to gas, moisture and light – aluminium foil is a suitable material.
(3) The inner layer – it is important that this layer is heat-sealable – polypropylene is a suitable material for the inner layer.

These layers are bonded together by high performance adhesives. Retort pouches must be able to withstand temperatures of 112–125°C for up to 120 min. They must also be strong enough to withstand processing operations and post-process handling.

### *Packaging of specific commodities*

#### *Cereal grains*

For intermediate bulk, woven polypropylene 'big bags' of 1–1.5 $m^3$ capacity may be used. The transport package is always some form of sack. Woven sacks of jute or similar fibres, or of woven PP, are usual in areas where bulk systems do not dominate, with contents usually in the range 50–100 kg. While both types of material can provide a sufficiently strong sack for effective containment, jute is generally regarded as easier to handle and stack, and more suitable for repeated use. The closely woven PP fabric presents somewhat greater resistance to water vapour transmission and airflow; in some circumstances this may be an advantage, but deterioration of maize in large stacks has been attributed to reduced air movement through convection following the use of PP sacks. Whatever the technical merits, the PP sacks are increasingly preferred on grounds of availability and lower cost. For certain markets, such as retail animal feed, multiwall paper sacks may be used; here the weight is likely to be 25 kg.

The retail market for cereals as grains is dominated by rice, for which a wide range of package weights, from 0.5 to 10 kg, may be found. Most packages are film bags, of LDPE or transparent laminates, generally produced on form-fill-seal machines. Some advertised brands and special types of rice are sold in cartons of up to 1 kg; in some instances the rice may be enclosed in separate portions in 'boil-in-the-bag' packs usually of perforated HDPE film. The demand for retail packs of other wholegrain cereals is limited; while film bags are customarily used, these are more likely to be filled semi-automatically.

*Cereal products*

Flours and meals, like raw cereal grains, are microbiologically stable at most ambient conditions. However, they are closer to being consumable foods, making hygiene more important. Consequently packages often provide more protection against contamination than those used for grains. The customary transport package is still a sack; although good quality jute sacks have often been used, cotton would be preferred. Woven PP, perhaps with a bonded liner of non-woven rayon fabric, is also used, but in industrialised countries multiwall paper sacks are customary. For example, valve sacks of 35 kg capacity are used for flour in the UK, while 25 kg open-mouth sacks with glued closure have been used in large numbers for food aid shipment of a range of blended and fortified cereal products.

Retail packaging of these products is often similar to that used for grains. However, in high-speed mechanised packing of flours, removal of air entrained in the flour is easier if a porous package is used, and consequently much flour is packed in paper bags of 0.5–3 kg.

Packages for cereal products which are further processed are variable, depending on the characteristics of the product. Dry pasta is stable, and may be wrapped in paper, bagged in plastic film, or cartoned. Crisp breakfast cereals require some protection against uptake of moisture from the atmosphere, and are often very fragile; the majority are therefore packed in a film bag enclosed in a carton. Bread and cakes sold at ambient conditions are generally short-life products, and a package is used mainly for hygiene and brand identification, although some restriction of moisture loss may improve shelf-life. Packages for the simpler products are usually wraps or bags of LDPE or PP film, often perforated.

*Pulses, nuts and seeds*

This group of products, whether raw or processed by shelling or hulling, are generally packed in the same ways as cereals at all levels. One exception is cashew kernels; because of their high value, these are protected in the wholesale pack by using an inert atmosphere, requiring either a tin or a strong high-barrier laminate bag as the package. Processed foods made from these products include nut butters. These require a package which will resist the oil and be appropriate for the pattern of use. Glass or plastic (especially PET) jars with screwed lids are very suitable, and plastic tubs may also be used; tins would be technically ideal but probably unacceptable in most markets. Snack foods are prepared from many of the seeds by roasting or frying and adding salt or other flavouring; these must be packed to retain their low water activity. Bags made of low permeability materials such as aluminised PET/LDPE or PVDC/PP/LDPE are typical. Where a longer shelf-life or a distinctive market position is required, tins or glass or PET jars may be used.

*Oils and fats*

Owing to their low surface tension, liquid oils can penetrate through very small gaps; package closures must be carefully chosen if the product is to be contained without any seepage. Packages should also preferably restrict the access of oxygen and light to avoid promoting oxidative rancidity. Mild steel drums are suitable for capacities of 25–200 litres; HDPE drums are also used. Drums in the 5–25 litre range, important in the catering market, are usually of either tinplate or HDPE. A greater variety of packages is found in the retail market. There is some use of tinplate cans (round or rectangular); sometimes plain seamed ends are used, requiring puncturing to remove the contents, but more frequently there is a spout and screw cap or other closure. Glass bottles with metal or plastic screw caps are used particularly for higher-valued oils. Plastic bottles are widely used; formerly these were of PVC, but PVC has now been largely replaced by PET.

Solid fats for consumption as foodstuffs have a melting point not far removed from ambient temperature. For large-scale packages they are likely to be treated as oils. For retail sale, sealed tins are commonly used when high ambient temperatures could be encountered and the availability of refrigeration cannot be assured. In cool climates or under refrigeration block fats are simply wrapped in greaseproof paper or in a paper/aluminium foil laminate. Softer fats formulated for spreading are more appropriately packed in plastic tubs, usually thermoformed from styrene co-polymers with a heat-sealed foil closure, a plastic press-on lid, or both.

*Spices*

While fresh spices such as chillies and ginger are treated as horticultural crops, the dried products are durable commodities. For transport and storage prior to processing, most spices are packed in sacks in the same way as cereals or oilseeds, despite their much higher value. One exception is vanilla; the pods would be damaged by loose packing into a sack, and this spice is traditionally packed in boxes. Packages used for retail sale of both whole and ground spices include lever-lid tins, glass jars with metal or plastic caps, moulded plastic pots, paperboard containers with metal ends, and plastic film bags. The rigid packages are used particularly for smaller quantities – say 10–100 g – while bags are usual for larger packs.

*Beverage crops*

Green coffee and cocoa beans are packed for transport and storage in sacks, in the same way as cereals. Owing to the critical importance of flavour with these crops, special attention may need to be given to the specification of the sack. Thus in some producing countries sisal sacks have been preferred to jute for coffee, because of their lower odour, and some jute sacks have been found unsuitable for cocoa because of their oil content. For tea, the 50 kg plywood chest with metal edges and foil lining, which was customary in most producing countries, has been replaced to a considerable extent by multiwall paper sacks incorporating a barrier layer of aluminium foil laminate, polypropylene film, or polyethylene. The paper sacks are, however, only considered satisfactory if shipped as pallet-loads in containers. In retail markets, coffee may be sold as roasted whole beans or roasted and ground. With both forms it is necessary to protect the product from oxygen, and packages usually comprise tins or flexible laminate bags, vacuum-packed or flushed with inert gas; roasted whole beans may require special attention to accommodate the release of carbon dioxide from the product. Most cocoa products sold at retail are beyond the scope of this discussion; the relatively small amount of cocoa powder marketed is packed in lined cartons, composite containers or tins. Tea is retailed in packs of about 100–500 g of loose tea or, increasingly, in packs of 20–200 infusion bags each containing around 2 g of tea. Paper bags (sometimes two-ply and perhaps incorporating a barrier ply of a flexible laminate), lined cartons and reclosable tins are customary packages for loose tea. Infusion bags are made of a porous paper, based on natural or synthetic fibres, and can be formed by heat-sealing or by folding and stapling; they may have a string and tag attached, to allow easy removal after infusion. Some packaging machines simultaneously also form a paper envelope around the infusion bag. The infusion bags are most appropriately packed in cartons, usually with a transparent film overwrap.

*Fresh horticultural produce*

Fresh horticultural produce differs from the commodity groups considered above in several respects; in particular, these products have relatively high respiration rates and are easily damaged. Packaging should therefore take into account the need for gas and heat exchange in addition to providing protection against impact, crushing, puncture and abrasion.

Intermediate bulk containers of about 250–500 kg are used in two contexts. For transport from the field and for long-term storage, for example of apples, pallet bins (box pallets) of wood or plastics are used. Commodities exported for processing or repacking may be shipped in fibreboard boxes on a pallet base.

Market packages include sacks, baskets and boxes. Sacks may be used to contain from about 10 kg to 50 kg of produce. They do not stack well, and provide very little mechanical protection to the produce. They are useful for produce which is fairly resistant to mechanical damage, for low-value produce, and for produce which has a low bulk density, to keep packaging costs as low as possible. Woven sacks, made of jute, sisal and other coarse natural fibres, or polypropylene tape are used principally in less industrialised countries. Open mesh sacks made of plastic tape or filament allow more visibility of the product, and are often preferred; in particular, this is the package of choice for bulb onions. Paper sacks are used principally for potatoes, for which exclusion of light is a particular advantage, and for root crops.

Baskets are employed mainly for local marketing of fruit and vegetables in less industrialised countries. They are usually cylindrical or conical, as a result of the method of manufacture; they are therefore difficult to stack, utilise space poorly, and may be inconvenient for arranging the contained produce. Stacking strength is never very good. Prevention of damage often depends on the use of a lining of a soft material, often leaves, grass or paper.

Boxes may be made from several alternative materials. In international trade, and in many domestic markets, corrugated fibreboard is the dominant material; among the many box designs used, the two-piece full telescope box (IFCC code 0320) has been especially popular for the important fruit trades. With the widespread adoption of palletised transport, the use of open packages has become acceptable, and fibreboard trays (especially machine-erected styles), which can be used to display the produce at the retailers, are now common. Folder styles have been customarily used for the smaller boxes needed for airfreighted off-season and exotic produce. Wooden boxes provide the most important alternative to fibreboard. Nailed boxes are used especially as multi-trip packages for internal trade, while the more lightly constructed wire-stitched boxes are more appropriate for export. In addition to natural wood, there is limited use of plywood, hardboard, and chipboard. A major advantage of wood is its retention of rigidity when wet, while the difficulty of disposal or of recycling the material has become an obstacle to its use in certain countries. Plastic boxes or crates made of PP or HDPE by injection-moulding are relatively expensive, and multiple use is essential. Conditions which allow their economic use for trade passing through a wholesale market are uncommon, but they are widely employed for deliveries of both loose and pre-packed produce to supermarkets.

Retail packages are found in several different categories, dependent both on the characteristics of the produce and the preferences of the individual retailer. In most instances the packs are intended to provide some restriction in the loss of moisture from the commodity while leaving the gas composition of the atmosphere in the pack unchanged; however a small proportion of packs are designed to modify the gas composition, with the objective, when combined with adequate temperature control, of reducing the rate of respiration and thereby increasing shelf-life. For items of produce which are purchased as single units and which are large enough to handle individually, and such items as bundles of leafy vegetables, direct film wraps are common. Close-fitting wraps may be applied using stretch film, heat-sealed or relying on the inherent adhesion of the film for closure, or using shrink film, which is applied loose, heat-sealed and shrunk by heat. Loose-fitting wraps are usually sealed with adhesive tape. Where a pack to contain a number of pieces is required, a tray or punnet may be used. Shallow trays are normally overwrapped with transparent film; deep punnets may be left uncovered, or covered using film, plastic netting or a lid of the same material as the punnet. For packs to hold a number of pieces of less fragile produce, bags of LDPE or PP film are suitable. These may be unsealed, for certain products such as lettuce and celery, or, more commonly, closed but with a few holes to allow gas exchange. Automatic machines can form and fill bags from reeled film, including printed film, at rates typically 40–60 packs per minute. These bags can be reliably sealed, and allow the possibility of modifying the atmosphere in the package. Paper bags (small sacks) are used for larger retail packs of potatoes. Net bags are commonly used for citrus fruits, nuts and Brussels sprouts and occasionally for other produce. Cartons are sometimes used for a few types of produce, notably garlic and cranberries. Small versions of market containers find some use for gift-packs of fruit, but these are larger than the usual retail pack.

### *Dried fruits and vegetables*

Traditional dried fruits such as raisins and prunes are shelf-stable; fibreboard or wooden boxes, with linings of paper or plastic film for hygiene, are used as large-scale packages. Other dried fruits, usually prepared as cut pieces, may be prepared at lower water activities and require a sealed package providing effective protection against water vapour uptake. For retail sale, the conventional products are usually offered in film bags or in lined or coated cartons. Higher moisture products, incorporating preservative and sold as 'ready to eat' or 'no soak' versions, and the low-moisture products both need more protective packaging to maintain quality, and flexible laminate bags are the usual package.

Most dried or dehydrated vegetables of commerce are produced as small cut pieces or as powders, and require moderate protection against moisture uptake. Typical trade packages contain 10–50 kg, and comprise a fibreboard box with a sealed liner bag of LDPE film or a flexible laminate; tins might be used for powders such as onion or garlic. At retail level, potato powder (instant mashed potato) is the most important product. To maintain quality, it is preferably packed in a nitrogen atmosphere and consequently a high-barrier package is needed, so tins or laminate pouches are usually employed. Pouches may be enclosed in a carton. Other single vegetables may be packed in film or laminate bags or in plastic or glass containers. Dry formulated foods incorporating dried vegetables, such as soup mixes, are usually packed in laminate pouches.

### *Juices and purees*

Juice concentrates and fruit and vegetable purees (notably tomato paste) require preservation even if they are to be further processed. The choice is between freezing, with transport under refrigeration, and heat treatment followed by aseptic filling into a package, using ambient transport. Drums of HDPE or steel (stainless or appropriately lined), usually about 200 litres capacity, are used for frozen product. Aseptic packing of heat-treated product into various types of package can be achieved with several proprietary systems. Alternatives include steel drums, sterilised *in situ* using steam, and flexible laminate bags, sterilised using radiation, enclosed in a fibreboard box or pallet bin.

The retail market for juices, and the related nectars and juice-based drinks, divides into long- and short-life sectors. For long-life products, aseptically filled laminated cartons are used in sizes from 150 ml to 1.5 litres, and glass bottles with metal (crown or twist-off) closures include a similar range. Cans are used particularly in small sizes, and there is limited application of plastic bottles. Short-life products, sold from refrigeration, are packed in cartons, plastic bottles (PET or HDPE) or glass. Tomato paste is sold canned, and in collapsible (aluminium) tubes, while lower viscosity products are packed in the same ways as long-life juices.

## Flour

The *Oxford Dictionary* gives the following definition of flour: 'Fine meal or powder made from grain after the bran has been sifted out, used in cooking.'

The western world generally thinks of flour as being made from wheat and used for bread, cake and biscuit manufacture. However, as the definition states, flour can be made from any cereal grain and may be used in a wide variety of ways. Cereal grains are used throughout the world as a major source of energy in the daily diet of both humans and animals. They include wheat, barley, rye, oats, maize, rice, sorghum, millet, fonio and teff and are used in a wide, and ever-increasing, variety of foodstuffs from simple, whole, cooked grains to complex processed foods. These include bread, biscuits, cakes, chapatis, tortillas, pasta, thick and thin porridges, baby foods, special diet foods, breakfast cereals, glucose drinks and alcoholic beverages.

In order to understand flour, its uses and nutritional role, it is necessary to examine the structure of cereal grains and how they are transformed from whole grains of low digestibility (for humans) into an attractive food product.

### Grain structure

Cereals are the seeds of cultivated grasses. The grain is the seed from which a new plant could grow if given the right conditions. It is therefore a food store for the emerging plant and is a rich source of nutrients. Each species of cereal has unique structural (morphological) and chemical properties which influence the way it is processed and utilised and affect the nutritional content of the food produced.

For instance sorghum grains are very similar in size to those of wheat and barley but are almost spherical in shape, whilst wheat and barley are more elongated and have a ventral crease. Maize has a much larger grain, which is flattened. All cereal grains consist of five main components, husk, bran, aleurone layer, endosperm and germ (Fig. 8.1).

The husk or hull is the external, fibrous part of the grain that protects it during its formation and has low digestibility because of its high cellulose content. In some grains such as wheat, maize, pearl millet, finger millet, sorghum and rye the husk is loose and is easily removed from the grain during threshing. Such grains, known as naked grains, may also be referred to as having a naked caryopsis. In other cereal species the husk fuses around the grain and is not removed by threshing and must be removed by a dehusking or milling process. These grains, such as rice, barley, some of the millets and oats, are said to have covered caryopses.

Beneath the hull are several different layers of cells, known as the bran, which makes up around 15% of the kernel weight. The outer layers, or pericarp, are fibrous and waxy and create a protective barrier against water and mould penetration. The innermost layer of the bran

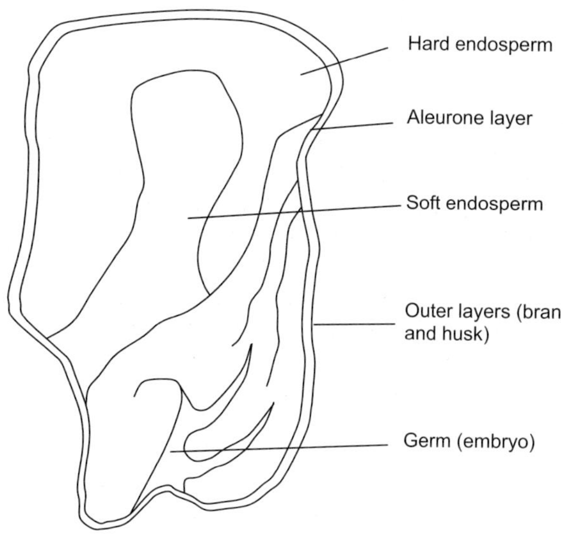


**Fig. 8.1** Structure of a maize grain.

is the testa or seed coat. It is in this layer that many of the pigments are located.

The aleurone layer lies between the bran and the floury part of the grain. This is a single layer of cells rich in fat, proteins and vitamins. In some milling processes the aleurone layer is removed as part of the bran while in others it is retained by the endosperm. It is not possible to isolate aleurone layers by milling.

The germ or embryo is the part of the grain from which a new plant could develop if given the right conditions (temperature and moisture). It is rich in protein, fat, vitamins and minerals. The germ of the maize kernel is very large and is situated at the base of the grain. It is therefore very easy to remove during processing and is an important source of vegetable oil. In wheat and barley the germ is very small but can still be removed by some milling processes and can therefore be sold as a by-product of the milling process for use as a food ingredient, in livestock feed or for oil extraction.

The endosperm or starchy part of the grain generally constitutes between 80% and 85% of the grain and is the energy store of the grain, providing nourishment for the germinating plant. It is also the most important part of the grain for the miller since it is from here that the flour is obtained. Some species and varieties have a starch structure which is more densely packed than others. Protein bodies situated between the individual starch granules determine the overall hardness of the grain. The starchy material in the centre of the grain is usually more loosely packed than that in the outer parts. This hard, outer layer helps to protect the grain from insect attack.

The endosperm in both sorghum and maize is usually a mixture of both hard and soft (vitreous and floury) starch. Dent maize has vitreous endosperm at the sides and back of the kernel (the softer endosperm in the middle collapses on drying, to give the grain its characteristic indentation). Flint maize has a hard endosperm which completely encases a small floury centre. Wheat and barley kernels consist of either vitreous or floury endosperm and can thus be classified as either hard or soft varieties. The harder wheats have a higher protein content and are more suitable for breadmaking while softer wheats, those with a lower protein content, are used for biscuit and cake manufacture.

### Nutritional content

The nutrients are unevenly distributed throughout the grain. Processing can therefore have a very significant effect on the nutritional value of the product. The nutritional content of the grain is affected both by intrinsic and extrinsic factors, varying from variety to variety due not only to genetic traits but also to variation in growing conditions such as climate, soil and agronomic practices. Reported data are, therefore, only a general indication of the nutritional characteristic of a species.

The four basic nutrients in the grain are protein, fat, carbohydrate (starch, sugars and fibre) and vitamins and minerals. The relative amounts of these nutrients is shown in Fig. 8.2 and Table 8.9. Each grain also contains water. If properly dried after harvest and given good storage conditions, the water content of grain is usually

**Table 8.9** Nutrient distribution throughout the grain.

| Fraction | Cereal | Soluble carbohydrates (%) | Protein (%) | Fat (%) | Sugar (%) | Fibre (%) | Minerals (%) |
|---|---|---|---|---|---|---|---|
| Endosperm | Wheat | 85.0 | 11.4 | 1.2 | 0.8 | 0.1 | 0.4 |
| | Maize | 86.4 | 9.4 | 0.8 | 0.6 | 0.1 | 0.3 |
| | Sorghum | 86.4 | 9.5 | 1.0 | 0.4 | 1.0 | 0.8 |
| | Rice | 88.9 | 9.8 | 0.5 | — | 0.3 | 0.6 |
| Germ | Wheat | 20.0 | 29.4 | 10.0 | 20.0 | 8.9 | 8.9 |
| | Maize | 19.0 | 18.8 | 34.5 | 10.8 | 4.6 | 10.1 |
| | Sorghum | 21.0 | 15.1 | 20.0 | 12.0 | 2.6 | 8.2 |
| | Rice | * | * | * | * | * | * |
| Bran + aleurone layers | Wheat | 1.0 | 11.1 | 3.5 | 0.3 | 13.5 | 6.1 |
| | Maize | 7.0 | 7.1 | 1.0 | 0.3 | 14.0 | 5.8 |
| | Sorghum | 5.0 | 8.9 | 5.5 | 0.3 | 8.6 | 2.4 |
| | Rice | 10.0 | 14.1 | 18.8 | 0.5 | 5.7 | 11.2 |
| Husk | Rice | 34.0 | 1.2 | 0.3 | 0.3 | 34.0 | 24.0 |

*Values for rice germ have been omitted as the germ constitutes less than 2% of the grain and is not used on its own as a food or feed product.

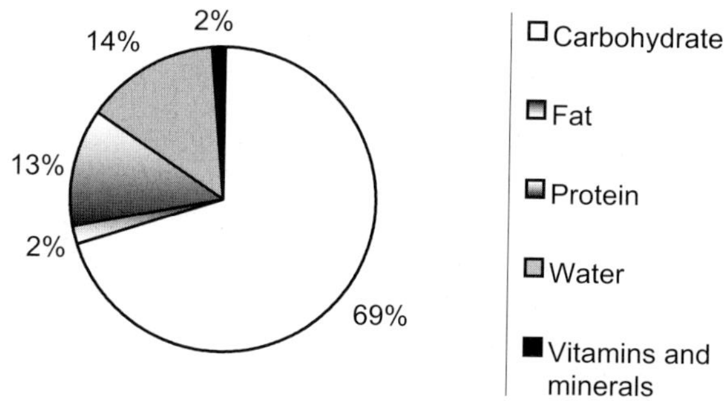


**Fig. 8.2** Typical chemical composition of a wheat grain.

between 7% and 14% and is therefore suitable for long-term storage.

### *Protein*

Protein occurs throughout the grain but is concentrated in the germ and the aleurone layer. Cereal protein is deficient in the essential amino acid, lysine. An essential amino acid is an amino acid which cannot be synthesised by the body but must be provided in the correct quantities by the diet. People, especially children, whose diet consists almost exclusively of cereals are likely to suffer from protein deficiency. Supplementing the diet with other protein sources can prevent this. Refined cereals, such as polished rice, have had almost all of the protein removed during processing.

### *Fat (oil)*

Fat from cereal germ is of good quality (unsaturated) and can be extracted for commercial use. This is only possible when the germ (or other fat-rich fraction) can easily be separated from the remainder of the grain, for example, corn (maize) oil and rice bran oil. Fat is also important as it is the carrier for the oil-soluble vitamins D, E and carotenoids as precursors of vitamin A.

### *Carbohydrate*

Carbohydrates can be broadly grouped into two types, soluble and insoluble. Generally, the insoluble fibres are located in the outer layers of the grain. Fibre, both soluble and insoluble, is very important in the diet. Societies which consume highly refined foods are more prone to cancers of the digestive system. For this reason many people now prefer to eat wholegrain products such as wholemeal flour and wholegrain breakfast cereals. The most important carbohydrate is starch. Starch is insoluble in water, but rapidly absorbs water when cooked (gelatinised). Cereals contain relatively small amounts of sugar.

### *Vitamins and minerals*

The majority of minerals (ash) are located in the germ, bran and husk as are most of the vitamins. Milling therefore reduces the vitamin and mineral content of the grain. It is for this reason that many countries fortify milled flour with vitamins of the B complex group to replace those lost during the milling process.

## Milling

The major operation in the food preparation chain of cereals is milling. Milling is carried out to make the grain more palatable, to improve digestibility and to increase variety, however it may also result in losses in both the overall quantity and the nutritional value of the product. Separation of parts of the grain from the final product alters the nutritional content; some nutrients are lost by the separation but this also has the effect of concentrating other nutrients.

Grain milling has been done since people first started to eat the seeds of wild grasses and cereals. The primary objectives of milling are to remove the indigestible parts of the grain from the edible, to expose the starchy interior and to reduce the particle size of the starch component.

From a consumer's perspective, milling of cereals:

- improves digestibility;
- reduces cooking time;
- adds variety to the diet;
- gives a range of different textures, flavours shapes and colours, and
- may reduce the nutritional content of the grain.

From the miller's perspective it widens the product range and adds value to the grain both in the primary product and by-products.

From the flour processor's perspective it enables different products to be made from the same cereal, and changes the functional and nutritional properties of the grain.

There is a wide difference in the perception of the importance of the milling process in the food production chain between people raised in the developed world and the rural poor in developing countries. In many cases it is fair to include the urban elite from developing countries in the former category.

When people in a developed country need flour, they simply buy it from a shop, and few people would recognise the efforts involved in producing a packet of enriched flour from the grain in the farmer's field to the shop.

Traditional milling practices are still dominant in many rural societies of the developing world; few techniques have been more resistant to change. They are usually linked to the farming system and are carried out by women who may devote 2–4 h of hard labour per day in hand pounding and grinding the daily requirement of flour for their family's needs. The continuance of these age-old techniques in rural areas is a strong indication that the operation is essential for the survival of man and that the technique applied is the best for the local food processing conditions. However with increased urbanisation most countries are faced with changing needs which will require new technologies to meet them. Mechanising the milling is a desirable step in reducing time and effort spent on food processing.

In the future the mechanism of rural, small-scale milling processes is likely to lead to conflicts of interest from the farm to national level. There is a tendency to underestimate the complexity by dealing only with the technical aspects of the mechanisation of milling. A narrow, technical approach runs the risk of leaving a number of socio-economic bottlenecks and issues unattended, which may limit the opportunities to promote sustainable processing systems.

Any changes in the techniques used for cereal processing will have implications, both positive and negative, in relation to:

- gender roles
- food preferences
- nutrition and health
- food security
- technology/productivity
- socio-economic factors
- income and/or costs, and
- sustainability.

### *Milling technologies*

The choice of milling technology depends on the cereal to be processed, the end use of the product and the economic viability of the enterprise.

### *Traditional technologies*

The earliest milling techniques were the saddle stone or mortar and pestle, both of which are still used today in many parts of the world. Greeks and Romans introduced man-powered rotating stone mills. From these the first animal-powered mills were developed as was the concept of 'dressing' (cutting grooves into) the stones to improve the rate of grinding. The mechanisation of milling continued with the introduction of watermills and windmills. Modern milling techniques were developed during the nineteenth century to feed the fast-growing urban population and emerging livestock industry (Ove-Jonsson *et al.* 1994).

The traditional, labour-intensive process, which is totally dependent upon the skill and strength of the operator, involves pounding the grain in a mortar and pestle, usually made from local hardwoods, to remove the bran layers (Plate 27). This is followed by winnowing and sieving to separate the fractions. The pounded grain may then be placed between two abrasive stones, the lower being fixed whilst the upper is moved backwards and forwards over it (Plate 28). This requires a lot of human energy but can produce a very fine, even flour.

Mechanised mills may be powered by diesel engines or electric motors. The earliest mills to be mechanised were developments of the hand-powered mills and produced a coarse meal which could be winnowed or sieved and reground to produce an extracted flour.

### *Stone mills*

Stone mills operate on the principle of grain being passed between two stones, one fixed, the other rotating over it to produce a grinding action. The first mechanised stone mills used granite stones over a metre wide which, because of their weight, were set in a horizontal position. Wind or water power was used to drive such stones. Nowadays the stones are often much smaller (20–60 cm in diameter), are usually set in a vertical position and are driven by electric or diesel power. The 'stones' are usually man-made and are intersected with groves to assist the grinding action and remove the flour from the abrasive surface. This type of mill produces a very fine wholemeal flour, particle size being determined by the gap between the stones, which is variable.

### *Plate mills*

Plate mills operate on a similar principle to stone mills: the grain is passed between two grinding surfaces, one fixed and the other driven by an external power source. In this case the millstones are replaced with grinding plates made of cast iron. The surface of the plate is again intersected with grooves and the plates are usually reversible. The product is a wholemeal flour, the fineness of the flour being determined by the gap between the plates and the depth and pattern of the grooves. Plate mills can be used to grind wet grains or oilseeds as the plates are easily cleaned and do not become clogged or blinded like stones or fine screens.

### *Hammer mills*

Hammer mills carry out size reduction of grain by impact. Grain is fed into a chamber containing rotating hammers fixed to a central shaft and surrounded by a perforated screen. The grain must be sufficiently reduced in size to pass through the screen and discharge from the chamber, thus, the smaller the screen aperture, the longer the residence time in the chamber and the finer the flour. The initial shattering of the grain occurs as the grain is struck by the rotating hammers; further impacts occur as grains hit each other, the sides of the chamber or the surface of the screen. The hammers may be either fixed or swinging and the impact surface of the hammer flat or blade-like. Particle size is determined by screen size and throughput varies with particle size, hardness and moisture content of the grain. Hammer mills produce a wholemeal flour, unless the grain has previously been decorticated. Considerable quantities of heat can be generated in a hammer mill, which results in an even distribution of the germ oil fraction throughout the flour.

### *Roller mills*

In the nineteenth century roller mills were introduced which mechanised every step of the separation process to produce a very fine, refined white flour. In general, roller milling equipment bears a high capital cost, requires a reliable supply of good-quality grain, a well-defined market for the flour and highly trained operators. However, there are some small, relatively low-cost, roller mills available for maize milling, which can also be used for other cereals, though the flour produced is not as well refined as that produced by commercial roller mills. Unlike the one-step mills previously described, roller milling gradually breaks down the grain, removes the bran and reduces the particle size of the flour in a series of carefully controlled operations.

Roller milling is a combination of shearing, scraping and crushing. The grain passes through a series of rolls each designed to yield a mixture of coarse, medium and fine particles. The fine particles are removed by sifting while the medium and coarse particles are transported to a further set of rolls for further separation. In simple roller mills there may only be two sets of rolls while in commercial roller mills the flour may pass through up to 20 different sets of rolls.

The rolls can be classified into two groups: the break system and the reduction system. The rolls in the break system consist of paired fluted (corrugated) metal rolls, rotating in opposite directions with a speed differential between the two rolls to create a shearing action at the point of contact. The gap between the rolls can be finely adjusted to allow precision grinding.

Grain is fed into the first set of break rolls at a carefully determined rate, the fluted rolls shear open the grain, usually along the crease, and unroll the bran coat so that each grain consists of a thick layer of endosperm adhering to a thin layer of bran. Some small fragments of endosperm become detached from the grain and a very small amount of fine flour is also removed. The various particles are separated from one another by a series of sieves. The bran coats are fed into the second set of break rolls, which have finer flutes and are set closer together than the first. More particles of endosperm are removed together with some flour. The process continues, the grain fractions passing through a succession of break rolls, each with finer flutes and set closer together than the previous set. The endosperm is progressively scraped from the bran until finally, after passing through the last break roll, no more endosperm can be extracted and the bran is discharged from the mill. At each break the fine flour is removed by sifting and the larger particles of endosperm are transported to the second stage of roller milling, the reduction rolls.

The endosperm then passes through up to 16 pairs of reduction rolls. These rolls are usually smooth, though in some mills the first one or two pairs may be finely fluted. The speed differential between the rolls is lower than in the break system, thus particle size reduction is achieved by crushing rather than shearing. Each pair of rolls has a smaller gap (nip) than the previous set. Flour produced at each stage is removed by sifting and larger endosperm particles pass on to the next set of rolls. Some small bran particles are inevitably carried over from the break rolls, however the slower smooth rolls flatten these particles, enabling them to be removed by sifting.

The objective in the milling of wheat into refined, white flour is to separate, as completely as possible, the endosperm from the bran and germ and thus improve the palatability, digestibility and extend the storage life of the flour. The wheat grain contains, on average, 82% of starchy endosperm. Theoretically, it should be possible to produce white flour of 82% extraction. In practice this is not possible to achieve due to the limitations of the milling process. Extraction rates of 75% white flour are possible, further increases in extraction rate will also contain bran, aleurone and germ and not be a truly 'white' flour. A good-quality white flour will, typically, have an extraction rate of 72–74%.

In order to achieve the best possible separation of bran and endosperm, grain for roller milling is usually conditioned before use. This involves adjusting the moisture content of the grain. As moisture content increases:

- bran becomes tougher and less brittle and does not shatter on impact and contaminate the flour;
- endosperm becomes more friable (crumbly), therefore requiring less energy to grind it;
- cohesion between bran and endosperm becomes stronger, making it more difficult to separate these two fractions; and
- fine particles agglomerate and may block screens.

There is an optimal moisture content for each variety of grain, at which the bran is toughened, the endosperm made more friable and separation of bran and endosperm can be achieved without causing blockages in the sieves. In general the optimum moisture content for roller milling wheat is 15–17.5% dependent on variety. Millers often use a mixed grist (mixture of varieties with different milling properties) to produce the required quality of flour. By conditioning each variety independently the miller ensures the maximum yield from his grain. Once cold water has been added to the grain it requires thorough mixing and a resting period of 24–72 h to achieve equilibrium. Few millers have storage capacity for this time, so many carry out warm conditioning of the grain at temperatures of approximately 40°C for 1.5 h.

### *Dehulling or pearling*

Some cereals, such as rice, barley and oats do not thresh clean of their husks. These must be removed prior to further processing. Other cereals such as sorghum have an unpalatable seed coat which adheres strongly to the endosperm and is not easily removed. There is a range of machinery available for these purposes. The resulting 'pearled grain' produced may be consumed as whole grains or undergo further processing into coarse meal or flour.

Abrasive disc dehullers utilising carborundum stones may be used to abrade the bran or hull from cereals. A series of abrasive discs, mounted on a rotating spindle, are encased in a milling chamber. Grain is fed into the chamber either batch-wise or on a continuous flow system. As the grain passes between the stones and casing the seed coat is gradually abraded. The bran, which is removed as a fine dust, is removed through an aspiration system or cyclone. Commercial sorghum dehullers may have up to 20 individual discs and are capable of processing up to 500 kg of grain per hour. The degree of abrasion is dependent on the residence time in the chamber, the speed of rotation, and the abrasiveness of discs (type and age).

### *Rice milling*

Rice mills vary in capacity from 250 kg to 5 tonnes per hour. Small-scale mills fall into two categories: steel hullers and rubber roll mills.

#### *Steel huller mills*

Steel huller mills, also known as Engleberg-type mills, are a one-pass operation with a throughput in the region of 250–450 kg per hour. The mill consists of a central, cylindrical, horizontal rotating steel shaft with a short Archimedean screw at the intake end and horizontal ribs along the rest of its length. This is encased in a cylindrical chamber slightly larger than the mill, the lower side of which is comprised of replaceable perforated screens. An adjustable steel blade protrudes into the chamber to create a variable gap between the blade and the horizontal ribs on the shaft.

The paddy grains are forced through the gap between the rotors and the horizontal ribs, which splits the husk open, allowing the separated rice kernels and husk to fall into the milling chamber. The smaller sized kernels pass easily through the gap between the blade and ribs. As they are forced along the length of the milling chamber the soft bran layers are removed by frictional forces between the metal components of the mill, other rice grains and residual husk.

#### *Small-scale rubber roll mills*

These are also a one-pass operation in that a batch of paddy is processed in a single machine. Within the machine are a number of processing components each with a dedicated purpose. Throughput is dependent on the size of the machine and the quality of the raw material but is generally in the range of 500–2500 kg per hour.

Paddy is fed into the mill and passes through a single pair of rolls, coated with a layer of rubber. The gap between the rolls is set in such a way that the husk is split open and removed with minimal damage to the rice kernel. The dehusked grain (brown rice) then passes to an integral polisher, usually an abrasive cylinder, where the bran and germ are removed. Large-scale mills (typically 2–4 tonnes per hour) operate on the same principle as the small rubber roll mills though there is a separate machine for each unit operation and extra equipment may be added to further improve product quality. Large-scale mills usually have additional components such as grain cleaners, destoners, broken grain separators and colour sorters to ensure production of a high-quality, uniform product.

The polished kernel can then be milled into flour if required.

### Wheat flour

Wheat flour is usually made from a mixture of different varieties of wheat (a 'grist') in order to achieve the desired product. Some varieties have hard grains, others soft. Hard wheats tend to be high in protein and break down into granular, free-flowing flour. Soft wheats have lower protein content and form finer flour though the particles are irregular and tend to stick together slightly. Hard wheats typically grow in regions where there are cold winters and hot summers (such as US, Canada and Australia) whereas soft wheats grow in more temperate climates such as Western Europe.

Wheat flour can be described as:

*Wholemeal*: 100% extraction rate. Wholemeal flour is simply the whole wheat grain ground into flour. The nutritional content of the flour is therefore the same as that of the whole grain, high in fibre and containing all natural vitamins and minerals. Because none of the germ has been removed, the fat content of wholemeal flour is higher than that of white flour and therefore wholemeal flour has a shorter shelf-life.

*Brown*: Brown flour has had some of the bran and germ removed and usually has an extraction rate of around 85%. It is more digestible than wholemeal flour and, because the germ has been partially removed, has a lower fat content and better storage qualities.

*White*: Usually around 75% extraction. Almost all of the bran and germ have been removed. The flour therefore consists mainly of starch and has had many of the naturally occurring nutrients removed. It is however very digestible and is preferred by many consumers.

Within the three broad groups of flour detailed above there are numerous types of flour for specific uses.

*Stoneground*: This flour is usually wholemeal and is milled in traditional stone mills. The flour particle size is very fine.

*Organic*: Any flour produced from grain grown without the use of artificial fertilisers or pesticides.

*Strong bread flour*: Made from hard wheats, this flour contains more protein (11–14%) than other flours. Two specific proteins, gliadin and glutenin, are needed to make bold, well-risen loaves. When water is added to the dough mix, the two proteins combine to form gluten, an elastic substance which traps the gas produced by the fermentation of the yeast and hardens during baking to form the structure of the loaf. Strong flour may be white, brown or wholemeal and is used for bread, pizza dough and puff pastry.

*Plain flour*: A general-purpose flour made from a mixture of hard and soft wheats and suitable for pastry, sauces and cakes. This flour would typically have a protein content of 7–10%.

*Self-raising flour*: Flour, usually white, which has had a raising agent, such as sodium hydrogen carbonate and calcium phosphate, added to it, making it suitable for cakes and biscuits.

*Multigrain and wheatgerm flours*: Pretreated whole grains, both of wheat and other cereals, may be added to flours to extend the product range by adding flavour and texture to the flour. Similarly wheatgerm, removed and isolated during the milling process, may be added back into the flour.

Table 8.10 shows the nutritional value of a selection of wheat flours.

**Table 8.10** The nutritional value of an average selection of wheat flours. Source: Food Standards Agency 2001.

| | Strong white | Plain white | Brown | Wholemeal |
|---|---|---|---|---|
| Extraction rate (%) | 75 | 75 | 85 | 100 |
| Protein (%) | 11.5 | 9.4 | 12.6 | 12.7 |
| Fat (%) | 1.4 | 1.4 | 1.8 | 2.2 |
| Carbohydrate (%) | 75.3 | 77.7 | 68.5 | 63.9 |
| Fibre (%) (Englyst) | 3.1 | 3.1 | 6.4 | 9.0 |
| Iron (mg/100 g) | 2.1 | 2.0 | 3.2 | 3.9 |
| Calcium (mg/100 g) | 140 | 140 | 130 | 38 |
| Thiamine (mg/100 g) | 0.32 | 0.32 | 0.39 | 0.47 |
| Niacin (mg/100 g) | 2.0 | 1.7 | 4.0 | 5.7 |

### Fortification

In some countries it is mandatory to fortify flour in order to replace some of the nutrients removed by processing. In the UK white and brown flours must be fortified with calcium, iron, thiamine and niacin. Wholemeal flour is not subject to these regulations as it already naturally contains the required amount of the nutrients. However, as the phytates in the bran, which bind calcium, have not been removed, wholemeal flour is deficient in this mineral. Wholemeal flour also contains folic acid.

#### *Flour improvers*

Bread-making flours often have flour improvers or 'flour treatment agents' added to them. The agent is usually vitamin C (ascorbic acid) which acts as an antioxidant and promotes cross-linking of the protein molecules, thus forming a stronger gluten matrix.

### Non-wheat flours

*Barley*: Barley is very low in gluten and cannot therefore be used for baking traditional fermented loaves.
*Buckwheat*: Buckwheat flour has no gluten but can be used for flatbreads and pancakes.
*Cornmeal (maize flour)*: Often eaten as a thick porridge or may be used to make soft breads and pancakes.
*Gram*: This is not a cereal flour but is made from pulses. It is often used in conjunction with wheat flour to give texture and flavour to flat breads such as chapatis.
*Oats*: Oatmeal can be eaten as a porridge or used to make biscuits.
*Rye*: Rye flour produces a very dense, dark loaf with a slightly sour taste.
*Sorghum*: Sorghum is broadly similar to wheat in its chemical composition. It is often consumed as a thick or thin porridge, sometimes fermented. It may also be used to make flatbreads.
*Spelt*: Spelt is an ancient form of wheat. The flour has a very distinctive flavour and texture and the gluten is easily digested. Spelt flour can, therefore be eaten by those who suffer from mild gluten intolerance.

## Oilseeds, oils and fats

### Origin and definition

Oils and fats are substances produced and used by plants and animals as an energy store. Plant seed oils are used by the plant during and after seed germination; some components of oils are essential to metabolic processes. Some plant fruits (notable the oil palm and the olive) are rich in oil. Fat is also contained in the milk of mammals and provides energy for offspring. When a fat is in the liquid form it is described as oil and when solid it is known as fat. Coconut oil melts between 22 and 24°C and so is usually a liquid oil in the tropics but a solid fat in temperate climates.

### Triglycerides and fatty acids

Oils and fats are obtained from a wide variety of plant and animal sources and each one has its own individual properties. They are, however, all of the same chemical type based on the structure of glycerol and fatty acids. Glycerol has a chemical formula that can be illustrated by the structure shown in Fig. 8.3. Each of the 'arms' of the glycerol molecule can combine with a fatty acid, to build up an oil molecule as illustrated.

The oil is composed of glycerol plus three fatty acid molecules and is therefore known as a triglyceride. All oils and fats are made up of a mixture of triglycerides.

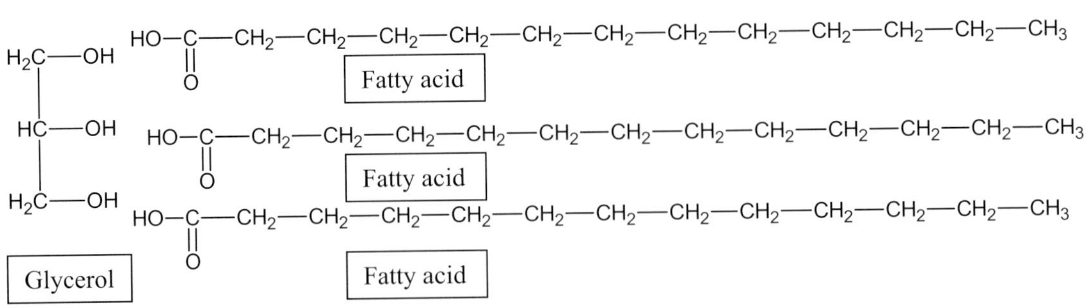


**Fig. 8.3** Chemical formula of glycerol showing how the 'arms' of the glycerol molecule can combine with a fatty acid.

A number of different fatty acids exist and to a large extent the character of a particular oil or fat depends on the actual fatty acids that are present in the individual triglyceride molecules. Some of the component fatty acids are longer or shorter than others and they can all combine with a glycerol 'arm'. Chemically, each arm of the glycerol molecule is an alcohol group. The alcohol group in the glycerol reacts with the acid group of the fatty acid to produce the triglyceride.

In addition, there are three different types of fatty acids: saturated, monounsaturated and polyunsaturated.

### *Saturated fatty acids*

In their simplest form, fatty acids are made up of a linear chain of carbon atoms linked to a group which provides the acidic properties. Such fatty acids are said to be saturated. The most common examples are lauric, myristic, palmitic and stearic acids and they differ according to the number of carbon atoms in their structure (Fig. 8.4). All these acids are solid at room temperature and their presence in high proportions in a triglyceride mixture is likely to make it solid.

### *Monounsaturated fatty acids*

Sometimes the carbon atom chain contains a double bond between two carbon atoms. These double bonds are centres of instability for the fatty acids in terms of its ability to react chemically with hydrogen, oxygen and other molecules. The reaction with hydrogen may stabilise the fatty acid but reactions with oxygen can induce rancidity in the oil. A fatty acid containing one double bond is said to be monounsaturated. The most common example is oleic acid which, like stearic acid, has 18 carbon atoms (Fig. 8.5).

### *Polyunsaturated fatty acids*

Some fatty acids have two or three double bonds and these are said to be polyunsaturated. Linoleic and linolenic acids are common examples of polyunsaturated fatty acids. They have the same number of carbon atoms as stearic acid, but linoleic acid has two double bonds and linolenic has three (Fig. 8.6).

Table 8.11 shows the relationship between the number of double bonds and melting point, in the common fatty acids including those with 18 carbon atoms (C18 acids). The presence of a double bond lowers the melting point.

$$\mathrm{HO{-}\underset{\|\atop O}{C}{-}CH_2{-}CH_2{-}CH_2{-}CH_2{-}CH_2{-}CH_2{-}CH_2{-}CH_2{-}CH_2{-}CH_2{\cdot}CH_2{-}CH_2{-}CH_2{-}CH_2{-}CH_2{-}CH_2{-}CH_3}$$

**Fig. 8.4** Simple fatty acid structure.

$$\mathrm{HO{-}\underset{\|\atop O}{C}{-}CH_2{-}CH_2{-}CH_2{-}CH_2{-}CH_2{-}CH_2{-}CH_2{-}CH_2{-}CH_2{-}CH_2{\cdot}CH{=}CH{-}CH_2{-}CH_2{-}CH_2{-}CH_2{-}CH_3}$$

**Fig. 8.5** Structure of oleic acid.

Linoleic acid (C18=2)

$$\mathrm{HO{-}\underset{\|\atop O}{C}{-}CH_2{-}CH_2{-}CH_2{-}CH_2{-}CH_2{-}CH_2{-}CH_2{\cdot}CH{=}CH{-}CH_2{\cdot}CH{=}CH{-}CH_2{-}CH_2{-}CH_2{-}CH_2{-}CH_3}$$

Linolenic acid (C18=3)

$$\mathrm{HO{-}\underset{\|\atop O}{C}{-}CH_2{-}CH_2{-}CH_2{-}CH_2{\cdot}CH{=}CH{-}CH_2{\cdot}CH{=}CH{-}CH_2{\cdot}CH{=}CH{-}CH_2{-}CH_2{-}CH_2{-}CH_2{-}CH_3}$$

**Fig. 8.6** Structure of polyunsaturated fatty acids.

**Table 8.11** Structure and melting points of some fatty acids.

| Fatty acid | Carbon atoms | Double bonds | Melting point (°C) |
|---|---|---|---|
| Lauric | 12 | 0 | 44 |
| Myristic | 14 | 0 | 54 |
| Palmitic | 16 | 0 | 63 |
| Stearic | 18 | 0 | 70 |
| Oleic | 18 | 1 | 13–16 |
| Linoleic | 18 | 2 | –5 |
| Linolenic | 18 | 3 | –11 |

Thus a triglyceride mixture containing a high proportion of monounsaturated or polyunsaturated fatty acids is likely to be liquid. It is important to note that the reactivity of a fatty acid increases with the number of double bonds.

### *Structure of triglycerides*

The triglycerides in Fig. 8.7 are just some of the many that are found in palm oil. The overall triglyceride composition is such that palm oil is liquid at tropical temperatures but solid in temperate climates.

Palmitic, Palmitic, Stearic (PPS – Melting point 62°C)

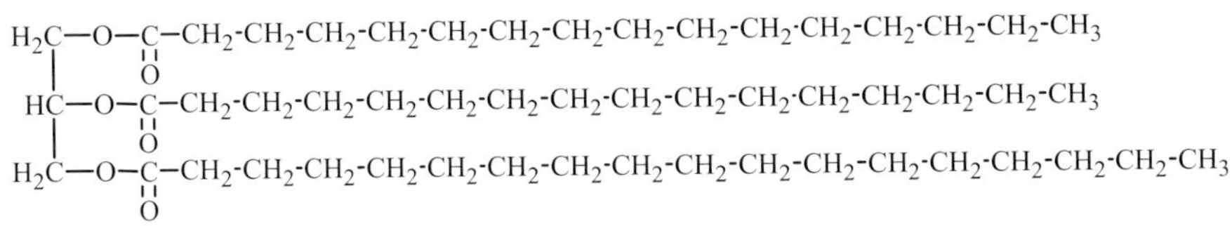

Stearic, Palmitic, Oleic (SPO – Melting point 39°C)

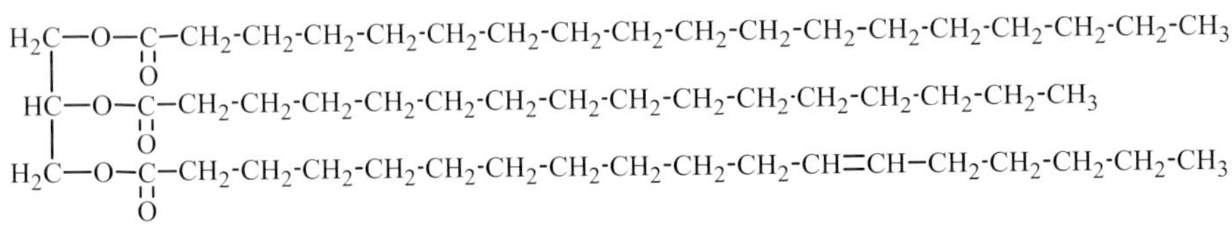

Palmitic, Oleic, Oleic (POO – Melting point 19°C)

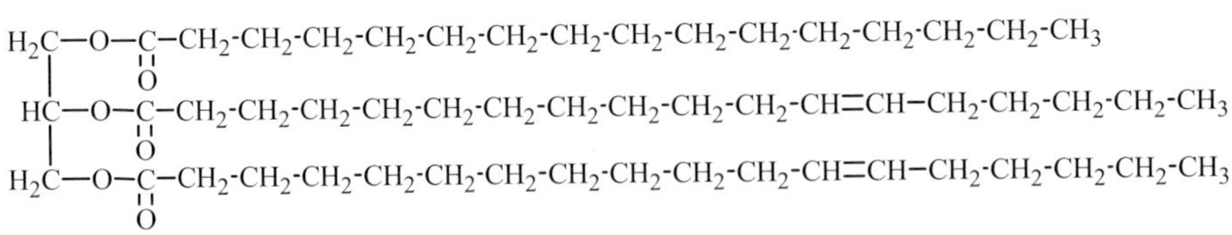

Oleic, Oleic, Oleic (OOO – Melting point 5.5°C)

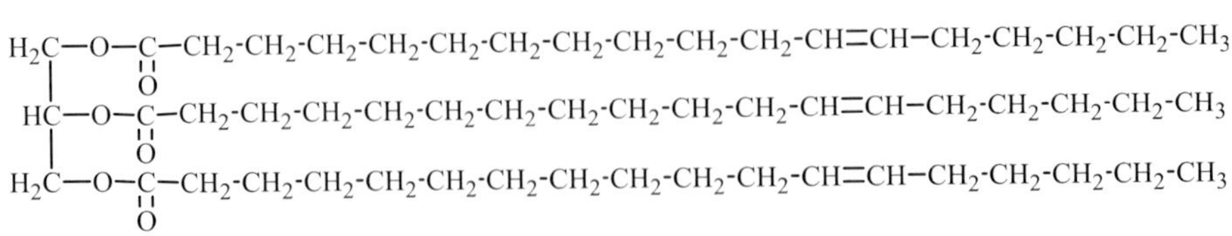

**Fig. 8.7** Structure of some triglycerides.

### *Fatty acid composition*

The fatty acid composition of an oil tends to be characteristic of the oilseed from which it is extracted. The fatty acid compositions and melting points of oils from different oilseeds are given in Table 8.12 together with the oil contents of the seeds.

### *Triglyceride deterioration*

Oils and fats can undergo deterioration under poor storage conditions. Oxidation (reaction of the fat with oxygen) occurs when the oil reacts with the oxygen present in the air. Hydrolysis (the splitting of a fatty acid from glycerol in the presence of water) occurs when oil reacts with moisture. Usually hydrolysis occurs in the presence of an enzyme which is either present naturally in the oilseed or is present in moulds or fungi that may grow on it. Oil deterioration normally occurs under poor storage conditions.

#### *Oxidation*

Oxygen in the air usually reacts with the triglyceride at a double bond on the fatty acid arm of the oil (Fig. 8.8).

Thus the fats with few double bonds are more stable than fats with a large number of double bonds. The process of oxidation in oils and fats is accelerated or catalysed by the presence of trace metals such as iron and copper.

#### *Hydrolysis*

Moisture reacts with oil, in the presence of lipase – an enzyme which is an active protein with the ability to split a fatty acid from a triglyceride molecule. Eventually glycerol and free fatty acid (FFA) will be obtained, but intermediate stages occur where only one or two of the fatty acids have been split away from the triglyceride to produce mono- and diglycerides as well as free fatty acids (Fig. 8.9).

Hydrolysis can occur both in oilseeds and in extracted oil. Hydrolysis in oilseeds is accelerated by the presence of moulds. It is therefore very important to dry oilseeds to a moisture level below which mould can grow – usually between 6% and 12% depending on the oilseed.

Free fatty acid content is used as a guide to the extent of deterioration in the quality of oils. Coconut oil from high-quality copra can have FFA levels in the region of 0.1% to 0.2% but levels as high as 5% can occur in oil from poor-quality copra. The active lipases present in oils can result in rapid increase in FFA unless the enzyme is deactivated. For example, the lipase in oil palm can rise to around 35% in 3 months at temperatures between 27 and 40°C (Nkapa *et al.* 1990). It is therefore essential that oil palm fruits are sterilised by heat-treatment immediately after harvest and before oil extraction to deactivate the lipase. FFA cause off-flavours in an oil, but they can be removed in the oil refining process.

## The composition of oilseeds

Oilseeds can be thought of as being mixtures of three components: oil, meal and water. Processing removes the bulk of the oil and the product remaining is oilcake. Oilcake contains the residual oil, all of the meal and some water. Since the meal is mainly composed of protein, carbohydrate and fibre, oilseed cakes in general are excellent materials for animal feeds. Table 8.13 shows the composition of copra (dried coconut kernel) both by weight, and by weight percentage terms, at various levels of moisture content.

Commercial copra is usually considered to be 'dry' when its moisture content is 6%. A typical oil content of dry copra is 65%, leaving a balance of 29% for the meal content.

The figures in Table 8.13 show the changes in oil and moisture content as a sample of fresh coconut kernel dries to copra. The figures are composed so that the weight of 'dry' copra is 100 units. Thus 117.5 kg of wet copra, containing 55.3% oil and 20% water, will dry down to 100 kg of 'dry' copra, containing 65% oil and 6% water. In other words, 100 kg of copra at 20% moisture only contains 55.3 kg oil whereas 100 kg of 'dry' copra will contain 65 kg.

Due to the variation of oil content in copra with moisture content, comparisons of oil contents are always made on a dry weight or 'moisture-free basis' (MFB). The oil content of copra (MFB) in Table 8.13 is 69.1%. The following formula may be used to convert oil content (calculated on an MFB basis) to oil content at any given moisture content:

$$\%\ \text{Oil (at X\% moisture)} = \frac{\%\ \text{Oil (MFB) multiplied by } (100 - \text{X})}{100}$$

**Table 8.12** Typical oil contents, melting points and oil composition of selected oilseeds.

| | | | | | | Rapeseed | | | | | |
|---|---|---|---|---|---|---|---|---|---|---|---|
| | | Copra | Palm kernels | Sunflower seed | Groundnut | High erucic | Low erucic | Cottonseed* | Sesame seed | Soybean | Oil palm |
| Oil content (%) | | 65–68 | 44–53 | 25–48 | 45–55 | — | 36–50 | 15–24 | 44–54 | | |
| Melting point (°C) | | 23–26 | 24–26 | −16 to −18 | −2 | −9 | −20 | −2 to 2 | −4 to 0 | −23 to −20 | 33–40 |
| Major fatty acid composition (%) | | | | | | | | | | | |
| Name | Structure** | | | | | | | | | | |
| Caproic | C6(0) | 0–1 | 0–1 | — | — | — | — | — | — | — | — |
| Caprylic | C8(0) | 3–15 | 2–5 | — | — | — | — | — | — | — | — |
| Capric | C10(0) | 6–15 | 3–5 | — | — | — | — | — | — | — | — |
| Lauric | C12(0) | 41–56 | 44–51 | — | — | — | — | — | — | — | — |
| Myristic | C14(0) | 13–23 | 15–17 | trace | trace | trace | trace | trace–2 | trace | trace | 1–1.5 |
| Palmitic | C16(0) | 4–12 | 7–10 | 3–10 | 6–16 | 1–6 | 2–6 | 17–29 | 7–12 | 7–12 | 42–47 |
| Palmitoleic | C16(1) | — | — | 0–1 | 0–1 | 0–3 | trace | 0–2 | trace | trace | — |
| Stearic | C18(0) | 1–5 | 2–3 | 1–10 | 1–7 | 0–3 | 1–3 | 1–4 | 3–6 | 2–6 | 4–5 |
| Oleic | C18(1) | 3–12 | 12–19 | 14–65 | 36–72 | 8–50 | 50–66 | 13–44 | 35–50 | 15–33 | 37–41 |
| Linoleic | C18(2) | 1–4 | 1–3 | 20–75 | 13–45 | 13–29 | 17–30 | 33–58 | 35–50 | 43–58 | 9–11 |
| Linolenic | C18(3) | trace | trace | trace | 0–1 | 5–16 | 6–14 | 0–2 | 0–1 | 5–11 | — |
| Arachidic | C20(0) | trace | trace | 0–1 | 1–3 | 0–3 | 0–1 | trace | 0–1 | 0–1 | — |
| Eicosenoic | C20(1) | trace | trace | trace | 0–2 | 3–15 | 1–4 | trace | trace | 0–1 | — |
| Behenic | C22(0) | — | — | 0–2 | 2–5 | 0–2 | trace | trace | trace | trace | — |
| Erucic | C22(1) | — | — | trace | trace | 5–60 | trace–5 | trace | — | — | — |
| Docosadienoic | C22(2) | — | — | — | — | 0–2 | — | — | — | — | — |
| Lignoceric | C24(0) | — | — | trace | 1–3 | trace | trace | trace | — | — | — |
| Tetracosenoic | C24(1) | — | — | — | — | 0–3 | trace | — | — | — | — |
| Specific sample composition | | | | | | | | | | | |
| Saturated | | 91 | 85 | 17 | 17 | 6 | — | 34 | 15 | 15 | 53 |
| Monounsaturated | | 7 | 13 | 29 | 61 | 86 | — | 26 | 40 | 25 | 38 |
| Polyunsaturated | | 2 | 2 | 52 | 22 | 8 | — | 40 | 45 | 60 | 9 |

*Cottonseed oil also contains up to 1% cyclopropenoid acids.
**Denotes the number of carbon atoms followed by the number of double bonds in parentheses.

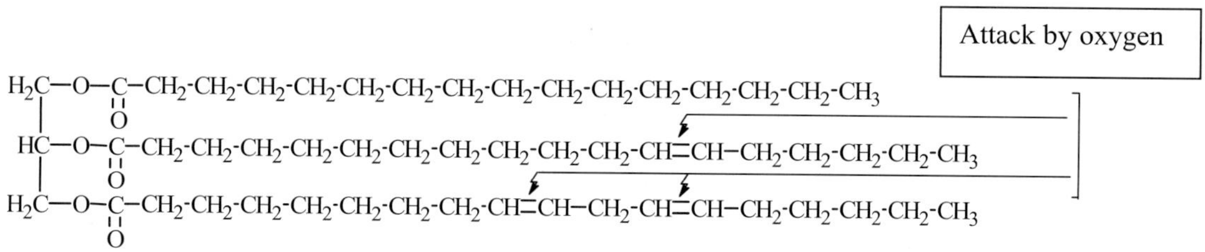


**Fig. 8.8** Oxidative breakdown of a triglyceride.

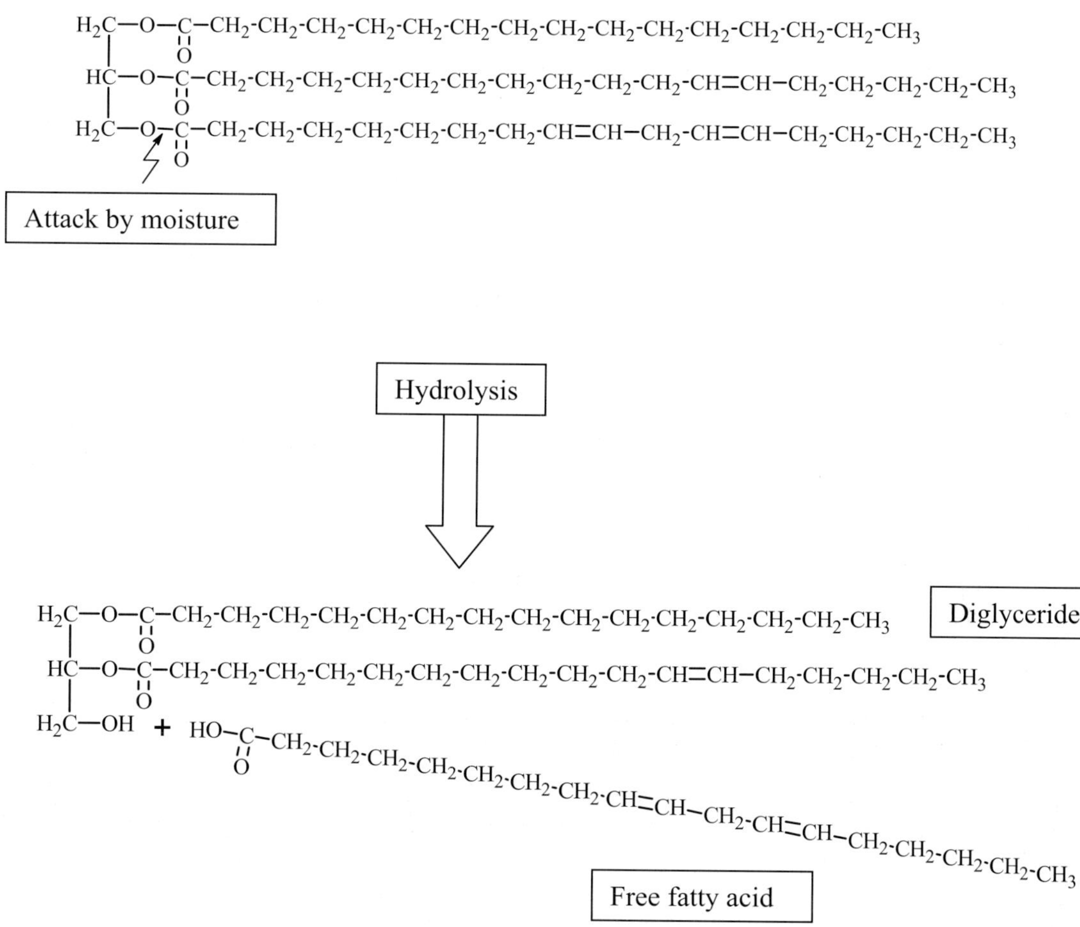


**Fig. 8.9** Hydrolytic breakdown of a triglyceride.

| | Composition by weight | | | | % composition | | |
|---|---|---|---|---|---|---|---|
| | Oil | Meal | Water | Total | Oil | Meal | Water |
| Copra at high moisture content | 65.0 | 29.0 | 23.5 | 117.5 | 55.3 | 24.7 | 20.0 |
| Underdried copra | 65.0 | 29.0 | 12.8 | 106.8 | 60.9 | 27.1 | 12.0 |
| Dry copra | 65.0 | 29.0 | 6.0 | 100.0 | 65.0 | 29.0 | 6.0 |
| Copra (moisture-free) | 65.0 | 29.0 | 0.0 | 94.0 | 69.1 | 30.9 | 0.0 |

**Table 8.13** Composition of copra at various levels of moisture based on the production of 100 kg of dry copra prior to oil extraction.

Thus if the oil content of groundnuts (MFB) is 50% and the moisture content of the nuts is 10%, then the oil content at 10% moisture content is:

$$\frac{50 \times (100 - 10)}{100} = 45\%$$

### *Extraction efficiency (EE)*

Extraction efficiency is the percentage of oil extracted in relation to the amount of oil present in the seed, and can be presented on a dry-weight basis. If 100 kg of sunflower seeds (oil content 32% MFB) are processed to yield 29 kg of the oil then the extraction efficiency is:

$$\frac{29 \times 100}{32} = 90.6\%$$

Extraction efficiency tends to be related to the oil content of the seed. It is difficult to achieve high levels of efficiency with low oil content seed using mechanical expellers since these machines are unable to generate sufficient pressures to release all the oil from the cellular structure of the seeds. However, large-scale expellers generally extract in excess of 90%. The oil extraction efficiency using solvent extraction (whereby the oil is removed from the crushed seed using a petroleum-based solvent – usually hexane) is around 98%. In small-scale expelling the EE is usually in the range 60–65% and rarely exceeds 80%.

## Oilseed milling

### *Pressing processes*

Oil extraction from oilseeds and nuts ('milling') is carried out using the following steps:

- the seeds are passed over magnetic separators to remove metal and sieved to remove dirt and stones;
- if required, the shells or hulls are removed in decorticating machines;
- the kernels are reduced in size to produce flakes or small pieces by grinding between grooved rollers or using hammer mills; and
- the crushed kernels are pressed in hydraulic or screw presses with or without heating, depending on the type of oil-bearing material and the oil quality required.

Oil expressed without heating can be of edible quality. These 'crude' (unrefined) oils are known as cold-pressed, or virgin oils. Virgin olive oil, for example, commands a high price, has a very distinctive flavour and odour, can have an intense green colour and is highly regarded as a salad oil. After mechanical pressing the residue is known as oilseed cake or oilcake. The term 'cake' refers to the hard block-like texture of the pressed oilseed residue as it exits the press. After solvent extraction the residue is known as meal because of its free-flowing properties. Oilcakes and meals are concentrated sources of high-quality protein and are generally used in compound livestock rations. Soybean meal is often used as a human food in texturised protein products.

### *Pressing machines*

Many different mechanical devices have been used for pressing. The Romans developed a screw press for the production of olive oil, and wedge presses were commonly used in China centuries ago. The stamper press invented in the seventeenth century was used almost exclusively in Europe for pressing oilseeds such as linseed until the early part of the nineteenth century, when the hydraulic press was developed. The yield of oil from the hydraulic press was considerably higher than that from earlier processing methods due to the higher pressures applied (approximately 400 kg cm$^{-2}$ is possible). The modern screw expeller has largely replaced the hydraulic presses because it is a continuous process, has greater capacity, requires less labour, and will generally remove more oil.

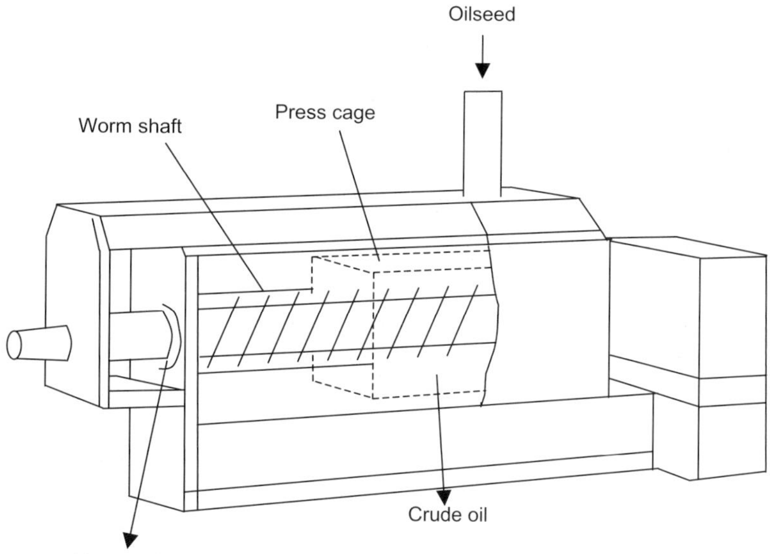


**Fig. 8.10** A screw expeller for continuous pressing of oil.

Continuous pressing using screw expellers is used widely to extract oil from oilseeds and nuts. The expeller (Fig. 8.10) consists of a screw (or worm) which rotates slowly inside a cylindrical barrel or cage along which narrow slots are provided for the oil to escape. The oilseed material is fed via a hopper into the barrel. The screw rotates and transports the oilseed along the barrel. Increasing pressure is exerted by decreasing the clearance between the screw shaft and the cage. This can be achieved by progressive increase in the shaft diameter; reducing the length of the screw flight in the direction of the axial movement; or by reducing the size of the outlet from the barrel end (the choke mechanism). Pressures vary from 700 kg $cm^{-2}$ to 2100 kg $cm^{-2}$. Oil flows out of the barrel through the slots due to the increase in pressure. The press-cake is discharged through the choke outlet.

Oilseeds are normally cleaned, dehulled, flaked, cooked and dried prior to expelling.

*Dehulling* (or decortication) of the fibrous outer hull of the oilseed is carried out to increase both the oil content and protein content of the material.

*Flaking* decreases the distance that the oil will have to travel to reach the particle surface and therefore facilitates oil release.

*Cooking* denatures the proteins, helps oil droplets coalesce and plasticises the flakes, making them less brittle so they do not disintegrate due to the shearing forces within the expeller. Flake disintegration reduces oil yield and produces a crude oil with a high 'foots' content (fine particles of oilseed debris). The moisture content of the oilseed during expelling depends upon the material and is adjusted during the cooking process by adding steam while thoroughly mixing the flaked oilseed material.

*Drying* – after cooking, excess moisture is removed from the flaked oilseed material by indirect steam heating.

Expellers can be used to extract oil from many oilseeds and nuts, and commercial processing plants are normally multipurpose. The process is simple and not capital-intensive compared to solvent extraction plants. The smallest solvent extraction plant processes 100–200 t per day. Expellers are available for much smaller capacities, from micro-expellers processing 40 kg per hour up to several hundreds of tonnes per day.

A well-maintained, powerful press will produce an oilcake of 3–5% residual oil content. However, the residual oil can be recovered using solvent extraction. This two-stage process of pre-pressing coupled with solvent extraction is common. Most oilseeds have an oil content of over 30% and are subjected commercially to pre-pressing followed by solvent extraction where throughput makes this a viable proposition. The high oil content must be reduced before solvent extraction can be carried out. Oilseeds with a low oil content, notably soybeans which have only 18–21% oil content, can be directly solvent extracted. The commercial value of the soybean meal is usually higher than that of the oil extracted. The quality of the meal is therefore of prime importance. Expelled soybean cake does not store well due to its relatively high oil content. Also high pressures cause high temperatures during expelling and may impair the nutritive value of the protein, especially that of lysine bio-availability. However, soybeans contain anti-nutritive factors including proteolytic enzyme inhibitors, haemagglutinins and saponins, which must be deactivated by heat treatment after solvent extraction. Solvent-extracted soybean meal has an oil content of below 0.5% (normally 0.1%) and stores well.

### *Solvent extraction*

Soybeans are given as an example of the process. Other oilseeds are treated in a similar way. The beans are weighed, unloaded and conveyed to large steel vertical storage silos. Pneumatic or mechanical conveying is used depending on the size of the processing plant. Mechanical conveyors and elevators are less costly and more suited to smaller plants. Oilseed quality is assessed by analysing representative samples for moisture, foreign matter, colour, broken beans, oil and protein content. Prior to extraction the beans are dried, tempered, cleaned, cracked, often dehulled, conditioned and flaked.

*Drying*: Soybeans must be dried to less than 10% moisture content before extraction if they are to be dehulled, since the fibrous hulls are separated more easily at low moisture content.

*Tempering*: After cooling, the dried soybeans are tempered (stored in vertical silos for 2–5 days) so that moisture equilibration is achieved.

*Cleaning*: Iron particles known as tramp iron are removed by electromagnetic or permanent magnetic separators attached to conveyors or chutes; or by revolving drum-type magnets. Stones, sand, dust and extraneous matter are removed by a two-stage vibrating screen seed cleaner. Cyclonic aspirators are used to remove dust and hull particles.

*Cracking*: Cracking is carried out in roller mills. Two or three pairs of rolls are normally used, mounted one on top of the other. The corrugations on the upper pair of rolls are coarser and deeper than those on the lower pairs. The corrugated rolls rotate in opposite directions at different speeds to shear the soybeans. The beans are reduced in size, the hulls are loosened and a vibrating screen is then used to separate the broken particles into hulls and soybean kernel pieces. The hulls are removed by aspiration. Soybean particles are subsequently flaked.

*Conditioning*: This process is carried out to increase the plasticity of the soybean particles so that when they are flattened into flakes they do not break up and produce fines. The soybean particles are heated to 65–70°C with a moisture content of about 11%.

*Flaking*: Flaking machines are similar to cracking machines but the rolls are smooth. Conditioned soybean particles are flattened between the rolls. The pressure between the rollers is adjusted to produce a suitable flake thickness of about 0.2–0.4 mm. One purpose of flaking is the same as for expelling – to reduce the distance that the solvent and the extract will have to travel in the process of extraction. Another is to increase the contact surface between the oilseed solvent (normally hexane), allowing the oil to be extracted more efficiently.

*Alternative processes*: The processes described above are conventional oil-mill operations. Other processes include:

- The 'Hot Dehulling (Popping) System' (Buhler-Miag Ltd). Soybeans at a moisture content of 13% are heated to 60°C. The beans are then transported on a fluidised bed of hot air, causing the soybean hulls to split. The hulls are separated by aspiration. The advantage of the process is its lower energy consumption, since the multiple heating and cooling, drying and humidification steps of conventional dehulling are not required. The short heat treatment time prevents protein denaturation.
- The 'Alcon Process' (Lurgi GmbH). Flaked soybeans are humidified and heated in a conditioner, then tempered for 15–20 minutes whereby the flakes are fused into more compact, porous granules. The granules are then dried and cooled before solvent extraction. The rate of percolation of solvent through the granules is tripled. The amount of solvent which remains in the spent granules is about 30% less than flakes so less energy is needed to remove it.

Other methods include extruding soybeans into pellets prior to solvent extraction.

#### *Principle of solvent extraction*

The extraction of oil from oilseeds by means of a solvent (also known as miscella) involves three steps:

- diffusion of the solvent into the solid;
- dissolution of the oil droplets in the solvent; and
- diffusion of the oil from the solid particle to the surrounding liquid.

The diffusion step takes the longest so oilseeds are rolled into thin flakes to increase the rate of extraction. An open, porous structure of the solid material is best because it facilitates diffusion as well as percolation.

A counter-current extraction process is carried out. Flakes with the lowest oil content are extracted first with fresh solvent. Newly arriving flakes with high oil content are extracted with solvent containing oil already extracted from earlier batches of soybean flakes.

*Types of extractors*

Two types of solvent extractors are in current use:

- Semi-continuous systems consist of several batch extractors connected in series. The miscella flows from one extractor to the next one in the series. When the oil is completely removed from the first extractor, the second basket receives fresh solvent while the first extractor is emptied and refilled with a batch of fresh flakes. This extractor then becomes the last unit to receive miscella. The process is repeated.
- Continuous extraction. Both the oilseeds and the solvent are fed into the extractor continuously. In the de Smet belt extractor, flaked soybeans are fed onto a moving belt via a hopper. The solvent is introduced at the spent flake discharge end (opposite the flake feeding side of the extractor) in a spray which percolates through the bed, giving the spent flakes a final wash and removing some oil. The resulting miscella is collected underneath the belt and resprayed onto the flakes at the next section in the direction opposite to belt movement. This process of miscella collection, pumping and spraying at the next section is repeated.

*Miscella distillation*

The oil content of the miscella after undergoing the extraction process is around 30% oil. About 2.5 tonnes of solvent needs to be recovered for every tonne of soybean processed. Flash evaporation, vacuum distillation and steam stripping are all used to recover the solvent.

*Removal of solvent from the meal*

The spent flakes contain about 35% solvent. The recovery of the solvent is critical since it determines the quality of the meal. The lower the temperature and contact time with heat the less the protein is denatured. The desolventiser-toaster consists of a vertical cylindrical stack of compartments fitted with stirrers or racks attached to a central vertical shaft. Spent flakes are fed at the top of the stack and direct or indirect steam heating is carried out. The meal is then dried and cooled.

While desolventising-toasting is the standard method for the manufacture of soybean oil meal for animal feeding, this process is not suitable for the production of meal with minimum protein denaturation destined for the production of soybean protein isolates, most concentrates and texturised products. For these products flash desolventising is used. In this process, the spent flakes which exit from the extractor are fluidised in a stream of superheated solvent vapours which provides the energy for the evaporation of solvent from the flakes.

**Processing of extracted oil**

The extent of refining steps applied to oils depends on their source, quality, and end-use. Many oils are used for edible purposes after simply clarifying by settling or filtering. Most cold-pressed oils (notably olive oil) can be used in food products without further processing. However, the demand for bland-tasting oils with a long shelf-life requires several refining steps to be carried out.

***Refining***

Components such as free fatty acids, waxes, colour bodies, mucilaginous materials, phospholipids, and carotenoids contribute undesirable properties in oils. Crude oil is refined in three main steps comprising neutralisation, bleaching and deodorising, to remove fatty acids, colour and off-flavours, respectively. Each of these processes involves heating the oil to a certain extent. The refining processes may be carried out in stages either in batches or continuously. The fatty acids are neutralised by mixing the oil with a solution of caustic soda (sodium hydroxide). The caustic soda solution reacts with the fatty acids to produce soap, which can be washed from the oil with water. The soap solution, known as soapstock, can be used for soap manufacture. The oil is then bleached by adding a powdered clay known as 'fuller's earth'. This operation removes coloured pigments and produces a light-coloured oil. Deodorisation is carried out by passing high-pressure steam through the oil under vacuum to remove the taints and odours present in the oil at the beginning of the refining process and those produced at the neutralising and bleaching stages.

Separation of high-melting-point glycerides, or stearin, usually requires very slow cooling in order to form crystals that are large enough to be removed by filtration or centrifuging. Palm oil is often winterised (subjected to progressive cooling which allows saturated fatty acids to crystallise preferentially for removal by filtration) for simultaneous production of a hard fat called stearin (high in stearic acid content for special uses such as vegetable shortening) and of liquid oil called olein.

Hydrogenation (a process of hydrogen addition to double bonds in the presence of a catalyst, normally nickel) is carried out to produce solid fats from liquid oils. Margarine and many shortenings are produced this way. Thus oleic or linoleic acids, which are normally liquid at room temperature, can be converted to the saturated stearic acid. During the hydrogenation process, isomerisation (rearrangement of the molecular structure) also takes place, producing *trans* fatty acids. The *trans* isomers have much higher melting points than the naturally occurring *cis* form. When *trans* fatty acids are formed there is also a migration of double bonds along the chain. Thus isomers of oleic acid may be formed with the double bond in any position from carbon atom 2 to carbon atom 17.

## Weaning foods

Wholesome food is essential for growth and development. Breastmilk is the best and safest food for young babies. However, as the child develops, milk alone cannot provide the nutrients and energy requirements to enable the child to grow and thrive. From the age of around 4–6 months other foods must gradually be introduced into the diet to complement the milk intake. This process is known as weaning, the process whereby a baby gradually becomes accustomed to solid food and the adult diet. During weaning the child's eating pattern is gradually changed from the consumption of only milk to the mixed diet eaten by the rest of the family. Weaning foods need to be nutritionally balanced and energy-dense in order to support the period of rapid growth that occurs in children of this age.

Protein energy malnutrition is a serious problem in pre-school children in many developing countries. If a child's food is insufficient in quantity, given at an inappropriate time, or is of poor nutritional value, the child will fail to grow and will become malnourished. Such children are more susceptible to other illnesses, suffer from stunted growth and possibly mental retardation. Weaning is a critical period in a child's life when malnutrition may occur.

As the child grows, other foods must be introduced to provide the range of nutrients necessary for healthy development. However, care must be taken that any additional ingredients do not affect the safety, preparation, functional properties or organoleptic characteristics of the product.

There are two possibilities for the development of weaning foods: at household level or for industrial production. The first is the enrichment or supplementation of traditional foods and the second is the development of a 'new food'. It is crucial that any weaning foods are prepared to the highest hygiene standards as they can be a major source of infection to a very vulnerable group.

When developing weaning foods, for either commercial or domestic purposes, local habits and taboos, current weaning practices and the availability of ingredients, current nutritional deficiencies, water supply and food processing technologies all need be considered.

Possible actions in preparation for the production or introduction of weaning foods might include:

- dietary surveys to determine the kinds of foods consumed, the amounts ingested, the frequency of consumption, as well as the general characteristics of the diet in terms of its major components, and their availability;
- seasonality of food supplies;
- collection of information related to the traditional dietary practices of the local population;
- understanding the food preparation processes used at the home level;
- understanding the function of foods in the diet both nutritionally and organoleptically;
- establishing the cost of locally available, or imported, ingredients;
- establishing local hygiene standards with particular emphasis on water supplies;
- considering the possibility of using by-products from existing food processing operations, for example, broken rice or bran products;
- taking into account the presence of naturally occurring anti-nutritional factors; and
- examining traditional weaning practices.

Home-prepared food can give babies all the nourishment they need providing suitable foods are selected and are freshly and hygienically prepared. Food given to young children should not contain salt or spices. It is important that weaning foods contain the correct proportions of digestible energy and protein.

It is vital that the food is thoroughly cooked to ensure that pathogens (bacteria and parasites) are destroyed and that all feeding utensils (including the child's hands) are scrupulously clean. For these reasons it is best that the child's food is kept separate from the rest of the family's – this will also allow the amount the child eats to be monitored. Also preparing the porridge just before feeding will give pathogenic micro-organisms less opportunity to grow (Plate 29).

In many countries young children are first fed on porridge made from cereal flour (e.g. rice, maize, sorghum, oats, barley), root crops (e.g. cassava, yam, cocoyam, potato, sweet potato) or starchy fruits (plantain, breadfruit, banana) and water. Indeed, most of the manufactured first-stage weaning foods are based upon cereals, as their bland taste and smooth consistency are acceptable to the child's developing palate. However these staple foods do not contain enough of the important nutrients. Root crops contain very little protein, cereals contain some protein but lack some of the essential amino acids. Most cereals contain negligible amounts of vitamin A or C.

When cereal flour is cooked in water the starch granules swell and gelatinise. Different cereals absorb lesser or greater amounts of water; for example, millet absorbs more than maize which absorbs more than rice. A very thick, sticky porridge is difficult for the child to eat, therefore it is often necessary to add more water to thin it down. However this also dilutes the energy and nutrient density of the porridge. Thin porridges may only contain about 0.5 kcal per ml, whereas breast milk contains 0.75 kcal per ml. Starchy root crops do not absorb as much water as cereal flour during cooking because they already contain a lot of water, so porridges (paps) made from these are also very bulky. Options for overcoming the problem of bulky foods for children include frequent feeding, enrichment (e.g. with legumes, oils and fats, fruit and vegetables) and pretreatment of the starch components to make the porridge thinner.

## Pretreatment of starch

### *Fermentation (souring)*

This breaks down the starch granules so that they absorb or retain less water. The porridge does not thicken on cooling, so more flour, and therefore more calories, can be added to the mixture. In addition to its thinning effect, fermentation also improves the keeping quality of the porridge as the increase in lactic acid helps to prevent the growth of harmful micro-organisms. A typical fermentation might include soaking grains in water for 12 h. The grains would be pounded and sieved to remove excess water and then left to ferment for 48 h.

### *Germination (malting)*

The terms sprouting, malting and germination are often used interchangeably to describe the process of soaking or steeping dry grains in water until they are saturated, followed by germination under controlled conditions. Most cereals and legumes can be germinated and then used to make food. These include the cereals (rice, wheat, maize, barley, sorghum, millet and rye) and the legumes (cowpea, faba beans, chickpea, winged bean, pigeon pea, etc.).

During germination a large variety of plant endogenous enzymes are produced or activated which degrade the starch, lipid and protein to easily available, water-soluble products. It can be appreciated therefore, that humans absorb the nutrients in germinated grains more easily than those in ungerminated grains. A typical germination might include soaking grain for 1–2 days in potable water. The grain is then drained, rinsed in clean water, spread on cloth or mesh (to allow the air to circulate) and covered with cloth or leaves to retain the moisture. When the grains have sprouted they are rinsed in clean water and then carefully sun-dried to below 10% moisture content.

Some of the advantages of germination include:

- reduction of viscosity, to enable a liquid consistency and increased energy density;
- decrease of tannin content in, for example, *Vicia faba* and red sorghum, due to complexing with proteins. Also, lectins are inactivated in *Phaseolus vulgaris*;
- increased availability of amino acids including lysine, tryptophan and methionine;
- phytic acid is partly degraded; however, this is not reflected in increased availability of minerals;
- oligosaccharides are gradually degraded; and
- germination is easy to carry out; it can be done without sophisticated equipment.

### *Roasting*

Roasting grain in a large pan, with frequent stirring to prevent burning, partially gelatinises the starch which is therefore unable to absorb so much water. Cooking time is reduced, as is the microbial load. This improves the safety of the product. Enzymes and anti-nutritional factors (e.g. cyanide in cassava) are inactivated, thus improving storage time. Taste and digestibility are also improved.

### Formulation of a weaning food

In order to identify suitable ingredients for a weaning food the following factors need to be considered:

- What foods are available locally?
- Can the family afford them?
- Are any foods prohibited on traditional or religious grounds (locally acceptable)?
- Are the foods considered to be suitable for a young child?
- Are they easy to prepare, using traditional methods?
- Are they palatable?

Once possible ingredients have been identified, the chemical and nutritional content of each needs to be examined to determine the formulation of the food. The daily requirements for young children are shown in Table 8.14. It is important that any nutritional changes due to processing, such as the loss of vitamins, are considered. The basic nutrients found, to a lesser or greater extent, in all food commodities comprise carbohydrate, protein, fat vitamins and minerals.

#### *Energy*

Energy is needed for growth but the total energy content of the infant's diet must be maintained within controlled limits. An insufficient energy intake could lead to failure to thrive, whereas an energy intake in excess of requirements may lead to obesity. The 'energy density' (amount of energy in a given quantity of food) is therefore important. For example, high fibre foods have a low energy density, while sugar and fat have a high energy density. An energy density of 38–48 kcal/100 ml is considered to be sufficient for most infants' needs.

#### *Carbohydrate*

As source of energy, carbohydrate intake must be controlled. If given in excess it may be converted into fat and stored as fat in the body. The type of carbohydrate included in the food is also important as babies need carbohydrate that is easily digestible. Sources of carbohydrate calories to be used in weaning food formulations include starchy foods, such as cassava, yams and plantain. Carbohydrates have an energy density of 4 kcal/g.

#### *Protein*

An adequate amount of protein must be provided in the diet, but too much protein should be avoided. Cereal grains can be considered to be protein sources, although the quality of the protein is low because of deficiencies in lysine and tryptophan (Table 8.15). Legumes are also an important source of protein and are often used to complement cereal proteins in the ratio 60–65% cereal to 25–35% legume. Animal products (meat, fish, milk) are the richest source of protein and must be included in the diet once breastfeeding is withdrawn. Proteins have

**Table 8.14** Daily requirements for energy, protein, fat, iron, iodine and vitamins for children of both sexes. Source: Burgess & King 1993.

| Age* (years) | Weight (kg) | Energy (kcal) | Protein (g) | Fat (g) Diet H | M | L | Iron (mg) | Iodine (μg) | Vit. A (RE) | Thiamine (mg) | Riboflavin (mg) | Niacin (mg) | Folate (μg) | Vit. C (mg) |
|---|---|---|---|---|---|---|---|---|---|---|---|---|---|---|
| 0–½ | 5.4 | 585 | 10 | — | — | — | — | — | 350 | — | — | — | 19 | 20 |
| ½–1 | 8.8 | 960 | 14 | — | 7 | 11 | 21 | 50 | 350 | 0.3 | 0.5 | 5.4 | 32 | 20 |
| 1–3 | 11.9 | 1250 | 14 | 35 | 5 | 7 | 13 | 70 | 400 | 0.5 | 0.8 | 9.0 | 40 | 20 |
| 3–5 | 15.9 | 1510 | 18 | 42 | 5 | 7 | 14 | 90 | 400 | 0.7 | 1.0 | 10.5 | 53 | 20 |
| 5–7 | 19.6 | 1710 | 20 | 48 | 7 | 10 | 19 | 90 | 400 | 0.8 | 1.1 | 12.1 | 65 | 20 |
| 7–10 | 25.9 | 1880 | 26 | 52 | 8 | 12 | 23 | 120 | 400 | 0.9 | 1.3 | 14.5 | 85 | 20 |

* 1–3 years means 1 year 0 month to 2 years 11 months.
— No value available. Assumption made that breastmilk covers needs.

**Table 8.15** Essential amino acid contents of some foods.

| | Deficient in | Source of |
|---|---|---|
| *Cereal grains* | | |
| Whole corn | Lysine, tryptophan | Methionine |
| Degerminated corn | Lysine, tryptophan | Methionine |
| Rice (milled, white) | Lysine, threonine | Methionine |
| Wheat flour | Lysine | — |
| Sorghum | Lysine | Methionine |
| *Cereal grain by-products* | | |
| Corn germ | Methionine | Lysine |
| Rice bran | Methionine | Lysine |
| Wheat middlings | Methionine | — |
| *Oilseeds* | | |
| Soybean protein | Methionine | Lysine |
| Cottonseed flour | Lysine | — |
| Sesame protein | Lysine | Methionine |
| Peanut | Lysine, methionine | — |
| *Other* | | |
| Yeast | Methionine | Lysine |
| Leaf protein | Methionine | Lysine |
| *Food legumes* | | |
| All kinds | Methionine | Lysine |

an energy density of 4 kcal/g. Weaning foods should contain around 15 g protein per 100 g portion.

All proteins are made from amino acids, the essential building blocks for muscle and tissue development. The body is able to synthesise many amino acids but eight essential amino acids cannot be synthesised and must be available in the diet if healthy growth and development are to be maintained. Some foods are deficient in some of these essential amino acids while others are a better source of them (deficiencies are relative to levels of amino acids found in milk protein) (see Table 8.15).

### *Fat*

Infants must receive an adequate amount of fat as it is the most important source of energy. Fat is also important to provide essential fatty acids and the fat-soluble vitamins, A, D, E and K vitamins.

Fat has an energy density of 9 kcal/g. No more than 10% fat should be included in a weaning food.

### *Vitamins and minerals*

Added vitamins and minerals may be needed to ensure an adequate intake or restore losses which may occur during processing. For children up to 6 months old sufficient vitamins are supplied from breastmilk. After that the diet needs to be supplemented with fruit and vegetables to provide the necessary vitamins and minerals. The function of some of the vitamins and minerals is shown in Table 8.16.

### *Fibre*

Unlike adults, infants do not benefit from a high fibre intake and therefore fibre levels must be controlled. Fibre can interfere with the absorption of essential minerals and because babies have a small stomach capacity it is difficult to consume sufficient quantities of fibre-rich foods to meet energy requirements. Weaning foods should therefore contain less than 5% fibre. Figure 8.11 shows how different foods can contribute to a well-balanced diet.

**Table 8.16** Function of some vitamins and minerals. Source: www.idfa.org.uk and Anon. 1998.

| Nutrient | Function | Source |
|---|---|---|
| Vitamin A | Essential for eye function, bone formation, the immune system, differentiation of tissues and for growth and reproduction | Retinol – in liver, fish, eggs and milk. Carotene – in dark green leafy vegetables, yellow or orange vegetables and fruit |
| Vitamin C | Essential for wound healing and for the development of connective tissue. It is essential for the absorption and metabolism of non-haem iron | Vegetables and fruit |
| Iron | A component of haemoglobin and many enzymes. It is essential for the supply of oxygen to the body | Meat, fish, eggs, pulses |
| Iodine | Essential for the thyroid hormones which control metabolic rate, the development of the brain and central nervous system | Plants and animals in areas where the salt is rich in iodine. Seafood is rich in iodine |
| Calcium | Essential for the development of bones and teeth | Milk and milk products, beans and peas |

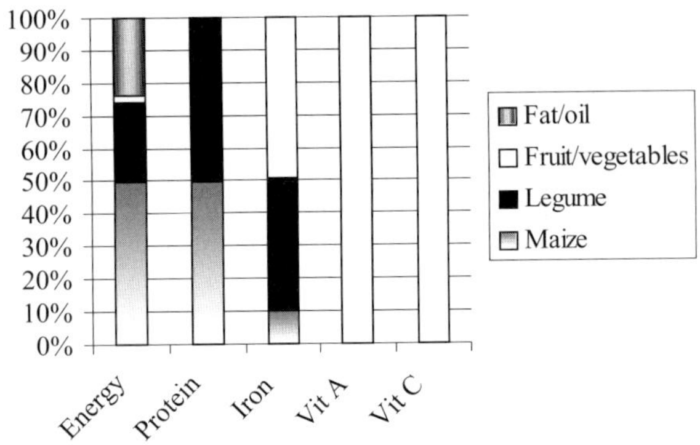


**Fig. 8.11** How different foods can contribute to a well-balanced weaning diet.

A typical weaning food might therefore contain:

- 80% cereal, such as maize, sorghum or millet (preferably roasted, germinated or fermented);
- 10% legumes, such as soybeans (roasted and peeled), chickpeas, lentils, beans and peas;
- 10% groundnuts; and
- supplements of fruit or vegetables.

### Small-scale production of weaning foods

While many mothers will simply prepare small quantities of weaning food for their own children, others may consider turning it into an income-generating activity. Whether the weaning food is to be prepared for home use or for small-scale retail sale, it is vital that all processes are carefully carried out under hygienic conditions, including storage of the raw materials (ingredients). It is essential that all raw materials are safely stored in a suitable, clean and dry environment. This is particularly important for cereals and groundnuts in order to prevent fungal growth and possible production of mycotoxins. If at all possible raw ingredients should not be stored for prolonged periods.

Before use the raw materials should be sieved or winnowed to remove any extraneous matter and ensure that the product is as clean as possible. Depending on the next processing step it might be advisable to wash the grain in potable water, if available.

The raw materials may be roasted, germinated or fermented before mixing and grinding (milling or pounding). A local custom-fee mill may be used or the mixture may be pounded in the traditional manner.

If the product is to be packaged or stored, careful consideration must be given to the type of packaging and storage conditions. As this is the final product, scrupulous attention must be paid to hygiene. The shelf-life of the product will be dependent on the type of packaging used. Any packaging should contain instructions for preparation and use of the product.

### Commercially produced weaning foods

Commercial weaning foods are expensive to produce. They must be specially formulated to meet nutritional requirements and produced to the highest possible standards of hygiene. Commercially available weaning foods fall into the following categories:

*Cereal-based weaning foods*: Usually based on precooked cereals with additional ingredients such as meat, vegetables or fruit. These are usually reconstituted with milk or water;

*Ready-to-eat baby foods*: Cooked and sterilised meals marketed in cans or jars;

*Dried baby foods*: Cooked meals which have been dried and powdered for reconstitution with milk or water; and

*Biscuits and rusks*: These may be used directly in the dried form or softened with water or milk.

Commercially produced weaning foods have advantages and disadvantages (Table 8.17).

### Legislation covering the manufacture of weaning foods

In Europe, manufactured weaning foods must comply

**Table 8.17** Advantages and disadvantages of commercially produced weaning foods.

| Advantages | Disadvantages |
|---|---|
| Quick to prepare | Expensive |
| May not need cooking | If contaminated water is used to reconstitute the weaning food it may not be heated sufficiently to destroy pathogens, and could pose a food safety risk |
| Hygienically prepared | If imported, supplies may be unreliable |
| Clean and safe | |
| Likely to be nutritionally balanced | |
| Palatable | |

with all relevant food legislation covering food hygiene, labelling, nutrition labelling, additives and contaminants. Legislation on additives prohibits the use of artificial preservatives, antioxidants, colourings and artificial sweeteners in commercial weaning foods.

In addition, weaning foods are a category of products contained within Council Directive relating to Foodstuffs intended for Particular Nutritional Uses (89/398/EEC), known as the PARNUTS Directive. Within this framework Directive, weaning foods are controlled by Commission Directive 96/5/EC of 16 February 1996 on Processed Cereal-Based Foods and Babyfoods for Infants and Young Children. Based on recommendations of the EU Scientific Committee for Food, this Directive sets standards for sodium, carbohydrate, fat and protein levels as well as defining which nutrients in the form of vitamins or minerals may be added to a product. This Directive has been implemented in UK law by the Processed Cereal Foods and Baby Foods for Young Children Regulations 1997 (SI No 2942) and amendment 1999 (SI 275).

As with all food products, baby food labels must list ingredients in descending order by weight, and foods intended for infants and young children must display a 'prescribed energy statement' which requires the declaration of protein, fat, carbohydrate and energy content. More detailed nutritional information is also given on the label relating specifically to saturated fat content, sugar, sodium and fibre. Other useful information may be provided, such as a statement indicating whether the product is gluten-free, suitable for vegetarians or contains added vitamin C.

In February 2000 new labelling regulations (SI No 1398, 1999 implementing EU Directive 97/4/EC) required all foods, including baby foods, to provide information on the percentage quantity of the main characterising ingredients in the product (Quantitative Ingredient Declaration – QUID).

EC Directive 1999/50/EC sets a maximum limit of 0.01 mg/kg for individual pesticides in baby foods and prohibits the use of certain pesticides in their manufacture. This Directive will soon be implemented in the UK and many products already comply. All commercial baby foods, however, must comply by 1 July 2002 (www.idfa.org.uk).

## Animal feeds

The farming of domestic livestock for the production of meat, milk and eggs has been steadily increasing due to demand from expanding world populations. Although many of the world's livestock are ruminants raised on extensive rangelands in developing countries, large populations of monogastric animals, such as pigs and poultry, are grown under more intensive conditions where the total package of nutrients for production is provided by the livestock farmer, rather than obtained through scavenging.

Statistics from the FAO illustrate the importance of livestock products within the global economy. For illustrative purposes, Table 8.18 presents world production levels for meat, eggs, milk and fishery products from aquaculture for the year 2000.

Much of the milk and ruminant meat is produced by grazing, though high-yielding cows require additional feed sources beyond grazing. Pig, poultry and aquaculture products are highly dependent on feed provided.

Expansion in knowledge of livestock nutrition, genetics and animal management has enabled feeding systems to be designed for optimum livestock performance according to a range of environmental, feed supply and production systems. Since feed is the main input into a livestock production system, it is important that feeds are formulated to provide appropriate levels of nutrients at least cost.

Many of the raw materials, which are the nutrient sources for livestock, are the by-products of agro-processing for refining cereal flours and extraction of oils from oilseeds for human consumption or industrial raw materials. Changes in the processing technologies for the manufacture of human foods will have a corresponding impact on the nutritional value of by-products. Changes in nutritional value of raw materials will influence their price and inclusion level in feed. To accommodate fluctuating raw material supply and price, most commercial livestock feeds are formulated on a least cost basis by computer.

**Table 8.18** World production of livestock products in 2000. Source: FAO. http://apps.fao.org/page/collections?subset=agriculture

| Livestock product | Metric tonnes |
|---|---|
| Beef and buffalo meat | 60 233 078 |
| Sheep and goat meat | 11 204 408 |
| Milk (all sources) | 568 486 839 |
| Pig meat | 90 909 402 |
| Poultry (chicken, duck, turkey) meat | 65 510 499 |
| Eggs | 54 727 606 |
| Aquaculture (fish and crustaceans) | 38 964 000 |

The process of animal feed manufacture is not just the blending of a series of raw materials to a fixed and proven formula, but is a dynamic relationship between:

- the nutritional requirements of the livestock to be fed according to their productive phase;
- the nutritional characteristics of the raw materials available;
- the physical, chemical and microbiological characteristics of the raw materials and their ability to flow, and changes due to the application of heat and moisture; and
- the consumers' requirements for feed in any specific formulation according to the feeding system. This is particularly important when feeding aquaculture species, which are feeding in a water environment.

### Feed nutrients

In general terms, the nutrients required for the growth of animals are the same as those needed by humans. Many come from the same agricultural raw materials. The most important nutrients are carbohydrates (which includes starches, sugars and fibres), fats, proteins, minerals, vitamins and water. Table 8.19 gives a summary of nutrients for livestock and from what sources they may come.

Animals can obtain energy for work, growth or reproduction from carbohydrates, fats and proteins, but the most cost-effective sources are carbohydrates and fats. Protein is primarily needed for tissue and muscle development. Fat has the highest energy value per unit weight, though starch is often the cheapest source of energy.

**Table 8.19** Summary of nutrients needed for livestock.

| Nutrients | From | Sources | |
|---|---|---|---|
| Energy | Carbohydrate | Cereals | |
| | Starch | Cassava | |
| | Sugar, fat | Oil, fat | |
| Non-essential amino-acids | Proteins | Cereals, vegetable meals, animal meals | |
| Essential amino-acids | Proteins | | |
| Lysine | | Vegetable meals | |
| Methionine | | Soya meal | |
| Cystine | | Soya meal | |
| Tryptophan | | Sunflower meal | |
| Threonine | | Animal meals | |
| Leucine | | Meat bone meal, fish meal, etc. | |
| Isoleucine | | Meat bone meal, fish meal, etc. | |
| Macro-elements | | | |
| Calcium | Chalk, calcium carbonate | | |
| Phosphorus | Dicalcium phosphate, tricalcium phosphate | | |
| Sodium | Salt | | |
| Micro-elements (ppm) | | Needed by pigs | Needed by chickens |
| Iron | | 100 | 40 |
| Zinc | | 100 | 40 |
| Copper | Sulphate or oxide | 10 | 2 |
| Manganese | | 40 | 70 |
| Cobalt | | 0.1 | 0.2 |
| Iodine | | 0.1 | 0.1 |
| Selenium | | 0.6 | 1 |
| Vitamins (ppm) | | | |
| A, D, E, K | | | |
| B1 | | 1 | 0.5 |
| B2 | | 4 | 4 |
| B3 | | | |
| Biotin | | 0.1 | — |
| Folic acid | | 0.5 | 0.2 |
| B12 | | 0.03 | 0.01 |

### Carbohydrates

Carbohydrates are polymers of sugars linked together in certain structures.

Sugars are produced in plant leaves from carbon dioxide by the process known as photosynthesis. There are many different ways in which plants link sugars together to form plant food for growth, structure and storage energy. Some sugars are linked together to make the fibres which enable the plant to be protected, to form pipes for moving liquids within the plant, for stem and leaf structure, and to protect the new seeds as they grow. However the chemical linking of the sugar molecules makes fibres undigestible to man and other monogastric animals.

The same sugars may be linked in a different polymer structure to form starch, particularly within growing seeds. In plants like wheat and barley these seeds will eventually become the starch-rich cereal grains for animal feed or for milling to bread flour. Some sugars will remain as simple sugars and be moved about the plant and used to give energy for plant growth.

Starches and sugars are easily digested by livestock and are usually the cheapest energy sources for animal production. If there is an excess in the diet beyond what is needed, the body converts the excess starch and sugar into fat. Fat is an important flavour component in meat, and is also a useful way for some animals to sustain their development during periods of feed shortage.

The fibre carbohydrates (celluloses) are poorly digested by pigs and poultry and will pass through the intestines without releasing any nutrients. High levels of fibre in pig and poultry diets will also accelerate the passage of feed through the intestine to the extent that there is insufficient time for other available feed nutrients to be digested and absorbed. However, some fibre is needed (about 3–7% of the diet) to stimulate muscle function in the intestinal wall and keep feed moving in the gut.

Fibre levels in pigs and poultry feeds are therefore kept low, particularly for young animals. Many oilseed processors decorticate or remove the hulls from the seeds prior to extraction, in order to prevent the fibrous hulls from diluting the desired level of protein in the oilseed meal.

Ruminant animals such as cattle and sheep have a complex stomach called the rumen. The rumen acts as a large fermentation tank containing millions of microorganisms, which have the ability to break down the plant fibres and convert them to volatile fatty acids which are absorbed by the rumen wall into the bloodstream to be used as sources of energy.

### Fats

Fats are also used within the animal body to provide energy for work, reproduction and growth. In plants, where they originate, they are part of the storage energy sources in certain seeds. For example, cereals store their seed energy as starch, while oilseeds store their seed energy as oil or fat.

Fats are composed of fatty acids and glycerol. Some fatty acids are referred to as essential, because they cannot be synthesised in the body and must be provided in the diet.

The storage quality of fats is very dependent upon their structure. Reactive linkages in the fat molecule are susceptible to reaction with oxygen, leading to oxidative rancidity, which gives a bitter and unpalatable taste to fats and oils, accelerates loss of fat-soluble vitamins, and at high levels can make the oils toxic. A second kind of rancidity occurs when the fatty acids are split from the parent fat, so releasing free fatty acids. This reaction is called hydrolysis, and causes the fat to taste soapy. Hydrolysed fats, although less appetising than whole fats, are readily digested by livestock. However, fatty acids are corrosive to metal pumps and pipework, and for feed manufacturing it is preferable to use fats with low acidity.

Fats are particularly important for livestock because they contain more than twice the energy concentration of starches. Some of the fatty acids are also essential for the working of body tissues for maintaining good health.

Monogastric animals fed high oil diets will have soft fat in the carcass. This is particularly undesirable for pig meat. Animals fed starch as the main energy source will tend to lay down a harder fat in the carcass. Since yellow carotenoid pigments (xanthophylls) are present in the fat fraction in maize, poultry feed diets high in yellow maize will have yellow carcass fat, and lay eggs with yellow pigmented yolks. In contrast, poultry fed diets based on white maize, or barley, and without additional sources of carotenoid pigments, will produce white-fleshed birds, and eggs with pale yellow yolks.

### Proteins

Proteins are long, nitrogen-based polymers made within plants and animal tissue. In plants they are part of the cell structures, particularly in leaves and seeds. In animals they form muscle tissue and give structure to skin, hair, blood and all the body organs.

The building blocks of proteins are amino acids, of which there are 23 important ones. In order to promote animal growth, sufficient protein must be provided in the diet to enable new tissue to be built from ingested protein.

In general terms, for pigs and poultry, proteins derived from animal muscle tissue contain the appropriate mixture of amino acids for building animal muscle and tissue. These materials are expensive; the alternative is to blend plant proteins to give a mixture of amino acids that will enable the body to digest the proteins and rebuild them into muscle or tissue proteins. Since the body cannot synthesise certain amino acids they must be included in the diet. These are known as essential amino acids, of which 21 are important in animal nutrition. Those which are most important for poultry and pig feeds are lysine, methionine and cystine. If these are absent from the diet, then muscle tissue cannot be made. Hence a diet can be high in protein but produce little tissue growth, because the important links in the muscle chains are missing. It is for this reason that fishmeal and meat and bone are added to the diet to provide the missing essential or limiting amino acids.

In ruminants, a high proportion of the proteins are broken down by microbes in the rumen and rebuilt into bacterial proteins. These bacteria then pass into the small and lower intestine to be digested (in a way similar to protein digestion in pigs and poultry), before being absorbed into the blood.

For higher levels of ruminant production, particularly milk, it is desirable to have a proportion of the protein fraction of the diet that is not degraded by bacterial action, but passes through the rumen to the small intestine for digestion. A degree of protection is given to protein through heat treatment.

### *Minerals*

Most of the minerals in feeds originate from the soil. They have been absorbed by the plants into their structure and are essential for good health and growth of both plants and animals.

The most important elements are calcium and phosphorus, which are the key minerals in bone structure. If there is an imbalance in the absorption of these two minerals, bones may become weak or even break. These two minerals are also essential for good teeth development, eggshell production in poultry and milk production in cows.

Although plants contain a reasonable level of phosphorus, as indicated by analysis, much is bound into substances of low digestibility. Since mineral sources of calcium and phosphorus are more readily digested than plant sources, feeds are supplemented with calcium phosphates, which provide both minerals, or limestone (calcium carbonate) as a source of calcium.

Iron is essential for the structure and functioning of blood (lack of iron causes anaemia); selenium is involved in vitamin E utilisation, and cobalt is part of the vitamin B12 molecule.

The animal body needs a wide range of trace minerals in order to maintain efficient cellular function. If present to excess, some can be toxic. Hence, care must be taken in the formulation of mineral premixes for inclusion in livestock feeds.

### *Vitamins*

Vitamins have essential functions in the body to maintain healthy tissue and metabolism. Vitamin A is essential for cell function and eyesight, vitamin D for bone structure, thiamine (vitamin B1) is necessary for the conversion of energy within the body, etc. As with minerals, an excess of certain vitamins can cause toxicity (particularly vitamin A).

### *Enzymes*

Enzymes are special proteins built up from amino acids. There are many different types of enzymes, each with a different and often very specialised function. They are the 'fixer' molecules in the body, the biological catalysts which make things work. For example, one set of enzymes will enable the leaf cells in a plant to convert carbon dioxide and water into sugar; another set will enable the glucose sugar to be converted into starch, while another enzyme is necessary to build the same sugars into complex fibres. While pigs and poultry produce enzymes for digesting starch, these same enzymes cannot digest the fibres made from the same sugars.

Without enzymes there can be no production or utilisation of nutrients for livestock life, growth or reproduction.

## Digestion of nutrients

Digestion is the process by which feed raw materials are broken down within the alimentary canal into simple substances which can pass through the wall of the intestine into the bloodstream. Once absorbed, the body rebuilds these simple substances into the tissues of the body, or the nutrients provide energy or keep the system operating efficiently (metabolism).

### *What happens to feed in the alimentary canal?*

*Pigs and poultry*

The basic stages are intake by mouth or beak, followed by muscular movement down the oesophagus to the stomach (of different types depending on the animal) where feed is acidified to coagulate the protein, and well mixed. On leaving the stomach the mix is made alkaline and prepared for true enzymatic digestion in the small intestine. Most of the enzymes are secreted into the small intestine from an organ situated just below the stomach, called the pancreas, while the liver secretes substances like detergents to help in the digestion and absorption of fats. After digestion the small nutrient molecules of sugars, amino acids and fatty acids, vitamins and soluble minerals are absorbed through the intestine wall into the bloodstream where they are moved to the appropriate site in the body for use. The undigested food passes down the intestine to regions where excess water is re-absorbed back into the body. In pigs and ruminants there may be some additional breakdown of fibre by bacteria, and possibly the absorption of vitamins from bacterial action. The undigested material then leaves the body for somebody to clear up, store and hopefully return to the land as nutrients to stimulate further plant growth.

*Cattle and other ruminants*

Cattle chew their feed excessively in the mouth before swallowing for activity in the rumen. The rumen is not an acid secreting, predigestion vessel like the true stomach of pigs, but is a fermentation system through which all feed must pass before entering the small intestine for normal enzymatic digestion.

Ruminant animals, through the enzyme activity of rumen micro-organisms, have the capability to degrade cellulose fibres to compounds (volatile fatty acids) capable of giving energy to the animal.

Proteins in raw materials are also degraded by bacteria and the nitrogen-based products of fermentation are built into bacterial protein. On leaving the rumen, these bacteria are digested in the small intestine by enzymatic processes similar to those of pigs and poultry.

Fats may be hydrolysed to fatty acids in the rumen, but high levels of fat inhibit microbial action and can disturb the effective functioning of the rumen. On leaving the rumen, fats will be digested in the small intestine and absorbed through the gut wall into the blood.

Undigested material will pass into the large intestine where water is reabsorbed into the body and secondary bacterial fermentation will produce some nutrients for absorption before the waste leaves the animal body.

## Feed and its relationship to animal production

Modern livestock species possess fairly well-defined genetic characteristics for growth and performance in terms of liveweight gain, meat, milk or eggs. Performance can be predicted with some degree of certainly on the assumption that dietary, management and housing conditions can be achieved.

The preceding sections have described the background information necessary for ration development and the planning of feeding regimes for livestock. This is a subject for specialists, but the following will serve to illustrate the stages in the planning of animal feeding systems.

## Rational feeding

### *Qualities of feed rations*

The target of rational feeding is to bring to each animal all the nutrients it needs, in the right proportion and at the right time. The feed and the daily ration must have six qualities at the very least:

- easily eaten
- complete – providing all the nutrients required for each animal
- well mixed
- well balanced
- sufficient, and
- adapted to the physiological phase of production.

This good-quality daily ration is necessary for success in breeding and production. Good feed is not the only criterion. It is also necessary to satisfy some other conditions including good buildings; providing animals with a certain comfort with more or less constant temperature, sufficient water and air and efficient manure removal and cleaning; disease-free animals and animals with good genetic potential.

### *Feed and ration*

A ration is the package of nutrients that an animal needs to meet its daily production target of, say, 650 g per day liveweight for a grower pig or 15 litres per day of milk from a dairy cow.

For the pig, all the nutrients for its daily ration may come from one source – the feed from the mill. To grow at 650 g per day, it will need a certain intake of energy, protein, etc. included in a certain weight of feed. The feed may contain all the right nutrients, but if insufficient is fed, then the production will not be achieved. In contrast, excess feed may be wasteful of nutrients, and may not result in cost-effective improvement in performance.

For the dairy cow, a large proportion of its daily nutrient needs for the production of 15 litres of milk per day will come from grass or hay, silage, etc. The balance of nutrients to meet its ration for production will come from compounded dairy meal (concentrate) prepared in the feedmill or on farm. Knowledge of the nutritional quality of grazing is as important as that of the milk concentrate.

## Raw materials and their processing

### *Raw materials as sources of feed nutrients*

Raw materials for feed processing generally fall within the following groups:

- cereals and their by-products;
- oilseeds and their by-products;
- animal protein sources; and
- miscellaneous – those not coming into these categories.

Many of the raw materials used for feed manufacture or 'compounding' are derived from other agroprocessing operations. The following sections illustrate the processes through which common raw materials are prepared for animal feed use. Many are the by-products of seeds (cereals or oilseeds) whose primary product is food for human use. Data on 'typical' nutritional content of processed materials are given as examples only. It is important to obtain the real analytical data for each raw material and from each supplier.

### *Analytical values*

The nutritional constituents of feed raw material specifications are determined by chemical analysis. The steps in the analysis chain are illustrated in Fig. 8.12. The most important components are moisture, protein, fat, fibre, ash, calcium, phosphorus and salt, though there could be many more including amino acids, anti-nutritional factors, microbial or other toxigenic contaminants. However, so far there are no fully internationally agreed methods for measuring their presence in feeds. When writing feed specifications it is important that the chemist defines which method is to be used for the specification. For example, the UK method for measuring fat in sunflower meal may be different from that used in its country of origin, say, Argentina. If sunflower meal is imported from Argentina at a specification of 5% fat content but on arrival it is found, using the UK method, to contain only 4% fat, is this difference commercially important? Is it a difference in analytical methods of the seller and the buyer, or is it a real difference that has financial implications for the value of the shipment? For trading purposes it is important to ask details of the

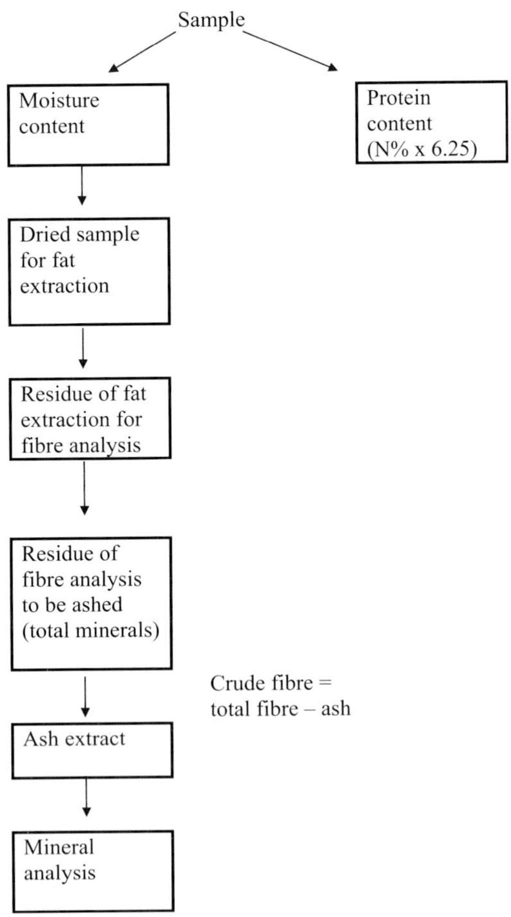


**Fig. 8.12** Steps in the analysis of the major nutrients in animal feed raw materials and compound feeds.

methods used, and if necessary to clearly state whose methods should be used, or to specify an independent laboratory which will analyse raw materials on behalf of the buyer and seller.

*Moisture*

Moisture (water) is one of the most difficult substances to measure because of the way in which it is chemically or physically present within a material. In feed materials it is normally measured as the weight loss during drying at a fixed temperature (for example, 105°C for 5 h). Other methods might require 100°C overnight in a vacuum oven. Different methods may give slightly different results for the same sample.

*Crude protein*

Proteins are composed of carbon, hydrogen, oxygen, nitrogen and sometimes sulphur. It is very difficult, though not impossible, to measure levels of true proteins, but to overcome the costs and difficulties we measure not protein but the level of protein nitrogen in the sample. Although there are many types of proteins almost all contain a similar proportion of nitrogen within the protein molecule. Each 100 g of protein contain approximately 16 g of nitrogen. Hence to determine the level of protein the level of nitrogen extracted from the protein is multiplied by 6.25 (i.e. 100/16). Interestingly, wheat flour millers use a factor of 5.7 rather than 6.25 since this factor is specifically representative of the amino acid protein in Canadian wheats (Kirk & Sawyer 1991).

However, other compounds present in some feed raw materials also contain nitrogen. The nitrogen in these compounds will also be extracted in the protein analysis, and therefore increase the level of apparent protein to a level which is greater than the true level of protein. This is why the method is called 'crude protein'.

*Crude fat*

Once a sample has been dried, the fat is extracted with a hot solvent (light petroleum spirit). After extraction the solvent is evaporated and what remains is fat. However, other substances are soluble in the solvent, such as plant pigments, and these are also present in the extracted fatty material. What is, therefore, measured is 'crude fat'. If the fat has become rancid due to exposure to oxygen or to fat breakdown to fatty acids, then these levels will also need to be measured independently since they may not have been extracted in the solvent extraction.

*Crude fibre*

In recent years there has been much discussion as to the usefulness of this analysis from a nutritional point of view, because the analytical result does not correlate with any real levels of animal performance. However, the method has been accepted and used as a measure of quality of feed raw materials, and is likely to remain in use for many years to come. Again, the word 'crude' fibre means that we are measuring a mixture of things and nothing very specific. What is being measured is the level of fibrous cell tissue which remains after moisture and fat extraction, after it has been reacted with weak acids and weak alkalis (to remove the starches, sugars and proteins). It is a measure of the level of potentially indigestible matter (fibre) in the raw material that is to form part of a diet for pigs or poultry. Ruminant animals like cattle and sheep have the ability, through the action of bacteria in the rumen, to partially break down the chemical components of fibres to volatile fatty acids for energy. Nevertheless, crude fibre is a useful indication of the maturity of a plant feed. The older the plant, the higher the fibre level. The analytical value for crude fibre is the total fibre less the quantity of minerals (ash) retained in the extracted fibre fraction

*Ash*

Ash is the total mineral matter in a raw material. It is the residue left after the material has been burnt at about 550°C in a special furnace. The ash contains all the calcium, phosphorus, trace minerals, etc. as well as undesirable contaminants like sand. If the ash is washed in acid, the soluble minerals dissolve and the sand (silica) remains. The ashing of a raw material is often the first stage to extracting the sample for the estimation of calcium, phosphorus or other nutrient minerals.

*Minerals and trace elements*

Calcium and phosphorus are the two important elements for bone structure. There are several methods the chemist can use to measure the level of these elements. Some require the complexing of these metals with compounds in solution, which give colour changes in relation to the quantity of metal present. Others use the colour and intensity of light emitted when a solution of the metal is burned in a very hot flame. These are called spectrophotometric and chromatographic (light and colour) methods. Each requires the comparison of results from the sample with those of standard solutions containing known amounts of the desired element.

*Salt (sodium chloride)*

Salt is extracted from feed with cold water and measured by complexing it with a silver compound or by using an electronic meter (chloride meter).

*Starch*

Starch is not always measured in feed raw materials, though it can be measured by converting the starch to glucose with acids or enzymes, and then measuring the sugar produced. The non-fibre carbohydrate fraction (most of which is starch) is often called nitrogen-free extract (NFE) and is equal to 100 – (% moisture + % protein + % fat + % fibre + % ash).

### Data sources on raw material composition

Commercial feed manufacturers will build up complete databases on the analytical composition of the raw materials used in formulations and relate these to the suppliers. For indicative formulations, the general nutritional composition of raw materials can be obtained from standard databases (e.g. Bo Gohl 1975; Food Standards Agency 2001).

### Cereals and their by-products

*Cereals*: Cereals are purchased unprocessed for use in the feedmill. They may have been partially dried with forced warm air to reduce the moisture level at harvest to that suitable for safe storage (14% moisture).

*Cereal by-products*: Cereal by-products are primarily derived from the extraction of flour or starch from grains (Table 8.20). The by-product for feed use can be relatively high or low in starch, depending on the method of cereal flour fractionation. They are valuable sources of plant fibre, protein, minerals and micro-nutrients suitable for all classes of livestock.

The processes by which these cereal by-products are produced are illustrated in Fig. 8.13 (milling of wheat and rye), Fig. 8.14 (milling of maize) and Fig. 8.15 (milling dried brewers' grains).

### Oilseeds and their by-products

*Oilseeds*

Oilseeds are valuable sources of both energy and protein and their inclusion in a feed formulation will be dependent upon their price relative to the price of alternative sources of oil and protein. Since oilcakes and meals are the major raw materials of interest to the feed mill, their processing will be examined before that of whole oilseeds. Figure 8.16 shows the general method of oilseed processing.

*Oilseed cakes and meals*

Oilseed meals are the by-product of the processing of oilseeds for their oil content. The seeds are normally low in starch content, but high in protein. Many are covered with an outer fibrous layer which may be removed before the oil is removed. The removal of the outer covering is called dehulling or decortication. Oil may be extracted by crushing (expelling) or by crushing followed by extraction with chemical solvents (often hexane). The resulting product is an oilseed meal for feed use. An oilseed 'cake' is the product from expelling, since it has a hard 'cake' texture as it leaves the expelling machine. When the cake is broken or ground it becomes a meal. Oilseeds which are solvent-extracted produce 'meals'

**Table 8.20** Sources of cereal by-products for the feed industry.

| Cereal | Primary food product | By-product | Process |
|---|---|---|---|
| Wheat | Flour for bread and baked goods | Wheat bran/wheat feed | Dry grain roller milling and sieving |
| Rye | Flour for bread and baked goods | Rye bran | Dry grain roller milling and sieving |
| Maize | Maize starch for food or industrial processing (e.g. paper) | Maize gluten feed, maize gluten meal, maize germ meal | Wet milling |
| Barley | Beer/alcohol production | Wet or dry brewers' grains | Wet fermentation of malted barley grain |

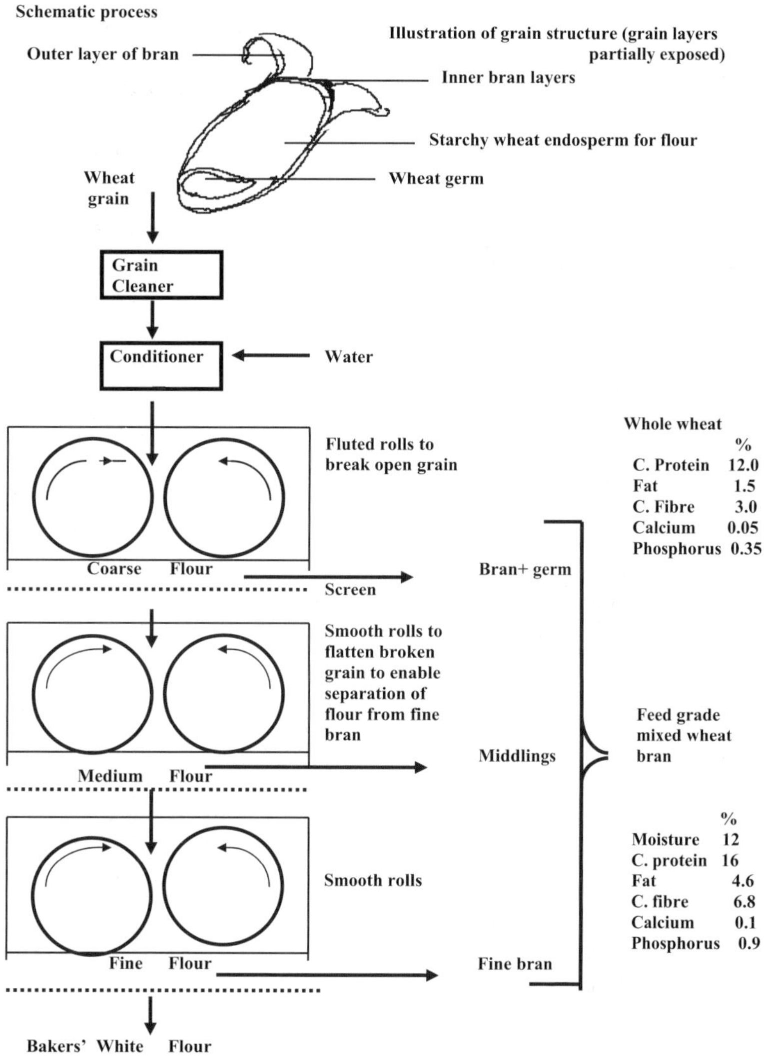


**Fig. 8.13** Milling of wheat and rye to separate the flour from the fibrous bran. This diagram shows only three pairs of rollers. In a modern roller mill there may be up to 20 pairs of rolls for the extraction of high quality flour. In most European mills the bran and germ are separated from the middlings and fine bran. Bran and germ are used in human food, and bran for horses. A mix of middlings and fine bran for feed use may be called wheatfeed.

as the product, though they may be ground to a smaller particle size before sale.

Decortication or dehulling may be only partial, since some fibrous material can aid the removal of oil during the expelling process. However, in most cases the removal of the fibre improves the efficiency of oil removal.

Within the expeller the oilseeds are subjected to high pressure as they are pushed forward by a worm shaft turning within a perforated press cage. As the oilseeds are compressed the oil cells are ruptured and the oil released through the perforated cage.

If the press is used as the only method of oil expelling, its throughput will be 20–25% of that when the machine is used for pre-expelling prior to solvent extraction. However, the important factor is to define the lowest capital and operating costs for oil extraction.

Oilseed meals processed by expelling may contain up to 15% oil in the meal, whereas solvent-extracted meals contain less than 5% oil. Some oilseeds contain anti-nutritive substances which must be removed if the meals are to be of optimum digestibility.

For the animal feed industry worldwide, the important oilseeds are soybean, cotton, sunflower, rapeseed,

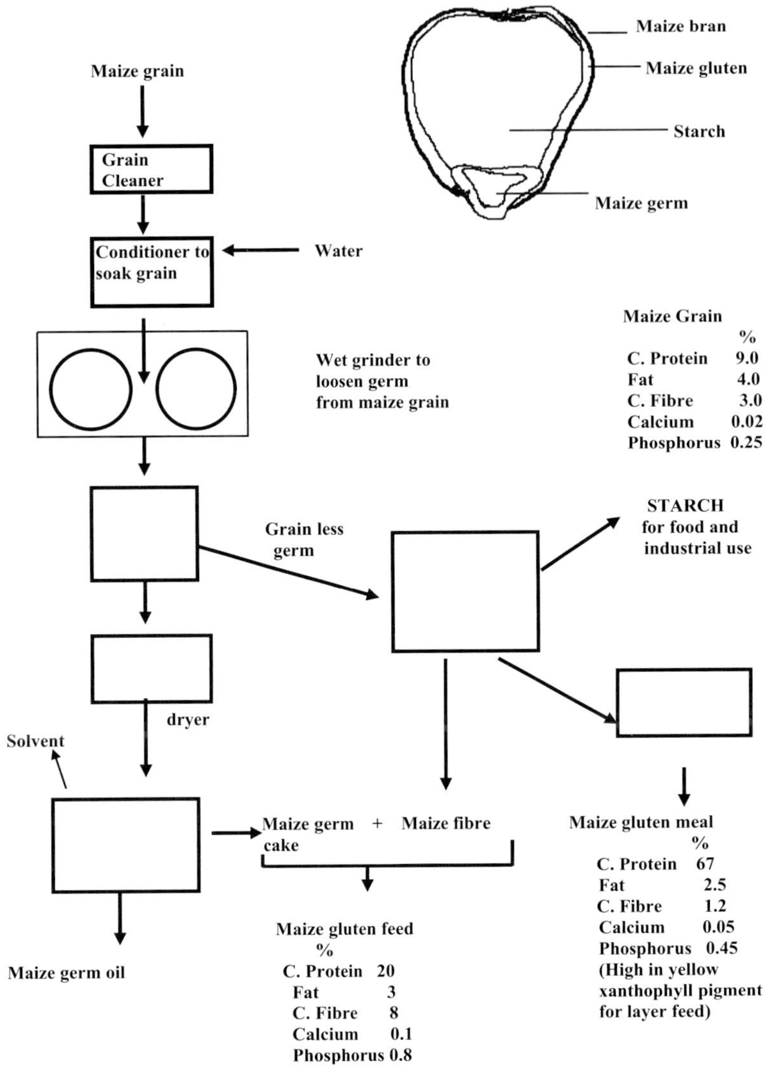


**Fig. 8.14** Milling of maize to separate the endosperm starch from the protein and fibre fractions of the grain. Whereas the bran by-products of wheat and rye are obtained from a dry milling process, maize by-products are derived from the wet milling of maize for starch extraction.

cotton seed, coconut, palm kernel, linseed, sesame and safflower.

### *Oilseed meals by type*

#### *Soybean meal*

Soybeans are dehulled before oil expelling or extraction. Different levels of hulls may be blended with the meal to give products of different protein, oil and fibre levels.

Raw soybeans contain anti-nutritive factors, which are not destroyed during pre-pressing by expelling or solvent extraction. After oil extraction it is therefore necessary to heat or 'toast' the extracted meal to inactivate the anti-nutritive factors – in particular, trypsin inhibitors. These inhibitors are proteins which preferentially attach themselves to the protein-digesting enzyme (trypsin) in the small intestine, preventing it from digesting the protein in the soybean meal. As a result the animal cannot utilise the soya protein for maximum growth. The heat treatment must be sufficient

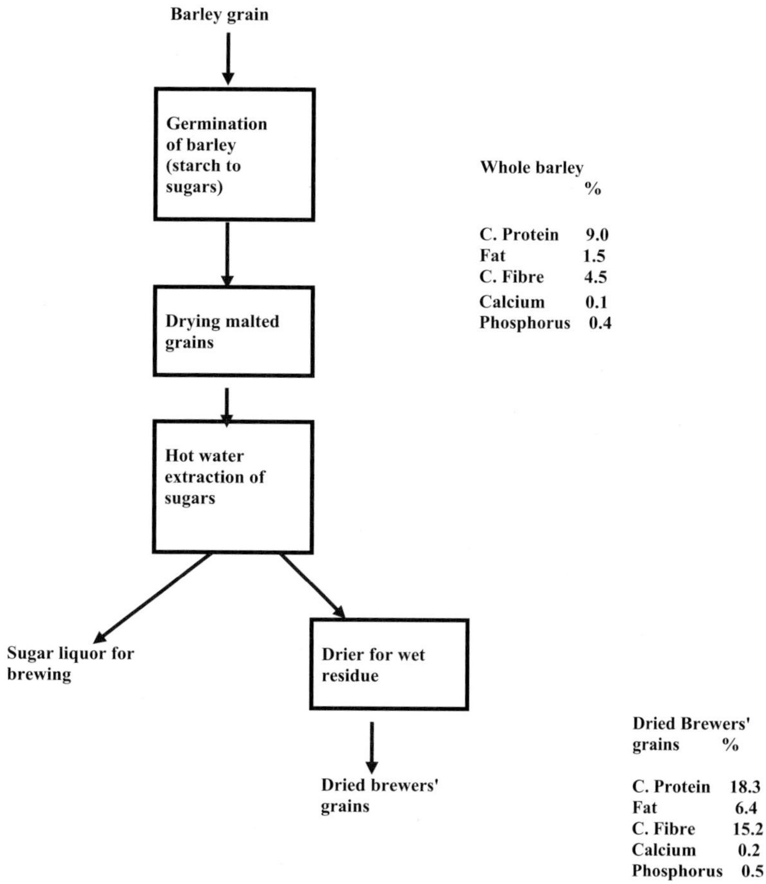


**Fig. 8.15** Milling of dried brewers' grains. The dried residue from barley fermentation for beer production. The barley grain is similar in structure to wheat. For brewing, the barley must be germinated to convert the starch to sugars (the malting process). The germination of the grains is stopped by heating, and the grains dried in a special oven to yield 'malted barley'. The dried grains are mixed with hot water to remove the sugars which will be converted to alcohol in the brewing process while the residues from hot washing are removed and dried for feed use – as dried brewers' grains.

to inactivate trypsin inhibitors, but not heat damage the soya protein.

In Europe the buying specification for soybean meal will include a level for trypsin inhibitor, and/or a level of urease activity. Urease is another enzyme present in soybeans which is also inhibited during heat treatment. However, in the laboratory it is easier to measure urease activity than trypsin inhibitor activity, so urease is used as an indicator of appropriate heat treatment. If the urease level is high, then heat treatment is insufficient. If the urease level is low, then this may indicate excessive heat treatment (i.e. excessive heat damage to the soya protein and a lower protein digestibility). Raw soybeans also contain blood-clotting agents (haemagglutinins), but these are also destroyed during toasting. Most commercial soya meals are extracted with hexane as the fat/oil solvent. Typical compositions are shown in Table 8.21.

### *Soybeans (whole processed)*

Although soybean meal is the most common form in which soya is fed to livestock, there has been a growing use of whole, full-fat soya in animal feeds. Soya oil is a very good source of energy for livestock and when present in whole soya may be a cheaper source of fat (energy) for animal feed diets than commercial feed grade fat.

The anti-nutritive factors of raw soya must be destroyed by heat treatment before the whole beans can be used in an animal feed.

There are several commercial methods for heat-treating whole soya beans: the two most common are extrusion processing and treatment with infra-red heat radiation. The purpose of both processes is to heat the beans to a temperature which is both sufficiently high to denature

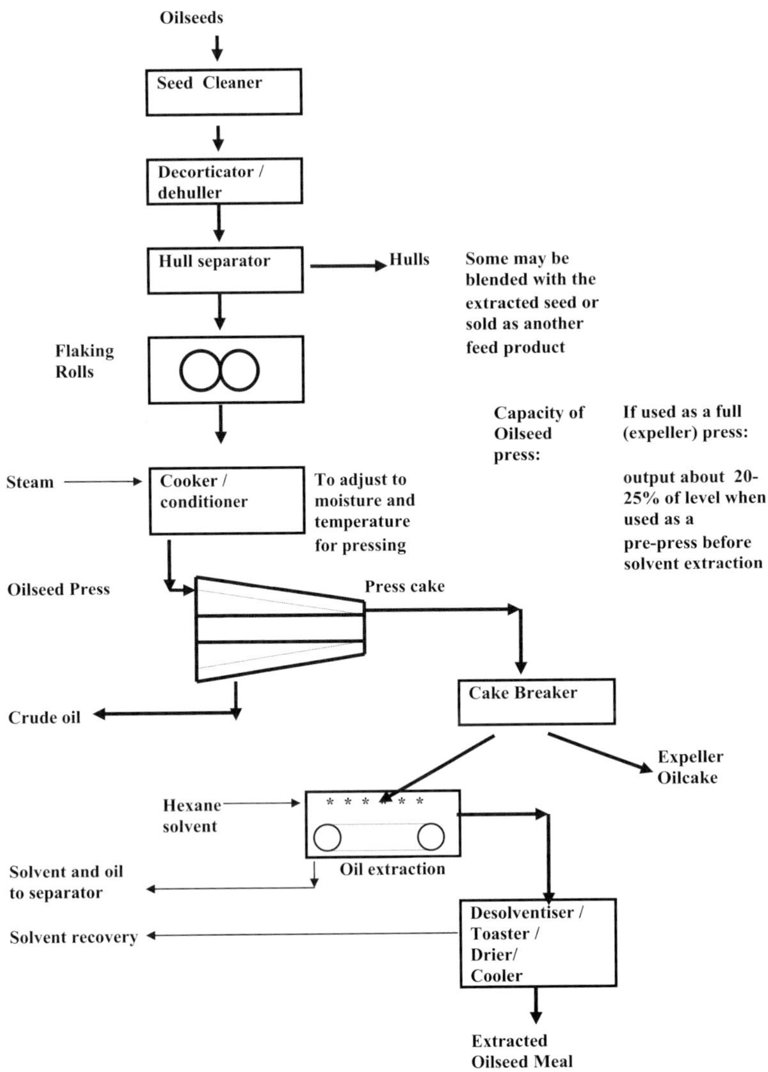


**Fig. 8.16** General method for oilseed processing.

**Table 8.21** Typical composition of soya meals.

| | Expeller | Solvent extracted |
|---|---|---|
| Crude protein | 40.3 | 45.0 |
| Fat | 5.4 | 1.0 |
| Crude fibre | 4.3 | 5.5 |
| Calcium | 0.1 | 0.6 |
| Phosphorus | 0.6 | 0.6 |
| Trypsin inhibitor | n.a. | 4–5 mg/g |
| Urease | n.a. | 0.1–0.3 mg N |

the trypsin inhibitors without damage to the soya protein, and also capable of rupturing the soya oil cells to enable the oil to be readily digested by livestock. However, the process must not cause the oil to be released from the processed meal during storage and handling. Products are known as 'full-fat soya' (see Table 8.22).

*Extrusion processing*: Whole raw soya beans are extruded using a machine of similar design to the oil expeller. Both machines have a worm shaft for applying pressure to the oilseed to rupture the oil cells, but the extruder is designed to retain the oil while

**Table 8.22** Typical analysis of full-fat soya.

| | Raw beans (%) | Heat-treated beans (%) |
|---|---|---|
| Moisture | 9.0 | 5.0* |
| Crude protein | 38.0 | 39.5 |
| Fat | 18.0 | 19.5 |
| Crude fibre | 5.0 | 5.0 |
| Calcium | 0.25 | 0.24 |
| Phosphorus | 0.5 | 0.5 |
| Available lysine | 2.4 | 2.3 |
| Trypsin inhibitor (mg/kg) | 21 | 1.5 |
| Urease activity (mg N/g min at 30°C) | >0.5 | 0.06 |

*On storage, this level will increase to about 9% as the product reaches equilibrium with the atmosphere.

cooking the protein under pressure. The pressure chamber of the extruder is called the barrel, and the cooked product is extruded through a pressure plate (die), into the atmosphere. As the product exits the extruder barrel into atmospheric conditions the oil is quickly reabsorbed into the cooling meal, to produce a product with dry free-flowing properties. Within the extruder barrel, soya may be heated by external steam or electric jackets, or on small extruders, simply by the frictional heat generated as the soya is forced through the machine. For the deactivation of the trypsin inhibitors, the soya enters the extruder at room temperature and moisture content and is extruded at about 150°C. Passage through the extruder takes about 30 seconds.

*Infra-red heat treatment*: In this process the beans are heated as they pass as a thin layer beneath special gas-fired, infra-red ceramic heaters. The beans are cleaned, but not conditioned with water. During heating the beans reach a temperature of 120°C within 135 seconds. The whole, hot beans are then held in a lagged residual hopper for up to 15 minutes to continue 'cooking', before being flaked using heavy-duty rollers and cooled for storage. The flaked soya would normally be ground at the feed mill to the desired particle size for feed inclusion.

The micronisation process is illustrated in Figs 8.17 and 8.18.

### *Sunflower meal*

Sunflower meal may be prepared as expeller or extracted meal with various levels of decortication depending upon the source of the product. The higher the level of hull inclusion, the higher the fibre level and the lower the protein level. The oil percentage is dependent upon the efficiency of the expeller or extraction process.

Sunflower meal does not appear to contain anti-nutritive factors. Toasting of the meal after oil removal is therefore not necessary. Analysis of sunflower meal is shown in Table 8.23.

### *Rapeseed meal*

Rapeseed meal has become of increasing importance within the European feed industry since it can be grown under the temperate European climate rather than the hotter climates needed for soya and sunflower.

The oils from rapeseed were originally used for industrial purposes (paint, lubricants, etc.) since the seeds and oils contained anti-nutritive factors – erucic acid found in the oil, and glucosinolates in the seed cell walls. However, new varieties of rapeseed yield oils with very low levels of these substances, and are suitable for salad and frying oils.

The glucosinolates (which can cause thyroid problems) are not destroyed during oilseed processing or toasting, but new varieties now contain very low levels of these compounds. Rapeseed has become a valuable alternative protein source for livestock, particularly cattle (Table 8.24). Most rapeseed is solvent extracted.

### ***Animal protein products***

Of all the raw materials available for feed use, animal proteins are often the most expensive because of the costs of the basic raw materials and processing, and their higher nutritional value as sources of the essential amino acids that plant proteins often lack. The two most common animal protein sources are fishmeal, meat meal and bonemeal. Others include blood meal, feather meal, and poultry by-product meal.

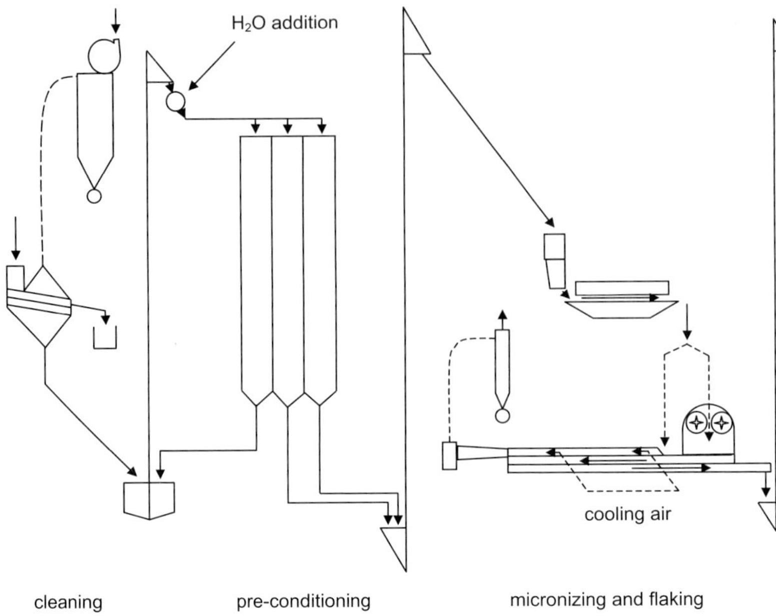


**Fig. 8.17** Flow diagram of an infra-red micronising system for cereals. Source: Micronizing Co. (UK) Ltd.

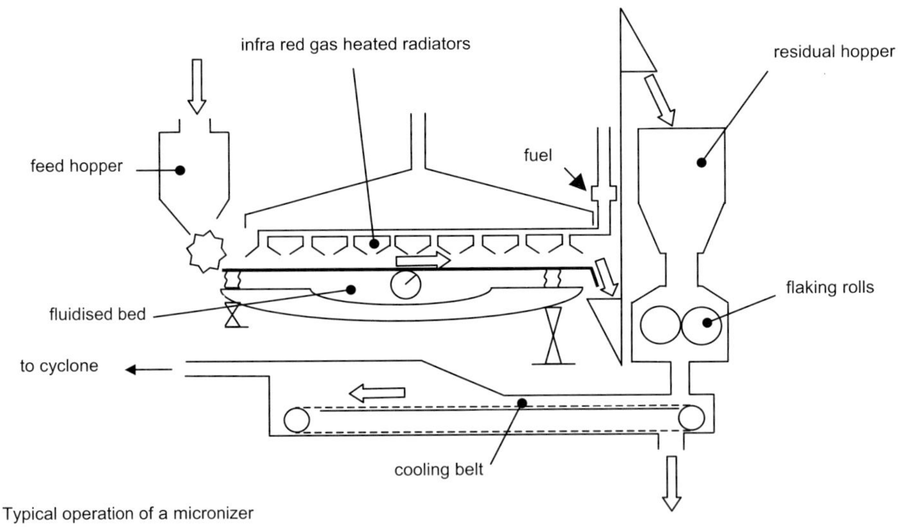


Typical operation of a micronizer

1. Whole soya beans are pre-cleaned of stones, metal, etc.
2. Beans (without water conditioning) are evenly cooked with infra red heat at 120°C on a vibratory conveyor.
3. Hot products are held in a residual lagged hopper for about 15 min.
4. Beans are flaked with a heavy duty roller mill to rupture the oil cell structure, and cooled to ambient temperature.

**Fig. 8.18** Flow diagram of an infra-red micronising system for soybeans. Source: Micronizing Co. (UK) Ltd.

**Table 8.23** Typical analysis of sunflower meal.

| | Undecorticated meal % | Decorticated | |
|---|---|---|---|
| | | Expeller % | Solvent extracted % |
| Crude protein | 18.5 | 37.2 | 37.0 |
| Fat | 7.2 | 13.7 | 1.0 |
| Crude fibre | 29.1 | 12.1 | 19.5 |
| Calcium | 0.4 | 0.27 | 0.4 |
| Phosphorus | 1.0 | 1.18 | 1.20 |

**Table 8.24** Typical analysis of rapeseed meal.

| | Expeller meal % | Solvent extraction % |
|---|---|---|
| Crude protein | 35.3 | 37.0 |
| Fat | 9.6 | 1.5 |
| Crude fibre | 12.0 | 13.0 |
| Calcium | 0.5 | 0.6 |
| Phosphorus | 0.7 | 1.0 |

Animal protein products can be sources of contamination with pathogenic bacteria. The processing conditions must be sufficient to cause virtual sterility of the final products but without undesirable heat damage to the proteins. Nevertheless, because animal proteins can be recontaminated, meals may be treated with an organic acid (formic acid) at about 1% inclusion level.

The processing of fishmeal, meat meal and bonemeals is illustrated in Figs 8.19 and 8.20.

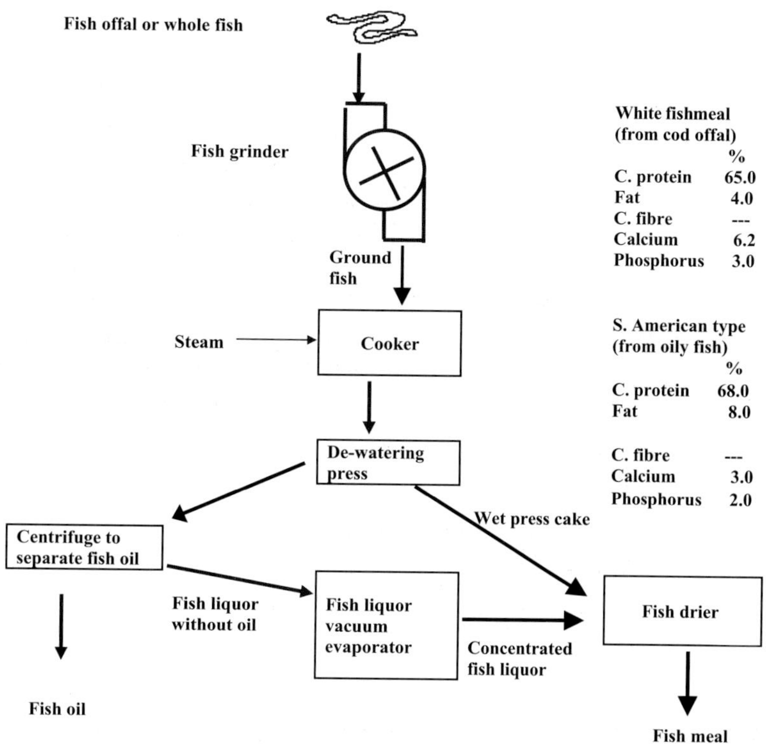


**Fig. 8.19** Processing of fishmeal.

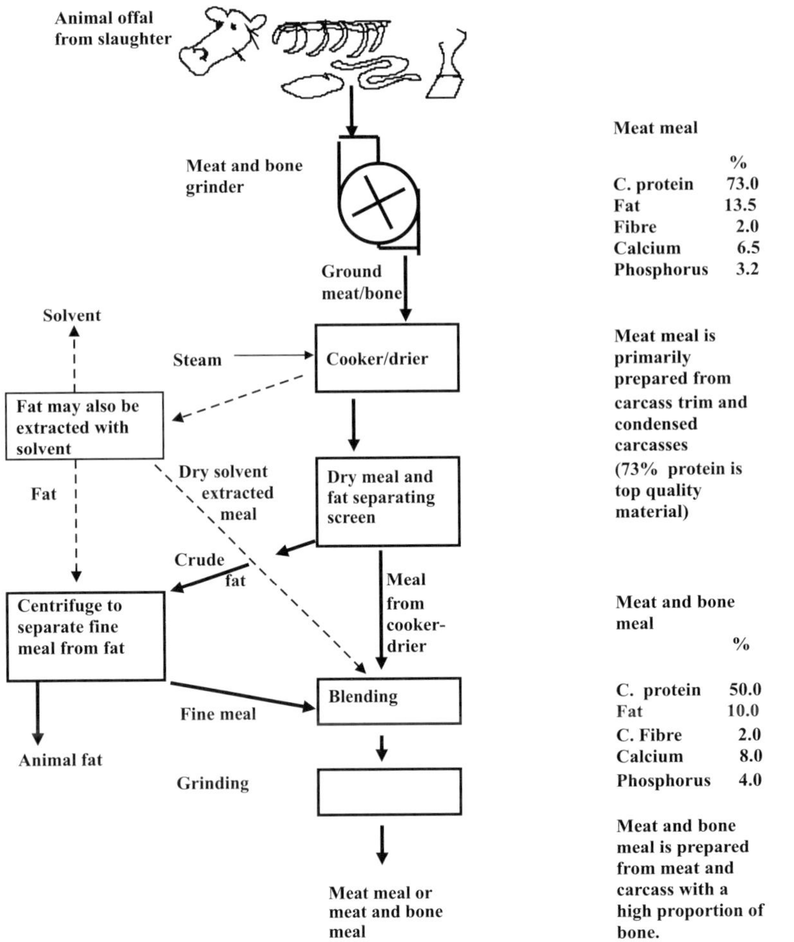


**Fig. 8.20** Processing of meat meal.

The composition of fishmeals is dependent upon the fish material from which they are prepared: fish offal or whole fish, and from oily fish (herring or anchovy type) or from white fish.

Fishmeals are good sources of essential amino acids, calcium and digestible phosphorus. However, they are often variable in composition and quality. The basic stages of fishmeal processing are maceration, cooking (to denature the protein and allow the water in the muscle tissue to be released), pressing and drying. The press liquor is normally concentrated and added back to the press-cake before drying. This process requires less energy than drying whole fish.

The composition of meat meal and bonemeal is dependent upon the operations of the slaughter processes from which the meat and bone offals are derived. Meat- and bonemeals are good sources of essential amino acids, calcium and digestible phosphorus. However, they are often variable in composition and quality. The processing of meat offals is sometimes referred to as 'rendering'. The basic stages of processing are maceration, cooking (to denature the protein and allow the water and fat in the muscle tissue to be released) and drying. The fat is extracted for food, feed or industrial use depending upon the rendering process and the quality of fat obtained.

The range of raw materials available for animal feed use is very extensive. For use in the feed mill almost all agroindustrial by-products require drying to less than 12% moisture content for safe storage. Products like yeast from the wood processing industry will be separated and dried in a process similar to that for dried brewers' grains.

Within the feed industry of Europe, products such as dried citrus pulp and dried sugar beet pulp have, in some countries, become particularly useful for cattle diets. Where cereal costs are high, for example in Holland and to some extent Germany, feed compounders have used large tonnages of tapioca – the sun-dried roots of the cassava plant – as a major source of starch for feed energy. Molasses from either sugar cane or sugar beet processing can be low-cost energy sources for livestock, but their inclusion level may need to be limited for poultry and pigs. Molasses contains high levels of potassium and other minerals, which can cause diarrhoea.

By-products from food processing may also be effectively used in compound feeds, if they are available in sufficient supply. Within Europe new companies have been established to purchase and blend wastes from bread, biscuit, and snackfood processors. These ingredients are carefully analysed and blended, and sold as a local ingredient for feed processors. Their price will be based on the nutritional value of the product in relation to other competitive sources. Raw materials which may be considered to be of little interest to one country's feed industry may, therefore, be of major interest to another, whose agricultural economy is different. Examples from the feed industries of Holland and the UK are presented in the next section to illustrate how different raw materials can be used to meet the desired nutrient requirements for animal feeds.

### Raw materials and feed formulation

The task of the nutrition specialist is to formulate feeds of appropriate nutritional quality, and at the most competitive price for a practical feed. This task is not an easy one when the quality of raw materials and their prices at the mill gate are constantly changing. This search for nutrients at least cost has resulted in very wide differences in the levels and types of raw materials included in feed formulations around the world, although the feeds are formulated to similar nutritional specifications.

The differences can be illustrated by examples of raw material usage and formulation types between feeds from The Netherlands and the UK. For The Netherlands, grains are expensive, and the industry has developed using imported tapioca (sun-dried cassava roots) as an important source of starch, and increased levels of plant proteins from grain by-products (e.g. maize gluten feed) and oilseed cakes and meals. On a grain equivalent basis, one tonne of cereal is approximately equivalent in nutrient terms to 750 kg of tapioca plus 250 kg of soybean meal.

Tables 8.25 and 8.26 illustrate the differences in typical raw material usage between these two countries. Although the raw material information was presented in 1987 to the UK Society of Feed Technologists, it demonstrates that fixed formulations are inappropriate for modern feed production.

However, new raw materials will always bring with them new problems of handling, grinding and pelleting in the mill. Processing technologists have, therefore, an important part to play in resolving these practical problems, without reducing final feed quality.

**Table 8.25** Raw material usage in The Netherlands and the UK feed industries. Based on 1986 national usage statistics.

| | The Netherlands | | UK | |
|---|---|---|---|---|
| Raw material | % of fraction | % of total | % of fraction | % of total |
| Cereals | | 14.1 | | 39.3 |
| of which | | | | |
| Maize | 47 | | 5 | |
| Wheat | 27 | | 75 | |
| Barley | 20 | | 17 | |
| Oats | 3 | | 2 | |
| Sorghum | 2 | | 1 | |
| Others | 1 | | — | |
| Wheat by-products | | 3.9 | | 9.3 |
| Tapioca | | 15.7 | | 0.6 |
| Maize gluten feed | | 10.0 | | 3.1 |
| Oilseed cakes and meals | | 22.2 | | 14.5 |
| Animal proteins | | 6.8 | | 5.0 |
| Oils and fats | | 1.9 | | 1.7 |
| Molasses | | 3.7 | | 4.5 |
| Other materials | | 21.7 | | 22.0 |

**Table 8.26** Standard dairy feed (for performance level of 1 kg feed for 2 kg milk).

| Raw materials | % Inclusion | |
|---|---|---|
| | The Netherlands* | UK |
| Cereals | – | 19.0 |
| Cereal by-products | – | 33.0 |
| Vegetable protein | 62.5 | 28.0 |
| Oil and fat† | 2.0 | 2.0 |
| Others‡ | 35.5 | 18.0 |

*For Dutch feeds, maize gluten feed could be used at up to 40% of the feed (depending on its price).
†Vegetable proteins include soybean hulls, palm kernel meal, citrus pulp, rapeseed meal.
‡'Others' includes molasses and sugar beet pulp.

### Examples of feed types

When considering these examples, note particularly the wide range of products and by-products used in feeds in The Netherlands, many of which are imported. In the UK cereal brans and wheatfeed are important cereal by-products for the feed industry. These are of limited availability in The Netherlands.

In Table 8.25, the raw material inclusion levels shown were applicable to the 1986 feed market, although a somewhat similar picture would be apparent in the 1990s. However, this may not remain the case, since new international agreements may result in a new opportunity for Europe to purchase its cereals at more competitive prices on the world market. If the economic conditions were favourable, The Netherlands could also become a cereal-based feed industry.

In The Netherlands cereal by-products are mainly maize gluten feed and maize gluten meal. Vegetable products include: green peas, whole heat treated soybeans, and sunflower meal. Animal proteins include: meatmeal, fishmeal, hydrolysed feather meal. Others include limestone, minerals and vitamins. In The Netherlands, maize and wheat are the main cereals. Maize is limited because of the market demand for a white-skinned bird. Tapioca may be used in broiler feeds up to 35%. Vegetable proteins include green peas, soybean meal and whole heat-treated soybeans. Animal proteins are the same as in layer feeds.

The above is an example only. Formulations will change according to prevailing prices and availability of raw materials (Nijweide 1987).

### The feed manufacturing process

The previous sections have described the steps in the development of a feed formulation which meets the nutritional requirements of the target animal. The manufacturing stage is to convert the formula into a palatable product for the animal to consume. The usual steps in the feed manufacturing process include the following:

- raw material intake, cleaning, storage and selection
- raw material grinding
- raw material weighing
- mixing of dry ingredients and addition of liquids
- pelleting and cooling of mixed feeds, and
- feed bagging, storage and despatch.

The actual flow line in the factory will vary according to the raw materials and products to be made. The following is an example from a European mill majoring in pig feeds.

Figure 8.21 presents the flow line for feed manufacture. This mill is computer-controlled and designed to weigh and premix the main raw materials before grinding. Capacities are shown in Table 8.27. Other mills may grind raw materials into storage bins and then weigh the ground materials into the mixer. The choice will depend upon the raw materials selected for feed manufacture, engineering design preferences, building limitations, and the products to be manufactured.

#### *Raw material reception*

Raw materials are delivered according to the production plan. On arrival they are visually checked for colour, odour and appearance. They must be at a moisture content which will prevent bio-deterioration during storage, and be free from physical, chemical or microbiological contaminants.

On arrival at the mill samples are taken for immediate analysis for moisture, bulk density, impurities, protein, starch and cellulose. Specialist analyses for vitamins, trace elements and aflatoxins may be undertaken at a service laboratory. In addition to raw materials, the laboratory also conducts selected analyses on finished products.

Before entering the storage system, raw materials arriving directly from field stores may be cleaned of stones, plastic, metal, etc. by screening. Magnets are fitted throughout the mill system to remove ferrous metal contamination from any source.

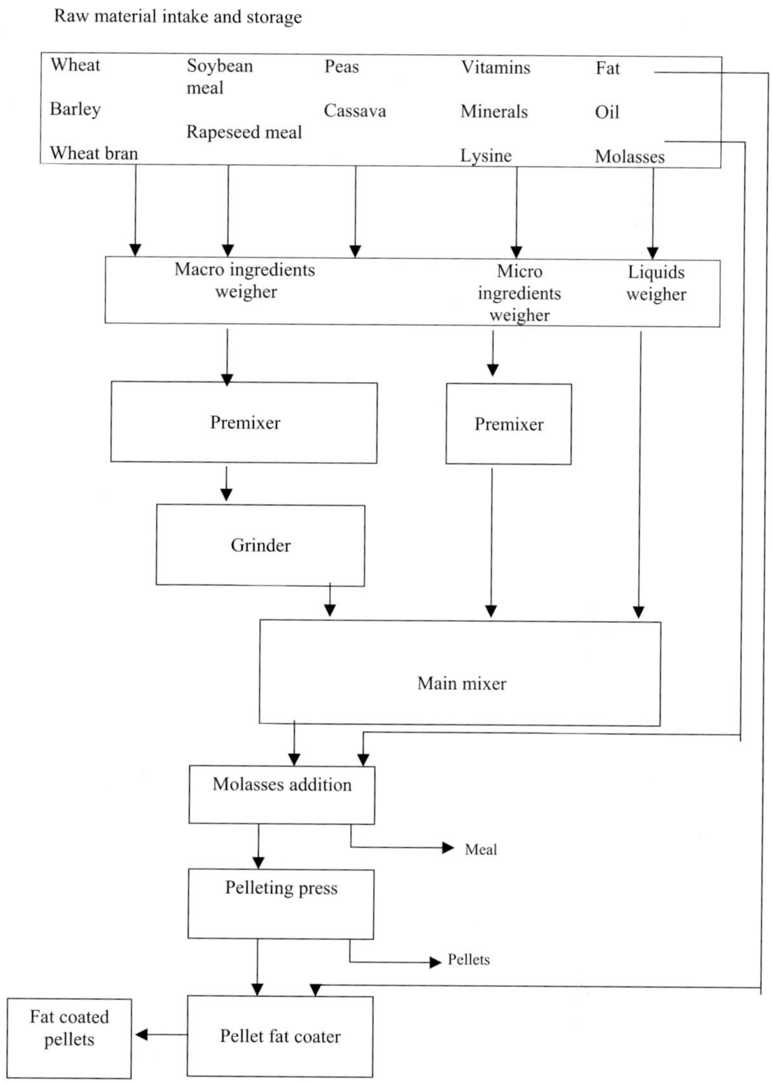


**Fig. 8.21** Flow diagram of a modern feed mill.

### *Weighing*

The accurate weighing of raw materials is the most critical step in ensuring that the feed formulation can meet the nutritional requirements of the animal. Modern weighing systems are based on electronic load cells, for the sequential addition of raw materials into the mixer.

### *Premixing*

The flow line of the illustrative mill uses a premixer before grinding. Some raw materials are malleable rather than rigid in nature, particularly if they contain high levels of fat. These materials can block the screens of a hammer mill if they are milled as a single commodity. Co-milling malleable materials with more fibrous or friable materials can prevent screen blocking.

In contrast some argue for grinders dedicated to materials of similar physical properties as a means of more effective control of particle size and energy consumption during grinding.

**Table 8.27** Example of mill capacities from a European mill majoring in pig feeds.

| Area | Capacity |
|---|---|
| Raw material reception | 500 t per hour |
| Total cereal storage capacity | 40 000 t |
| Raw material intake availability | operating 23 hours per day |
| Feed production capacity | 60 t per hour |
| Raw material silos | 52 silos with total capacity 2000 t |
| Final product silos | 52 silos with total capacity 1000 t |
| Weighers | 7 |
| Grinders | 2 of 200 kW (270 hp) with combined throughput approx. 60 t per hour |
| Mixer | 12 000 litres with combined output of 60 t per hour |
| Pellet presses | 5 presses with combined output capacity of 70 t per hour |
| Pellet crumblers | 5 |
| Fat coating line | 1 |

The mixer preferred by the feed industry is the horizontal mixer fitted with a contra-spiral mixing scroll and emptying by a bomb door. These mixers are claimed to accurately mix 125 parts in 1 million in 6 minutes. The efficiency of mixing is dependent upon many factors including range in particle sizes, presence of liquids, and density. Tests to determine the coefficient of variability of any test ingredient can be undertaken to determine the optimum mixing time for a particular product.

### *Grinding*

The function of the grinder is to reduce the particle size of the raw materials to a spectrum of particle sizes most suitable to the needs of the animal to be fed. In general terms, the smaller the animal, the finer must be the particle size. The most commonly used grinder in the feed industry is the hammer mill (Fig. 8.22).

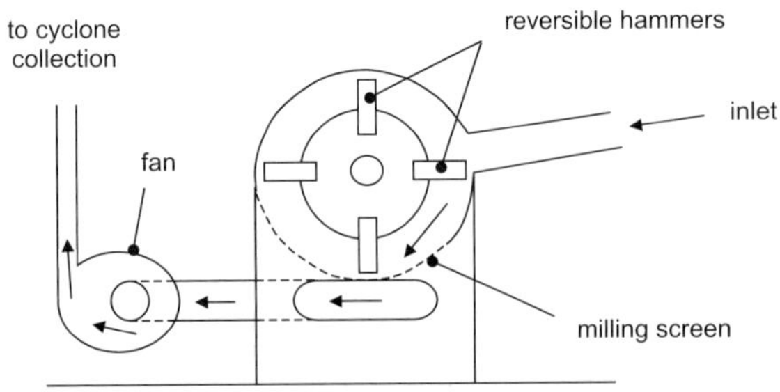


**Fig. 8.22** Typical hammer mill.

The efficiency of grinding is influenced by numerous factors including the physical characteristics of the raw materials, dimensions of the milling chamber, tip velocity of the hammers in the mill, level of wear in the mill, airflow, and screen thickness. Examples of mill performance are illustrated in Table 8.28.

The energy expended in grinding causes the temperature of the raw material to increase by up to 20°C. Air discharge will help to reduce the heat in the product, but care is needed to prevent condensation of moisture which was released from the material during milling.

Milling utilises large quantities of air within the milling chamber and for product removal to storage or other parts of the mill. Air and product are separated from each other by the use of cyclones or filters.

### *Main mixer*

The function of the main mixer is to produce the final homogeneous blend of all the dry raw materials which have undergone a two-part premix step. Premix A contains cereals, oilseed cakes, and other protein and energy sources, and Premix B contain the vitamins, minerals and amino acids which are in powder form.

Although some mills attempt to add liquids such as molasses in the main mixer, this is often undertaken in a high-speed continuous molasses mixer (which may have steam injection) prior to the meal entering the pelleting press.

### *Pelleting press*

Pelleting is the method by which the mixed meal from the molasses mixer (or having by-passed the molasses

**Table 8.28** Typical characteristics and performances from two example hammer mills.

| Parameter | Characteristics | |
|---|---|---|
| Motor size | 75 kW (100 hp) | 185 kW (250 hp) |
| Speed (rpm) | 3000 | 3000 |
| No. of beaters (hammers) | 6 rows of 10 | 6 rows of 24 |
| Screen size (aperture diameter; mm) | 2.5–5.0 | 2.5–5.0 |
| Throughput (tonnes per hour) | | |
| Barley | 3–10 | 11–31 |
| Maize | 7–18 | 27–58 |

mixer) is compressed into hard pellets. There are several reasons for pelleting feed:

- to prevent segregation of feed mixtures of widely differing particle size;
- to enable feed mixtures to flow in mechanical feeding systems for animals. Mixed feed in meal form may have poor flow characteristics, and can bridge in storage bins;
- to increase the bulk density of the feed and thus improve its transportation characteristics;
- to reduce the energy expended by animals in feeding; and
- to prevent livestock segregating feed ingredients of good nutritional value but low palatability from the feed ration.

When feed meals are to be pelleted they are softened (conditioned) before pelleting by the injection of dry steam as it passes through the molasses mixer or conditioner. Feed entering the rotating pelleter die head is forced into the tapered holes in the die by a pair of rollers. Steam, together with the frictional heat generated when feed meal is compressed through the pelleter die head, can raise the pellet temperature to 80–85°C. Temperatures above this will cause steam expansion within the pellet, resulting in pellet expansion and loss of structure.

Steam conditioning will increase the moisture content of the meal and resulting pellets from approximately 13% moisture content to 16–18% moisture. At this moisture level pellets will quickly deteriorate and become mouldy. Pellets must therefore be quickly cooled by forced air to reduce the temperature to ambient and the moisture level to about 13% for safe storage.

The hardness (kg pressure to crack the pellet) and durability (resistance to breakage during transportation and handling) of the pellets is dependent on the raw materials present in the feed, particle size, conditioning temperatures and compression pressure within the die head. Thicker dies with a long hole for pellet formation will give more time for the pellet to form, but will require more energy for pellet production.

The presence of fat in the feed mix is an important factor in pellet stability. When fat is released from raw materials such as oilcakes during steam conditioning, it acts as a lubricant between particles so preventing their adhesion under compression. To overcome this problem, feed mixes which require pelleted feed with more than about 3.5% fat are usually prepared by spraying liquid fat to the outer surface of the pellet in a fat-coating machine. The fat is absorbed into the outer surface of the pellet. The base pellet has sufficient rigidity to withstand handling and transport without crumbling to a powder.

## Further reading

Fellows, P.J. (2000) *Food Processing Technology: Principles and Practice*. Second Edition. Woodhead Publishing, Cambridge, UK.

Hui, Y.H. (Ed.) (1992) *Encyclopaedia of Food Science and Technology*. 4 volumes. John Wiley & Sons, New York NY, USA.

MacRae, R., Robinson, R.K. & Sadler, M.J. (Eds) (1983) *Encyclopaedia of Food Science, Food Technology, and Nutrition*, 8 vols. Academic Press, London, UK.

Rosengarten, F. (1984) *The Book of Edible Nuts*. Walker, New York NY, USA.

Salunkhe, D.K., Chavan, J.K., Adsule, R.N. & Kadamet, S.S. (1992) *World Oilseeds: Chemistry, Technology, and Utilisation*. Aspen Publishers, New York NY, USA.

Woodroof, J.G. (1979) *Tree Nuts: Production, Processing, Products*. Second Edition. AVI Publishing, Westport CT, USA.

## References

Anon. (1998) Small-scale production of weaning foods. *Agrodok-Series* **22**. Agromisa. Netherlands.

Arthey, D. & Dennis, C. (Eds) (1991) *Vegetable Processing*. John Wiley & Sons, New York NY, USA.

Bo Gohl, B. (1975) *Tropical Feeds: Feeds Information Summaries and Nutrition Values*. Food and Agriculture Organization of the United Nations, Rome, Italy.

Burgess, A. & King, F.S. (1993) *Nutrition in Developing Countries.* Oxford University Press, Oxford, UK.

FAO. http://apps.fao.org/page/collections?subset=agriculture.

Food Standards Agency (UK) (Eds) (2001) McCance & Widdowson's *The Composition of Foods.* Sixth Edition. The Royal Society of Chemistry, Cambridge, UK.

Kirk, R.S. & Sawyer, R. (1991) *Pearson's Composition and Analysis of Foods.* Longman Scientific and Technical, London, UK.

Nijweide, R.J. (1987) *Raw materials, evaluation, source and usage; the influence of Rotterdam on raw material usage by a Dutch feed compounder and consequential problems and their resolution.* Society of Feed Technologists, UK.

Nkapa, N.N., Osanu, F.C. & Arowolo, T.A. (1990) Effects of packaging materials on storage stability of crude palm oil. *Journal of the American Oil Chemists Society*, **67**, 259–263.

Ove-Jonsson, L., Dandy, D., Wellings, J.K. & Bokalders, V. (1994) *Small-scale Milling: a Guide for Development Workers.* Intermediate Technology Publications, London, UK.

Rocha, T., Lebert, A. & Marty-Audouin, C. (1993) Effect of pre-treatments and drying conditions on drying rate and colour retention of basil. *Lebensmittel Wissenschaft und Technologie*, **26**, 456–463.

# Chapter 9
# Food Systems

**P. S. Hindmarsh, H. M. Kindness, A. E. Gordon and D. J. Walker**

## Access to food

Many commentators agree with the World Bank, which defines household food security as:

> 'Access by all people at all times to enough food for an active, healthy life. Its essential elements are the availability of food and the ability to acquire it. Food insecurity, in turn, is the lack of access to enough food. There are two kinds of food insecurity: chronic and transitory. Chronic food insecurity is a continuously inadequate diet caused by the inability to acquire food. Transitory food insecurity is a temporary decline in a household's access to enough food.'
>
> *World Bank (1986)*

A complementary definition that brings in a national dimension has been proposed.

> 'A country and people are food secure when their food system operates efficiently in such a way as to remove the fear that there will not be enough to eat. In particular, food security will be achieved when the poor and vulnerable, particularly women, children and those living in marginal areas, have secure access to the food they want. Food security will be achieved when equitable growth ensures that these groups have sustainable livelihoods; in the meantime and in addition, however, food security requires the efficient and equitable operation of the food system.'
>
> *Maxwell (1989)*

Access to food is central to these definitions of food security. The key question in this chapter is: to what extent does storage at the household level, and at the national level, influence access to food grains, the staple food of millions in the tropics?

All individuals have a certain degree of command over food (Sen 1981). The command is determined by the entitlement set of the individual. Sen distinguished between production entitlements (food which is grown) and exchange entitlements (food which is purchased in cash or kind). Using these categories, it can be seen that at the household level the balance between production and exchange entitlements will determine the importance of storage in relation to food access.

Rice, maize, wheat, sorghum, millet and barley are critically important components in rural diets, providing more than 60% of energy intake. Cereal production is normally confined to one, or exceptionally two or three, harvests a year. Storage is used to even out fluctuations between supply and demand, whether these are intra-year, inter-year, demand-induced, weather-induced, due to transport delay, and so on. The need for storage is a function of the variability of these factors and the level of protection from the effects of fluctuation one desires.

For example, in situations where households rely upon agricultural production as their main livelihood activity, production is sufficient to meet consumption needs. Where access to grain markets is very poor (due to remoteness from roads or lack of a nearby marketing outlet), then on-farm grain storage is important for household food security. In contrast, where production is low and households derive a large proportion of their livelihoods from non-agricultural activities, such as trading, and where grain markets are easily accessible and well-functioning, on-farm storage may be much less important. In the latter case, storage at the local market, or at district, regional or national centres, can take on a high level of importance for household food security, as these sources may be critical in ensuring a constant supply at an affordable price – 'access at all times'.

## Characteristics of storage systems

The possession of a food store in almost every rural household in Africa and Asia is evidence of the importance of this asset to rural livelihoods. Efficient and effective storage facilities contribute to indicators of improved livelihoods for poor people, which include greater income, increased wellbeing, reduced vulnerability, improved food security and sustainable use of natural resources.

Traditional farm storage systems have evolved over long periods to provide maximum protection against deterioration caused by rain and ground moisture, to provide a barrier to pest attack and to deter theft. Most are well adapted to their environment and losses to stored food are generally low, often below 5% of grain weight over a storage season. Higher losses occur when the equilibrium of the traditional post-harvest system is disrupted (Boxall *et al.* 1997). Important examples are the introduction and increased production of hybrid cereal varieties, and the spread of the larger grain borer (LGB) in sub-Saharan Africa.

New, introduced crop varieties that do not possess the good storage characteristics of the traditional varieties are often more susceptible to pest attack. If they mature earlier in the season they may give rise to handling and drying problems. The newer high-yielding varieties grown by more progressive small-scale farmers may produce grain in quantities exceeding the traditional storage capacity. In the past, these farmers, and larger-scale commercial farmers, have depended upon parastatal grain marketing organisations and grain traders to take the surplus soon after harvest.

## Credit and storage

Work in the tropics has shown that many farmers sell a proportion of their production at low prices, straight after harvest. Households have pressing cash needs – such as school and hospital fees and other non-food expenditures – which are met through crop sale proceeds. Unfortunately these same farmers are often forced to buy grain later on in the season when prices are higher. They need to purchase grain either in cash or kind (through labouring on someone else's farm, for example). This seasonal pattern of prices and purchases has a negative impact on the welfare of the households involved.

Furthermore, there is inadequate recognition of the important role and value to a nation's food security of the farmer who grows and stores sufficient food to feed his family. Much work has been done attempting to reduce losses in farm storage. Effective and efficient techniques are available, yet uptake of recommendations has been slow, except where spectacular losses occur such as those caused by LGB and by rodents, or by farmers wanting to produce only the higher yielding hybrids. The reason for this may be a combination of:

- the perception by farmers that storage losses are generally low or unavoidable;
- a lack of market price incentives for storage as a result of government policy;
- a scarcity of credit to withhold grain from the market after harvest.

Attention is now being given to schemes permitting the producer to receive cash for the crop at harvest, retain title to the production, and benefit from price rises later in the season. Schemes, such as warehousing inventory credit (Box 9.1), are initially attractive to traders and larger producers. However, they could be of importance in rural grain deficit areas, where traders export production, soon after harvest. The benefits of improving rural storage facilities in these areas look attractive when the costs are compared with the high value of replacement stock from outside. The overall benefits would be to increase farm incomes, improve rural food security and decrease costs to the public sector.

### Box 9.1 The case for warehouse receipts in Africa.

Most African grain marketing systems are largely liberalised, and the state is generally less involved in marketing than in Eastern Europe. However, there are serious problems to be addressed.

- In Ghana, Zambia, Tanzania and Uganda, wholesale maize prices in the lean season typically rise to 70% or more than prices at harvest time, while the costs of storage and financing are typically less than 25% (figures are in real terms). A notable exception to this is South Africa where seasonal price fluctuations approximate carry costs in most years.

- Farmers do not have the information to enable them to price their grain when dealing with traders.
- Undifferentiated prices at the farm gate provide inadequate rewards for grain quality.
- There is little credit available for smallholders to purchase inputs, except indirectly where these are supplied through (often unsustainable) public schemes. Fears of default and subsidised public supplies drive private companies to seek cash and barter terms.

One of the major attractions of warehouse receipts is that they can facilitate improvements in many areas, notably:

- financing, where the receipt gives the lender high quality collateral;
- grading, by a trained warehouse operator;
- trading, using receipts to document delivery; and
- contract enforcement, with the warehouse guaranteeing performance by the seller.

In Africa, warehouse receipts have mostly been used in conjunction with international trade, to cover situations where the lender knows there is an international sales contract for the stored goods. The level of usage for domestically traded commodities is growing but, except in South Africa, it is mainly confined to multinational or foreign-owned companies, with centralised stocks being managed or overseen by international inspection companies. Involvement of local traders and farmers is minimal. Few companies issue warehouse receipts acceptable to the banks.

Coulter and Norvell (1998) tried to discover how warehouse receipts can be popularised and their use generalised, and how more companies can get involved in the storage function. The authors studied various cases from around the world, and identified three models for analysis.

(1) The regulated elevator-company model. Elevator companies are grain merchants, trading companies and farmers' co-operatives, registered with and overseen by government agricultural authorities. By law, they are required to open up their stores to third parties, be they farmers or other trading companies. Government also establishes an official grading system and a state or state-licensed inspection company certifies the quality and grades of grain handled. Elevator companies are normally close to the farmers they serve, and buy from them, offering a variety of contracts that may be hedged on futures and options markets (when such facilities are available). Negotiable warehouse receipts may also be used as delivery mechanisms in commodity exchanges. To be licensed, elevator companies must satisfy net worth and professional requirements, they must be regularly inspected and submit audited accounts. Stocks must be insured, and elevator performance has to be underwritten – this is normally achieved by requiring companies to purchase bonds, or by requiring the entire industry to subscribe to an indemnity fund.

(2) The general warehousing model. General warehouses are dedicated to the storage function, but store all kinds of goods, not simply agricultural products. They sometimes carry out other service functions such as freight forwarding, but tend to be barred from trading on their own account, to avoid conflicts of interest between service and trading functions. They often engage in 'field warehousing', i.e. taking control of the stores of farmers, traders or manufacturers, and issuing warehouse receipts used to raise bank loans. They are licensed and overseen by public authorities dealing with trade or monetary matters, who normally place particular emphasis on capital adequacy requirements.

(3) The private trader model. In countries without warehousing legislation, private traders will sometimes offer services similar to those described in the licensed elevator-company model, including the storage of grain on behalf of farmers. Where such services do not exist, government, development finance institutions and donors may collaborate to provide a framework of incentives to encourage private traders.

The regulated elevator company model has been furthest developed in the United States. In the grain sector, there were 11 000 elevators with storage capacity of 212 million tonnes of grain, all regulated by federal and/or state authorities. When the oversight system was established in 1916, farmers made considerable use of warehouse receipts to raise loans, but with the development of other financing alternatives, they are now mainly used as a delivery mechanism on commodity exchanges. However, this regulated model is generally acknowledged to have had a crucial role in creating a relatively transparent and efficient system, involving narrow trading margins.

Many of the features of this system have been adopted in South Africa, but here it is the commodity exchange, SAFEX, rather than government, which oversees the warehouses, and only those warehouses used as SAFEX delivery locations. SAFEX futures contracts have rendered warehouse receipts highly liquid and made them much more attractive to lenders.

The public warehousing model is widely used in Code Law countries (countries with codified laws based on the Napoleonic model, to be contrasted with Common Law countries whose legal systems are based on English law), particularly in Latin America, and it has recently found favour in Eastern European countries that have resurrected pre-war legislation. Performance has been varied.

- In Hungary, three warehousing companies, with public storage subsidies, now issue receipts worth more than $100 million or 10% of total grain production. Field warehousing accounts for over 80% of the total.
- In Argentina, ten companies have contributed to an expansion of agricultural lending resulting from economic stabilisation and liberalisation in the 1990s. The value of warrants for all goods including grains now exceeds $1 billion.
- In Colombia, by contrast, liberalisation has weakened the general warehousing companies, which were formerly the beneficiaries of a system of preferential agricultural credits. They have been heavily regulated by the monetary authorities, but still enjoy a reputation for financial probity.
- In Brazil, warehouse receipts have by and large performed poorly, notwithstanding long experience under an Act of 1903. The sector has become over-dependent on public storage contracts, and hence on political patronage. Systems of public oversight have been very weak.

The private trader model was recently implemented by Cargill in the Krasnodar area of Russia, with financial support from USAID. The scheme has, so far, only had a limited impact, but this can be attributed to policy factors and the lack of a storage structure rather than to the concept *per se*. On the one hand public-sector intervention has flattened the seasonal price curve, greatly diminishing incentives for private storage. On the other hand, there is a local disincentive: Krasnodar harvests relatively early, and so prices are likely to fall during part of the storage period.

This would not be a major problem in African countries, most of which now have strong 'carry' structures.

## Conclusion

Warehouse receipts have a crucial role to play in improving the performance of African grain markets, and should be prioritised by policy-makers. Brazil's 100-year experience with warehouse receipts underlines a fundamental consideration in the design of appropriate systems. There is a range of political pressures that tend to undermine the credibility of warehouse receipts and, given the political sensitivity of agriculture and food supplies, these pressures are particularly strong in developing countries.

The authors' view, in the current context, is that an ideal system will consist of an elevator-company model regulated by a professional licensing agency insulated from day-to-day political pressures. Paradoxically, such models depend on political processes for their design and ratification, and assistance in enforcing overseers' decisions. Such an approach has in effect been achieved in

South Africa, with SAFEX taking over the role normally assigned to government. In the absence of a business-led solution, countries wishing to institute a proper elevator-company model should consider appointing an independent and internationally reputed body with proven expertise in this area, to license and oversee warehouses, and enforce grain standards. Such an approach would be most effective if implemented on a regional basis. In the case of Southern Africa, this would facilitate links to commodity exchanges and help create a more transparent and competitive regional grain market.

A less ambitious solution would be to institute the private company model. This might require companies to tender for government warehousing facilities and formulate specific commitments regarding the development of warehouse receipt systems (Coulter and Norvell 1998).

## The case for food security reserves

National food security reserves are generally considered to be stocks of staple food grains held at strategic locations for release in response to an emergency. In fact, the number of countries holding discrete, separately managed reserves of food grains specifically for emergency relief purposes is small, and many countries rely on either import and export arrangements or national stockholdings with multiple functions (Box 9.2).

It will be seen from the examples in Box 9.2 that some countries were simultaneously operating stockholding strategies in several of the categories. They were also often balancing the demands of their varied obligations from the same stock inventory.

### Box 9.2 Strategic famine relief stocks.

Emergency reserves, often established and maintained with donor assistance, and subject to strict release, replenishment and recycling criteria. Aimed specifically at defined recipient groups and often calculated on the basis of a proportion of the population receiving free food for a predetermined period of time in response to local or national food crises. Size and locations of reserves are often heavily dependent on logistical criteria. Examples can be found in Ethiopia, Mali and Tanzania.

#### Operational buffer stocks

Set up to prevent supply surges and to regulate routine food distribution flows. The concept was commonly applied in trend surplus countries, where local production is heavily concentrated seasonally, but regular distribution programmes (e.g. ration shops, budget groups) demand even flows throughout the year. The concept was also valid for structurally deficit countries where programmed food aid delivery to recipients is dependent upon timely arrival of food shipments. Regulated flow prevents surges in market supply, depressed prices and consequent disincentives for local producers, as in Bangladesh, India and Indonesia.

#### Price stabilisation stocks

The framework used by countries operating a classical system of floor and ceiling support prices for producers and consumers, respectively. The size of the price band between the two levels is crucial in determining whether private trade can flourish or will be squeezed out by the narrowness of the operating margin. Other key issues are coping with bumper harvests, stockholding costs, pan-territorial or pan-seasonal pricing, and heavy transport costs. Price stabilisation is found in India, Malawi, Indonesia and South Africa.

#### Inter-annual stocks

Essentially a facility whereby production, surplus to immediate in-country requirements in one year, is retained and carried forward in anticipation of stock run-down in subsequent years. This method has been used in India, Indonesia, Malawi, Zambia and Zimbabwe. Major issues are production variability, opportunity costs of stockholding, import/export parities, and logistics.

In a study of food security reserves in southern Africa, (Anon. 1993), the authors concluded that national plans for establishing reserves, where these existed, had been based on rough and subjective criteria with minimal analytical foundation. For example:

- holding reserve stocks was just one of a range of options to improve national food security, but alternative options were rarely taken into consideration when the optimal size of a physical reserve was estimated;
- national policies had rarely been based on a rigorous analytical assessment of the likely needs which such a reserve might be required to meet, or on the trade-off between cost and the degree of food security provided;
- how to determine the optimum size of a reserve was often seen as subsidiary to the institutional and political issues surrounding financial and management rules for stock drawdown and replenishment.

Many of the attempts to establish national strategic grain reserves either failed at the outset, or became associated in one way or another with a reputation for inefficiency and unsupportable overhead costs. Some of the parastatal grain marketing agencies were charged with responsibility for operating the food security reserves, but management operations were often characterised by:

- non-transparent accounting procedures;
- failure to observe specified release and recycling procedures
- high operating costs
- non-replacement of released grain
- counterpart fund irregularities
- high physical loss levels and poor storage management, and
- subsequent loss of donor confidence.

Often, the agencies charged with the management responsibility for a national strategic reserve were given neither the appropriate funding for the job nor the essential managerial and technological skills with which to discharge this demanding new mandate. Major problems were encountered, including:

- unsuitability of locally procured and donated grain for reserve storage;
- unsuitability of donated grain for recycling on local markets;
- inadequate storage facilities and funding to maintain quality of reserve stocks;
- political pressure to release grain in non-mandated situations;
- inadequate funding for stock replacement;
- unclear differentiation between reserve stocks and normal operational stocks;
- conflicting demands of government and donors.

By the mid-1990s, experience suggested that few countries would be able to sustain a strong case for establishing and maintaining strategic, famine relief reserves. Successful grain market liberalisation presupposed the proliferation of private sector storage, theoretically reducing the burden on centralised stockholding. Better and faster commercial access to international grain markets could be expected to markedly affect the historical lead times for grain importation in response to impending or identified food crises. The evolving sophistication and application of early warning system outputs to food production would enhance predictive capability in a manner which could well erode the rationale for in-country food security reserves.

Given the rather undistinguished history of attempts to establish and maintain strategic grain reserves, it is not surprising that policy analysts have stressed the greater relevance and cost-effectiveness of alternative approaches to national food security. Several of these approaches have taken on a new dimension with the widespread adoption of grain market liberalisation policies in many developing countries. These largely centre on the 'trade versus stocks' arguments and include:

- liberalised input supply and output procurement at producer level;
- provision of incentives to encourage on-farm and private sector storage;
- provision of market information to all levels of the trade;
- improved physical access and market infrastructure development;
- development of cross-border trade;
- removal of import/export restrictions;
- leasing of parastatal facilities to private sector; and
- moves to import parity pricing for local grain crops.

As we enter the twenty-first century, deregulation in cross-border grain trading has gained momentum as the role of marketing boards and national grain agencies has been systematically reviewed and reduced. Centralised monopolies in the major grain exporting nations such as Australia and Canada are being overhauled to include the private sector, while the importing activities of government agencies in major end-user markets such as Indonesia and Bangladesh have been opened up to licensed private trade. Major state stockholding commitments to service food rationing schemes for the urban and rural poor have been reduced or eliminated in Bangladesh, and even the future of the world's biggest public food distribution system in India, with 16 million recipients, is currently under review.

Those few countries which have elected to maintain a discrete, famine relief focused reserve have either decided to retain an existing, reorganised parastatal agency with a reduced mandate relating specifically to that task, or have created new agencies, largely in an attempt to restore waning donor confidence and support.

Attempts to establish a rationale for a physical, food security reserve need to accept that a historical analysis of real international prices for food grains will show a steadily declining trend over time. Policy analysts will, therefore, regard stockpiled grain as a wasting asset when comparing national investment options. Decisions should be based on an assessment of full economic cost, balanced by the extra degree of food security insurance obtained. A thorough screening of the basic criteria should take place, namely:

- precise role of reserve
- level of food security required
- identification of beneficiaries
- historical risk of stock-out
- lead times between decision to import and arrival of grain
- status of food staple production
- production variability
- correlation between domestic and international shortfalls in food staple, and
- import and export parity price relationships.

## Economies of scale in grain storage

Although store rehabilitation or the provision of new storage capacity is a regular component of storage development and food security programmes, the interaction of how much capacity, its location and the most appropriate type is rarely tackled satisfactorily. The minimum information required is quite large:

- the location of existing storage (including private and farm-held stocks);
- warehouse capacity based upon correct storage management procedures;
- storage losses;
- time series data on regional production and demand;
- transport availability and costs; and
- marketing system structure and responsiveness.

Economic theory suggests that storage costs per unit of grain decrease as the size of storage facilities increases. These economies of scale would be due to lower investment costs per unit stored – storage volume increases as a cube, and construction costs as a square, of the linear dimensions – and more efficient use of grain handling and protection equipment. In reality, the picture regarding economies of scale in grain storage in developing countries is mixed. There are numerous examples where the construction of large warehouses or silos has led to inefficient use of resources.

A careful assessment is needed of the actual requirements and purpose of the facility. High storage costs are often due to over-capacity in comparison to actual demand for storage space. In addition, the availability and cost of construction material, as well as relevant building skills, will influence the type and size of a store.

When deciding about the construction of one, large, warehouse or several smaller ones, it is important to take into account both investment and operating costs. Although the case may look positive as far as capital costs are concerned, larger operations tend to be less flexible. The higher local and foreign exchange cost of increased transport associated with centralised storage limits the degree of centralisation. In fact, centralised stocks incur higher performance risks because of small transport fleet size, lack of spare parts due to lack of foreign exchange, poor road maintenance and long response times. Relief work in Ethiopia would attest to this. In addition, existing idle storage capacity, because of the sunk costs, can influence location as well.

However, with correctly located centralised stores it may be possible to decrease the total size of the storage needed. As long as the variability in production (or other food security variable) is not regionally correlated and transport mechanisms exist, one smaller central store can offer the same food security protection as two regional stores. The need to recycle inter-year buffer stocks to control quality and quantity losses must not be overlooked. A flow of grain in and out of the pool is necessary even when the stock is not called upon for food security protection. In practice, this tends to limit decentralisation of stocks and highlights the need for intake quality standards and knowledge of domestic markets.

The issue of operating costs and the right choice of technology is an important one, in that a number of factors such as availability and cost of labour need to be considered. Modern, large-scale stores tend to be equipped with capital-intensive technology for grain handling and protection (e.g. elevators, fumigation facilities), resulting in lower unit costs if used efficiently. This obviously requires well-trained staff, inexpensive power (preferably electricity), and easy access to spare parts. On the other hand, bag handling may be more cost-efficient than bulk handling if labour is relatively abundant and cheap.

The high cost of grain sacks for bagged storage is focusing attention on bulk grain storage. Bulk storage is more capital- and technology-intensive with higher fixed costs, but offers high-speed handling with low variable costs (no sacks and labour). The investment is more sensitive to the rate of storage utilisation and interest rates. Bulk handling is economically justified at bottlenecks in the marketing chain, where large volumes of grain have to be handled at great speed, e.g. ports, railway terminals, large mills (Coulter & Magrath 1994). Among the different bulk handling scenarios, bunker storage has the lowest construction costs (as low as \$2 per tonne), but potentially high operating costs (Newman 1994). According to the same source and based on 1992 prices, larger tank stores (i.e. >5000 tonnes) cost in the range of \$25–35 per tonne of capacity, whereas smaller tanks (1000–2000 tonnes) may cost \$40 per tonne. Construction costs of tall hopper-bottom silos can be of the order of \$75–100 per tonne capacity.

The changeover from bag to bulk storage is not simply a technical decision but involves the social issues of displacing labour and increasing capital intensity in countries where unemployment or under-employment is large and capital savings low.

As far as storage economics is concerned, it is important to see stores as part of an overall operation. If this is efficiently run then the likelihood of achieving economies of scale from large stores and the use of more capital-intensive technology is greater. In addition, storage costs must be compared to storage benefits, requiring financial calculations. The 'annualised capital cost' method is more appropriate if benefits are either difficult to quantify, or if they are the same for all scenarios. In addition to this cost-effectiveness analysis, discounted cash flow analyses are particularly useful if both costs and benefits of an operation can be valued without too many difficulties.

As far as operations such as inventory credit schemes are concerned, economies of scale are a key consideration, both in the running of warehouses, and in their certification and inspection by an outside agency. In this case, the requirement for economy of scale concerns management costs, rather than storage costs *per se*.

## Changes in marketing infrastructure

Food storage needs to be seen in the context of the policy environment.

During the 1970s and early 1980s, the main emphasis for many tropical countries was to encourage local production through guaranteed purchasing, inter-regional trade bans, pan-territorial and pan-seasonal pricing and, to satisfy consumers, by subsidising sales. This led to many problems, some of which were:

- production of surpluses for which no market had been identified;
- production of surpluses in areas distant from consumers with consequently large transport costs for both inputs and production;
- entrenchment of parastatals as storers of first not last resort; and
- low-quality stocks because of guaranteed purchasing without inbound quality control.

Since the early 1980s, structural adjustment programmes have led to market liberalisation in many developing countries. As far as agricultural markets in industrialised countries (e.g. European Union) are concerned, some liberalisation efforts have taken place during the 1990s partly as a result of the GATT agreement, partly as a result of budgetary constraints. Nevertheless, subsidies still play an important part in the agriculture of the European Union and North America.

Changes in the way that grain markets operate have increased the need for on-farm storage of grains for many of the world's rural poor, particularly those living in semi-arid areas. The effects of market liberalisation, involving the replacement of parastatal monopolies with private trading activities, the removal of subsidies on inputs, and the abolition of pan-territorial output pricing, have benefited farmers in the most favourable locations, but disadvantaged poorer farmers in more remote regions and villages. In the worst affected areas, farmers have retreated into subsistence production, because private marketing services are either too unpredictable or offer prices too low to justify production for the market.

Major weaknesses in the way the marketing systems function are:

- poor and costly mechanisms for financing trade;
- performance and payment risks associated with transactions, and a lack of effective mechanisms for resolving disputes;
- lack of forward contracting; and

- lack of standardised quality and grading.

## Market access

This section on market access is largely based on Kindness and Gordon (2001).

There is general recognition of the role that improved market access can play in enhancing rural development and poverty reduction, though there is limited information on the extent and nature of the problem and of options for developing appropriate mechanisms that would enable rural communities to take advantage of the market opportunities commonly found in urban and semi-urban centres. For this reason, the discussion below focuses on the important factors that influence access to markets for small-scale producers in rural areas of developing countries. The influence of globalisation and liberalisation on market access is also considered.

### *Marketing constraints*

Farmers with access to markets have sufficient information and the physical, financial and social means to purchase inputs and food, and sell agricultural produce on favourable terms. The consequence of lack of market access is a low volume of buying and selling transactions. Transactions that take place are usually on unfavourable terms for farmers. This decreases farmers' incomes, removes production incentives (thereby reducing yields of food and cash crops) and leads to food insecurity.

Small-scale producers in rural communities face considerable marketing constraints, including:

- Inadequate orientation of small-scale production to the demands of consumers, where market orientation refers to the degree to which producers respond to these demands. Inadequate orientation can be attributed, in part, to the failure of market information systems to provide producers with information on current market trends and consumer preferences. Producers are unaware of marketing opportunities and are therefore not producing the right products for the market.
- Inability to take advantage of seasonal price fluctuations. Small-scale, resource-poor farmers tend to market their surpluses soon after harvest when markets are over-supplied and prices are low. To be able to take advantage of seasonal price fluctuations, producers and traders must store the produce and have sufficient liquidity to pay for consumption, inputs and debt. Delaying the sale of produce is a risk that entails storage and credit costs.
- Marketing costs may be prohibitively high for small producers who are marketing their goods individually and in small quantities. For small volumes of trade, the fixed costs of marketing can form a high proportion of total transaction costs per unit.
- In remoter areas there are likely to be relatively few traders purchasing farm produce or selling farm inputs. This limits the choice and therefore limits competition between traders, contributing to lower prices for produce and high prices for inputs.
- Moral hazard: 'Moral hazard arises when an individual takes an action to maximise his own welfare that is to the detriment of others in situations where informational problems prevent the assignment to the individual of the full damage caused by his action' (Hoff *et al.* 1993). With respect to marketing, traders may deliberately adjust prices against information-poor producers who are unaware of prevailing market prices.
- Transportation difficulties: The greatest transportation difficulties facing rural communities are local, in moving goods (inputs, produce) to and from fields or to storage facilities or local markets. The commonest means of transport rely on humans using headloads, bicycles and handcarts, or carts drawn by animals. Use of motorised transport is usually neither practical nor realistic. Infrastructure and vehicles are too expensive. Notwithstanding, transportation on foot is also expensive (in terms of time lost) and leads to losses due to lower carrying capacities (Sieber 1997).
- Large distances must be travelled to reach markets. The cost of this increases pressure to sell upon arrival at the market; this may lead to farmers undercutting one another to dispose of their produce.
- Sparse populations: The land required to support a household is relatively large. This reduces the possibility of achieving economies of scale in marketing, and increases the distances between the household and the farmer's fields.
- High production costs and low productivity, for example, due to climatic variability, poor soils and expensive inputs: In avoiding these risks farming systems tend to have a relatively large percentage of resources devoted to providing food for the household, reducing the surplus available for marketing.

### *Changes in marketing infrastructure in the context of liberalisation*

In recent years developing country economies have

been through structural adjustment programmes which have entailed some degree of market liberalisation. With respect to agricultural marketing, the effect has been to reform or abolish state monopolies, which previously dominated the sector. Highly centralised systems were characterised by low, pan-seasonal and pan-territorial prices. These prices were justified by their protection of urban consumers, but provided little incentive to producers and traders to invest in marketing infrastructure or services (Mittendorf 1988). Producers often had no option but to join marketing co-operatives, which had responsibility for purchasing and marketing all farm produce. Such co-operatives were typically grossly inefficient and a drain on the national budget. However, the withdrawal of the state and the growth in the role of the private sector has had a mixed impact. In remoter areas, where the costs of marketing are high, the withdrawal of the state has often left a void which the private sector has not filled (Carney 1995; Conway & Tyler 1995).

State intervention was justified to prevent exploitation by private traders of small farmers, to stabilise domestic prices and to protect urban consumers. These policies resulted in artificially low prices for agricultural produce. Prices today in theory more accurately reflect market demand than during pre-liberalisation times, though this assumes that the private sector is relatively well developed and well informed. A feature of the withdrawal of the state in price-setting is that greater seasonal and annual fluctuations of prices may occur. This increases both the challenges and the potential profitability of marketing. Quality of produce is of increasing significance in an open, competitive marketplace, with prices varying according to the quality of the product and the degree of processing (i.e. the amount of value added).

### *Improving market access*

A holistic approach is required to improve market access for farming communities in remote areas. The main elements include provision of an enabling economic environment for marketing activity, public and private investment in transportation, improved provision of related rural services (including market information systems, research and extension, and rural finance) and institutional development, including community-based and non-governmental organisations. Institutional issues need to be considered in the context of the ongoing decentralisation efforts in many less developed countries, which are highly significant for improved market access.

There are two key requirements for the promotion of marketing: increasing market opportunities, for example through globalisation, economic liberalisation and market orientation, and better access to those opportunities. Increasing market access requires a reduction in the costs involved in marketing activities. The most critical marketing constraints for remote communities are transportation costs and information costs, and the route to increasing market access involves lowering these costs.

### *Transportation*

Transportation costs are high due to, *inter alia*, large distances, weak infrastructure and the small volumes of produce transported. Costs can be reduced by optimising the provision of transport infrastructure, that is, by providing infrastructure that is appropriate to users' needs and means, and which can be constructed and maintained at low cost. The need for close consultation with affected communities, and instilling a sense of ownership, suggests that a decentralised authority may be best placed to oversee infrastructure provision at the local level, while private contractors carry out the actual construction and maintenance work.

The key to improving rural transport infrastructure and transportation services is the development of low-cost solutions (Sieber 1997). Government funds for investment are limited, as is the potential for mobilising local funds. The private sector is unlikely to be interested in investing in either infrastructure that is essentially a public good or transport services in areas where the population is relatively sparsely distributed and infrastructure is poor. This highlights the need to look for alternative approaches, such as those provided by intermediate means of transport (IMT). IMT is defined as:

> '...those means of transport which are intermediate in terms of initial cost and transport characteristics...between the traditional methods of walking and headloading and conventional motor vehicles [and] intermediate in time, i.e. they are a stage in the process of developing a traditional to a modern transport system.'
>
> *Howe (1994), quoted in Sieber (1997)*

The majority of goods in rural communities are carried on paths and tracks, and the most prevalent form of transportation is by headloads. Estimates of the costs of transport indicate that headloading is expensive when compared with other forms of IMT (Sieber 1997).

Relatively small investments can transform paths into tracks suitable for donkey and ox-drawn carts. Data show that shifting from headloads to animal traction can significantly reduce costs per unit transported – by 60% for donkey carts and 90% for ox carts. Animal-drawn carts are most suitable for carrying large loads over short to medium distances. Motor vehicles are cheaper if operated on good roads over long distances and with high rates of use.

Thus investment in roads may not only be costly but also have a limited impact on the needs of rural communities. Roads in Ghana that were upgraded from earth to gravel had very little knock-on effect on producer prices, but footpaths converted into roads had substantial benefits, up to one hundred times greater (Hine 1993).

The main constraint to IMT for resource-poor households is the initial capital expenditure. Appropriate credit schemes would be needed. Overall, a broader approach is required to transportation and marketing for rural communities. Although roads and motorised transport clearly have an important role, especially further up the marketing chain when volumes of trade increase, consideration needs to be given at the same time to local constraints, for which roads may not be the answer.

If local transport systems allow the accumulation of marketable surpluses in sufficient quantities, whether at periodic or fixed markets, private traders will be encouraged to participate in marketing. Such economies of scale may be achieved through marketing groups at the local level. However, remote areas are less attractive to the private sector as costs of providing services are higher, the risks involved are greater and the purchasing power of potential clients is lower.

### *Market information*

Market information is a vital component of the marketing system. Lack of information limits market access by increasing transaction costs and risk. For remote communities, knowledge is required to inform production and marketing decisions. From a marketing perspective, farmers need to know where they can sell their produce or purchase inputs. In addition, they need to know when to sell in order to maximise returns. Information must be reliable, easy to interpret, up-to-date and broadcast to all areas. This requires a market information system that collates the relevant data, puts it into an understandable form and disseminates it widely, accurately and rapidly.

> 'Up-to-date, or current, market information enables farmers to negotiate with traders from a position of greater strength. It also facilitates spatial distribution of products from rural areas to towns and between markets. Well-analysed historical market information enables farmers to make planning decisions, including those related to new crops. It also permits traders [and producers] to make better decisions regarding the viability of intra-, and perhaps, inter-seasonal storage.'
>
> *Shepherd (1997)*

Producers are disadvantaged by information asymmetries. For these to be reduced, all parties to the exchange must have access to the information, and so market information can be considered a public good to be provided by the state. However, the state's record in the provision of market information has been poor, being:

> 'unsustainable and [they]... failed to provide commercially useful advice, confining themselves to the gathering of, frequently unused, data'.
>
> *Robbins (1998)*

Traders are put off from dealing with small farmers in remote areas by the high identification and screening costs. An appropriate market information system will improve the market orientation of producers, and enable them to develop marketing strategies. Such a system needs to be sympathetic to the needs of communities with respect to the type of information and the media used to communicate the information. Radio has proved particularly effective and appropriate as a vehicle for dissemination of agricultural information (Robbins 1998), and newer dissemination technologies, such as the World Wide Web and mobile telephony, will no doubt also be made use of. However, there is a danger of lack of equitable access to newer dissemination methods, particularly since those most in need of information are those least likely to be able to afford the cost of accessing it electronically.

### *Unresolved issues and areas for research*

The key dilemma in discussions of market access is the question of what objectives are realistic and appropriate. More work is needed to identify and analyse experiences with remote and disadvantaged communities, to study the approaches used and the balance between social and commercial objectives. Some outstanding issues are:

- Rural transportation policy in less developed countries, especially the role of decentralised authorities, and the extent of public/private participation.
- Examples of successful self-help groups involved in infrastructure services, such as construction and maintenance of roads and markets. Information is lacking on successful experiences with marketing groups, particularly those linked to the private sector.
- Means of transportation: Are communities adequately served by public transport and/or private transport? Are there any policies or programmes for the promotion of intermediate means of transport? Research is needed on IMT.
- Market information systems: What exists and is it appropriate? This issue could be investigated in the wider context of agricultural knowledge and information systems.
- Co-ordination between governments and donors on market access related issues.
- To what extent do policies related to market access take account of issues such as rural poverty alleviation, gender and the environment?

Following economic reforms and the official embracing of the market economy, much is said about the need for farmers to be more market oriented and to treat farming as a business. However, more information is needed to establish exactly what is required and identify positive experiences in farmers' access to markets, as individuals and in groups. Furthermore, the impact and sustainability of alternative intervention strategies have not been investigated. Analysis of impacts needs to take the wider benefits of improved market access into consideration, rather than focus narrowly on marketing-related results.

### *Future prospects for marketing in developing countries*

Private sector participation is vital. Improving the infrastructure, information systems and institutional aspects of marketing will lead to increased volumes of trade, and encourage deeper penetration of private traders into more remote areas. The forging of linkages between the private sector and communities can be enhanced through group actions at the local level.

Globalisation is at the forefront of any discussion of markets. All countries are affected by changes in consumer demand, trading practices and regimes and new technology. Yet in much of the developing world, particularly in Africa, many rural populations have not adjusted to, or benefited from, the market reforms implemented in the 1980s and 1990s, let alone acquired the capacity to compete in a globalised marketplace. Indeed, rural communities often faced worse access to services and markets, and greater uncertainty, in the wake of state withdrawal and the reluctance or inability of the private sector to fill the gap.

The process of adjustment continues. In many areas there is gradual development of markets, sometimes boosted by improvements in infrastructure or knock-on effects from new export opportunities. In many countries there is more emphasis on dialogue, participation and partnership, with NGOs and community-based organisations (CBOs) increasingly playing key roles in this debate. The evolving orientation of these organisations and the new participatory paradigm signal the way for constructive arrangements to link farmers and markets. The greatest need for this is in areas where the task is most difficult, that is, in the poorest communities disadvantaged through geographical and economic isolation. Improving access to markets in these areas must be a key objective for poverty reduction at the beginning of the twenty-first century.

## Seed security

A commonly accepted definition of seed security is:

> 'the continuous availability and access to sufficient quantities of seed of adequate genetic and physical quality under both normal and emergency conditions'.

Farmers in the more marginal areas of the world are highly seed insecure, even in times of peace and stability. However, while food security has, justifiably, had an extremely high profile for many years, there is a significant perception in the seed sector that seed security should receive more prominent recognition in its own right.

Over the last 20 years, many developing countries have established national laboratories to provide seed testing and certification services. They have commonly been linked with seed improvement programmes aimed at developing high-yielding varieties. Associated seed multiplication projects have assisted in promoting and distributing these varieties. Although the improved varieties frequently require significantly greater inputs such as fresh seed, fertiliser and pesticides, the impact on crop production in many developing countries has been substantial. An additional move towards the use of certified seed of improved varieties has been achieved by the proactive marketing by the multinational seed companies, which have received a boost in recent years from the agricultural sector liberalisation process, i.e. privatisation, of the near market seed sector. This is especially the situation in many African countries.

Prior to liberalisation, government agencies commonly had responsibility for making improved seed widely available. Privatisation of national seed companies may have been a two-edged sword, in that the companies mostly gave up any role they might have had in serving the poorest farmers, and in supplying seeds of secondary crops or crops with low multiplication rates. Even major food crops such as sorghum and millet are frequently neglected as companies seek to improve their financial profitability.

Notwithstanding the impact of the formal seed sector, either government or private, and wide variations between crops and countries, it is widely accepted that more than 80% of seed for food crops in developing countries is still produced and retained on the farm from one season to the next. The role of the traditional farmer and the informal seed sector as a whole has not been well supported and is not adequately understood. In many developing countries the traditional seed sector is only just beginning to receive appropriate recognition. The informal seed sector will continue to be a major consideration in the agriculture of developing countries for the foreseeable future. Hence, a much better understanding of farmer seed management and the related decision-making processes is required.

Farmers are accepted as playing a leading role in the *in situ* conservation of agricultural biodiversity. However, how farmers value biodiversity and how they perceive their role in its management are two issues that have received little investigation. While some attention is given to the seeds of the major food crops, it should be noted that seeds of secondary food crops, herbs, spices, medicinal plants, tropical trees, pasture and forage crops are poorly researched.

International opinion is that past and present efforts have focused on the normal, i.e. non-emergency, situation and that it was now important to raise the profile of seed security in emergency situations. The result will be a more holistic understanding of seed security, which in turn could lead to more effective national and donor seed policies.

However, a study by van Wijk (1996) of the effects of the international convention on Trade-Related Intellectual Property Rights (part of GATT) and more comprehensive plant variety protection indicated that on-farm seed saving activities could be restricted. More recent international trade developments only serve to underline the threat to the small-scale farmer.

### Seed management by traditional farmers

Literature surveys indicate that much of the available information on small-scale farmer seed practices is anecdotal and that farmer seed management is a neglected area (Wright *et al.* 1994). Field surveys of farmer seed have been conducted with national collaborators in Ghana, Malawi and Tanzania by Wright and Tyler (1994) and Wright *et al.* (1995). Farmer practices were described and the effect of their use on seed quality determined using germination tests. The results indicated that farmer practices were generally sufficiently effective to ensure adequate germination.

Subsequently, however, more in-depth survey of farmer practices in Ghana and Zambia (Tripp *et al.* 1998a, b) revealed that farmers paid much less attention to seed selection than had been presumed. Studies in Ghana show that farmers and traders most often extract seed from those fruits that cannot be sold because they are rotten or damaged by insect pests: rotten fruit can spread seed-borne diseases (Orchard & Suglo 1999). It was apparent that farmers' selection methods had resulted in a rather heterogeneous mixture of varieties, with a gradual adaptation towards local conditions imposed by stress factors as influenced by climate, soil fertility, disease pressure, etc. It was found that productivity could be easily increased five-fold or more following pure line selection of some of the local land-races. This raises questions about the assumption that farmers function effectively as plant breeders through their seed selection activities. Farmers also paid little attention to testing the quality of their seed, possibly because they did not perceive it as an issue. Additionally, there were serious problems with the protection of legume seeds from insect infestation. Hence, there are opportunities for interventions to improve seed selection and seed quality but these must be viewed in a multidisciplinary context.

The issue for many small farmers around the world is not so much one of the availability or quality of seed *per se* but of the variety of seed which is accessible. Farmers have specific needs to plant varieties which maximise the potential of their holdings while fulfilling the varied demands of the household and marketplace. As modern varieties are promoted, many of the traditional varieties are being displaced and, in many cases, lost. The implications of this loss are far-reaching; not only are these varieties highly adapted to the local conditions and provide relatively stable yields under marginal conditions, but they also contain the genetic resource base that could be accessed by future generations of plant breeders.

There are specific problems associated with the supply of vegetable seeds in Africa. Imported seeds commonly come from sources that are not inspected by regulatory authorities and hence, on occasion, introduce seed-transmissible pathogens. Additionally, they are frequently from varieties that do not breed true to type. The growing demand for seeds of both 'western' and 'indigenous' vegetables in west and central Africa requires an increased availability of appropriate and good-quality seed.

To view the issue of farmer preference in perspective it will be necessary to obtain a better understanding of farmers' selection criteria, i.e. why do farmers adopt or abandon varieties given a range of agronomic, socio-economic, biodiversity, conservation and cultural factors? These investigations should not be limited to the main staples but should also include the secondary or locally important food crops on which local communities can be dependant.

Studies of vegetable farmers in Ghana revealed that commercial-oriented farmers considered yield, size of fruit and disease resistance to be equally important characteristics for tomato (Tables 9.1 and 9.2) (Orchard & Suglo 1999). Post-harvest quality attributes were often seen to be overlapping, especially resistance to physical damage, storage life, flavour and the amount of pulp in tomatoes. The respondents found it hard to think of all the different criteria separately as their decisions are made by evaluating the variety as a whole. While adequate yield is necessary for farmers to accept varieties, high-yielding varieties are rejected by traders if they do not have the required post-harvest qualities, particularly fruit size and firmness at maturity. Traders were not interested in yield or fruit size but particularly concerned that fruits should be mature (nearly at full red stage) but firm to withstand transportation to the market.

Farmers have to take into consideration traders' preferences, as well as important crop criteria such as length of harvest period, early maturing, disease and physiological disorders.

Women frequently play an important role in on-farm seed management and seed exchange systems, but gender and other social issues tend to be poorly documented and understood.

Seed marketing is one of the weakest links in seed security and it is the informal sector that is normally most effective in servicing the needs of farmers in marginal areas with often difficult access. In many countries local food markets also serve as sources of seed.

### Seed security in emergencies

Most disaster prevention and relief programmes focus on food relief and neglect the issue of seed security. In the absence of organised responses, relief efforts commonly revert to introducing modern varieties, which require expensive external inputs, are not adapted to the climate

**Table 9.1** Selection criteria for tomatoes.

| Criteria | Urban-market oriented farmers* | Home consumption/ local market farmers |
|---|---|---|
| Yield | ✓✓ | ✓ |
| Size of fruit | ✓✓ | ✓ |
| Resistance to pests and diseases | ✓✓ | ✓✓ |
| Resistance to heat/sun | ✓ | ✓ |
| Resistance to lack of rain | ✓ | ✓✓ |
| Does not crack or fall | ✓✓ | |
| Long harvest period | ✓✓ | ✓✓ |
| Early maturing | ✓ | ✓ |
| Does not spoil when left on plant | ✓ | |
| Grows on less fertile land | ✓ | ✓ |
| Resistance to physical damage | ✓✓ | ✓ |
| Stores well | ✓✓ | ✓ |
| Food in it | ✓✓ | ✓✓ |
| Flavour | ✓ | ✓✓ |

✓ Important, ✓✓ Very important
* Commercial, urban market-oriented farmers will also take into account trader and consumer preferences to ensure they can sell their produce.

**Table 9.2** Traders' criteria.

| Criteria | Total number of times cited |
|---|---|
| Lasts long (does not rot quickly) | 64 |
| Strong/firm/hard | 53 |
| Food in it (not watery) | 43 |
| Shape (resembles others) | 33 |
| Ripens properly after harvest | 19 |
| Colour | 19 |
| Farmers like it | 16 |
| Good for certain dishes | 15 |
| Size | 15 |
| Does not crack | 10 |
| Good even when rotten | 6 |
| Can get infested | 5 |
| Seed content | 5 |

and are vulnerable to various biotic and abiotic stresses. Use of this type of seed subjects the farmers to greater risks. In many cases, farmers lacking the necessary varieties may plant any seed from the marketplace or food relief stocks, thereby exacerbating their problems.

Emergencies can have very complex effects on seed security, especially when combined with market imperfections such as misplaced subsidies, for example, groundnut seed produced with subsidies in Malawi in 1994 was mainly used for food and processing.

While national and regional formal sectors have the potential to meet emergency demands for a narrow range of varieties, the only source of seeds of traditional food crops is the informal sector. However, the informal sector is highly susceptible to disruption during droughts or civil disruption. In any emergency, effective support needs to be provided to protect the local agricultural crop biodiversity and to use this as a base for regenerating seed for distribution to deficit farmers.

There is growing NGO involvement in seed security during emergencies. Non-governmental organisations adopt variable approaches to seed security. However, despite notable success, it should be noted that the technical competence of NGOs in this sector cannot be assumed.

As a result of these concerns, thinking in some international development agencies is focusing on the following topics in seed security:

- development of a seed security strategy and framework procedures at regional and global level in order to guide better emergency interventions;
- designing mechanisms to assist countries in disaster-prone areas in implementing seed security strategies and policies, for example, development of an early warning system on seed security in order to encourage interventions to protect local land-races in the regions affected by agricultural calamities;
- development of inter-country co-operative programmes and networks with a view to motivating the formal and informal seed sector to preserve the genetic material at risk, and to multiply it for redistribution to farmers of disaster-prone areas; and
- harmonisation and rationalisation of regional seed regulations, varietal testing and release mechanisms and quality standards to facilitate rapid cross-border responses.

## Food aid

The donation of food aid is an age-old practice that has existed throughout human history. It is provided by families, communities and governments for those affected by poverty, war, civil unrest and the consequences of natural disasters such as drought and flooding.

After World War II, the provision of food aid was established on an international scale. However, notwithstanding the humanitarian considerations, there is no doubt that food aid has its roots firmly in the disposal of agricultural surpluses. This aspect, together with the significant political implications, continues to influence food aid policy and practice.

In 1954, under Public Law 480, the US government authorised government-to-government donations of food for famine and other relief programmes to nations regardless of their political systems.

In 1960, the US sponsored a resolution in the UN General Assembly calling upon the Food and Agriculture Organization (FAO) to assist in making the large international food surpluses available for relief and development. This resolution led to the foundation in 1963 of the World Food Programme (WFP). The WFP is a joint subsidiary body of the United Nations and the FAO, and is the United Nations' agency responsible for providing food aid. While it was originally tasked to support longer-term rehabilitation, reconstruction, and development through the provision of food aid, its role has developed and WFP is now also a key operator in emergency relief provision. Subsequent to the creation of the WFP many nations have formalised their contributions to the international food aid system.

In 1999, WFP supplied 3.4 million tonnes of food aid to help feed 89 million people in 82 countries. WFP typically supplies around one-third of all food aid and is therefore a significant player in this sector. Individual donor nations supply WFP with cash for the procurement of food aid or they provide the food itself. Other supplies of food aid are made by national bilateral programmes and by national or international non-governmental organisations. WFP delivers 95% of all international multilateral food aid. Additionally, WFP has been steadily increasing the amount of food it purchases from developing countries, and in 1997 it reached a peak of 1.3 million tonnes.

Some donor nations make food aid available direct to the receiving states in the form of bilateral food aid. Members of the European Union will also contribute to the food aid supplied by the European Commission.

The full extent of food aid is not known with any precision. Provision at local community level is not documented and provision at a formal level is unclear because of uncertainties arising from a lack of clarity regarding the definition of the term food aid.

However, over the past decade, FAO statistics indicate that the formal supply of food aid commodities at an international level has averaged the equivalent of over 10 million tonnes of cereal per annum. There are wide fluctuations around this mean, ranging from 15.2 million tonnes in 1992/93 to 5.6 million tonnes in 1996/97. These fluctuations do not arise totally from changes in need due to the incidence of natural disasters or civil unrest that lead to hunger and food insecurity. They are just as influenced, if not more so, by the need to dispose of agricultural surpluses, the level of political interest and the newsworthiness of large-scale disasters.

### Where food aid goes

Over the past decade, sub-Saharan Africa and Asia have normally received around one-third of all food aid. However, in 1999 the proportion of the global total that went to sub-Saharan Africa fell to 18%, while deliveries to Europe and the CIS countries increased from between 10% and 20% to a high of 40%.

### What are the main commodities?

Cereal grains comprise approximately 85% by volume and 50% by value of food aid. The bulk of this is wheat and there are significant quantities of maize and rice. Other cereals such as barley, millet and sorghum are supplied in much smaller quantities. Whether the cereals are shipped in bulk or in bags will depend on the specific circumstances, although bulk shipment has become much more common in recent years. Bulk shipments not destined for processing near the port area will be bagged as they leave the ship for further distribution.

Other components are supplied in much smaller quantities but have much greater economic unit costs and supply essential contributions to the diet.

Pulses, such as beans, lentils, peas and soya, comprise around 8% by volume. Their major role is the provision of vegetable protein in food aid diets. Vegetable oil is a major food aid component which, while only 4% of the volume, provides an essential component in the dietary provision. Canned meat and fish, less than 1% by volume, provide important animal proteins. Dairy products, less than 1% by volume, are supplied as dried skimmed milk. Other items will depend on dietary needs and preferences, but could include dried dates, ghee, high-energy biscuits, salt and sugar.

Some food aid situations require the cereals to be part processed, for example, bulgur wheat, maize meal, pasta, rolled oats and wheat flour. Bulgur wheat is wheat that has been soaked in water, boiled or steamed, and then dried. The gelatinous grain is subsequently pearled, cracked and screened to remove most of the bran. Bulgur wheat is comparable to parboiled rice in nutritive value, with similar cooking properties. Processed commodities can have a greater interaction between enzymes and grain constituents, and are much more susceptible to insect infestation than are the whole grains. Processing also results in more exacting packaging requirements, for example, multiwall paper sacks.

Blended foods are important contributions to balanced diets, especially for weaning infants, children and the elderly. Various fortified and blended foods have been formulated using cereals, pulses, dried skimmed milk and vegetable oils. One of the first was corn (maize) soya milk (CSM) which comprises processed gelatinised maize meal, defatted toasted soya flour, refined deodorised stabilised soya oil, non-fat spray dried milk, mineral premix and vitamin antioxidant premix. Other blended foods include wheat soya blend, corn soya blend, wheat soya milk and soya fortified bulgur wheat.

Some emergency rations are made of compact dry foods (only 5% moisture) pressed into vacuum-packed bars. They contain high-quality protein, fat and carbohydrates with added vitamins and minerals. Others are freeze-dried products which have long shelf-life and low transport costs, but their expense would normally limit their use in food aid operations.

### Quality assurance issues

Recipients of food aid are frequently under stress or disadvantaged. Hence, it is important that they receive food aid commodities that match local preferences, are easy to cook given the existing constraints, and are in good condition.

Matching supply of food aid bèneficiary preferences can be problematic. The system is still very much supply-driven and some local staple foods are difficult to procure or supply. Many communities have a staple that is little traded on the world market, for example, millet, yellow maize and certain types of rice. Others have root or tuber staples that are not durable and therefore difficult to transport. It is inevitable that some recipients will receive food aid commodities with which they are unfamiliar and, in some instances, they might be unsure how to cook.

Cereals traded in temperate areas of the world will commonly have a moisture content of 15–16%. In cool climates these levels of water will rarely cause any difficulty during storage. However, in tropical or subtropical climates it is necessary to have a lower moisture content, e.g. 12–13%, to ensure that the grains will store for several months without spoilage. Because food aid is procured from normal trading systems it is inevitable that some food aid cereals will be delivered with moisture contents in excess of the safe storage maximum for local conditions. The result will be reduced shelf-life and possible heating of the stacks leading to damage and loss.

Food aid, by definition, is delivered to areas or countries under stress in some form. Therefore, it is often difficult to ensure that adequate warehouse management competence is in place prior to its arrival. Facilities and logistics will frequently be suboptimal with the result that there is a significant potential for larger than normal levels of insect and water damage. Aid agencies need to focus much more on quality assurance issues than do normal commodity trading organisations.

## Further reading


Demery, L. & Addison, T. (1987) Food insecurity and adjustment policies in sub-Saharan Africa: a review of the evidence. *Development Policy Review*, **5**, 177–196.

Donaldson, G. (1984) Food security and the role of the grain pool. *American Journal of Agricultural Economics*, **May**.

Grant, I.F. & Sear, C. (Eds) (1999) *Decision Tools for Sustainable Development*. Natural Resources Institute, University of Greenwich, Chatham, UK.

Huddleston, B. & Konandreas, P. (1981) *Insurance approach to food security: simulation of benefits for 1970/71 to 1975/76 and for 1978 to 1982. Food Security for Developing Countries*. Westview Press, Boulder CO, USA.

Josling, T. & Baricrello, R. (1984) The role of developed countries since the World Food Conference. *Food Policy*, November.

Kasalu, E. & Johnson, W.I.R. (1988) Policy issues for wheat in Zambia. Planning Division seminar, Ministry of Agriculture and Water Development, Zambia.

Kirkpatrick, C. & Diakosawas, D. (1985) Food insecurity and the foreign exchange constraint in developing countries: the case of sub-Saharan Africa. *Journal of Modern African Studies*, **23**.

Koester, U. (1984) Regional co-operation among developing countries to improve food security. *Quarterly Journal of International Agriculture*, **23,** 99–115.

Monke, E. & Salam, S. (1986) Trade policies and variability in international grain markets. *Food Policy*, August.


## References


Anon. (1993) *Review of FSRs*. NRI Report No. R1897 (S). Natural Resources Institute, University of Greenwich, Chatham, UK.

Boxall, R., Golob, P. & Taylor, R. (1997) *Pest Management in Farm Granaries*. Natural Resources Institute, University of Greenwich, Chatham, UK.

Carney, D. (1995) Changing public and private roles in agricultural service provision: a literature survey. *ODI Working Paper* **81**. Overseas Development Institute, London, UK.

Conway, J.A. & Tyler, P. (1995) *Practical Implications of Grain Market Liberalisation in Southern Africa*. Proceedings of a workshop, 18–20 October 1994, Harare, Zimbabwe. Natural Resources Institute, Chatham, UK.

Coulter, J.P. & Magrath, P. (1994) Economics of grain handling and storage in developing countries. In: *Grain Storage Techniques – evolution and trends in developing countries*. Agricultural Services Bulletin **109**, United Nations Food and Agriculture Organization, Rome, Italy.

Coulter, J. & Norvell, N. (1998) The role of warehousing in Africa – lessons for implementation from four continents. Partners for Development Summit, November 1998, Lyon, France.

Hine, J. (1993) Transport and marketing priorities to improve food security in Ghana and the rest of Africa. In: H.-U. Thimm & H. Hahn (Eds) *Regional Food Security and Rural Infrastructure, Volume 1*. Symposium Geissen Rauischholzhausen, Germany, 3–6 May 1993. Schriften des Zentrums für Regionale Entwicklungsforschung der Justus Liebig Universitat Giessen, Giessen, Germany.

Hoff, K., Braverman, A. & Stiglitz, J. (Eds) (1993) *The Economics of Rural Organisation: Theory and Practice*. World Bank, Washington DC, USA.

Kindness, H. & Gordon, A. (2001) *Agricultural marketing in developing countries: the role of NGOs and CBOs*. Policy Series **13**. Natural Resources Institute, University of Greenwich, Chatham, UK.

Maxwell, S. (1989) Food insecurity in North Sudan. *Discussion Paper* **262**, Institute of Development Studies, University of Sussex, Brighton, UK.

Mittendorf, H.J. (1988) *Improving Agricultural Physical Marketing Infrastructure in Africa Through More Self-help.* FAO Occasional Paper. United Nations Food and Agriculture Organization, Rome, Italy.

Newman, C. (1994) Bulk storage. In: *Grain Storage Techniques – Evolution and Trends in Developing Countries.* Agricultural Services Bulletin **109**, United Nations Food and Agriculture Organization, Rome, Italy.

Orchard, J.E. & Suglo, E.K.J. (1999) *Integrated food crops systems project: enhancing smallholder livelihoods through reducing costs and adding value to agricultural production.* Final Technical Report. Natural Resources Institute, University of Greenwich, Chatham, UK.

Robbins, P. (1998) *Review of market information systems in Botswana, Ethiopia, Ghana and Zimbabwe.* CTA, CMIS, London, UK.

Sen, A. (1981) *Poverty and Famine: an Essay on Entitlement and Deprivation.* Clarendon Press, Oxford, UK.

Shepherd, A. (1997) *Market Information Services: Theory and Practice.* United Nations Food and Agriculture Organization, Rome, Italy.

Sieber, N. (1997) *Economics of appropriate agricultural transport – a broader approach towards on-farm and market transport.* East Africa Regional Project Planning Workshop, 2–7 November 1997, Thika, Kenya. Agricultural Rural Transport Research Project, Department for International Development, UK.

Tripp, R., Miti, F., Mukumbuta, S., Walker, D.J & Zulu, M. (1998a) *Farmers' seed management in Zambia – A study of sorghum, cowpea and groundnut seed in Southern and Western Provinces.* NRI Bulletin **76**. Natural Resources Institute, University of Greenwich, Chatham, UK.

Tripp, R., Walker, D.J., Apau, A., Dankyi, A.A. & Delimini, L.L. (1998b) *Farmers' seed management in Ghana – A study of maize and cowpea seeds in Brong-Ahafo and Volta Regions.* NRI Bulletin **68**. Natural Resources Institute, University of Greenwich, Chatham, UK.

Van Wijk, J. (1996) How does stronger protection of intellectual property rights affect seed supply? Early evidence of impact. *Natural Resource Perspectives* **13**, Overseas Development Institute, London, UK.

World Bank (1986) *Poverty and Hunger: Issues and options for food security in developing countries.* World Bank, Washington DC, USA.

Wright, M. & Tyler, P. (1994) *Traditional seed-saving practices in northern Ghana and central Malawi.* NRI Report R2102 (S). Natural Resources Institute, University of Greenwich, Chatham, UK.

Wright, M., Donaldson, T., Cromwell, E. & New, J. (1994) *The retention and care of seeds by small-scale farmers.* NRI Report R2103. Natural Resources Institute, University of Greenwich, Chatham, UK.

Wright, M., Delimini, L., Luhanga, J., Mushi, C. & Tsini, H (1995) The quality of farmer saved seed in Ghana, Malawi and Tanzania. Unpublished Research Report. Natural Resources Institute, University of Greenwich, Chatham, UK.

### *Website*

World Food Programme. http://www.wfp.org.

# Chapter 10
# Applied Research and Dissemination

**H. C. Coote, N. K. Marsland, I. M. Wilson, S. Abeyasekera and U. K. Kleih**

## Extension methods and technology transfer in less developed countries

Concern is frequently voiced by governments, donors and researchers about post-harvest crop loss:

> 'There is a substantial amount of food loss during post-harvest due to backward storage and food processing. In order to curb this problem research will be conducted to reduce both pre- and post-harvest losses.'
>
> *Government of Ethiopia*

> 'Farmers in sub-Saharan Africa lose a sizeable proportion of their harvested grain to insect pests during storage...Reducing storage losses will enhance household food security, as well as increasing farmers' control over both the scale and timing of their grain marketing.'
>
> *DFID-funded research project*

To what extent these fears and concerns reflect the reality is discussed elsewhere in this book. The extent to which extension resources are directed towards assisting farmers to reduce post-harvest losses and the best ways of providing such assistance, including the introduction of new or improved technologies, are considered below.

### Technology choice and identification of needs

The first step in the identification of appropriate technology should be the assessment of the needs of potential users. The rate of adoption of technologies at farm level, to reduce or prevent post-harvest losses such as new storage or chemical pest control, has often been poor (Compton 1992). Technology transfer may have failed because the technology was promoted on the basis of false assumptions. It has often been incorrectly assumed that storage ranked high among farmers' priorities. However, improved storage can involve substantial costs and risks as well as potential benefits. Storage competes with other activities and an understanding is required of where storage fits into the farming system and the household economy, as well as the external policy situation, in order to assess the need for interventions and the probability of uptake. In field research into post-harvest issues farmers may identify pre-harvest production problems which are of much greater concern to them (Marsland & Golob 1996).

For smallholder farm families the main purpose in storing grains is to ensure household food supplies. In addition, on-farm storage provides a form of saving, to cover future cash need through sale, or for barter exchange or gift-giving. Grain is also stored for seed and as inputs into household enterprises such as beer brewing, or the preparation of cooked food (Proctor 1994). When grain marketing was largely in the hands of the state and crop prices were controlled there were often few financial advantages from delaying grain sales. Immediate cash needs after the harvest for other investments, debt repayment and social obligations also mitigate against storage. Farmers may use traditional storage methods that only allow products to be safely stored for short periods due to the likelihood of spoilage.

The need for economic and social analysis in the planning and design of storage interventions has become more widely recognised. This has stemmed from the realisation that any 'improvements' in storage will only be attractive to farmers, traders or governments if the perceived benefits substantially outweigh the costs. Technical superiority is generally not sufficient. Farmers and traders may tolerate quite high storage losses before undertaking complex or expensive changes to their storage systems. Local, established storage systems are usually well-adapted to local conditions, and losses from grain storage are already low and acceptable to farmers (Compton 1992). In some countries the government has actively discouraged on-farm storage in the past and this may also affect how farmers currently think about storage. Other post-harvest investments which may increase grain quality may not be profitable as higher quality has rarely been reflected in a higher price, and therefore provides no incentive for adoption of new technologies.

Regarding the perceived need for storage, there is often a difference between the technically estimated loss and farmers' perceptions of losses, particularly the way in which these are reflected in decision-making concerning crop disposal. Technical losses are relatively straightforward to measure, although researchers in Malawi found 'existing methods of loss assessment are too crude or too technically complex' (Marsland & Golob 1996). The same fieldwork ascertained that people's perceptions of losses have a seasonal dimension and are also related to the use of the crop:

> 'Cultural celebrations such as weddings and festivals take place soon after harvest, a lot of food is consumed and some is wasted, thrown away for dogs and pigs. More generally, from harvest in May until up to about August, people consume possibly more than is required; there is a popular belief that people are not satisfied if there are leftovers on the plate after a meal. As the year progresses, more and more people run out of food, theft increases and pest attack increases on the diminishing stock of stored grains and legumes. Farmers' perceptions of loss also change; they become more sensitive to the levels of loss.
>
> 'In addition, it appears that perceptions of loss are also influenced by end use. For example, in Chimango village (Lilongwe district), the perception of loss was greater if maize was used for sale than if it was used for consumption, because, even if weeviled, farmers argued that they could still make *nsima* (maize porridge) (although it would not taste as good), whereas weight loss would mean lower market prices. In addition, it was noted that if the germ is eaten then the crop will not germinate, thus leading to a big loss in terms of seed. However, it can still be eaten. If the endosperm is attacked then the farmers noted that it could still be used as seed and food. On this basis, even though weight loss might be the same, the farmers perceived the germ loss as more important than the endosperm loss. The key point is that perceptions are related to the functional uses of grain.'
>
> *Marsland & Golob (1996)*

The changing external environment may result in the need for increased storage of grain at various stages in the marketing chain. Six factors have been identified where there may be new impetus for seeking new storage technologies for conserving grain safely on-farm (Coulter & Magrath 1994).

(1) Increasing urban demand plus large increases in intensive animal production that create large markets for feed grains.
(2) Market deregulation has removed the guaranteed market for grain surplus immediately after harvest, which previously relieved the producer of storage and quality maintenance problems, and increased the role of the private sector in marketing grain including storage. The removal of pan-seasonal prices may make it more profitable for farmers to store grain for selling later in the season.
(3) Changes in the farming system, such as the adoption of higher yielding crops which tend to have poorer storage durability (the very qualities which lead to higher yields also make the grain more attractive to pests).
(4) Short duration varieties have enabled increased cropping intensity, which can give rise to further storage problems when one of the harvests occurs in the wet season, making it difficult for farmers to dry the grain sufficiently for storage.
(5) High yields also require storage or sale of larger quantities of grain within a shorter space of time, which in itself may cause problems and encourage farmers to sell at harvest, in order to free up the labour for field preparation of the next crop.
(6) A major change in the incidence of pests can prompt demand for new storage technologies, for example, the spread of the larger grain borer (*Prostephanus truncatus*), a severe pest of stored grain, in sub-Saharan Africa.

Identifying the likely increase in interest in new storage technologies will not necessarily lead to more government resources for extension in this area. There is ongoing debate on who should pay for extension and the appropriate roles of private, voluntary and public sectors in funding and delivering agricultural extension services (Picciatto & Anderson 1997). Changes in extension needs and practices are likely to arise from market liberalisation. It is hoped that more open and liberalised agricultural markets will bring the knowledge and skills of private agribusiness to farmers, without necessarily involving public sector intermediaries. Providing commercial incentives for farmers to store may be followed by an increase in availability of commercial storage structures for sale, with information on how to use them and the benefits they will bring.

### Extension issues

Agricultural extension, 'as the organised exchange of information and the purposive transfer of skills' (Nagel 1997), has tended to concentrate on pre-harvest activities, particularly crop production and management. The value of extension is often given as the increase in crop yields due to the provision of extension (Evenson 1995). There is relatively little literature on extension and technology transfer in the post-harvest sector compared to pre-harvest activities, or information about extension programmes covering post-harvest activities (a rare exception is the Bureau of Post-Harvest Research and Extension in the Philippines). This may have been due to the preoccupation of extension services with crop-oriented extension programmes which promoted the uptake of high-yielding, fertiliser-responsive crop varieties, and food shortages which forced output prices high enough to make the use of the new technologies viable (Picciatto & Anderson 1997).

State regulation of marketing and operation of pan-territorial (country-wide price) and pan-seasonal (same price year round) pricing regimes did not provide an incentive for farmers to store with a view to speculation. There was also little reason for scarce extension resources to be devoted to this sector since storage was in the hands of parastatal organisations. The crops that were stored were mainly those used for home consumption, often the preserve of women.

Much has been written about the failure of extension services to address the needs of particular groups and this might be a further reason why little attention was paid to on-farm storage, particularly in Africa (Spurling 1995; Nagel 1997; Jiggins *et al.* 2000). In a rapid rural appraisal of post-harvest issues in Malawi, men and women were asked who was responsible for the various activities and who made the decisions for maize. These are summarised in Table 10.1.

Similarly, much has been written on the evolution of extension methods and their success, or otherwise, in providing farmers with 'efficient, effective and appropriate technology, training and information' (Jiggins *et al.* 2000) over the past few decades. Extension services have largely failed in meeting these objectives for the vast majority of rural inhabitants. Reasons for failure put forward by Nagel (1997) include:

- The contradictory nature of extension goals, e.g. public interest implies serving both farmers and the urban population, securing subsistence production and promoting cash crops, reaching the masses of rural households and serving the needs of specific groups;
- The hierarchical and highly bureaucratic way in which services are organised hampers a full realisation of their potential – priority-setting for research is rarely based on extension field evaluations because the system does not foster critical upward communication;
- The way in which technical knowledge is transformed into field messages leads to distorted and outdated information;

**Table 10.1** Responsibilities for harvest and post-harvest activities for maize in Malawi. Source: Marsland & Golob 1996.

| Activity | Responsibility |
|---|---|
| Drying, while still planted | Children (to scare off monkeys) |
| Stooking | Men |
| Transporting from field to home | Women and men |
| Drying | Women |
| Threshing/shelling | Women |
| Winnowing | Women |
| Bagging | Women |
| Filling *nkokwe* (granary) | Men and women |
| Pest control | Men |
| Removal from *nkokwe* | Women |
| Processing (soaking, pounding/milling) | Women |
| Marketing | Men usually |

- Extension has never been a purely educational activity – non-educational activities have included supervising credit repayment, policing disease control measures, organising 'voluntary' community work;
- Ministry-based extension unable to reach a majority of its potential clientele; and
- Extension workers selecting the more responsive section of their clientele, to fulfil their production plans, improve their job satisfaction or because of prejudice against certain target groups.

Storage advice is said to be generally available from agricultural extension services, usually as one or two basic messages or packages, but it may be inappropriate, inconvenient or too expensive for some farmers. Table 10.2 gives some examples of such packages. However, farmers need advice that is tailored to their individual and specific situations. In particular, recommendations assume that farmers are able to plan how they will manage their stored grain from the start of the season. Smallholders tend to make a series of decisions on how to manage their crop, in sequence, throughout the season based on the options and constraints at each stage. Effective storage loss reduction programmes involve farmers in the analysis of their storage problems and in identifying appropriate storage systems and storage management techniques (Boxall *et al.* 1997).

One of the first real attempts to introduce post-harvest extension was associated with attacks by the larger grain borer (LGB) beetle. This severe pest of farm-stored maize and cassava 'created an urgent extension need' which included the introduction of a major change of post-harvest practice, the shelling of maize before storage, in Eastern and Central Africa (Golob & Eisendrath 1996). It took a major effort by extension staff in Tanzania to convince farmers to modify their storage structures to grain stores, to shell the maize shortly after harvest and to add an insecticide dust to the raw grain. Indigenous grain storage methods were inadequate to deal with an introduced pest. Prior to the start of the LGB programme of the Food and Agriculture Organization, extension officers also had had no formal training in on-farm crop storage. One of the main tasks of the FAO programme was to give training to the extension officers so that they could face farmers, who demanded a great deal of information about the huge problem of protecting grain supplies, without feeling inadequate (Golob & Eisendrath 1996).

### Evolution of extension methods

In the early years after independence, agricultural extension services tended to imitate their colonial antecedents and were geared towards the production and export of crops. Extension systems were largely desk-bound, with one of the main dissemination methods of research results being farm demonstration programmes. In the late 1960s and early 1970s the focus of agricultural extension shifted towards technology diffusion of high-yielding crop varieties. In order to support this system an innovative organisational method for extension delivery was devised – the training and visit (T&V) system. T&V dominated agricultural extension in South Asia and Africa for over 20 years. Its strength was to concentrate on fixed visits at regular intervals with contact farmers who were expected to pass on information to farmers with similar problems. Regular sessions were held for extension workers to receive training and discuss issues. However, implementation often proved difficult. The contact farmer concept, with its two-step flow of information from the extension worker to the farmer and from there to other farmers, frequently failed. T&V used a top-down approach that left few possibilities for participation and initiative, both by farmers and village extension workers. Standardised messages passed on were often of little relevance to local conditions. 'Once T&V was

**Table 10.2** Recommended storage packages in Ghana and Togo. Source: Boxall *et al.* 1997.

| Medium and large-scale maize farmers (Ghana) | Small-scale farmers (Togo) |
|---|---|
| Grow a recommended high-yielding variety of maize | Grow a recommended variety of maize with good husk cover |
| Dehusk cobs at harvest | Harvest at the recommended time |
| Spray cobs with an approved insecticide | Select only cobs with good husk cover |
| Store in a narrow ventilated crib | Clean the store before putting in new maize |
| Shell when cobs are dry | Smoke the store when weevil infestation is seen |
| Mix with an approved insecticide | |
| Store in sacks | |

extended to less favoured regions it became clear that technology of the green revolution showing quick and visible results was not available' (Nagel 1997).

The integrated rural development project approach, prevalent in the late 1970s and early 1980s, aimed at influencing the entire rural development process, and at alleviating mass rural poverty by simultaneously improving the utilisation of natural resources and of human potential through targeting the whole population. Extension was only one area of intervention, though often crucial. However, evaluations of integrated rural development programmes have shown that few attained the goal of mass poverty alleviation due to disregard of the target group principle, lack of due consideration of economic and institutional conditions, and lack of compatible technical solutions (Nagel 1997).

Farmer field schools have been operating for over 15 years in Asia, as a way of introducing problem-oriented training and development of sustainable solutions. Farmer field schools were established primarily to introduce integrated pest management in rice production (Braun *et al.* 2000). The field schools adopted a participatory approach whereby the extension agent facilitated farmers to think about and apply their own knowledge. This style of approach – with a fundamental change in the role of the extension agent – is gaining ground in extension. The extension agent is no longer seen as the expert who has all the information and technical solutions.

> 'The clients' own knowledge and ingenuity, individually and collectively, are recognised as a major resource; solutions to local problems are to be developed in partnership between agents and clients.'
>
> *Swanson* et al. *(1997)*

There is strong argument for farmers' participation and the greater use of local expertise for programme development, capacity building, cost-effectiveness and familiarity with local content (Rivera 1996), which is the case both for pre- and post-harvest activities.

Whatever the topic addressed, extension tends to be expensive and there are never enough resources to reach even a tiny proportion of the rural population. Increasingly, debates on extension are focused on who is to be served, who will pay and who will deliver (Rivera 1996). In Europe, little farm management advice comes free. However, in developing countries it is likely that small-scale and subsistence farmers, whose requirements include needs assessment, organisational skills and simple technologies, will need to be publicly funded. As a response to limited state resources and interest an increasing role is being taken by NGOs (and contract farming schemes for cash crops) in providing extension to smallholders, including post-harvest issues such as household food security, storage and marketing. In Francophone West Africa some non-governmental organisations have encouraged villages to create 'cereal banks' – food supply mechanisms for remote villages involving the storage of cereals immediately after harvest and selling grain back to villagers in periods of shortage.

Where the knowledge being transferred is embedded in, or is closely associated with, market goods (plantation crops, tractors, hybrid seed) it is considered that delivery of advisory services is best left to the private sector, within an appropriate regulatory framework. Where the technology or practice being promoted is associated with a 'toll' good (farm management or marketing information), provision of extension is best undertaken by a mix of public and private organisations. With a 'common-pool' good it is critical to link extension closely to co-operative or voluntary action. It is only where market and participation failures are high, for example where subsistence farming dominates as in sub-Saharan Africa, or where social conditions preclude voluntary action, that a pure public-sector approach to agricultural extension is considered desirable and justifiable (Picciato & Anderson 1997).

The increasing use of participatory methods leads to farmers becoming more confident in putting forward their needs. An example, from Zimbabwe, relates to the revival of *zunde ramambo* – the communal production of grain for disadvantaged members of the community. This has led to requests for technical assistance in store construction and management (Stathers 2000).

### Steps in technology transfer

Where farmers are convinced of the need to adopt new, or upgrade their current technology, the profitability of a particular storage investment needs to be shown using a range of input and output prices and discount or interest rates, using techniques such as discounted cash flow or annualised costs. This should be followed (Coulter & Magrath 1994) by:

- test-marketing, supported by a delivery structure
- installation of prototypes (if appropriate)
- monitoring growth of sales
- analysis of market penetration

- investigation of reasons for non-adoption, and
- changing/modifying the technology, if necessary.

Work on the behavioural aspects of technology adoption by farmers in the US indicates that technology and change will most likely be implemented when the benefits will be quickly realised (usually within 12–18 months), the tools for implementation are readily available in the local marketplace, the risk of implementation can be diminished and when the change or new technology can be comfortably integrated into other basic ongoing aspects of daily life (Barao 1992).

One of the most successful storage technologies to date is the use of insecticides. Insecticides can easily be integrated into existing storage systems and often give a high rate of return. The main constraints on increased use include the availability of appropriate insecticides at the right time, stability of the formulations used, farmer training in the correct types and correct use of insecticide and cost, which sometimes renders their usage uneconomical (Coulter & Magrath 1994). However, farmers do voice concerns about using chemicals with food grains.

> 'Farmers' fears of mixing synthetic chemical pesticides with their food have highlighted the need for alternative grain protection methods such as diatomaceous earths [inert dusts].'
>
> *Stathers (2000)*

The use of these is currently being researched in Zimbabwe, with smallholder farmers.

### Participatory rural appraisal and survey techniques

The foregoing section indicates the importance of ascertaining the reasons why people store, and the systems within which storage occurs in order to estimate how the benefits and costs of innovations are likely to be assessed by the intended users of the technology. Extension services are often provided at group level. A participatory approach, involving groups or individuals in deciding which storage management strategy to adopt, may be more time-consuming than promoting a single extension message. However, it is considered likely to be more effective and result in a higher adoption rate of recommendations (Boxall *et al.* 1997).

A participatory rural appraisal (PRA) approach can be used to identify opportunities for improvements and assisting farmers to find appropriate solutions. The PRA methodology stresses the inclusion of all stakeholders, that is, all those affected by or involved in a particular issue. This may include taking care to interview a representative sample of men and women farmers and traders, including significant minority groups likely to have an important role in crop storage (e.g. larger mechanised farmers). Out of this activity should come an assessment of whether any improvements are worth considering in greater depth, and a list and description of those ideas or concepts suited to particular groups of farmers.

Participatory methods are now widely used to obtain an overview of issues, ascertain the key players and enable target groups to be involved in a process of change. A considerable body of literature and websites exists on this topic, both on the philosophy and on methodologies and tools. A number of techniques taken from PRA are used in farming systems' research and participatory market research. These involve working with farmers and traders to gain an understanding of their needs and aspirations. For example, a rapid loss assessment method for estimating storage losses in maize and cassava was developed in Togo that attempted to incorporate farmer criteria in defining categories of loss. As the measurement occurs in the field, rather than at a laboratory, results can be discussed with farmers on the spot (Compton 1992). Such methods could usefully be integrated into post-harvest technology projects.

A key PRA tool of particular use for discussing existing and new practices and technologies is matrix scoring. This involves tabulating alternative technologies (on a horizontal axis) against a range of criteria used in their selection (on a vertical axis). The criteria should be developed in discussion with key informants knowledgeable about on-farm storage. Each technology can be scored or ranked by groups in terms of their perception of its performance against each criterion. If some criteria appear to be much more important for the villagers than others then they can be given more weight by multiplying their scores by a weighting factor (e.g. × 2). An example is shown in Table 10.3.

In this example the metal bin has the highest score. However, if 'low construction costs' were of greatest importance to the villagers and this criterion was given a double weighting, by multiplying its scores by two, then the traditional bin would come out top, with 31 points, compared to 28 and 24, respectively, for the other two constructions. As an alternative to scoring, data may be ranked against each criterion. The method can rapidly elicit information on why participants give priority to

**Table 10.3** Matrix scoring for three grain storage structures.

| | Metal bin | Improved crib | Traditional crib |
|---|---|---|---|
| Durability of structure | ******** | ***** | ** |
| Ease of handling | ****** | ****** | **** |
| Peace of mind | ***** | *** | *** |
| Low construction costs | * | *** | ********** |
| Does not attract pests | ******* | **** | ** |
| Total score | 27 | 21 | 21 |

certain criteria. However, ranking only conveys an order of preference, not the degree of like or dislike, as ranked data for different criteria cannot be added up (Coulter & Magrath 1994).

Participatory methods also encourage the involvement of a multidisciplinary team. Agricultural research is often organised on the basis of scientific disciplines rather than commodity programmes or farming systems. This has meant that it is difficult to realise the needs of the farmers, who often require a complete package of recommendations addressing various constraints, both pre- and post-harvest.

Other data collection methods involve use of sample surveys (based on a questionnaire) to get an overview of numbers and make inferences about wider needs. Again, the techniques are widely described in the literature (see, for example, Casley & Lury 1987). Increasingly, a mix of these two methods is used. The size of the target population has a bearing on the importance of external validity but it plays less of a role if the target population is small (e.g. a small number of villages in the case of an NGO-led development project). On the other hand, projects covering entire regions or countries depend on results representative of these areas. Formal sample survey work has probably the most to gain from informal participatory approaches in the area of credibility and objectivity, whereas informal work (if it is to be generalised) can borrow from formal methods to improve external validity (Marsland *et al.* 2000). These issues are discussed in greater detail on page 448.

## Farmer participation in assessing post-harvest needs

Needs assessment is essentially an agricultural research planning exercise and, in the case of technological interventions, it should address farmers' needs and constraints in such a way that there is a greater chance that the interventions will be readily adopted.

A common criticism of post-harvest research is that technical innovation has been high, but adoption has been poor. However, research and development activities in this area are complex because of the interaction of technical constraints with the social and economic contexts of those involved. When research and development activities are undertaken, there needs to be a clear understanding of these complexities so that improved technologies will be adopted.

Historically, agricultural problems tended to be assessed within the narrow focus of subject disciplines. Field workers announced their interests and set boundaries for their questioning. As an agricultural research planning exercise it was faulty in that the wider context of the problem was often not addressed and its place in the hierarchy of wider agricultural constraints was not identified. Participatory needs assessments (PNAs) undertaken in Africa and elsewhere indicate, for example, that farmers have a clear list of reasons for choosing particular storage structures and storage protection methods. Work in Northern Ghana and in Zimbabwe demonstrated that PNA using ranking and scoring are useful tools for uncovering opportunities and constraints in relation to storage structures and storage protection. The findings of the research are proving useful in guiding project implementation and the thrust of extension efforts (Marsland 1997). This should have a positive impact on uptake of technical research outputs. Box 10.1 gives an example of how needs assessment exercises are leading to a more holistic understanding of the place of post-harvest research in farming systems in Malawi.

In Ghana, the market price for cowpea, free from insect damage, increases considerably 3–4 months after harvest. However, there is little likelihood of producers taking up safe storage methods if there is a shortage of cowpea seed and little chance of producing marketable surpluses. Cereal crops such as sorghum and maize are also prone to storage losses, but farmers may have other more pressing concerns. For example, participatory needs assessment done during 1996 in several villages

**Box 10.1 The relationship between monitoring and evaluation instruments and project level logical frameworks. Source: Marsland & Golob 1996.**

A rapid rural appraisal (RRA) of household food security constraints was conducted in six villages in three agro-ecological zones in Malawi's Central Region. The study focused on post-production problems and issues, which can be divided into those occurring: before harvest; at harvest; during storage; during processing; and marketing.

Farmer groups identified post-production issues as constraints to food security, but production problems, including input prices, field pests and diseases, and land access and land productivity were of greater concern. The RRA identified post-production problems, which, if addressed, would make a significant impact on the wellbeing of rural communities. Many of the problems were interrelated and could not easily be solved in isolation. The key constraints were as follows:

- Drying cereals and legumes (the introduction of improved drying cribs, tried in several African countries with varying degrees of success, may be a possible intervention here).
- Difficulties in transporting ripened and partly or fully dried crops from the field to the household.
- Controlling insect pests of cereals and legumes, particularly the larger grain borer, which has been found throughout the country. Low cost, sustainable solutions are required to complement on-going GTZ work on biological control. NRI research on botanical and inert dust insecticides conducted at Chitedze Research Station, Lilongwe, is appropriate.
- Researchers do not understand farmers' perceptions of losses and how these are reflected in decision-making concerning crop disposal. Moreover, existing methods of loss assessment used in Malawi are either too crude or too technically complex to be used as effective instruments by research or extension fieldworkers.
- A decline in the availability of natural materials required for store construction (there may be a need to identify alternative methods of storage if alternative construction materials are not available).
- Growers of onions require improved storage technologies to prolong shelf life.
- New varieties of perishables, such as tomatoes and cabbages, are required to change the growing season and avoid gluts. Alternatively, growers need to diversify out of these crops to other vegetables for which new markets may have to be found.
- The lack of credit facilities for farmers requires research into the transfer of ideas from inventory credit schemes to the Malawian context.

An integrated approach is required, which builds on the results of the study and would:

- deepen understanding of identified constraints;
- identify any additional constraints;
- introduce appropriate techniques to overcome the constraints;
- initiate trials to assess efficacy of the solutions;
- assess the socio-economic viability of improved technologies;
- encourage uptake of technologies using resources of the government extension service and NGOs; and
- monitor and evaluate uptake and efficacy.

in northern Ghana found that farmers view production-related problems as being more important than storage problems (Table 10.4). The linkages between storage and non-storage issues means that storage projects may fail to achieve storage outcomes if they (the projects) do not acknowledge these linkages. Figure 10.1 also illustrates these points.

## A methodological framework for combining quantitative and qualitative survey methods

Qualitative survey methods started to gain prominence in development projects during the 1980s, primarily in response to the drawbacks of questionnaire-type surveys, which were considered time-consuming, expensive and not suitable for providing in-depth understanding of an

**Table 10.4** Participatory needs assessment in Ghana. Source: Marsland 1997.

| Constraint | Type | Overall ranking | Regions where not mentioned[a] |
|---|---|---|---|
| Access to labour-saving technology for land preparation[b] | P | 1 | |
| Cost or availability of fertiliser | P | 2 | |
| Seeds uprooted by birds and rodents | P | 3 | Upper East Region |
| Marketing problems[c] | M | 4 | Upper East Region |
| Weeds[d] | P | 5 | |
| Storage pests | S | 6 | |
| Poor rainfall[e] | P | 7 | |

P = Production (farming system) constraint, S = Storage constraint, M = Marketing constraint
a This column gives only a very crude indication of regional differences in priorities.
b Usually either tractors or bullocks.
c Includes early sale, low prices, low bargaining power in relation to middlemen, transport problems.
d Usually *Striga* spp.
e This was mentioned in only one village in the Northern Region where it was ranked 6th. It was ranked highest in the Upper West Region.

issue (Chambers 1983, 1994; Pretty *et al.* 1995). This led to a polarisation in collection and analysis of information with 'traditional', quantitative techniques on the one hand, and qualitative methods, on the other.

The result of this polarisation of approaches and the associated shortcomings was that the users of information were often dissatisfied with the quality of data and the resulting analytical conclusions. At the same time, it was recognised that there are areas or interfaces where the two types of approach can benefit from each other, leading in turn to improved quality of information which is required for intelligent decision-making at the various stages of field-based research projects and programmes.

During the second half of the 1990s, attempts were made to highlight the complementarity of the two types of approach, for example, in relation to poverty assessments in Africa (Carvalho & White 1997). Other work by Mukherjee (1995) examined the pros and cons of each type of approach and the potential for synergy in a general development context. In the field of renewable natural resources research it was realised that, while some research practitioners were combining methods as a matter of course when conducting field research, experiences were often not documented. Moreover, several avenues of potential remained untapped.

This section is based on work by Marsland *et al.* (2000) and addresses in general terms the basic question: 'Given a set of information objectives on the one hand, and constraints such as time, money and expertise on the other, which combinations of qualitative and quantitative approaches will be optimal?'

### Practical aspects of the selection of survey techniques

In order to work out the most appropriate combinations of methods for a given task, it is necessary to consider both objectives and constraints.

#### *Objectives*

Investigation of a problem or phenomenon may be seen as the overall goal of data collection. Researchers need to decide:

*What characteristics* (e.g. precision, scope of extrapolating from findings) the information ought to have;
*For whom* is the information being collected? (e.g. project managers, policy makers, etc.);
*Degree of participation* – in most (many) research activities there will be objectives which relate to how information is collected and analysed; and
*Training objectives* – there may be training objectives attached to the collection and analysis of information guiding the choice of methods.

#### *Constraints*

An important point to note in this context is that objectives interact with each other: having one objective will affect the extent to which other objectives can be achieved. In this sense, one objective can become a constraint to the achievement of another. This is because resources of time and money and expertise are limited.

These resources will often shape the parameters of a fieldwork just as much as objectives.

*Time*: One of the reasons why informal methods came into greater use in the 1970s and 1980s was that practitioners and managers were fed up with the excessive time taken to conduct, analyse and disseminate sample surveys. While in practice it is not possible to say unequivocally that participatory exercises are quicker than sample surveys – everything depends on the particular circumstances including expertise, logistics and institutional constraints – it appears that informal work is quicker than formal more often than not. Certainly, this is the tentative conclusion of Mukherjee (1995) who notes that 'On balance … by and large … PRA method takes relatively less time'.

In most project situations, time is at least as important as cost per day. For many project managers, the quicker turn-around time of informal work is a powerful argument for undertaking such work. It is important to compare like with like in terms of quality and quantity of coverage: a weak sample may be a false economy.

*Cost*: Received wisdom has it that sample surveys are expensive and that participatory and rapid rural appraisal (PRA/RRA) exercises are cheap. However, as Mukherjee (1995) notes: 'It is not easy to arrive at a relatively simple comparison of cost for the two methods [sample surveys and PRA].' There are a host of factors to be considered in this regard which can influence both actual cost and inputed cost for undertaking conventional survey or PRA-type studies. As a consequence, it is not possible to say categorically that one type or collection of methods will automatically be more expensive than another type or collection, thus cost *per se* cannot be reliably used in a blueprint sense to select methods. Each case needs to be taken on its merits.

*Expertise*: As a general statement, informal survey work requires a greater array of skills per researcher than formal work, and formal work requires a greater number of people to undertake the research process. In addition, the need for a degree of multidisciplinarity is greater in informal work, which derives much of its internal consistency from 'triangulation', including that achieved by the debate between investigators from different disciplines. For informal work, the interviewer normally will need to be highly skilled in interview techniques, and often to be familiar with a range of instruments. He or she will probably also be required to analyse the data at high speed, much of it in the field itself. Characteristically, in formal work a number of different individuals will be involved in the task of research design, training of enumerators, data collection, design of data entry programmes, analysis and write-up.

### *Trustworthiness of information*

The value of information depends on its trustworthiness. Here it is argued that the trustworthiness of information will be greater if quantitative and qualitative approaches to data collection and analysis are combined rather than used separately. The following four tests of trustworthiness can be discerned:

*Internal validity or credibility*. The key question here is: how confident can we be about the 'truth' of the findings?

*External validity or transferability*: Can we apply these findings to other contexts or with other groups of people?

*Reliability or dependability*: Would the findings be repeated if the inquiry were replicated with the same or similar subjects in the same or similar context?

*Objectivity or confirmability*: How can we be certain that the findings have been determined by the subjects and context of the inquiry, rather than the biases, motivations and perspectives of the investigators?

Internal and external validity, reliability and objectivity are the terms used in conventional scientific research. Credibility, transferability, dependability and confirmability are the terms put forward by Pretty (1993), after Lincoln and Guba (1985), to describe the equivalent criteria implicitly and routinely used in much participatory field research.

Obviously, the size of the target population has a bearing on the importance of these criteria for a particular study. For example, external validity plays less of a role if the target population is small (e.g. a small number of villages in the case of an NGO-led development project). On the other hand, research projects covering entire regions or countries depend on results representative of these areas. Overall, formal work has probably most to gain from informal in the area of credibility and objectivity, whereas informal work (if it is to be generalised) can borrow from formal methods to improve external validity.

### *Types of combinations*

Merging is one way of combining qualitative and quantitative approaches. It consists of swapping tools and attitudes from one tradition to the other. In addition to merging, there are two other types of combining: sequencing and concurrent use of tools and attitudes. If they are to lead to integrated conclusions, sequenced and concurrent combinations should be followed by a synthesis of the information collected. Box 10.2 illustrates the differences between the different types of combinations with some examples.

## Box 10.2 Types of qualitative and quantitative combinations that may be used in sample surveys and experiments. Source: Marsland *et al.* 2000.

**Type A: Swapping tools and attitudes: 'Merging'**

- Thinking about sampling in designing enquiry based on qualitative methods
- Coding responses to open-ended questions from qualitative enquiries.
- Using statistical techniques to analyse unbalanced data sets and binary, categorical and ranked data sets, arising from participatory enquiry:
  – creating frequency tables from coded data;
  – modelling binary and categorical data generated from ranking and scoring exercises.
- Using mapping to generate village sampling frames for: questionnaire surveys; type 2 or type 3 on-farm trials.
- Using attitudes from participatory methods, e.g. to reduce the non-sampling error in questionnaire surveys or farmer-researcher misunderstandings in on-farm trials.

**Type B: 'Sequencing'**

- Using participatory techniques in exploratory studies to set up hypotheses, which can then be tested through questionnaire based sample surveys, or via on-farm trials.
- Choosing a random sample and conducting a short questionnaire survey to gain information on key variables which are then investigated in-depth by participatory enquiry.

**Type C: Concurrent use of tools and methods from the different traditions: 'Mixed Suite'**

*Concurrent use of:*

- Survey of statistically selected sample members, using pre-coded questionnaires to determine target population characteristics of a qualitative (e.g. opinions on a new technology) or quantitative (e.g. crop production) nature.
- Setting up scientific experiments (on-station or type 1 trials) to study the effects of specific interventions in a controlled environment (e.g. on-station or 'contract' research).
- Using aerial photographs, GIS.

*along with:*

- Participatory enquiry for attitudes, beliefs and perceptions of the target population.
- Type 3 trials.

Note: Type 1 on-farm trials are those designed and managed by researchers. Type 2 trials are designed by researchers but managed by farmers. Type 3 trials are designed and managed by farmers and monitored by researchers (Coe & Franzel 1997).

Within a particular applied research or development project dealing with the sustainability of livelihoods, any mixture of these types of combination can be used. Of them all, sequencing has probably been the most widely practised in the past. While aspects of types A, B and C have undoubtedly been used in the field for some time, it is only relatively recently that examples have been documented and disseminated widely (see e.g. PLA Notes 28; Carvalho & White 1997). The latter paper stresses the importance of synthesising of information obtained through combinations of survey techniques.

### *Combinations, objectives, trustworthiness and researcher-researched relationships*

Figure 10.1 shows how combinations of survey instruments form part of a continuum in relation to the objectives of a given research project. The different types of combinations need to be seen in relation to the different

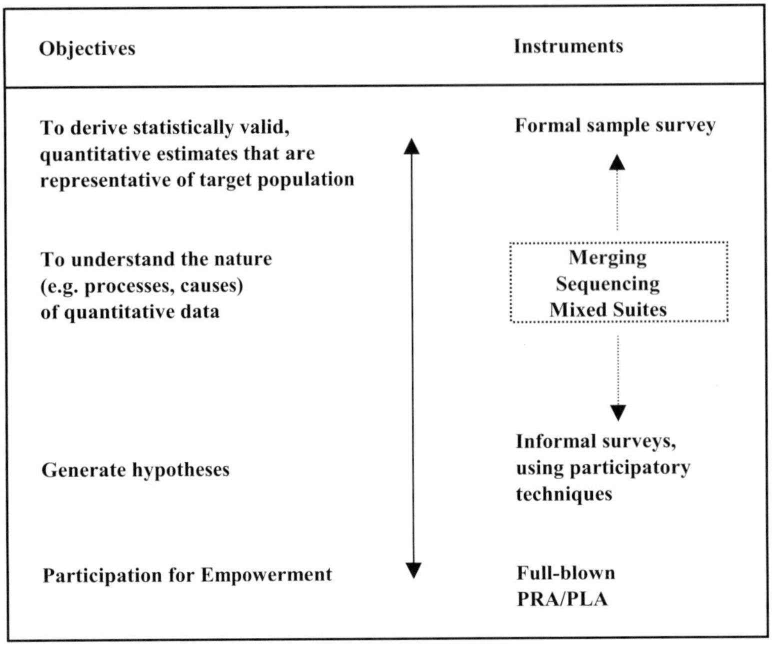


**Fig. 10.1** Continuum of objectives and combinations of instruments.

stages of the research process where they can be applied. Although this discussion focuses on survey techniques, it is important not to lose sight of the other stages leading to a research output.

Table 10.5 presents types of formal and informal combinations at the various stages of the research cycle, and their relationship to aspects of trustworthiness. The latter will be enhanced as a result of 'examining, explaining, confirming, refuting, and/or enriching information from one approach with that from another' (Carvalho & White 1997).

Type A and C approaches are illustrated with examples below. For a more complete description see Marsland *et al.* (2000).

## Type A: Swapping tools and attitudes: merging

### *Informal contributions to formal approaches*

*Informalising and contextualising interviews in surveys and experiments*: Including semi-structured interviewing in a structured questionnaire format can improve the quality of data generated due to increased flexibility and openness, allowing the questionnaire as a whole to adapt better to particular local environments (Ziche 1990). This adaptation ranges from contextualising of questionnaires through use of appropriate locally specific vocabulary, to being better able to deal with certain types of information within a questionnaire format. To some extent qualitative response is routinely incorporated in many questionnaires, with the inclusion of open-ended questions. The addition of a checklist of points or hints for probing on particular issues takes this process one step further and introduces a greater degree of interaction on the part of the interviewee. Summarising any substantial number of such responses requires a careful coding exercise.

*Using maps to create village sampling frames*: Once villages in a region are chosen for a study, based on (say) agro-ecological conditions, social mapping can be used to generate a list of households, together with their physical locations within a village. This can then be used as a sampling frame in sample selection. In a 1993 study, India's National Council of Applied Economic Research (NCAER) found that social mapping compared favourably with standard household listings often employed in sample surveys.

*Using qualitative understanding to inform classification procedures*: Cluster analysis is a technique commonly applied to quantitative data by statisticians. Based on a survey, it entails agglomerating the respondents into groups on the basis of 'similarity' with respect to responses to some set of survey questions. The starting point is a choice of 'cluster seeds' to which others are

**Table 10.5** Types of formal and informal combinations and their relationship to aspects of trustworthiness.

| Stage in information cycle | Type of combination | Explanation/example | Function: relationship to elements of trustworthiness |
|---|---|---|---|
| Design | Merging | Formal sampling procedures for informal work | Reduced sampling error: better external validity for informal work |
| | | Informal attitudes for formal work | Reduced non-sampling error: better internal validity for formal work |
| | | Use of social mapping for formal work | Reduced time and cost for household listing and sampling |
| | Concurrent | Correct use of different instruments for different variables within the same survey/experiment | Better internal validity for qualitative variables – belief, motivations, etc. – alongside better external validity for quantitative variables – rates, proportions, etc.<br>'Enriching': the outputs of different informal and formal instruments adding value to each other by explaining different aspects of an issue |
| Data collection | Sequential | Analysis of informal outputs feeding into the design of formal instruments, i.e. using informal studies to 'map out' key issues and approaches to be explored further in formal work, e.g. using informal work to generate hypotheses to be tested in formal work | 'Enriching' |
| Analysis | Sequential | Analysis of formal outputs with informal approaches, e.g. testing null hypotheses, investigating unexpected outcomes | 'Refuting': where one set of methods disproves a hypothesis generated by another set of methods<br>'Confirming': where one set of methods confirms a hypothesis generated by another set of methods<br>'Explaining': where one set of methods sheds light on unexpected findings derived from another set of methods |
| | Merging | Applying statistics to categorical and unbalanced data sets | Improved credibility of analytical conclusions from informal work |
| | | Coding responses from informal work | Enhances possibilities for aggregation, thus facilitating generalisation<br>Enhances possibilities for stratification of sample for subsequent sample survey |
| Synthesis | Merging | Blending the analytical outputs from informal and formal work into one set of policy recommendations | Higher quality policy recommendations |

then joined in the process of cluster formation. If these seed respondents – core members of groups – have been studied intensively and are well understood through qualitative work, clusters formed on the basis of similarity to the seeds will have an understandable character. Ideally, seed respondents are prototypical of what could become effective strata or recommendation domains.

### *Formal contributions to informal approaches*

In some instances, researchers have found it necessary to incorporate more structure into a previously unstructured exercise. For example, one general conclusion of the IIRR/CIP-funded review of participatory monitoring and evaluation was that:

> 'with the emphasis on participation and learning processes, much of the PM & E experiences started off using qualitative and semi-structured methodologies. However, there is an emerging recognition of the need to build into current participatory methodologies some of the quantitative tools to provide

for better triangulation of information and greater acceptability of the results when endorsed as inputs to policy. This includes paying greater attention to establishing baseline data to more systematically monitor progress and facilitate ante and post evaluation procedures.'

*UPWARD (1997)*

*Sampling and stratification*: Pretty (1993) argues for the trustworthiness of participatory inquiry, citing the four characteristics of credibility, transferability, dependability and confirmability. It is interesting and important to note, however, that the case for transferability (equivalent to external validity in structured research) appears to be considerably weaker than the one he makes for the other characteristics (Pretty 1993). It is perhaps in the question of transferability that the most obvious 'Achilles heel' of informal research lies, at least insofar as its practitioners try to generalise their findings in much the same way as sample surveys. Effective and statistically based methods of sampling are needed if the domain of validity of research conclusions is to be extended.

Many issues have to be considered in the sample selection process if results are to be generalised to a wider population. Some important issues are (a) a clear identification of the recommendation domain; (b) the use of secondary data and relevant grey literature in assessing the availability of a suitable sampling frame; (c) where a sampling frame is unavailable, evaluating the feasibility of adopting a hierarchical sampling procedure so that sampling frames can be built up for just selected units in the hierarchy; (d) clearly defining the sampling units most appropriate for study objectives; (e) methods to be used in sample selection, in particular, including an element of randomness in the procedure; (f) being open to the possibility of post-stratification at the data analysis stage; (g) sample size considerations. Wilson (2000) gives more detailed consideration to these elements.

*Applying statistical analysis to unbalanced, binary, categorical and ranked data sets*: During the 1990s, practitioners of informal surveys and PRA type work in developing countries had started to recognise the potential for applying modern statistical methods to unconventional data sets. Martin and Sherington (1996) and Abeyasekera (2000) among others have outlined some of the ways in which statistical techniques can play a useful role for such data.

One starting point is coding open-ended questions from informal work. This is common in questionnaire work. What is less common is coding of information collected informally. Certain types of information collected during informal work can be coded readily, and others with rather more careful thought.

*ANOVA*: The principal method for the statistical analysis of data from on-farm participatory trials is the analysis of variance (ANOVA). The power of the method lies in its ability to 'disentangle', 'correct' or, in a loose sense, 'explain' the effects of one or more factors (e.g. new technologies) on response variables such as results from participatory scoring exercises (Abeyasekera 2000).

*Generalised linear models for binary data*: Farmers' preference ranking of factors as 'good' or 'poor' allows the resulting binary data to be analysed via a generalised linear modelling approach to determine factors which affect their preference. In particular, the dependence of preference ranking on ethnic groups is demonstrated (Martin & Sherington 1996).

*Multi-level models*: A recent set of statistical developments that extends the idea of general linear models to multi-level models which explicitly acknowledge and model hierarchical information, as found for instance where some data are at community level, some at household and some at individual level.

*Qualitative residuals*: A general idea which runs through regression and ANOVA modelling as well as generalised and multi-level modelling is that of the 'residual', the difference between the observed result and that suggested by the model fitted.

*Ranking and scoring*: Ranking and scoring data arise from activities where precise numerical measurement is inappropriate, including a range of qualitative work, some of it participatory. Ranking entails an ordering, for example, between a set of crop varieties in terms of cooking characteristics. For the same task, scoring would entail assessing each variety separately on a fixed scale, say a four-point scale with values 1, 2, 3, and 4. Simple scoring and ranking data can be analysed very straightforwardly (see Box 10.3), but where the study has more structure, statistical methods can be used to correct for respondent grouping factors, e.g. respondent's ethnic group and gender. In a substantial number of cases, scoring data can be treated by relatively standard statistical methods, so the results can be modelled and simultaneously corrected for a range of 'explanatory factors', even when these occur in an unbalanced fashion (Abeyasekera 2000).

*Bayesian statistics*: Bayesian statistics is based on the notion of subjective probability or degree of belief. Briefly, the Bayesian paradigm consists of modelling

## Box 10.3 Example of analysing ranked data from a study in Tanzania. Source: Marsland *et al.* 1999.

The larger grain borer (LGB), *Prostephanus truncatus*, was first reported in Africa in 1981. The beetle, a severe pest of farm-stored maize and dried cassava was initially a major problem to farmers in western Tanzania.

The two principal objectives of the study were:

- to assess the role played by *P. truncatus* in determining changes in production, storage, and marketing of the maize and cassava crop during the period between the time of the establishment of the beetle and today; and
- to assess the factors determining the role played by *P. truncatus* in these stages of the maize and cassava commodity system, in particular the impact of the insecticide treatment.

In order to achieve these objectives, a combination of sample survey and rapid rural appraisal (RRA) techniques was required.

In pursuing one component of the above objectives, attempts were made to apply statistics to the ranking data derived from the RRA exercises. Chi square tests and variants thereof were used to test for changes in rankings of the importance of *P. truncatus* when farmers compared the situation at the time of establishment of the pest with the situation at the time of the survey (i.e. 1998). As an example, in one exercise farmer groups were asked to rank the importance of the pest in comparison to all other storage problems (a) at the time of establishment and (b) for the present day. The ranks were then compared and analysed using McNemar's test (Siegal & Castellan 1998). The following table illustrates how ranking data for the past and present can be summarised.

| | | Present | |
|---|---|---|---|
| | | Rank = 1 | Rank > 1 |
| Past | Rank = 1 | 24 | 13 |
| | Rank > 1 | 2 | 4 |

The cells representing no change give no information about how the ranking of LGB has changed over the years. Only the bottom left and top right cells give information about change. McNemar's test (sign test in this case) can be used to test the null hypothesis of no change in attitude. The test gives a *P*-value of 0.0045, which indicates strong evidence for rejecting the null hypothesis. It is clear from the table that there was a significant increase in the ranking, giving significant evidence for a reduction in the role of LGB as a storage problem.

beliefs before observing data, by prior probabilities, and using Bayes's theorem to combine information from observations with the prior distribution to obtain a posterior distribution.

*Procedural aspects of applying statistical analysis to qualitative data sets*: A compromise needs to be struck so that informal data can be analysed by using statistical techniques. Some of the flexibility inherent to RRA/PRA exercises needs to be given up in favour of a minimum of rigour, making the data suitable for cross-site analysis. Nevertheless, if well blended into the exercise, this can be done without seriously restricting participation.

A number of aspects should be taken into account during survey design and data collection which employs statistical analysis to qualitative data sets, particularly if the research is to lead to generalisable results:

- the study group needs to be adequately large and representative of the target population;
- there has to be an element of randomness in the selection of the study units;
- the format of the data collection tool should remain the same throughout the survey (e.g. use of the same format of matrix throughout the exercise; use of a uniform scoring system);
- well-defined consistent recording of information so that, for example, results from individual PRA practitioners can be coded in a coherent way and put together for analysis; and
- clear and complete recording of meta-data (i.e. details of where and how the information was collected), so that information summaries can be based on a clear-cut rationale and have proper support for any claim to generalisability.

### Type C: Concurrent uses of tools

#### *Survey work*

NCAER (1993) found several benefits in using informal and formal techniques together in its evaluation of 'India's National Programme on Improved Chullah'. The NCAER experience concerned a geographically broadly spread sample in which a questionnaire was used to collect quantitative or quantifiable information on a limited number of variables. Other mainly qualitative data were collected through RRA/PRA methods from a smaller sample, spread across fewer villages picked from all regions. The questionnaire results provided 'representativeness', while the RRA/PRA work provided 'contextual linkages for explaining behavioural patterns, [and] … additional in-depth qualitative data which could be helpful during analysis and report writing stages' (NCAER 1993). Overall, 'The blending of the two approaches can lead to a more reliable data base'; in other words there was a definite 'trustworthiness payoff' (Box 10.4).

Schoonmaker-Freudenberger (1996) makes precisely this point, arguing that we should not attempt to extrapolate from PRAs, but instead use the findings to stimulate…

> '…a more accurate debate about a policy issue by identifying the diversity of local conditions. By combining PRA with questionnaires or remote sensing techniques which capture broader spatial information, one can derive "an attractive combination of range and depth of information".'

> **Box 10.4 Combinations of broad, formal survey and narrow, in-depth study.**
>
> It often makes sense to think of a combination of a broad shallow study which provides good 'representativeness' and one or more deep narrow studies which provide the depth referred to above. This combination may be thought of as providing a table or platform supporting the research conclusions. When such a combination of studies is planned, it is of course desirable that the sampling structure be organised so that effective merging of conclusions can follow. This implies that the in-depth studies are planned with special attention to how their selection relates to the broad shallow study. For more information see Wilson (2000).

Table 10.6 shows the concurrent use of both PRA exercises and formal household questionnaires.

#### *Experimental work*

Another type of concurrent combination is that which involves detailed scientific measurements on the one side

**Table 10.6** Concurrent use of research tools: LGB study. Source: Marsland *et al.* 1999.

| Thematic area | Research approach |
|---|---|
| Changes in role of crop production in household food security strategies comparing 1985 with 1998 | RRA (groups of men and women – some single-gender groups – ranking strategies for 1985 and 1998) |
| Changes in farmers' perceptions of the importance of maize and cassava, comparing 1985 with 1998 | RRA (groups of men and women – some single-gender groups – ranking both crops against all other crops for 1985 and 1998) |
| Influence of *P. truncatus* on production, storage and marketing outcomes:<br>• production levels<br>• role of *P. truncatus* in maize and cassava harvests<br>• role of *P. truncatus* in the choice of maize and cassava varieties<br>• role of *P. truncatus* in the duration of storage and volume of sales at farm level | Household sample questionnaire |
| Is *P. truncatus* still regarded as a problem?<br>• *P. truncatus* in the context of major agricultural problems<br>• *P. truncatus* in the context of other storage problems | RRA (groups of men and women – some single-gender groups – ranking strategies for1985 and 1998) |
| Coping strategies for *P. truncatus*:<br>• Actellic Super Dust perceptions<br>• Storage operations and structures | Household sample questionnaire |

and informal investigations of perceptions, beliefs and attitudes on the other. An example of this is the qualitative and quantitative sorghum loss work conducted in India. This seeks to compare detailed laboratory-based analysis of mycotoxins, pest damage of stored sorghum with farmers' perceptions of the importance of losses (R.J. Hodges, pers. comm.).

### Conclusions

There are a variety of ways in which qualitative and quantitative methods may be combined to improve the trustworthiness of survey and experiment findings. Several combinations are already known to practitioners in the field, while others have not yet found practical expression. It is clear that the choice of particular instruments and combinations will be conditioned not only by the extent to which they improve trustworthiness, but also by time, money, expertise and other factors which can act as constraints to the process of data collection and analysis. Clearly, all information objectives need to be resourced, and, in many cases, the types of instruments used will be as much – or more – a reflection of resource constraints as they are of objectives.

Both objectives and resource constraints have implications for the selection of survey teams. Aside from the typical multidisciplinary combination of social and natural science inputs, there is a need to consider inputs from statisticians, especially in more complex cases.

Case study exercises have shown that it is important that survey teams are sufficiently trained and familiar with approaches and have been provided with sufficient resources to achieve their targets. Supervision can be a problem, in particular if exercised over long distances without direct contact. Unforeseen circumstances can push a relatively inexperienced survey team to the limits of its capabilities. If in doubt about the experience of the team and the tasks expected, it may be more appropriate to choose a less demanding survey design.

Well-synthesised survey results are required so that decisions can be taken by project leaders or policy decision-makers. A unified set of recommendations should reflect a balanced use of tools, which ultimately led to more trustworthy information. Aside from swapping tools for the collection and analysis of data (i.e. merging of techniques), findings obtained through the use of one approach can be confirmed, enriched, or refuted by research results obtained from the concurrent or sequenced use of the other approach.

## Monitoring investment and the evaluation of impact

Three basic questions need to be addressed in designing and implementing a good monitoring and evaluation (M&E) system:

(1) What to monitor;
(2) Methods and instruments used to collect information and develop indicators; and
(3) Who is to do the monitoring.

> 'Monitoring is a continuous assessment both of the functioning of project activities in the context of implementation schedules and of the use of project inputs by targeted populations in the context of design expectations. It is an internal project activity, an essential part of good management practice and therefore an integral part of day-to-day management. Evaluation is a periodic assessment of the relevance, performance, efficiency and impact of the project in the context of its stated objectives. It usually involves comparisons requiring information from outside the project – in time, area or population.'
>
> *Casley & Kumar (1987)*

In order to track progress against indicators, project monitoring and evaluation systems need to employ a range of instruments. Certain types of instrument are generic to good M&E.

### The monitoring system

Monitoring systems should consist of three distinct but inter-related parts: financial and physical monitoring, beneficiary assessments and diagnostic studies.

The financial and physical part of the monitoring system measures the extent to which the research or development interventions are being delivered on time, within budget and to the target group. Beneficiary assessments gauge beneficiaries' perceptions of the relevance and quality of the process of events leading up to the delivery of interventions and the degree to which the interventions are being used. The role of diagnostic studies is to try and understand why and how beneficiaries are not satisfied and not reacting in the way that was previously anticipated.

Typically, the role of research or development interventions in the early phases of a project is to create the conditions which will allow the chain of events to occur – the intervention model (i.e. completion of activities will achieve certain results which will in turn contribute to the purpose). Thus, the initial emphasis will be monitoring the delivery system (financial and physical monitoring). Once the delivery system is proven to be working in accordance with expectations, monitoring of beneficiary reactions can begin (beneficiary assessments and, if necessary, diagnostic studies).

### The management information system

A management information system (MIS) is intended to draw all three components of the monitoring system together. Records of financial and physical data must be linked with beneficiary assessments. Without follow-up diagnostic studies, the monitoring system cannot fulfil its fundamental purpose of improving management decision-making.

In making judgements about the scope and nature of an MIS, answers to the following questions are critical:

- What is the minimum information requirement for effective management?
- What are the timing requirements for each information item?
- Are staffing responsibilities for generating, handling and disseminating monitoring information clearly delineated?
- In what form should the information be conveyed?
- How will the information be used?

### Evaluation

There are typically four types of evaluation: appraisal, mid-term review, end-of-project review and ex-post review.

The appraisal, or evaluation of the idea, is undertaken before the financing arrangements are implemented, to help translate a good idea into a viable project. During the implementation stage of a project, a mid-term review – often an Output to Purpose Review, in the terminology of log framework analysis (Anon. 1989, 1993) – is undertaken to help improve project performance during the rest of the implementation period. When the project finishes, an end-of-project evaluation is undertaken. This assesses project performance. An ex-post evaluation may be done a few years after the project to assess the project's lasting impact.

Although not evaluations in themselves, baseline surveys are undertaken to establish a starting point from which changes induced by the project can be evaluated subsequently. Information collected by these surveys should relate to the specific changes that are expected to occur by the end of the project and beyond. They are therefore relevant to the higher levels of the log frame, the purpose and goal.

Evaluation is a periodic activity, in contrast to monitoring which is a continuous activity (Casley & Kumar 1987). In addition, the primary consumers of evaluation information are normally not directly involved in day-to-day project management (although management will be important consumers for, especially, mid-term evaluations). Thus, if the post-harvest department of a Ministry of Agriculture is undertaking a project, in all likelihood the central government and the donor agencies will be most concerned with evaluation. These bodies will normally take responsibility for organising the evaluation. Non-project staff, for example, international consultants or donors, will normally undertake the evaluation itself.

#### *Links between monitoring and evaluation*

Monitoring and evaluation are distinct functions with different objectives and are executed by different groups (Fig. 10.2). However, there are important links between the two functions. First, it is usual for evaluations to make use of monitoring information. Indeed the job of evaluation is facilitated significantly by high-quality

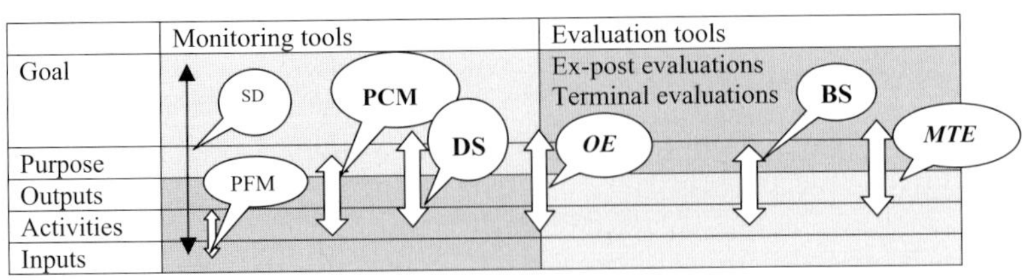


SD = Secondary Data
PCM = Participant Contact Monitoring
OE = On-going Evaluation
BS = Baseline Studies (including Participatory Needs Assessments)
PFM = Physical Delivery and Financial Monitoring
DS = Diagnostic Studies
MTE = Mid-Term Evaluation

**Fig. 10.2** Links between monitoring and evaluation.

monitoring information. Second, there is one type of activity, not mentioned above, which actually straddles the monitoring and the evaluation function. This activity is known as ongoing evaluation. These evaluations are modestly conceived enquiries into specific areas not routinely covered by the three components of the MIS. Ongoing evaluation tries to answer three types of question:

(1) What effects are beginning to emerge from the use, and repeated use, of project interventions?
(2) Is the project having consequences that were not intended or anticipated in its design?
(3) Does the intervention model upon which the project is based remain valid in a changing environment?

**Example of a qualitative evaluation: impact of a larger grain borer risk minimisation campaign in Tanzania**

Evaluating the impact of a technology should normally take place some years after introduction. Box 10.5 shows how, through using a combination of participatory and questionnaire-based methods, farmers' views on the impact of LGB in the farming system in Tanzania were gauged 15 years after the implementation of a nationwide prevention programme.

## Box 10.5 Evaluation of impact of control regimes for larger grain borer on farmers. Source: Golob *et al.* 1999.

The larger grain borer, *Prostephanus truncatus*, has been established in East Africa for more than 15 years. During that time, recommendations have been introduced to enable farmers to combat this most destructive pest of stored maize and dried cassava. The initial control method introduced was to shell maize, treat with an insecticidal mixture, and store it in a suitable container. This procedure represented a major change for families who traditionally stored maize on the cob. Subsequently, other methods have been tried including spraying cobs with insecticide, and the release (in Kenya) of a biological control agent, *Teretrius nigrescens*, a histerid beetle. No specific methods have been developed for cassava; farmers have simply been advised to leave the tubers in the ground for as long as possible, to reduce the storage period. Other than an evaluation of the rate of uptake of recommendations there has not been any assessment of the efficacy of treatments used to protect maize against LGB, nor the extent to which farmers are coping with this pest.

A qualitative evaluation of the impact of the recommendations was conducted in seven districts in LGB-infested areas in Tanzania in 1997. The districts were spread over six regions: Arusha, Iringa, Kilimanjaro, Morogoro, Rukwa and Tabora. The surveys were designed to determine the extent of the pest problem, the strategies farmers are using to overcome it, and the constraints they face. A total of 390 randomly selected farmers were interviewed – between 47 and 79 per district. In addition to the individual interviews, group discussions were held to put the LGB, and other storage problems, into a general agricultural perspective and to compare the current situation to that which occurred when the pest was first introduced.

The results of the evaluation present something of a puzzle. With the exception of a minority of farmers in the Tabora and Kilimanjaro survey areas, LGB does not appear to have influenced production, storage and marketing outcomes to any significant degree. It is true that there is considerable evidence that farmers in some areas have changed their storage behaviour. There has been a much more widespread use of Actellic Super Dust (ASD) and, in Morogoro, Actellic EC. There has been an increase in the incidence of shelling maize, and a decrease in crib and platform storage. However, the changes have by no means been universal, and there are significant concerns about cost, availability and efficacy of ASD, and cost and availability of sacks for storage. In these circumstances, it is legitimate to wonder why LGB has not had more of an impact on production, storage and marketing outcomes. One possible answer is that farmers have been overestimating the difficulties that they are encountering with pesticides and sacks. However, problems with adulteration of pesticides, and their cost and availability, are well known. Another possible answer is that a combination of good protection measures in the past, together with the effects of the droughts of the last decade, has prevented the build-up of LGB in the villages, so that it no longer causes significant damage, even to poorly protected grain.

In these circumstances, there have to be legitimate concerns over the prospects for the future. If adequate protection measures are not taken, LGB populations will soon increase, and losses will rise. When farmers have access to reliable insecticides and adequate storage methods, the indications are that they will use them if they can afford to. The implication is, therefore, that measures should be taken to tighten up the regulation of pesticides and ensure that they are more widely available. In addition, the question of cost should be considered. If it is not tenable to reduce cost, then this argues for an increased emphasis on low cost (perhaps botanical) protectants and on integrating other, individually less effective, measures to achieve adequate control. A further benefit of naturally occurring pesticides is greater environmental sustainability.

These issues merit further investigation.

This second evaluation is an example of an economic benefit study. Consultants were commissioned by the major funding agency (UK Department for International Development) to conduct an evaluation of selected agricultural research themes. One of the themes examined was research on LGB control in Ghana and Tanzania. The consultants' report is summarised in Box 10.6.

## The impact of post-harvest research on household food security

The World Food Summit in November 1996 set the target of reducing the number of under-nourished people in the world by half by 2015. This means not only that food production would have to grow by 4% each year for the next two decades, but also, as DFID (1997) recognised, access by poor people to sufficient safe food and water will be critical.

Goletti and Wolff (1998) provide a useful account of the impact of post-harvest research and the critical role that post-harvest systems play in meeting the overall

### Box 10.6 Cost–benefit of research to manage the larger grain borer in Africa. Source: DTZ Pieda 1998.

The objective of the evaluation was to identify, describe and quantify the benefits achieved that are directly attributable to the research carried out and to assess the value for money achieved. The review focused on economic benefits achieved, but also highlighted issues relevant to poverty, equity, the environment, and institutional effects on partner organisations.

The study showed that research on LGB has had considerable impact, both in Tanzania and Ghana. In Tanzania, there is evidence that the work has had a lasting impact well beyond the duration of the project. The available data on impact, combined with a number of assumptions, suggest total gross savings of some £21.5m. This is well in excess of the DFID expenditure in Tanzania (£0.8m), giving a net impact of £20.7m. Indeed, the savings also compare favourably with the total expenditure on Tanzanian projects by all aid organisations (£6.4m at 1998 prices).

In Ghana, estimates of the savings resulting from the DFID-funded research have also been made. It is unfortunate that no follow-up survey was conducted in Ghana, as the lack of any survey evidence has meant that the estimates of impact are based on very approximate assessments of uptake. These have suggested that savings equivalent to £3.7m at 1998 prices have been achieved in Ghana through the reduction of losses in stored maize. Set against the expenditure on research in Ghana of £2m (at 1998 prices), this gives a net benefit of around £1.7m. Taking into account the contribution of the Ghanaian government, the net benefit is approximately £1.6m.

Total expenditure on LGB research has been of the order of £15.6m at 1998 prices, of which £6.1m has been funded by DFID or its predecessor, ODA. Thus, the benefits from the Tanzanian project alone entirely offset total LGB research expenditure, including that incurred by organisations other than the DFID. The table below summarises the impact achieved.

**Summary of benefits**

| | £ million (1998 prices) |
|---|---|
| Gross savings – Tanzania and Ghana | 25.2 |
| *Expenditure on LGB research and extension* | |
| DFID expenditure in Tanzania and Ghana | 2.9 |
| DFID expenditure worldwide | 6.1 |
| All aid organisations in Tanzania and Ghana | 8.5 |
| All aid organisations worldwide | 15.6 |
| *Net benefits on basis of* | |
| DFID expenditure in Tanzania and Ghana | 22.3 |
| DFID expenditure worldwide | 19.1 |
| All aid organisations in Tanzania and Ghana | 16.7 |
| All aid organisations worldwide | 9.6 |

In the longer term, the work will have ongoing impact in the countries affected. If the LGB problem spreads further within Ghana and Tanzania and/or to other African countries, the annual savings would be still greater as the same results could be achieved with a lower level of expenditure.

The benefits achieved are substantial. The monetary value of the benefits is considerable and will work towards achieving the ultimate goal of poverty elimination. In addition to the monetary value of the impact, there will also have been benefits to nutrition, health and education (higher incomes will enable families to pay school fees). These factors will help support and sustain rural communities and provide a stronger basis for continued development.

goals of food security, poverty alleviation and sustainable agriculture.

As people live further away from where food is produced, they increasingly rely on smooth transport, storage, processing and marketing systems to give them access to a secure food supply. The reduced time for food preparation and the increased demand for processed food fuels the need to develop healthy, affordable food products and appropriate processing. Research gives rural people choices for processing their food crops and for using their time and effort more efficiently in supporting the needs of the family. Beneficiaries are rural women, many of whom provide the major labour inputs for harvesting and processing, while male members of the household seek alternative employment outside the village community.

Alternative rural income sources are essential to limit rural–urban migration, and post-harvest activities, such as processing and marketing, can provide much-needed employment. Research on policies, institutions and technologies to strengthen the development of rural agroenterprises make a direct, positive contribution. More liberal international trade systems and an increasing orientation of developing countries toward export markets are important sources of economic growth. Participation in international markets requires relatively sophisticated marketing, information and transportation networks. Successful competition requires quality control and product standardisation. This trend towards improved infrastructure and communication networks opens up new market opportunities for resource-poor farmers.

Three reasons justifying an increased commitment to post-harvest are suggested:

*Effect on poverty*: Post-harvest research provides income-earning opportunities for the rural poor and time-saving processed foods to the urban poor. Reduced wastage during storage reduces food and income losses for farmers. In the case of tropical fruit, improved storage technology opens up new markets, creates income opportunities and reduces poverty. In addition, processed convenience foods reduce the amount of time spent, mainly by urban women, in preparing meals. Improved processing frees up time for other activities such as wage work, further contributing to poverty reduction.

*Effect on food security and health*: Improved storage technologies, such as biological pest control or controlled atmosphere storage, reduce post-harvest food losses. Reducing losses increases the amount of food available for consumption. For example, control of the larger grain borer reduces losses in on-farm storage for smallholders, and thus enhances food security; the reduction of cyanide potential in cassava has an important effect on food safety, since a significant proportion of consumers suffer from cyanide-related diseases; control of rodents makes a considerable contribution to minimising the spread of transmissible disease.

*Effect on sustainable use of resources*: Alternative pest control mechanisms for grain storage reduce the need for pesticides, which reduces pollution, minimises accidents with pollutants and lowers pesticide residues in food consumed by humans. The reduction of post-harvest food losses in itself contributes to sustainability. Reducing waste of already produced food is more sustainable than increasing production to compensate for post-harvest losses. Increasing production leads to more intensive farming or to an expansion of the area under cultivation, both of which may have negative effects on the environment, especially when poor rural households tend to farm in fragile ecosystems or on marginal land. Value-adding opportunities generate income for improving welfare and provide farmers with the finance needed for investment in resource-enhancing technologies.

Goletti and Wolff (1998) conclude that as the significant contribution of post-harvest technology to goals such as poverty reduction, food security and sustainability becomes clear, and in the light of high rates of return, the very skewed allocation of funds to production versus post-harvest topics cannot be justified. Since, so far, relatively little has been invested in post-harvest research, there is potential for large impacts as constraints and bottlenecks are removed. It would thus be desirable to re-examine current funding priorities and to allocate a larger proportion of resources to the post-harvest area.

## Suggested reading

Davis, R. (1997) Combining rapid appraisal with quantitative methods: an example from Mauritania. *PLA Notes*, **28**.

Floyd, S. (1998) *When is quantitative data collection appropriate in farmer participatory research and development? Who should analyse the data, and how?* Natural Resources Institute, University of Greenwich, Chatham, UK.

Galpin, M., Dorward, P. & Shepherd, D. (2000) *Participatory Farm Management Methods for Agricultural Research and Extension Needs Assessment: A Manual.* Departments of Agriculture and Agricultural Extension and Rural Development, University of Reading, Reading, UK.

Hagmann, J., Chuma, E. & Gundani, O (1995) Integrating formal research into a participatory process. *ILEIA Newsletter*, **11(2)**, 12–13.

Leach, M. & Kamanga, J. (1997) Shotgun wedding or happy marriage? Integrating PRA and sample surveys in Malawi. *PLA Notes*, **28**.

McCracken, J.A., Pretty, J.N. & Conway, G.R. (1987) *An Introduction to Rapid Rural Appraisal for Agricultural Development.* International Institute for Environment and Development, London, UK.

Poate, C.D. & Daplyn, P.F. (1993) *Data for Agrarian Development.* Cambridge University Press, Cambridge, UK.

Shanley, P., Galvao, J. & Leda, L. (1997) Limits and strengths of local participation: a case study in Eastern Amazonia. *PLA Notes*, **28**.

## References

Abbot, J. & Guijt, I. (1997) Methodological complementarity: creativity and compromise. *PLA Notes* **28**, 27–32.

Abeyasekera, S. (2000) *Analysis approaches in participatory work involving ranks and scores.* Theme paper from DFID Project R7033 Methodological Framework Integrating Qualitative and Quantitative Approaches to Socio-Economic Survey Work. Statistical Services Centre, University of Reading, Reading, UK.

Abeyasekera, S. & Lawson-McDowall, J. (2000) *Computerising and Analysing Qualitative Information from a Case Study Concerning Activity Diaries.* Case study from DFID Project R7033 Methodological Framework Integrating Qualitative and Quantitative Approaches to Socio-Economic Survey Work. Statistical Services Centre, University of Reading, Reading, UK.

Anon. (1989) *The Logical Planning Framework Approach (LFA): Handbook for Objectives-oriented Project Planning.* Norwegian Agency for Development Co-operation, Oslo, Norway.

Anon. (1993) *The Logical Framework.* Project Analysis Department. International Development Bank, Washington DC, USA.

Barao, S.M. (1992) Behavioral aspects of technology adoption. *Journal of Extension*, **30**. http://www.joe.org/joe/1992summer/a4.html.

Boxall, R., Golob, P. & Taylor, R. (1997) *Pest Management in Farm Granaries, with Special Reference to Africa.* Natural Resources Institute, University of Greenwich, Chatham, UK.

Braun, A.R., Thiele, G. & Fernàndez, M. (2000) *Farmer Field Schools and Local Agricultural Research Committees: complementary platforms for integrated decision-making in sustainable agriculture.* ODI Agricultural Research and Extension Network Paper **105**. Overseas Development Institute, London, UK.

Carvalho, S. & White, H. (1997) *Combining the Quantitative and Qualitative Approaches to Poverty Measurement and Analysis: the practice and the potential.* World Bank Technical Paper **366**, Washington DC, USA.

Casley, D.J. & Kumar, K. (1987) *Project Monitoring and Evaluation in Agriculture.* World Bank, Washington DC, USA.

Casley, D.J. & Kumar, K. (1988) *The Collection, Analysis and Use of Monitoring and Evaluation Data.* World Bank, Washington DC, USA.

Casley, D.J. & Lury, D.A. (1987) *Data Collection in Developing Countries.* Oxford University Press, Oxford, UK.

Chambers, R (1983) *Rural Development: Putting the Last First.* Longman, London, UK.

Chambers, R. (1994) *Paradigm shifts and the practice of participatory research and development.* Institute for Development Studies, University of Sussex, Brighton, UK.

Coe, R. & Franzel, S. (1997) *Designing on-farm experiments: lecture notes.* International Centre for Research on Agroforestry, Nairobi, Kenya.

Compton, J.A.F. (1992) *Reducing Losses in Small Farm Grain Storage in the Tropics.* Natural Resources Institute, University of Greenwich, Chatham, UK.

Coulter, J. & Magrath, P. (1994) Economics of grain handling and storage in developing countries. *FAO Agricultural Services Bulletin*, **109.** United Nations Food and Agriculture Organization, Rome, Italy.

DFID (1997) *Eliminating World Poverty: a Challenge for the*

*21st Century*. DFID White Paper. Department for International Development, London, UK.

DTZ Pieda (1998) *Control of the Larger Grain Borer in Tanzania and Ghana*. DTZ Pieda Consulting, Edinburgh, UK.

Evenson, R. (1995) Extension, technology and efficiency in agriculture in sub-Saharan Africa. In: S.A. Brett (Ed.) *Achieving greater impact from research impacts in Africa.* Proceedings of the Workshop Developing African Agriculture, 26–30 September 1995. Centre for Applied Studies in International Relations, Global 2000, Sasakawa Africa Association.

Goletti, F. & Wolff, C. (1998) *The Impact of Post-harvest Research*. Markets and Structural Studies Discussion Paper **29**. International Food Policy Research Institute, Washington, DC, USA.

Golob, P. & Eisendrath, S. (1996) Tanzania: new pest, new technology, new extension approach. *The Rural Extension Bulletin*, **9,** 12–16.

Golob, P., Marsland, N., Nyambo, B., *et al.* (1999) Coping strategies employed by farmers against the Larger Grain Borer in East Africa. In: Jin Zuxun, Liang Quan, Liang Yong-sheng, Tan Xianchang & Guan Lianhua (Eds) *Proceedings of the 7th International Working Conference on Stored Product Protection*, Volume 11. Sichuan Publishing House of Science and Technology, Beijing, China.

Jiggins, J., Samanta, R.K. & Olawoye, J.E. (2000) Improving women farmers' access to extension services. United Nations Food and Agriculture Organization, Rome, Italy.

Lincoln, Y.S. & Guba, E.G. (1985) *Naturalistic Enquiry.* Sage, Beverly Hills CA, USA.

Marsland, N. (1997) *Bugs, beans and dawadawa: perceptions of commodity systems in Northern Ghana.* Presentation made at Overseas Development Institute, London, UK.

Marsland, N. & Golob, P. (1996) *Rapid rural appraisal of post-maturity issues in the central region of Malawi.* Natural Resources Institute, University of Greenwich, Chatham, UK.

Marsland, N., Golob, P. & Abeyasekera, S. (1999) *Larger Grain Borer Coping Strategies.* Case study for DFID Project R7033 Methodological Framework Integrating Qualitative and Quantitative Approaches to Socio-Economic Survey Work. Natural Resources Institute, University of Greenwich, Chatham, UK.

Marsland N., Wilson I., Abeyasekera S. & Kleih U. (2000) *A Methodological Framework for Combining Quantitative and Qualitative Survey Methods.* Natural Resources Institute, University of Greenwich, Chatham, UK.

Martin, A. & Sherington, J. (1996) *Participatory research methods – implementation, effectiveness and institutional linkages.* Paper prepared for SEM Workshop, 28–30 April 1996. Overseas Development Administration, London, UK.

Mukherjee, N. (1995) *Participatory Rural Appraisal and Questionnaire Survey.* Concept, New Delhi, India.

Nagel, U.J. (1997) Alternative approaches to organising extension. In: B.E. Swanson, R.P. Bentz & A.J. Sofranko (Eds) *Improving Agricultural Extension: A Reference Manual.* United Nations Food and Agriculture Organization, Rome, Italy.

NCAER (1993) *Comparative Study of Sample Survey and Participatory Rural Appraisal Methodologies with Special Reference to Evaluation of National Programme on Improved Chullah.* National Council of Applied Economic Research, New Delhi, India.

Picciatto, R. & Anderson, J.R. (1997) Reconsidering agricultural extension. *The World Bank Research Observer*, **21**, 249–259.

Pretty, J.N. (1993) *Participatory Enquiry for Sustainable Agriculture.* International Institute for Environment and Development, London, UK.

Pretty, J.N., Guijt, I., Thompson, J. & Scoones, I. (1995) *Participatory Learning and Action: a trainer's guide.* International Institute for Environment and Development, London, UK.

Proctor, D.L. (Ed.) (1994) *Grain Storage Techniques: evolution and trends in developing countries. FAO Agricultural Services Bulletin* **109.** United Nations Food and Agriculture Organization, Rome, Italy.

Rivera, W.M. (1996) Agricultural extension in transition worldwide: structural, financial and managerial strategies for improving agricultural extension. *Public Administration and Development*, **16,** 151–161.

Schoonmaker-Freudenberger, K. (1996) The use of RRA to inform policy: observations from Madagascar and Guinea. *PLA Notes*, **27**, 28–33.

Shah, P. (1993) *Questionnaires: participatory, reliable and interesting?* A note prepared for IIED/IDS Workshop on Alternatives to Questionnaire Surveys. International Institute for Environment and Development, London, UK.

Siegal, S. & Castellan, N.J. (1998) *Nonparametric Statistics for the Behavioral Sciences.* Second Edition. McGraw-Hill, New York NY, USA.

Spurling, D. (1995) *Integrating gender into extension services: rural women in the Sahel. Findings, Africa Region.* World Bank, Washington DC, USA.

Stathers, T. E. (2000) Project summary completion report. DFID Crop Post Harvest Programme Annual Report, 2001. Natural Resources International, University of Greenwich, Chatham, UK.

Swanson, B.E., Bentz, R.P. & Sofranko, A.J. (1997) *Improving Agricultural Extension: A Reference Manual.* United Nations Food and Agriculture Organization, Rome, Italy.

UPWARD (1997) *Participatory Monitoring and Evaluation: The Asian Experience.* Los Baños, Laguna, Philippines.

Wilson, I.M. (2000) *Sampling and Qualitative Research.* Theme paper produced for DFID Project R7033 Methodological Framework Integrating Qualitative and Quantitative Approaches to Socio-Economic Survey Work. Natural Resources Institute, University of Greenwich, Chatham, UK.

Ziche (1990) Some critical aspects in gathering socio-economic data in rural areas of non-Western societies. In: B. Hoeper (Ed.) *Qualitative versus Quantitative Approaches in Applied Empirical Research in Rural Development.* German foundation for International Development, Bonn, Germany.

# Chapter 11
# Trade and International Agreements

**P. Greenhalgh, J. A. Conway, M. E. Blowfield, R. Black and L. Nicolaides**

The special nature of the agricultural sector has led governments, at both national and international levels, to intervene in the market. Many justifications have been provided: the need for food security; the cyclical nature of production and prices; agriculture's vital importance to both employment and export earnings; food safety issues; and the need to provide a regulatory framework to facilitate trade are just some of the arguments propounded. The vital importance of agriculture is reflected in the value of all agricultural exports, which in the late 1990s exceeded \$300 bn per year. Table 11.1 provides details of the export values of major agricultural commodities (excluding fish and forestry products), many of which are exported from developing countries. During the second half of the twentieth century, when many developing countries achieved independence, they have been the principal advocates of international intervention in commodity markets. Meanwhile, many governments in industrialised countries have intervened to support their domestic agricultural sectors, particularly in the European Union, Japan and the US. The trading arrangements that have arisen from these interventions have taken a wide variety of forms and it is only possible to deal with a limited number of aspects.

The second section of this chapter discusses and differentiates between the various international and regional organisations that have a direct impact on trade including WTO, GATT, NAFTA, EU, APEC, while mention is made of various UN bodies that play a role in trade including the FAO, UNCTAD, CFC and ITC. Prominence is given to the World Trade Organization (WTO) which since its establishment in 1995 has had a growing impact on agricultural trade, an influence that will almost certainly increase. The WTO is now at the forefront of globalisation and liberalisation movements (WTO 1998), which over the past two decades have created many opportunities, including an expansion of world trade and economic development.[1] However, there are many risks involved, of both an economic and political nature. Globalisation can be detrimental to weak or badly governed countries and sectors, which can become more vulnerable to external price shocks and to the potentially destabilising effects of large and rapid capital flows. Strong arguments can still be made for continued government intervention especially in the provision of public goods, to improve equity and to overcome market failure. However, global events and international agreements are increasingly influencing the markets for many agricultural commodities rather than actions by specific governments and commodity producers.

The next section discusses the role of specific international commodity organisations, which in contrast to the WTO have seen a diminution of their role. Some of these bodies have incorporated economic clauses that have directly affected trade (e.g. coffee, cocoa, rubber) while others are more oriented to study and research groups (sugar, wheat, tropical timber, cotton, jute, olive oil, and the various FAO inter-governmental commodity groups – bananas, citrus fruit, fish, hard fibres, hides and skins, meat, oils, oilseeds and fats, rice and tea).

Alongside international interventions in agricultural production and trade, the individual sectors themselves have developed organisations aimed at facilitating their development. The fourth section discusses the role of various trade associations, which have been established to represent the stakeholders in the specific commodity sector (such as the CAL – Cocoa Association of London, CMA – Cocoa Merchants of America, AFCC, GAFTA, FOSFA, and the Edible Nuts Association). Often these

**Table 11.1** Value of exports of major agricultural products in 1999 ($ bn). Source: FAO 2000.

| | World total | | | Developing countries | | | Developed countries | | |
|---|---|---|---|---|---|---|---|---|---|
| | 1998 | 1999 | % change | 1998 | 1999 | % change | 1998 | 1999 | % change |
| Beverage crops | 18.0 | 12.4 | −31.2 | 16.7 | 12.4 | −25.8 | 1.3 | — | — |
| Cocoa | 3.2 | 2.2 | −31.2 | 2.8 | 2.2 | −22.0 | 0.4 | — | — |
| Coffee | 11.6 | 7.4 | −36.4 | 11.1 | 7.4 | −33.1 | 0.6 | — | — |
| Tea | 3.2 | 2.8 | −12.6 | 2.8 | 2.8 | −1.3 | 0.4 | — | — |
| Sugar | 10.2 | 7.3 | −27.8 | 6.4 | 5.0 | −21.2 | 3.8 | 2.3 | −39.2 |
| Bananas | 3.5 | 3.0 | −15.0 | 3.3 | 2.8 | −15.0 | 0.2 | 0.2 | −14.8 |
| Citrus | 5.0 | 4.6 | −6.7 | 2.6 | 2.4 | −5.8 | 2.4 | 2.2 | −7.7 |
| Cereals | 34.3 | 30.4 | −11.3 | 12.3 | 10.7 | −13.2 | 22.0 | 19.8 | −10.2 |
| Wheat | 14.7 | 13.5 | −8.1 | 2.1 | 2.0 | −6.0 | 12.6 | 11.5 | −8.5 |
| Rice | 9.2 | 7.4 | −19.3 | 7.4 | 6.0 | −19.7 | 1.8 | 1.5 | −17.4 |
| Coarse grains | 10.4 | 9.5 | −8.8 | 2.7 | 2.7 | −1.1 | 7.7 | 6.8 | −11.5 |
| Cassava | 0.6 | 0.7 | 22.4 | 0.6 | 0.7 | 15.6 | 0.0 | 0.0 | — |
| Meat | 24.9 | 25.0 | 0.3 | 6.5 | 6.1 | −6.7 | 18.4 | 18.5 | 0.7 |
| Bovine meat | 10.0 | 11.5 | 14.2 | 2.6 | 2.5 | −3.0 | 7.5 | 8.5 | 13.4 |
| Ovine meat | 1.4 | 1.4 | −2.2 | 0.1 | 0.1 | −8.0 | 1.3 | 1.3 | −1.8 |
| Pig meat | 6.3 | 5.9 | −5.8 | 1.0 | 0.8 | −19.9 | 5.3 | 5.1 | −2.8 |
| Poultry meat | 6.4 | 5.4 | −15.0 | 2.5 | 2.4 | −4.6 | 3.9 | 3.1 | −19.2 |
| Other meat | 0.7 | 0.7 | 0.7 | 0.2 | 0.2 | −11.0 | 0.5 | 0.5 | 8.4 |
| Milk & milk products | 12.8 | 11.2 | −12.6 | 1.3 | 1.2 | −4.3 | 11.6 | 10.0 | −13.5 |
| Butter | 1.6 | 1.2 | −23.7 | 0.1 | 0.1 | −16.4 | 1.5 | 1.1 | −24.3 |
| Cheese | 4.0 | 3.1 | −22.1 | 0.2 | 0.2 | −5.2 | 3.8 | 2.9 | −23.1 |
| Powder & other products | 7.3 | 6.9 | −5.1 | 0.9 | 0.9 | −2.5 | 6.3 | 6.0 | −5.4 |
| Oils, oilseeds & meals | 42.5 | 36.8 | −13.2 | 23.8 | 20.8 | −12.5 | 18.6 | 16.0 | −14.2 |
| Oilseeds | 13.3 | 12.4 | −6.8 | 4.8 | 4.2 | −12.0 | 8.5 | 8.2 | −3.9 |
| Oils and fats | 21.8 | 18.9 | −13.1 | 14.0 | 12.7 | −9.3 | 7.8 | 6.3 | −19.8 |
| Cakes and meals | 7.4 | 5.5 | −25.3 | 5.1 | 4.0 | −21.8 | 2.3 | 1.6 | −32.7 |
| Agricultural raw materials | 15.9 | 14.7 | −7.2 | 6.8 | 6.0 | −12.3 | 9.1 | 8.5 | −6.9 |
| Cotton | 8.3 | 7.5 | −10.5 | 2.3 | 2.4 | 4.4 | 6.0 | 5.0 | −16.2 |
| Jute | 0.1 | 0.1 | −13.0 | 0.1 | 0.1 | −15.5 | 0.0 | 0.0 | — |
| Hard fibres | 0.4 | 0.4 | 3.2 | 0.3 | 0.4 | 4.3 | 0.0 | 0.0 | −10.8 |
| Natural rubber | 3.6 | 3.0 | −17.3 | 3.5 | 2.7 | −24.8 | 0.1 | — | — |
| Hides and skins | 3.5 | 3.8 | 10.1 | 0.5 | 0.5 | −11.7 | 3.0 | 3.4 | 13.9 |
| Total of the above | 167.6 | 146.2 | −12.8 | 80.2 | 68.1 | −15.1 | 87.4 | 77.4 | −11.4 |
| All agricultural products* | 308.6 | — | — | 132.5 | — | — | 176.1 | — | — |
| Forestry products | 132.3 | 139.5 | 5.0 | 19.9 | 21.0 | 5.6 | 109.5 | 115.0 | 5.0 |
| Fishery products | 48.9 | — | — | — | — | — | — | — | — |

Note: Export values for 1999 are preliminary estimates, derived on the basis of estimated changes in trade volumes form 1998 and in world market prices. 1998 trade data are from FAOSTAT, except for hard fibres and cassava. The value of exports for developed countries and the world exclude intra-trade of the EU. Oils and fats exclude butter and fish oil. Meals and cakes exclude fishmeal. Wheat includes flour in wheat equivalent. Beverage crops, cotton and rubber in 1999 do not include re-exports. Hides and skins in 1999 do not exclude intra-EU exports. Hard fibre values are FAO estimates and include processed products. Cassava values are FAO estimates. Export values are f.o.b.

*These include all agricultural products reported in FAOSTAT (the trade data are not available for 1999; for 1997 the corresponding values were: $332 bn for the world, $143 bn for developing, and $189 bn for developed countries).

organisations create and police the contracts under which much of the trade is undertaken.

The final section outlines various commercial practices including export documentation and procedures, cargo surveying, microbiological stability, microclimate effects in shipping containers, litigation and food safety.

## The WTO and other international and regional organisations

### The World Trade Organization

The WTO was formally established in 1995 following

the successful conclusion of the Uruguay Round (UR) of trade talks, which included agriculture for the first time.[2] In just a few short years, the WTO has emerged as one of the world's most influential and important international organisations. There is little doubt that the WTO's impact on commodity production and trade will intensify over the coming decades as the adoption of the UR agreements and possible new trade round agreements lead to substantial economic gains from the stimulation of international trade and capital flows.[3] Over the past two decades there have been unprecedented changes in the structure of global economic relations, especially in trade and finance. A reduction of trade barriers, improved transport and communications, the globalisation of corporate production, and the growing integration of financial markets have facilitated these developments. The links between trade, financial flows and investment are inextricably mixed[4] and the WTO is becoming increasingly involved in trying to regulate these relationships. However, there are still many countries and economic sectors, especially in agriculture, that have not become globalised. While membership of the WTO has led to some liberalisation of a country's economy there are still many countries where the transition period is quite long and considerable restrictions remain.

The establishment of the WTO followed seven years of protracted multilateral negotiations under the Uruguay Round of trade talks. The UR agreement signed in April 1994 led to a considerable overhaul of the GATT (General Agreement on Tariffs and Trade) system, which had been in existence since 1948 and had seen enormous changes in economic relationships over the previous decades. The GATT was established with the aim of overseeing international trade in goods and services and gradually liberated trade through a series of rounds – Geneva, Torquay, Annecy, Kennedy, Tokyo – and the most recent, the Uruguay Round. Discussions are currently under way to begin another round (the Millennium Round) with the aim of further lowering tariffs and other impediments to trade – and agricultural trade is likely to take centre stage in these negotiations.

The dominant aim of the various rounds was to encourage freer trade. The rounds have led to a considerable decline in tariff levels; elaborate machinery was established for obtaining tariff concessions via multilateral trade negotiations leading to the registration of the concessions thus achieved; known as tariff bindings.[5] In addition, GATT prohibited quantitative restrictions on trade although there are a number of exceptions, usually relating to agricultural and fisheries products. Over the years, the areas covered and the issues involved have become much more complex, and the number of contracting nations has increased from the original 22 GATT members to the 136 WTO members in early 2001.

The UR agreement brought agriculture under GATT disciplines for the first time. The fiendishly complicated agricultural agreement did three things. First, with a few temporary exceptions, it converted all non-tariff barriers and unbound tariffs into bound tariffs. It was agreed that these tariffs had to be cut by an unweighted average of 36% between 1995 and 2000. Second, it prohibited new export subsidies and cut existing ones. Third, it began to tackle the domestic agricultural subsidies, which, in effect, protect farmers against foreign competition in much the same way as tariffs (Josling 1998).

The UR covers 19 agreements in all (listed in Box 11.1), the most notable being the establishment of WTO, which provides the institutional framework that encompassed the whole of GATT. Never before had a GATT round covered so many different aspects of international trade: from agriculture to intellectual property rights; from sanitary standards to import licensing; and from textiles to investment. Thus, the WTO is much wider than GATT; in addition to trade in goods the WTO has expanded to new areas including services (GATS), intellectual property rights (TRIPs), investment (TRIMs), government procurement and increasingly environmental issues. Hence, the legal and regulatory framework of the WTO embodies the GATT, GATS, accords on TRIPs and TRIMs, plurilateral agreements on government procurement, meat and dairy, civil aviation plus trade policy review and dispute settlement mechanisms. In addition, there are improved rules and disciplines to prevent abuses of the system, including unfair trade practices. 'Safeguard' and anti-dumping measures still remain the key instruments but these have been further refined to ensure fair application and prevent abuse. Each of these agreements can have some impact on the agricultural sector. Thus, the WTO 'dispute settlement mechanism' enables countries to settle their trade differences rapidly, and reduces the uncertainties and disruption for business operators. Indeed the dispute settlement procedure has been described by a former head of the WTO (Renato Ruggiero) as the WTO's most individual contribution to the stability of the global economy. The new system is stronger, more automatic and more credible than its GATT predecessor and is based on equitable, fast, effective and mutually acceptable principles, with clearly defined rules and procedures and timetables for completing a case.

## Box 11.1 List of multilateral agreements of the World Trade Organization.

**Annex 1a: Multilateral agreements on trade in goods**

General Agreement on Tariffs and Trade 1994 – GATT
- Understanding on the Interpretation of Article II: 1 (b) of GATT 1994
- Understanding on the Interpretation of Article XVII of GATT 1994
- Understanding on Balance-of-Payments Provisions of GATT 1994
- Understanding on the Interpretation of Article XXIV of GATT 1994
- Understanding in Respect of Waivers of Obligations under the GATT 1994
- Understanding on the Interpretation of Article XXVII of GATT 1994

Marrakech Protocol to the GATT 1994
Agreement on Agriculture
Agreement on the Application of Sanitary and Phytosanitary Measures
Agreement on Textiles and Clothing
Agreement on Technical Barriers to Trade
Agreement on Trade-Related Investment Measures (TRIMs)
Agreement on Implementation of Article VI of the GATT 1994
Agreement on Implementation of Article VII of the GATT 1994
Agreement on Pre-Shipment Inspection
Agreement on Rules of Origin
Agreement on Import Licensing Procedures
Agreement on Subsidies and Countervailing Measures (ASCM)
Agreement on Safeguards

**Annex 1b: General agreement on trade in services (GATS)**

**Annex 1c: Agreement on trade-related aspects of intellectual property rights (TRIPs)**

**Annex 2: Understanding on rules and procedures governing the settlement of disputes**

**Annex 3: Trade policy review mechanism**

**Annex 4: Plurilateral trade agreements**

In contrast to previous GATT Rounds, new members must agree to all 19 individual Agreements (except the Bovine Meat Agreement) before they can become new WTO members. The four major GATT principles have been incorporated into the WTO, namely:

*Most favoured nation (MFN)*: Trade must be conducted on the basis of non-discrimination – no country is to give special trading advantages to others or to discriminate against it – exceptions to this rule are allowed only under special circumstances.

*National treatment*: If a country chooses to give protection to a domestic industry, this protection should be provided through a customs tariff and not through other commercial barriers.

*Transparency*: This ensures that traders are not handicapped because of a lack of information on regulations or taxation. In addition, the 'binding' of tariff levels negotiated among contracting parties provides a stable and predictable basis for trade: when a country becomes a WTO member, it agrees to fix its tariffs, which can only be raised if compensation is negotiated under special terms.

*Presumption of fair trade*: underpinned by procedures for the imposition of anti-dumping or countervailing duties.

There are a number of other features of the WTO, which are relevant to international trade:

- consultation, conciliation and dispute settlement is fundamental;
- exemption: a country may, when its economic or trade circumstances warrant it, seek exemption from certain WTO obligations;
- restrictions: there is a certain general prohibition on the use of quantitative restrictions, for example, import quotas, except when a country's balance of payments so warrants; and

- regional trading arrangements: the WTO permits the establishment of regional trade arrangements such as the EU, NAFTA and ASEAN, provided certain criteria are met. The criteria are meant to ensure that such trade arrangements do not raise barriers to world trade.

Currently, the WTO has the mandate of its members to function as administrator on WTO trade agreements, handle trade disputes, monitor national trade policies and act as a forum for trade negotiations. In addition to the countries that are signatories to the Agreements, including the Triad of US, the EU and Japan, a number of other countries have applied to join including China, Russia and Saudi Arabia. However, there has been growing discontent, particularly among developing countries, that feel increasingly threatened by the path and pace of globalisation and liberalisation, which are undermining their ability to exercise economic and political sovereignty. One major concern is the distribution of the gains, the bulk of which are benefiting the Triad and multinational companies, while the economically and politically under-leveraged nations are shouldering the heavy burden of integration into the world economy. Another major concern is the environmental implications of WTO decisions and agreements, some of which are seen as being detrimental to the environment. The issue of trade and environment was not included in the UR negotiations although the preamble to the WTO Agreement includes direct references to the objective of sustainable development and the need to protect and preserve the environment. Certainly the issue of the relationship between trade and the environment has been moving rapidly up the WTO agenda. As discussed in detail elsewhere in this chapter, the new Agreements on Technical Barriers to Trade and on Sanitary and Phytosanitary Measures explicitly take into account the use by governments of measures to protect human, animal and plant life and health, and the environment. The Agreement on Agriculture exempts direct payments under environmental programmes from WTO Members' commitments to reduce domestic support for agricultural production, subject to certain conditions.

Many countries dependent on agriculture are facing erosion of their economic leverage within the multilateral WTO negotiations. They face the stark choice of either opening up to liberalised trade and investment regimes and braving the difficulties that this brings, or facing the risks of not joining or leaving the WTO and of being side-stepped by foreign investment and multinational companies, leading to lower rates of growth and a widening of the gap in GDP and per capita incomes. Adoption of the liberalisation approach involves a painful readjustment process and a growing inability to decide national economic policies. Moreover, as global competition intensifies both in overseas markets and for sources of external finance, there is growing pressure to adapt and adopt international standards meted out by the international institutions (i.e. the World Bank/IMF and WTO) in order to become a player or maintain their international competitiveness. However, they may lack the human and physical capacity to absorb the demands of technology and competition, thus compromising their growth prospects. Alternatively, if they walk away from liberalisation and WTO membership, they will distance themselves even further from global economic links.

The new Millennium Round provides a vital opportunity for commodity producing and exporting countries to effect improvements in the rules governing agricultural trade and investment. A dominant objective of many countries is to reduce the substantial agricultural trade barriers, subsidies and tariff escalation that still exist in major importing countries, particularly in the EU, US and Japan. Some of the policies being suggested to assist in trade reform include:

- strengthening producing countries' capacities to negotiate a fairer deal – which will necessitate a substantial increase in resources to facilitate their ability to analyse the economic and social impacts of a range of policy options; this is particularly true of a country's ability to protect its plant varieties under TRIPs;
- food aid donations should be limited to where they are absolutely necessary;
- supporting 'fair trade' and 'ethical trade' initiatives; and
- promoting 'smallholder-friendly' quality assurance schemes to assist in meeting stringent sanitary and phytosanitary standards.

Trade policy should be viewed within a wider development context; trade liberalisation alone will not help the poor as the low take-up of the EU's Lomé provisions illustrated. To exploit the benefits of trade, many agricultural producing countries need better economic governance and reforms to attract investment as well as better trade opportunities.

### Regional trading blocks

Since the end of World War II there has been a proliferation of regional trading blocs, aiming to protect geo-economic niches. Over 150 such agreements have been notified to the WTO and GATT – more than half of which have been set up since 1990 (some of which are revisions of previous deals). This deluge of 'alphabet soup' has deeply divided economists – some seeing them as 'building blocks' for international trade negotiations, others as 'stumbling blocks' discriminating between members and the rest of the world.[6] Some of the major ones include:

- EU – the European Union, which can be regarded as the grandfather of the current groups; this has made several agreements with other bodies, along with EFTA (the European Free Trade Area) it has formed the European Economic Area – it also has Europe agreements with North Africa and Central Europe, each of which incorporates some agreements on trade in agricultural goods;
- NAFTA – North American Free Trade Area – involving Canada, Mexico and the US;
- Mercosur – covering several South American states; and
- APEC – the Asia Pacific Economic Co-operation – which has a grand goal of 'free trade in the Pacific' by 2020.

Besides regional trading blocks there are a number of United Nations organisations that are involved in agricultural production and trade. The Common Fund for Commodities (CFC) remains one of the very few international bodies focused upon support for commodity trade development. The CFC was established in 1989 in Amsterdam to provide commodity finance for projects submitted by approved bodies – which include all international organisations. As originally conceived the CFC would have access to some $6 bn to finance price stabilisation operations for a dozen or so commodities. However, the shift in the political and economic environment against market intervention has led to the CFC operations being more modest. Its 'first account' has never been used as intended for market interventions, although some funds have been released in the late 1990s specifically aimed at helping countries improve commodity marketing and risk management. The CFC's activities have focused mainly on the 'second account' which funds commodity research and development and measures aimed at encouraging diversification, improving quality, expanding local processing, developing new uses and pest and disease control. CFC project identification, implementation and management have become increasingly important activities for commodity organisations. For example, the International Coffee Organization (ICO) in 2000 had seven projects under way totalling $31m and a further four projects valued at $25m in the pipeline. The CFC's agenda has become increasingly poverty-focused in line with those of most major donors. In recent years it has also increasingly sought to finance projects that explore more novel approaches, for example, with respect to trade finance and ethical trade. CFC operations have been constrained by the international commodity bodies through which it is obliged to operate. Over the past decade, a key role for commodity organisations has become the accessing of funds from the CFC.

Almost all efforts regarding international commodity arrangements took place under the auspices of the United Nations, in particular, since the mid-1960s, the United Nations Committee on Trade and Development (UNCTAD), which is a permanent UN body based in Geneva. There have been numerous UNCTAD meetings since the 1960s and considerable increased expectations regarding international commodity control measures, but these expectations have invariably failed to be met. Concerning the future role of UNCTAD, the organisation's strength is seen in 'measurement' in the sense of 'devoting resources to collecting, verifying, and publishing trade policy and trade data. Binswanger and Lutz (2000) argue for a wider role:

> 'UNCTAD should define its functions and work programme in partnership with the WTO, FAO, IMF and World Bank. These should include: providing a forum for developing countries on trade and related issues; maintaining trade-related databases and providing information; undertaking high-quality analyses; providing technical assistance in norms and standards and in dispute settlement; advocating better market access in industrial countries; helping to build coalitions and seeking common developing country positions in multilateral trade negotiations.'

The Food and Agricultural Organization of the UN (FAO) is closely involved in agricultural production and trade. It is responsible for organising a number of intergovernmental groups on a range of commodities including jute, kenaf and allied fibres, hard fibres, bananas, tea, citrus fruits, rice, grains, oils and oilseeds.

Also worth noting are various financial compensation schemes, which are commodity related, introduced by the EU as part of is development project. Stabex, the most popular system, was introduced in 1975 as part of the first Lomé Convention. Stabex is an export earning stabilisation scheme between EU members and approximately 70 African, Caribbean and Pacific(ACP) developing countries. The International Monetary Fund in 1981 implemented its compensatory financing facility (CFF) whereby excesses in cereal import costs were accepted as a new reason for drawing down from the CFF.

## International commodity agreements

At the end of World War II many ideas emerged regarding commodity control schemes, some of which were incorporated into the 1947 agreement establishing GATT. As a result, internationally agreed procedures and principles were established for the creation and operation of international commodity agreements.[7] It was anticipated that a large number of international commodity agreements would be established, but in the half century since 1948 only a limited number of fully-fledged international control agreements have been concluded – and often operated only fitfully. These included agreements on international trade in wheat, sugar, coffee, cocoa, rubber and grains – each of which led to the establishment of specific commodity bodies, e.g. the International Cocoa, Coffee and Sugar Organizations. Other agreements have been concluded relating to jute, olive oil, wool and timber, but these have been more in the form of study groups and have not aimed to affect prices and markets.

In the early days, GATT did not actively promote or support commodity agreements but rather left the negotiation and organisation of such agreements to various UN bodies. Indeed, various GATT members, notably the US, believed that commodity agreements were not consistent with the GATT system, since they frequently involved measures such as price controls and export restrictions. This was particularly the case with agreements that aimed to raise prices above their long-term trend, as opposed to stabilising fluctuations around the long-term trend.[8] Prior to the adoption of Part IV, GATT had not been active in the various multilateral activities concerning international commodity trade (Dam 1970). Beginning in 1955, GATT had initiated annual reviews in commodity trends and arrangements had been established for consultation on commodity problems. Although Part IV encouraged GATT to be more proactive with regard to commodity agreements, particularly with regard to temperate agricultural products of wheat and meat, there was a tendency to leave actual responsibility for negotiations agreements to UN bodies, particularly the UN Committee on Trade and Development.

Since the end of World War II, several agricultural commodities have been covered by an international commodity agreement (ICA). In each of these agreements both producing (exporting) and consuming (importing) countries were members, and the various agreements have permitted the imposition of some form of market controls such as buffer stocks, export controls and indicative price bands. These agreements had several objectives including reduced price volatility, higher prices and greater revenue stability for producers, assisting export diversification and improving market access, but invariably the aim of such control was to maintain export prices and hence revenues.[9] From the early 1990s a key role for ICAs has become the accessing of funds from the CFC. While ICAs have had some limited success in these economic objectives, most of the market control efforts have ended in failure (see, for example, Gilbert 1995). For example, the International Coffee Organization, which has existed since 1962, was able to impose controls on a large proportion of unprocessed coffee entering international trade, since producing members accounted for well over 90% of global coffee production, while consuming members accounted for approximately three-quarters of consumption. If coffee prices fell below a certain level, the ICO imposed export quotas on shipment from coffee-producing countries to those consuming countries that were ICO members. Moreover, it was the consuming member countries, which included the EU, the US and Japan, that monitored these exports. For a variety of reasons the export quota system collapsed in 1989 following a failure to renew the economic clauses of the International Coffee Agreement.

While international commodity organisations have continued to survive, the 1990s saw the demise of their market intervention activities – with the exception of rubber. At the start of the new millennium, co-operation among commodity trading states is chiefly concerned with providing reliable statistical data and analysis to improve market transparency to facilitate trade. Each commodity agreement provides for regular inter-governmental meetings, at which officers from the various sectors' trade associations are often in attendance. If the present hostility to market intervention measures ends then these meetings could act as the springboard for renewing market intervention and stabilisation measures. As regards the future role of ICAs and international commodity bodies (ICBs), it is important to distinguish between long-term and short-term price stability (Maizels 2000). The long-term decline in real prices is the result of excess supply. This calls for production strategies being better informed by production and demand forecasts, based on which producer countries can decide their longer-term production plans for a certain commodity while consumption is stimulated and price volatility reduced. However, to date, the International Cocoa Organization's Production Management Plan based on this strategy has failed. The initial failure arises from producers being simply unable to exert control over output, because production is prey to disease, pests and weather and output cannot be fine-tuned. However, the application of the approach is still at an early stage; it must be linked closely with diversification strategies, and it remains to be seen to what extent it can contribute to more balanced commodity markets.

Currently several ICA and Study Groups (ISG) are being, or will shortly be, renegotiated. In these renegotiations there is a need to stress greater poverty focus of their actions (e.g. practices to improve bargaining power of smallholders including development of farmers' associations, improved access to credit and risk management). There is also the need for greater involvement and/or membership of these organisations from the private sector and international bodies such as the IMF, WB, WTO, UNCTAD, CFC and FAO.

The ICBs should provide a business forum for private sector participants of a commodity chain. This would include operators from the exporting as well as importing countries, including representatives of workers and smallholders (e.g. farmer associations and trades unions). In the context of poverty reduction, it appears important that the latter groups are given a more prominent voice in the commodity system. Given the importance of some of the internationally traded commodities (e.g. coffee) and the number of people depending on them in one way or another, such business forums are indispensable. They can also serve a lobbying function for the industry. At the same time, ICBs need to improve their operational efficiency, including more transparency regarding decision-making and budgetary matters. ICBs should expand their role in identifying best practices, co-ordinating research and development and sustainable management initiatives. Two major functions should continue, namely, the provision of market data and improvement of market transparency and a continued role regarding project identification, implementation and monitoring.

Alongside ICAs other international commodity policies exist. Among these are the commodity protocols under the EU's Lomé Convention – these include protocols on bananas, sugar, rum, beef and veal. While these have had significant welfare gains through income transfers for specific countries, the overall success of these protocols is questionable.

## Trade associations

Various trade associations have been established to represent the stakeholders in the specific commodity sector. Although their functions differ somewhat, most of them are internationally recognised non-profit associations which, among other things, promote, protect and regulate the specific commodities with which they are involved. This often includes the administration of the major physical trading contracts, the publication of market rules and collection of market data as well as the provision of a panel of arbitrators to deal with disputes that have not otherwise been settled amicably. They invariably aim to establish uniformity in commercial practices, especially with regard to contracts, bills of lading, freight, insurance contracts and other trade documents. In addition, they will disseminate information to both members and the general public as well as represent members in discussions with governments, both at a national and international level.

Some of the major associations are listed below and include:

*Cocoa*: the Cocoa Association of London (CAL); the Cocoa Merchants of America (CMA);

*Coffee*: there are a large number of national and international association in the coffee sector, such as the Coffee Trade Federation (CTF in the UK), Deutscher Kaffee Verband in Germany, Committee of European Coffee Associations (CECE), Association of Coffee Producing Countries (ACPC);

*Vegetable oils and seed*: GAFTA (Grains and Feed Trade Association), FOSFA (Federation of Oils, Seeds and Fats Association Ltd), NSA (National Sunflower Association) and NSPA (National Soybean Processors Association);

*Others*: Edible Nuts Association.

These associations were created to represent members' interests – and depending on the particular organisation, a wide range of disciplines are represented in the group including producers, traders, importers, exporters, agents, brokers, processors, futures markets, warehousers, freight forwarders, port authorities, banks, and insurance companies. Often these organisations create and police the contracts under which much of the trade is undertaken.

## Commercial practices

Cereal grains and their food or feed derivatives, together with oilseeds and their derivatives, dominate international trade in durable agricultural commodities. At a global level, the five major exporting nations – United States, Canada, European Union, Australia and Argentina – similarly dominate this trade although several other countries are major players in the international export trade in some of these commodities, e.g. rice. A simplified breakdown of a World Grain Review for 1999 serves to illustrate the major features of the international trade picture (Table 11.2).

The great majority of these commodities are shipped in bulk in purpose-built dry bulk carriers of the Panamax or Handysize categories, typically with a capacity of 20 000–65 000 tonnes. Processed or semi-processed food products such as wheat flour and some rice products, however, are commonly shipped in bags, either as containerised or general cargo on liner or tramp-type vessels. Bagged cargo forms a very small percentage of global trade in durable agricultural commodities. Higher value commodities, such as edible nuts, cocoa beans, coffee, and so on are nowadays invariably shipped in containers.

### Export documentation and procedures

Whereas commodities traded domestically are commonly governed by contract terms specific to a particular country, international trade normally takes place under terms evolved by the major trade associations, e.g. GAFTA and FOSFA. These 'standard' contracts define clearly the respective obligations of buyer and seller and are widely accepted internationally. Within the standard contract both buyer and seller will agree on specific standards of quality with which a contracted consignment must comply, or, commonly, will stipulate a widely traded and recognised grade of produce. Grades set by the United States Department of Agriculture (USDA) for internally traded cereals and oilseeds are commonly used for this purpose.

Having located an appropriate carrier (shipping company) which makes available a suitable vessel at the export point, the seller provides all essential information relating to the contracted cargo contained in Bills of Lading which are issued to the carrier's representative on completion of loading in the export port. This is the point at which the master of the vessel or his chief officer may decide to register any reservations concerning the loaded condition of the cargo in the form of an endorsement to the Bills of Lading. Under the maritime guidelines as interpreted by the Hague Visby Rules, the carrier's representatives are not expected to exercise any objective judgements on the quality of the loaded cargo except where there is clear visual evidence that it is not of a condition considered normal by accepted trade parameters.

Guidance to shippers and carriers on the characteristics of shipped commodities, including precautions to be taken, is available in trade manuals and in information based on a series of 'Carefully to Carry' booklets produced by the London-based Protection and Indemnity (P&I) Club in the 1960s and 1970s. P&I Clubs are basically insurance syndicates to which almost all commercial freight vessel owners belong. Under the International Grain Code administered by the International Maritime Organization, all ships carrying grain must comply with the requirements of the Code and

**Table 11.2** Major features of the international grain trade. Source: *World Grain*, November 1999.

| Commodity | Global quantity traded (million tonnes) | Major exporters | Major importers |
|---|---|---|---|
| Wheat | 100.9 | USA, Canada, Australia, EU, Argentina | Egypt, Japan, Brazil, Algeria, South Korea |
| Wheat flours | 10.2 | EU, Turkey, Argentina, USA, Japan | CIS, Yemen, Algeria, Libya |
| Maize (corn) | 62.9 | USA, Argentina, China, Hungary, South Africa | Japan, South Korea, Taiwan, Mexico, Egypt |
| Sorghum | 6.9 | USA, Argentina | Mexico, Japan |
| Barley | 12.7 | EU, Australia, Canada, Turkey, Russia | Saudi Arabia, Japan, China, USA, Iran, Libya |
| Soybeans | 40.4 | USA, Brazil, Argentina, Paraguay | Netherlands, Japan, Mexico, Germany, Spain |
| Rice | 27.4 | Thailand, India, Vietnam, China, USA | Indonesia, Bangladesh, Philippines, Brazil |

carry a document of authorisation issued under it. More detailed information on the structures and procedures underpinning the international shipment of grain are given in Sewell (1999).

Either at the request of the buyer, as a requirement by the importing country, or as part of routine procedure in the exporting country, an official phytosanitary certificate may be issued by an exporting authority confirming the generic description of the cargo and voyage details, together with a record of pre-shipment pesticide application, if any. A statement attesting to freedom from scheduled weed seeds, moulds or fungi, and 'substantial' freedom from living insect pests may also be given.

## The role of cargo/marine surveyors

It is common for the contracting parties to commission the services of either an official inspection agency or a commercial superintending company to provide information both on the pre-loading condition of the ship's stowage areas as well as on the condition of the intended cargo. In the US all export consignments are inspected, sampled and analysed prior to clearance for loading by the Federal Grain Inspection Service (FGIS) which also certifies the vessel as fit for carriage. Certificates attesting to these clearances are normally issued prior to departure of the vessel. In other countries, commercial surveyors are engaged to provide these services and certification.

Whereas the FGIS procedures follow clearly defined rules in an attempt to arrive at representative findings concerning the pre-loaded quality of a commodity shipment, the *modus operandi* of commercial companies engaged in this work and their capability to provide objectively derived information is not always clear. The trade appears to be content with a system which provides one set of quality determinations purporting to represent the average condition of an entire cargo, often consisting of many thousands of tonnes. The determinations may be based on samples drawn from different storage sites in a port area or possibly from some considerable distance removed. Consequently, a degree of heterogeneity in loaded commodity quality can be considered to be the rule rather than the exception in this trade. Pre-loading quality certificates give no insight into the range of loaded qualities making up a cargo and tell us little concerning the propensity of such a cargo to undergo a sea voyage safely. It is also remarkable how many of the certificates record average quality parameters determined either exactly at, or fractionally below, the stipulated contractual maxima.

## Microbiological stability and biodeterioration

As noted above, commodities moving in international trade do so under contractual conditions which include generic or specific description of required 'quality' parameters. These descriptions are often 'grades' of produce or 'quality standards' as commonly used in internal trade within the major producing and trading nations for those commodities. Hence they will have evolved to reflect specific conditions of climate, logistics, post-harvest technology, storage periods, delivery times, etc. pertinent to trading in that country and found to be acceptable as a trading platform by the various market players. In many cases, a blend of inherent or acquired quality characteristics may be incorporated into a 'grade' of commodity which is considered acceptable in-country by a trade typified by short storage periods, modern aeration facilities, rapid transit times, low ambient temperatures, etc., whereas that same blend of characteristics would sound warning bells to a storage technologist experienced in biodeterioration. Where these 'grades' or 'quality standards' are transposed from their country or region of origin and applied to an international trade working to a totally different set of operating parameters, the potential for biodeterioration to occur is apparent.

The concept of the use of equilibrium relative humidity (ERH) or the water activity ($a_w$) of a product as a measure of its storage potential, and the need to relate these determinations to moisture content for practical purposes, are well established. It is, for example, widely accepted that mould growth, from spores ubiquitously present in traded commodities, is not significant at ERH levels below 70% or at $a_w$ values below 0.62. A considerable body of scientific literature exists which records data on the laboratory-derived moisture content/relative humidity equilibria for a range of commodities under specific temperature regimes. The literature on these relationships for cereals, oilseeds and legumes, spices and beverages has been well reviewed by Gough and Lippiatt (1978).

In a later publication, Gough (1996) extends the use of laboratory data to explain the relationships between physical condition and hazards to food cargo quality. The crucial physical conditions governing the microbiological stability of a shipped commodity are the temperature of the cargo and that of the surrounding air, the moisture content of the cargo and the relative humidities of the interparticulate air and of the surrounding air.

Although insect infestation may contribute directly or indirectly to the biodeterioration of some shipped commodities, it is microbiological instability exploited by mould species which leads to the most common manifestation of cargo spoilage in bulk shipments. All cereals and oilseeds, in any part of the world and in all climatic regions, when taken from the harvesting field for conditioning and entry into national or international trade, are ubiquitously infected with a wide range of field and storage species of moulds and fungi. Conditioning processes, for example, mechanical drying, will kill off the actively growing and sporulating stages of these organisms and reduce the moisture content of the commodity to such a level that the micro-environment is no longer suitable for further development of surviving stages. The more heat-resistant dormant spores of some species, however, may be present in large numbers and some will always be present in any traded consignment of raw cereals or oilseeds. These spores are microscopic in size and cannot be detected without considerable magnification. They cannot be detected either by cargo surveyors in routine inspection or sampling of cargoes or ships' crew at point of loading. These fungal spores will remain dormant and undetected throughout a normal voyage unless the specific micro-environmental conditions necessary for their development and multiplication occur.

In a paper summarising experiences in the mould spoilage of cereals during transportation by sea from Latin America to Europe, Mossel (1988) describes a predictive model showing how shipments of cereals could be protected against mould growth during sea voyages. For example, model predictions proposed that at a true loaded moisture content of 13.3%, maize cargoes would transport safely, that is, without significant mould development, for up to 15 weeks at temperatures not exceeding 25°C. Experience with deteriorated cargoes arriving at Rotterdam within this shipment time showed that these had invariably been loaded at average moisture contents considerably above this threshold value. The author comments upon the practice of reliance on an average moisture content figure at loading, pointing out that moisture determinations can vary tremendously even within consignments considered by the trade to be homogeneous.

Some indication of the extent to which scientific data on moisture content equilibria for 'safe' storage is at variance both with international standards and with national grades often used in commercial contracts is shown in Table 11.3.

### Micro-climatic conditions, the role of ventilation and the 'ship's sweat' phenomenon

Throughout the sea carriage of bulk agricultural commodities, the temperature of the headspace air above the commodity in the cargo holds and its water carrying capacity, i.e. the microclimatic conditions, are heavily influenced by the temperature and moisture content of the loaded cargo. Vessels contracted for the carriage of such commodities are equipped with ventilation facilities consisting of various permutations and models of manually closable vents which allow passage of ambient air directly from the deck into the upper part of each cargo hold. The passage of air, with the vent in the open position, is either natural or fan-assisted. Ventilation allows ambient air to enter the hold and displace the air rising from within the cargo and gathered in the free or headspace above it. Air displaced from the headspace during ventilation is evacuated to the deck outside the hold. The underlying concept of such ventilation is to introduce ambient air into the headspace only when the air to be introduced is at a lower relative humidity than the air to be displaced. In this manner, the storage microenvironment at the cargo surface is stabilised or even improved to the extent that moisture is given up from the cargo surface to the ventilating air. Introduction of

**Table 11.3** Definitions of 'safe' moisture content (% wet basis).

| Commodity | Laboratory data* maximum levels for safe storage | Maximum tolerance | | |
|---|---|---|---|---|
| | | EC | ISO | National grade |
| Maize | 13.5 | 14.5 | 15.5 | 15.5 (Thai grade 2) |
| Wheat | 13.5 | 14.5 | 15.5 | |
| Soyabeans | 11.7–12.5 | | | 14.0 (US no. 2) |
| Milled rice | 13.0 | 15.0 | 15.0 | 15.0 (US grades) |

*Sources: Hall 1970; Gough & Lippiatt 1978.

ventilating air at a higher relative humidity than that of the air in the headspace could raise moisture levels in the surface layers of cargo to the point at which microbiological activity could be stimulated and subsequent damage could occur. Based on these principles, a ship's crew applying internationally-accepted best practice will ventilate cargo holds selectively after having established the critical ambient and within-hold climatic conditions using approved equipment and applying psychrometric conversion tables.

Ventilation of cargo holds as described, whether natural or fan-assisted, is intended to retard the development of a micro-climate which could be deleterious to the surface of a cargo loaded in sound condition and, as such, can only influence climatic conditions in the headspace of the cargo hold. Such a ventilation system, however effectively applied, is neither designed for, nor is capable of, exercising any conditioning effect whatsoever on the cargo bulk below the surface layer.

One spoilage phenomenon which a ship' s ventilation system can mitigate or even avoid is the condensation of moisture on the hold superstructure known in the shipping trade as 'ship's sweat'. This can occur when warm, moisture-laden air in the headspace of a cargo hold comes into contact with areas of the ship' s superstructure which are cooled by ambient conditions to a temperature which is known as the 'dew point'. Headspace moisture condensing under dew-point conditions on the underside of the deck plating and hatch covers or beams will fall as liquid water or run down the wings or shell plating onto the surface of the cargo, leading to characteristic caking, heating and mould growth. By regular monitoring of dew-point calculations and selective ventilation, a ship's crew can prevent the build-up of warm, moisture-laden air in the headspace and consequent condensation to a considerable extent, even under conditions favourable for 'ship's sweat'.

Taking actual data from a ship's log records, it is possible to construct a dew-point profile for a cargo of milled white rice shipped from Vietnam to West Africa via the South African Cape during the month of August, i.e. during the South African winter in the southern hemisphere. This cargo arrived at destination with severe mould spoilage and sustained considerable losses. The ship's records give air and sea temperatures during the voyage. The extremes are summarised in Table 11.4 and Fig. 11.1.

**Table 11.4** Extremes of air and sea temperatures sustained during a voyage.

| | Ho Chi Minh (Vietnam) | Cape Town (South Africa) | Lomé (Togo) |
|---|---|---|---|
| Air temperature (°C) | 29 | 15 | 24 |
| Sea temperature (°C) | 30 | 6 | 26 |

### *Commodities travelling between climatic zones*

As indicated above, barring unforeseen events such as significant rain or seawater ingress into cargo holds, throughout a sea voyage the microclimatic regime within a ship's hold will be heavily dominated by the combination of climatic characteristics present at point of loading within the commodity cargo.

In detailing the relationship between changes in physical conditions and hazards to food cargo quality, Gough (1996) points out that both the thermal conductivity and moisture conductivity of bulk food and feed commodities are very low. Such cargoes are thus poor conductors of heat and the movement of moisture through the commodity by isothermal diffusion, in the absence of temperature gradients, is very slow. Conversely, the movement of moisture by non-isothermal diffusion, when temperature gradients are present in the commodity, can be significant.

These changes and their practical significance have been measured in a series of instrumented studies involving commercial shipments of cereals from temperate to tropical regions (Boxall & Bisbrown 1989; Boxall & Gough 1989, 1993). These trials involved shipments of maize from Canada to Brazil and Angola, and shipments of wheat from the UK to Sudan and Bangladesh. Whereas surface grain temperatures closely mirrored ambient temperatures, rising by as much as 25°C during a typical voyage of 6–7 weeks, these temperature changes did not extend beyond 200 mm into the cargo. Similarly, only at the grain surface and close to the sides of the hold were any measurable changes in moisture content recorded and these only just exceeded a 1% rise at the maximum. In none of the cargoes studied did physical conditions within the stowed commodities beyond these peripheral areas change appreciably throughout the voyages.

Considerable biodeterioration can and does arise when cargoes loaded at commercially acceptable and commonly traded moisture contents in temperate zones are discharged and stored in tropical ports. This is because exposure to the higher temperatures at discharge leads to a rapid increase in the ERH of the commodity to a level above which a range of fungal species can develop.

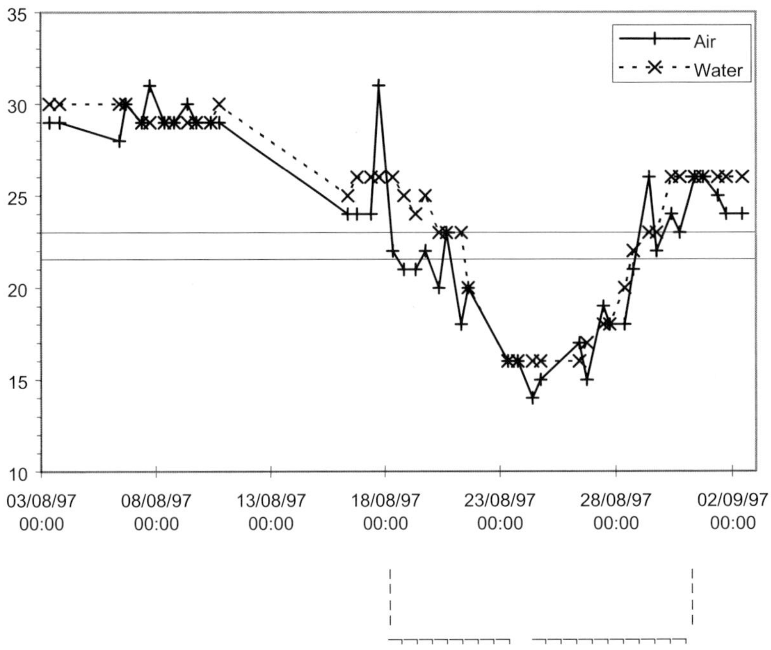


**Fig. 11.1** Air and water temperatures in a ship's hold on a voyage from Vietnam to West Africa. Dashed lines show the time period during which dew-point temperatures were achieved for a rice cargo loaded at 30°C and 14% moisture. During this period warm water vapour rising from the stow would condense on the cooler ship's superstructure and subsequently fall as liquid water precipitation onto the cargo surface and down the bulkheads/wings.

### Common litigation issues

Many of the cases of biodeterioration which pit ship owners' representatives against cargo interests hinge on the actual loaded quality of the shipped commodity and its propensity to carry safely at that level under normal, predictable shipping conditions. That propensity has been commonly referred to by the shipping, legal and insurance industries as the 'inherent vice' of the commodity. 'Inherent vice' is often used in an attempt to explain away a particular phenomenon of biodeterioration of an otherwise durable commodity. By inference we are to believe that the very nature of the commodity predisposes it to spoilage of some kind and possibly also that this characteristic exonerates one or more of the parties involved from responsibility for any such spoilage. In most cases of this type an experienced food technologist would be looking not for any 'inherent vice' but rather for any 'acquired characteristic' which might more readily explain the form of biodeterioration experienced. Often such a characteristic, for example, excessive water in the loaded cargo, may have led directly or indirectly to the spoilage and would normally have been within the remit of the shipper of the commodity to control or influence.

Of the millions of tonnes of cereals, grain pulses and oilseeds shipped annually, a considerable proportion of these cargoes is loaded at quality levels which threaten the microbiological stability of the consignments. Microbiological instability leads to the classic cases of biodeterioration characterised by cargo heating, moisture migration, caking, discoloration, mould formation, etc. For most shipments this may be of little consequence as voyages are short, extreme climatic regimes are not encountered, ventilation slows things down, discharge is rapid and the visual evidence of the onset of biodeterioration is masked by the mixing which occurs during bulk evacuation.

## Ethical trade

This section draws heavily on Blowfield (1999).

### Motivation for ethical trade

Despite problems of definition, ethical trade has captured the imagination of both the public and the business community. According to the US Fair Trade Federation, the fairtrade market accounts for \$400m in retail sales each year in Europe and the US. The global organic market is estimated to be worth \$11 bn, with organic imports from developing countries worth \$500m. According to some estimates, certified timber products will account for 15%

of the timber market in certain European countries by 1999 (Blowfield *et al.* 1998).

Although these are relatively small market shares, they are growing rapidly and advocacy organisations cite several surveys showing strong consumer support (Mintel 1997). More significantly, media pressure and NGO campaigns have generated consumer interest in the conditions under which certain items such as garments, cut flowers, coffee and vegetables are produced in developing countries. This has led to the growth in interest in ethical sourcing, particularly among European retailers, and in the near future many items on supermarket shelves are likely to be from sources that comply with certain basic ethical standards (Blowfield *et al.* 1998).

Similar pressure has also been applied on companies less dependent on consumer image, but equally dependent on shareholder perception. Extractive companies such as Shell and Rio Tinto, for instance, which have been criticised for negative social and environmental impacts, now publish annual statements on social and environmental performance. Organisations in the US such as the Investor Responsibility Research Center and the Council on Economic Priorities provide shareholders with regular updates on the ethical performance of major companies.

The reasons for shareholder and consumer concern vary. Concerns about sustainability are probably less important than not wanting to be party to processes that cause harm to the environment, other people, animals or oneself. For some people, the need for greater business responsibility is a result of globalisation whereby a company's economic activities are frequently beyond the reach of any one national regulatory system. This has caused

> 'a notable shift in power from the provident state to privatized and multinational enterprise, concomitant with a waning influence of governments and local workers' organisations'.
>
> *ILO (1998)*

In a world where the largest corporation has sales greater than the GDP of the fourth most populous country, there is a strong feeling that companies have both the responsibility and the power to behave ethically in ways that are outside the realm of free market forces (Varley 1998).

There may also be a darker motivation, one that is often cited by governments and entrepreneurs in developing countries. Interest groups that feel threatened by competition from the South may see environmental and social standards as a means of protecting their jobs or companies (Varley 1998). This has resulted in a number of cases being brought before the World Trade Organization and led to the development of the WTO's Code of Good Practice of the Technical Barriers to Trade Agreement which offers guidelines on how voluntary standards can be used without being interpreted as a non-tariff barrier (ILO 1998).

### Approaches to ethical trade

The broad scope of ethical trade and the plethora of initiatives throw up various possibilities for categorisation. One way is to use the categories employed by organisations directly involved in the trade: this would distinguish, for instance, ethical sourcing from fairtrade (Murphy 1997; Lake 1998), and the trade in organic produce from that in responsibly produced timber. This form of categorisation appeals to those who feel comfortable with established sectoral segmentation, or who wish to maintain a distinct identity (e.g. in the marketplace or when seeking donor funding). However, it does not help identify common ground shared by different approaches, and furthermore goes against a trend for more cross-sectoral debate being promoted by supporters within various movements.[10]

It might also be possible to categorise approaches according to the stage in the production–marketing chain where they seek to intervene. In theory this would distinguish initiatives that focus on trading from those that are primarily concerned with conditions of production. Fairtrade seeks to challenge conventional wisdom about trading relations by establishing a partnership between producer and buyer based on long-term commitment, stable prices and greater producer involvement in marketing. In contrast, sustainable fisheries or environmentally and socially responsible cut flower initiatives are about managing production, and do not address trading relations *per se*. Even ethical sourcing, despite sometimes being called ethical trading, is about managing the social conditions of production, not examining trading relations.

Approaches to ethical trade can be divided into two categories: enterprise initiatives and labelling initiatives. Enterprise initiatives are typically standards used by a company to measure the environmental and/or social performance of the company and its suppliers or subcontractors. Labelling initiatives are independently managed standards that companies may seek to comply with, and as a consequence earn the right to use a label or mark as proof of compliance. Such a simple categorisation is not without its problems and does not show, for instance, the differences that exist between the ethical issues facing a mining company and a supermarket, or the different challenges facing developers of standards for organic production compared to developers of fairtrade marks. However, it does highlight the shared strengths and weaknesses, opportunities and challenges of different approaches – aspects that are often rendered opaque when viewed from other vantage points.

## Enterprise initiatives

Without examining the reasons why, the fact is that a growing number of companies, particularly large corporations, are adopting policies relevant to ethical trade and taking part in joint initiatives (Table 11.5). A growing number of smaller companies, known as alternative trade organisations, are developing approaches to fairtrade, typically working with small producers in developing countries. Most of these companies are based in Europe or North America, although some would argue that this is not because such companies are more ethical than their Japanese or Korean equivalents, but because it is easier to profile ethical initiatives as a distinct approach in European and North American companies which place greater emphasis on targets and quantification rather than on overarching creeds or philosophies (ILO 1998).

Enterprise initiatives draw heavily on accepted management systems. No commercial enterprise today contests the importance of financial recording and reporting, both as a management tool and as a way of demonstrating performance. Yet auditing only became a legal requirement in the nineteenth century (Zadek *et al*. 1997), and the idea that a company's financial health was more than a private concern was disputed until the collapse of American and European markets in 1929 (Ranganathan 1998).

Today, large companies are very familiar with systems for managing quality, safety, hazard and other aspects of their business, sometimes to fulfil legal requirements, but also because such systems can deliver direct and indirect benefits to the company. Schemes and tools such as the International Standards Organization's series of standards for quality assurance (ISO 9000) and Hazard Analysis Critical Control Point share basic steps: the setting of a clear and detailed policy in the relevant area, allocating managerial resources for effective dissemination of the policy, and acquiring adequate tools for monitoring, reporting and taking corrective action.

Given this familiarity, it is not surprising that the mechanisms for accounting, auditing and reporting on social and environmental aspects of a company's operations are very similar to the ones already employed for other aspects of its business. While some people may be surprised at how rapidly companies have adopted the language of social and environmental responsibility, it is more accurate to say that these companies have simply increased their lexicon and applied it to a grammar which they had already mastered.

## Social and environmental labelling

Labelling is used as a means of communicating information about the social or environmental conditions surrounding the production of goods or provision of a service. Labelling gained notoriety when numerous companies sought market advantage by labelling their products environmentally 'friendly' in the 1980s. There were no fixed standards for demonstrating this 'eco-friendliness' and many products were shown to be no less harmful than their non-labelled equivalents.

Despite this adverse publicity, a number of labelling organisations have become well established. They offer an independent service, setting the standard for a particular sector or commodity, and overseeing the development accreditation and certification processes. By doing so, they address concerns about enterprise initiatives defining their own standards and not subjecting their performance to independent monitoring and verification. They also help reduce customer confusion

**Table 11.5** Joint enterprise initiatives.

| Initiative (country) | Sector | Examples of participating organisations |
|---|---|---|
| 95 Buyers' Group (UK) | Forestry | Habitat, B&Q, World Wide Fund for Nature |
| IFOAM (international) | Organic agriculture | Growers, traders, consultancy companies, general public |
| Horticulture Producers' Council | Horticulture | Various commercial growers |
| Apparel Industry Partnership | Apparel | Reebok, Liz Claiborne, Robert Kennedy Memorial |

about what a company means by ethical, and in some instances are promoted as offering market advantage (e.g. fairtrade and organic labels allowing companies to charge an ethical premium).

There are labels addressing both social and environmental performance, and increasingly labelling organisations' standards attempt to combine the two (Table 11.6). As the examples in Table 11.6 show, labelling initiatives operate in different ways. Some are membership organisations engaging a wide range of stakeholders in developing the label, but there are also private companies promoting labels for use in the business community and by government. Initiatives such as the FSC function as a custodian body that sets basic standards which can then be developed by accreditation agencies, and these agencies approve auditing or certification bodies (Forest Stewardship Council 1998). In other initiatives, auditors are either accredited directly by the custodian body (e.g. Fairtrade Labelling Organisation International) or that body carries out the auditing itself (e.g. Kenya Flower Council, Milieu Project Sierteelt).

Although some labelling initiatives are international, there are also national ones (e.g. the Child-Friendly Company programme in Brazil, the Kenya Flower Council) and ones that are specifically intended for use in developing countries (e.g. Rugmark, fairtrade labels). In the case of Milieu Project Sierteelt, environmental sections of its standard are international, but the social chapter applies only to developing countries.

In contrast with most company codes of practice, there are several labelling initiatives that have emerged from developing countries rather than the North. The Kenya Flower Council code of practice has been favourably received by a number of UK multiple retailers; the Child-Friendly Company programme was developed by

**Table 11.6** Examples of social and environmental labels and their scope.

| Programme | Comments |
|---|---|
| Forest Stewardship Council (FSC) | Developing international standards for responsible forest management. Membership organisation comprising commercial forestry companies, retailers, processors and environmental NGOs. Attempting to incorporate environmental and social issues. Auditing by accredited independent auditors. |
| Rugmark | Label for hand-knotted carpets produced in developing countries. Managed by membership organisation comprising religious and secular NGOs. Main focus is child labour. Auditing by member NGOs. |
| Fairtrade Labelling Organisation International (FLO) | Custodian body for fairtrade standards for developing country producers and Western buyers in products such as tea, coffee, bananas and cocoa. Managed by three European fairtrade labelling bodies (Max Havelaar, Transfair and Fairtrade Foundation). Main focus is social issues, but provides non-compulsory guidelines on some environmental aspects. Auditing by fairtrade labelling bodies. |
| Kenya Flower Council (KFC) | Cut flower industry association managing label for Kenyan growers. Managed by members within the flower industry. Main focus is environmental impact, but has recently included a social chapter. Auditing by KFC, although selective independent verification likely soon. |
| Social Accountability 8000 (SA 8000) | International standard for working conditions. Managed by Council on Economic Priorities Accreditation Agency. Sole focus is social issues, primarily in manufacturing. Auditing by accredited independent auditors. |
| ISO 14000 series of environmental standards | International standard for environmental management. Managed by International Organisation for Standardisation and recognised national bodies (e.g. British Standards Institute). Sole focus is environmental issues. Auditing by accredited independent auditors. |
| International Federation of Organic Agriculture Movements (IFOAM) | Custodian body for organic standards for agriculture and other sectors. Membership organisation of organic growers, buyers, retailers, processors and NGOs. Main foci are soil management and health issues, but has non-compulsory guidelines on social issues. Auditing by accredited independent auditors. |
| Milieu Project Sierteelt (MPS) | International standard for flower production. Private company. Main focus is environmental issues on flower farms, but recently introduced social chapter for developing countries. Auditing by MPS with some work subcontracted to country offices of SGS. |

the Abrinq Foundation, established by members of the Brazil's Association of Toy Manufacturers; and the FSC is based in Mexico.

#### *Strengths and weaknesses of labelling initiatives*

Labels help set common standards for certain sectors or commodities, and help prevent confusion among consumers. They can reduce the costs of ethical trade by creating greater economies of scale, and can access a greater body of expertise than could be afforded by most individual companies. Labelling initiatives can foster forums to ensure greater representation in the development and implementation of standards, and are seen as more independent than enterprise initiatives. This independence has been questioned in cases where the label is controlled by an industry association or an organisation with close links to a particular company or industry.[11]

#### *Independent monitoring and verification*

Non-governmental organisations and, increasingly, sourcing companies would like to see more independent auditing in relation to labelling initiatives. Some initiatives such as those of IFOAM, FSC and ISO already have accredited auditing organisations, and the Council on Economic Priorities Accreditation Agency is training auditors in the use of SA8000. But auditing remains costly and of dubious quality, especially in developing countries (Blowfield *et al.* 1998). The meaning of independence in relation to some labelling initiatives can also be questioned. Labelling initiatives are often financially dependent on revenues from companies using the labels, and labelling organisations must strike a balance between the need to generate income and the need to maintain label credibility. This is most problematic where there is as yet a small demand for labels, or competing standards for the same industry. Similar considerations apply to auditing organisations, particularly where the auditing organisation issues the certificate or right to use the label (Ghazali & Simula 1996). As yet there is a lack of legislation or industry accepted practices for establishing labels, although some organisations use ISO standards for auditors.

#### *Procedural versus performance approaches*

Criticisms about the implementation, criteria and indicators, and scope of enterprise initiatives apply equally to labelling initiatives. A further criticism that applies to both but which is best understood in the context of labelling initiatives is that of approach. There are two types of approach towards developing ethical standards: one that measures performance against a specified set of criteria and indicators, and a second that assesses achievement by adopting procedures that will lead to improved performance. The procedural approach has been adopted for a number of international labelling initiatives such as ISO 14000.

Performance and procedural approaches each have their respective strengths and weaknesses. Performance approaches such as Milieu Project Sierteelt have the advantage of setting clear standards that observers can understand, but are often criticised for being too static, and for encouraging performance to be measured on a pass-or-fail basis. Procedural approaches are more dynamic and many value progress towards a goal rather than simply its achievement. However, exclusively procedural schemes can allow organisations to set goals that are too low, and may be too complicated and reliant on a paper trail to be implemented by small producers.

An increasingly common solution to this is to use a performance-based approach as the starting point but then to introduce elements of a procedural approach to encourage continual improvement (e.g. SA 8000). There is a growing consensus that the object of any standard or code of practice is to encourage improvement in all companies, not to fail certain companies for not being as good as others. A 'fail' is only the sanction of last resort. The logic behind this is to encourage as many companies as possible to adopt ethical standards, and to avoid a situation where 'good' companies adopt standards but are then criticised for not being 'perfect', while 'bad' companies carry on regardless because they refuse to participate in labelling initiatives.[12]

#### *Power relations – whose ethics?*

Just as power relations affect enterprise initiatives, similar forces apply to labelling initiatives. Some labelling initiatives offer more opportunity for stakeholder representation than initiatives generated by a single company, but this is not always the case. In part this is because labelling initiatives are voluntary, and some have argued that to become more rigorous stronger legislation is required (Welford 1995). Some labelling initiatives such as IFOAM have already informed legislation, but the move from voluntary principles to legal requirement is not without its problems, as the recent experience of trying to set legal standards for organic agriculture in the US has shown. Suspicion of possible WTO rulings on non-tariff barriers have also hindered the process of pushing for legislation on social and environmental standards (Blowfield *et al.* 1998).

### Future agendas

The challenges facing ethical trade should not detract from the general mood of optimism that these approaches will make a valuable contribution to North–South relations. This stems not least from the fact that ethical trade in its various forms is succeeding in creating new partnerships that bring business, civil society and government together, and that within these partnerships the challenges are being recognised and addressed. Companies, business associations, trades unions, NGOs and academics are working both separately and through a burgeoning number of networks and common forums. However, this enthusiasm is not always matched by resources. Few companies other than alternative trade organisations have full-time dedicated staff working in this field, and even high-profile initiatives such as the Ethical Trading Initiative may have insufficient human resources. Furthermore, there are many initiatives in developing countries that depend on grants or below-market rate loans from donor organisations, and whose long-term viability is open to question (Blowfield *et al.* 1998). Resolving problems with human resources and financial viability is fundamental to ethical trade's future.

Codes of practice have proved a useful entry point for companies wishing to engage in ethical trade, but there is considerable opportunity to make both company codes and those related to labelling initiatives more effective. Criteria and indicators need to be more complete and more relevant to reflect the norms, values and priorities of employees, homeworkers, small producers and outgrowers as well as to reflect diverse environmental conditions. It is becoming apparent that a code of practice is only as useful as its implementation, and there is a need for more investment in building models of best practice. This would include mechanisms for continual improvement of the code and systems of monitoring and verification, but would also need to ensure that codes do not overlook the needs of certain primary stakeholders (e.g. homeworkers, outgrowers) or result in certain types of producer being excluded from income-generating opportunities.

The limitations of codes of practice need to be recognised, particularly where they attempt to be universal or where a company has significant impacts away from the locus of production. Rather than have codes of practice as the focus of a company's ethical policy, the interests of primary stakeholders might be better served by having codes as an element of a more holistic ethical strategy. Such a strategy should be a corporate strategy, not, as is normally the case, with codes focused on a particular aspect of the company's operations (e.g. treatment of workers; environmental management practices within a factory). For instance, current codes of practice for horticulture apply to on-farm and in some cases pack-house conditions, and in some countries growers have been strongly criticised for over-use of chemical inputs on farms. But ongoing work with the Ghana pineapple industry shows that part of the reason for the degree of chemical use is to meet consumer demand for a golden fruit, and to ripen fruit at short notice for European wholesalers who seem unable to predict market demand.[13] In such circumstances, consumer education and more stable markets appear a prerequisite for ethical trade, but neither aspect is addressed by codes of practice.

These types of issue have been high on the agenda of the fairtrade movement for some time (EFTA 1998). Price stability, long-term buyer–producer relations and greater emphasis on partnership are also features of best practice among many multiple retailers (Blowfield *et al.* 1998). The standards developed by multiples as enterprise initiatives are now forming the basis for industry standards, where individual codes of practice are encouraged to meet a common base standard. An example of this is EUREP, a nascent European standard for agriculture, which builds on national standards such as the Assured Produce scheme in the UK that in turn came from individual multiple retailers' standards.

The development of common standards is likely to increase, although who bears the cost of implementing them remains a subject for debate. Some multiple retailers are likely to continue auditing their suppliers themselves, but some would like to use third-party auditors paid for by producers. It remains to be seen whether this will significantly increase the cost to consumers and whether consumers will on a routine basis be prepared to pay higher prices for ethically produced goods, particularly when ethical standards remain voluntary, and 'non-ethical' but cheaper goods are still available. If consumers prove unwilling, then it is possible that producers will improve their social and environmental performance but prefer to invest in labour-saving technology, thus generating less employment (Varley 1998). However, at present, information on the cost implications of implementing ethical trade are largely anecdotal, and there is a need for further research before conclusions can be drawn.

In conclusion, ethical trade is developing in a range of different sectors, but that many of the lessons learnt and challenges faced are not sector-specific. Intersectoral linkages for exchanging learning about successes and failures are weak, and as a result there is much duplication and 'reinventing of wheels'. Practitioners have various reasons for wishing to maintain the distinct identity of separate approaches to ethical trade, but in the medium term this is likely to confuse consumers and businesses interested in applying ethical trade principles. Nowhere is this more evident than in the segregation of social and environmental approaches to ethical trade, as most consumers assume that ethically produced goods have a positive human and ecological impact. Furthermore, without a conscious merging of environmental and social standards, ethically trade is unlikely to become the step towards the development of sustainable business and trade that early proponents hoped for. As Ranganathan (1998) has stated:

> 'At present the corporate sustainability measurement "picture" is broken into several pieces [and] there is little evidence to suggest that the separate pieces will be brought together in the near term and assembled into a coherent picture of business sustainability.'

Regardless of whether codes or other instruments are used to measure and manage ethical trade, there is a need for greater stakeholder participation. It is ironic, given that ethical trade is primarily described in terms of its benefits for developing country people, that the participatory development approaches common in international development have not been widely used in designing, implementing or evaluating ethical trade. Trades unions argue that codes of practice are only of use where labour organisations are outlawed or weak, and that the aims of ethical trade will best be achieved through freedom of association, strong labour laws and a workforce educated in techniques such as collective bargaining.[14] This would be one way of increasing participation, but not all companies are comfortable with trades unions' role in ethical trade (Greenhouse 1998). Moreover union representation only addresses social issues, and unions are not always able to represent the interests of all types of worker.

Alternative trade organisations such as GEPA and Oxfam Fairtrade argue strongly that participation is more important than any code of practice, and that for companies with limited resources labelling initiatives are an unnecessary cost and less important than developing a credible brand. Indeed, they can reduce the resources available to invest in building the capacity of producer organisations to organise and negotiate effectively. ATO brands are one example of the emic[15] approach also being explored by companies such as B&Q and the extractive industries. Fear of adverse media coverage has made companies wary of allowing outsiders to learn more about these approaches and how they foster participation, but the development of ethical trade is being hampered by an absence of independently assessed examples of best practice. The strengths and weaknesses of fairtrade labels and brands were assessed by Murray (1998).

Without greater participation, and particularly a shift in decision-making from the North to the South, ethical trade will at best be paternalistic and at worst harmful to those it is intended to benefit. Indicators of achievement will reflect the values and concerns of Northern companies and consumers, and the instruments employed will remain those that Northern stakeholders understand rather than those best able to identify the concerns and aspirations of people in the South. One of the successes of ethical trade is that it has opened the North's eyes to conditions in the South, but it will require courage in the South's capacity and ability to define its own ethical goals if Northern companies, civil society organisations and consumers are to relinquish the control over the ethical trade movement that they currently exert.

## Phytosanitary agreements, requirements and standards

### International Plant Protection Convention and International Phytosanitary Certificate

Before 1995, international agreement in plant quarantine was governed by the International Plant Protection Convention (IPPC) which came into effect in 1952 and was subsequently revised in 1979. The latest revision of the IPPC (1997) recognises the current principles governing plant quarantine in relation to international trade established under the World Trade Organization in 1995 (available at the FAO website http://fao.org/ag/agpp/pq/default.htm). Apart from the fact that signatory to the original IPPC was not universal, its implementation was not monitored or policed because the responsible international authority, the Food and Agricultural Organization (FAO) of the United Nations, had no funds to operate a Secretariat. Furthermore, the only international standard under IPPC (or any other convention) applicable to phytosanitary control was the International Phytosanitary Certificate. With the possible exception of the potato ring rot bacterium (*Clavibacter michiganensis* subsp. *sepodonicus*), no plant pathogens or invertebrate pests were universally accepted as organisms of quarantine significance. This contrasts with the situation in human and animal health with the recognition of such diseases as smallpox and yellow fever as requiring internationally agreed vaccination certificates and allowing official action to be taken in default.

The following definitions are used in this account, taken from the *Glossary of Phytosanitary Terms* of the FAO:

*Official*: established, authorised or performed by a National Plant Protection Organisation;

*Pest*: any species, strain or biotype of plant, animal or pathogenic agent injurious to plants or plant products (includes fungi, bacteria, viruses, phytoplasmas, nematodes, insects, molluscs and vertebrates);

*Plant quarantine*: all activities designed to prevent the introduction and spread of quarantine pests or to ensure their official control;

*Phytosanitary measure*: any legislation, regulation or official procedure having the purpose to prevent the introduction and/or spread of pests;

*Quarantine pest*: a pest of potential economic importance to the area endangered thereby and not yet present there, or present but not widely distributed and being officially controlled; and

*Regulated non-quarantine pest*: a non-quarantine pest whose presence in plants for planting affects the intended use of those plants with an economically unacceptable impact and which is therefore regulated within the territory of the importing contracting party.

The guiding principle for phytosanitary inspectors and decision-makers was the precautionary principle as applied in health and environmental protection (Griffin 2000; Smith 2000). On this basis, restrictive phytosanitary measures such as refusal of entry, treatment or detention under post-entry quarantine could be imposed on any consignment of plants or other plant-derived material as a precaution to avoid the possibility of introducing an exotic organism. Undoubtedly, this allowed the use of plant quarantine as a trade barrier, ostensibly as a phytosanitary precaution but actually as protectionist device. In addition, the International Phytosanitary Certificate was vaguely worded and allowed the exporting authorities to evade their responsibilities for ensuring the phytosanitary integrity of exported consignments. The IPPC as an international convention is theoretically binding on all signatory countries but signature is not obligatory, even on member nations of the United Nations, and adherence by signatory countries has rarely if ever been challenged let alone enforced. Prior to 1995, the phytosanitary authorities could and did take the view that the importer of the plant material under consideration had to satisfy them that the material was 'pest-free' before allowing entry, or that it could be detained indefinitely while exhaustive investigations could be carried out. From 1 January 1995, phytosanitary authorities (in the developed world at least) have been under an obligation to assist trade by only imposing restrictive measures when there was firm evidence of a pest or pest risk. The reason for this dramatic change is a set of rules and agreements adopted by the World Trade Organization.

## Plant quarantine and WTO (technical barriers to trade)

Since its inception, the General Agreement on Tariffs and Trade had been evolving ways of gradually liberalising international trade by abolishing restrictive trade barriers. The process of trade liberalisation made a quantum leap during the Uruguay Round of GATT negotiations. This culminated in the formation on 1 January 1995 of the World Trade Organization to replace GATT. Several new international agreements concluded under the Uruguay Round came into effect on this date; these have had a dramatic effect on international trade by eliminating technical barriers to trade and subjecting decision-making imports to objective, 'scientific' criteria. The Agreement on Technical Barriers to Trade (http://www.wto.org/wto/english/tratop_e/tbt_e/tbt_e.htm) underlies the Agreement on the Application of Sanitary and Phytosanitary Measures (henceforth referred to simply as SPS) that has had a revolutionary impact on international plant quarantine (http://www.wto.org/english/tratop_e/sps_e/sps_e.htm). In contrast to the IPPC, which until recently had no teeth, SPS is enforced and is automatically binding on all members of the WTO (virtually all nations of the world).

### *Agreement on Application of Sanitary and Phytosanitary Measures*

SPS covers both sanitary measures (human and animal health) and phytosanitary measures in international trade. In the text of SPS, three core principles or doctrines governing the ability of the relevant authorities to impose restrictive measures are laid down:

(1) Measures are to be 'scientifically justified' – this implies the use of pest risk analysis as the basis for decision-making;
(2) 'Transparency' – publication and dissemination of phytosanitary prohibitions, restrictions and requirements and subject to request, explanation of the rationale for these measures; and
(3) 'Harmonisation' – measures to be based wherever possible on international (and regional) standards, guidelines and recommendations developed within the framework of the IPPC.

Each of these principles is elaborated in more detail in *Principles of Plant Quarantine as Related to International Trade*, which was the first of the international standards for phytosanitary measures from the Secretariat of the IPPC.

Article 10 of SPS allows delayed compliance with requirements imposed on 'developing country Members' (of WTO) subject to specific request. Article 14 made special provision for 'least-developed country Members' and for 'other developing country Members'. These provisions have now lapsed and do not appear to have been extended. Article 9 provides for technical assistance for these Members although little assistance of this sort has yet to be provided by the WTO itself (Box 11.2).

## Box 11.2 Articles 9, 10 and 14 of the WTO Agreement on Application of Sanitary and Phytosanitary Measures

### Article 10 – Special and Differential Treatment

(1) In the preparation and application of sanitary or phytosanitary measures, Members shall take account of the special needs of developing country Members, and in particular of the least-developed country Members.
(2) Where the appropriate level of sanitary or phytosanitary protection allows scope for the phased introduction of new sanitary or phytosanitary measures, longer time-frames for compliance should be accorded on products of interest to developing country Members so as to maintain opportunities for their exports.
(3) With a view to ensuring that developing country Members are able to comply with the provisions of this Agreement, the Committee is enabled to grant to such countries, upon request, specified, time-limited exceptions in whole or in part from obligations under this Agreement, taking into account their financial, trade and development needs.
(4) Members should encourage and facilitate the active participation of developing country Members in the relevant international organizations.

### Article 9 – Technical Assistance

(1) Members agree to facilitate the provision of technical assistance to other Members, especially developing country Members, either bilaterally or through the appropriate international organizations. Such assistance may be, *inter alia*, in the areas of processing technologies, research and infrastructure, including in the establishment of national regulatory bodies, and may take the form of advice, credits, donations and grants, including for the purpose of seeking technical expertise, training and equipment to allow such countries to adjust to, and comply with, sanitary or phytosanitary measures necessary to achieve the appropriate level of sanitary or phytosanitary protection in their export markets.
(2) Where substantial investments are required in order for an exporting developing country Member to fulfil the sanitary or phytosanitary requirements of an importing Member, the latter shall consider providing such technical assistance as will permit the developing country Member to maintain and expand its market access opportunities for the product involved.

**Article 14 – Final Provisions**

The least-developed country Members may delay application of the provisions of this Agreement for a period of five years following the date of entry into force of the WTO Agreement with respect to their sanitary or phytosanitary measures affecting importation or imported products. Other developing country Members may delay application of the provisions of this Agreement, other than paragraph 8 of Article 5 and Article 7, for two years following the date of entry into force of the WTO Agreement with respect to their existing sanitary or phytosanitary measures affecting importation or imported products, where such application is prevented by a lack of technical expertise, technical infrastructure or resources.

### *Implementation of SPS in 1995 without phytosanitary standards*

The combination of sanitary and phytosanitary measures in SPS and abandonment of the precautionary principle for plant and animal/human health is highly revealing because of the difference in the status of standards in the two domains. As mentioned above there was no accepted list of what pests could be regarded internationally as of quarantine concern and indeed no definition of the concept of quarantine pest in relation to plant quarantine. These came later with FAO's International Standards for Phytosanitary Measures (ISPMs). It is true to say that the response of the international phytosanitary community, led by FAO, to SPS was reactive and that the post-WTO revision of the IPPC was also a reaction.

### *SPS covering biological and chemical factors*

SPS covers both chemical and biological factors potentially contaminating food of animal and plant origin and live animals and plants and it also covers the use of chemicals in the production of food. Pre-existing standards such as the CODEX Alimentarius on chemical residues in food are supported within SPS and are applicable grounds for decision-making. The abandonment of the precautionary principle was tested in the domain of human health with the conflict between the European Union (EU) and the United States and Canada over the issue of exports from the latter of beef from animals treated with growth promoting hormones. The EU lost this dispute in effect because the WTO Dispute Settlement Panel ruled that the precautionary principle could not be applied – there had to be scientific proof of harm to European citizens eating hormone-treated beef.

### *International Standards for Phytosanitary Measures since 1995*

As a reaction to SPS, FAO firstly sought funds from member countries to establish a Secretariat of the IPPC which would enable the convention to be implemented properly as an item of international law. Secondly, a major revision to the text of the IPPC was negotiated to bring it in line with the provisions of the SPS. Thirdly, an Interim Commission on Phytosanitary Measures (ICPM) was convened to produce the standards that had hitherto been lacking in phytosanitary affairs. The most important of these for the purposes of this chapter are:

ISPM 1: Principles of Plant Quarantine as Related to International Trade
ISPM 2: Guidelines for Pest Risk Analysis (original version)
ISPM 5: Glossary of Phytosanitary Terms
ISPM 7: Export Certification
ISPM 11: Pest Risk Analysis for Quarantine Pests

There are other ISPMs that will be considered later with the above. All the ISPMs may be downloaded from the IPPC website (http://www.fao.org/ag/agp/agpp/PQ/Default.htm).

The *Glossary of Phytosanitary Terms* contains formal and official definitions of such important concepts as quarantine pest. Relevant definitions are also given in each of the ISPMs.

### *European Union*

The other major event in the 1990s affecting international plant quarantine was the further integration of the economies of the members of the European Community by establishment of the single market for goods and services. This process was part of the process embodied in the Maastricht Treaty of 1992 establishing the European Union (EU). Internal borders were abolished from January 1993 (from June 1993 for plants and plant products); for plants and plant products this meant that the national laws of member countries had to follow the EC Plant Health Directive (2000/29/EC replacing 77/1993/

EEC). In principle, at least, plants and plant products were to move freely between member countries with no control or inspections. The borders of the EC were to be the points of entry from non-member countries ('third countries'). Most significantly, the quarantine pest lists of individual member countries were replaced by a single list of pests and restricted commodities operating on the imports from third countries. Henceforth, countries of northern Europe traditionally importing fruit and vegetables from tropical and subtropical countries had to take account of the quarantine concerns of the warmer parts of the EC. For practical reasons, the theory of 'no internal borders' had to be modified by a number of regulatory inventions, especially Plant Passports and Protected Zones.

### *Role of phytosanitary services in free-trade era*

The role of phytosanitary services has changed considerably since the advent of the WTO. Phytosanitary services have responsibility for ensuring the phytosanitary integrity of the territories for which they are responsible but they do this within the overall context of also providing assistance to international trade.

Another role has emerged for phytosanitary services in recent years, either by accident or design. Such services are responsible for approval of importation of all planting materials (germplasm) after consideration of the risks of importing pests potentially infesting the material. If the germplasm is a genetically modified plant (or transgenic plant), then in fact the risk of damage to the environment or to the health of the population also has to be taken into account. At the moment it seems that the phytosanitary services bear responsibility for the risk assessment and decision-making. The same considerations apply to biological control agents, whether genetically modified or not. Further consideration of this is beyond the scope of this section but raises the question of the different criteria of risk assessment that appear to apply to considerations of human health and environmental protection on the one hand and plant health on the other (Griffin 2000; Smith 2000).

## International phytosanitary requirements and standards

### *Pest risk analysis*

Pest risk analysis (PRA) is pivotal to plant quarantine in the free trade era, providing the means for decision-making based on 'scientific principles' and therefore consistent with SPS. However, the difficulties with PRA as with risk analysis in other fields (e.g. safety of nuclear power plants or aircraft) are: (a) devising a consistent and formal framework for the analysis, (b) acquiring sufficient and relevant data and (c) making a meaningful quantitative or probabilistic assessment of the risks. PRA is bound up with the official definition of quarantine pest – the economic damage perspective; in fact quarantine pests cannot be nominated without PRA. According to the basic theory, PRA is divided into three stages: stage 1, initiation; stage 2, pest risk analysis (formerly referred to as assessment); and stage 3, pest risk management (Royer & Kubicek 1993). The descriptions below are based on the original international standard (ISPM 2) and the current guidelines (ISPM 11):

- area endangered (endangered area): an area where ecological factors favour the establishment of a pest whose presence in the area will result in economically important loss;
- pathway: any means that allows the entry or spread of a pest;
- pest risk analysis: the process of evaluating biological or other scientific and economic evidence to determine whether a pest should be regulated and the strength of any phytosanitary measures to be taken against it;
- pest risk assessment: determination of whether a pest is a quarantine pest and evaluation of its introduction potential; and
- pest risk management: the decision-making process of reducing the risk of introduction of a quarantine pest.

### *Guidelines for pest risk analysis*

*Stage 1: Initiation* (Fig. 11.2): PRA is initiated in several ways according to circumstances or events. Pest-initiated PRA is begun when a new pest threat is identified or when the regulatory body wishes to formulate or revise its list of quarantine pests and consequent commodity-based restrictions. Pathway-initiated PRA is appropriate for identifying the risks associated with an existing or proposed trading link. Finally, a new category was adopted in of ISPM 11: PRA initiated by policy.

Considering pest-initiated PRA, stage 1 starts with the collection and evaluation of relevant information about the pest or pests being considered including their precise taxonomic status, the definition of the area under consideration and identification of those pests that are potential quarantine pests, i.e. they could be a threat to the PRA

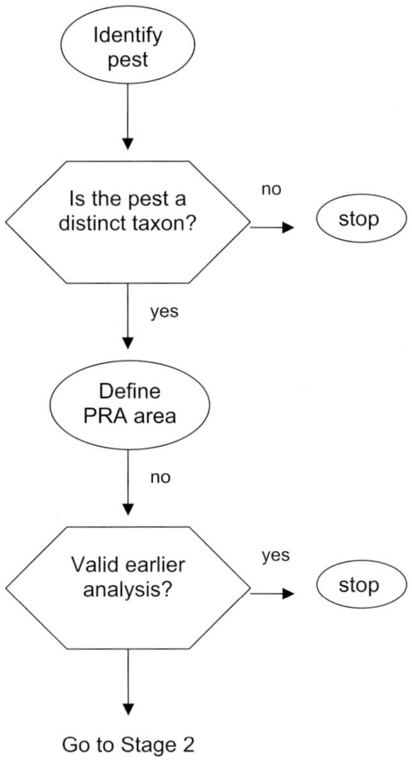


**Fig. 11.2** Stage 1 of pest risk analysis scheme. Source: EPPO (http://www.eppo.org).

area and have not been the subject of a previous PRA. These pests are then taken forward into stage 2.

*Stage 2: Pest risk assessment* (Fig. 11.3): Stage 2 is the most contentious stage of PRA, supposedly resulting in the quantitative assessment of the risk of a pest ultimately causing economic damage within the PRA area. The risk is assessed firstly by categorising the pest in terms of identity, geographical and regulatory factors and the potential for establishment and causing economic/environmental effects. If this preliminary assessment indicates that the pest has the potential to be a quarantine pest, the analysis should continue with detailed assessment of the risks:

- assessment of the probability of introduction and spread; and
- assessment of the potential economic consequences (including environmental impacts).

Provided the geographical and regulatory indicators have been satisfied, i.e. the pest is a quarantine pest, the overall risk is then carried forward into stage 3. Risk may be scored as a percentage probability or as a value on a scale (scales of 1–3, 1–5 and 1–9 have been variously adopted).

The difficulty with ISPM 2 and ISPM 11 is that little or no guidance is given as to how the available data are used to derive the quantitative risk assessment for each criterion and, even more significantly, it is left open how the overall assessment is made. Some current regional schemes specify summation of individual scores, others leave it to expert judgement.

*Stage 3: Pest risk management:* In stage 3 the various options available for managing or minimising the risk to an acceptable level of risk (ALR) are considered. IPSM 2 lists a number of options, which are not mutually exclusive:

- inclusion in a list of prohibited pests;
- phytosanitary inspection and certification prior to export;
- definition of requirements to be satisfied before export (e.g. treatment, origin from pest free area, growing season inspection, certification scheme);
- inspection at entry, inspection station or, if appropriate, at place of destination;
- detention in post-entry quarantine;
- post-entry measures (restrictions on use of commodities, control measures); and
- prohibition of entry of specific commodities from specific origins.

The efficacy and impact of the options in achieving reduction of the risk to an acceptable level should be considered. In general, the measures should have minimal impact. Unfortunately, the ALR may be highly contentious and a matter of conflict between the phytosanitary authorities of the exporting country and those of the importing country.

### *More detailed guidelines for PRA*

Some countries have officially adopted the ISPM 2 or ISPM 11, such as the Australian Quarantine and Inspection Service (http://www.aqis.gov.au). OIRSA is an example of a regional plant protection organisation that does the same (http://ns1.oirsa.org.sv/Castellano/DI02/Dtsv.htm); others apparently use their own scheme without publishing it (e.g. APHIS in the US). The European and Mediterranean Plant Protection Organization (EPPO) is one of the few regional plant protection organisations to publish their own scheme (on the web); this is of a similar level of detail to ISPM 11(http:

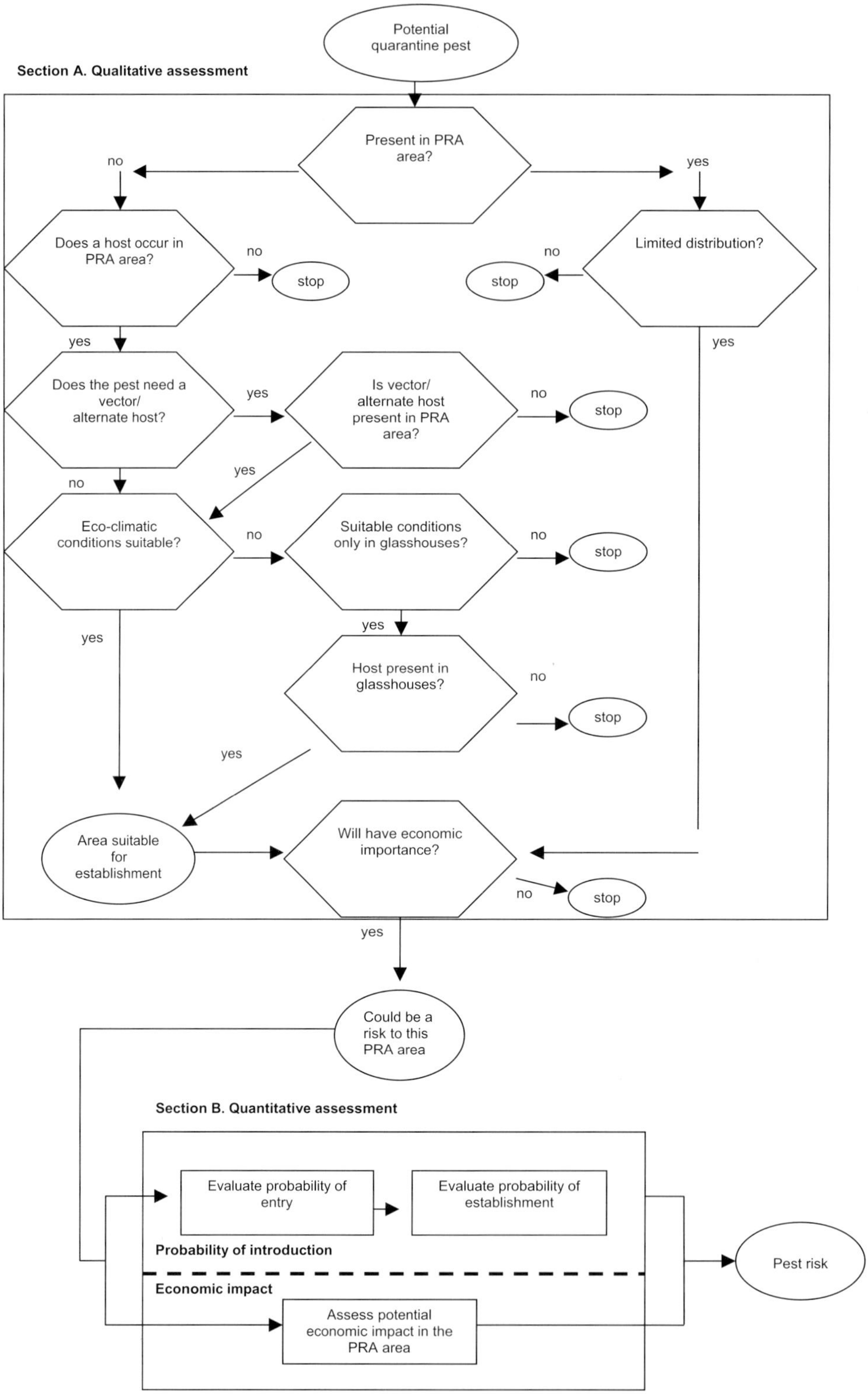


**Fig. 11.3** Stage 2 of pest risk analysis scheme. Source: EPPO (http://www.eppo.org).

//www.eppo.org/quarantine/pra/pra.html. However, the EPPO scheme only covers stages 1 and 2 of PRA (Box 11.3).

## Box 11.3 Extract from the pest risk analysis scheme of the European Plant Protection Organization

A key feature of the EPPO scheme is the addition of a preliminary evaluation to risk assessment before the full quantitative assessment is carried out. In this way, it may be possible to make a clear decision immediately one way or the other. In particular, if high scores are immediately given for certain important questions, or a high risk is immediately identified for one or more important pathways or important hosts, it may be superfluous to search for information for and reply to other questions, or to consider other pathways or hosts. Expert judgement will be used to decide this, and the preliminary assessment will thus provide guidance on the information which will be needed for the full assessment. On the other hand, it can quickly be obvious in section A that a particular pest does not have all the essential characteristics for being a quarantine pest, so that there is no purpose in continuing with a full assessment.

In a full assessment, there are a number of questions to be answered with a score of 1–9 that break up the risk criteria into more manageable elements. The scheme also indicates a number of questions that are more important than others. The assessor should first consider the quality and quantity of the information used to answer the questions, and give an overall judgement of how reliable the pest risk assessment can be considered. If other relevant information is available that has not been considered, this should be noted.

By the means of his choice, the assessor should attempt to make a separate estimate of the probability of introduction of the pest and its probable level of economic impact. As explained in the introduction, these estimates cannot, on the basis of the procedure used in the scheme, be expressed in absolute units. The numerical scores may be combined, weighted and averaged in appropriate ways that may enable the assessor who uses them consistently to make useful comparisons between pests, pathways and hosts. No particular mode of calculation is specifically recommended by EPPO. Certain questions have been identified as more important than others, and the assessor should take due account of this.

The assessor may then combine his estimates of probability of introduction and probable economic impact to formulate a single estimate of pest risk. This may usefully be compared with one or several reference levels of risk to decide whether the pest should be considered to be a quarantine pest, so that phytosanitary measures should be taken against it.

Finally, the scores given in answer to the different sections (particularly that on pathways) may be used again in pest risk management.

From EPPO (www.eppo.org)

The detailed assessment in the EPPO scheme takes the form of a questionnaire which has to be followed through from start to finish. In fact, this questionnaire represents a flowchart with several branch points, depending on the answers to particular questions. This is to avoid the user coming to a dead end when the risk in a particular category is low or non-existent. The end result is a complicated scheme in which it might be difficult to take an overall view.

An alternative approach adopted in Tanzania and other countries in East Africa is to use a spreadsheet to record the scores for each pest risk criterion and to calculate the overall risk by whatever means appropriate (Abdallah & Black 1998). Different pests are presented in different rows of the spreadsheet. A linked worksheet may be used to record the scores for component risk elements of each main criterion; the score for each criterion, again calculated by an appropriate formula, then appears automatically in the main spreadsheet. The advantage of this approach is that all assessments can be seen at one time and the effects of chains of scores (for risk elements or risk criteria) can also be seen.

Another approach is to use 'mind mapping' to plot the risk criteria and analyse them into risk elements at successively lower levels (Zhu *et al.* 2000) (Fig. 11.4). With this approach, a stage is reached where a question can be asked that can be answered by reference to a short statement or data element from a pest data sheet. Software is

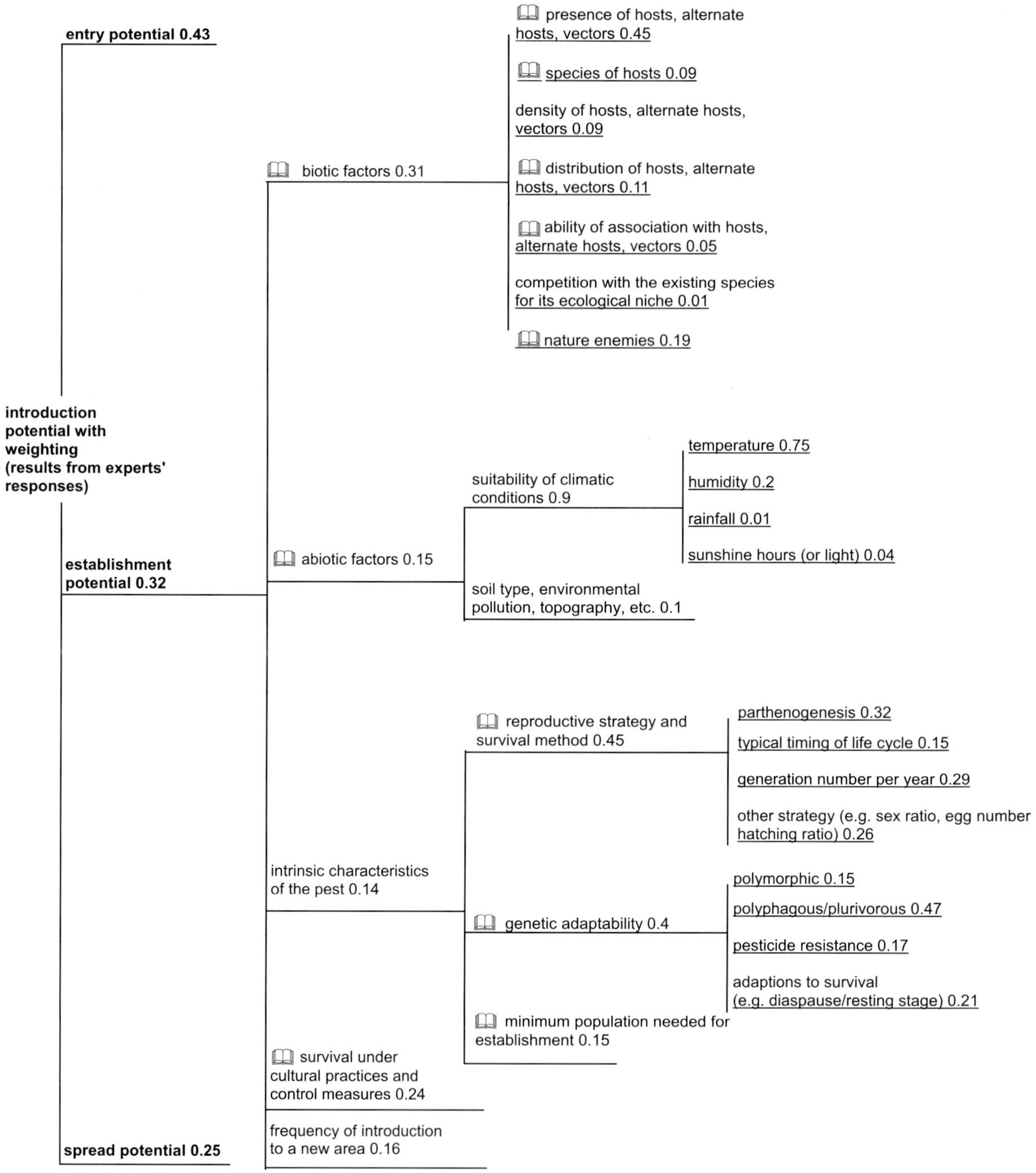


**Fig. 11.4** An example of a mindmap showing the breakdown of part of introduction potential into quantifiable risk elements.

available commercially for mind mapping that can reveal the question, answer and other information for each ultimate risk element. The advantage of this approach is that, within the constraints of the general guidelines, the PRA practitioner has freedom to construct a 'risk map' appropriate to the particular case being examined.

The EPPO scheme (and ISPM 11) recognise that need to express uncertainty:

> 'It is important to document the areas of uncertainty and the degree of uncertainty in the assessment, and to indicate where expert judgement has been used. This is necessary for transparency and may also be useful for identifying and prioritizing research needs.'
>
> *ISPM 11*

In the methods developed by Zhu *et al.* (2000), uncertainty is expressed by intervals or fuzzy numbers and fuzzy arithmetic used to calculate overall scores for risk.

The other concept now recognised as essential is expert judgement, particularly in the overall assessment of risk from the scores of individual criteria or elements. Hard-line exponents of the 'scientific basis' doctrine of the WTO would say that this is subjectivity by another name but exponents of expert judgement reply that PRA is not possible without the judgement of the PRA practitioner, based on his or her experience (Smith 2000). This relates to the debate on the role of the precautionary principle in plant quarantine and world trade generally (Black 2000; Griffin 2000). The legal basis for decisions based on the precautionary principle has also been examined. See, for example, the website recording viewpoints on the precautionary principle (http://www.cid.harvard.edu/cidbiotech/comments/) where David Hegwood notes:

> 'A recent report from the Organization of Economic Cooperation and Development (OECD) concluded: "There is general agreement within OECD countries that precaution has been and should remain an essential element of risk analysis". The report goes on to note, however, that OECD countries have differences over how to incorporate the precautionary principle into food safety decisions. It is safe to assume these differences extend to other types of risk assessments as well.'

Every step of the PRA process and every decision should be fully documented in order to satisfy the WTO requirement for transparency. The PRA should be backed by full data on pests, host plants, commodities and pathways.

## Permits

It is necessary to discuss the question of import permits (import licences) in relation to PRA. The requirement for import permit or import licence is regarded as a trade barrier by the European Union and other plant quarantine authorities, not surprisingly because the import permit system has often been abused to protect domestic fruit and vegetable production. However, the use of import permits should not be dismissed in less well-resourced countries because an application for an import permit gives the appropriate authority an opportunity to assess risks and determine measures before the items are ordered from abroad (in ideal situations). Their phytosanitary inspectors do not generally have the training or information and diagnostic resources to do PRA and make appropriate decisions at the point of entry on material of this sort. In a hot climate lack of facilities means that expensively obtained germplasm could deteriorate rapidly if detained for further investigation or inspection. France recognises this in its derogation allowing its overseas departments to require import permits (Arrĕté de 3 Decembre 1991, Ministère de l'Agriculture).

Regrettably, the requirement for an import permit in many southern countries has often been ignored or evaded (either through ignorance or arrogance) by influential bodies or individuals such as development projects, government departments or prestigious consultants with more power than the ministry of agriculture. The result is either an embarrassing confrontation at the point of entry or in departmental offices or, all too frequently, the importer bullies a hapless inspector into submission (Black & Sweetmore 1995). The risk analysis and consequent decisions surrounding an application for an import permit should be fully documented. A suggested standard form for an import permit and the application for it are given in Figs 11.5 and 11.6, respectively.

### *Export certification*

The requirement for a phytosanitary certificate to accompany most exports of plants and plant products, and the model of the phytosanitary certificate in the Annex to the International Plant Protection is the only pre-WTO international phytosanitary standard. Unfortunately, poor export certification and issue of phytosanitary certificates without due inspection is one of the biggest problems in international plant quarantine. Consignments arrive all too frequently, visibly infested with invertebrate quarantine pests and other pests. The certified health of germplasm (i.e. freedom from more cryptic plant pathogens) is often viewed with suspicion. Frequently offending countries are threatened with blacklisting or imports are subject to extra inspections.

Ministry of Agriculture
<ANYCOUNTRY>

PERMIT TO IMPORT PLANT MATERIAL
(Regulation X of the Plant Protection Regulations)

Name of importer.......................................Permit No...............................................

Address.................................................Date..............................................

---

Plant species (variety) ..............................................................................................

Type of plant material ..............................................................................................

Quantity ..............................................................................................

Origin ..............................................................................................

Destination/Location where plants will be grown ...................................................................
.........................................................................

Point of entry ..............................................................................................

Expected date of arrival ..............................................................................................

---

Permission is granted to import the above specified plant material on the condition that:

(a) the consignment is free from soil;

(b) the consignment is accompanied by a phytosanitary certificate;

(c) the consignment will be subject to inspection by a Plant Quarantine Officer, and released if found to conform with the phytosanitary regulations;

(d) specific condition of entry...........................................................................
.........................................................................
.........................................................................
.........................................................................
.........................................................................
.........................................................................
.........................................................................

---

Place of issue:..............................................................................................

Name of authorized officer:..............................................................................................

Signature:..............................................................................................

**Fig. 11.5** Suggested form for an import permit.

Ministry of Agriculture
<ANYCOUNTRY>
Plant Protection Act

APPLICATION FOR AN IMPORT PERMIT

Type of plant material (common name): ..........................................................................

To the Chief Agricultural Officer
Ministry of Agriculture
<anycountry>

I,
..................................................................................................................................

of.................................................................................................................................
hereby apply under Regulation X of the Plant Protection Regulations, for a permit to import into <anycountry> the following plant material:

Genus or species: ..............................................................................................

Variety: ..............................................................................................................

Quantity: ............................................................................................................

Name and address of sender: ..........................................................................

Locality where grown: .......................................................................................

Reason for importation: ....................................................................................

Nature of any treatment given, or proposed to be given, prior to despatch to <anycountry>

..................................................................................................................................

Proposed date of despatch: ..............................................................................

Expected date of arrival: ..................................................................................

Dated this...................day of................., 20

Signed:........................
(Importer)

**Fig. 11.6** Suggested form for application for an import permit.

This situation arises partly because of the poor enforcement of the IPPC and partly from the wording of the vital certifying clause in the model certificate (Fig. 11.7) but mainly because of lack of international guidelines for the conduct of inspections and documentation. The Export Certification System (ISPM 7) now provides the necessary standards for the process (http://www.fao.org/ag/agp/agpp/PQ/Default.htm).

MODEL PHYTOSANITARY CERTIFICATE
(to be typed or printed in block letters)

Plant Protection Organisation No……………………………………

of…………………………………………………..

To Plant Protection Organization(s)

of…………………………………………………………..…………………………………………...…

DESCRIPTION OF CONSIGNMENT

Name and address of exporter……………………………………………………………….…………

Name and address of consignee……………………………………………………….……………………..

Number and description of packages…………………………………………………….………………..

Distinguishing marks……………………………………………………………………………..………..…

Place of origin…………………………………………………………………………………..…………

Means of conveyance……………………………………………………………………..……………..

Declared point of entry…………………………………………………………………….…..……………

Name of produce and quantity declared…………………………………………………………...…..

Botanical name of plants……………………………………………………………………………………

---

This is to certify that the plants or plant products described above have been inspected by appropriate procedures and are considered to be free from quarantine pests, and practically free from other injurious pests; and that they are considered to conform with the current phytosanitary regulations of the importing country.

---

DESTINATION AND/OR DISINFECTION TREATMENT

Date……………………………Treatment………………………………………………………...………..

Chemical (active ingredient)………………… Duration of exposure……………………….………………..

Concentration……………………………….Additional information………………………………….……..

…………………………………………………………………………………………………………….….

Additional declaration……………………………………………………………………………………….

……………………………………………………………………………………………………………..…

Place of issue…………………………………………………..…..

(Stamp of organisation) Name of authorized officer……………………………….…….

Date……………………… ……………………………
(Signature)

---

No financial liability with respect to this certificate shall attach to………………………………(name of Plant Protection Organization)…………..or to any of its officers or representatives.

---

Optional clause

**Fig. 11.7** Model phytosanitary certificate (in accordance with current text of International Plant Protection Convention).

***Basic elements of ISPM 7***

The basic elements are:

- ascertaining the relevant phytosanitary requirements of the importing country (including import permits if required);
- verification of compliance – inspections/examinations in relation to these specific phytosanitary requirements; and
- issuing phytosanitary certificate.

Requirements for a certification system to fulfil these functions comprise the legal authority and management responsibility.

Inspections need to be carried out in relation to specified quarantine pests and regulated non-quarantine pests. A checklist of requirements is shown in Box 11.4. There are also draft guidelines for phytosanitary certificates on the same website that provide more detail about issuing the certificates.

The Model Phytosanitary Certificate itself was revised in the 1997 revision of the IPPC. The certifying clause now reads:

> 'This is to certify that the plants, plant products or other regulated articles described herein have been inspected and/or tested according to appropriate official procedures and are considered to be free from the quarantine pests specified by the importing contracting party and to conform with the current phytosanitary requirements of the importing contracting party, including those for regulated non-quarantine pests. They are deemed to be practically free from other pests.'*
>
> * Optional clause

The reference to regulated non-quarantine pests is discussed below. Additional declaration can be used as a certificate of origin to show that the consignment comes from an area free of a certain pest. This is especially important for germplasm.

### Box 11.4 Checklist of requirements for export certification.

- What are the requirements of the importing country?
- What type of inspections are required in relation to requirements?
- Are laboratory examinations needed?
- Are resources available to carry out the inspection programme – human (trained) and physical?
- What sampling system is required?
  - relate to tolerances in importing country;
  - if zero tolerance, base on sample size used in import inspections.

Is management of the process competent? (Management is all about quality assurance, including record keeping.)

Written instructions must be issued to inspection personnel covering:

- sampling and inspection procedures;
- authorisation to inspect, issue and sign certificates;
- maintaining security (authenticity of certificates, identity of consignments, secure delivery of certificates);
- record keeping and consignment traceability; and
- completing certificates, additional declarations, certified alterations.

***Import inspections***

It is clear that import inspections should be based on procedures to detect quarantine pests that might be associated with the particular commodities and that these quarantine pests have been determined by PRA. On the basis of risk analysis, a particular type of commodity may be prohibited if it originates in a particular country or it might be subject to treatment (e.g. fumigation), detailed inspections or diagnostic tests, post-entry quarantine or special documentary requirements (e.g. certificate of germplasm health). Inspections should involve three components:

*Documentary check*: phytosanitary certificate, shipping documentation, certificate of origin, import permit;

*Physical check*: whether the number of containers, boxes or other physical forms agree with the documents and that they have not been tampered with; and

*Actual inspection for pests*: the inspector should have information (and training) on specific pests for each type of commodity (including packaging and dunnage) and particular countries of origin as appropriate.

Close liaison with customs and with the port or airport authorities as appropriate is necessary. The quarantine inspectors should have a handbook which summarises quarantine pest lists and the restrictions in practical terms as well as proper standing orders for their duties.

### Quarantine vs quality; 'regulated non-quarantine pest'

A problem frequently encountered by quarantine inspectors at points of entry is plant products such as fruit, vegetables, flour and grain infested with insects or contaminated by moulds that are not quarantine pests but organisms already present in the country and in fact ubiquitous. This applies to common grain beetles such as *Tribolium* spp or common fungi such as *Aspergillus, Penicillium* or *Alternaria* spp. The produce may be so heavily contaminated as to be unfit for human (or animal) consumption and in the case of moulds (or green potatoes) may actually be toxic. In fact, the laws of some countries in the past have listed such organisms as quarantine organisms. In the WTO era such material cannot be excluded on quarantine grounds because the contaminating organisms in no way fit the definition of quarantine pest. However, there may be other laws under health or quality standards or consumer protection which provide a basis for refusal of entry and the quarantine inspectors may have standing orders that allow them to take action on non-quarantine laws.

So far this discussion of quality versus quarantine has concerned plant products for consumption. Another aspect of quality is very important in phytosanitary matters and that is the health (i.e. freedom from plant pathogens) of plant germplasm. Safeguarding the quality of imported and domestically produced germplasm is important for a country's farmers. Some viruses, fungi and bacteria potentially contaminating germplasm will be quarantine pests for the country concerned but others will not be. The revised text of the IPPC and the new model phytosanitary certificate contained therein recognises the category of regulated non-quarantine pest specifically to allow the phytosanitary authorities to control the importation of germplasm from a disease point of view.

### Pest-free areas

If a quarantine pest for a particular endangered area (normally a national territory or a state or province in a federation) is detected within that area, export markets for any produce susceptible to that crop (e.g. mango fruit in the case of a species of fruit fly known to infest mango) are likely to be suspended. The presence of a declared quarantine pest is acceptable grounds for exclusion of susceptible produce by an importing country and to be consistent this requirement would normally be enforced. This would apply even if there is no significant infestation of the crop in question and limited damage. However, in some circumstances it is in the economic interests of both the exporter and importer to maintain the trade in a situation where the pest has not yet entered the particular area of crop production. The way round this predicament is to establish and maintain a pest-free area in which the pest in question does not occur and from which it can be excluded by phytosanitary measures and/or natural means. For example, the quarantine pest may have been detected (and indeed become established) in the southern part of a country which produces, say, citrus for domestic consumption. The northern part of the country may be separated from the south by a natural barrier such as a mountain range or desert across which the pest cannot move and in which, say papaya, is grown for export. Provided that transport of potential hosts of the pest (e.g. citrus fruit) from the south to the north is monitored and controlled, the north could be maintained as a pest-free area. Then, with the agreement of both parties to the trade, exports of papaya could continue.

Evidence for the status of a pest-free area (and indeed the status of the pest in the original area of outbreak) is provided by surveillance, which is an official process of collecting and recording data on pest occurrence or absence. Surveillance may be done by surveys, monitoring or other means. Surveys of trees or fields of an annual crop are done to record infection by a pathogen or infestation by other relatively non-mobile pests. Monitoring usually involves trapping for more mobile pests using pheromones or other attractant chemicals. Further details of procedures for surveillance may be found in Guidelines for Surveillance (ISPM 6) at the IPPC website (http://www.fao.org/ag/agp/agpp/PQ/Default.htm).

## International co-operation

### *FAO and IPPC*

The Plant Protection Service of FAO (AGPP) was always supposed to have a central role in international plant quarantine and in the pre-WTO era it provided guidance and information in important areas and co-ordinated the work of regional plant protection in the developing world. However, the IPPC was not an effective item in international law because FAO had no resources to monitor compliance. With the advent of the WTO, the Secretariat of the IPPC has been established and the IPPC has been revised to take account of international trade rules. International standards are being developed in a consistent manner for the first time.

### *Regional plant protection organisations and harmonisation*

Regional plant protection organisations (RPPOs) developed as bodies to co-ordinate plant protection and quarantine activities and facilitate information exchange. All regions currently have an RPPO in various states of activity. One important concept developed by RPPOs were A1 and A2 pests, from which member countries could select pests of most relevance to their circumstances. Although RPPOs were recognised in the 1979 version of the IPPC, they had no formal role in the implementation of the IPPC, being restricted to information exchange (clauses 1 and 2 of Article IX of the revised text of the IPPC). The revised text now has a much greater role in international plant quarantine, as declared in clauses 3 and 4 of Article IX, particularly in development of international standards (including PRA) and harmonisation (Box 11.5).

A1 and A2 pest lists are lists of pests for regional purposes from which members of the region may select potential quarantine pests (subject to appropriate hosts and climatic conditions). A1 pests are not present in any of the countries of the region but their introduction into the region is potentially very damaging because of the presence of important host crops and suitable climate for growth and reproduction. A2 pests are already present in some of the countries of the region but their further spread is of concern to several of the remaining countries not yet affected.

Regional plant protection organisations are:

APPPC (Asia and Pacific Plant Protection Commission)
CAN (Communidad Andina de Naciones – Bolivia, Colombia, Ecuador, Peru and Venezuela) http://www.comunidadandina.org/
COSAVE (Comité regional de sanidad vegetal del Cono Sur – Argentina, Brazil, Chile, Paraguay, Uruguay) http://www.cosave.org.py/
CPPC (Caribbean Plant Protection Commission)
EPPO (European and Mediterranean Plant Protection Organization) http://www.eppo.org

## Box 11.5 Article IX of the Sanitary and Phytosanitary Measures (of the IPPC).

**Regional plant protection organizations**

(1) The contracting parties undertake to cooperate with one another in establishing regional plant protection organizations in appropriate areas.
(2) The regional plant protection organizations shall function as the coordinating bodies in the areas covered, shall participate in various activities to achieve the objectives of this Convention and, where appropriate, shall gather and disseminate information.
(3) The regional plant protection organizations shall cooperate with the Secretary in achieving the objectives of the Convention and, where appropriate, cooperate with the Secretary and the Commission in developing international standards.
(4) The Secretary will convene regular Technical Consultations of representatives of regional plant protection organizations to:
    (a) promote the development and use of relevant international standards for phytosanitary measures; and
    (b) encourage inter-regional cooperation in promoting harmonized phytosanitary measures for controlling pests and in preventing their spread and/or introduction.

EU (European Union – the EU functions in some ways like a national plant protection organisation, in others like an RPPO)
IAPSC (Inter-African Phytosanitary Council)
NAPPO (North American Plant Protection Organization) www.nappo.org
NEPPO (Near East Plant Protection Organization)
OIRSA (Organisation Regional Internacional Sanidad Agropecuaria – Central America) http://www.oirsa.org.sv/
PPPO (Pacific Plant Protection Organization)

The EPPO website has further contact details for RPPOs.

### *EPPO and the EU*

There is a unique relationship between the EU and EPPO. Since the demise of the Soviet Union, the EU is the only supra-national body that provides the laws of a group of independent countries (since 1993). The EU Office of Veterinary and Phytosanitary Inspections and Control (under Directorate General VI of the European Commission) is largely regulatory and relies heavily for technical advice on EPPO as well as on experts from the member countries. The EU commissions information resources from EPPO and adopts EPPO schemes for PRA and so on. However, it must be remembered that EPPO's membership is much larger than that of the EU, currently covering all of Western and Eastern Europe, many countries of the former Soviet Union, most of North Africa and some countries of the Middle East. In terms of quarantine pests, therefore, EPPO has much wider concerns than the EU.

### *Diplomatic baggage (Article 36 of Vienna Convention 1961)*

It is commonly assumed that diplomatic personnel are able to ignore the phytosanitary laws of their host country and import or export any sort of plant or plant materials. In fact, this is no more the case than that diplomats are free to import arms, explosives or other threats to national security, as stated quite clearly in Article 36 of the Vienna Convention 1961 – Diplomatic Relations (Box 11.6). Needless to say, it is a different matter as to whether a quarantine inspector or customs officer has the courage to confront a diplomat or diplomat family member suspected of contravening the national phytosanitary regulations or indeed carrying a bunch of cut flowers or a buttonhole of orchids that are important pathways for quarantine pests.

**Box 11.6 Article 36 of the Vienna Convention 1961 – Diplomatic Relations.**

(1) The receiving State shall, in accordance with such laws and regulations as it may adopt, permit entry of and grant exemption from all customs duties, taxes, and related charges other than charges for storage, cartage and similar services, on:
  (a) articles for the official use of the mission;
  (b) articles for the personal use of a diplomatic agent or members of his family forming part of his household, including articles intended for his establishment.

(2) The personal baggage of a diplomatic agent shall be exempt from inspection, unless there are serious grounds for presuming that it contains articles not covered by the exemptions mentioned in paragraph 1 of this Article, or articles the import or export of which is prohibited by the law or controlled by the quarantine regulations of the receiving State. Such inspection shall be conducted only in the presence of the diplomatic agent or of his authorized representative.

## Plant health in the European Union

### *Abolition of internal borders*

Whereas some countries have traditionally had customs union (e.g. members of the former Soviet Union), the abolition of internal borders between member states of the EU in 1993 when these countries had very diverse political, administrative–legal and commercial cultures was a dramatic step in phytosanitary matters. It meant that the national laws of the member countries had to be consistent with the EC Plant Health Directive (a Directive is of indirect effect, to be interpreted and implemented in a national law, in contrast with a EC Regulation which has direct effect.). An important consequence was that those countries which had points of entry from non-EU countries ('third countries') were responsible for the phytosanitary security of the whole of the EU. This included countries with international airports receiving flights from third countries (practically all EU members), countries with sea or river ports with international traffic, and countries with land border crossings with third countries (such as Finland with Russia and Germany with Poland and the Czech Republic).

### *EC Plant Health Directive 2000/29/EC*

The Directive has been subject to many amendments since the first version in 1993. The requirements of the EU are mostly found in the Annexes. These are lists of quarantine pests for the EU and the consequent list of requirements in relation to plants and plant products.

Details of the Plant Health Directive text may be found at http://europa.eu.int/(requires registration). However, this site is not very user-friendly and it is better to search for specific provisions rather than search for the consolidated version of the directive. The text of the laws of the European Union may also be accessed through the EPPO's electronic information service (http://eppo.org):

Annex 1 lists pests prohibited in the EU;
Annex II lists pests prohibited when present on certain plants;
Annex III lists prohibited plants and plant products;
Annex IV lists the special requirements that must be met before plants and plant products can be imported; and
Annex V lists the plants and plant products that must be officially inspected before they can be imported (i.e. these plants and plant products must be accompanied by a phytosanitary certificate).

In the United Kingdom, the EC Plant Health Directive is implemented by means of the Plant Health (Great Britain) Order 1993 and the Plant Health (Forestry) (Great Britain) Order 1993. The former covers all agricultural and horticultural material and forestry products; the latter covers wood and bark. *The Plant Health Guide for Importers* is a useful guide to the EC Plant Health Directive from the United Kingdom standpoint (MAFF 1998a). There is also a booklet, *A Grower's Guide to Passporting* from the same source (MAFF 1998b).

### *Phytosanitary certificates and plant passports*

As mentioned previously, a major change in northern Europe introduced by the EC Directive is the requirement for many fruits of tropical or subtropical origin to be accompanied by a Phytosanitary Certificate and thus be subject to inspection on entry into the EU. This applies to citrus (*Citrus*, *Fortunella*, *Poncirus* and their hybrids) and more exotic fruits like mango (*Mangifera*), Sharon fruit (*Diospyros*), guava (*Psidium*) and soursop, sweetsop, custard apple and so on (*Annona*).

The plant health authorities of the member states recognised that the removal of barriers to trade within the EU created a risk of pest introductions on material moving from an area where the pest was no threat to an area where the pest could be economically damaging (e.g. citrus pests on citrus fruit imported into northern Europe being transported to markets in southern Europe). The response was to introduce Plant Passports which control the movement of specified materials within the EU. The Plant Passport differs from the Phytosanitary Certificate in a number of respects (Table 11.7).

Third-country exporters do not need to concern themselves with Plant Passports. They may require a Phytosanitary Certificate for entry of their goods into the EU but, once admitted, further movement within the EU is of no concern to them.

### *Protected zones and buffer zones*

Further control of movement of plant and plant products within the EU is provided by protected zones and buffer zones. A protected zone is an area where an otherwise well-distributed and established pest is absent and whose plant production would be economically affected by the pest. Examples of protected zones are shown in Box 11.7.

A buffer zone is an area surrounding a protected zone, which is maintained free of the pest in question. Should the pest be detected in this area, attempts will be made to eradicate it. The pest in other areas is not subject to any official action.

### *Derogations*

Because some pests are regarded as very serious threats to the agriculture and forestry of the EU and because some pathways would provide a ready opportunity for entry, certain plants and plant materials are prohibited (Annex III of the EC Plant Health Directive). In some cases the prohibition is universal, for example, plants

**Table 11.7** Differences between a phytosanitary certificate and a plant passport.

| Phytosanitary certificate | Plant passport |
|---|---|
| Issued by national plant health authority (government) | Issued by private grower or merchant sending material to another member state |
| Requirement of International Plant Protection Convention | Issuing organisation licensed by national plant protection authority under EC Plant Health Directive and subject to regular quality control checks |
| Required by third-country exporters to EU | Not required by third-country exporters to EU |

### Box 11.7 Protected zone for *Erwinia amylovora* (cause of fireblight disease of pome fruit).

The following countries are protected zones: Austria, Spain, Finland, Ireland, Italy, Portugal, parts of France and Northern Ireland, Isle of Man and the Channel Islands.

It is forbidden to move host plants of *Erwinia amylovora* (various genera of Rosaceae) imported from third countries (other than those in which *E. amylovora* is EU-recognised not to occur) into protected zones for *E. amylovora.* It is similarly forbidden to move cut branches and pollen of these species into the protected zones.

Relevant plant material originating within the EU and moved to a protected zone must be accompanied by a Plant Passport affirming that the material meets the requirements of the protected zones.

of *Citrus, Fortunella, Poncirus*, etc., and plants of *Vitis* (grapevine) other than fruit (the definition of plants in the Directive is wide-ranging, including plants, fruit, cut flowers and foliage as well as plants in tissue culture). In other cases there is selective prohibition on plants from certain countries, for example, plants of *Phoenix* (date palm) other than fruit and seed from Algeria and Morocco. In isolated cases the economic need to import prohibited plants into a specific member state is held to outweigh the pest risk and the European Commission allows a derogation (exemption) from the regulations. Derogations apply to imports of seed potato from Canada into Greece, Italy, Portugal and Spain (a Community Decision as Document 399D0751). There is a general derogation allowing the overseas Departments of France to require import permits. The United Kingdom has been seeking a derogation against the prohibition of lime leaves (*Citrus hystrix*) which are an essential component of Thai cooking.

## Trade disputes involving phytosanitary matters

### *Disputes under SPS*

Article 11 of SPS implements the Dispute Settlement Understanding of WTO that in turns applies Articles XXII and XXIII of GATT 1994. A member country of WTO may therefore initiate a dispute if it feels that its rights to free trade have been undermined by a phytosanitary measure of another member country of WTO, for example, imposition of restrictions that allegedly are not based on scientific evidence and risk analysis. A dispute is heard by the Dispute Settlement Body (DSB), which first establishes a Panel to meet the parties, gather evidence, hear the case and report to the DSB. There is also an Appellate Body to hear appeals. Article 11 also allows member countries to settle disputes under the aegis of other international bodies.

### *Role of the secretariat of the IPPC*

Notwithstanding, at all stages parties are encouraged to consult with one another and try to settle out of court. In this regard, WTO has formally commissioned the Secretariat of the IPPC to conduct a pre-hearing process that will encourage the parties to settle the matter without resort to the costly and time-consuming DSB.

### *Future of the dispute settlement process*

Opponents of WTO point to lack of involvement of developing countries in dispute settlements, either originators or defendants, as being plainly unfair. In fact, there is evidence of growing third-world involvement in the dispute settlement process, while many lawyers argue that the WTO system is basically fair although the proceedings could be more transparent; the problem rather is lack of resources for the poorest countries to enable them to participate. The problem starts with the lack of permanent representation at WTO in Geneva by many of the poorest countries of WTO, together with a lack of specialised legal counsel and resources to employ expert witnesses. A partial solution to this problem is to allow third parties such as non-governmental organisations (NGOs) to gather and submit evidence on behalf of parties to the dispute (so-called *amicus* briefs). This appears to be adopted in principle following the recent shrimp-turtle case, although ironically in this case the NGO submitted an *amicus* brief on behalf of the United States, not the developing countries with which it was in dispute. A more controversial approach is to allow third parties to initiate disputes on behalf of poor member countries or their citizens. This *locus standi* principle has been accepted in some national courts.

The beef hormone case was the first major dispute under SPS. Essentially the United States prevailed over the EU on the grounds that applying the precautionary principle to agricultural products in order to safeguard the health of European citizens runs contrary to the principle of scientifically based risk assessment. In spite of the ruling of the DSB in this case, the role of the precautionary principle in health and environmental protection and more directly in phytosanitary matters is still being debated by technical experts and lawyers. There are other disputes on the horizon concerning transgenic crops that relate more directly to phytosanitary matters because many phytosanitary authorities have control over imports of plant germplasm.

## Information resources for plant quarantine decision-making

Much of the background information on phytosanitary agreements, standards and requirements and the more detailed technical information needed for decision-making in plant quarantine is available electronically, either from Internet web pages or from information products such as floppy disks or CD-ROMs to run on the user's computer. This part of the chapter is a guide to these resources as well as to paper-based information.

### *Treaties, agreements and international organisations*

*World Trade Organisation*: WTO's site map is at http://www.wto.org/english/info_e/site2_e.htm. As well as the legal texts (the text of the SPS Agreement is also available through IPPC), there are various explanatory texts. For example, SPS is explained at http://www.wto.org/english/tratop_e/sps_e/sps_e.htm

*FAO and IPPC*: The home page of the IPPC has already been referred to several times in this text (http://www.fao.org/ag/agp/agpp/PQ/Default.htm). Another useful starting point is FAO's Plant Protection Service (http://www.fao.org/ag/agp/agpp/) from which one can navigate to IPPC. These websites provide the text of the IPPC, SPS and the ISPMs.

*Regional plant protection organisations and PRA schemes*: The RPPOs and their websites were listed earlier. EPPO deserves special mention because this organisation is unique to date in publishing full details of its PRA scheme (http://www.eppo.org/QUARANTINE/quarantine.html#pestrisk). Other RPPOs make reference to their PRA scheme but do not publish such full details.

### *Information resources for PRA and decision-making*

References are provided to information resources that can be used in PRA and plant quarantine decision-making, that is, data on pest taxonomy, distribution, economic effects and associated crop data.

*FAO*: The FAO in the past produced a number of information resources that are now no longer available. The journal *FAO Plant Protection Bulletin* was the only official means of notifying outbreaks of plant pests and diseases, albeit two or three years after the event, but this journal was last issued in 1994. The latest product was the *Global Plant and Pest Information System* (GPPIS), an Internet-based database of plant quarantine-related information driven from a CD-ROM. This was launched in 1998. However, FAO now distances itself from such products because of fears of potential legal liability associated with providing information that may be used for PRA. The GPPIS concept has been taken up in the ECOPORT system of the University of Florida (http://www.ecoport.org/) with some information on pests, including a range of colour photographs of the organism and the symptoms it causes. However, sensitivity about plant quarantine is probably the reason why such essential information as distribution is lacking.

*CAB International*: CAB International (formerly Commonwealth Agricultural Bureaux) is a major source of information for plant quarantine. Many readers will be familiar with the series of *CAB International Abstracts* that are now largely accessed online or on CD-ROM by subscribers. CABPEST-CD was a pioneer in information processing for crop protection. The flagship of CAB International in this field is the *Crop Protection Compendium* (CPC), a multi-media CD-ROM product that provides comprehensive information of relevance to plant quarantine. Annual updates are available with further details available from CAB International (http://pest.cabweb.org/cpc/cpchp.htm). From the 2001 edition onwards, the CPC will be significantly enhanced with respect to plant quarantine (including pest risk analysis).

CAB International also produces several publications in conjunction with EPPO (see below) including the book *Quarantine Pests for Europe* and distribution maps of pests and diseases. All CAB International information products are produced commercially although there are special prices for *bona fide* public sector institutions in developing countries, and some development agencies may sponsor copies of the CPC and other products.

*EPPO*: Although centred on the information needs of the EPPO region, EPPO is the only provider of information of global significance essentially free of charge.

*Quarantine Pests for Europe* is based on pest data sheets that are fundamental to EPPO's information service. The data sheets are updated regularly and are available on request through the electronic information service but also form the basis for the EPPO PQR database and the EPPO Reporting Service.

A brief description of these products is given below. Further information and some downloads are available from the EPPO website (http://www.eppo.org):

*EPPO PQR* database may be downloaded from the website while the twice-yearly updates are also available on diskette for an annual subscription of Ffr.250 (useful for laptops);

*EPPO Reporting Service* is published monthly and contains news of outbreaks, definitive distribution records and some technical information on diagnostic methods. *EPPO Reporting Service* may be obtained electronically through the email server;

Also available from the email server are plant health regulations of all EPPO members and some other countries, pest data sheets, EPPO plant quarantine standards, etc.;

*Bulletin OEPP/EPPO Bulletin* is a journal specialising in diagnostic and survey methods particular relevant to European plant protection and quarantine.

## Food safety and HACCP

Food safety is a global issue of increasing concern for governments, food producers, food processors and handlers, as well as consumers. Safe wholesome food supplies play a key role in ensuring the health of populations worldwide and are achieved by improving knowledge of the causal agents of food-borne illness. This provides information on how to control these agents and, ultimately, reduces the occurrence of sources of food hazards that can result in morbidity and mortality. The design and implementation of food safety management systems (FSMS), combining good practices (GP) and the Hazard Analysis and Critical Control Point (HACCP) concept, will ensure the provision of safe food. HACCP is also a legal requirement for all major trading blocks of the world.

Increasing globalisation of economic activity and expanding world trade in food, fresh and preserved, have made food safety a global issue. Today, food businesses are obtaining their raw materials from regions around the world. However, to trade with member countries of the WTO, producers have to meet specified food safety standards. Lack of adequate local regulations is often a hindrance for export. Hence, potential exporting countries or regions have endeavoured to develop regulatory frameworks. This has impacted on producers and processors of food in developing countries, as the technical ability of small and medium enterprises to meet new standards required for export is limited.

Food safety regulations affect consumer health, food habits and food productivity in all parts of the world. In this dynamic field it is important for all stakeholders to understand issues of food safety, hazards of food contamination, and strategies that can affect the viability of food enterprises and the ability of these enterprises to meet with the requirements of new standards.

The World Trade Organization governs international trade through a number of agreements negotiated by WTO member countries. The two main agreements associated with world trade are the Agreement on Technical Barriers to Trade (TBT) and the SPS.

### The role of CODEX Alimentarius

The CODEX Alimentarius standards, guidelines and recommendations are the international reference under the TBT and SPS agreements. CODEX has prepared and published over 230 food commodity standards and more than 40 hygiene and technical codes of practice. Providing that local food industries, with support from the public health sector, follow the recommendations of CODEX, the food that they produce and process is safe for consumption and acceptable for export. CODEX includes requirements to support all areas of food control, from the farm to the fork. Examples of these documents covering the area of food safety are the food hygiene basic texts, pesticides and horticulture produce and grains.

### Good practice

Good practices are an essential first step to ensure that all food is produced under safe and hygienic conditions. All stakeholders in the production and processing chain have a responsibility to ensure that food is safe and suitable for consumption. Good practices lay a firm foundation for the subsequent development of a food safety management system. They will start on the farm or on the fish farm as good agricultural or good aquacultural practices (GAP). Then, if the food moves into a processing plant, good manufacturing practice (GMP) will be followed. Last but not least, food will be kept under good storage practice (GSP) and transported under good transportation practice (GTP). Good practices lay the foundations for the establishment of a preventative food safety management system, by applying the HACCP concept to the food system.

### Hazard Analysis and Critical Control Point

The concept of hazard analysis critical control point (HACCP) analysis was first introduced during the mid-1960s when a fail-safe method with zero tolerance was required for the production of food for the US space programme.

HACCP has been successfully applied in the control of safety in low-acid canned foods in the US and many food companies in Europe and the US have adopted the approach. Increasingly, regulatory bodies have recognised the usefulness of this concept and it has been incorporated into legislative requirements by both the EU and the US Food and Drug Administration.

Until the introduction of HACCP, end product testing, using measurements of physical properties, microbiological testing, chemical analyses and organoleptic evaluations were used as a means of assessing food safety. A number of limitations to this approach were recognised:

- the problems associated with the design and implementation of appropriate sampling plans
- the time required to obtain results
- the cost

The HACCP approach to safety assurance moves away from testing of the final product and instead emphasises control of the commodity chain. Control is taken out of the laboratory and into the processing environment. HACCP provides a structured and critical approach to the control of identified hazards and involves:

- identification and description of the product and its intended use – assessment of hazards associated with all stages of product handling and processing;
- production and identification of Critical Control Points (CCPs) at which the identified hazards must be controlled;
- specification of the critical limits for the identified hazards;
- monitoring the CCPs supported by a corrective action that should be taken if the CCP is moving out of control;
- verification; and
- documentation and records.

A thorough understanding of the whole process is required in order to identify the most appropriate means of monitoring CCPs. Tests where results are obtained quickly are paramount (for example, measurement of pH level instead of counting for bacteria which produce acid), while for other stages visual or sensory evaluation may be required, such as the quality rating of fresh fish. It is therefore important to assemble a team of specialists who can each look at the whole process from the point of view of their own area of expertise and who can contribute to the overall HACCP plan.

Food safety is the principal aim when applying the HACCP concept to a process. The technique was originally developed for control of microbiological hazards but it can just as easily be applied to other areas such as chemicals and foreign bodies. More recently, quality control points (QCPs; control points designed to safeguard quality criteria of a product), as well as safeguards to protect against economic fraud, have been included in integrated plans. It is recommended that all safety CCPs have been considered, identified and are correctly monitored before QCPs are incorporated into such systems.

There are a number of factors outside the handlers' and processors' control that can affect the quality of food. For example, vegetable production can involve large mechanised farms down to small family units. Hazardous practices, such as the use of raw sewage as fertiliser, or allowing animals carrying bacteria, viruses and parasites to forage among crops, together with inadequate processing and storage facilities, may increase risks associated with the products. All these factors must be considered when drawing up quality systems.

#### Risk analysis

Risk analysis can be broken down into three components:

*Risk assessment* – the scientific evaluation of known or potential adverse health effects resulting from human exposure to food-borne hazards. This process is divided into identification of hazards, hazard characterisation, exposure assessment and risk characterisation. (The definition of risk assessment includes quantitative risk assessment, which emphasises reliance on numerical expressions of risk and also the qualitative expressions of risk, as well as indication of the attendant uncertainties.)

*Risk management* – the process of weighing policy alternatives to accept, minimise or reduce assesses risks and to select and implement appropriate actions.

*Risk communication* – an interactive process of exchange of information and opinion on risk among risk assessors, risk managers and other interested parties.

Risk analysis is already carried out within the CODEX system by a number of different bodies that work together to prepare draft standards, guidelines and recommendations for consideration by the CODEX Alimentarius Commission (CAC). These bodies cover areas such as food additives and contaminants, pesticide residues, residues of veterinary drugs in foods, food hygiene, meat hygiene, food import and export inspection and certification systems.

### Food safety management systems (FSMS)

The design of appropriate food safety management systems effectively combines the practical aspects of good practices and HACCP within a structure of a quality management system such as ISO 9000–2000. Such a structure ensures that the FSMS can be easily checked and maintained and includes a requirement for traceability of product. Such systems can be audited by customers to be confident that the outputs of the FSMS comply with their specifications.

### EU food law

EU food law is divided into two sets of directives. The vertical directives cover proteinaceaous foods, meat and meat products, fish and fishery products and milk and milk products. The horizontal directives cover all other commodities and include areas such as compositional aspects and maximum residue levels for identified pesticides.

At the moment the requirement for a food safety management system (FSMS) based on the HACCP concept is included in both the vertical and horizontal directives. There are specific instructions in the vertical directives on how FSMS should be achieved for meat, fish, milk and their associated products: Directive 91/493/EEC – *Hygiene fishery products* (Article 6, controls based on HACCP), Directive 92/5/EEC – *Hygiene meat products* (Article 7, controls based on HACCP) and Directive 92/46/EEC – *Hygiene milk* (Article 14, controls based on HACCP). To ensure that all other commodities meet with safety requirements, the *Hygiene of Foodstuffs Directive* 93/43/EEC was published in 1993, where Article 3 makes HACCP obligatory for all foods.

The vertical directives and the horizontal general hygiene directives are currently being revised by the EU to standardise the approach in the design and implementation of FSMS, incorporating the HACCP concept. The EU Commission's objective is to introduce a 'farm to fork' approach to managing food safety, with all areas of production, including primary production, covered. Within the revised *General Hygiene Directive* the responsibility to produce safe food moves away from the enforcement agencies to the food producer and there will be an added requirement for traceability for all food ingredients.

The EU Commission published a Green Paper on General Principles of Food Law (Document COM (97) 176 final) in 1997. Included in this was a requirement that business operators should be responsible for food safety at all stages of the food system through the implementation of HACCP: an adoption of the 'farm to fork' approach which would be reinforced by official controls.

In 2000, the Commission presented a White Paper on Food Safety which includes an Action Plan on Food Safety. The requirements of the white paper will lead to better coherence and transparency of food legislation for all member countries, and the role of stakeholders at all stages in the food chain is better defined. The Commission wants to keep EU food law dynamic, so a series of activities, such as risk assessments, will be launched for this area. It is planned to include the outputs of such activities in future legislation.

#### *Transition of EU food law into food law in England and Wales*

The *General Hygiene of Food Directive* (93/43/EEC) has been combined with the Water in Food Production provisions of the *Drinking Water Directive* (80/778/EEC) in British Food Law and published *as The Food Safety (General Food Hygiene) Regulations 1995*. These regulations require that adequate food safety procedures must be identified, implemented, maintained and reviewed on the basis of HACCP principles, which are listed in the directive. The regulations are of a general nature and include detailed hygiene requirements covering premises and practices.

In the past, departments of the British government that had responsibility for the enforcement of these laws were the Ministry of Agriculture, Food and Fisheries, the Department of Health (DoH) and the Scottish, Welsh and Northern Ireland Offices. Responsibilities for different areas of legislation were overseen by the department concerned. For example, the DoH has responsibility for the safety of food with regard to the consumer; it dealt with health aspects of the environment and food. However, the Food Standards Agency was formed in 2000, and the Department for Environment, Food and Rural Affairs in 2001 (to replace MAFF), as the British government and consumers felt that government departments overseeing food safety issues in the UK were not sufficiently independent. Food law is enforced at local levels by environmental health officers (EHOs).

The network of local authority co-ordination bodies (LACOTS) have responsibility to monitor food and trading standards at local government level. The Port Health Authority monitors food imported into the UK, whilst the Meat Hygiene Service deals with the safety and quality aspects of meat and meat products produced locally, as well as those imported into the country.

To assist the British food industry implement the requirements of national and European legislative requirements national trade associations have prepared codes of practice to act as layman's guidelines for the various food sectors, e.g. catering sector and fresh fruit consortium.

Major supermarket chains within the UK have always been a driving force behind the implementation of safety and quality management systems, so that they can demonstrate 'due diligence' in the production of safe and wholesome food, a requirement of the *Food Safety Act 1990*. The supermarket systems are a combination of the requirements for good manufacturing practice, HACCP and a quality management system based on the requirements of the ISO 9000 series. The supermarket chains require that all suppliers conform to these requirements and have their own documented systems. The suppliers are required to carry out self-audits to demonstrate that their own system is working. The supermarket technical officers or a third-party assessor will also carry out third-party audits to demonstrate that the supplier has a safety and quality management system in place, i.e. they are doing what they say they are doing.

National bodies recognised as champions in the education of food producers and inspectors are the Certified Institute of Environmental Health (CIEH) and The Royal Institute of Public Health (RIPH). Both bodies have a series of accredited courses that cover the areas of basic, intermediate and advanced hygiene and food safety, as well as HACCP. Following a request from the British food industry the RIPH developed a standard for the training of the principles of HACCP, which was published in 1996. This standard is now becoming accepted worldwide. The RIPH is currently developing a standard for an advanced HACCP course. All courses have been designed to train British food workers to a level commensurate with their duties. It is a requirement of EU legislation that all the work force should have an understanding of how food poisoning occurs, basic food hygiene and personal hygiene in relation to food handling and temperature abuse. Supervisors and managers will also be required to have an understanding of hazard analysis and risk assessment, how to collate and manage a documented system and be able to verify that the system is in use.

A case study outlining these points is given below.

## Case study – appropriate food safety systems for small and medium enterprises in Central America and the Caribbean

The UK Natural Resources Institute has been working with research institutes and government organisations around the world to establish centres of excellence in the area of food safety and quality management (FS&QM). A six-year programme from 1993 to 1999, funded by the British Department for International Development, was established to strengthen the agro-processing sector of Central America and the Caribbean. A centre of excellence for food safety and quality management was established in Costa Rica to provide support to small- to medium-scale enterprises (SMEs) of the region. Scientists from the Centro Naçional de Ciencia y Tecnologia de Alimentos (CITA) attended a professional development programme based at the NRI facilities in the UK, and included visits to food producers, processors and major retailers. Teams undertook training in all aspects of safety and quality management and participated in courses on general hygiene, Hazard Analysis and Critical Control Point and quality management systems such as ISO 9000, 1994. Concomitant programmes in the establishment of laboratory routine to comply with good laboratory practice required in ISO 17025 were also run.

In Costa Rica, meat and meat products are distributed to a range of consumers through a variety of outlets including local markets, supermarkets and export markets. The latter expect and demand quality attributes that will result in the rejection of product, and hence economic losses to the livestock and meat industries, if the specified quality attributes are not met. Increasingly, as the quality consciousness of the consumer improves, the internal markets are also very likely to impose standards of quality that will mean rejection of meat produce that fails to meet the minimum standards. Opportunities for adding value will also be lost if suppliers of meat and meat products do not adopt a more quality-conscious attitude.

A prototype project was established with the national meat product sector. Twenty volunteer enterprises were invited to participate in a workshop that established a baseline in understanding of good practices which needed to be in place before hazard analysis can be introduced. Participating industries received regular support from the CITA FS&QM team at all stages of the programme, which involved the following key components:

- a preliminary project workshop to ensure the active participation of the key members of the meat industry;
- training in good practices, HACCP and the BS EN ISO 9000, 1994 series for key members of the meat industry;
- identification and surveillance of collaborating meat processors to evaluate the efficacy of the new quality systems; and
- an end-of-project workshop to discuss and disseminate the project outputs.

### The Hazard Analysis and Critical Control Point system

The HACCP concept is a systematic approach to hazard identification, assessment and control. However, it is costly and inappropriate to attempt to transfer methods directly from developed countries to developing country meat industries and, to date, such attempts have met with little success. However, since HACCP is an approach and not a prescriptive system, the concept can be used to develop tailor-made systems applicable to the production and marketing of meat products in developing countries. Such systems should incorporate control systems which combine both good practices and HACCP. Good practice is a basic and subjective approach which addresses environmental conditions and the control of working procedures. However, when combined with the systematic approach used in the HACCP concept, its application results in a significant improvement in quality and a reduction in related foodborne illness. Training in HACCP and GP was provided both in Costa Rica and in the UK, and included first-hand experience of how the British food industry follows and applies the legislative requirements of the EU and the US.

### Collaborating small- to medium-scale meat processors

Baseline surveys of seven representative SMEs were performed in 1993, using a statistically based diagnostic approach developed by the CITA FS&QM group. The handling and manufacturing protocols followed by each processor were evaluated using raw material condition, the process, the product, hygiene of the plant and operations, equipment condition and maintenance and quality management system.

### Appropriate safety and quality management systems

After the baseline survey of the SMEs had been completed, flow diagrams describing the process utilised by each manufacturer were prepared and the critical control points identified. A simplified process flow diagram for salami, a popular meat product manufactured in Costa Rica, is shown in Table 11.8. Those constraints which were contributing to the non-enforcement of GP (for example, cost and lack of understanding) were also identified. An appropriate safety and quality system was then designed and commissioned for each SME, ensuring that the system ensured the safety of the product.

**Table 11.8** A simplified process flow showing potential hazards for the production of salami.

| Process | Potential hazards | Control measures |
|---|---|---|
| *Raw materials* | | |
| Minced beef (0–5°C) | Delays | Avoid delays<br>Chill (0–5°C) |
| Chopped pork and fat (–20°C) | Delays | Avoid delays<br>Maintain in deep freeze (–20°C) |
| Dry raw materials | Incorrectly stored | Store all dry materials in a dry material store |
| *Processing* | | |
| Weigh ingredients | Incorrect weights<br>Out-of-date ingredients | Calibrate scales<br>Adequate training<br>Operate stock rotation |
| Prepare emulsions | Delays<br>Faulty machinery | Avoid delays<br>Maintenance of equipment |
| Stuff into casings | Poor filling<br>Inadequate hygiene | Increased supervision<br>Adequate training |
| Cook (internal temperature 78°C for 3 hours, rh 100%) | Incorrect times and temperatures | Check and monitor process<br>Calibrated recording |
| Cool (15 minutes to 40°C) | Contaminated water | Chlorinated chilling water |
| Chill to 5°C | Delays<br>Cross-contamination | Avoid delays<br>Good hygienic practices |
| Store (0–5°C) | Incorrect storage | Temperature control<br>Maintain records |
| Distribute (3–5°C) | Poor temperature control during transportation | Specify the use of refrigerated trucks |
| Retail (3–5°C) | Inadequate temperature control<br>Poor hygiene practices | Maintenance of equipment<br>Adequate training |

**Follow-up surveillance**

After a four-year period, the surveillance of the collaborating meat processors was repeated in order to evaluate the efficacy of the new quality systems. The results of the surveys showed a clear improvement in the six protocols, namely condition of raw material, process, product quality, hygiene, equipment and quality (Table 11.8). The overall mean performance of the SMEs had increased as a result of an increase in the performance associated with each of the protocols.

**End-of-project workshop**

The experiences of the participating SMEs were reported and discussed at an end-of-project workshop which enabled the results and conclusions, including the benefits of the improved quality systems, to be circulated to those members of the meat industry, as well as other industries, who had not directly participated in the project. One of the participating enterprises, La Unica, had improved operations to an extent that they were able to establish an export business for countries of the region.

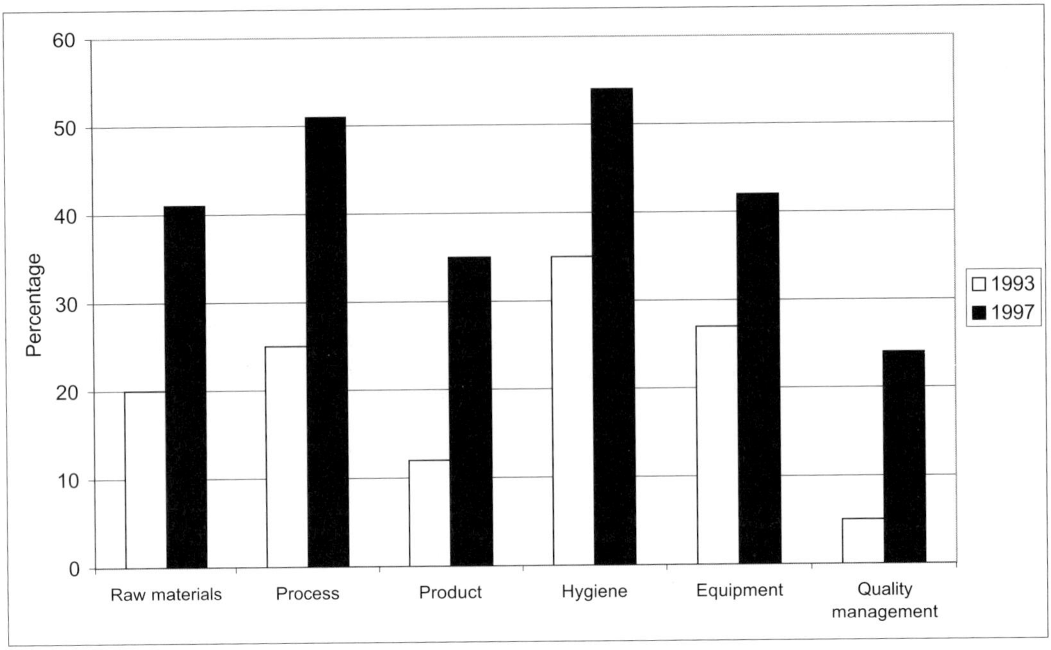


**Fig. 11.8** Performance indicators of six processors before and after introduction of appropriate safety and quality systems.

## Notes

1 For a detailed discussion of globalisation and liberalisation see DFID White Paper (2000) *Eliminating World Poverty: Making Globalisation Work for the Poor*, which also contains detailed references.
2 Detailed up-to-date information on the WTO can be found on its website http://www.wto.org.
3 One recent study estimated the gains from the UR at $171 bn, although 'the dollar gains are concentrated in developed countries, especially the United States, the European Union and Japan' (Harrison *et al.* 1997).
4 Approximately one-third of the $6.1 trillion total for world trade in goods and services in 1995 was trade within companies.
5 In the UR agreement there are 22 500 pages listing individual countries' commitments on specific categories of goods and services, which are legally binding. They include commitments to cut and 'bind' their customs duty rates on imports of goods. In some cases, tariffs are being cut to zero. There has been a significant increase in the number of 'bound' tariffs – duty rates that are committed in the WTO and are difficult to raise.
6 See, for example, Anon. (1998).
7 For a discussion of the various principles and procedures see Rowe (1965).
8 Some agreements attempted to achieve both.
9 For an excellent overview of these agreements see Rowe (1965).
10 See, for instance, Bo van Elzakker's presentation on bridge building at the 12th Scientific Conference of the International Federation of Organic Agriculture Movements, Argentina, November 1998, and work on incorporating social indicators into standards for responsible forest management by Prabhu *et al.* (1996).
11 Such criticisms have been made in interviews of the Kenya Flower Council and the MSC when the author was conducting consultancy work on the flower industry and fisheries.
12 This comment was made by Sharon Cohen, Vice President of Human Rights, Reebok, at a presentation to the American Bar Association, New York, September 1998.

13 Natural Resources and Ethical Trade Programme 'Ethical Trade and Horticulture project'; http://www.nri.org/NRET/nret.htm.

14 Dwight Justice in presentation to the Ethical Trading Initiative workshop, London, 2 December 1998; Mark Hankin, American Centre for International Labour Solidarity, pers. comm.

15 Emic is the direct approach to measuring similarities in a domain. The indirect approach is called etic. The words were coined by Pike in 1958. Emic is used in anthropology and sociology to measure items on a series of attributes, and then correlate each item's profile across these attributes.

## Further reading


Adams, R.P. & Adams, J.E. (Eds) (1998) *Conservation and Utilization of African Plants*. Missouri Botanical Garden Press, St Louis MI, USA.

Buckley, J. (Ed.) (1996) *Guide to World Commodity Markets*. Seventh Edition. London, UK.

CABI/EPPO (1997) *Quarantine Pests for Europe*. Second Edition. CAB International, Wallingford, UK.

Dillon, M. & Griffith, C. (1996) *How to HACCP*. Second Edition. M.D. Associates, Lincolnshire, UK.

Ebbles, D. (Ed.) (1993) *Plant Health and the European Single Market*. Monograph **54**. British Crop Protection Council, Farnham, UK.

FAO (1993) Guidelines for the application of the Hazard Analysis Critical Control Point (HACCP) System. CODEX Alimentarius Commission Vol. 1, Supplement 1, Section 7.5. United Nations Food and Agriculture Organization, Rome, Italy.

FAO (1995) *Application of risk analysis to food standards issues*. Report of the Joint FAO/WHO Expert Consultation. United Nations Food and Agriculture Organization, Rome, Italy.

FAO (1997) Food hygiene basic texts. Joint FAO/WHO Food Standards Programme. CODEX Alimentarius Commission. United Nations Food and Agriculture Organization, Rome, Italy.

FAO/IPGRI (undated) Guidelines for the safe movement of germplasm. http://www.fao.org.

ICMSF (1988) *Micro-organisms in Foods 4 – Application of the Hazard Analysis Critical Control Point (HACCP) System to Ensure Microbiological Safety and Quality.* International Commission on Microbiological Specifications for Foods. Blackwell Academic and Professional, London, UK.

Jackson, J.H. (1998) *The World Trade Organization: Constitution and Jurisprudence*. Royal Institute for International Affairs, London, UK.

Josling, T. (1998) *Agricultural Trade Policy: Completing the Reform*. Institute for International Economics, Washington DC, USA.

Krueger, A.O. (Ed.) (1997) *The WTO as an International Organization*. University of Chicago Press, Chicago IL, USA.

Maizels, A. (1992) *Commodities in Crisis*. Clarendon Press, Oxford, UK.

Martin, W. & Winters, L.A. (1996) *The Uruguay Round and the Developing Countries*. World Bank, Washington DC, USA.

Mortimore, S. & Wallace, C. (1994) *HACCP: A Practical Approach*. Chapman & Hall, London, UK.

Nicolaides, L., Coker, R., Wareing, P.W. & Nagler, M. (1998) *Food Safety in Developing Countries*. Natural Resources Institute, University of Greenwich, London, UK.

NRET (1998) *Ethical trade and export horticulture in sub-Saharan Africa: the development of tools for ethical trading of horticultural exports by resource poor groups*. Natural Resources and Ethical Trade Working Paper **1**. Natural Resources Institute, University of Greenwich, Chatham, UK.

WHO (1996) *Guidelines for strengthening a national food safety programme*. World Health Organization, Geneva, Switzerland.

WHO (1997) *Food safety and globalisation of trade in food – a challenge to the public health sector.* World Health Organization, Geneva, Switzerland.


## References


Abdallah, R. & Black, R. (1998) Safe movement of germplasm between countries. In: R.P. Adams & J.E. Adams (Eds) *Conservation and Utilisation of African Plants*. Missouri Botanical Garden Press, St Louis MO, USA.

Anon. (1998) A survey of world trade. *The Economist* 3 October.

Binswanger, H. & Lutz, E. (2000) Agricultural trade barriers, trade negotiations and the interests of developing countries. Paper prepared for UNCTAD X: High Level Round Table on Trade and Development: Directions for the Twenty-first century, 12 February 2000, Bangkok, Thailand.

Black, R. (2000) Can phytosanitary services in African countries meet the challenges of globalisation? In: *Proceedings of the Brighton Crop Protection Conference, Pests and Diseases*, pp. 1167–1174. British Crop Protection Council, Farnham, UK.

Black, R. & Sweetmore, A. (1995) *Plant Quarantine: A Primer for Development Workers*. Natural Resources Institute, University of Greenwich, Chatham, UK.

Blowfield, M.E. (1999) Ethical trade: a review of developments and issues. *Third World Quarterly*, **20 (4)**, 753–770.

Blowfield, M.E., Malins, A., Nelson, V., Maynard, W.B., Gallat, S. & Robinson, D. (1998) *Ethical Trade and Sustainable Rural Livelihoods*. Natural Resources Institute, University of Greenwich, Chatham, UK.

Boxall, R.A & Gough, M.C. (1989) *Investigation of technical problems associated with the distribution of food grain from temperate to tropical regions. Food aid wheat to Sudan*. Report No. 2. Natural Resources Institute, Chatham, UK.

Boxall, R.A & Gough, M.C. (1993) Moisture problems in food aid grain. *World Grain* January/February, 6–11.

Boxall, R.A. & Bisbrown, A.J.K. (1989) *Investigation of technical problems associated with the distribution of food grain from temperate to tropical regions. Food aid wheat to Bangladesh*. Report No. 1. Natural Resources Institute, Chatham, UK.

Dam, K.W. (1970) *The GATT: Law and International*

*Economic Organization.* University of Chicago Press, Chicago IL, USA.

DFID (2000) *Eliminating World Poverty: Making Globalisation Work for the Poor.* White Paper on Development. Department for International Development, London, UK.

EFTA (1998) From fair trade to responsible consumption: the power of the citizens of Europe to change the conditions for North/South relations. European Fair Trade Association, Maastricht, The Netherlands.

FAO (2000) *Commodity Market Review 1999–2000.* Commodity and Trade Division, United Nations Food and Agriculture Organization, Rome, Italy.

Forest Stewardship Council (1998) *Group Certification: FSC Guidelines for Certification Bodies.* Doc. 3.6.1. 31 July 1998. http://www.fscoax.org.

Ghazali, B.H. & Simula, M. (1996) Study on the Development and Formulation and Implementation of Certification Schemes for All Internationally Traded Timber and Timber Products. International Timber Trade Organization, Manila, Philippines.

Gilbert, C.L. (1995) International commodity agreements: an obituary notice. GEI Working Paper **2**, ESRC, London, UK.

Gough, M.C. (1996) Marine transport of some tropical food cargoes to cold climates: maintenance of quality. *NRI Bulletin,* **69**. Natural Resources Institute, University of Greenwich, Chatham, UK.

Gough, M.C. & Lippiatt, G.A. (1978) Moisture humidity equilibria of tropical stored produce. *Tropical Stored Products Information* **35**, 15–29.

Greenhouse, S. (1998) Plan to curtail sweatshops rejected by union. *New York Times*, 5 November.

Griffin, R.L. (2000) The precautionary approach and phytosanitary measures. In: *Proceedings of the Brighton Crop Protection Conference, Pests and Diseases*, pp. 1153–1158. British Crop Protection Council, Farnham, UK.

Hall, D.W. (1970) *Handling and storage of food grains in tropical and sub-tropical areas.* FAO Agricultural Development Paper No. 90.

Harrison, G., Rutherford, T. & Tarr, D. (1997) Trade policy options for Chile: a quantitative evaluation. Policy Research Paper **1783**, The World Bank, Washington DC, USA.

ILO (1998) *Overview of Global Developments and Office Activities Concerning Codes of Conduct, Social Labelling and Other Private Sector Initiatives Addressing Labour Issues.* International Labour Organization, Geneva, Switzerland.

Josling, T. (1998) *Agricultural Trade Policy: Completing the Reform.* Institute for International Economics, Washington DC, USA.

Lake, R. (1998) Fair trade and ethical trade: distinct but complementary. Discussion note for ETI communications task group. Traidcraft, Gateshead, UK.

MAFF (1998a) *The Plant Health Guide for Importers* (PB 1881). Ministry of Agriculture, Fisheries and Food, London, UK.

MAFF (1998b) *A Grower's Guide to Plant Passporting.* (PB2160). Ministry of Agriculture, Fisheries and Food, London, UK.

Maizels, A. (2000) Economic dependence on commodities. Paper prepared for UNCTAD X: High level Round Table on Trade and Development: Directions for the Twenty-first Century, 12 February 2000, Bangkok, Thailand.

Mintel (1997) *Organic and Ethical Foods.* Market Survey, Mintel Market Intelligence, London, UK.

Mossel, D.A.A. (1988) Mould spoilage of cereals during transportation by sea from Latin America to Europe – mechanisms, impact and management. *International Journal of Food Microbiology*, **7(3)**, 205–209.

Murphy, D.F. (1997) *Responsible international trade: key issues and best practice.* New Academy of Business Briefing Paper **5** New Academy of Business, London, UK.

Murray, R. (1998) Ethical marks and ethical brands. Unpublished Report. TWIN Trading, London, UK.

Prabhu, R., Colfer, J.P. & Venkateswarlu, P. (1996) *Testing criteria and indicators for the sustainable management of forests: phase 1 final report.* Centre for International Forest Research, Bogor, Indonesia.

Ranganathan, J. (1998) *Sustainability Rulers: Measuring Corporate, Environmental and Social Performance.* World Resources Institute, Washington DC, USA.

Rowe, J.W.F. (1965) *Primary Commodities in International Trade.* Cambridge University Press, Cambridge, UK.

Royer, M.H. & Kubicek, Q. (1993) Pest risk assessment theory. In: D. Ebbles (Ed.) *Plant Health and the European Single Market.* Monograph **54**. British Crop Protection Council, Farnham, UK.

Sewell, T. (1999) *Grain – Carriage by Sea.* Third Edition. Lloyd's of London Press, London, UK.

Smith, I.M. (2000) New constraints to plant quarantine. In: *Proceedings of the Brighton Crop Protection Conference, Pests and Diseases*, pp. 1147–1152. British Crop Protection Council, Farnham, UK.

Varley, P. (Ed.) (1998) The sweatshop quandary: corporate responsibility on the global frontier. Investor Responsibility Research Centre, Washington DC, USA.

Welford, R. (1995) *Environmental Strategy and Sustainable Development: The Corporate Challenge for the 21st Century.* Routledge, London, UK.

WTO (1998) *Trading into the Future: An Interactive Guide.* World Trade Organization, Geneva, Switzerland.

Zadek, S., Pruzan, P. & Evans, R. (1997) *Building Corporate Accountability: Emerging Practices in Social and Ethical Accounting, Auditing and Reporting.* New Economics Foundation, London, UK.

Zhu, L., Holt, J. & Black, R. (2000) New approaches for pest risk analysis for plant quarantine. In: *Proceedings of the Brighton Crop Protection Conference, Pests and Diseases*, pp. 495–498. British Crop Protection Council, Farnham, UK.

# Glossary

**Abscission** The organised shedding of plant parts. It occurs in a zone where hydrolytic enzymes reduce cell adhesion.

**Absorb/absorption** The uptake of moisture (water) into a hygroscopic material (such as grain).

**Acervulus** (pl. **acervuli**) An erupting cushiony mass of hyphae with conidiophores and conidia.

**Acquired qualities** Qualities that are additional to the normal complement of intrinsic qualities, the result of changes in intrinsic qualities or the consequence of certain intrinsic qualities being lost or not attained.

**Active ingredient** The biologically active component of a formulated pesticide.

**Adjunct** Non-essential additive.

**Adjuvant** A compound added to a pesticide to increase its efficacy.

**Aerobe** An organism needing free oxygen for growth.

**Aleurone layer** The outermost, highly differentiated proteinaceous layer(s) of the endosperm, which secretes digestive enzymes during germination.

**Ambient** Referring to the outside air in the atmosphere as opposed to that in the store.

**Amino acid** The 'building blocks' from which proteins are produced, containing an $NH_2$ group.

**Amylases ($\alpha$ and $\beta$)** Enzymes which break down starch.

**Anaerobe** An organism able to grow in the absence of oxygen.

**Anamorph** The asexual or imperfect state of a fungus.

**Angle of repose** 'Natural' angle between the sloping surface of, for example, a heap of grain and the flat surface on which it is placed.

**Antinutritive factors (ANFs)** In cereals – those components which prevent the uptake of certain nutrients, breakdown by the body's enzyme system or make eating a food unpleasant.

**Apothecium** (pl. **apothecia**) A cup-shaped spore-bearing structure.

**Appressorium** (pl. **appressoria**) A swelling on a fungal cell formed before penetration of the host.

**Arbitrage** The movement of supplies between one place and another until prices come into balance, taking into account transport and financing costs.

**Ascospore** A fungal spore produced in an ascus.

**Ascus** (pl. **asci**) A sac-like cell usually containing eight ascopores; characteristic of the Ascomycota.

**Bill of lading** Statement that specified goods have been loaded onto a ship or other transport vehicle.

**Blanching** Heating perishable produce in boiling water or steam to inactivate naturally-occurring enzymes, that would otherwise cause undesirable changes during subsequent storage.

**Bonemeal** Animal food made from finely crushed bone.

**Bract** A reduced or modified leaf subtending a flower or floral axis.

**Brewers' grains** The grains of barley which have undergone malting and hot washing, and are used in animal feed formulations.

**BSE** Bovine spongiform encephalopathy, widely known as 'mad cow disease', is a chronic, degenerative disease affecting the central nervous system of cattle. BSE is caused by a transmissible agent which has not been fully characterised.

**Budget** A detailed quantitative statement of a plan for a farm or marketing enterprise together with the costs involved and the returns expected.

**Buffer stock** Stocks – usually government controlled – built up by purchasing when supplies are ample and prices low, and sold when supplies become scarcer and prices are higher, with the goal of stabilising prices.

**Bulk density** The density (kg/m$^3$) of the grain mass, i.e. of the grains and air spaces between them.

**Bulk handling** The product is moved or held loose as opposed to in sacks or other packaging.

**Bulk sample** See **Composite sample**.

**C&F** Cost and Freight – a contract term by which the seller is responsible for charges incurred up to the port of final destination, but is responsible for loss and damage only until he/she delivers the commodity into the custody of the shipowner at the port of shipment or FOB point.

**Calibration** The comparison of a meter's reading to that which is known to be correct in order to determine the meter's accuracy. In the case of grain moisture meters, the comparison is normally to samples analysed using a laboratory oven.

**Calorie** A unit of heat. The energy content of food is expressed in terms of calories or kilojoules, where one calorie is the unit of energy that is required to raise the temperature of one gram of water by 1°C.

**Calyx** A collective term for all of the sepals, the outer leaves, of a flower.

**Capital** Goods which have not been used up including land, equipment, livestock and money.

**Capital investment** Money spent on equipment, stock or on improvement which has a life of more than one year and which adds to the productive capacity of an enterprise.

**Carbohydrate** Any member of a large class of chemical compounds that includes sugars, starches, cellulose, and related compounds. Carbohydrates are produced naturally by green plants from carbon dioxide and water.

**Cartel** A group of firms which collaborate on output pricing and other policies so as to restrict competition.

**Caryopsis** A dry, one-seeded fruit characteristic of the grasses. Cereal grains are caryopses.

**Cellulase** An enzyme that degrades cellulose.

**Cellulose** A straight chain polysaccharide composed of glucose molecules. Cellulose forms the chief constituent of plant cell walls and is the world's most abundant organic compound.

**Chilling** Reducing the temperature of a product to one favourable for conservation during transport, but above freezing point (usually around 0°C to 5°C).

**CIF** Cost, Insurance, Freight – a contract term (similar to C&F) except that the seller provides insurance on the commodity (but usually only the minimum necessary) on terms that are current in the trade up to the final destination.

**Coenocyte** A mass of protoplasm with many nuclei, i.e. the fungal cell is non-septate.

**Cold chain** An organised sequence of refrigerated assembly, storage, transport and sales display facilities.

**Commodity** The material being stored, i.e. grain, fresh or preserved produce.

**Common goods** When the consumption of benefits is subtractive (as with private goods) but the feasibility of exclusion is difficult, then the output is known as a common good.

**Composite sample** Sample consisting of mixture of primary samples (may be referred to as a bulk sample).

**Condensation** Formation of moisture on a surface, deposited by the air. Occurs when saturated air (i.e. 100% r.h.) is further cooled, resulting in the air having to release some of its moisture.

**Conidium** (pl. **conidia**) An asexual fungal spore.

**Consignment** Dispatch of produce without prior agreement on the price to a commission agent who will sell it and return the proceeds less his commission and expenses.

**Contaminant** (As distinct from foreign matter) – a substance that cannot be readily removed from grain.

**Contingency allowance** Allowance to cover unexpected events, e.g. a drought resulting in severe losses of cattle or crops.

**Co-operative** An enterprise established on the basis of equal participation by its members in its capitalisation, management, profits and losses.

**Covalent bond** A chemical bond by which two or more atoms are held together by the sharing of their electrons.

**Crude fat** A total measure of the fat content from all components of the material under study.

**Crude fibre** A total measure of the fibre content from all components of the material under study.

**Crude protein** A total measure of the protein content from all components of the material under study.

**Cultivar** A variety of a cultivated plant.

**Cyclical price variations** Recurring movements in average prices over a period of years, due to changes in demand or supply conditions and delayed production response in the same time period.

**Decortication** Removal of the outer covering of oilseeds.

**Dehulling** Removal of the outer covering of oilseeds.

**Dehumidify** Removal of moisture from the air.

**Density** General term used to indicate the mass of a commodity for a given volume (kg/m$^3$).

**Desorb/desorption** The release of moisture (water) from a hygroscopic material.

**Dew point** The temperature to which a sample of air is cooled to become saturated – further cooling will result in moisture being lost from the air.

**Disorder** A harmful deviation from normal growth not caused by a pathogen.

**Dockage** A term used primarily in the US and Canada to describe non-grain material that can and must be removed from grain using approved cleaning equipment, before grain can be assigned a grade.

**Dry basis (d.b.)** One method of calculating the moisture content of grain – ratio of weight of water to the dry weight of the grain (see **Wet basis**).

**Dry-bulb temperature** Actual temperature of the air.

**D-value** A measure of the heat resistance of a micro-organism. It is the time in minutes at a given temperature needed to destroy one log cycle (90%) of the target micro-organism.

**Endogenous** Processes, substances or organs that arise within a plant.

**Endosperm** Tissue containing reserve food substances in seeds or grains.

**Endo-β-gluconase** An enzyme involved in carbohydrate metabolism.

**Enthalpy** The heat content of a substance.

**Entrepreneur** A person or enterprise that is willing to initiate a marketing operation, bringing together the necessary resources and taking the risk of success or failure.

**Enzyme** A protein which acts as a biological catalyst.

**Equilibrium relative humidity (ERH)** For a sample of grain, at a fixed moisture content, the ERH would be reached when moisture in the air is in equilibrium with the grain, i.e. there is no net movement of moisture from or to the grain.

**Essential amino acid** An amino acid which has an important role to play within a living organism but which cannot be produced by that organism and must therefore be ingested in the diet.

**Eukaryote** An organism whose cells have a distinct nucleus.

**Exogenous** Resulting from causes external to an organism. In plants, developed on the outside of an organ or at or near the surface.

**Expeller** Machine used to crush oilseeds to yield oil.

**External economies** Advantages gained by an enterprise or country through easy access to supplies, markets, skilled labour, financial, research and service facilities or other productive resources that are already available. An enterprise benefits from these when it locates in an area where they are already available.

**Facultative parasite** A parasite able to live as a saprobe and to grown on artificial media.

**FAQ** Fair Average Quality.

**FAS** Free Alongside Ship – a contract term by which the seller is responsible for all the charges up to the point alongside ship.

**Fat** An organic compound composed of glycerol and fatty acids.

**Fatty acid** One of a series of acids with the general formula $C_nH_{2n}O_2$, of which some members are found in natural fats.

**Feasibility of exclusion** A term used to indicate whether it is easy or difficult to control access to a good or service.

**Feed** A formulation of various raw materials given to livestock to provide nutrients and minerals, but which may not necessarily contain all the essential ingredients in the correct proportions.

**File sample** Sample set aside in the laboratory to be analysed at a later date in case of dispute over the first analytical results.

**Fishmeal** Animal food made from finely crushed fish.

**Floret** One of the small flowers in a larger group of flowers, for example sunflower.

**FOB** Free on Board – a contract term by which the seller is responsible for all charges incurred and all loss and damage until the commodity is placed on board the vessel or other carrier.

**Foreign matter** Any matter mixed with grain that does not answer to the type description of the grain itself.

**Forma specialis** An infraspecific taxon characterised by its physiology rather than its morphology.

**Frass** Insect waste consisting of excrement, faeces, cast skins and sometimes also dead insect bodies.

**Fructose** A sugar commonly found in fruits.

**Full-fat soya** Refers to products which are made from whole soybeans which have been processed to rupture their internal oil-containing cells.

**Fumigant** A chemical that is toxic in its volatile form.

**Gall** A localised swelling caused by proliferation of plant tissues.

**Gametangium** (pl. **gametangia**) Fungal cell containing gametes, the site of meiosis.

**Germ** The part of the grain which develops into the young plant. It includes the scutellum and embryo.

**Glass transition temperature** The temperature boundary above which a material is flexible (the atoms can move) but below which the material is inflexible, being typically glassy and brittle (i.e. the atoms cannot move).

**Glucose** A sugar commonly found in fruit, animals and blood.

**Glume** In the grasses, one of a pair of bracts found at the base of the spikelet.

**Gluten** The water-insoluble protein complex extractable from cereal grains.

**Glycerol** (= **glycerine**) A sweet, viscous liquid compound; a molecule of which forms the 'backbone' of a fat molecule.

**Grading** Sorting products by size or other quality features in accordance with their appearance, physical condition or market value.

**Gram-negative** A bacterium that stains red, i.e. it takes up a red counterstain, but not Gram's stain.

**Gram-positive** A bacterium that stains dark blue, i.e. it takes up Gram's stain.

**Grist** Mixture of milled malt and cereals.

**Gross margin** The gross income from a business activity less the variable costs of that activity.

**Haemagglutinins** Blood-clotting agents present in raw soybeans.

**Halophile** A salt-loving organism that tolerates saline conditions.

**Haustorium** (pl. **haustoria**) A type of fungal hypha within a host cell that absorbs food.

**Hemicellulose** A group of compounds in plant cell walls that form part of the matrix in which cellulose is embedded.

**Hermetic** Airtight storage.

**Hydrolysis** Splitting of a compound by interaction with water.

**Hygrometer** An instrument used to measure the relative humidity of air.

**Hygroscopic** A material that is capable of absorbing and desorbing water (i.e. taking in and releasing moisture).

**Hymenium** (pl. **hymenia**) In the Basidiomycota, a spore-bearing layer of a fruiting body.

**Hypha** (pl. **hyphae**) A thread of fungal mycelium.

**Hysteresis** The slight difference in an isotherm depending on whether the grain moisture contents were being increased or decreased at the time that the ERHs were being recorded.

**Imperfect state** The state of a fungus in which asexual or no spores are produced; anamorph.

**Impurities** All elements which are considered undesirable in a sample or batch of grain.

**Inflation** An increase in the supply of money in relation to the goods and services available and, in consequence, a decline in its value.

**Informal sector** Small trading enterprises without much investment in fixed facilities, e.g. street hawkers.

**Infrastructure** Facilities and services, e.g. stores, roads, transport, banking and communications services.

**Intergranular** Relating to the air spaces between the grain, for example, intergranular air is the air between the grains.

**Intrinsic qualities** Qualities possessed by the whole, unblemished fresh commodity.

**Inventory** A record of stocks and supplies in hand.

**Inventory credit** Obtaining finance against stocks of a stored commodity held in a bonded warehouse.

**Isotherm** The relationship between the m.c. of the grain (in equilibrium with the surrounding air) and the r.h. of the air at a constant temperature – usually depicted by a graph.

**Kilning** Term used for drying of malted grains.

**Kunitz trypsin inhibitor** A protein which inhibits the action of the enzyme trypsin.

**Laboratory sample** Sample extracted from a bulk sample for analysis in the laboratory – see also **Submitted sample**.

**Lignin** A complex polyphenol that anchors and stiffens plant cell walls.

**Limiting amino acid** That (essential) amino acid which is present in the diet, but in insufficient quantity so as to limit the rate of protein production.

**Lipase** An enzyme that breaks down lipids.

**Lodicule** Small scales below the stamens in the florets of most grasses.

**Lysine** An amino acid.

**Macro-nutrient** A substance providing nourishment which is required in relatively large quantities.

**Maltase** An enzyme which hydrolyses the sugar maltose.

**Malted barley** Barley which has undergone the malting process, including drying.

**Malting barley** Barley which is germinated (malted) for use in brewing and distilling.

**Malting process** Germination of barley to convert the starch to sugars.

**Market liberalisation** Opening up of markets to the private sector.

**Marketing** Identification and meeting of a customer's need.

**Marketing board** A body set up by government action to guide marketing operations or to undertake them directly like a public marketing enterprise.

**Marketing channels** The sequence of enterprises and markets by which produce is moved from producer to consumer.

**Marketing organisation** A marketing enterprise or body undertaking control, support, promotion or other marketing-related functions; in a group sense the whole set of such enterprises and bodies involved in marketing and their relationship to each other.

**Marketing system** Marketing enterprises, channels, facilities and support services taken as a whole.

**Meat meal** Animal food produced from finely crushed meat.

**Mesophile** Organisms that grow at temperatures from 25°C to 40°C, i.e. between psychrophilic and thermophilic organisms.

**Methionine** An essential amino acid.

**Micro-nutrient** A substance providing nourishment which is required in relatively small quantities.

**Middlings** Grade of commodities, such as flour, of secondary quality or fineness.

**Mineral** Element or compound occurring naturally as a product of inorganic processes.

**Moisture** Alternative name for water. Usually relates to water contained on or in a commodity, whereas water is used to describe 'free' water, i.e. rainfall, puddles, etc.

**Moisture capacity** A measure of how much moisture an object could hold.

**Moisture content (m.c.)** A measure of how much moisture an object actually holds. Usually expressed as a percentage (see **Dry basis** and **Wet basis**).

**Moisture diffusion/diffusivity** A measure of how easily moisture can move through an object.

**Mycelium** (pl. **mycelia**) A mass of agglomerated fungal cells or hyphae.

**Non-essential amino acid** An amino acid which can be produced within the body and therefore does not need to be ingested as part of the diet.

**Obligate parasite** Organism only capable of living as a parasite.

**Offal** Parts cut off (as waste) in dressing carcass of animal killed for food.

**Oilcake (oilseed cake)** The product of expelling oilseeds to yield oil.

**Oilseed** Seed of a legume which is grown for its oil content.

**Oilseed meal** The product of expelling and solvent extraction of oilseeds, or the result of grinding oilseed cake after expelling.

**Oligopoly** Where only a few sellers of a certain product or service exist so that each will be affected substantially by a change in policy on the part of another.

**Oligopsony** Where there are only a few buyers so that each will be affected substantially by a change in policy on the part of another.

**Oligosaccharide** Compounds in which monosaccharides are joined by glycosidic linkages.

**Oosphere** In fungi, the female sex cell.

**Oospore** A resting fungal spore from a fertilised oosphere.

**Operating costs** Variable costs plus overhead costs.

**Operating profit** Gross income less operating costs.

**Opportunity costs** Income foregone by keeping a given set of resources out of the most profitable alternative use that would be practicable.

**Organelle** A distinct structure, usually bounded by a membrane, within a cell that has a specialised function.

**Organoleptic properties** Those properties of a foodstuff or food ingredient which are perceived in the mouth during consumption. The properties include taste, mouth-feel, consistency, texture, chewability, stickiness, etc.

**Osmophile** An organism able to grow in high solute concentrations.

**Ostiole** (adj. **ostiolate**) A pore through which spores are liberated from a pycnidium or other fungal fruiting body.

**Overhead (fixed) costs** Costs which do not vary greatly with changes in the volume of produce handled, or of services provided.

**Parameter** In economics, any factor which has an important effect on operating profit (e.g. yield, price, direct cost).

**Parastatal** An autonomous state-sponsored organisation.

**Pasteurisation** The heating of every particle of milk or milk product to a specific temperature for a specified period of time without allowing recontamination of that milk or milk product during the heat treatment process, followed by rapid cooling.

**Pectolysis** Enzymatic breakdown of pectin, which is a mixture of polysaccharides found in the cell walls of fruits.

**Peltier pump** A thermoelectric heat pump that is cold on one side and hot on the other. This results from the passage of a DC current through semiconductor junctions assembled between and bonded to two ceramic plates. Heat is 'pumped' from the cold side to the hot side.

**PER** The PER or protein efficiency ratio represents the gain in body weight of a growing animal fed a test protein divided by the grams of protein consumed.

**Perfect state** The state of a fungal life cycle during which spores are formed from nuclear fusion or parthenogenesis; teleomorph.

**Perithecium** (pl. **perithecia**) In the fungi, a globe- or flask-shaped container that contains asci.

**Peritrichous** Bacterial or fungal cells with hairs or flagellae all over the surface.

**Petiole** The stalk by which a leaf is attached to the rest of the plant.

**Pheromone** Chemical produced by one member of a species that has a physiological effect on members of the same species. Sex pheromones attract males to females or vice versa. Aggregation pheromones attract both sex.

**Phytic acid** Myo-inositol hexakisphosphate – a storage form of phosphorus in seeds which forms insoluble complexes with many minerals.

**Plasmodesmata** Bridges of cytoplasm that connect adjacent plant or animal cells.

**Polymer** A large macromolecule composed of small sub-units. For example, starch is a polymer of glucose, a lipid a polymer made up of fatty acid attached to a glycerol molecule.

**Polyphenol** The name used to describe many different forms of phenolic compounds. Lignins and tannins are common types of polyphenols. Polyphenol compounds are capable of binding to proteins and to some complex carbohydrates.

**Primary sample** Samples collected from primary units.

**Private good** When the feasibility of exclusion is easy and consumption is subtractive, an output is described as a private good. For example, if a tree can be protected from outsiders it is likely that most of its products can be classed as private goods.

**Productivity** The return in terms of marketing output or services from the resources applied by a marketing enterprise or service organisation. This is commonly defined in terms of return to capital invested or labour units employed.

**Prokaryote** An organism whose cells do not have a distinct nucleus.

**Protease** An enzyme which breaks down proteins.

**Protein** A class of organic compounds composed of amino acids, containing carbon, hydrogen, oxygen, nitrogen (and sometimes sulphur); forming an important part of all living organisms and the essential nitrogenous constituent of food for animals.

**Psychrometric chart** A chart illustrating the various properties of air (such as dry-bulb and wet-bulb temperatures, r.h.s, absolute humidities, etc.) at a given atmospheric pressure.

**Psychrophile** An organism that thrives in cool conditions. Optimum growth is typically at 15°C to 20°C but some growth may be possible near freezing.

**Psychrotroph** Micro-organism that has a minimum growth temperature below that commonly found in refrigerators (5°C) but an optimal growth temperature around room temperature.

**Pycnidium** (pl. **pycnidia**) A flask-shaped fungal organ which is lined with cells that produce conidia.

**Rachis** The axis that carries the flower or leaflets (in compound leaves).

**Relative humidity (r.h.)** Measure of the quantity of moisture held by air, expressed as a percentage of what the air could hold at that temperature. Defined as the ratio of the partial pressure to the saturation pressure.

**Respiration** The intake of oxygen and release of carbon dioxide by living objects to enable the oxidation of food materials to produce energy.

**Safe moisture content** That moisture content with a corresponding ERH of 70% or below, i.e. the lower limit for mould growth.

**Saprobe** An organism that feeds by absorbing dead organic matter.

**Saturated air** Air that is unable to pick up any more moisture, i.e. at 100% r.h.

**Saturated fat** Fat that has no double bonds within its structure.

**Sclerotium** A fungal resting body composed of a hard dark mass of hyphae, from which mycelia or fruiting bodies may emerge.

**Scutellum** Applied to grains – the cellular layer in the shape of a shield which separates the grain embryo from the endosperm.

**Seasonal price variations** Movements in average prices within one year generated by the bunching of output during a limited production season or seasons.

**Secondary market** Features sales by wholesalers to retailers and other wholesalers.

**Septum** (pl. **septa**) A fungal cell wall.

**Silo** A vertical store in which grain is held in bulk, as opposed to in sacks.

**Specific density** The density (kg/$m^3$) of the individual grains, i.e. not including any air spaces between the grains.

**Spikelet** The basic unit of the flowering branch in grasses, some reeds and sedges that consists of an axis, two bracts or glumes and one or more florets.

**Sporangiospore** A fungal spore contained in a sporangium.

**Sporangium** (pl. **sporangia**) A fungal organ that contains spores not derived from the containing structure.

**Stabilisation** Limiting fluctuations in supplies and prices, but not necessarily eliminating them entirely.

**Standard** An established specification of size and other quality factors.

**Standardisation** Arranging for products, or containers, weights and measures, etc., to conform to an established set of size, shape, or form specifications.

**Starch** A carbohydrate occurring widely in plants, especially cereals and potatoes, and which forms an important part of the diet of humans and livestock.

**Sterilisation** The destruction of all micro-organisms in an environment.

**Store** The building in which a commodity is to be stored.

**Stowage factor** The quantity of space taken up by 1 tonne of the commodity ($m^3$/kg); reciprocal of the bulk density.

**Stroma** (pl. **stromata**) A mass of fungal hyphae which may incorporate host tissue and from which spores may arise.

**Submitted sample** Sample extracted from a bulk sample for analysis in the laboratory; see also **Laboratory sample**.

**Sudanophilic** A fat that takes up Sudan red stain.

**Tannins** Tannins (or polyphenols) are known to bind iron and also inhibit some enzymes.

**Teleomorph** The sexual or perfect state of a fungus.

**Tender** An offer to buy or supply a designated quantity of a product in response to an announcement inviting such offers.

**Thallus** (pl. **thalli**) The vegetative fungal body.

**Thermal capacity** A measure of the ability of an object to store heat.

**Thermal conductivity** A measure of the ability of an object to conduct heat.

**Thermophile** A micro-organism whose optimum growth temperature is warmer than that seen with most micro-organisms. Such organisms can grow at temperatures above those that denature proteins in most organisms (typically 50°C), and sometimes up to the temperature of boiling water.

**Thiamin** Vitamin B1 – vital for normal development, growth, reproduction, healthy skin and hair, blood production and immune function.

**Toast** The process of heating soybeans to inactivate their antinutritive factors.

**Toll goods** When the feasibility of exclusion is relatively easy (as with a private good) but consumption of benefits is joint rather than subtractive, then the outcome is called a toll good. Parks and game reserves are examples of toll goods or services.

**Trace element** An element that is required in only tiny quantities.

**Transnational** An enterprise operating in several countries at the same time.

**Trypsin inhibitors** Proteins found in soybeans and other pulses, which prevent the digestion of soybean proteins in the intestine.

**Tryptophan** An essential amino acid which the body cannot produce and so has to be part of the diet.

**Unsaturated fat** A fat that has one or more double bonds within its structure.

**Urease** Enzyme which helps in the chemical breakdown of nitrogenous compounds.

**Variable (direct) costs** Costs which vary as the level of production varies.

**Viability** When related to a seed indicates the ability of the seed to germinate and develop into a new plant.

**Vigour** A measure of viability.

**Vitamin** One of a number of substances essential for growth and nutrition, occurring in certain foodstuffs or produced synthetically.

**Void ratio** Proportion of volume of air (i.e. the volume of the air spaces between the grains) to the total volume of a bulk of grain, expressed as a percentage.

**Warehousing** Provision of a guaranteed storage service in return for a fee. Under a bonded warehousing system a negotiable certificate of storage can be provided under government supervision. This is acceptable collateral for credit from a bank (see **Inventory credit**).

**Water activity** The water activity ($a_w$) is the ratio of the water vapour pressure of the food to the water vapour pressure of pure water under the same conditions. It is expressed as a fraction.

**Wet basis (w.b.)** One method of calculating the moisture content of grain – ratio of weight of water to the total weight of the grain (see **Dry basis**).

**Wet-bulb temperature** The temperature indicated on the wet bulb of a wet-and-dry thermometer (lower than the dry-bulb temperature since water evaporating from the wet bulb cools the thermometer slightly).

**Wheat feed** A mix of middlings and fine bran for feed use.

**Wholesaler** An enterprise which sells in relatively large quantities to retailers or other merchants rather than to consumers.

**Working sample** A subdivision of a submitted sample which may be too large for analysis.

**Xanthophyll** A yellow caratenoid pigment found in plants and egg yolk, usually associated with chlorophyll, and forming the yellow and brown colour of autumn leaves.

**Xerophile** An organism able to grow in very dry conditions.

**Z-value** The temperature required to change the D-value by one logarithm.

**Zoospore** A motile fungal sporangiospore.

**Zygote** A fungal cell in which two gametes of opposite sex have fused.

# Appendix 1
# Some Important Post-harvest Pests

**Table A1.1** Insect, mammal and miscellaneous pests associated with stored products. No list of this type can be exhaustive.

| Scientific name | Common name | Family | Products affected | Comment |
|---|---|---|---|---|
| *Acanthoscelides obtectus* (Say) | Bean bruchid | Bruchidae | Pulses | *Phaseolus* mostly |
| *Acanthoscelides zeteki* (Kingsolver) | Bruchid | | Pigeon pea | West Indies |
| *Acheta domesticus* (L.) | Cricket | Gryllidae | Polyphagous | Occasionally found indoors |
| *Achroia grisella* (F.) | Lesser wax moth, honey moth | Pyralidae | Pollen, etc. in beehives | Cosmopolitan, prefer abandoned or hives with weakened colonies |
| *Aglossa* spp | | Pyralidae | Oil palm | West Africa |
| *Aglossa caprealis* Hübner | The murky meal caterpillar | Pyralidae | Spilled grain, clover, hay, lucerne | On damp produce; minor importance |
| *Agriotes* spp | Wireworms | Elateridae | Potato | Eat holes in tubers |
| *Agrotis segetum* (Denis & Schiffermüller) | Cutworm | Noctuidae | Potato | Larvae feed on tubers |
| *Ahasverus advena* (Waltl) | Foreign grain beetle | Silvanidae | Small grains, cereal products, cocoa beans | Scavenger mostly |
| *Alphitobius diaperinus* (Panzer) | Lesser mealworm | Tenebrionidae | Grains, cereal products, animal feeds | Omnivorous; prefers damp conditions |
| *Alphitobius laevigatus* (F.) | Black fungus beetle | | Grains, cereal products, groundnuts | Omnivorous |
| *Alphitobius viator* (Mulsant & Goddart) | | | Ginger, chillies, bone | Mainly West Africa, uncommon |
| Alphitophagus bifasciatus (Say) | Two-banded fungus beetle | Tenebrionidae | Flour, copra, palm kernels and animal products | Uncommon scavenger |
| *Anastrepha fracterculus* (Weid) | South American fruit fly | Tephritidae | Citrus, mango and other fruits | |
| *Anastrepha ludens* (Loew) | Mexican fruit fly | | Grapefruit, oranges | |
| *Anastrepha obliqua* (Macquart) | West Indian fruit fly | | Mango | |
| *Anastrepha suspensa* (Loew) | Caribbean fruit fly | | Guava and other fruits | |
| *Anarsia lineatella* Zeller | Peach twig borer | Gelechiidae | Peaches | Larvae inside fruits |
| *Anobium punctatum* Degeer | Furniture beetle | Anobiidae | Wood | Larval tunnelling |
| *Anthicus floralis* (L.) | Ant beetle | Anthicidae | Polyphagous plant feeders | Cosmopolitan |
| *Anthicus formicarius* (Goeze) | Ant beetle | Anthicidae | Polyphagous plant feeders | Cosmopolitan |
| *Anthrenus* spp | Carpet beetles | Dermestidae | Wool, fur and feathers | Adults and larvae feed on keratin |
| *Anthrenus coloratus* Reitter | | | Wool | Southern Europe, North Africa, South Asia and US |
| *Anthrenu flavipes* (Le Conte) | Furniture carpet beetle | | Wool | Tropical and subtropical |
| *Anthrenus museorum* L. | Museum beetle | | Wool | Holarctic and Australasia |
| *Anthrenus verbasci* (L.) | Varied carpet beetle | | Wool | Cosmopolitan |
| *Aonidiella* spp | Armoured scales | Diaspididae | Citrus | On surface of fruits |
| *Apate* spp | False powder post beetle | Bostrichidae | Wood, cassava | Minor pest |
| *Apion* spp | Seed weevils | Apionidae | Pulses | Legume seeds |
| *Araecerus fasciculatus* (Degeer) | Coffee bean weevil, nutmeg weevil | Anthribidae | Cassava, nuts, coffee beans, cocoa beans, spices | Larvae feed on tubers; serious pest of coffee |
| *Archips argyrospilus* (Walker) | Fruit tree leaf roller | Tortricidae | Nuts | Larvae in kernels |
| *Arenipses sabella* (Hampson) | Greater date moth | Pyralidae | Dried fruits | Pupates in stored produce |
| *Argyresthia* spp | Fruit tortrix moth | Tortricidae | Apple | Larvae in fruits |

| | | | | |
|---|---|---|---|---|
| *Attagenus* spp | Carpet beetles | Dermestidae | Dried meat, skins and hides, wool, fur and feathers | Adults and larvae feed |
| *Attagenus cyphonoides* Reitter | | | Wool, skins, fur, grain residues | Cosmopolitan |
| *Attagenus fasciatus* Thunberg | | | Wool, skins, fur | In warm climates and heated buildings |
| *Attagenus gloriosae* (F.) | | | Groundnuts, maize | Uganda |
| *Attagenus pellio* (L.) | Two-spotted carpet beetle | | Wool, skins | Larval feeding |
| *Attagenus piceus* (Olivier) | | | Rotten beans | Kenya |
| *Attagenus smirnovi* Zhantiev | | | Grain residues | Russia, northern Europe in heated stores; Kenya and Ethiopia |
| *Attagenus unicolor* (Brahm) | Black carpet beetle | | Grain and cereal products, animal products | Larval feeding |
| *Aulacorthum solani* (Kaltenbach) | Potato aphid | Aphididae | Potato | Infest eyes in storage |
| *Bactrocera tryoni* (Froggatt) | Queensland fruit fly | Tephritidae | Tropical fruits | |
| *Bactrocera dorsalis* (Hendel) | Oriental fruit fly | | Apple, mango, papaya and others | |
| *Bactrocera jarvisi* (Tryon) | Jarvis's fruit fly | | Mango | |
| *Balanogastris kolae* (Desbrochers) | | Cuculionidae | Kola nuts | |
| *Blaps polychresta* Förster | Egyptian beetle | Tenebrionidae | Feed on faeces of rats, mice and rabbits | Presence is likely result of heavy rodent infestation |
| *Blaps mucronata* Latreille | Cellar beetle | Tenebrionidae | Polyphagous | Adults omnivorous but larvae show preference for damaged grain |
| *Blapstinus* spp | Darkling beetle | Tenebrionidae | Dried fruits | Adults and larvae |
| *Blatella germanica* (L.) | German cockroach | Blattellidae | Polyphagous | Ubiquitous scavengers |
| *Blatta orientalis* L. | Oriental cockroach | Blattidae | Polyphagous | Ubiquitous scavengers |
| *Bostrychoplites cornutus* (Olivier) | False powder post beetle | Bostrichidae | Wood, cassava, sweet potato | Minor pest, East Africa |
| *Brachycorynella asparagi* (Mordvilko) | Asparagus aphid | Aphididae | Asparagus spears | |
| *Brachypeplus gabonensis* (Murray) | | Nitidulidae | Maize | Nigeria and Cameroon |
| *Brachypeplus pilosellus* Murray | | | Maize cobs, cocoa beans and castor beans | Nigeria and Uganda |
| *Bruchidius* spp | Pea and bean bruchids | Bruchidae | Pulses | Minor pests in store |
| *Bruchus* spp | Pea and bean bruchids | Bruchidae | Pulses | Mainly in the field |
| *Calliphora* spp and others | Blowflies | Calliphoridae | Fresh meat, dried fish and fishmeal | Larvae eat flesh, adults feed on surface |
| *Callosobruchus analis* (F.) | Bruchid | Bruchidae | Green gram, cowpea, white soybean | Larval feeding; major pest in India and Indonesia |
| *Callosobruchus chinensis* L. | | | Chickpea, lentil, green gram, adzuki bean and cowpea | Larval feeding; major pest; origin in tropical Asia |
| *Callosobruchus maculatus* (F.) | | | Lentil, green gram, and cowpea | Larval feeding; major pest; origin in Africa |
| *Callosobruchus phaseoli* (Chevrolate) | | | Grams and dolichos bean | In Africa, South Asia and Central and Southern America |
| *Callosobruchus rhodesianus* (Pic) | | | Cowpea | Mainly southern Africa; very tolerant of cold temperatures |
| *Callosobruchus subinnotatus* (Pic) | | | Bambara groundnut | West Africa |
| *Callosobruchus theobromae* (L.) | | | Black soybeans | Mainly a field pest of pigeon pea in India or groundnut in Nigeria |
| *Carpophilus* spp | Dried-fruit beetles, sap beetles | Nitidulidae | Small grains, dried fruits, cocoa beans | On mouldy grain |

**Table A1.1** (*Continued.*)

| Scientific name | Common name | Family | Products affected | Comment |
|---|---|---|---|---|
| *Carpophilus binotatus* Murray | | | Maize cobs | Nigeria |
| *Carpophilus dimidiatus* (F.) | Corn sap beetle | | Maize, oilseeds, cocoa, nuts | Probably the commonest *Carpophilus* sp. |
| *Carpophilus freemani* Dobson | Freeman sap beetle | | Cereals, nuts | Uncommon; Americas, Africa, South Asia |
| *Carpophilus fumatus* (Boheman) | | | Maize and sesame | Mainly a field pest in Africa |
| *Carpophilus hemipterus* (L.) | Dried-fruit beetle | | Maize, dried fruits | On mouldy grain |
| *Carpophilus humeralis* (F.) | | | Maize, cassava, over-ripe tomatoes | |
| *Carpophilus ligneus* Murray | | | Dried fruit, cocoa, cereals, nuts, oilseeds | Cosmopolitan but uncommon |
| *Carpophilus maculatus* (Erichson) | | | Maize, nuts, oilseeds | Africa, Pacific, Caribbean, Asia |
| *Carpophilus mutilatus* Erichson | Confused sap beetle | | Dried fruits, cereals and oilseeds | Tropical and subtropical |
| *Carpophilus obsoletus* Erichson | | | Copra, oilseeds, maize | Relatively common |
| *Carpophilus pilosellus* Motschulsky (= *Carpophilus halli* Dobson) | | | Rice, oilseeds, nuts, maize, dried fruit | Widespread in tropics on mouldy produce |
| *Carpophilus zeaphilus* Dobson | | | Maize | In East Africa and Nigeria |
| *Caryedon serratus* Olivier | Groundnut beetle | Bruchidae | Legumes | Groundnuts mostly, serious in West Africa |
| *Caryedon sudanensis* Southgate | | | Senna pods | Northeast Africa |
| *Cathartosilvanus vulgaris* (Grouvelle) | Flat bark beetle | Silvanidae | Brazil nuts, cocoa beans | South America |
| *Cathartus quadricollis* (Guérin) | Square-necked grain beetle | Silvanidae | Maize, copra, cocoa beans | Widespread in tropics |
| *Catolethrus fallax* Boheman | | Curculionidae | Maize | Central America only |
| *Catolethrus longulus* Boheman | | | Maize | Central America only |
| *Catorama herbarium* (Gorham) | | Ptinidae | Books, maps, paper | Uganda |
| *Caulophilus oryzae* Gyllenhal | Broad-nosed grain weevil | Curculionidae | Maize, ginger | North and South America |
| *Ceratitis* spp | Fruit flies | Tephritidae | Citrus, peaches | Larvae inside fruit |
| *Cernuella* spp | Small banded snails | Helicellidae | Small grains | Contaminants on harvested grain |
| *Cernuella virgata* Da Costa | Small banded snail | Helicellidae | Pulses | Contaminants of peas for freezing |
| *Ceroplastes sinensis* Del Guercio | Pink waxy scale | Coccidae | Spices | Scales found on bay leaves |
| *Ceutorhynchus pleurostigma* Marsham | Turnip gall weevil | Curculionidae | Turnips and mangels | Larvae in root gall |
| *Chalcophora japonica* Gory | Pine jewel beetle | Buprestidae | Wood | Larvae in *Pinus* spp |
| *Chlorophorus annularis* F. | Bamboo longhorn | Cerambycidae | Bamboo | Larvae bore stem internodes |
| *Citripestis sagittiferella* (Moore) | Citrus fruit borer | Pyralidae | Citrus | Caterpillars in fruits; Malaysia and Indonesia |
| *Coccus hesperidum* L. | Soft brown scale | Coccidae | Spices | Scales found on bay leaves |
| *Coccotrypes dactyliperda* F. | Date stone borer | Scolytidae | Dried fruits | Larvae in unripe fruits, adults emerge in store |
| *Coelopalorus foveicollis* Blair | | Tenebrionidae | Wide range of commodities, especially those with high moisture contents | Minor pest in India, Orient, East Africa, Trinidad |
| *Colpterus* spp | | Nitidulidae | Maize, barley | US |

| | | | | |
|---|---|---|---|---|
| *Columba* spp | Pigeons | Columbidae | Small grains | Of sporadic importance |
| *Contarinia sorghicola* (Coquillett) | Sorghum midge | Cecidomyiidae | Small grains | Sorghum only |
| *Coptotermes* spp | Wet-wood termites | Termitidae | Small grains, wood, sacking | Foraging workers remove grains; in damp wood or hessian |
| *Corcyra cephalonica* (Stainton) | Rice moth | Pyralidae | Small grains, dried fruits, nuts, cocoa beans, chocolate, spices | Humid tropics; nut shell must be damaged to allow access to kernel |
| *Cossonus suturalis* (Boheman) | | Curculionidae | Sweet potato | Zanzibar |
| *Cryptoblabes gnidiella* (Milliere) | Christmas berry webworm, honeydew moth | Pyralidae | Citrus | In fruits from Mediterranean |
| *Cryptolestes* spp | Red grain beetles | Cucujidae | Maize, small grains, cereal products, cassava, cocoa beans | Secondary pests; on cassava chips |
| *Cryptolestes capensis* (Waltl) | | | Small grains, cereal products, oilseeds | North Africa and Europe, able to survive dry conditions |
| *Cryptolestes cornutus* Thomas & Zimmerman | | | Dried chilli peppers | Thailand |
| *Cryptolestes ferrugineus* (Stephens) | | | Small grains, cereal products, oilseeds | Cosmopolitan |
| *Cryptolestes klapperichi* Lefkovitch | | | Nutmegs, cassava, dried chillies | Southeast Asia |
| *Cryptolestes pusilloides* (Steel & Howe) | | | Small grains, cereal products, oilseeds | Australia, South America, southern and East Africa |
| *Cryptolestes pusillus* (Schönherr) | | | Small grains, cereal products, oilseeds | More common in humid tropics |
| *Cryptolestes turcicus* (Grouvelle) | | | Milled cereals | US |
| *Cryptolestes ugandae* Steel & Howe | | | Cassava, cereals, groundnuts | Tropical Africa |
| *Cryptophagus* spp | Fungus beetles, silky grain beetles | Cryptophagidae | Maize, small grains, nuts, dried fruit | On mouldy grain |
| *Cryptophagus subfumatus* Kraatz | Silky grain beetle | | | Walnut, dried fruit in Europe; grain in North America |
| *Cryptophilus integer* (Heer) | Lizard beetle | Languridae | Wide range of products | Europe, US, Africa; uncommon |
| *Cryptophlebia leucotreta* Meyrick | False codling moth | Tortricidae | Citrus | Caterpillar bores into fruits |
| *Cryptophlebia ombrodelta* (Lower) | Macadamia nut borer | | Nuts | Larvae in kernels |
| *Cryptotermes brevis* (Walker) | Dry-wood termites | Kalotermetidae | Wood | Uncommon |
| *Ctenopseustis obliquana* (Walker) | Brown-headed leafroller | Tortricidae | Fruit | |
| *Curculio nucum* (L.) and others | Hazelnut weevils | Curculionidae | Nuts | Larval emergence holes in shell |
| *Cybocephalus* sp. | | Nitidulidae | Cassava | Tanzania |
| *Cydia molesta* (Busck) | Oriental fruit moth | Tortricidae | Peaches | Larvae in fruits |
| *Cydia nigricana* (F.) | Pea moth | | Pulses | Larvae feed on peas |
| *Cydia pomonella* L. | Codling moth | | Apple | Caterpillars in fruits, may pupate in store |
| *Cydia prunivora* Walsh | Lesser appleworm | | Apple | Caterpillars in fruits, may pupate in store |
| *Cylas* spp | Sweet potato weevils | Apionidae | Sweet potato | Larvae and adults in tunnels |
| *Cynaeus angustus* (Le Conte) | Larger black flour beetle | Tenebrionidae | Cereals | Mexico and US |
| *Dacus* spp | Fruit flies | Tephritidae | Citrus, peaches | Larvae inside fruits |
| *Delia antiqua* (Meigen) | Onion fly | Anthomyiidae | Onion | Larvae inside bulbs |
| *Dermestes ater* Degeer | Black larder beetle | Dermestidae | Copra | Mostly larval damage |
| *Dermestes carnivorus* F. | | | Dried meat, fish, hides | Adults and larvae feed |
| *Dermestes frischii* Kugelann | | | Dried meat, fish, hides | Adults and larvae feed |

**Table A1.1** (*Continued.*)

| Scientific name | Common name | Family | Products affected | Comment |
|---|---|---|---|---|
| *Dermestes haemorrhoidalis* Kuester | | | Dried meat, fish, hides | Adults and larvae feed |
| *Dermestes lardarius* L. | Larder beetle | | Dried meat, fish, hides | Adults and larvae feed |
| *Dermestes maculatus* Degeer | Hide beetle, leather beetle | | Dried meat, fish, hides | Adults and larvae feed |
| *Dermestes peruvianus* Laporte | Hide beetle, leather beetle | | Dried meat, fish, hides | Adults and larvae feed |
| *Derocerus reticulatum* (Müller) and other species | Slugs | Limacidae | Potato | Tunnel in tubers |
| *Dinoderus distinctus* (Lesne) | Bamboo borer | Bostrichidae | Cassava | |
| *Dinoderus minutus* (F.) | Bamboo borer | | Cassava, bamboo | Adults bore into tubers and bamboo stems; reported also on maize, sorghum, sweet potato |
| *Doloessa viridis* Zeller | Bark borer | Pyralidae | Copra, cocoa beans | Larvae are minor pests |
| *Drosophila* spp | Small fruit flies | Drosophilidae | Citrus, peaches, pineapple, soy sauce | On ripe fruits |
| *Dysides* spp | False powder post beetle | Bostrichidae | Wood, cassava | Minor pest |
| *Dysmicoccus brevipes* (Cockerell) | Pineapple mealybug | Pseudococcidae | Pineapple | Bugs infest fruit |
| *Ectenostoma* spp | | Cistelidae | Beans | East Africa |
| *Ectomyelois ceratoniae* Zeller | Locust bean moth, carob moth | Pyralidae | Citrus, nuts | Larvae in fruits; nut shell must be damaged to allow access to kernel |
| *Elasmolomus sordidus* (F.) | Seedbug | Lygaeidae | Oilseeds | Suck out seed contents |
| *Endrosis sarcitrella* L. | White-shouldered house moth | Oecophoridae | Small grains, cereal products, sacking, wool, fur and feathers | Secondary pest, damage from larval feeding |
| *Ephestia* spp | Warehouse moths | Pyralidae | Polyphagous | Larvae feed on broken grains; secondary pests |
| *Ephestia calidella* Guénée | Dried fruit moth | | Dried fruits, cork | Pupates in stored produce |
| *Ephestia cautella* (Walker) | Tropical warehouse moth | | Small grains, cereal products, dried fruits, nuts, cocoa beans | More tropical; nut shell must be damaged to allow access to kernel |
| *Ephestia elutella* (Hübner) | Warehouse moth | | Small grains, cereal products, dried fruits, cocoa beans, tobacco | More temperate |
| *Ephestia figuliella* Gregson | Raisin moth | | Dried fruits | Pupates in stored produce |
| *Ephestia kuehniella* Zeller | Mediterranean flour moth | | Cereal products | Subtropical |
| *Ephystris chersaea* Meyrick | | Gelechiidae | Sorghum | Uganda |
| *Epitragus sallaei* Champion | | Tenebrionidae | Cereals | Minor pest in Nicaragua |
| *Epurea ocularis* (Kraatz) | | Nitidulidae | Coffee beans | Tanzania |
| *Etiella zinckenella* (Treitschke) | Pea pod borer; lima bean pod borer | Pyralidae | Pulses | Larvae in pods of various legumes |
| *Euborellia annulipes* (Lucas) | Ring-legged earwig | Carcinophoridae | Polyphagous | Mainly a field pest, tropical |
| *Eumerus* spp | Bulb flies | Syrphidae | Potato, ginger, onion, carrot, flower bulbs | Larvae in roots, tubers and bulbs (Lilliaceae mostly) |
| *Euophryum* spp | Wood boring weevil | Curculionidae | Wood | Larvae feed on decaying softwood |
| *Euxestoxenus rotundus* (Arrow) | | Colydiidae | Maize | Uganda |
| *Euzophera bigella* Zeller | Quince moth | Pyralidae | Peaches | Larvae in flesh or stone of fruit |
| *Euzophera osseatella* Treitschke | Eggplant stem borer | | Potato | Occasional pest on tubers |

| | | | | |
|---|---|---|---|---|
| *Euzophera sagax* Meyrick | | | Tea seed, coffee | Uganda |
| *Fannia canicularis* L. | Lesser housefly | Fanniidae | Bean curd, soy sauce | Maggots in curd |
| *Forficula auricularia* L. | European earwig | Forficulidae | Polyphagous | Temperate |
| *Gibbium psyllioides* (de Czenpinski) | Shiny spider beetle | Ptinidae | Store residues | Scavengers |
| *Glischrochilus* spp | Picnic beetles | Nitidulidae | Wheat, maize | Canada, US |
| *Gnatocerus cornutus* (F.) | Broad-horned flour beetle | Tenebrionidae | Small grains, cereal products | Secondary pest, found on many farinaceous materials |
| *Gnatocerus maxillosus* (F.) | Narrow-horned flour beetle | Tenebrionidae | Maize | In field and store |
| *Gonocephalum* spp | | Tenebrionidae | Cereals and palm kernels | Africa and Asia |
| *Grapholitha glycinivorella* Matsumura | Soya bean pod borer | Tortricidae | Pulses | Larvae in pods of soybean |
| *Haptoncus luteolus* (Erichson) | | Nitidulidae | Decaying oranges | Uganda |
| *Hapsifera horridella* Walker | | Tineidae | Dried sweet potato | Uganda |
| *Henoticus californicus* (Mann) | | Cryptophagidae | Dried fruit, cocoa beans, spices | Favours damp conditions |
| *Henoticus serratus* Gyllenhal | | Cryptophagidae | Wheat | Favours damp conditions; temperate |
| *Heterobostrychus aequalis* Waterhouse | Black borer | Bostrichidae | Wood, oilseeds, pulses, cassava | Southeast Asia |
| *Heterobostrychus brunneus* (Murray) | Black borer | | Potato, coffee beans, cassava | Sub-Saharan Africa |
| *Himatismus villosus* (Haag-Rutenberg) (= *Curimosphena villosa* (Haag-Rutenberg)) | | Tenebrionidae | Groundnuts, wheat, chillies | Africa and Middle East |
| *Hofmannophila pseudospretella* (Stainton) | Brown house moth | Oecophoridae | Small grains, cereal products, pulses, dried fruits, cork, book bindings, dried seaweed, sacking, wool, fur and feathers | Secondary pest, damage from larval feeding |
| *Holoparamecus depressus* Curtis (and other genera) | | Merophysiidae (and Lathridiidae) | Mould feeders, uncommon in the tropics | |
| *Hylotrupes bajulus* L. | House longhorn | Cerambycidae | Wood | Domestic species in pine timber |
| *Hypophloeus* sp. | | Tenebrionidae | Cowpeas | |
| *Hypothenemus hampei* Ferrari | Coffee berry borer | Scolytidae | Coffee | Adults bore cherries |
| *Hypothenemus obscurus* F. | | | Coffee | |
| *Hypothenemus eruditus* Westwood | | | Maize | Nigeria |
| *Iridiomyrmex* spp | House and meat ants | Formicidae | Fresh and dried meat | In industrial premises |
| *Lasiodactylus* sp. | | Nitidulidae | Tea seed | Kenya |
| *Lasiodactylus tibialis* (Boheman) | | | Cloves | Madagascar |
| *Lasioderma serricorne* F. | Tobacco beetle | Anobiidae | Maize, small grains, cereal products, dried fruits, oilseeds, cocoa beans, tobacco, spices | Larvae damage grain and fruits; cosmopolitan and polyphagous |
| *Laspeyresia caryana* (Fitch) | Hickory shuckworm | Tortricidae | Nuts | Larvae in kernels |
| *Latheticus oryzae* Waterhouse | Long-headed flour beetle | Tenebrionidae | Small grains, cereal products, oilseeds | Secondary pest, found on most farinaceous products |
| *Lepinotus* spp | Booklice | Psocidae | Cereal products | Mouldy produce |
| *Lepisma sacharina* L. | Silverfish | Lepismatidae | Cereal products, book bindings | Scavenger |

**Table A1.1** (*Continued.*)

| Scientific name | Common name | Family | Products affected | Comment |
|---|---|---|---|---|
| *Leptacinus parampunctatus* (Gylhenhal) | | Staphylinidae | Barley | Kenya |
| *Leptophloeus* spp | Flat grain beetle | Cucujidae | Cereal products | Adults and larvae feed |
| *Leucophaea maderae* (F.) | Madeira cockroach | Blaberidae | Polyphagous | Scavengers |
| *Liposcelis* spp | Booklice | Psocidae | Cereal products | Mouldy produce |
| *Litargus balteatus* Le Conte | Hairy fungus beetle | Mycetophagidae | Cosmopolitan | Temperate; damp produce |
| *Lonchaea aristella* Beck | Fig fly | Lonchaeidae | Dried figs | |
| *Lophocateres pusillus* Klug | Siamese grain beetle | Lophocateridae | Small grains, pulses, cassava | Tropical minor pest; on cassava mostly in SE Asia |
| *Lucilla* spp | Greenbottles, etc. | Calliphoridae | Fresh and dried meat, dried fish and fishmeal | Larvae eat flesh, adults feed on surface |
| *Lyctus africanus* (Lesne) | Powder-post beetle | Lyctidae | Cassava, wood, groundnuts | Larvae bore into tubers; larvae in plywood |
| *Lyctus brunneus* (Stephens) | Powder-post beetle | | Sweet potato, cassava | |
| *Macrotermes* spp | Termites | Termitidae | Small grains, wood | Foraging workers remove grains |
| *Marava arachidis* Yersin | Earwig | Labiidae | Polyphagous | Tropical |
| *Martyringa xeraula* Meyrick | Grain moth | Oecophoridae | Small grains | Japan |
| *Maruca testulalis* Geyer | Mung moth | Pyralidae | Pulses | Larvae in pods of various legumes |
| *Melanogromyza* spp | Pulse pod flies | Agromyzidae | Pulses | Inside pods |
| *Melissopus latiferreanus* Walsingham | Filbertworm | Tortricidae | Walnuts, hazelnuts | |
| *Melolontha* spp | Chafer grubs | Scarabaeidae | Potato | Eat holes in tubers |
| *Merodon* spp | Large bulb flies | Syrphidae | Onion, flower bulbs | Usually a single larva |
| *Mezium affine* Boieldieu | Northern spider beetle | Ptinidae | Store residues | Scavenger |
| *Mezium americanum* (Laporte) | American spider beetle | | Grain, cayenne | Scavenger, worldwide |
| *Minthea rugicollis* (Walker) | Powder-post beetle | Lyctidae | Cassava, wood | Larvae bore into tubers |
| *Monanus concinnulus* (Walker) | Flat bark beetle | Silvanidae | Illipe nuts, maize, rice, cocoa beans | Cosmopolitan |
| *Monochamus* spp | Pine sawer | Cerambycidae | Wood | Larvae in pine timber |
| *Monomorium pharaonis* (L.) | Pharoah ants | Formicidae | Cereal products | Polyphagous |
| *Monopsis* spp | | Tineidae | Skins and hides, wool, fur and feathers | Caterpillars eat keratin |
| *Murmidius ovalis* (Beck) | Minute bark beetle | Cerylonidae | Various stored products | Associated with mouldy produce |
| *Murmidius segregatus* Waterhouse | Minute bark beetle | Cerylonidae | Various stored products | Associated with mouldy produce |
| *Mus musculus* (L.) | House mouse | Muridae | Maize, small grains, cereal products, meat, dried fish, cheese | Primary pests |
| *Musca domestica* (L.) | House fly | Muscidae | Fresh and dried meat, dried fish and fishmeal | Larvae eat flesh, adults feed on surface |
| *Mussidia nigrivenella* Ragonot | | Pyralidae | Maize | West Africa only |
| *Mycetaea hirta* (Marsham) | | Endomychidae | Cereals | On damp mouldy produce |

| | | | | |
|---|---|---|---|---|
| *Mycetophila* spp | Mushroom midges | Mycetophilidae | Mushrooms | Part of the mushroom maggot complex |
| *Myzus persicae* (Sulzer) | Peach-potato aphid | Aphididae | Potato | Infest eyes in storage |
| *Nauphoeta cinerea* (Olivier) | Lobster cockroach | Blaberidae | Fishmeal | |
| *Necrobia* spp | | Cleridae | Bonemeal, cheese, meat | Protein feeders |
| *Necrobia ruficollis* (F.) | Red-necked bacon beetle | | Bacon and ham | Larvae burrow into fatty parts, adults feed on surface |
| *Necrobia rufipes* Degeer | Copra beetle | | Oilseeds, copra, cocoa beans, spices, dried meat and fish, bacon | Most damage done by larvae |
| *Nemapogon* spp | Fungus moths | Tineidae | Mushrooms | Larvae feed on fruiting bodies |
| *Nemapogon granella* (L.) | Corn moth | Tineidae | Small grains, dried fruits, nuts, mushrooms | Temperate; primary pest; larvae polyphagous |
| *Niptus hololeucus* (Faldermann) | Golden spider beetle | Ptinidae | Store residues | Scavengers |
| *Odontotermes* spp | Termites | Termitidae | Small grains | Foraging workers remove grains |
| *Oligota parva* Kraatz | Rove beetle | Staphylinidae | | Widespread scavengers |
| *Oryzaephilus gibbosus* Aitken | | Silvanidae | Coconuts, Gambian groundnuts | East Africa |
| *Oryzaephilus mercator* (Fauvel) | Merchant grain beetle | Silvanidae | Oilseeds, cocoa beans | Most damage done by larvae |
| *Oryzaephilus surinamensis* (L.) | Saw-toothed grain beetle | Silvanidae | Maize, small grains, dried fruits, copra, spices | Cosmopolitan on damaged grain and fruit |
| *Pagiocerus frontalis* (F.) | | Scolytidae | Maize | Reaches pest status only in the high Andes |
| *Palembus dermestoides* (Fairmaire) | | Tenebrionidae | Groundnuts, flour | Southeast Asia |
| *Palembus ivoirensis* (Ardoin) | | | Maize cobs | Nigeria |
| *Palembus ocularis* (Casey) | | | Maize cobs | Nigeria |
| *Palorus ficicola* (Wollaston) | | Tenebrionidae | Cereals, rice, flours, nuts | Minor tropical pest |
| *Palorus genalis* Blair | | | Sago flour, rice, groundnut cake, beans | Asia, East and West Africa |
| *Palorus laesicollis* (Fairmaire) | | | Damaged maize, cereal residues | Highlands of Kenya and Ethiopia |
| *Palorus ratzeburgii* Wissman | Small-eyed flour beetle | | Wheat and oilseeds | Adults and larvae on damaged commodities |
| *Palorus subdepressus* (Wollaston) | Depressed flour beetle | | Wheat and oilseeds | Adults and larvae on damaged commodities |
| *Pantomorus cervinus* (Bohemon) | Fuller rose beetle | Curculionidae | Lemons | |
| *Paralipsa gularis* (Zeller) | Stored nut moth | Pyralidae | Dried fruits, nuts | Pupates in stored produce; nut shell must be damaged to allow access to kernel |
| *Paramyelois transitella* (Walker) | Navel orange worm | Pyralidae | Citrus, nuts | Larvae in fruits; nut shell must be damaged to allow access to kernel |
| *Passer* spp | Sparrows | Ploceidae | Small grains, cereal | Invade grain stores |
| *Pectinophora gossypiella* Saunders | Pink bollworm | Gelechidae | Cotton seed | |
| *Pentarthrum* spp | Wood boring weevil | Curculionidae | Wood | Larvae feed on decaying softwood |
| *Periplaneta americana* L. | American cockroach | Blattidae | Polyphagous | Ubiquitous scavengers |
| *Periplaneta australasiae* F. | Australian cockroach | Blattidae | Polyphagous | Ubiquitous scavengers |
| *Peristeptus marginalis* Gebien | | Tenebrionidae | Bran | Kenya |
| *Pharaxonotha kirschii* Reitter | | Languriidae | Wheat, beans, tubers and maize | Minor pests; Americas |
| *Phradonoma* spp | | Dermestidae | Oilseeds, coffee beans, bonemeal | Most damage done by larvae |
| *Phthorimaea operculella* (Zeller) | Potato tuber moth | Gelichiidae | Potato, tobacco | Larvae bore into tubers and mine tobacco leaves |

**Table A1.1** (*Continued.*)

| Scientific name | Common name | Family | Products affected | Comment |
|---|---|---|---|---|
| *Piezotrachelus varium* (Wagner) | | Apoinidae | Pulses | Occasional pests of stored products |
| *Piophila casei* (L.) | Cheese skipper | Piophilidae | Dried meat and fish, bacon and ham, cheese | Maggots feed in meat and fat |
| *Planococcus* spp and others | Citrus mealybugs | Pseudococcidae | Citrus | Infest eye and base of stalk |
| *Planolestes* spp | Flat grain beetle | Cucujidae | Cereal products | Adults and larvae feed |
| *Platydema* sp. | | Tenebrionidae | Maize | Uganda |
| *Platypus penetralis* (Sampson) | | Platypodidae | Cassava, rubber | East Africa |
| *Ploceus* spp | Weaver birds | Ploceidae | Small grains | Invade grain stores |
| *Plodia interpunctella* (Hübner) | Indian meal moth | Pyralidae | Maize, small grains, cereal products, Brazil nuts, chocolate, spices | Larvae feed on broken grains; secondary pests; mostly subtropic |
| *Polycompsitis pycnosaris* Meyrick | | Tineidae | Old cotton seed | Uganda |
| *Pnyxia scabiei* (Hopkins) | Potato scab gnat | Sciaridae | Potato | Larvae on tubers |
| *Prays endocarpa* Meyer | Citrus rind borer | Plutellidae | Citrus | Larvae mine rind (SE Asia and US) |
| *Prionoryctes* spp | Yam beetles | Scarabaeidae | Yams | Adults and larvae tunnel into tubers |
| *Promalactis inonisema* Butler | Cotton seedworm | Oecophoridae | Cotton seed | Japan |
| *Prostephanus truncatus* (Horn) | Larger grain borer | Bostrichidae | Maize, cassava | Adults and larvae feed on maize grains or bore into cassava tubers |
| *Pseudeurostus hilleri* Reitter | Spider beetle | Ptinidae | Barley, sorghum and store residues | Scavengers |
| *Psila rosae* (F.) | Carrot fly | Psilidae | Carrot | Larvae in roots |
| *Psyllipsocus* spp | Booklice | Psocidae | Cereal products | Mouldy produce |
| *Ptinus clavipes* Panzer | Brown spider beetle | Ptinidae | Store residues | Scavengers |
| *Ptinus fur* (L.) | White-marked spider beetle | Ptinidae | Store residues | Scavengers |
| *Ptinus pusillus* Sturm | Spider beetle | Ptinidae | Store residues | Scavengers |
| *Ptinus tectus* Boieldieu | Australian spider beetle | | Cereal products, coffee beans | More temperate; polyphagous |
| *Pyralis* spp | Meal moths | Pyralidae | Maize, potato | Larvae feed on maize broken grains; secondary pests; occasionally bore into potato tubers |
| *Pyralis farinalis* L. | | | Small grains, cereal products, straw dunnage | Cosmopolitan, damage from larval feeding |
| *Pyralis manihotalis* Guénée | Grey pyralid | | Cassava, dried meat, bonemeal | Larvae eat tubers |
| *Pyralis pictalis* (Curtis) | Painted meal moth | | Cereal products | Damage from larval feeding |
| *Pyroderces rileyi* (Walsingham) | Pink scavenger caterpillar | Cosmopterygidae | Maize | US only |
| *Quadraspidiotus perniciosus* (Comstock) | San José scale | Diaspididae | Apples | |
| *Rattus* spp | Rats | Muridae | Maize, small grains, cereal products, meat, dried fish, cheese | Primary pests |
| *Rhagoletis* spp | Fruit flies | Tephritidae | Apples and other fruits | Larvae in fruits |
| *Rhopalosiphoninus latysiphon* (Davidson) | Bulb and potato aphid | Aphididae | Potato | Infest eyes in storage |

| | | | | |
|---|---|---|---|---|
| *Rhyzopertha dominica* (F.) | Lesser grain borer | Bostrichidae | Maize, cereal products, cassava | Adults and larvae feed on grains or cassava tubers; subtropical |
| *Sarcophaga* spp | Flesh flies | Sarcophagidae | Soy sauce, fresh meat, dried fish | Larviparous, larvae deposited in soya fluid or on meat |
| *Sciara* spp | Mushroom flies | Sciaridae and other families | Mushroom | Part of the mushroom maggot complex |
| *Scrobipalpopsis solanivora* Povolny | Potato tuber moth | Gelechidae | Irish potato | Central America |
| *Setomorpha rutella* Zeller | Tropical tobacco moth | Tineidae | Cocoa beans, tobacco | West Africa |
| *Silvanophloeus* sp. | | Cucujidae | Maize | Uganda |
| *Silvanus* spp | | Silvanidae | Various | Mostly tropical |
| *Sinoxylon* spp | False powder post beetle | Bostrichidae | Wood, cassava | Minor pest |
| *Sipalinus* spp | Pine weevils | Curculionidae | Wood | Larvae in pine timber |
| *Sitophagus hololeptoides* Laporte | | Tenebrionidae | Maize and nuts | Cosmopolitan except in Asia |
| *Sitophilus granarius* L. | Grain weevil | Curculionidae | Small grains | Temperate; primary pest |
| *Sitophilus linearis* (Herbst) | Tamarind weevil | | Tamarind pods | Tropical |
| *Sitophilus oryzae* (L.) | Rice weevil | | Maize, small grains | Major primary pest |
| *Sitophilus zeamais* Motschulsky | Maize weevil | | Maize | Major primary pest |
| *Sitotroga cerealella* (Olivier) | Angoumois grain moth | Gelechiidae | Small grains | Moves from field to store |
| *Solenopsis xyloni* McCook | Southern fire ant | Formicidae | Almonds | US |
| *Sophronica ventralis* (Aurivillius) | | Lamiidae | Coffee, cotton seed | East Africa, Brazil |
| *Specularius* spp | | Bruchidae | Pulses | Mostly in the field |
| *Spermophagus subfasciatus* (Boheman) | Bruchid | Bruchidae | Groundnuts | East Africa |
| *Stathmopoda auriferella* (Walker) | | Helionidae | Sorghum | Nigeria |
| *Stegobium paniceum* (L.) | Biscuit beetle | Anobiidae | Cereal products, coriander seeds, chocolate, spices | Cosmopolitan and polyphagous; larval feeding |
| *Stelidota* spp | | Nitidulidae | Maize, nuts, cocoa beans | Central and South America |
| *Stephanoderes hampei* (Ferrari) | | Scolytidae | Coffee seeds | East Africa, Far East |
| *Stephanopachys* spp | False powder post beetle | Bostrichidae | Wood, cassava | Minor pest |
| *Supella longipalpa* (F.) | Brown-banded cockroach | Blatellidae | Polyphagous | Scavengers |
| *Tapinoma* spp and others | Sugar ants | Formicidae | Pineapple | Attracted by sap and found in tinned fruits |
| *Tenebrio guineensis* Imhoff | | Tenebrionidae | Cereals | Africa; occasionally found |
| *Tenebrio molitor* L. | Yellow mealworm | | Grain and cereal products | Polyphagous but adults and larvae favour cereals |
| *Tenebrio obscurus* F. | Dark mealworm | | Grain and cereal products | |
| *Tenebroides mauritanicus* (L.) | Cadelle | Trogossitidae | Small grains, oilseeds | Larvae are minor pests, widespread |
| Termites | | Termetidae etc. | Maize | Foraging workers remove grains |
| *Thaneroclerus buqueti* Lefèvre | | Cleridae | Cassava, sweet potato | East Africa |
| *Theba* spp | Small banded snails | Helicidae | Small grains | Contaminants in harvested grain |
| *Thermobia domestica* (Packard) | Firebrat | Lepismatidae | Starchy material | Scavengers |
| *Thorictodes heydeni* Reitter | | Dermestidae | Broken grains, oilseed cake, pulses | Adults small (1.5 mm long), tropical and subtropical |
| *Tinea* spp | Clothes moths | Tineidae | Sacking, skins and hides, wool, fur, feathers | Caterpillars polyphagous on plant and animal material |
| *Tinea pellionella* L. | Case-bearing clothes moth | | Wool, fur and feathers | Caterpillars feed on keratin |

**Table A1.1** (*Continued.*)

| Scientific name | Common name | Family | Products affected | Comment |
|---|---|---|---|---|
| *Tineola* spp | Clothes moth | Tineidae | Sacking, skins, hides, dried fish and fishmeal, wool, fur and feathers | Caterpillars polyphagous on plant and animal material |
| *Tineola bisselliella* (Hummel) | Webbing clothes moth | | Hides | Tropics |
| *Tipnus unicolor* (Piller & Mitterpacher) | Spider beetle | Ptinidae | Store residues | Scavengers |
| *Tribolium anaphe* Hinton | Flour beetle | Tenebrionidae | Cotton seed, cocoa beans and palm kernels | Africa |
| *Tribolium audax* Halstead | American black flour beetle | | Cereals | North America, Europe, Egypt |
| *Tribolium castaneum* (Herbst) | Red flour beetle | | Maize, cereal products, dried fruits, coffee, cocoa beans | Secondary and serious pest |
| *Tribolium confusum* J. du Val | Confused flour beetle | | As above | Secondary and serious pest; less frequent in the tropics |
| *Tribolium destructor* Uyttenboogaart | Confused flour beetle | | As above | Secondary pest; in cooler tropics |
| *Tribolium madens* (Charpentier) | European black flour beetle | | | |
| *Trichophaga* spp | Tapestry moths | Tineidae | Wool, fur, feathers | Prefer coarse fibres and hairs |
| *Trigonogenius globulus* Solier | Spider beetle | Ptinidae | Store residues | Scavengers |
| *Trigonogenius particularis* Pic | Spider beetle | | Store residues | Kenya |
| *Trogium* spp | Booklice | Psocidae | Cereal products | Mouldy produce |
| *Trogoderma glabrum* (Herbst) | Glabrous carpet beetle | Dermestidae | Maize, cereal products | Europe, US and Mexico |
| *Trogoderma granarium* Everts | Khapra beetle | | Maize, small grains, oilseeds, cocoa beans | Primary pest; larvae |
| *Trogoderma inclusum* Le Conte | Larger cabinet beetle | | Polyphagous | Scavenger, minor pest |
| *Trogoderma variabile* Ballion | Warehouse beetle | | Maize, pulses | North America |
| *Typhaea stercorea* (L.) | Hairy fungus beetle | Mycetophagidae | Small grains | Tropical mainly; on mouldy grain |
| *Urophorus humeralis* (F.) | Pineapple sap beetle | Nitidulidae | Maize, pineapple | On damaged grains; attracted by sap and found in tinned pineapple |
| *Urophorus nitidus* Murray | | | Dried banana | Very uncommon, West Africa |
| *Xestobium rufovillosum* Degeer | Death watch beetle | Anobiidae | Wood | Larvae in hardwoods |
| *Xylion* spp | False powder post beetle | Bostrichidae | Wood, cassava | Minor pest |
| *Xylocopa* spp | Carpenter bee | Anthophoridae | Wood | Breeding tunnels in beams and rafters in tropics |
| *Xylocopa iridipennis* Le Peletier | Bamboo carpenter bee | | Bamboo | Adults bite holes in internodes |
| *Xylopertha* spp | | Bostrichidae | Bamboo, eucalyptus | East Africa |
| *Xyloperthella* spp | False powder post beetle | Bostrichidae | Wood, cassava | Minor pest |
| *Xyloperthodes nitidipennis* (Murray) | | Bostrichidae | Cassava | East Africa |
| *Zabrotes subfasciatus* (Boheman) | Mexican bean beetle | Bruchidae | Pulses | *Phaseolus* and others |
| Zygaemoides monstrosus (Pascoe) | | Anthribidae | Coffee berries, cotton seed | |

**Table A1.2** Mites (Acarina) associated with stored products.

| Scientific name | Common name | Family | Product | Comment |
|---|---|---|---|---|
| *Acarus gracilis* Hughes | Flour mite | Acaridae | Small grains, cereal products, nuts, dried fish and fishmeal, cheese | Uncommon |
| *Acarus farris* (Oudemans) | Flour mite | | As above | Uncommon |
| *Acarus immobilis* Griffiths | Flour mite | | As above | Uncommon |
| *Acarus siro siro* L. | Flour mite | | As above | Abundant in UK grain stores but uncommon in tropics |
| *Aeroglyphus* spp | | Glycyphagidae | Various stored products | Occasionally found in temperate areas |
| *Aleuroglyphus ovatus* (Troupeau) | | Acaridae | Small grains, cereal products | Cosmopolitan |
| *Austroglycyphagus geniculatus* (Vitzthum) | | Glycyphagidae | Various | Occasionally found in Europe and Africa |
| *Blomia tropicalis* Bronswijk Cock & Oshima | | Glycyphagidae | Might feed on stored products | Tropical Asia and South America; house dust allergen |
| *Caloglyphus berlesei* (Michael) | | Acaridae | Various | Widespread in tropics, on damp produce |
| *Caloglyphus hughesi* (Samšiňák) | | | Various | Widespread in tropics |
| *Caloglyphus oudemansi* (Zachvatkin) | | | Various | Widespread in tropics, on damp produce |
| *Carpoglyphus lactis* L. and others | | Carpoglyphidae | Various | In cool climates |
| *Chortoglyphus* spp | | Chortoglyphidae | Various | Mainly temperate |
| *Coproglyphus* spp | | Glycyphagidae | Various | Occasionally found in temperate areas |
| *Ctenoglyphus* spp | | Glycyphagidae | Various | Occasionally found in temperate areas |
| *Dermatophagoides* spp | | Pyroglyphidae | Mainly in house dust | Widespread |
| *Glycyphagus* spp | House mite | Glycyphagidae | Dried fruits, cheese | Cosmopolitan |
| *Glycyphagus destructor* (Schrank) | Grain mite | | Small grains, cereal products, straw dunnage | Abundant in UK grain stores |
| *Glycyphagus domesticus* (Degeer) | House mite | | Tobacco | Infestation by all stages |
| *Glycyphagus michaeli* Oudemans | House mite | | Cereals | Mainly in temperate areas |
| *Gohieria* spp | | Glycyphagidae | Various | Occasionally found in temperate areas |
| *Kleemania* spp | | Ameroseiidae | Fungus feeders | Two species found in grain stores in Indonesia |
| *Kleemania plumigera* Oudemans | | | | Temperate regions |
| *Kleemania plumosus* (Oudemans) | | | | Temperate regions |
| *Histiostoma feroniarum* (Dufour) | | Anoetidae | Damp produce | Filter feeder |
| *Lardoglyphus angelinae* Olsen | | Glycyphagidae | Skins and hides, dried fish, fishmeal, bonemeal | Protein feeders; Hong Kong |
| *Lardoglyphus konoi* (Sasa and Asanuma) | | | As above | Protein feeders; Europe, India and Far East |
| *Lardoglyphus zacheri* Oudemans | | | As above | Protein feeders; South America, Australia and Europe |
| *Leiodinychus krameri* (G. & R. Canestrini) and other species | | Uropopidae | Cereals, cereal products, cassava, copra and pulses | Cosmopolitan; favour damp produce |
| *Michaelopus gallegoi* (Portus & Gomez) | | Acaridae | Various foodstuffs | Spain and Indonesia |

**Table A1.2** (*Continued.*)

| Scientific name | Common name | Family | Product | Comment |
|---|---|---|---|---|
| *Neognathus* spp | | Caligonellidae | Various | Two species from grain stores in Indonesia |
| *Phyllocoptruta oleivora* Ashmead | Citrus rust mite | Eriophyidae | Citrus | Mites infest skin |
| *Rhizoglyphus callae* Oudemans | Bulb mite | Acaridae | Cereal spillage, flower bulbs | Favours damp conditions; UK and Ethiopia |
| *Rhizoglyphus echinopus* (Fumouze & Robin) | Bulb mite | | Onion, mushrooms, flower bulbs | Mites between scales |
| *Stenotarsonemus laticeps* | Bulb scale mite | Tarsonemidae | Flower bulbs | Between bulb scales |
| *Suidasia nesbitti* Hughes | Dust mite | Acaridae | Cereal products | Cosmopolitan but less common in tropics |
| *Suidasia pontifica* Oudemans | Dust mite | | Cereals and beans | Cosmopolitan in tropics |
| *Tarsonemus* spp | Glossy grain mites | Tarsonemidae | Cereal residues, dust | Smallest mite pests (0.1–0.2 mm) |
| *Tetranychus urticae* Koch | Two-spotted spider mite | Tetranychidae | Cut flowers | |
| *Thyreophagus entomophagus* (Laboulbène & Robin) | | Acaridae | Various foodstuffs | Northern temperate regions |
| *Tyroborus* spp | | Acaridae | Omnivorous | Less common in tropics |
| *Tyrolichus* spp | | Acaridae | Omnivorous | Less common in tropics |
| *Tyrophagus brevicrinatus* Robertson | Cheese and mould mites | Acaridae | Omnivorous | Less common, often in mixed populations with *T. putrescentiae* |
| *Tyrophagus palmarum* Oudemans | | | Omnivorous | Less common, often in mixed populations with *T. putrescentiae* |
| *Tyrophagus perniciosus* Zachvatkin | | | Omnivorous | Less common, often in mixed populations with *T. putrescentiae* |
| *Tyrophagus putrescentiae* (Schrank) | | | Omnivorous | Commonest mite pest of stored food in tropics; fungus feeder |
| *Tyrophagus tropicus* Robertson | | | Omnivorous | Less common, often in mixed populations with *T. putrescentiae* |

**Table A1.3** Predators and parasites of stored product pests.

| Scientific name | Family | Prey | Comment |
|---|---|---|---|
| *Acarophenax tribolii* Newstead & Duvall | Pyemotidae | *Tribolium* spp | Ectoparasite |
| *Acaropsis docta* Berlese | Cheyletidae | *Trogoderma granarium* | |
| *Acaropsis sollers* Rohdendorf | Cheyletidae | Acarid mites and possibly small insect larvae | Temperate regions |
| *Acaropsella volgini* (Gerson) | Cheyletidae | Acarid mites and possibly small insect larvae | In small numbers in tropics |
| *Adelina tribolii* Bhatia | Coccidia | *Trogoderma granarium* and *Tribolium* spp | |
| *Amblyseius* spp | Phytoseiidae | Probably on mites and fungi | Not common |
| *Amphibolus venator* (Klug) | Reduviidae | *Trogoderma granarium* | |
| *Androlaelaps casalis* (Berlese) | Dermanyssidae | Acarid and gamasine mites, young *Tribolium* larvae | Requires high humidity for successful development |
| *Anisopteromalus calandrae* (Howard) | Pteromalidae | Moth and beetle larvae | Commonest parasitic wasp |
| *Antrocephalus mitys* (Walker) | Chalcididae | Moth larvae | |
| *Apanteles* spp | Braconidae | Moth larvae | Rarely found |
| *Blattisocius dentriticus* (Berlese) | Ascidae | Acarid mites, especially *Tyrophagus putrescentiae* | Widespread |
| *Blattisocius keegani* Fox | | Storage beetles and mites | Abundant in tropical stores |
| *Blattisocius quadridentatus* Haines | | *Ahasverus advena* | On chocolate from Jamaica |
| *Blattisocius tarsalis* (Berlese) | | Predominantly moth eggs, but also other insect eggs and mites | May cause significant natural control of *Ephestia cautella* |
| *Bracon brevicornis* Wesmael | Braconidae | Pyralid warehouse moths | Occasionally found |
| *Bracon hebetor* Say | | As above | Common in tropics and subtropics |
| *Bruchophagus* spp | Eurytomidae | *Lasioderma serricorne* and *Stegobium paniceum* | |
| *Caudacheles* sp. | Cheyletidae | Acarid mites and possibly small insect larvae | In small numbers in tropics |
| *Cephalonomia gallicola* (Ashmead) | Bethylidae | *Lasioderma serricorne*, *Stegobium paniceum* and Ptinidae | Specifically associated with storage; may be overlooked because of their small size |
| *Cephalonomia tarsalis* (Ashmead) | | Particularly on *Oryzaephilus* spp | Worldwide; may be overlooked because of their small size |
| *Cephalonomia meridionalis* Brèthes | | *Oryzaephilus* spp | As above |
| *Cephalonomia waterstoni* Gahan | | Particularly on *Cryptolestes* spp | As above |
| *Cerocephala dinoderi* Gahan | Pteromalidae | Moth and beetle larvae | Rare |
| *Chelacheles* sp. | Cheyletidae | Acarid mites and possibly small insect larvae | In small numbers in tropics |
| *Cheletomorpha lepidopterorum* (Shaw) | Cheyletidae | Acarid mites and possibly small insect larvae | Cosmopolitan but uncommon in tropical storage |
| *Cheletonata* sp. | Cheyletidae | Acarid mites and possibly small insect larvae | In small numbers in tropics |
| *Cheyletus* spp | Cheyletidae | Eggs of *Lasioderma serricorne* | |
| *Cheyletus aversor* Rohdendorf | | Acarid mites, psocoptera and probably small insect larvae | Rare |
| *Cheyletus eruditus* (Schrank) | | As above | Predominantly temperate |
| *Cheyletu. fortis* Oudemans | | As above | May be a variety of *C. malaccensis* |
| *Cheyletus trouessarti* Oudemans | | As above | Rare |
| *Cheyletus malaccensis* Oudemans | | As above | Dominant cheyletid in tropical stores |

**Table A1.3** *(Continued.)*

| Scientific name | Family | Prey | Comment |
|---|---|---|---|
| *Choetospila elegans* Westwood | Pteromalidae | Moth and beetle larvae | Common, and often found with *Anisopteromalus calendrae* |
| *Chortoglyphus gracilipes* Banks | Glycyphagidae | Eggs of *Lasioderma serricorne* | |
| *Cunaxa setirostris* (Hermann) | Cunaxidae | Acarid mites and possibly small insect larvae | Cosmopolitan but uncommon, red in colour |
| *Diadegma chrysostictos* (Gmelin) | Ichneumonidae | Moth larvae | More common in temperate regions |
| *Dinarmus colemani* (Crawford) | Pteromalidae | Moth and beetle larvae | Rare |
| *Dinarmus basalis* (Rondani) (= *D. laticeps* (Ashmead)) | | Bruchids, especially *Callosobruchus* spp | Tropics |
| *Euchalcidia caryobori* Hanna | Chalcididae | Moth larvae | |
| *Eucheyletia taurica* Volgin | Cheyletidae | Acarid mites and possibly small insect larvae | Temperate regions |
| *Eulaelaps stabularis* (Koch) | Dermanyssidae | Arthropods | |
| *Evania appendigaster* (L.) | Evaniidae | Cockroach oothecae | In unhygienic stores |
| Ground beetles | Carabidae | Generalised predators | Not important in controlling pests |
| *Haemogamasus pontiger* Berlese | Dermanyssidae | Astigmatan mites and young insect larvae | Probably more common in temperate stores |
| *Holepyris hawaiiensis* (Ashmead) | Bethylidae | Larvae of Phycinae (*Ephestia* spp and *Plodia interpunctella*) | Widespread; may be overlooked because of their small size |
| *Holepyris sylvanidis* Brèthes (= *Rhabdepyris zeae* Turner & Waterston) | | Mainly on *Tribolium* spp | Cosmopolitan; may be overlooked because of their small size |
| *Hypoaspis* spp | Dermanyssidae | Astigmatan mites | |
| *Israelius carthami* Richards | Bethylidae | *Lasioderma serricorne* | May be overlooked because of their small size |
| *Ker* sp. | Cheyletidae | Acarid mites and possibly small insect larvae | In small numbers in tropics |
| *Lariophagus distinguendus* (Förster) | Pteromalidae | Beetles, especially *Lasioderma* spp and *Stegobium* spp | Infrequent in tropics |
| *Lasioseius* spp | Ascidae | Mites and nematodes | |
| *Machrocheles matrius* (Hull) | Macrochelidae | Insect and fungus feeders | |
| *Machrocheles muscaedomestica* (Scopoli) | | As above | |
| *Machrocheles robustulus* (Berlese) | | As above | |
| *Mattesia dispora* (Weiser) | Gregarinidae | *Cryptolestes* spp | Spores of this protozoan are spread by *Cephalonomia waterstoni* Gahan |
| *Melicheres agilis* Hering | Ascidae | *Acarus siro* and *Carpophagus lactis* | Europe |

| | | | |
|---|---|---|---|
| *Monieziella angusta* Banks | Acaridae | Eggs of *Lasioderma serricorne* | |
| *Neoseiulus* spp | Phytoseiidae | Probably on mites and fungi | Not common |
| *Nodele* sp. | Cheyletidae | Acarid mites and possibly small insect larvae | In small numbers in tropics |
| *Paratilus carus* (Newman) | Cleridae | Anobids | |
| *Phanerotoma* spp | Braconidae | Moth larvae | Rare |
| *Plastanoxus munroi* Richards | Bethylidae | Moths and beetles | Frequent in tropics |
| *Plastanoxus westwoodi* (Kieffer) | | As above | As above |
| *Proctolaelaps* spp | Ascidae | Possibly acarids | On mouldy produce in temperate stores |
| *Pronematus* spp | Tydeidae | Small acarids | Two species found in rice stores in Indonesia |
| *Pteromalus semotus* (Walker) (= *Habrocytus cerealellae* Ashmead) | Pteromalidae | Usually *Sitotroga cerealella* | Cosmopolitan |
| *Pyemotes anobii* Krczal | Pyemotidae | Beetle and moth pests | Ectoparasites; probably allergenic |
| *Pyemotes beckeri* Krczal | | As above | As above |
| *Pyemotes herfsi* Oudemans | | As above | As above |
| *Pyemotes tritici* (Lagreze-Fossat & Montagne) | | | Ectoparasites; can cause severe dermatitis in people handling stored products |
| *Rhagidia* spp | Bdellidae | Eggs of *Lasioderma serricorne* | |
| *Seiulus* spp | Phytoseiidae | Eggs of *Lasioderma serricorne* | |
| *Spinibdella* sp. | Bdellidae | Mite pests | Grain and flour stores in UK |
| *Synopeas* spp | Platygastridae | *Trogoderma granarium* | |
| Trichogrammatidae | | Eggs | Often overlooked because of their very small size (0.5 mm) |
| *Tarsostenus univattus* (Rossi) | Cleridae | Anobids | |
| *Tenebroides mauritanicus* (L.) | Trogossitidae | *Lasioderma* spp | |
| *Teretrius* (= *Teretriosoma*) *nigrescens* (Lewis) | Histeridae | *Prostephanus truncatus* | Introduced into Africa as a biocontrol agent |
| *Thanasimus formicarius* (L.) | Cleridae | Anobids | |
| *Thanoclerus buqueti* (Lefèvre) | Cleridae | *Lasioderma* spp | |
| *Tydeus australensis* Baker | Tydeidae | Small acarids | Indonesia |
| *Tydeus interruptus* Thor | | Small acarids | Northern Europe |
| *Typhlodromus* spp | Phytoseiidae | Probably on mites and fungi | Not common |
| *Venturia canescens* (Gravenhorst) | Ichneumonidae | Moth larvae | More common in temperate regions |
| *Zeteticontus insularis* (Howard) | Encyrtidae | Various | Uncommon |
| *Zeteticontus laeviscutum* (Thomson) | | Various | Uncommon |
| *Zeteticontus scutellatus* (Howard) | | Various | Uncommon |
| *Zeteticontus utilis* Noyes | | Larvae of *Carpophilus hemipterus* and *C. mutilatus* | Uncommon |

# Appendix 2
# Some Important Post-harvest Pathogens

**Table A2.1** Fungi.

| Pathogen genus and important species[1] | Family and morphs[2] | Biology and appearance | Type of damage caused | Crops attacked[1] |
|---|---|---|---|---|
| Pythium, *P. aphanidermatum* (Edson) Fitzpatrick, *P. ultimum* Trow | Pythiaceae | Mycelium of slender, coenocytic, intracellular and intercellular hyphae. No haustoria. Asexual sporangia remain attached to the hyphae and give rise to zoospores or germ tubes. Zoospores are produced in a bubble-like vesicle that is attached to the sporangium by a long tube. Sexual reproduction produces oospores.[3] | Wet rot | Banana, beans, beets, broccoli, Brussels sprouts, cabbage, carrot, celery, chayote, chicory, cucumber, eggplant, endive, garlic, ginger, leek, lettuce, melons, onion, parsley, parsnip, pea, peppers, potato, pumpkin, rhubarb, shallot, squash, strawberry, swede, sweet potato, tannia, taro, tomato, turnip, watermelon, yam |
| Phytophthora, *P. cactorum* (Lebert & Cohn) J. Schröter, *P. capsici* Leonian, *P. nicotianae* var. *parasitica* (Dastur) G.M. Waterhouse, *P. palmivora* (E.J. Butler) E.J. Butler, *P. porri* Foister | Pythiaceae | Similar to *Pythium* except that zoospores are produced inside the sporangium and the vesicle is usually absent.[3] | Blight | Apple, asparagus, avocado, banana, beans, breadfruit, cabbage, carrot, cassava, chayote, citrus, cucumber, durian, eggplant, feijoa, fennel, garlic, guava, leek, lettuce, mango, marrow, melons, okra, onion, papaya, parsley, passion fruit, pea, pear, peppers, potato, pumpkin, rhubarb, sapodilla, shallot, spring onion, squash, strawberry, swede, sweet potato, sweet sop, tannia, taro, tomato, watermelon |
| Rhizopus, *R. oryzae* Went & Prins. Geerl., *R. stolonifer* (Ehrenb.) Vuill. | Mucoraceae | Sporangiospores released from a sporangium germinate by germ tube that develops into a fluffy, branched, aerial mycelium. *Rhizopus* is heterothallic. Sexual reproduction occurs when zygotes develop from gametangia resulting from the meeting of compatible thalli. Zygotes release a germ sporangium, that produces sporangiospores.[3] | Wet rot | Apple, apricot, avocado, banana, beans, beet, Brussels sprouts, cabbage, carrot, cassava, cauliflower, cherry, citrus, eggplant, fig, garlic, grape, guava, Jerusalem artichoke, litchis, mango, melons, nectarine, papaya, passion fruit, pea, peach, pear, peppers, persimmon, pineapple, plum, pomegranate, potato, pumpkin, raspberry, strawberry, sweet potato, tomato, watermelon, yam |
| Fusarium, *F. moniliforme* J. Sheld., *F. oxysporum* E.F. Sm. & Swingle and its special forms, *F. solani* (Mart.) Sacc. | Anamorphic Hypocreaceae, teleomorphs *Gibberella, Nectria* | Fusaria have fusoid, curved, septate microconidia in slimy masses (sporodochia) on branched conidiophores. Mycelia and spores are usually brightly coloured. | Dry rot | Apple, asparagus, avocado, banana, beans, cabbage, Cape gooseberry, carrot, cassava, celery, chayotes, citrus, cucumber, eggplant, fig, garlic, ginger, guava, Jerusalem artichoke, melon, okra, onion, papaya, parsnip, passion fruit, pea, pear, peppers, pineapple, potato, squash, sweet potato, tannia, taro, tomato, watermelons, yam |

| | | | | |
|---|---|---|---|---|
| Alternaria, *A. alternata* (Fr.) Keissl., *A. brassicae* (Berk.) Sacc., *A. tenuissima* (Kunze) Wiltshire | Anamorphic Pleosporaceae, teleomorph *Lewia* | Conidia are large and multicellular, having longitudinal and transverse septa. A conidium is formed at the end of the hypha, with new conidia budding from the apex of the previous one, so that long chains are formed. Hyphae and conidia are dark coloured.[3] | Blotch, blight, leaf spot | Apple, apricot, avocado, bean, blueberry, Cape gooseberry, carambola, cauliflower, chayote, cherry, citrus, cucumber, eggplant, fig, gooseberry, grape, mango, melon, nectarine, papaya, passion fruit, pea, peach, pear, pepper, persimmon, plum, potato, raspberry, squash, strawberry, tomato, watermelon |
| Phoma, *P. exigua* Sacc. and its varieties, *P. lingam* (Tode) Desm. | Anamorphic Pleosporaceae, teleomorph *Pleospora* | *Phoma* is characterised by small, ostiolate pycnidia sunk in the host tissue. The conidiophores are very short and the conidia are hyaline and spherical.[3] Conidia are dispersed by water and air currents, and the fungus can also be spread on or in diseased tubers.[1] | Dry rot | Asparagus, bean, beet, broccoli, Brussels sprout, cabbage, carrot, celeriac, celery, cauliflower, chayote, eggplant, fennel, guava, kiwifruit, melon, okra, papaya, parsnip, pea, pepper, persimmon, pomegranate, potato, raspberry, swede, sweet potato, tamarillo, tomato, turnip, watermelon |
| Sclerotinia, *S. sclerotiorum* (Lib.) de Bary | Sclerotiniaceae | Over-wintering sclerotia on mummified fruits germinate to produce stalked apothecia in the following spring. Germinating ascospores give rise to a mycelium that is the invasive stage of the fungus. Conidia may be formed in some species. | Brown and soft rot | Apple, artichoke, asparagus, banana, bean, beet, Brussels sprout, cabbage, carrot, cauliflower, celery, broccoli, chicory, citrus, cucumber, eggplant, endive, faba bean, fennel, garlic, Jerusalem artichoke, leek, lettuce, melon, onion, parsley, parsnip, pea, pear, pepper, potato, pumpkin, radish, squash, strawberry, swede, sweet potato, tomato, turnip, watermelon |
| Monilinia, *M. fructicola* (G. Winter) Honey, *M. fructigena* Honey, *M. laxa* (Aderh. & Ruhland) Honey | Sclerotiniaceae | As for *Sclerotinia.* | Brown and soft rot | Apple, apricot, blueberry, cherry, nectarine, peach, pear, plum |
| Botrytis, *B. cinerea* Pers. | Anamorphic Sclerotiniaceae, teleomorph *Botryotinia* | The fungus persists in the field as sclerotia or mycelium in soil or on infected plant debris. Sclerotia germinate to produce conidiophores with conidia. The conidia geminate and give rise to a mycelium that may remain quiescent in the flower parts but subsequently invades the fruits. Conidia on fruits are rather large, oval or spherical and formed in bunches on the ends of erect conidiophores, resembling bunches of grapes.[3] | Soft rot and mould | Apple, apricot, artichoke, asparagus, bean, blueberry, broccoli, Brussels sprout, cabbage, carrot, cauliflower, celery, cherry, citrus, cucumber, eggplant, endive, faba bean, feijoa, fennel, fig, garlic, grape, Jerusalem artichoke, kiwifruit, leek, lettuce, marrow, melon, nectarine, onion, parsnip, pea, peach, pear, pepper, persimmon, plum, pomegranate, potato, pumpkin, rhubarb, raspberry, squash, strawberry, sweet potato, taro, tomato, turnip, yam |

**Table A2.1** (*Continued.*)

| Pathogen genus and important species[1] | Family and morphs[2] | Biology and appearance | Type of damage caused | Crops attacked[1] |
|---|---|---|---|---|
| Penicillium, *P. cyclopium* Westling, *P. expansum* Link | Anamorphic Trichocomaceae, teleomorphs *Eupenicillium, Talaromyces* | The mycelium produces simple, long erect conidiophores that branch in a characteristic brush-like manner. The multiple branching gives rise to long chains of conidia. The coloured conidia are responsible for the green, blue or yellow tinges of the colonies.[3] | Mould | Apple, apricot, asparagus, avocado, carrot, cherry, citrus, cucumber, fig, garlic, grape, Jerusalem artichoke, kiwifruit, mango, melon, nectarine, onion, papaya, passion fruit, peach, pear, persimmon, pineapple, plum, pomegranate, raspberry, strawberry, sweet potato, tannia, taro, tomato, yam |
| Cladosporium, *C. herbarum* (Pers.) Link | Anamorphic Mycosphaerellaceae, teleomorphs *Mycosphaerella, Venturia* | The fungus survives as spores in the soil. The spores are two-celled, dark-coloured conidia that are dispersed by rain and wind. Some *Cladosporium* species can directly penetrate their hosts, others gain entry through wounds.[1] | Scab | Apple, apricot, banana, bean, cabbage, Cape gooseberry, carambola, cauliflower, cherry, citrus, cucumber, date, eggplant, fig, grape, leek, marrow, melon, nectarine, onion, papaya, passion fruit, pea, peach, pear, pepper, persimmon, plum, pomegranate, raspberry, squash, tomato, watermelon |
| Botryosphaeria, *B. ribis* Grossenb. & Duggar | Botryosphaeriaceae | Spores may arise from sexual (ascospores from perithecia) or asexual reproduction (conidia from pycnidia). The conidia are water-borne but the ascospores may be dispersed by air currents. The fungus gains entry through wounds and lenticles.[1] | | Apple, avocado, banana, citrus, grape, kiwifruit, peach, pear |
| Phomopsis, *P. mali* (Schulzer & Sacc.) Died. | Anamorphic Valsaceae, teleomorph *Diaporthe* | *Phomopsis,* like *Phoma*, is characterised by small, ostiolate pycnidia sunk in the host tissue. However, *Phomopsis* produces two types of pycnidiospore. The small round conidia are indistinguishable from those of *Phoma*, but the other type is elongated and sometimes bent at the tip.[3] | Dry rot | Apple, asparagus, avocado, bean, blueberry, citrus, cucumber, eggplant, grape, guava, kiwifruit, mango, melon, papaya, peach, plum, pomegranate, sapodilla, strawberry, sweet potato, sweet sop, watermelon |

| | | | | |
|---|---|---|---|---|
| Rhizoctonia, *R. carotae* Rader, *R. solani* J.G. Kühn *Ceratobasidium* | Anamorphic Corticiaceae, teleomorphs *Thanatephorus, Tricharina*, etc. | The fungus survives as sclerotia and mycelium in soils and plant debris. Mycelia on aerial parts of the host may give rise to the sexual stage (a hymenium with basidiospores). Sclerotia can be formed at any time, and may develop in store on the harvested produce, if the temperature and relative humidity are suitable.[1] | Scab | Bean, beet, cabbage, carrot, cauliflower, celery, cucumber, eggplant, Jerusalem artichoke, lettuce, melon, okra, pepper, potato, pumpkin, radish, squash, swede, sweet potato, strawberry, tannia, taro, tomato, turnip, watermelon, yam |
| Sclerotium, *S. cepivorum* Berk *S. rolfsii* Sacc. | Mitosporic fungi (*S. rolfsii* teleomorph is *Corticium*) | Sclerotia survive in the soil for many years. Mycelia produced by sclerotia infect the host, which may be harvested in good condition but decay during storage.[1] | Soft rot | Bean, beet, carrot, cassava, cauliflower, chive, eggplant, endive, garlic, ginger, guava, Jerusalem artichoke, leek, lettuce, melon, onion, parsnip, pea, pear, pepper, potato, pumpkin, rhubarb, shallot, spring onion, squash, sweet potato, tannia, taro, tomato, watermelon, yam |
| Lasiodiplodia, *L. theobromae* (Pat.) Griffiths & Maubl. (syn. *Botryodiplodia theobromae*) | Mitosporic fungi | *Lasiodiplodia* is a common soil inhabitant, with asexual conidia as the important infection stages. Infected produce may have many small black pycnidia on the surface. On citrus, the conidia are washed down the branches onto the developing fruits.[1] | Dry and soft rot | Apple, avocado, banana, cassava, carambola, citrus, cucumber, eggplant, ginger, grape, guava, litchi, loquat, mango, mangosteen, melon, onion, papaya, peach, pear, pepper, pineapple, pumpkin, rambutan, sour sop, squash, sweet potato, sweet sop, tannia, taro, tomato, watermelon, yam |

1 Snowdon, A.L. (1990, 1991) *A Colour Atlas of Post-Harvest Diseases & Disorders of Fruits & Vegetables.* Volumes 1 and 2. Wolfe Scientific, London, UK.
2 Kirk, P.M., Cannon, P.F., David, J.C. & Stalpers, J.A. (2001) *Ainsworth and Bisby's Dictionary of the Fungi*. Ninth Edition. CAB International, Wallingford, UK.
3 Alexopoulos, C.J., Mims, C.W. & Blackwell, M. (1995) *Introductory Mycology*. Fourth Edition. John Wiley & Sons, Chichester, UK.

**Table A2.2** Bacteria.

| Pathogen genus and important species[1] | Family and morphs | Biology and appearance | Type of damage caused | Crops attacked[1] |
|---|---|---|---|---|
| Erwinia, *Pectobacterium carotovorum* ssp. *atrosepticum* (= *E. carotovora* ssp. *atroseptica*), *Pectobacterium carotovorum* ssp. *carotovorum* (= *E. carotovora* ssp. *carotovora*) | Enterobacteriaceae | *Erwinia* are small, motile Gram-negative rods. They are facultatively anaerobic, being able to ferment carbohydrates in the absence of oxygen. The flagella are peritrichous. The carotovora group is fast-growing and pectolytic. | Wet rot | Asparagus, avocado, beet, broccoli, Brussels sprout, cabbage, carrot, cassava, cauliflower, celery, cucumber, fennel, garlic, ginger, guava, Jerusalem artichoke, leek, lettuce, mango, melon, onion, pea, papaya, parsley, parsnip, peppers, pineapple, potato, pumpkin, radish, rhubarb, squash, sweet potato, tannia, taro, tomato, turnip, watermelon, yam |
| Pseudomonas, *P. marginalis, P. syringae* and their pathovars, *Ralstonia solanacearum* (= *P. solanacearum*) | Pseudomonadaceae | Pseudomonads are motile, Gram-negative rods that can only utilise glucose oxidatively. They have one or several polar flagella. *Pseudomonas* may produce water-soluble, greenish fluorescent pigments on King's B medium. Non-fluorescent species have large, refractile sudanophilic inclusions. | Wet rot | Apple, asparagus, avocado, bean, broccoli, Brussels sprout, cabbage, carrot, cauliflower, celery, chicory, cucumber, fennel, ginger, Jerusalem artichoke, leek, lettuce, marrow, melon, okra, onion, passion fruit, parsnip, pea, pineapple, potato, rhubarb, spinach, squash, tomato, turnip, watermelon |
| Xanthomonas, *X. campestris* and its pathovars | Xanthomonadaceae | Xanthomonads are strictly aerobic, Gram-negative rods with a single polar flagellum. They produce a water-insoluble yellow pigment. | Wet rot | Apricot, artichoke, bean, broccoli, Brussels sprout, cabbage, carrot, cauliflower, cherry, lettuce, mango, melon, nectarine, passion fruit, pea, peach, pepper, plum, pumpkin, radish, swede, tomato, turnip, watermelon |

1 Snowdon, A.L. (1990, 1991) *A Colour Atlas of Post-Harvest Diseases & Disorders of Fruits & Vegetables.* Volumes 1 and 2. Wolfe Scientific, London, UK.

# Appendix 3
# Some Plants of Post-harvest Concern

**Table A3.1**

| Common name | Latin name | Family |
|---|---|---|
| Adzuki beans | *Phaseolus angularis* (Willd.) Wight | Leguminosae |
| Apple | *Malus* spp | Rosaceae |
| Asparagus | *Asparagus officinalis* L. | Asparagaceae |
| Aubergine | *Solanum melongena* L. | Solanaceae |
| Avocado | *Persea americana* Miller | Lauraceae |
| Bambara groundnut | *Vigna* (= *Voandzeia*) *subterranea* (L.) Verdc. | Leguminosae |
| Banana | *Musa* spp | Musaceae |
| Barley | *Hordeum vulgare* L. | Gramineae |
| Beetroot | *Beta vulgaris* L. | Chenopodiaceae |
| Black gram | *Vigna mungo* (L.) Hepper | Leguminosae |
| Blueberry | *Vaccinum corymbosum* L. | Ericaceae |
| Breadfruit | *Artocarpus altilis* (Parkinson) Fosb. | Moraceae |
| Broad bean (= fava bean) | *Vicia faba* L. | Leguminosae |
| Broccoli | *Brassica oleracea* var. *botrytis* L. | Cruciferae |
| Brussels sprout | *Brassica oleracea* var. *gemmifera* L. | Cruciferae |
| Butter bean (= lima bean) | *Phaseolus lunatus* L. | Leguminosae |
| Cabbage | *Brassica oleracea* var. *capitata* L. | Cruciferae |
| Cantaloupe | *Cucumis melo* L. | Cucurbitaceae |
| Cape gooseberry | *Physalis peruviana* L. | Solanaceae |
| Carambola (caramba) | *Averrhoa carambola* L. | Oxalidaceae |
| Carrot | *Daucus carota* L. | Umbelliferae |
| Cassava | *Manihot esculenta* Crantz | Euphorbiaceae |
| Cauliflower | *Brassica oleracea* var. *botrytis* L. | Cruciferae |
| Celeriac | *Apium graveolens* var. *rapaceum* (Miller) DC | Umbelliferae |
| Celery | *Apium graveolens* var. *dulce* Miller | Umbelliferae |
| Chayote | *Sechium edule* Jacq. | Cucurbitaceae |
| Cherimoya | *Annona cherimolia* Miller | Annonaceae |
| Cherry | *Prunus* spp | Rosaceae |
| Chick pea | *Cicer arietinum* L. | Leguminosae |
| Chicory | *Cichorium intybus* L. | Compositae |
| Chillies | *Capsicum frutescens* L. | Solanaceae |
| Chive | *Allium schoenoprasum* L. | Amaryllidaceae |
| Cocoa | *Theobroma cacao* L. | Sterculiaceae |
| Coconut | *Cocos nucifera* L. | Palmae |
| Coffee | *Coffea arabica* L. | Rubiaceae |
| Corn | *Zea mays* L. | Gramineae |
| Cotton (Egyptian) | *Gossypium barbadense* L. | Malvaceae |
| Cotton (upland) | *Gossypium hirsutum* L., *G. herbaceum* L., *G. barbadense* L. | Malvaceae |
| Cucumber | *Cucumis sativus* L. | Cucurbitaceae |

**Table A3.1** (*Continued.*)

| Common name | Latin name | Family |
|---|---|---|
| Custard apple | *Annona reticulata* L. | Annonaceae |
| Durian | *Durio zibethinus* Murray | Bombacaceae |
| Eggplant | *Solanum melongena* L. | Solanaceae |
| Endive | *Cichorium endiva* L. | Compositae |
| Feijoa | *Acca sellowiana* (O. Berg) Burret | Myrtaceae |
| Fennel | *Foeniculum vulgare* Miller | Umbelliferae |
| Fig | *Ficus carica* L. | Moraceae |
| Fonio | *Digitaria exilis* Stapf and *D. iburua* Stapf | Graminae |
| Ginger | *Zingiber officinale* Roscoe | Zingiberaceae |
| Globe artichoke | *Cynara scolymus* L. | Compositae |
| Gram | *Cicer arietinum* L. | Leguminosae |
| Granadilla | *Passiflora ligularis* Juss. | Passifloraceae |
| Grape | *Vitis vinifera* L. | Vitaceae |
| Grapefruit | *Citrus paradisi* Macfad. | Rutaceae |
| Green gram | *Vigna radiata* (L.) Wilczek | Leguminosae |
| Groundnut (= peanut) | *Arachis hypogaea* L. | Leguminosae |
| Guava | *Psidium guajava* L. | Myrtaceae |
| Guava (strawberry) | *Psidium cattleianum* Sabine | Myrtaceae |
| Jackfruit | *Artocarpus heterophyllus* Lam. | Moraceae |
| Jerusalem artichoke | *Helianthus tuberosus* L. | Compositae |
| Jujube | *Ziziphus jujuba* Miller | Rhamnaceae |
| Jute | *Corchorus* spp | Tiliaceae |
| Karela | *Momordica charantia* L. | Cucurbitaceae |
| Kenaf | *Hibiscus cannabinus* L. | Malvaceae |
| Kidney beans (= common bean) | *Phaseolus vulgaris* L. | Leguminosae |
| Leek | *Allium ampelprasum* var. *porrum* L. | Alliaceae |
| Lemon | *Citrus limon* (L.) Burm. f. | Rutaceae |
| Lentil | *Lens culinaris* Medikus | Leguminosae |
| Lettuce | *Lactuca sativa* L. | Compositae |
| Lime | *Citrus aurantifolia* (Christm.) Swingle | Rutaceae |
| Lime leaves (Kaffir or Keiffer) | *Citrus hystrix* L. | Rutaceae |
| Litchi | *Litchi chinensis* Sonn. with (Blumea) | Sapindaceae |
| Loquat | *Eriobotrya japonica* (Thunb.) Lindley | Rosaceae |
| Maize | *Zea mays* L. | Gramineae |
| Mammee apple | *Mammea americana* L. | Guttiferae |
| Mandarin | *Citrus reticulata* Blanco | Rutaceae |
| Mangetout | *Pisum sativum* var. *macrocarpon* Ser. | Leguminosae |
| Mango | *Mangifera indica* L. | Anacardiaceae |
| Mangosteen | *Garcinia mangostana* L. | Guttiferae |
| Marrow | *Cucurbita pepo* L. | Cucurbitaceae |
| Melon (canteloupe) | *Cucumis melo* L. | Cucurbitaceae |
| Melon (water) | *Citrullus lanatus* (Thunb.) Matsum. & Nakai | Cucurbitaceae |
| Millet (brown corn) | *Panicum miliaceum* L. | Gramineae |
| Millet (common) | *Panicum miliaceum* L. | Gramineae |
| Millet (pearl) | *Pennisetum glaucum* (L.) R. Br. | Gramineae |
| Millet (finger) | *Eleusine coracana* ssp. *africana* (Kenn.-O'Byrne) S. Phillips | Gramineae |
| Nectarine | *Prunus persica* Miller | Rosaceae |
| Oats | *Avena sativa* L. | Gramineae |
| Okra | *Abelmoschus esculentus* L. | Malvaceae |
| Onion | *Allium cepa* L. | Alliaceae |
| Orange | *Citrus sinensis* (L.) Osbeck | Rutaceae |
| Papaya/Pawpaw | *Carica papaya* L. | Caricaceae |
| Parsnip | *Pastinaca sativa* L. | Umbelliferae |

**Table A3.1** (*Continued.*)

| Common name | Latin name | Family |
|---|---|---|
| Passion fruit (purple) | *Passiflora edulis* Sims | Passifloraceae |
| Peach | *Prunus persica* (L.) Batsch | Rosaceae |
| Peanut | *Arachis hypogaea* L. | Leguminosae |
| Pear | *Pyrus communis* L. | Rosaceae |
| Peas | *Pisum sativum* L. | Leguminosae |
| Persimmon | *Diospyros kaki* L. f. | Ebenaceae |
| Pigeon pea | *Cajanus cajan* (L.) Millsp. | Leguminosae |
| Pineapple | *Ananas comosus* (L.) Merr. | Bromeliaceae |
| Plantain | *Musa paradisiaca* L. | Musaceae |
| Plum | *Prunus salicina* Lindley | Rosaceae |
| Pomegranate | *Punica granatum* L. | Punicaceae |
| Pomelo | *Citrus maxima* (Burm.) Merr. | Rutaceae |
| Potato | *Solanum tuberosum* L. | Solanaceae |
| Pumpkin | *Cucurbita maxima* Duchesne ex Lam. | Cucurbitaceae |
| Purple granadilla | *Passiflora edulis* Sims | Passifloraceae |
| Pyrethrum | *Tanacetum cinerariifolium* (Trev.) Schultz-Bip. (= *Chrysanthemum cinerariaefolium*) | Compositae |
| Radish | *Raphanus sativus* L. | Cruciferae |
| Rambutan | *Nephelium lappaceum* L. | Sapindaceae |
| Red squill | *Drimia maritima* (L.) Stearn | Hyacinthaceae |
| Rhubarb | *Rheum* × *hybridum* Murray | Polygonaceae |
| Rice | *Oryza sativa* L. | Gramineae |
| Sapodilla | *Manikara zapota* (L.) P. Royen | Sapotaceae |
| Sapote | *Pouteria sapota* (Jacq.) H. Moore & Stearn | Sapotaceae |
| Sisal | *Agave sisalana* Perrine | Agavaceae |
| Sorghum | *Sorghum bicolor* (L.) Moench | Gramineae |
| Soursop | *Annona muricata* L. | Annoncaeae |
| Soybean | *Glycine max* (L.) Merrill | Leguminosae |
| Squash | *Cucurbita* spp | Cucurbitaceae |
| Strawberry | *Fragaria vesca* L. | Rosaceae |
| Sugar cane | *Saccharum officinarum* L. | Gramineae |
| Swede | *Brassica napus* L. (Napobrassica group) | Brassicaceae |
| Sweet pepper | *Capsicum annuum* L. | Solanaceae |
| Sweet potato | *Ipomoea batatas* (L.) Lam. | Convolvulaceae |
| Sweetsop | *Annona squamosa* L. | Annonaceae |
| Tamarillo | *Cyphomandra betacea* Cav. | Solanaceae |
| Tamarind | *Tamarindus indica* L. | Leguminosae |
| Tangerine | *Citrus reticulata* Blanco | Rutaceae |
| Tannia | *Xanthosoma sagittifolium* L. | Araceae |
| Taro | *Colocasia esculenta* (L.) Schott | Araceae |
| Tea | *Camellia sinensis* (L.) Kuntze | Theaceae |
| Teff | *Eragrostis tef* (Zucc.) Trotter | Graminae |
| Tobacco | *Nicotiana tabacum* L. | Solanaceae |
| Tomato | *Lycopersicon esculentum* Miller | Solanaceae |
| Tomatillo | *Physalis philadelphica* Lam. | Solanaceae |
| Turnip | *Brassica rapa* L. | Cruciferae |
| Wheat | *Triticum aestivum* L. | Gramineae |
| Yam | *Dioscorea alata* L. (and other species) | Dioscoreaceae |

# Index